PROTECTIVE RELAYING
THEORY AND APPLICATIONS

PROTECTIVE RELAYING
THEORY AND APPLICATIONS
Second Edition, Revised and Expanded

WALTER A. ELMORE
Consulting Engineer
Blue Ridge, Virginia

MARCEL DEKKER, INC.   NEW YORK · BASEL

FIRST INDIAN REPRINT 2005

Although great care has been taken to provide accurate and current information, neither the author(s) not the publisher, nor anyone else associated with this publication, shall be liable for any loss, damage, or liability directly or indirectly caused or alleged to be caused by this book. The material contained herein is not intended to provide specific advice or recommendations for any specific situation.

Library of Congress Cataloging-in-Publication Data
A catalog record for this book is available from the Library of Congress.

ISBN: 0-8247-0972-1

Headquarters
Marcel Dekker, Inc. 270 Madison Avenue, New York, NY 10016, U.S.A.
tel: 212-696-9000; fax: 212-685-4540

World Wide Web
http://www.dekker.com

Printed and bound by Replika Press Pvt. Ltd., India

FOR SALE IN THE INDIAN SUBCONTINENT ONLY.

Preface

Continuous change in protective relaying has been caused by two different influences. One is the fact that the requirements imposed by power systems are in a constant state of change, and our understanding of the basic concepts has sharpened considerably over the years. The other is that the means of implementing the fundamental concepts of fault location and removal and system restoration are constantly growing more sophisticated.

It is primarily because of these changing constraints that this text has been revised and expanded. It began with contributions from two giants of the industry, J. Lewis Blackburn and George D. Rockefeller. From the nucleus of their extensive analyses and writings, and the desire to cover each new contingency with new relaying concepts, this volume has evolved. New solutions to age-old problems have become apparent as greater experience has been gained. No problem is without benefit in the solution found.

This new edition weeds out those relaying concepts that have run their course and have been replaced by more perceptive methods of implementation using new solid-state or microprocessor-based devices.

No single technological breakthrough has been more influential in generating change than the microprocessor. Initially, the methods of translating a collection of instantaneous samples of sine waves into useful current, direction, and impedance measurements were not obvious. Diligent analysis and extensive testing allowed these useful functions to be obtained and to be applied to the desired protective functions. This text attempts to describe, in the simplest possible terms, the manner in which these digital measurements are accomplished in present-day devices.

In addition to those already mentioned, huge contributions were made in the development and refinement of the concepts described in this book by Hung Jen Li, Walter Hinman, Roger Ray, James Crockett, Herb Lensner, Al Regotti, Fernando Calero, Eric Udren, James Greene, Liancheng Wang, Elmo Price, Solveig Ward, John McGowan, and Cliff Downs. Some of these names may not be immediately recognizable, but all have made an impact with their thoughtful, accurate, well-reasoned writings, and they all deserve the gratitude of the industry for the wealth of knowledge they have contributed to this book. I am keenly aware of the high quality of the technical offerings of these people, and I am particularly grateful for the warmth and depth of their friendship.

Walter A. Elmore

Contents

13 Backup Protection   **323**
Revised by E. D. Price

1

Introduction and General Philosophies

Revised by: **W. A. ELMORE**

1 INTRODUCTION

Relays are compact analog, digital, and numerical devices that are connected throughout the power system to detect intolerable or unwanted conditions within an assigned area. They are, in effect, a form of active insurance designed to maintain a high degree of service continuity and limit equipment damage. They are "silent sentinels." Although protective relays will be the main emphasis of this book, other types of relays applied on a more limited basis or used as part of a total protective relay system will also be covered.

2 CLASSIFICATION OF RELAYS

Relays can be divided into six functional categories:

Protective relays. Detect defective lines, defective apparatus, or other dangerous or intolerable conditions. These relays generally trip one or more circuit breaker, but may also be used to sound an alarm.

Monitoring relays. Verify conditions on the power system or in the protection system. These relays include fault detectors, alarm units, channel-monitoring relays, synchronism verification, and network phasing. Power system conditions that do not involve opening circuit breakers during faults can be monitored by verification relays.

Reclosing relays. Establish a closing sequence for a circuit breaker following tripping by protective relays.

Regulating relays. Are activated when an operating parameter deviates from predetermined limits. Regulating relays function through supplementary equipment to restore the quantity to the prescribed limits.

Auxiliary relays. Operate in response to the opening or closing of the operating circuit to supplement another relay or device. These include timers, contact-multiplier relays, sealing units, isolating relays, lockout relays, closing relays, and trip relays.

Synchronizing (or *synchronism check*) *relays.* Assure that proper conditions exist for interconnecting two sections of a power system.

Many modern relays contain several varieties of these functions. In addition to these functional categories, relays may be classified by input, operating principle or structure, and performance characteristic. The following are some of the classifications and definitions described in ANSI/IEEE Standard C37.90 (see also ANSI/IEEE C37.100 "Definitions for Power Switchgear"):

Inputs
Current
Voltage
Power
Pressure
Frequency
Temperature
Flow
Vibration

Operating Principle or Structures
Current balance
Percentage
Multirestraint
Product
Solid state
Static
Microprocessor
Electromechanical
Thermal

Performance Characteristics
Differential
Distance
Directional overcurrent
Inverse time
Definite time
Undervoltage
Overvoltage
Ground or phase
High or low speed
Pilot
    Phase comparison
    Directional comparison
    Current differential

A separate volume, *Pilot Protective Relaying*, covers pilot systems (those relaying functions that involve a communications channel between stations.

2.1 Analog/Digital/Numerical

Solid-state (and static) relays are further categorized under one of the following designations.

2.1.1 Analog

Analog relays are those in which the measured quantities are converted into lower voltage but similar signals, which are then combined or compared directly to reference values in level detectors to produce the desired output (e.g., SA-1 SOQ, SI-T, LCB, circuit shield relays).

2.1.2 Digital

Digital relays are those in which the measured ac quantities are manipulated in analog form and subsequently converted into square-wave (binary) voltages. Logic circuits or microprocessors compare the phase relationships of the square waves to make a trip decision (e.g., SKD-T, REZ-1).

2.1.3 Numerical

Numerical relays are those in which the measured ac quantities are sequentially sampled and converted into numeric data form. A microprocessor performs mathematical and/or logical operations on the data to make trip decisions (e.g., MDAR, MSOC, DPU, TPU, REL-356, REL-350, REL-512).

3 PROTECTIVE RELAYING SYSTEMS AND THEIR DESIGN

Technically, most relays are small systems within themselves. Throughout this book, however, the term *system* will be used to indicate a combination of relays of the same or different types. Properly speaking, the protective relaying system includes circuit breakers and current transformers (ct's) as well as relays. Relays, ct's, and circuit breakers must function together. There is little or no value in applying one without the other.

Protective relays or systems are not required to function during normal power system operation, but must be immediately available to handle intolerable system conditions and avoid serious outages and damage. Thus, the true operating life of these relays can be on the order of a few seconds, even though they are connected in a system for many years. In practice, the relays operate far more during testing and maintenance than in response to adverse service conditions.

In theory, a relay system should be able to respond to an infinite number of abnormalities that can possibly occur within the power system. In practice, the relay engineer must arrive at a compromise based on the four factors that influence any relay application:

Economics. Initial, operating, and maintenance
Available measures of fault or troubles. Fault magnitudes and location of current transformers and voltage transformers
Operating practices. Conformity to standards and accepted practices, ensuring efficient system operation
Previous experience. History and anticipation of the types of trouble likely to be encountered within the system

The third and fourth considerations are perhaps better expressed as the "personality of the system and the relay engineer."

Since it is simply not feasible to design a protective relaying system capable of handling any potential problem, compromises must be made. In general, only those problems that, according to past experience, are likely to occur receive primary consideration. Naturally, this makes relaying somewhat of an art. Different relay engineers will, using sound logic, design significantly different protective systems for essentially the same power system. As a result, there is little standardization in protective relaying. Not only may the type of relaying system vary, but so will the extent of the protective coverage. Too much protection is almost as bad as too little.

Nonetheless, protective relaying is a highly specialized technology requiring an in-depth understanding of the power system as a whole. The relay engineer must know not only the technology of the abnormal, but have a basic understanding of all the system components and their operation in the system. Relaying, then, is a "vertical" speciality requiring a "horizontal" viewpoint. This horizontal, or total system, concept of relaying includes fault protection and the performance of the protection system during abnormal system operation such as severe overloads, generation deficiency, out-of-step conditions, and so forth. Although these areas are vitally important to the relay engineer, his or her concern has not always been fully appreciated or shared by colleagues. For this reason, close and continued communication between the planning, relay design, and operation departments is essential. Frequent reviews of protective systems should be mandatory, since power systems grow and operating conditions change.

A complex relaying system may result from poor system design or the economic need to use fewer circuit breakers. Considerable savings may be realized by using fewer circuit breakers and a more complex relay system. Such systems usually involve design compromises requiring careful evaluation if acceptable protection is to be maintained. It should be recognized that the exercise of the very best relaying application principles can never compensate for the absence of a needed circuit breaker.

3.1 Design Criteria

The application logic of protective relays divides the power system into several zones, each requiring its own group of relays. In all cases, the four design criteria listed below are common to any well-designed and efficient protective system or system segment. Since it is impractical to satisfy fully all these design criteria simultaneously, the necessary compromises must be evaluated on the basis of comparative risks.

3.1.1 Reliability

System reliability consists of two elements: dependability *and* security. Dependability is the degree of certainty of correct operation in response to system trouble, whereas security is the degree of certainty that a relay will not operate incorrectly. Unfortunately, these two aspects of reliability tend to counter one another; increasing security tends to decrease dependability and vice versa. In general, however, modern relaying systems are highly reliable and provide a practical compromise between security and dependability. The continuous supervision made possible by numerical techniques affords improvement in both dependability and security. Protective relay systems must perform correctly under adverse system and environmental conditions.

Dependability can be checked relatively easily in the laboratory or during installation by simulated tests or a staged fault. Security, on the other hand, is much more difficult to check. A true test of system security would have to measure response to an almost infinite variety of potential transients and counterfeit trouble indications in the power system and its environment. A secure system is usually the result of a good background in design, combined with extensive model power system or EMTP (electromagnetic transient program) testing, and can only be confirmed in the power system itself and its environment.

3.1.2 Speed

Relays that could anticipate a fault are utopian. But, even if available, they would doubtlessly raise the question of whether or not the fault or trouble really required a trip-out. The development of faster relays must always be measured against the increased probability of more unwanted or unexplained operations. Time is an excellent criterion for distinguishing between real and counterfeit trouble.

Applied to a relay, high speed indicates that the operating time usually does not exceed 50 ms (three cycles on a 60-Hz base). The term *instantaneous* indicates that no delay is purposely introduced in the operation. In practice, the terms *high speed* and *instantaneous* are frequently used interchangeably.

3.1.3 Performance vs. Economics

Relays having a clearly defined zone of protection provide better selectivity but generally cost more. High-speed relays offer greater service continuity by reducing fault damage and hazards to personnel, but also have a higher initial cost. The higher performance and cost cannot always be justified. Consequently, both low- and high-speed relays are used to protect power systems. Both types have high reliability records. Records on protective relay operations consistently show 99.5% and better relay performance.

3.1.4 Simplicity

As in any other engineering discipline, simplicity in a protective relay system is always the hallmark of good design. The simplest relay system, however, is not always the most economical. As previously indicated, major economies may be possible with a complex relay system that uses a minimum number of circuit breakers. Other factors being equal, simplicity of design improves system reliability—if only because there are fewer elements that can malfunction.

3.2 Factors Influencing Relay Performance

Relay performance is generally classed as (1) correct, (2) no conclusion, or (3) incorrect. Incorrect operation may be either failure to trip or false tripping. The cause of incorrect operation may be (1) poor application, (2) incorrect settings, (3) personnel error, or (4) equipment malfunction. Equipment that can cause an incorrect operation includes current transformers, voltage transformers, breakers, cable and wiring, relays, channels, or station batteries.

Incorrect tripping of circuit breakers not associated with the trouble area is often as disastrous as a failure to trip. Hence, special care must be taken in both application and installation to ensure against this.

"No conclusion" is the last resort when no evidence is available for a correct or incorrect operation. Quite often this is a personnel involvement.

3.3 Zones of Protection

The general philosophy of relay applications is to divide the power system into zones that can be protected adequately with fault recognition and removal producing disconnection of a minimum amount of the system.

The power system is divided into protective zones for

1. Generators
2. Transformers
3. Buses

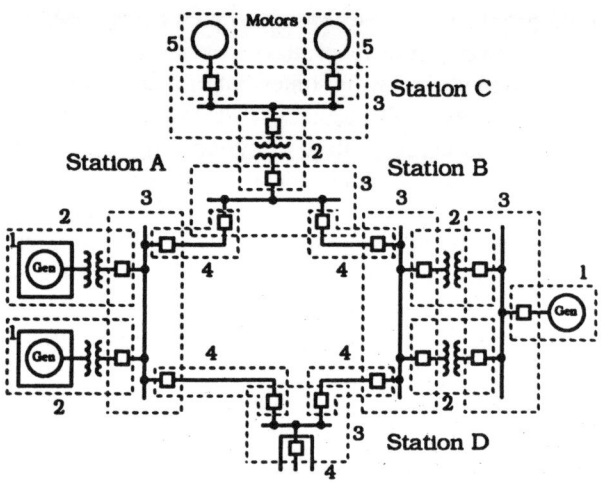

Figure 1-1 A typical system and its zones of protection.

4. Transmission and distribution circuits
5. Motors

A typical power system and its zones of protection are shown in Figure 1-1. The location of the current transformers supplying the relay or relay system defines the edge of the protective zone. The purpose of the protective system is to provide the first line of protection within the guidelines outlined above. Since failures do occur, however, some form of backup protection is provided to trip out the adjacent breakers or zones surrounding the trouble area.

Protection in each zone is overlapped to avoid the possibility of unprotected areas. This overlap is accomplished by connecting the relays to current transformers, as shown in Figure 1-2a. It shows the connection for "dead tank" breakers, and Figure 1-2b the "live tank" breakers commonly used with EHV circuits. Any trouble in the small area between the current transformers will operate both zone A and B relays and trip all breakers in the two zones. In Figure 1-2a, this small area represents the breaker, and in Figure 1-2b the current transformer, which is generally not part of the breaker.

4 APPLYING PROTECTIVE RELAYS

The first step in applying protective relays is to state the protection problem accurately. Although developing a clear, accurate statement of the problem can often be the most difficult part, the time spent will pay dividends—particularly when assistance from others is

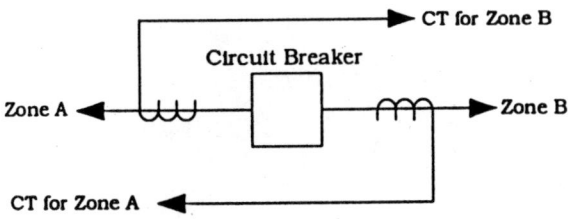

a) Dead Tank Breaker and Breakers With Separate Current Transformers on Both Sides of Breakers

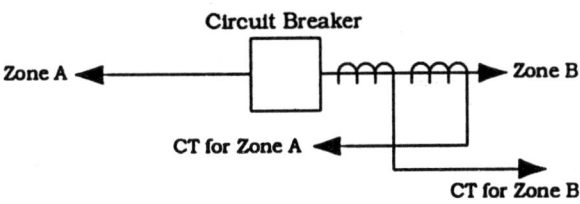

b) Live Tank and Breakers With Separate Current Transformers on One Side Only

Figure 1-2 The principle of overlapping protection around a circuit breaker.

desired. Information on the following associated or supporting areas is necessary:

System configuration
Existing system protection and any known deficiencies
Existing operating procedures and practices and possible future expansions
Degree of protection required
Fault study
Maximum load and current transformer locations and ratios
Voltage transformer locations, connections, and ratios
Impedance of lines, transformers, and generators

4.1 System Configuration

System configuration is represented by a single-line diagram showing the area of the system involved in the protection application. This diagram should show in detail the location of the breakers; bus arrangements; taps on lines and their capacity; location and size of the generation; location, size, and connections of the power transformers and capacitors; location and ratio of ct's and vt's; and system frequency.

Transformer connections are particularly important. For ground relaying, the location of all ground "sources" must also be known.

4.2 Existing System Protection and Procedures

The existing protective equipment and reasons for the desired change(s) should be outlined. Deficiencies in the present relaying system are a valuable guide to improvements. New installations should be so specified. As new relay systems will often be required to operate with or utilize parts of the existing relaying, details on these existing systems are important.

Whenever possible, changes in system protection should conform with existing operating procedures and practices. Exceptions to standard procedures tend to increase the risk of personnel error and may disrupt the efficient operation of the system. Anticipated system expansions can also greatly influence the choice of protection.

4.3 Degree of Protection Required

To determine the degree of protection required, the general type of protection being considered should be outlined, together with the system conditions or operating procedures and practices that will influence the final choice. These data will provide answers to the following types of questions. Is pilot, high-, medium-, or slow-speed relaying required? Is simultaneous tripping of all breakers of a transmission line required? Is instantaneous reclosing needed? Are generator neutral-to-ground faults to be detected?

4.4 Fault Study

An adequate fault study is necessary in almost all relay applications. Three-phase faults, line-to-ground faults, and line-end faults should all be included in the study. Line-end fault (fault on the line side of an open breaker) data are important in cases where one breaker may operate before another. For ground-relaying, the fault study should include zero sequence currents and voltages and negative sequence currents and voltages. These quantities are easily obtained during the course of a fault study and are often extremely useful in solving a difficult relaying problem.

4.5 Maximum Loads, Transformer Data, and Impedances

Maximum loads, current and voltage transformer connections, ratios and locations, and dc voltage are required for proper relay application. Maximum loads should be consistent with the fault data and based on the same system conditions. Line and transformer impedances, transformer connections, and grounding methods should also be known. Phase sequence should be specified if three-line connection drawings are involved.

Obviously, not all the above data are necessary in every application. It is desirable, however, to review the system with respect to the above points and, wherever applicable, compile the necessary data.

In any event, no amount of data can ensure a successful relay application unless the protection problems are first defined. In fact, the application problem is essentially solved when the available measures for distinguishing between tolerable and intolerable conditions can be identified and specified.

5 RELAYS AND APPLICATION DATA

Connected to the power system through the current and voltage transformers, protective relays are wired into the control circuit to trip the proper circuit breakers. In the following discussion, typical connections for relays mounted on conventional switchboards and for rack-mounted solid-state relays will be used to illustrate the standard application practices and techniques.

5.1 Switchboard Relays

Many relays are supplied in a rectangular case that is permanently mounted on a switchboard located in the substation control house. The relay chassis, in some implementations, slides into the case and can be conveniently removed for testing and maintenance. The case is usually mounted flush and permanently wired to the input and control circuits. In the Flexitest case, the electrical connections are made through small, front-accessible, knife-blade switches. A typical switchboard relay is shown in Figure 1-3; its corresponding internal schematic is shown in Figure 1-4. While the example shown is an electromechanical relay, many solid-state relays are in the Flexitest case for switchboard mounting.

Figure 1-3 A typical switchboard type relay. (The CR directional time overcurrent relay in the Flexitest case.)

The important designations in the ac schematic for the relay, such as that illustrated in Figure 1-5, are

Phase rotation
Tripping direction
Current and voltage transformer polarities

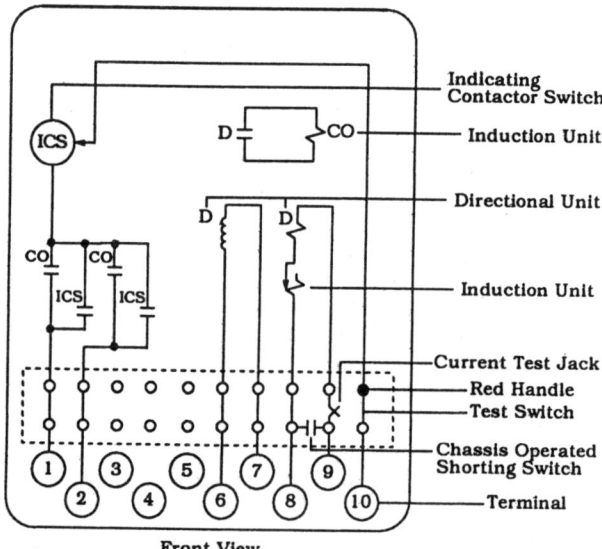

Figure 1-4 Typical internal schematic for a switchboard-mounted relay. (The circuit shown is for the CR directional time overcurrent relay of Figure 1-3.)

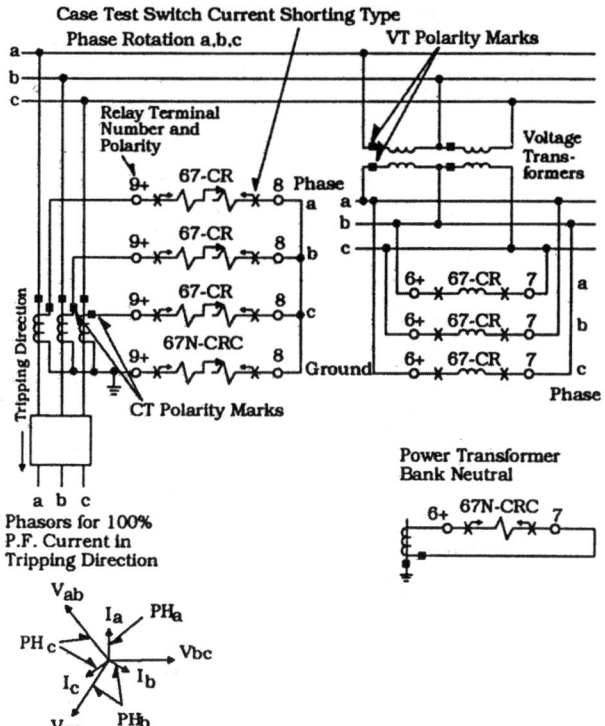

Figure 1-5 Typical ac schematic for a switchboard-mounted relay. (The connections are for the CR phase and CRC ground directional time overcurrent relay of Figure 1-3.)

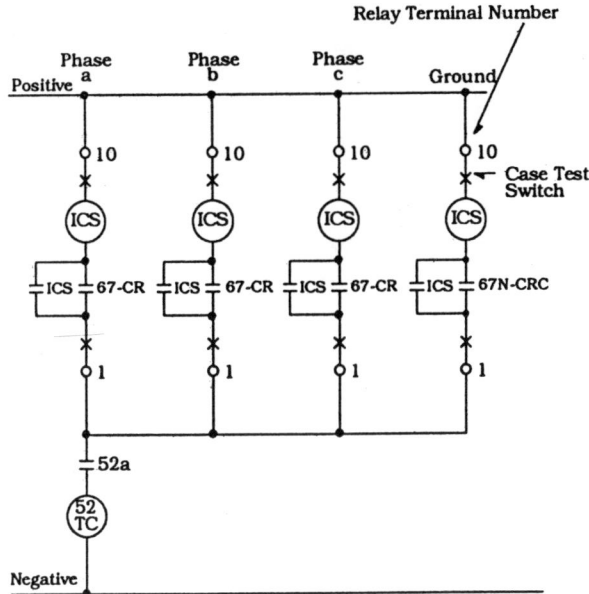

Figure 1-6 Typical dc schematic for a switchboard-mounted relay. (The connections are for three phase type CR and one CRC ground directional time overcurrent relays of Figure 1-3 applied to trip a circuit breaker.)

the breaker. Line voltage cannot be used directly since, of course, it may be quite low during fault conditions.

5.2 Rack-Mounted Relays

Solid-state and microprocessor relays are usually rack-mounted (Fig. 1-8). Since these relays involve more complex and sophisticated circuitry, different levels of information are required to understand their operation. A block diagram provides understanding of the basic process. Figure 1-9 is a block diagram for the MDAR microprocessor relay. Detailed logic diagrams plus ac and dc schematics are also required for a complete view of the action to be expected from these relays.

Relay polarity and terminal numbers
Phasor diagram

All these designations are required for a directional relay. In other applications, some may not apply. In accordance with convention, all relay contacts are shown in the position they assume when the relay is deenergized.

A typical control circuit is shown in Figure 1-6. Three phase relays and one ground relay are shown protecting this circuit. Any one could trip the associated circuit breaker to isolate the trouble or fault area. A station battery, either 125 Vdc or 250 Vdc, is commonly used for tripping. Lower-voltage batteries are not recommended for tripping service when long trip leads are involved.

In small stations where a battery cannot be justified, tripping energy is obtained from a capacitor trip device. This device is simply a capacitor charged, through a rectifier, by the ac line voltage. An example of this arrangement is presented in Figure 1-7. When the relay contacts close, the discharge of the energy in the capacitor through the trip coil is sufficient to trip

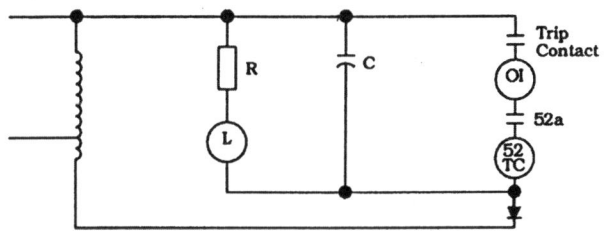

Figure 1-7 Typical capacitor trip device schematic.

Figure 1-8 A typical rack type relay. (The SBFU static circuit breaker failure relay.)

6 CIRCUIT-BREAKER CONTROL

Complete tripping and closing circuits for circuit breakers are complex. A typical circuit diagram is shown in Figure 1-10. In this diagram, the protective relay circuits, such as that shown in Figure 1-6, are abbreviated to a single contact marked "prot relays." While the trip circuits must be energized from a source available during a fault (usually the station battery), the closing circuits may be operated on ac. Such breakers have control circuits similar to those shown in Figure 1-10, except that the 52X, 52Y, and 52CC circuits are arranged for ac operation.

The scheme shown includes red light supervision of the trip coil, 52X/52Y antipump control, and low-pressure and latch checks that most breakers contain in some form.

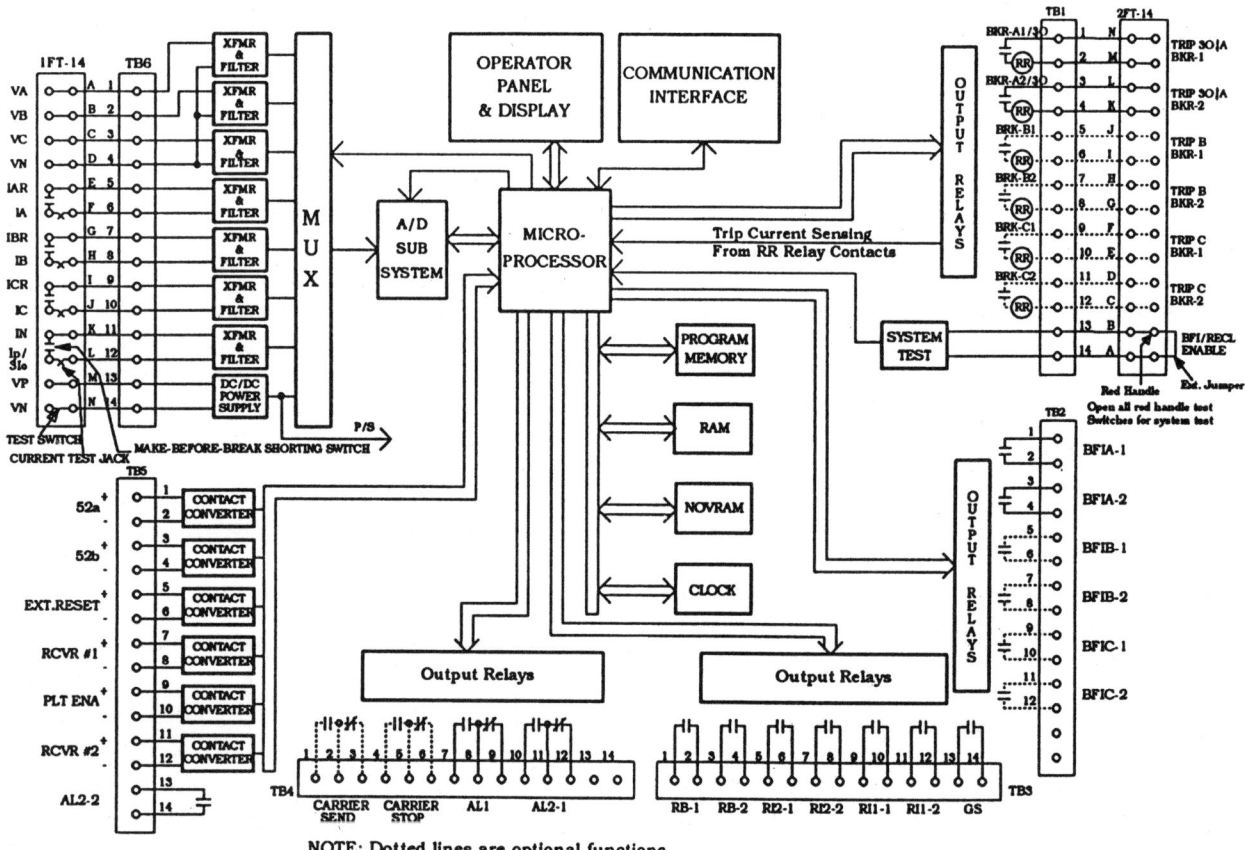

NOTE: Dotted lines are optional functions.

Figure 1-9 Block diagram of MDAR relay.

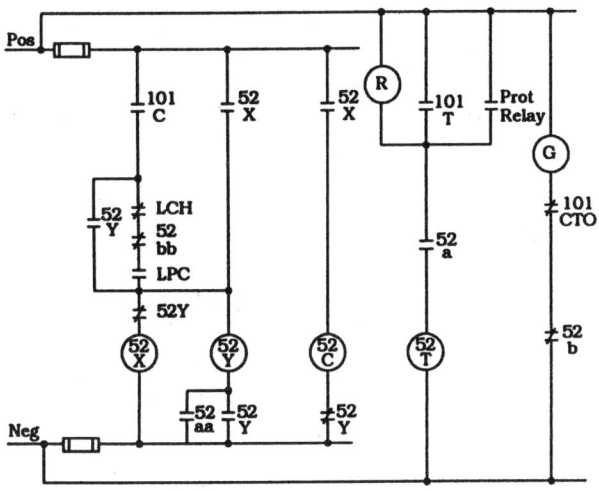

Figure 1-10 A typical control circuit schematic for a circuit breaker showing the tripping and closing circuits.

7 COMPARISON OF SYMBOLS

Various symbols are used throughout the world to represent elements of the power system. Table 1-1 compiles a few of the differences.

Table 1-1 Comparison of Symbols

Element	U.S. practice	European practice
Normally open contact		
Normally closed contact		
Form C		
Breaker		
Fault		
Current transformer		
Transformer		
Phase designations (typical)	A,B,C (preferred) 1, 2, 3	RST
Component designations (positive, negative, zero)	1, 2, 0	1, 2, 0
Current	I	I
Voltage	V	U

2

Technical Tools of the Relay Engineer: Phasors, Polarity, and Symmetrical Components

Revised by: **W. A. ELMORE**

1 INTRODUCTION

In addition to a general knowledge of electrical power systems, the relay engineer must have a good working understanding of phasors, polarity, and symmetrical components, including voltage and current phasors during fault conditions. These technical tools are used for application, analysis, checking, and testing of protective relays and relay systems.

2 PHASORS

A *phasor* is a complex number used to represent electrical quantities. Originally called vectors, the quantities were renamed to avoid confusion with space vectors. A phasor rotates with the passage of time and represents a sinusoidal quantity. A vector is stationary in space.

In relaying, phasors and phasor diagrams are used both to aid in applying and connecting relays and for the analysis of relay operation after faults.

Phasor diagrams must be accompanied by a circuit diagram. If not, then such a circuit diagram must be obvious or assumed in order to interpret the phasor diagram. The phasor diagram shows only the *magnitude* and *relative phase angle* of the currents and voltages, whereas the circuit diagram illustrates only the *location, direction*, and *polarity* of the currents and voltages. These distinctions are important. Confusion generally results when the circuit diagram is omitted or the two diagrams are combined.

There are several systems and many variations of phasor notation in use. The system outlined below is standard with most relay manufacturers.

2.1 Circuit Diagram Notation for Current and Flux

The reference direction for the current or flux can be indicated by (1) an identified directional arrow in the circuit diagram, as shown in Figure 2-1, or (2) the double subscript method, such as I_{ab}, defined as the current flowing from terminal a to terminal b, as in Figure 2-2.

In all cases, the directional arrow or double subscript indicates the actual or assumed direction of current (or flux) flow through the circuit during the positive half-cycle of the ac wave.

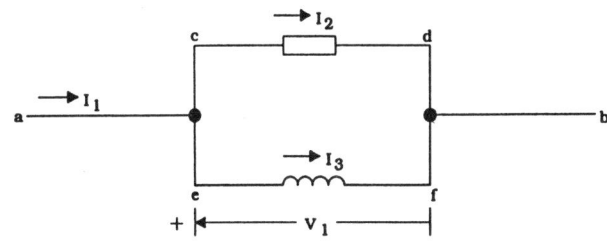

Figure 2-1 Reference circuit diagram illustrating single subscript notation.

11

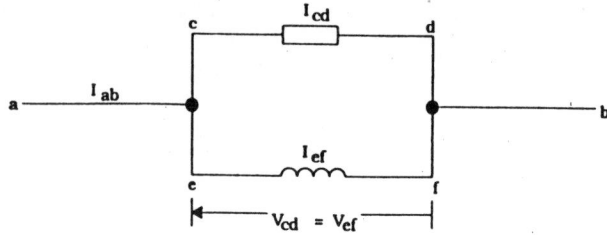

Figure 2-2 Reference circuit diagram illustrating double subscript notation. (Current arrows not required but are usually shown in practice.)

2.2 Circuit Diagram Notation for Voltage

The relative polarity of an ac voltage may be shown in the circuit diagram by (1) a + mark at one end of the locating arrow (Fig. 2-1) or (2) the double subscript notation (Fig. 2-2). In either case, the meaning of the notation must be clearly understood. Failure to properly define notation is the basis for much confusion among students and engineers.

The notation used in this text is defined as follows:

  The letter "V" is used to designate voltages. For simplicity, *only voltage drops are used*. In this sense, a generator rise is considered a negative drop. Some users assign the letter "E" to generated voltage. In much of the world, "U" is used for voltage.

  If locating arrows are used for voltage in the circuit diagram with a single subscript notation, the + mark at one end indicates the terminal of actual or assumed positive potential relative to the other in the half-cycle.

  If double subscript notation is used, the order of the subscripts indicates the actual or assumed direction of the voltage drop when the voltage is in the positive half-cycle.

Thus, the voltage between terminals a and b may be written as either V_{ab} or E_{ab}. Voltage V_{ab} or E_{ab} is positive if terminal a is at a higher potential than terminal b when the ac wave is in the positive half-cycle. During the negative half-cycle of the ac wave, V_{ab} or E_{ab} is negative, and the actual drop for that half-cycle is from terminal b to terminal a.

2.3 Phasor Notation

Figure 2.3a demonstrates the relationship between a phasor and the sinusoid it represents. At a chosen time (in this instance at the time at which the phasor has advanced to 30°), the instantaneous value of the sinusoid is the projection on the vertical of the point of the phasor.

Phasors must be referred to some reference frame. The most common reference frame consists of the axis of real quantities x and the axis of imaginary quantities y, as shown in Figure 2-3b. The axes are fixed in the plane, and the phasors rotate, since they are sinusoidal quantities. (The convention for positive rotation is counterclockwise.) The phasor diagram therefore represents the various phasors at any given common instant of time.

Theoretically, the length of a phasor is proportional to its maximum value, with its projections on the real and imaginary axes representing its real and imaginary components at that instant. By arbitrary convention, however, the phasor diagram is constructed on the

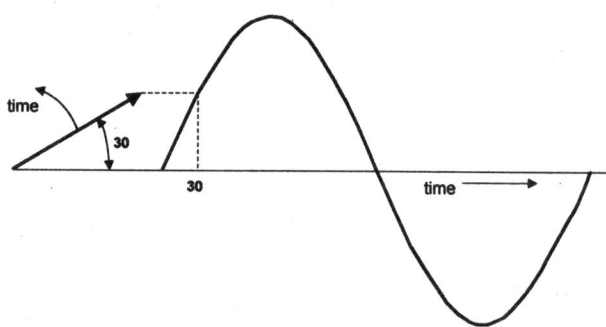

Figure 2-3a Phasor generation of sinusoid.

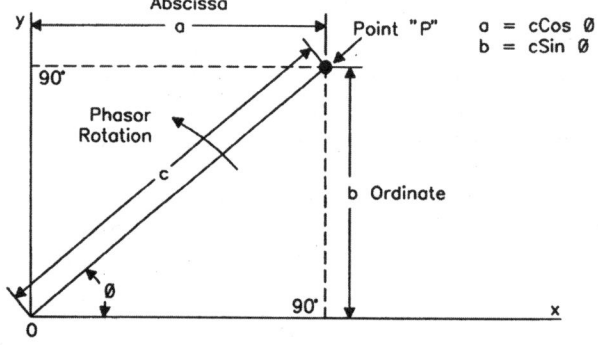

Cartesian Coordinates – Ø Measured Counter-clockwise is Positive

Figure 2-3b Reference axis and nomenclature for phasors.

basis of rms values, which are used much more frequently than maximum values. The phasor diagram indicates angular relationships under the chosen conditions, normal or abnormal.

For reference and review, the various forms, for representation of point P in Figure 2-3b are as follows:

Rectangular form	Complex form	Exponential form	Polar form	Phasor form	
$a + jb$	$= \lvert c\rvert(\cos\theta + j\sin\theta)$	$= \lvert c\rvert\varepsilon^{j\theta}$	$= \lvert c\rvert\angle\theta°$	$= c$	(2-1)
$a - jb$	$= \lvert c\rvert(\cos\theta - j\sin\theta)$	$= \lvert c\rvert\varepsilon^{-j\theta}$	$= \lvert c\rvert\angle-\theta°$	$= \hat{c}$	(2-2)

where

    a = real value
    b = imaginary value
    $\lvert c\rvert$ = modulus or absolute value (magnitude)
    θ = argument or amplitude (relative position)

If c is a phasor, then $\hat{c}$ is its conjugate. Thus, if

$$c = a + jb$$

then

$$\hat{c} = a - jb$$

Some references use c^* to represent conjugate.

The absolute value of the phasor is $\lvert c\rvert$:

$$\lvert c\rvert = \sqrt{a^2 + b^2} \qquad (2\text{-}3)$$

By adding Eqs. (2-1) and (2-2), we obtain

$$a = \frac{1}{2}(c + \hat{c}) \qquad (2\text{-}4)$$

Substracting Eqs. (2-1) and (2-2) yields

$$jb = \frac{1}{2}(c - \hat{c}) \qquad (2\text{-}5)$$

In addition to the use of a single term such as c for a phasor, $\dot{c}, \bar{c},$ and $\vec{c}$ have also been used.

2.3.1 Multiplication Law

The absolute value of a phasor product is the product of the absolute values of its components, and the argument is the sum of the component arguments:

$$EI = \lvert E\rvert \times \lvert I\rvert\angle(\theta_1 + \theta_2) \qquad (2\text{-}6)$$

or

$$
\begin{aligned}
E\hat{I} &= \lvert E\rvert\varepsilon^{j\theta_1} \times \lvert I\rvert\varepsilon^{-j\theta_2} \\
&= \lvert E\rvert \times \lvert I\rvert\angle(\theta_1 - \theta_2)
\end{aligned} \qquad (2\text{-}7)
$$

2.3.2 Division Law

The division law is the inverse of multiplication:

$$\frac{E}{I} = \frac{\lvert E\rvert\varepsilon^{j\theta_1}}{\lvert I\rvert\varepsilon^{j\theta_2}} = \frac{\lvert E\rvert}{\lvert I\rvert}\angle(\theta_1 - \theta_2) \qquad (2\text{-}8)$$

2.3.3 Powers of Complex Numbers

The product of a phasor times its conjugate is

$$(\lvert I\rvert\varepsilon^{j\theta})^{\eta} = \lvert I\rvert^{\eta}\varepsilon^{j\eta\theta} \qquad (2\text{-}9)$$

Thus, I^2 equals $\lvert I\rvert^2\varepsilon^{j2\theta}$:

$$\sqrt[\eta]{\lvert I\rvert\varepsilon^{j\theta}} = \sqrt[\eta]{\lvert I\rvert}\left(\varepsilon^{\frac{j\theta}{\eta}}\right) \qquad (2\text{-}10)$$

The product of a phasor times its conjugate is

$$
\begin{aligned}
I\hat{I} &= \lvert I\rvert\varepsilon^{j\theta} \times \lvert I\rvert\varepsilon^{-j\theta} \\
&= \lvert I\rvert^2\varepsilon^{j(\theta-\theta)} \\
&= \lvert I\rvert^2
\end{aligned} \qquad (2\text{-}11)
$$

Other reference axes used frequently are shown in Figure 2-4. Their application will be covered in later chapters.

2.4 Phasor Diagram Notation

In Figure 2-5, the phasors all originate from a common origin. This method is preferred. In an alternative method, shown in Figure 2-6, the voltage phasors are moved away from a common origin to illustrate the phasor addition of voltages in series (closed system). Although this diagram notation can be useful, it is not

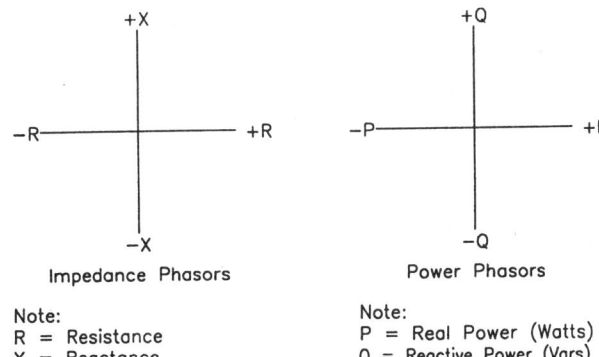

Figure 2-4 Other reference axes for phasors used in relaying and power systems.

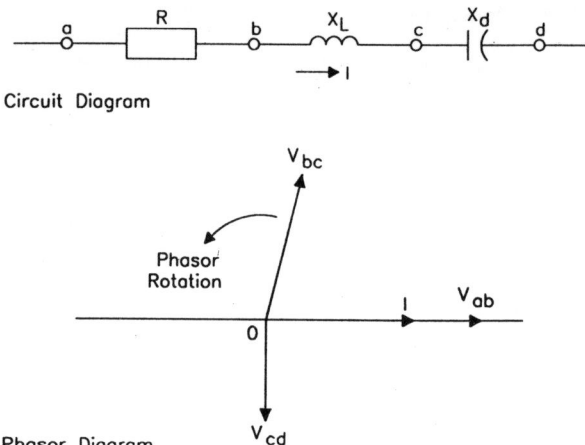

Circuit Diagram

Phasor Diagram

Figure 2-5 Open-type phasor diagram for the basic elements (resistor, reactor, and capacitor) connected in series.

generally recommended since it often promotes confusion by combining the circuit and phasor diagrams.

Notation for three-phase systems varies considerably in the United States; the phases are labeled a, b, c or A, B, C or 1, 2, 3. In other countries, the corresponding phase designation of r, s, t is frequently used.

The letter designations are preferred and used here to avoid possible confusion with symmetrical components notation. A typical three-phase system, with its separate circuit and phasor diagrams, is shown in Figure 2-7. The alternative closed-system phasor diagram is shown in Figure 2-8. With this type of diagram, one tends to label the three corners of the triangle a, b, and c—thereby combining the circuit and phasor diagrams. The resulting confusion is apparent when one notes that, with a at the top corner and b at the lower right corner, the voltage drop from a to b would indicate the opposite arrow from that shown on V_{ab}.

However, when it is considered that, *always*, $V_{ab} = V_{an} + V_{nb}$, it is evident that $V_{ab} = V_{an} - V_{bn}$.

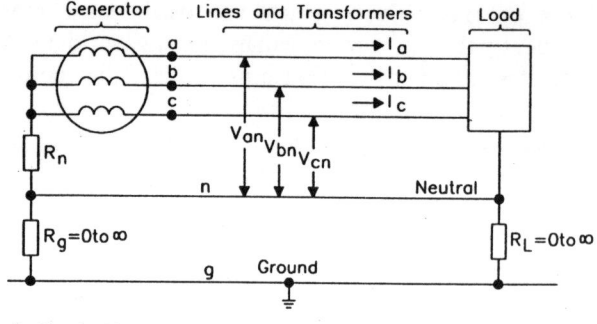

a) Circuit Diagram

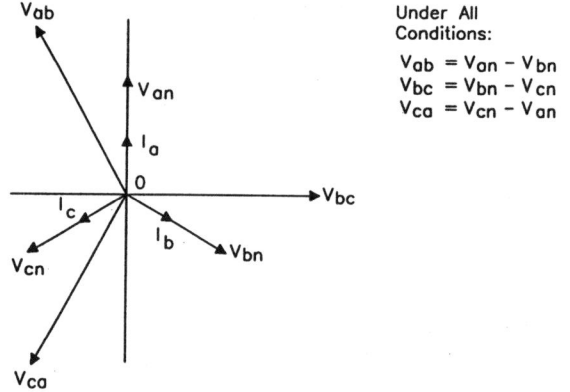

Under All Conditions:

$$V_{ab} = V_{an} - V_{bn}$$
$$V_{bc} = V_{bn} - V_{cn}$$
$$V_{ca} = V_{cn} - V_{an}$$

b) Phasor Diagram (Open System)

Figure 2-7 Designation of the voltages and currents in a three-phase power system.

Similarly, $V_{bc} = V_{bn} - V_{cn}$, and $V_{ca} = V_{cn} - V_{an}$. The associated phasors are shown in Figure 2-7.

Neutral (n) and ground (g) are often incorrectly interchanged. They are not the same. The voltage from n to g is zero when no zero sequence voltage exists. With zero sequence current flowing, there will be a voltage between neutral and ground, $V_{ng} = V_o$.

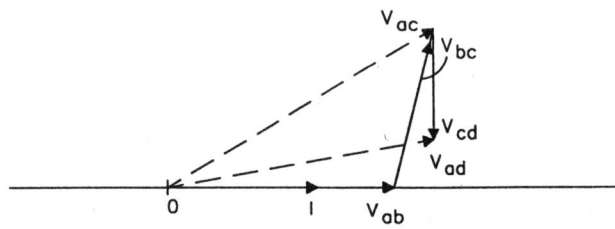

Figure 2-6 Alternative closed-type phasor diagram for the basic circuit of Figure 2-5.

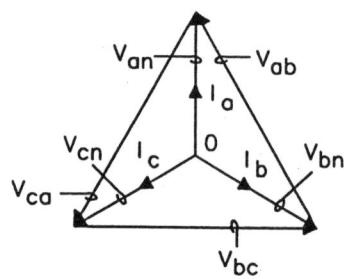

Figure 2-8 Alternative closed system phasor diagram for the three-phase power system of Figure 2-7.

Ground impedance (R_g or R_L) resulting in a rise in station ground potential can be an important factor in relaying. This will be considered in later chapters.

According to ANSI/IEEE Standard 100, "the neutral point of a system is that point which has the same potential as the point of junction of a group of equal nonreactive resistances if connected at their free ends to the appropriate main terminals or lines of the system."

2.5 Phase Rotation vs. Phasor Rotation

Phase rotation, or preferably *phase sequence*, is the order in which successive phase phasors reach their positive maximum values. Phasor rotation is, by international convention, counterclockwise in direction. Phase sequence is the order in which the phasors pass a fixed point.

All standard relay diagrams are for phase rotation a, b, c. It is not uncommon for power systems to have one or more voltage levels with a, c, b rotation; then specific diagrams must be made accordingly. The connection can be changed from one rotation to the other by completely interchanging all b and c connections.

3 POLARITY IN RELAY CIRCUITS

3.1 Polarity of Transformers

The polarity indications shown in Figures 2-9 and 2-10 apply for both current and voltage transformers, or any type of transformer with either subtractive or additive polarity.

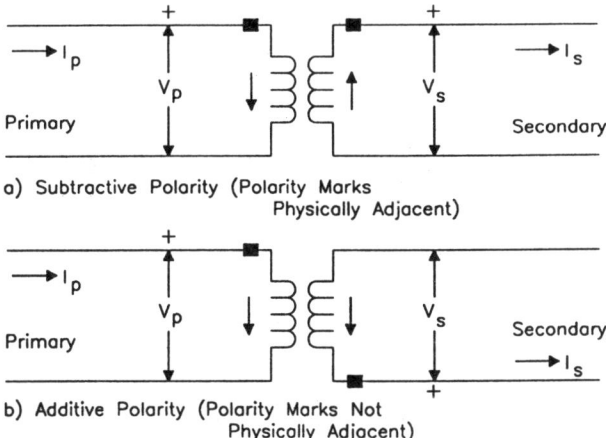

a) Subtractive Polarity (Polarity Marks Physically Adjacent)

b) Additive Polarity (Polarity Marks Not Physically Adjacent)

Figure 2-9 Polarity and circuit diagram for transformers.

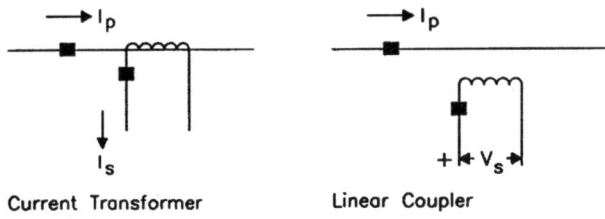

Current Transformer   Linear Coupler

Figure 2-10 Polarity and circuit diagram for conventional representation of current and linear coupler transformer.

The polarity marks X or —■— indicate

The current flowing out at the polarity-marked terminal on the secondary side is essentially in phase with the current flowing in at the polarity-marked terminal on the primary side.

The voltage drop from the polarity-marked to the non-polarity-marked terminal on the primary side is essentially in phase with the voltage drop from the polarity-marked to the non-polarity-marked terminals on the secondary side.

The expression "essentially in phase" allows for the small phase-angle error.

3.2 Polarity of Protective Relays

Polarity is always associated with directional-type relay units, such as those indicating the direction of power flow. Other protective relays, such as distance types, may also have polarity markings associated with their operation. Relay polarity is indicated on the schematic or wiring diagrams by a small + mark above or near the terminal symbol or relay winding. Two such marks are necessary; a mark on one winding alone has no meaning.

Typical polarity markings for a directional unit are shown in Figure 2-11. In this example, the markings indicate that the relay will operate when the voltage drop from polarity to nonpolarity in the voltage coil is in phase with the current flow from polarity to nonpolarity in the current coil. This applies irrespective of the maximum sensitivity angle of the relay. Of course, the levels must be above the relay pickup quantities for the relay to operate.

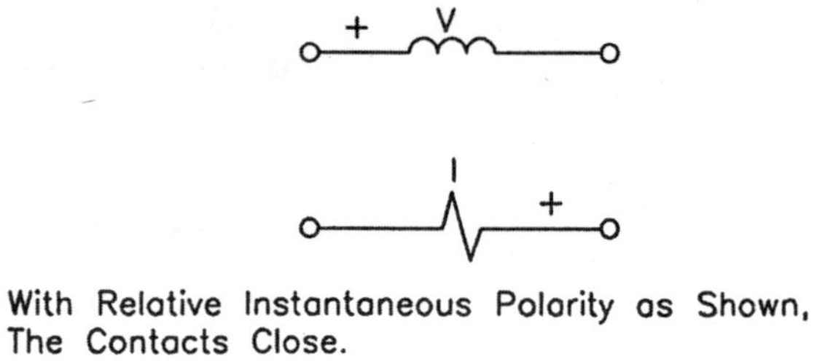

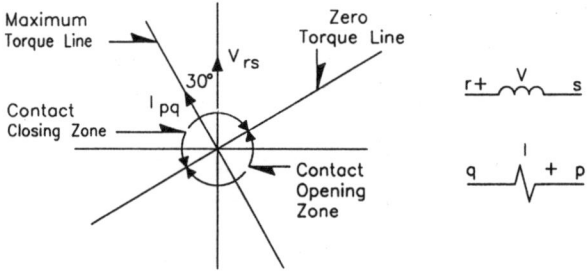

With Relative Instantaneous Polarity as Shown,
The Contacts Close.

Figure 2-11 Polarity markings for protective relays.

3.3 Characteristics of Directional Relays

Directional units are often used to supervise the action of fault responsive devices such as overcurrent units. The primary function of the directional units is to limit relay operation to a specified direction. These highly sensitive units operate on load in the tripping direction.

Directional units can conveniently serve to illustrate the practical application of phasors and polarity. In addition to polarity, these units have a phase-angle characteristic that must be understood if they are to be properly connected to the power system. The characteristics discussed below are among the most common.

3.3.1 Cylinder-Type Directional Unit

As shown in Figure 2-12, the cylinder-type unit has maximum torque when I, flowing in the relay winding from polarity to nonpolarity, leads V drop from polarity to nonpolarity by 30°. The relay minimum pickup values are normally specified at this maximum torque angle. As current I_{pq} lags or leads this maximum torque position, more current is required (at a constant voltage) to produce the same torque. Theoretically, at 120° lead or 60° lag, no torque results from any current magnitude. In practice, however, this

zero torque line is a zone of no operation and not a thin line through the origin, as commonly drawn.

3.3.2 Ground Directional Unit

As shown in Figure 2-13, the ground directional unit usually has a characteristic of maximum torque when I flowing from polarity to nonpolarity lags V drop from polarity to nonpolarity by 60°. Although this characteristic may be inherent in the unit's design, an auxiliary phase shifter is generally required in analog relays.

3.3.3 Watt-Type Directional Unit

The characteristic of the watt-type unit is as shown in Figure 2-14. It has maximum sensitivity when relay current and voltage are in phase.

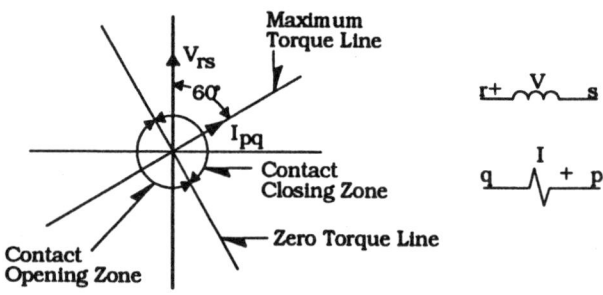

Figure 2-13 Phase-angle characteristics of a ground directional relay unit.

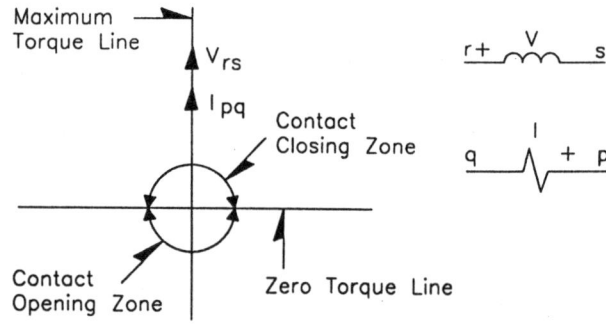

Figure 2-14 Phase-angle characteristics of a watt-type directional relay unit.

Figure 2-12 Phase-angle characteristics of the cylinder-type directional relay unit.

3.4 Connections of Directional Units to Three-Phase Power Systems

The relay unit's individual characteristic, as discussed so far, is the characteristic that would be measured on a single-phase test. Faults on three-phase power systems can, however, produce various relations between the voltages and currents. To ensure correct relay operation, it is necessary to select the proper quantities to apply to the directional units. For all faults in the operating zone of the relay, the fault current and voltage should produce an operating condition as close to maximum sensitivity as possible. Fault current generally lags its unity power factor position by 20 to 85°, depending on the system voltage and characteristics.

Four types of directional element connections (Fig. 2-15) have been used for many years. The proper

a) Available Quantities

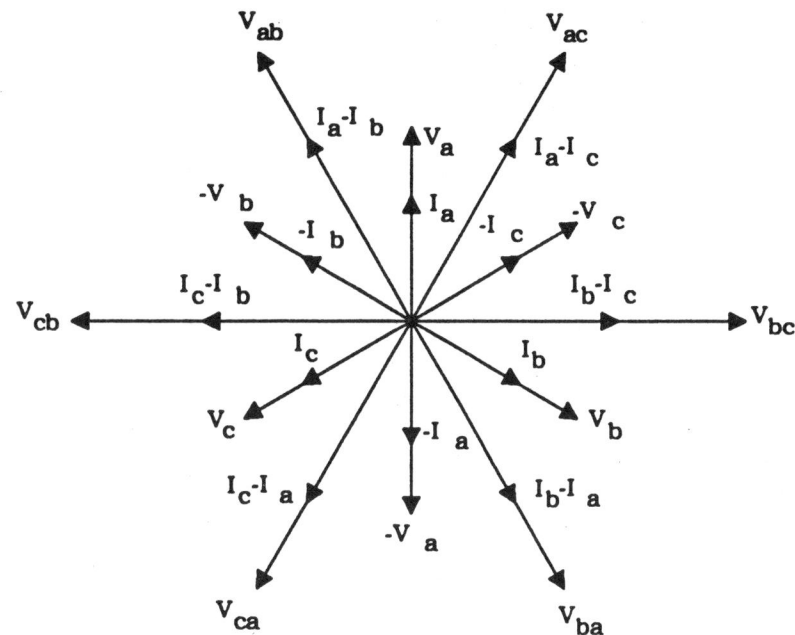

b) Available Connections

Connection +		Unit Type	Phase A		Phase B		Phase C		Maximum Torque When *
			I	V	I	V	I	V	
1	30°	Watt	I_a	V_{ac}	I_b	V_{ba}	I_c	V_{cb}	I lags 30°
2	60° △	Watt	$I_a - I_b$	V_{ac}	$I_b\text{-}I_c$	V_{ba}	$I_c\text{-}I_a$	V_{cb}	I lags 60°
3	60° ⅄	Watt	I_a	$-V_{cg}$	I_b	$-V_{ag}$	I_c	$-V_{bg}$	I lags 60°
4	90°	Cylinder	I_a	V_{bc}	I_b	V_{ca}	I_c	V_{ab}	I lags 60°

* Maximum Torque When "I" Lags its 1.0 p.f. Position by The Angle Indicated.

+ Connection is Defined as Angle by which 1.0 p.f. "I" Leads Voltage Applied to Relay

Figure 2-15 Directional element connections.

system quantities are selected to yield the best operation, considering the phase-angle characteristic of the directional unit. A study of these connections reveals that none is perfect. All will provide incorrect operation under some fault conditions. These conditions are, moreover, different for each connection. Fortunately, the probability of such fault conditions occurring in most power systems is usually very low.

For phase directional measurements, the standard 90° connection is the one best suited to most power systems. Here, the system quantities applied to the relay are 90° apart at unity power factor, balanced current. With this connection, maximum sensitivity can occur at various angles, depending on relay design, as in connection 4. The 90° connection is one standard for phase relays. The 90° angle is that between the unity power factor current and the voltage applied to the relay.

Some experts use a dual numbered system to describe the relationship of the system quantities and to identify the nature of the relaying unit itself. For example, the 90–60° connection is one in which the unity power-factor current applied to the relay and flowing in the relay trip direction leads the voltage applied to the relay by 90°. The nature of the relay referred to is such that the maximum sensitivity occurs when the system current lags its unity power phase position by 60°. The relay has its maximum sensitivity in this case when the current applied to it (into the polarity marker and out nonpolarity) leads the voltage applied to it (voltage drop polarity to nonpolarity) by 30°. Since this is somewhat confusing, it is recommended that the system quantities that are applied to the relay be defined independent of the characteristics of the relay, and that the characteristics of the relay be described independent of the system quantities with which it is used.

Figure 2-16 is a composite circuit diagram illustrating the "phase-a" connections for these four connections that have been used over the years, together with the connection for a ground directional relay. The phasor diagrams are shown in Figure 2-15a for the phase relays and Figure 2-17 for a commonly used ground relay.

4 FAULTS ON POWER SYSTEMS

A fault-proof power system is neither practical nor economical. Modern power systems, constructed with as high an insulation level as practical, have sufficient flexibility so that one or more components may be out of service with minimum interruption of service. In addition to insulation failure, faults can result from electrical, mechanical, and thermal failure or any combination of these.

4.1 Fault Types and Causes

To ensure adequate protection, the conditions existing on a system during faults must be clearly understood. These abnormal conditions provide the discriminating means for relay operation. The major types and causes of failure are listed in Table 2-1.

Relays must operate for several types of faults:

Three-phase (a-b-c, a-b-c-g)
Phase-to-phase (a-b, b-c, c-a)
Two-phase-to-ground (a-b-g, b-c-g, c-a-g)
Phase-to-ground (a-g, b-g, c-g)

Unless preceded by or caused by a fault, open circuits on power systems occur infrequently. Consequently, very few relay systems are designed specifically to provide open-circuit protection. One exception is in the lower-voltage areas, where a fuse can be open. Another is in EHV, where breakers are equipped with independent pole mechanisms.

Simultaneous faults in two parts of the system are generally impossible to relay properly under all conditions. If both simultaneous faults are in the relays' operating zone, at least one set of relays is likely to operate, with the subsequent sequential operation of other relays seeing the faults. When faults appear both internal and external simultaneously, some relays have difficulty determining whether to trip or not. Fortunately, simultaneous faults do not happen very

Table 2-1 Major Types and Causes of Failures

Type	Cause
Insulation	Design defects or errors
	Improper manufacturing
	Improper installation
	Aging insulation
	Contamination
Electrical	Lightning surges
	Switching surges
	Dynamic overvoltages
Thermal	Coolant failure
	Overcurrent
	Overvoltage
	Ambient temperatures
Mechanical	Overcurrent forces
	Earthquake
	Foreign object impact
	Snow or ice

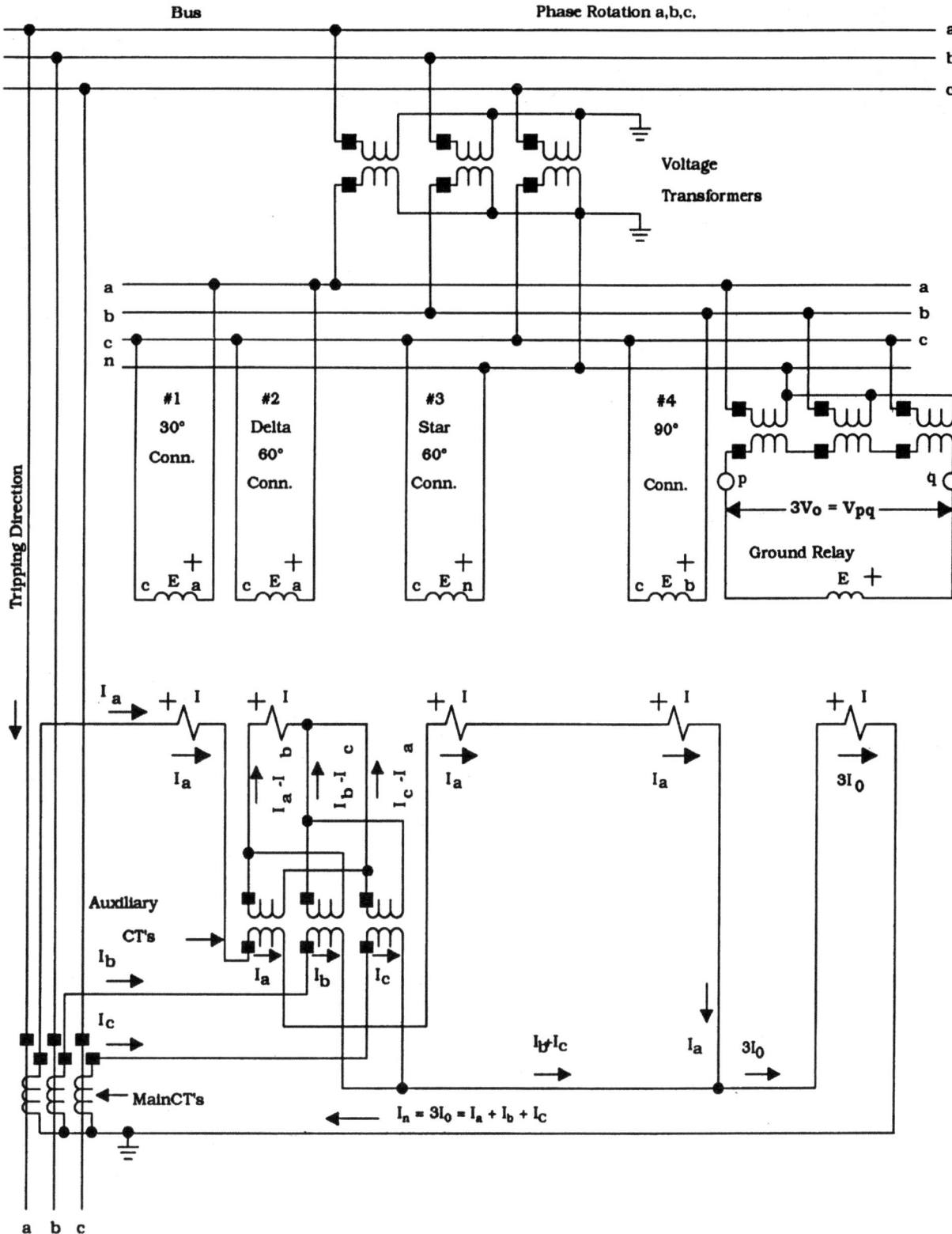

Figure 2-16 Directional unit connections (phase "a" only) for four types of connections plus the ground directional relay connections.

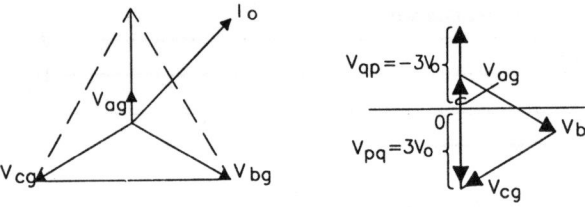

Figure 2-17 Phasor diagram for the ground directional relay connection shown in Figure 2-16. (Phase "a"-to-ground fault is assumed on a solidly grounded system.)

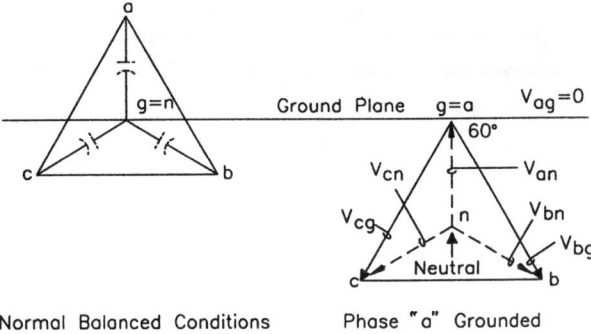

Normal Balanced Conditions   *Phase "a" Grounded*

Figure 2-18 Voltage plot for a solid phase "a"-to-ground fault on an ungrounded system.

often and are not a significant cause of incorrect operations.

4.2 Characteristics of Faults

4.2.1 Fault Angles

The power factor, or angle of the fault current, is determined for phase faults by the nature of the source and connected circuits up to the fault location and, for ground faults, by the type of system grounding as well. The current will have an angle of 80 to 85° lag for a phase fault at or near generator units. The angle will be less out in the system, where lines are involved.

Typical open-wire transmission line angles are as follows:

7.2 to 23 kV: 20 to 45° lag
23 to 69 kV: 45 to 75° lag
69 to 230 kV: 60 to 80° lag
230 kV and up: 75 to 85° lag

At these voltage levels, the currents for phase faults will have the angles shown where the line impedance predominates. If the transformer and generator impedances predominate, the fault angles will be higher. Systems with cables will have lower angles if the cable impedance is a large part of the total impedance to the fault.

4.2.2 System Grounding

System grounding significantly affects both the magnitude and angle of ground faults. There are three classes of grounding: ungrounded (isolated neutral), impedance-grounded (resistance or reactance), and effectively grounded (neutral solidly grounded). An ungrounded system is connected to ground through the natural shunt capacitance, as illustrated in Figure 2-18 (see also Chap. 7). In addition to load, small (usually negligible) charging currents flow normally.

In a symmetrical system, where the three capacitances to ground are equal, g equals n. If phase a is grounded, the triangle shifts as shown in Figure 2-18. Consequently, V_{bg} and V_{cg} become approximately $\sqrt{3}$ times their normal value. In contrast, a ground on one phase of a solidly grounded radial system will result in a large phase and ground fault current, but little or no increase in voltage on the unfaulted phases (Fig. 2-19).

4.2.3 Fault Resistance

Unless the fault is solid, an arc whose resistance varies with the arc length and magnitude of the fault current is usually drawn through air. Several studies indicate that for currents in excess of 100 A the voltage across the arc is nearly constant at an average of approximately 440 V/ft.

Arc resistance is seldom an important factor in phase faults except at low system voltages. The arc does not elongate sufficiently for the phase spacings involved to decrease the current flow materially. In addition, the arc resistance is at right angles to the reactance and, hence, may not greatly increase the total impedance that limits the fault current.

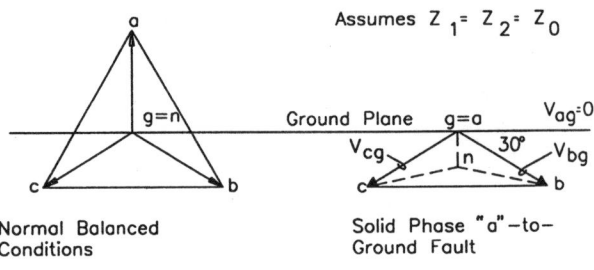

Normal Balanced Conditions   *Solid Phase "a"-to-Ground Fault*

Figure 2-19 Voltage plot for a solid phase "a"-to-ground fault on a solidly grounded system.

For ground faults, arc resistance may be an important factor because of the longer arcs that can occur. Also, the relatively high tower footing resistance may appreciably limit the fault current.

Arc resistance is discussed in more detail in Chapter 12.

4:2.4 Distortion of Phases During Faults

The phasor diagrams in Figure 2-20 illustrate the effect of faults on the system voltages and currents. The diagrams shown are for effectively grounded systems. In all cases, the dotted or uncollapsed voltage triangle exists in the source (the generator) and the maximum collapse occurs at the fault location. The voltage at

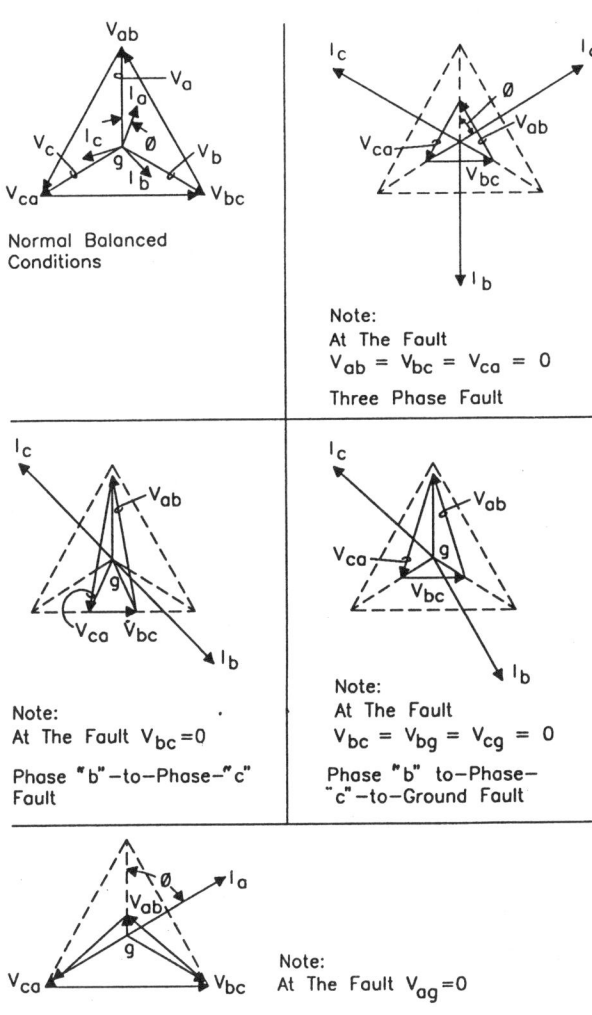

Figure 2-20 Phasor diagrams for the various types of faults occurring on a typical power system.

other locations will be between these extremes, depending on the point of measurement.

5 SYMMETRICAL COMPONENTS

Relay application requires a knowledge of system conditions during faults, including the magnitude, direction, and distribution of fault currents, and often the voltages at the relay locations for various operating conditions. Among the operating conditions to be considered are maximum and minimum generation, selected lines out, line-end faults with the adjacent breaker open, and so forth. With this information, the relay engineer can select the proper relays and settings to protect all parts of the power system in a minimum amount of time. Three-phase fault data are used for the application and setting of phase relays and single-phase-to-ground fault data for ground relays.

The method of symmetrical components is the foundation for obtaining and understanding fault data on three-phase power systems. Formulated by Dr. C. L. Fortescue in a classic AIEE paper in 1918, the symmetrical components method was given its first practical application to system fault analysis by C. F. Wagner and R. D. Evans in the late 1920s and early 1930s. W. A. Lewis and E. L. Harder added measurably to its development in the 1930s.

Today, fault studies are commonly made with the digital computer and can be updated rapidly in response to system changes. Manual calculations are practical only for simple cases.

A knowledge of symmetrical components is important in both making a study and understanding the data obtained. It is also extremely valuable in analyzing faults and relay operations. A number of protective relays are based on symmetrical components, so the method must be understood in order to apply these relays successfully.

In short, the method of symmetrical components is one of the relay engineer's most powerful technical tools. Although the method and mathematics are quite simple, the practical value lies in the ability to think and visualize in symmetrical components. This skill requires practice and experience.

5.1 Basic Concepts

The method of symmetrical components consists of reducing any unbalanced three-phase system of phasors into three balanced or symmetrical systems:

the positive, negative, and zero sequence components. This reduction can be performed in terms of current, voltage, impedance, and so on.

The positive sequence components consist of three phasors equal in magnitude and 120° out of phase (Fig. 2-21a). The negative sequence components are three phasors equal in magnitude, displaced 120° with a phase sequence opposite to that of the positive sequence (Fig. 2-21b). The zero sequence components consist of three phasors equal in magnitude and in phase (Fig. 2-21c). *Note all phasors rotate in a counterclockwise direction.*

In the following discussion, the subscript 1 will identify the positive sequence component, the subscript 2 the negative sequence component, and the subscript 0 the zero sequence component. For example, V_{a1} is the positive sequence component of phase-a voltage, V_{b2} the negative sequence component of phase-b voltage, and V_{c0} the zero sequence component of phase-c voltage. *All components are phasor quantities, rotating counterclockwise.*

Since the three phasors in any set are always equal in magnitude, the three sets can be expressed in terms of one phasor. For convenience, the phase-a phasor is used as a reference. Thus,

Positive sequence	Negative sequence	Zero sequence	
$V_{a1} = V_{a1}$	$V_{a2} = V_{a2}$	$V_{a0} = V_{a0}$	(2-12)
$V_{b1} = a^2 V_{a1}$	$V_{b2} = a V_{a2}$	$V_{b0} = V_{a0}$	
$V_{c1} = a V_{a1}$	$V_{c2} = a^2 V_{a2}$	$V_{c0} = V_{a0}$	

The coefficients a and a^2 are operators that, when multiplied with a phasor, result in a counterclockwise angular shift of 120 and 240°, respectively, with no

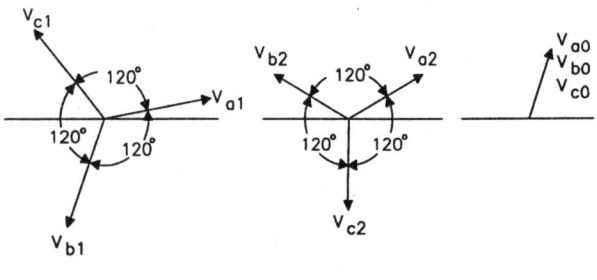

a) Positive   b) Negative   c) Zero

Figure 2-21   Sequence components of voltages.

change in magnitude:

$$a = 1\angle 120°$$
$$\quad = -0.5 + j0.866 \tag{2-13}$$
$$a^2 = 1\angle 240°$$
$$\quad = -0.5 - j0.866 \tag{2-14}$$
$$a^3 = 1\angle 360°$$
$$\quad = 1.0 + j0 \tag{2-15}$$

From these equations, useful combinations can be derived

$$1 + a + a^2 = 0$$
$$1 - a^2 = \sqrt{3}\angle 30° \tag{2-16}$$

or

$$a^2 - 1 = \sqrt{3}\angle 210°$$
$$a - 1 = \sqrt{3}\angle 150° \tag{2-17}$$

or

$$1 - a = \sqrt{3}\angle -30°$$
$$a^2 - a = \sqrt{3}\angle 270° \tag{2-18}$$

or

$$a - a^2 = \sqrt{3}\angle 90° \tag{2-19}$$

Any three-phase system of phasors will always be the sum of the three components:

$$V_a = V_{a1} + V_{a2} + V_{a0} \tag{2-20}$$
$$V_b = V_{b1} + V_{b2} + V_{b0}$$
$$\quad = a^2 V_{a1} + a V_{a2} + V_{a0} \tag{2-21}$$
$$V_c = V_{c1} + V_{c2} + V_{c0}$$
$$\quad = a V_{a1} + a^2 V_{a2} + V_{a0} \tag{2-22}$$

Since phase a has been chosen as a reference, the subscripts are often dropped for convenience. Thus,

$$V_a = V_1 + V_2 + V_0$$

and

$$I_a = I_1 + I_2 + I_0$$
$$V_b = a^2 V_1 + a V_2 + V_0 \tag{2-23}$$

and

$$I_b = a^2 I_1 + a I_2 + I_0$$
$$V_c = a V_1 + a^2 V_2 + V_0 \tag{2-24}$$

and

$$I_c = aI_1 + a^2I_2 + I_0 \qquad (2\text{-}25)$$

Quantities $V_1, V_2, V_0, I_1, I_2,$ and $I_0,$ can always be assumed to be the phase-a components. Note that the b and c components always exist, as indicated by Eq. (2-12). Note that dropping the phase subscripts should be done with great care. Where any possibility of misunderstanding can occur, the additional effort of using the double subscripts will be rewarded.

Equations (2-20) to (2-22) can be solved to yield the sequence components for a general set of three-phase phasors:

$$V_{a1} = \frac{1}{3}(V_{ag} + aV_{bg} + a^2V_{cg})$$

and

$$I_{a1} = \frac{1}{3}(I_a + aI_b + a^2I_c)$$

$$V_{a2} = \frac{1}{3}(V_{ag} + a^2V_{bg} + aV_{cg}) \qquad (2\text{-}26)$$

and

$$I_{a2} = \frac{1}{3}(I_a + a^2I_b + aI_c)$$

$$V_{a0} = \frac{1}{3}(V_{ag} + V_{bg} + V_{cg}) \qquad (2\text{-}27)$$

and

$$I_0 = \frac{1}{3}(I_a + I_b + I_c) \qquad (2\text{-}28)$$

A sequence component cannot exist in only one phase. If any sequence component exists by measurement or calculation in one phase, it exists in all three phases, as shown in Eq. (2-12) and Figure 2-21.

5.2 System Neutral

Figure 2-22 describes the definition of power-system neutral and contrasts it with ground. Neutral is established by connecting together the terminals of three equal resistances as shown with each of the other resistor terminals connected to one of the phases. We can thus write

$$V_{ag} = V_{an} + V_{ng}$$
$$V_{bg} = V_{bn} + V_{ng}$$
$$V_{cg} = V_{cn} + V_{ng} \qquad (2\text{-}29)$$

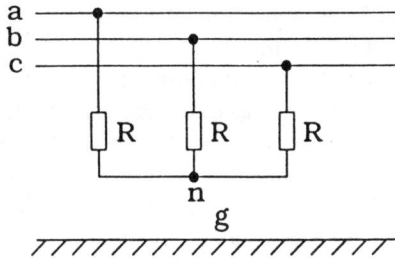

Figure 2-22 Power system neutral.

From Eq. (2-28),

$$V_0 = \frac{1}{3}(V_{ag} + V_{bg} + V_{cg})$$

Substituting Eq. (2-29), we obtain

$$V_0 = \frac{1}{3}(V_{an} + V_{ng} + V_{bn} + V_{ng} + V_{cn} + V_{ng})$$

Since $V_{an} + V_{bn} + V_{cn} = 0,$

$$V_0 = \frac{1}{3}(3V_{ng})$$

$$V_0 = V_{ng}$$

Neutral and ground are distinctly independent and differ in voltage by V_0.

Grounding and its influence on relaying are discussed in Chapters 7 and 12.

5.3 Sequences in a Three-Phase Power System

Several important assumptions are made to greatly simplify the use of symmetrical components in practical circumstances. Interconnections of the three sequence networks allow any series or shunt discontinuity to be investigated. For the rest of the power-system network, it is assumed that the impedances in the individual phases are equal and the generator phase voltages are equal in magnitude and displaced 120° from one another.

Based on this premise, in the symmetrical part of the system, positive sequence current flow produces only positive sequence voltage drops, negative sequence current flow produces only negative sequence voltage drops, and zero sequence current flow produces only zero sequence voltage drops. For an unsymmetrical system, interaction occurs between components. For a particular series or shunt discontinuity being represented, the interconnection of the networks produces the required interaction.

Any circuit that is not continuously transposed will have impedances in the individual phases that differ. This fact is generally ignored in making calculations because of the immense simplification that results. From a practical viewpoint, ignoring this effect, in general, has no appreciable influence.

5.4 Sequence Impedances

Quantities Z_1, Z_2, and Z_0 are the system impedances to the flow of positive, negative, and zero sequence currents, respectively. Except in the area of a fault or general unbalance, each sequence impedance is considered to be the same in all three phases of the symmetrical system. A brief review of these quantities is given below for synchronous machinery, transformers, and transmission lines.

5.4.1 Synchronous Machinery

Three different positive sequence reactance values are specified. X_d'' indicates the subtransient reactance, X_d' the transient reactance, and X_d the synchronous reactance. These direct-axis values are necessary for calculating the short-circuit current value at different times after the short circuit occurs. Since the subtransient reactance values give the highest initial current value, they are generally used in system short-circuit calculations for high-speed relay application. The transient reactance value is used for stability consideration and slow-speed relay application.

The unsaturated synchronous reactance is used for sustained fault-current calculation since the voltage is reduced below saturation during faults near the unit. Since this generator reactance is invariably greater than 100%, the sustained fault current will be less than the machine rated load current unless the voltage regulator boosts the field substantially.

The negative sequence reactance of a turbine generator is generally equal to the subtransient X_d'' reactance. X_2 for a salient-pole generator is much higher. The flow of negative sequence current of opposite phase rotation through the machine stator winding produces a double frequency component in the rotor. As a result, the average of the subtransient direct-axis reactance and the subtransient quadrature-axis reactance gives a good approximation of negative sequence reactance.

The zero sequence reactance is much less than the others, producing a phase-to-ground fault current magnitude $[3/(x_1 + x_2 + x_0)]$ greater than the three-phase fault current magnitude $(1/x_1)$. Since the machine is braced for only three-phase fault current magnitude, it is seldom possible or desirable to ground the neutral solidly.

The armature winding resistance is small enough to be neglected in calculating short-circuit currents. This resistance is, however, important in determining the dc time constant of an asymmetrical short-circuit current.

Typical reactance values for synchronous machinery are available from the manufacturer or handbooks. However, actual design values should be used when available.

5.4.2 Transformers

The positive and negative sequence reactances of all transformers are identical. Values are available from the nameplate. The zero sequence reactance is either equal to the other two sequence reactances or infinite except for the three-phase, core-type transformers. In effect, the magnetic circuit design of the latter units gives them the effect of an additional closed delta winding. The resistance of the windings is very small and neglected in short-circuit calculations.

The sequence circuits for a number of transformer banks are shown in Figure 2-23. The impedances indicated are the equivalent leakage impedances between the windings involved. For two-winding transformers, the total leakage impedance Z_{LH} is measured from the L winding, with the H winding short-circuited. Z_{HL} is measured from the H winding with the L winding shorted. Except for a 1:1 transformer ratio, the impedances have different values in ohms. On a per unit basis, however, Z_{LH} equals Z_{HL}.

For three-winding and autotransformer banks, there are three leakage impedances:

Impedance	Winding measured from	Shorted winding	Open winding
$Z_{HM}(Z_{HL})$	H	M(L)	L(T)
$Z_{HL}(Z_{HT})$	H	L(T)	M(L)
$Z_{ML}(Z_{LT})$	M(L)	L(T)	H

Both winding conventions shown above are in common use. In the first convention, the windings are labeled H (high), L (low), M (medium); in the second H (high), L (low), and T (tertiary). Unfortunately, the L winding in the second convention is

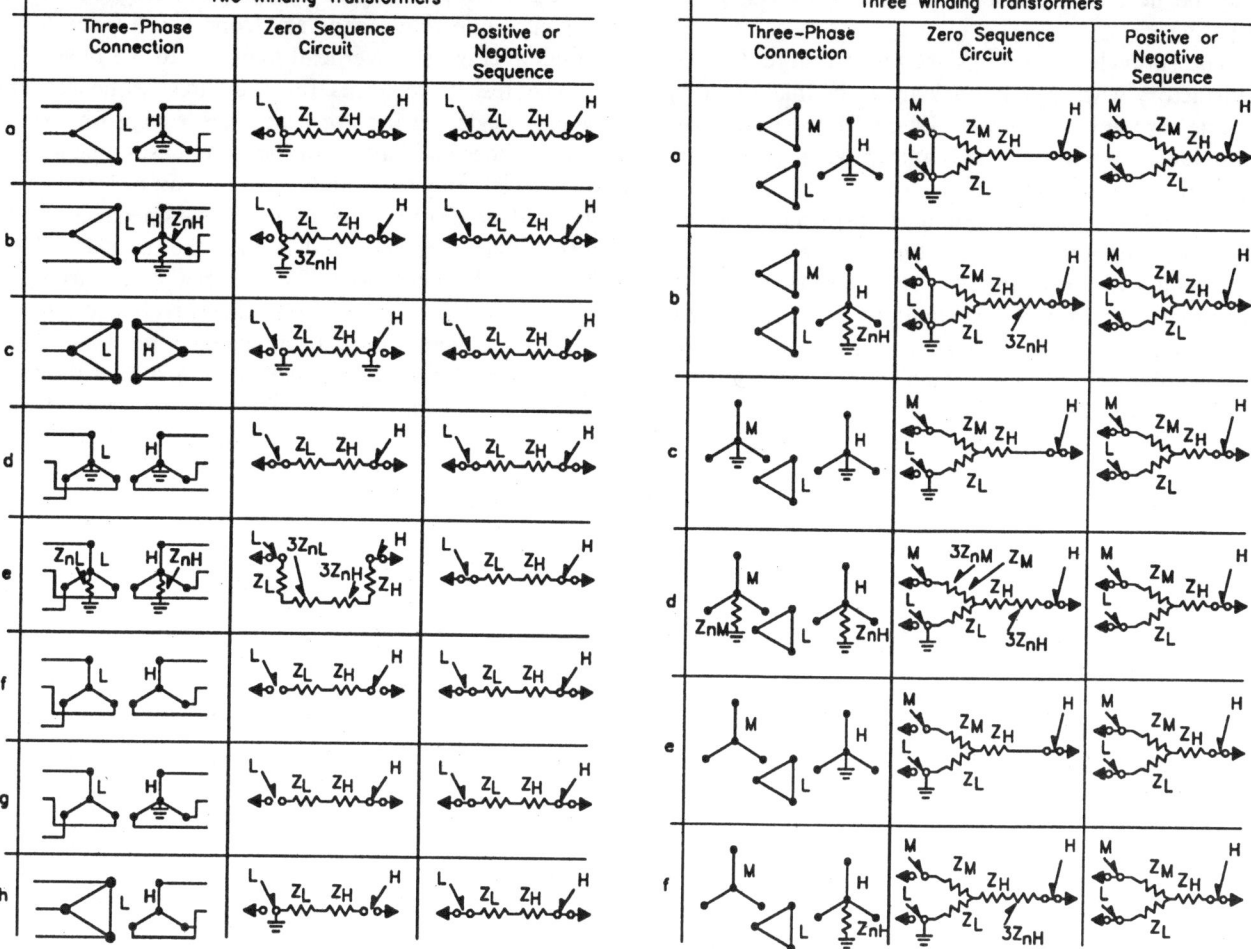

Figure 2-23 Equivalent positive, negative, and zero sequence circuits for some common and theoretical connections for two- and three-winding transformers.

equivalent to M in the first. The tertiary winding voltage is generally the lowest.

On a common kVA base, the equivalent wye leakage impedances are obtained from the following equations:

$$Z_H = \frac{1}{2}(Z_{HM} + Z_{HL} - Z_{ML})$$

or

$$Z_H = \frac{1}{2}(Z_{HL} + Z_{HT} - Z_{LT})$$

$$Z_M = \frac{1}{2}(Z_{HM} + Z_{ML} - Z_{HL})$$

or

$$Z_L = \frac{1}{2}(Z_{HL} + Z_{LT} - Z_{HT})$$

$$Z_L = \frac{1}{2}(Z_{HL} + Z_{ML} - Z_{HM})$$

or

$$Z_T = \frac{1}{2}(Z_{HT} + Z_{LT} - Z_{HL})$$

As a check, Z_H plus Z_M equals Z_{HM}, and so on. The wye is a mathematical equivalent valid for current and voltage calculations external to the transformer bank. The junction point of the wye has no physical significance. One equivalent branch, usually $Z_M(Z_L)$, may be negative. On some autotransformers, Z_H is negative.

The equivalent diagrams shown in Figure 2-23 are satisfactory when calculations are to be made relative to one segment of a power system. However, a more complex representation is required when phase currents and voltages are to be determined at points in the system having an intervening transformer between them and the point of discontinuity being examined. For delta-wye transformers, a 30° phase shift must be accommodated. For ANSI standard transformers, the *high-voltage* phase-to-ground voltage *leads* the *low-voltage* phase-to-ground voltage by 30°, irrespective of which side the delta or wye is on. This phase shift may be included in the equivalent per unit diagram by showing a 1 ∠30°:1 ratio for it.

The phase shift in the negative sequence network for the delta-wye transformer is the same amount, but in the opposite direction, to that in the positive sequence network. The phase shift then, for an ANSI standard transformer, would be 1 ∠ − 30°:1 in the negative sequence per unit diagram.

The phase shift must be used in all the combinations of Figure 2-23 where a wye and delta winding coexist. This effect is extremely important when consideration is being given to the behavior of devices on both sides of such a transformer.

5.4.3 Transmission Lines

For transmission lines, the positive and negative sequence reactances are the same. As a rule of thumb, the 60-Hz reactance is roughly $0.8\,\Omega/\text{mi}$ for single conductor overhead lines and $0.6\,\Omega/\text{mi}$ for bundled overhead lines.

The zero sequence impedance is always different from the positive and negative sequence impedances. It is a loop impedance (conductor plus earth and/or ground wire return), in contrast to the one-way impedance for a positive and negative sequence. Zero sequence impedance can vary from 2 to 6 times X_1; a rough average for overhead lines is 3 to 3.5 times X_1.

The resistance terms for the three sequences are usually neglected for overhead lines, except for lower-voltage lines and cables. In the latter cases, line angles of 30 to 60° may exist, and resistance can be significant. A good compromise is to use the impedance value rather than reactance and neglect the angular difference in fault calculations. This yields a lower current to assure that the relay will be set sensitively enough.

Zero sequence mutual impedance resulting from paralleled lines can be as high as 50 to 70% of the zero sequence self-impedance. This mutual impedance becomes an increasingly important factor as more lines are crowded into common rights of way.

5.5 Sequence Networks

With the system assumed to be balanced or symmetrical to the point of unbalance or fault, the three sequence components are independent and do not react with each other. Thus, three network diagrams are required to separate the three sequence components for individual consideration: one for positive, one for negative, and one for zero sequence. These sequence network diagrams consist of one phase to neutral of the power system, showing all the component parts relevant to the problem under consideration. Typical diagrams are illustrated in Figures 2-24 through 2-26.

The positive sequence network (Fig. 2-24) must show both the generator voltages and impedances of the generators, transformers, and lines. Balanced loads may be shown from any bus to the neutral bus. Generally, however, balanced loads are neglected. Compared to the system low-impedance high-angle quantities, they have a much higher impedance at a

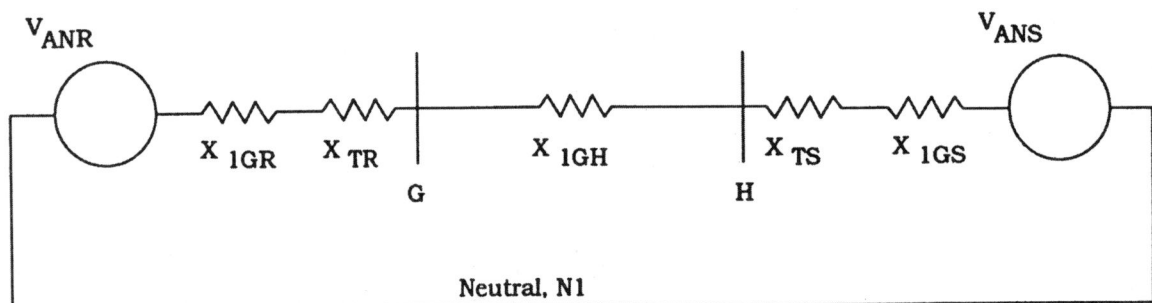

Figure 2-24 Example system and positive sequence network.

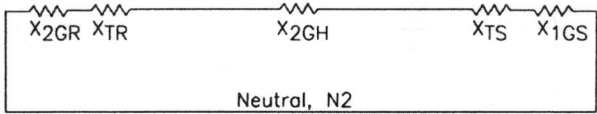

Figure 2-25 Negative sequence network for example system.

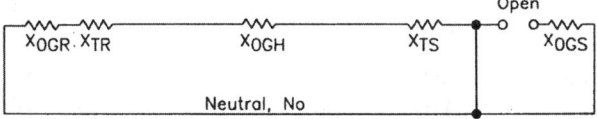

Figure 2-26 Zero sequence network for example system.

very low angle. In short, balanced loads complicate the calculations and generally do not affect the fault currents significantly.

With two exceptions, the negative sequence network (Fig. 2-25) will be a duplicate of the positive sequence network: (1) There will be no generator voltages, since synchronous machines generate a positive sequence only, and (2) the negative sequence reactance of synchronous machinery may be different from the positive, as previously described. For all practical calculations involving faults or discontinuities remote from the generating plant, however, X_1 is assumed to be equal to X_2.

The zero sequence network (Fig. 2-26) is quite different from the other two. First of all, it has no voltage: Rotating machinery does not produce zero sequence voltage. Also, the transformer connections require special consideration and grounding impedances must be included. Figure 2-23 shows the zero sequence circuits for many transformers.

A three-line system diagram is usually not required to determine the zero sequence network, but if a question arises as to the flow of zero sequence currents, the three-line diagram can be useful. From this three-phase system diagram, the zero sequence network requirements can be resolved by determining whether or not equal and in-phase currents can exist in each of the three phases. If the zero sequence current component can flow, the zero sequence network must reflect its path.

For simplicity, Figure 2-27 shows the generators solidly grounded. In practice, however, solid grounding is used only in very special cases.

5.6 Sequence Network Connections and Voltages

The current flow direction and voltage connections illustrated in Figure 2-28 must be followed for Eqs. (2-29), (2-30), and (2-31) to apply. Current reference direction in any circuit element must be the same in all three networks to avoid confusion. Current flow in one or more of the networks may reverse for some types of unbalances, particularly if the networks are complex. Reverse flow should be treated as a negative current to ensure that it will be properly subtracted when determining the phase currents.

Each sequence network is, of course, a per unit diagram representing one of the three phases of the symmetrical power system. Therefore, a resistor (reactor, impedance) connected between the system neutral and ground, as shown in Figure 2-28, must be multiplied by 3 as indicated. In the system, $3I_0$ flows through R; in the zero sequence network, however, I_0 flows through 3R, producing an equivalent voltage drop.

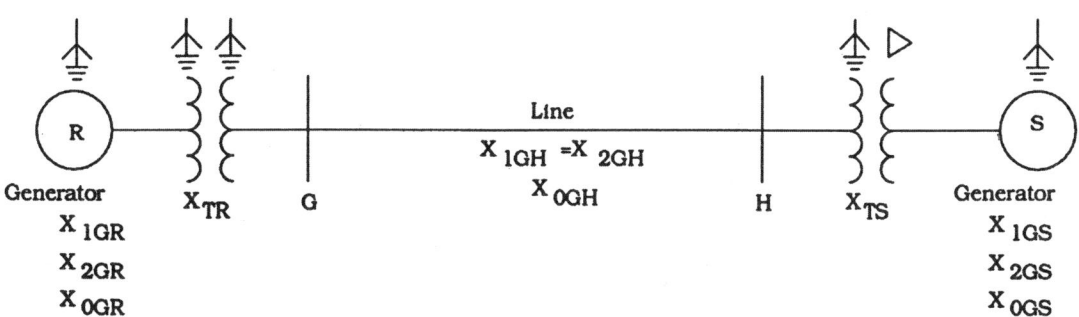

Figure 2-27 Example system (generators shown solidly grounded for simplification).

In the positive sequence network, the voltage drop at any point in the network is:

$$V_1 = V_{an} - \Sigma I_1 Z_1$$

Where $\Sigma I_1 Z_1$ is the phasor sum of the $I_1 Z_1$ drops from the neutral or zero potential bus (N_1) to the point where the voltage is to be determined.

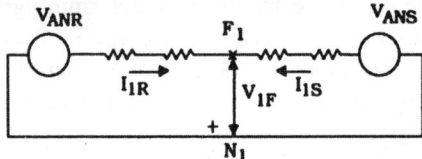

In the negative sequence network, the voltage drop at any point in the network is:

$$V_2 = 0 - \Sigma I_2 Z_2$$

Where $\Sigma I_2 Z_2$ is the phasor sum of the $I_2 Z_2$ drops from the neutral or zero potential bus (N_2) to the point where the voltage is to be determined.

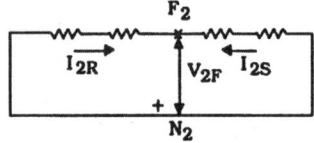

In the zero sequence network, the voltage drop at any point in the network is:

$$V_0 = -\Sigma I_0 Z_0$$

Where $\Sigma I_0 Z_0$ is the phasor sum of the $I_0 Z_0$ drops from the zero potential bus (N_0) to the point where the voltage is to be determined.

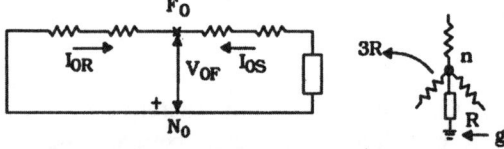

Figure 2-28 Sequence network connections and voltages.

5.7 Network Connections for Fault and General Unbalances

The sequence networks can be interconnected at a point of discontinuity, such as a fault. In such areas, negative and zero sequence voltages are generated, as previously described. Sequence network connections for various types of common faults are shown in Figures 2-29 through 2-32. From the three-phase diagrams of the fault area, the sequence network connections representing the fault can be derived. These diagrams do not show fault impedance, and fault studies do not include this effect except in very rare cases. The single-sequence impedance Z_1, Z_2, Z_0 (practically equivalent to X_1, X_2, X_0) shown in the

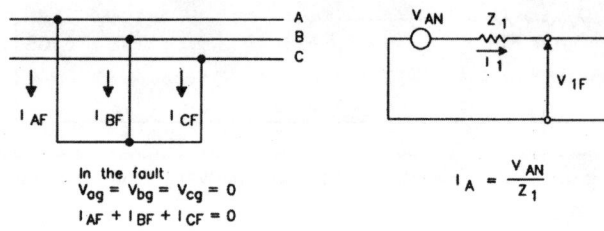

Figure 2-29 Three-phase fault and its network connection.

figures is the net impedance between the neutral bus and selected fault location. Based on zero load, all generated voltages (V_{AN}) are equal and in phase.

Since the three-phase fault is balanced, symmetrical components are not required for this calculation. However, since the positive sequence network represents the system, the network can be connected as shown in Figure 2-29 to represent the fault.

For a phase-a-to-ground fault, the three networks are connected in series (Fig. 2-30). Figure 2-31 illustrates a phase-b-c-to-ground fault and its sequence network interconnection. The phase-b-to-phase-c fault and its sequence connections are shown in Figure 2-32.

Fault studies normally include only three-phase faults and single-phase-to-ground faults. Three-phase faults are the most severe phase faults, whereas single-phase-to-ground faults are the most common. Studies of the latter faults provide useful information for ground relaying.

A fundamental study of both series and shunt unbalances was made by E. L. Harder in 1937. The shunt unbalances summarized in Figure 2-33 are taken from Harder's study. Note that all the faults shown in Figures 2-29 through 2-32 are also represented in Figure 2-33.

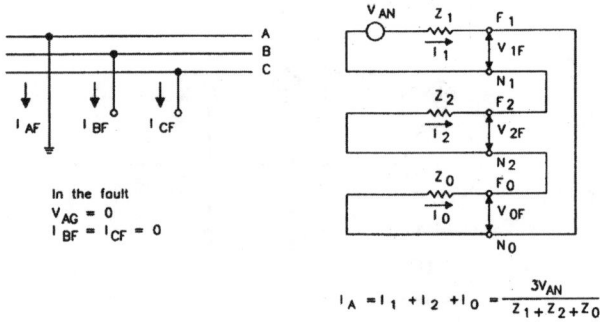

Figure 2-30 Phase-to-ground fault and its sequence network connections.

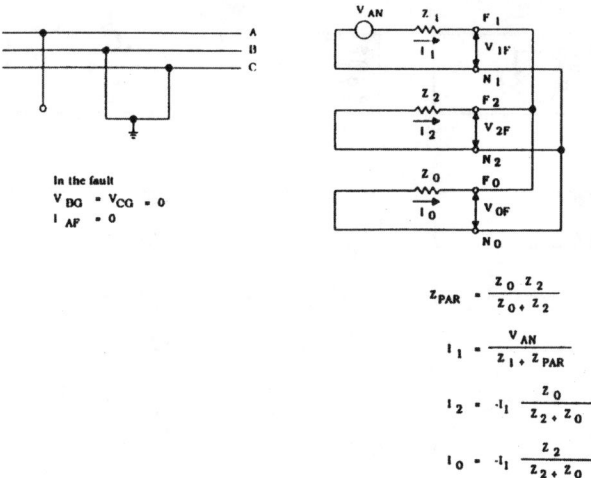

Figure 2-31 Double phase-to-ground fault and its sequence network connections.

In Figure 2-33, the entire symmetrical power system up to a point x of the shunt connection is represented by a rectangular box. Inside the topmost box for each shunt condition is a four-line representation of the shunt to be connected to the system at point x. The three lower boxes for each shunt condition are the positive, negative, and zero sequence representations of the shunt.

The sequence connections for the series unbalances, such as open phases and unbalanced series impedances, are shown in Figure 2-34. As before, these diagrams are taken from E. L. Harder's study. Here again, the diagrams inside the topmost box for each series condition represent the area under study, from point x on the diagrams left to point y on the right. The power system represented by the box is open between x and y to insert the circuits shown inside the box. Points x and y can be any distance apart, as long as there is no tap or other system connection between them. The

positive, negative, and zero sequence interconnections for the discontinuity shown in the top box are illustrated in the three boxes below it.

Simultaneous faults require two sets of interconnections from either Figures 2-33 or 2-34 or both. As shown in Figure 2-35, ideal or perfect transformers can be used to isolate the two restrictions. Perfect transformers are 100% efficient and have ratios of 1:1, 1:a, 1:a^2.

It is sometimes necessary to use two transformers as shown in Figure 2-35f. In this case, the first transformer (ratios $1:\varepsilon^{-j30°}$, $1:\varepsilon^{j30°}$, and 1:1) represents the wye-delta transformer, and the second transformer (ratios 1:a^2, 1:a, 1:1) represents the b-to-neutral fault. These can be replaced by an equivalent transformer with ratios $1:\varepsilon^{-j150°}$, $1:\varepsilon^{j150°}$, and 1:1.

Figure 2-35a, for example, represents an open phase-a conductor with a simultaneous fault to ground on the x side. The sequence networks are connected for the open conductor according to Figure 2-34j, with three 1:1 perfect transformers providing the restrictions required by Figure 2-33f. The manual calculations required, which involve the solution of simultaneous equations, may be quite tedious.

5.8 Sequence Network Reduction

When manual calculations are performed, the complete system networks are reduced to the single impedance values of Figures 2-29 through 2-32.

To simplify this reduction, with negligible effect on the results, the following basic assumptions are sometimes made:

All generated voltages are equal and in phase.
All resistance is neglected, or the reactance of machines and transformers is added directly with line impedances.
All shunt reactances are neglected, including loads, charging, and magnetizing reactances.
All mutual reactances are neglected, except on parallel lines.

By using these assumptions, the positive sequence network can be drawn with a single-source voltage V_{an} connected to the generator impedances by a bus.

If voltages are different, either the voltages must be retained in the network or Thevenin theorem and superposition must be used to reduce the network and calculate fault currents and voltages. Note that for the series unbalances of Figure 2-34, a difference in

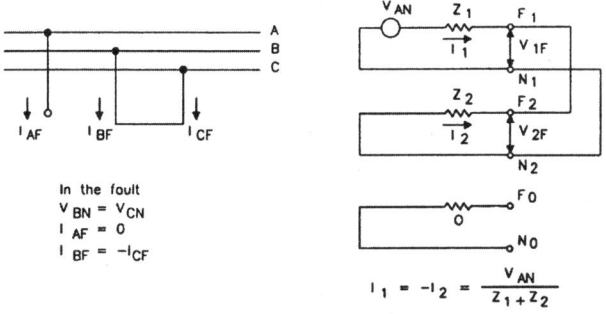

Figure 2-32 Phase-to-phase fault and its sequence network connections.

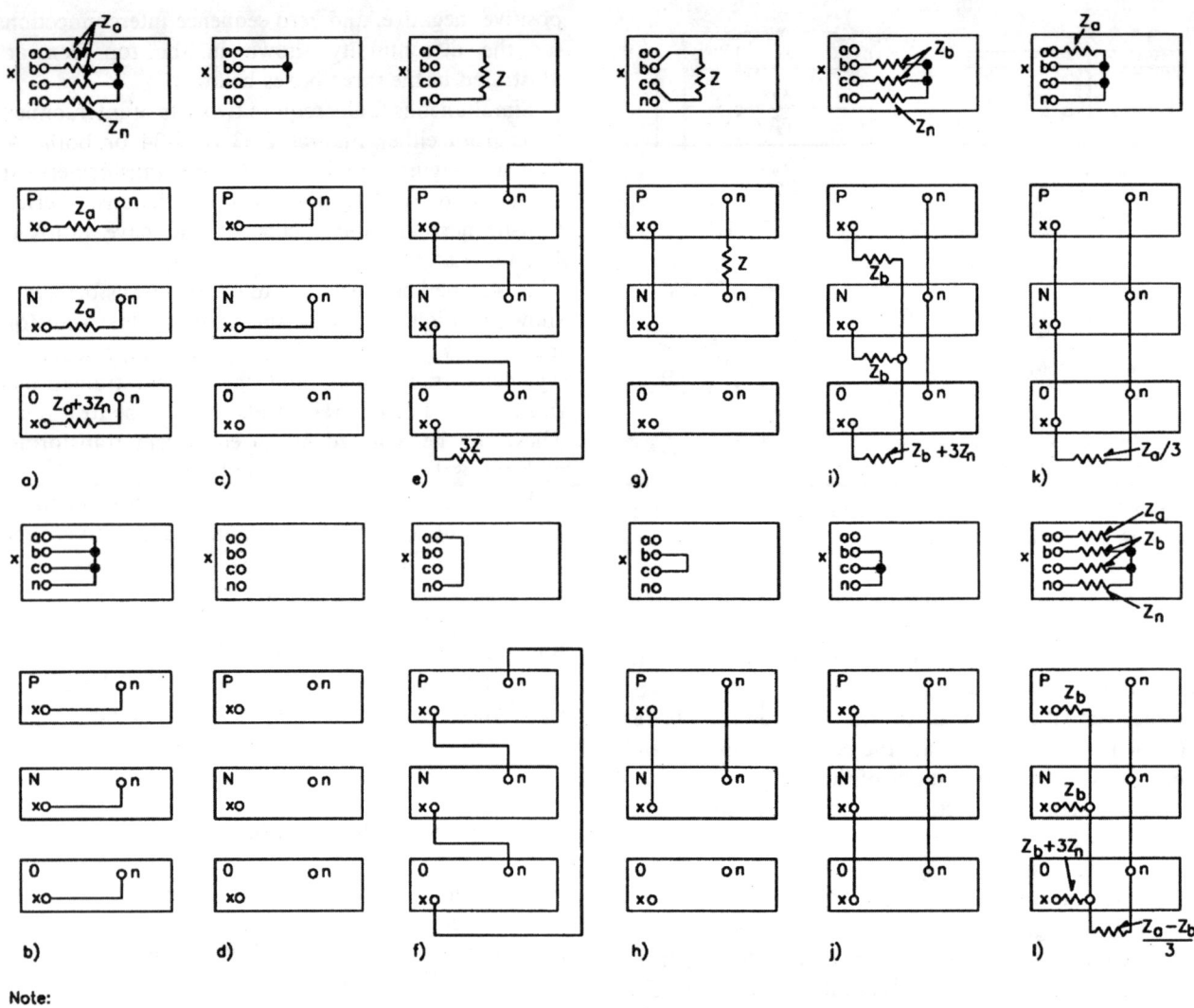

Figure 2-33 Sequence network interconnections for shunt balanced and unbalanced conditions.

Note:

a) Balanced load or three—line—to—ground fault with impedances.
b) A three—line—to—ground fault.
c) A three—phase fault.
d) A shunt circuit open.
e) A line—to—ground fault through an impedance.
f) A line—to—ground fault.

g) A line—to—line fault through impedance.
h) A line—to—line fault.
i) A two—line—to—ground fault with impedance.
j) A two—line—to—ground fault.
k) A three—line—to—ground fault with impedance in phase a.
l) Unbalanced load or three—line—to—ground fault with impedance.

voltage—either magnitude, phase angle, or both—is required for current to flow.

The single-sequence impedances Z_1, Z_2 and Z_0 of Figures 2-29 through 2-32 will be different for each fault location because of the different network reductions. During the network reduction, the distribution of currents in the various branches should be calculated, both as a check and to determine the current flow through the relays involved in a fault. These distribution factors are calculated with the assumption that 1

per unit current flows in these single-sequence impedances at the fault or point of discontinuity.

Network reduction calculations for the system of Figure 2-24 are illustrated in Figures 2-36, 2-37, and 2-38. In these figures, X_1, X_2, and X_0 are the impedances between the neutral bus and the fault at bus G. I_{1R}, I_{1L}, I_{2R}, I_{2L}, I_{0R}, I_{0L} are the per unit distribution factors. I_1, I_2, and I_0 are all assumed to be equal to 1 per unit.

Analog or digital studies should be tailored to produce outputs that allow each branch current in

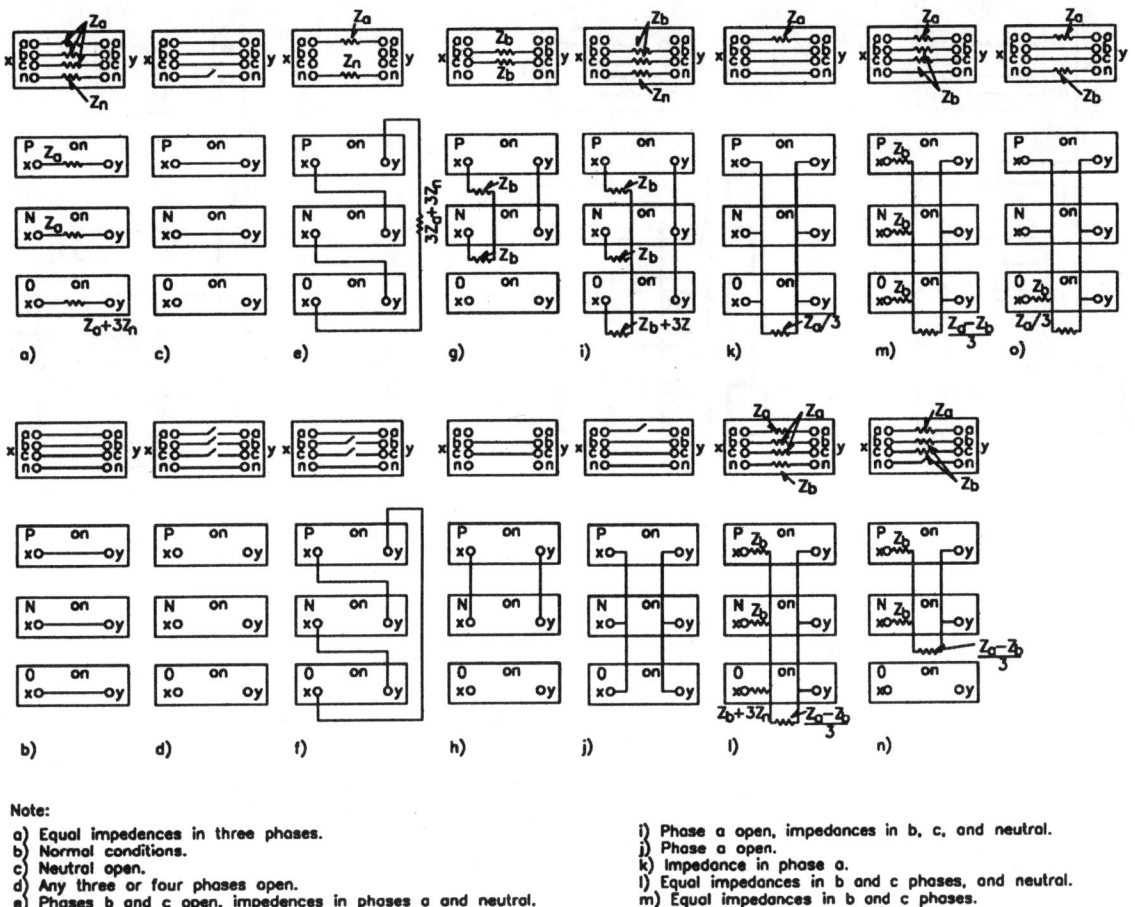

Figure 2-34 Sequence network interconnections for series balanced and unbalanced conditions.

Note:

a) Equal impedances in three phases.
b) Normal conditions.
c) Neutral open.
d) Any three or four phases open.
e) Phases b and c open, impedances in phases a and neutral.
f) Phases b and c open.
g) Phases a and neutral open, impedance in b and c.
h) Phases a and neutral open.

i) Phase a open, impedances in b, c, and neutral.
j) Phase a open.
k) Impedance in phase a.
l) Equal impedances in b and c phases, and neutral.
m) Equal impedances in b and c phases.
n) Equal impedances in b and c phases, neutral open.
o) Impedances in phase a and neutral.

each network to be identified. For single-phase-to-ground faults, $3I_0$ is required for relays.

When using the computer for sequence network reduction, the impedance data are input for the positive and zero sequence networks, along with bus and fault node points. The network is then solved for three-phase and single-phase-to-ground faults. Tabulated printed data are provided for phase-a fault current and three-phase fault voltages, along with the corresponding $3I_0$, $3V_0$ values for the phase-to-ground fault. I_2 and V_2 values should also be obtained for negative sequence relays.

These voltage and current values are needed for not only faults near the relay, but also those several buses or lines away. Among the operating conditions normally considered are maximum and minimum generation, selected lines out of service, and line-end faults where the adjacent breaker is open. This information allows the

correct relay types and settings to be selected in a minimal amount of time for the entire power system.

The following steps must be performed for calculating fault currents and voltages:

Obtain a complete single-line diagram for the entire system, including generators, transformers, and transmission lines, along with the positive, negative, and zero sequence impedances for each component.

Prepare a single-line impedance diagram from the system diagram or establish the nodes in a digital study for the positive, negative, and zero sequence networks.

Reduce the impedance values of all network branches to a common base. Values may be expressed as per unit on a common kVA base or as ohms impedance on a common voltage base.

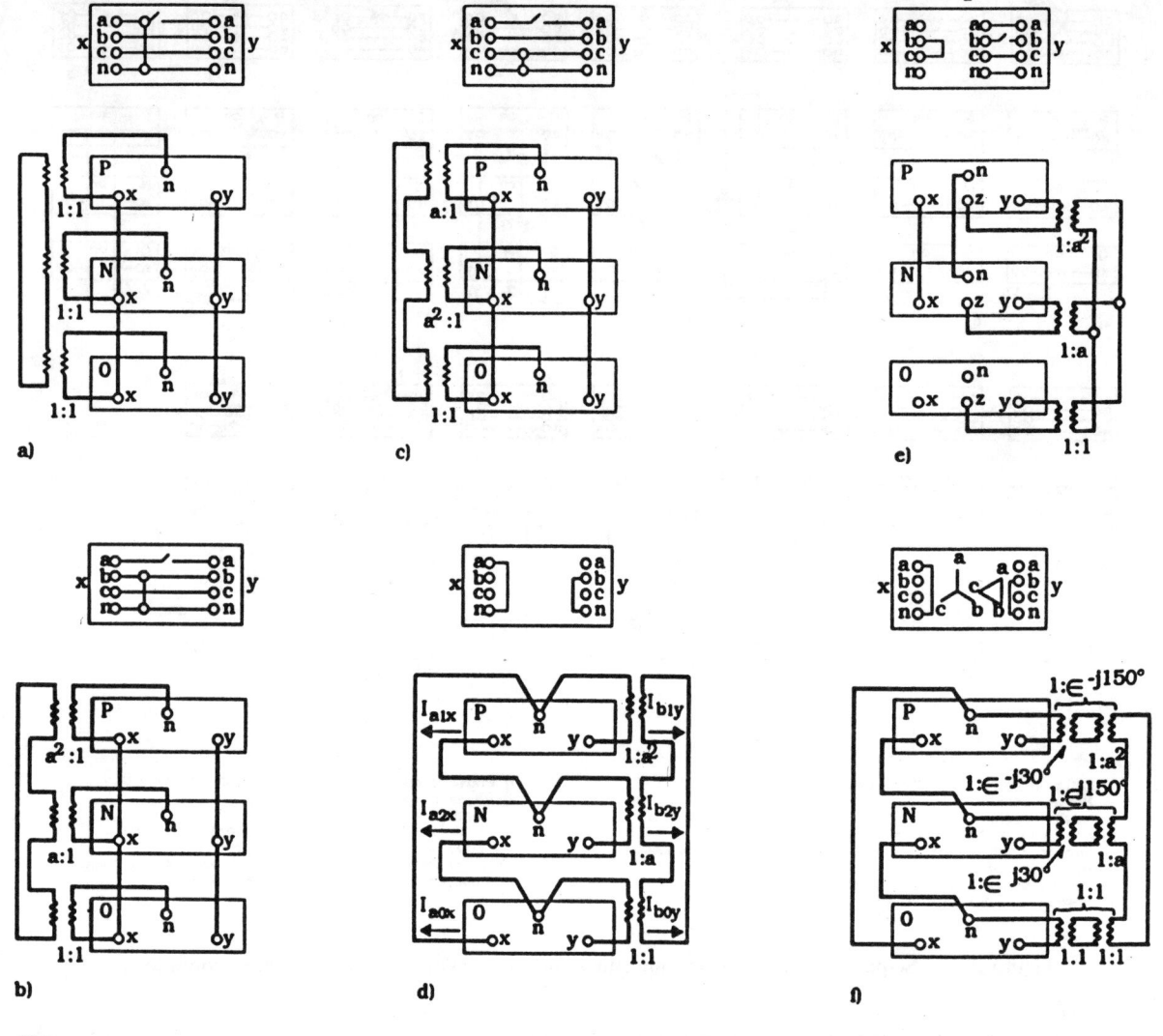

Figure 2-35 Representations for simultaneous unbalances.

Note:

a) One phase open and a fault to ground.
b) Phase a open and b-phase-to-ground fault.
c) Phase a open and c-phase-to-neutral fault.

d) Phase a-to-ground fault at x and a b-phase-to-ground fault at y.
e) A b-to-c fault at x, and b phase open z to y.
f) Phase a-to-neutral fault at x, phase b-to-neutral fault on other side of star-delta transformer bank at y x is taken as the reference point.

Obtain, or have the computer obtain, the equivalent single impedance of each sequence network, current distribution factors, and equivalent source voltage for the positive-phase sequence network. All quantities must be referred to the proper base.

Interconnect the networks or utilize the computer program to represent the fault type involved, and calculate the total fault current at the fault.

Determine the current distribution and voltages as required in the system. Total fault current is

seldom of use as relays generally see a fraction of that current except for radial circuits.

5.9   Example of Fault Calculation on a Loop-Type Power System

For the typical loop system shown in Figure 2-39, the generator units at stations D, S, or E could each be combinations of several machines. Alternatively, they

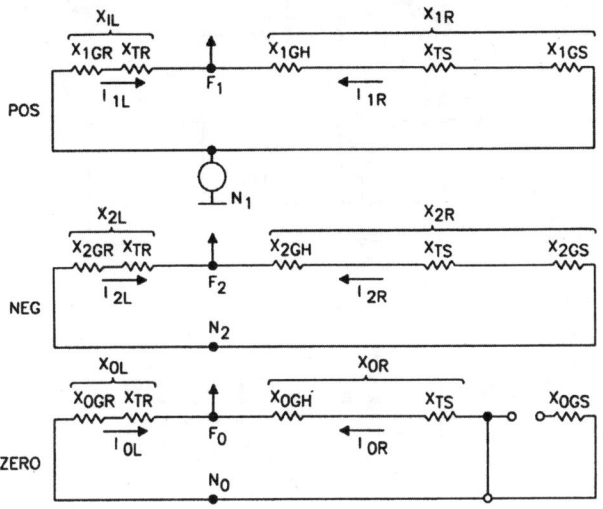

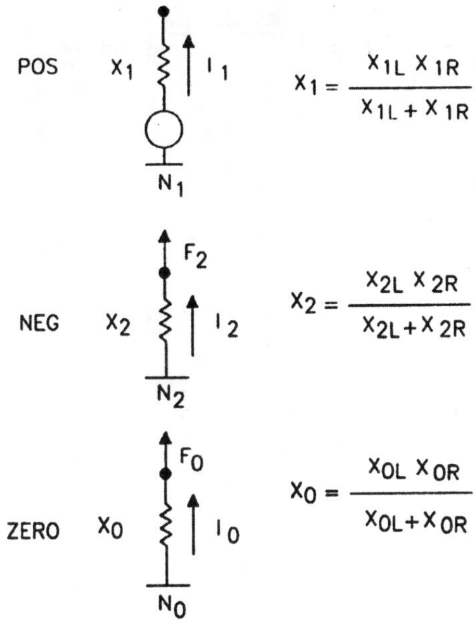

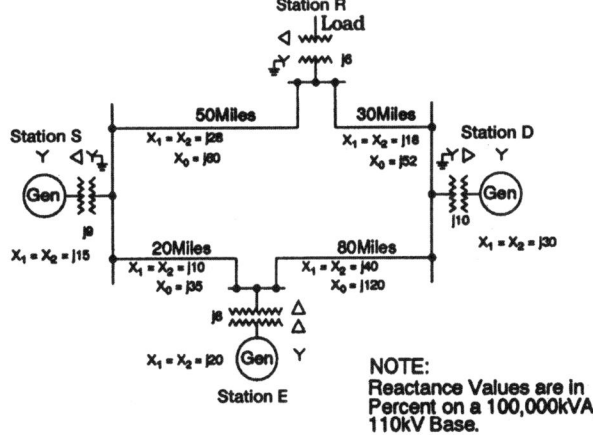

Figure 2-38 Final network reduction for fault at bus G in Figure 2-24.

Figure 2-36 Network reduction for example system (Figure 2-24) fault at bus G.

loops, at least one must be converted to wye-equivalent in order to reduce the networks. After one loop is chosen (arbitrarily), the equivalent X, Y, Z branches for an equivalent wye are dotted in as shown in Figures 2-40 and 2-41.

The X, Y, Z conversion from delta to wye-equivalent is a simple process: The X branch of the wye-equivalent is the product of the two delta reactances on either side divided by the sum of the three delta impedances. The same relation applies to

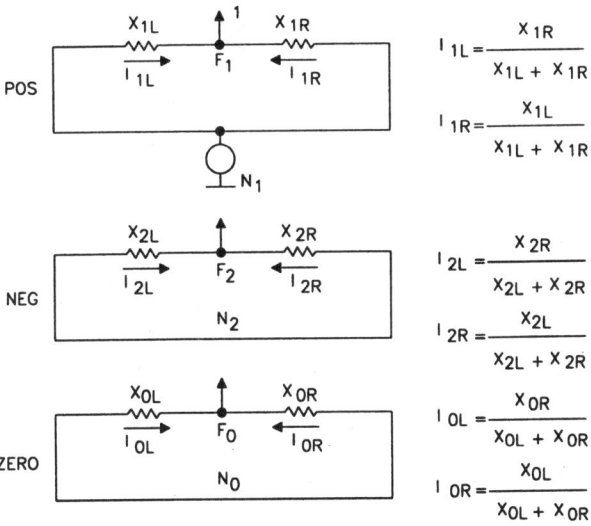

Figure 2-37 Network reduction and current distribution.

could represent the equivalent of a complex system up to the bus. All the impedances have been reduced to a common base, as indicated in the diagram. The positive sequence network for this system is shown in Figure 2-40, the zero sequence network in Figure 2-41. The negative sequence network is equal to Figure 2-40, except that V_{an} is not present.

To perform this sample calculation of a phase-to-ground fault on the bus at station D, the networks must be reduced to a single reactance value between the neutral bus and fault point. Of the several delta

Figure 2-39 Single line diagram for a typical loop-type power system.

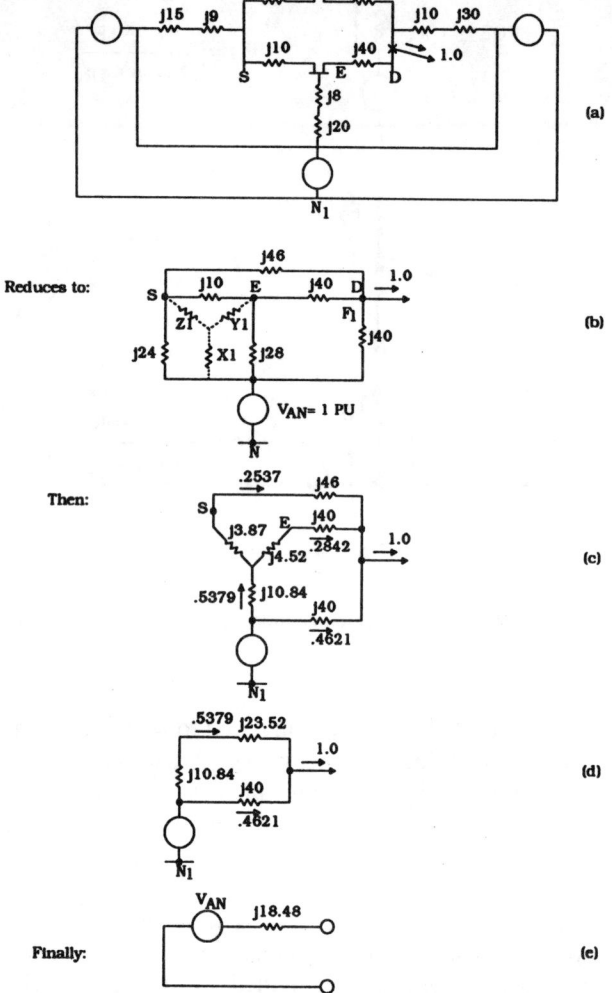

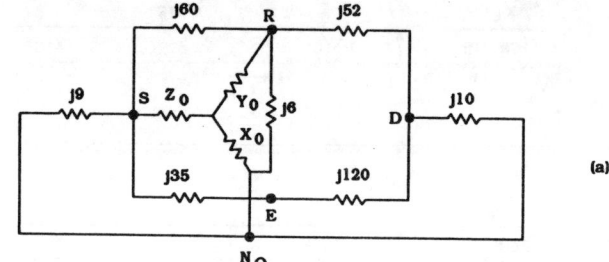

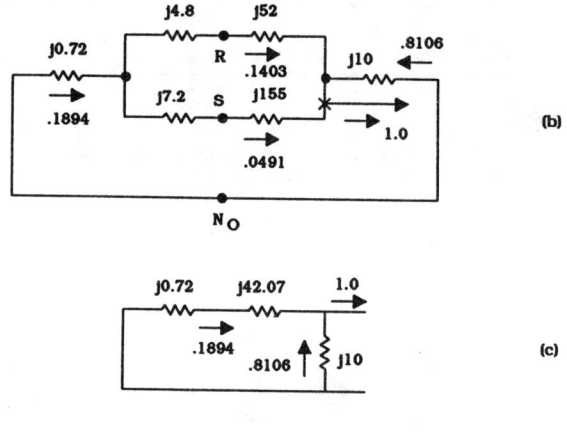

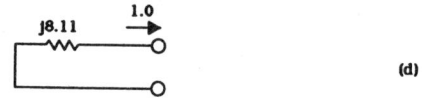

Figure 2-40 Positive sequence network reduction for the system of Figure 2-39.

Figure 2-41 Zero sequence network reduction for the system of Figure 2-39.

the Y and Z branches. Thus, in Figures 2-40 and 2-41, the networks are reduced as follows:

Positive and negative sequence networks	Zero sequence network
$X_1 = \dfrac{24 \times 28}{62} = j10.84$	$X_0 = \dfrac{9 \times 6}{75} = j0.72$
$Y_1 = \dfrac{28 \times 10}{62} = j4.52$	$Y_0 = \dfrac{6 \times 60}{75} = j4.8$
$Z_1 = \dfrac{24 \times 10}{62} = j3.87$	$Z_0 = \dfrac{9 \times 60}{75} = j7.2$

The networks now reduce to the simpler forms shown in Figure 2-40c. Since the two upper branches of each network are in parallel, they can be reduced as follows:

Positive and negative sequence networks	Zero sequence network
0.4716 0.5284	0.2594 0.7406
$\dfrac{44.52 \times 49.87}{94.39}$	$\dfrac{56.8 \times 162.2}{219.0}$
$= 23.52$	$= 42.07$

These reductions are shown in Figures 2-40d and 2-41c. The remaining branches are in parallel and can also be reduced:

Positive and negative sequence networks	Zero sequence network

$$X_1 = X_2 = \overset{0.4621 \quad 0.5379}{\frac{34.36 \times 40}{74.36}}$$

$$= j18.48\%$$

$$X_0 = \overset{0.8106 \quad 0.1894}{\frac{42.79 \times 10}{52.79}}$$

$$= j8.11\%$$

The numbers written above the equations are the distribution factors for the parallel circuits. These factors are expressed as the ratio of each term in the numerator and denominator. Determining these factors provides a convenient check on the calculations, since the sum of the two fractions must be 1.

Distribution factors can be determined by working back through the reduction. The factors should be written on the diagrams as shown in Figure 2-42.

The distribution factors for the upper parallel branches of Figure 2-40c are determined as follows:

Positive sequence network

44.52% branch : $0.5284 \times 0.5379 = 0.2842$

49.87% branch : $0.4716 \times 0.5379 = \underline{0.2537}$

$$0.5379 \text{ (check)}$$

The distribution factors in the zero sequence network are

Zero sequence network

162.2% branch : $0.2594 \times 0.1894 = 0.0491$

56.8% branch : $0.7406 \times 0.1895 = \underline{0.1403}$

$$0.1894 \text{ (check)}$$

In turn, these distribution factors are added to the diagram, as shown in Figure 2-42b.

The delta current distribution factors are obtained from the X, Y, Z equivalents. The conversion technique is straightforward: The voltage drop across two of the wye branches is equivalent to the drop across the delta branch. Calculating from Figure 2-40c, we obtain

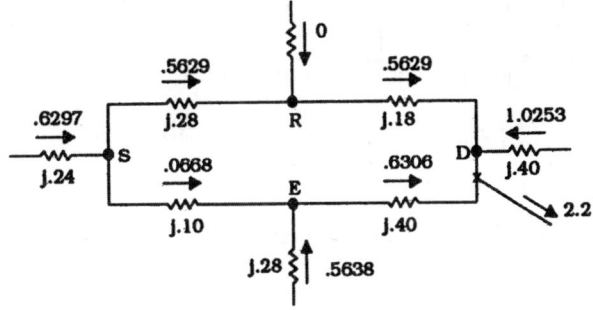

(a) Positive and Negative Sequence Current

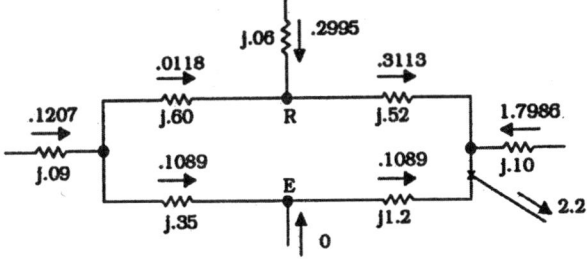

(b) Zero Sequence Current

Figure 2-42 Per unit current distribution for AG fault at D.

Positive sequence network

$$\frac{0.5379 \times j10.84 + 0.2842 \times j4.52}{j28} = 0.2541$$

$$\frac{0.5379 \times j10.84 + 0.2537 \times j3.87}{j24} = 0.2838$$

$$\frac{-0.2537 + j3.87 + 0.2842 \times j4.52}{j10} = 0.0301$$

Zero sequence network

$$\frac{0.1894 \times j0.72 + 0.1403 \times j4.52}{j6} = 0.1350$$

$$\frac{0.1894 \times j0.72 + 0.0491 \times j7.20}{j24} = 0.0544$$

$$\frac{-0.0491 + j7.20 + 0.1403 \times j4.80}{j60} = 0.0053$$

Figure 2-42 shows the complete per unit distribution for the original network of Figure 2-39.

The three networks are connected in series for the phase-to-ground fault (Fig. 2-28). For convenience, the sequence currents are calculated in per unit values:

$I_1 = I_2 = I_0$

$$= \frac{j1.0}{j0.1848 + j0.1848 + j0.0811}$$

$$= \frac{1.0}{0.4507}$$

$$= 2.22 \text{ p.u.}$$

The 100% (1 p.u. base) current is

$$I_B = \frac{\text{kVA base}}{\sqrt{3} \text{ kV}}$$

$$= \frac{100,000}{\sqrt{3} \times 110}$$

$$= 524.86 \text{ A at } 110 \text{ kV}$$

$I_1 = I_2 = I_0$

$$= 2.22 \times 524.86$$

$$= 1164.55 \text{ A at } 110 \text{ kV}$$

The current flowing in each branch of the networks can now be determined by multiplying the actual fault current by the distribution factor. These currents may be expressed in either per unit or ampere values. Currents in the fault are calculated for each phase as follows:

$$I_a = 3I_1 = 3I_2 = 3I_0 = 6.66 \text{ p.u.}$$

or

$$I_a = 3493.66 \text{ A } 110 \text{ kV}$$

$$I_b = (a^2 I_1 + a I_2 + I_0)$$

$$= (-I_1 + I_0) = 0$$

$$I_c = (a I_1 + a^2 I_2 + I_0)$$

$$= (-I_1 + I_0) = 0$$

For each branch, the per unit positive, negative, and zero sequence currents can then be used to determine the individual phase currents by using Eqs. (2-23), (2-24), and (2-25). These are recorded in Figure 2-43.

Next, the sequence and phase voltages at each bus are determined as in Figure 2-28. It is convenient to calculate the voltages in per unit values. Note that the impedances listed in Figure 2-39 appear in percent, rather than ohms, and may be converted easily to per unit.

In the following calculations, the values in parentheses are volts, converted from the per unit values for the 110-kV system of Figure 2-39:

$$V_{\text{line-to-neutral}} = 1.0 \text{ p.u.}$$

$$= \frac{110,000 \text{ V}}{\sqrt{3}}$$

$$= 63,508.53 \text{ V}$$

From Figure 2-42, first the sequence and phase voltages are calculated at bus S:

$$V_1 = j1.0 - 0.6297 \times j0.24$$

$$= j1.0 - j0.1511$$

$$= j0.8489 \text{ p.u. } (53,912.39 \text{ V})$$

$$V_2 = 0 - 0.6297 \times j0.24$$

$$= -j0.1511 \text{ p.u. } (9596.14 \text{ V})$$

$$V_0 = 0 - 0.1207 \times j0.09$$

$$= -j0.0109 \text{ p.u. } (692.24 \text{ V})$$

$$V_{ag} = V_1 + V_2 + V_0$$

$$= j0.6869 \text{ p.u. } (43,624.01 \text{ V})$$

$$V_{bg} = a^2 V_1 + a V_2 + V_0$$

$$= 0.8489 \angle -30° + 0.1511 \angle +30° - j0.0109$$

$$= 0.7352 - j0.4245 + 0.1309 + j0.0756 - j0.0109$$

$$= 0.8661 - j0.3598$$

$$= 0.9379 \angle -22.56° \text{ p.u. } (59,594.65 \text{ V})$$

$$V_{cg} = a V_1 + a^2 V_2 + V_0$$

$$= 0.8489 \angle 210° + 0.1511 \angle 150° - j0.0109$$

$$= -0.7352 - j0.4245 - 0.1309 + j0.0756 - j0.0109$$

$$= -0.8661 - j0.3598$$

$$= 0.9379 \angle 202.56° \text{ p.u. } (59,564.65 \text{ V})$$

Next, the sequence and phase voltages are calculated at bus D, the fault location:

$$V_1 = j1.0 - 1.0253 \times j0.40$$

$$= j1.0 - j0.4101$$

$$= j0.5899 \text{ p.u. } (37,463.68 \text{ V})$$

$$V_2 = 0 - 1.0253 \times j0.40$$

$$= -j0.4101 \text{ p.u. } (26,044.85 \text{ V})$$

$$V_0 = 0 - 1.7986 \times j0.1$$

$$= -j0.1798 \text{ p.u. } (11,418.83 \text{ V})$$

$$V_{ag} = 0$$

$$V_{bg} = 0.5899 \angle -30° + 0.4101 \angle 30° - j0.1798$$

$$= 0.5109 - j0.2950 + 0.3552 + j0.2051 - j0.1798$$

$$= 0.8661 - j0.2697$$

$$= 0.9071 \angle -17.30° \text{ p.u. } (57,608.59 \text{ V})$$

$$V_{cg} = 0.5899 \angle 210° + 0.4101 \angle 150° - j0.1798$$

$$= 0.5109 - j0.2950 - 0.3552 + j0.2051 - j0.1798$$

$$= -0.8661 - j0.2697$$

$$= 0.9071 \angle 197.30° \text{ p.u. } (57,608.59 \text{ V})$$

Similarly, the sequence and phase voltages can be

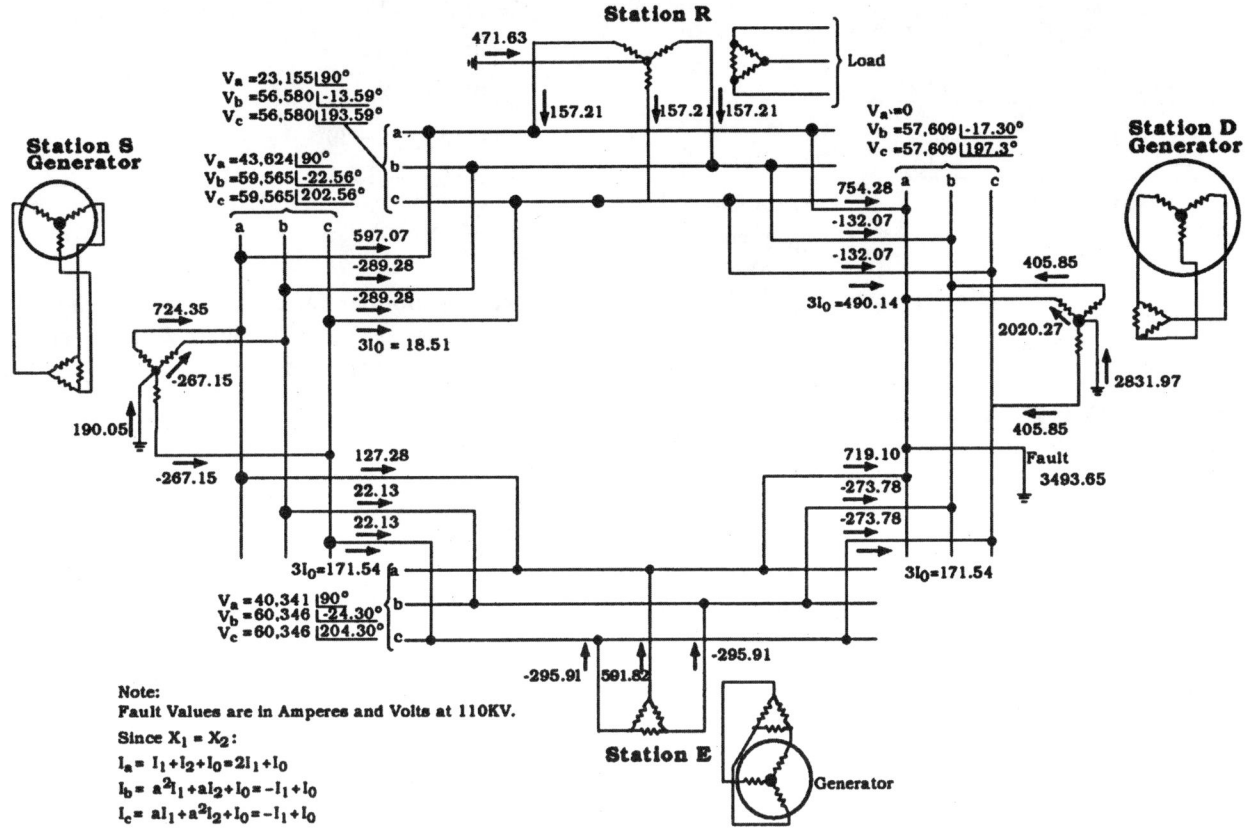

Figure 2-43 Current and voltage distribution for a single phase-to-ground fault at bus "D" of the system of Figure 2-37.

calculated at bus E:

$$V_{ag} = j0.6352 \text{ p.u. } (40,340.62\,\text{V})$$
$$V_{bg} = 0.9502\angle -24.30° \text{ p.u. } (60,345.80\,\text{V})$$
$$V_{cg} = 0.9502\angle 204.30° \text{ p.u. } (60,345.80\,\text{V})$$

Finally, the voltages are calculated at bus R:

$$V_{ag} = j0.3646 \text{ p.u. } (23,155.21\,\text{V})$$
$$V_{bg} = 0.8909\angle 13.59° \text{ p.u. } (56,579.75\,\text{V})$$
$$V_{cg} = 0.8909\angle 193.59° \text{ p.u. } (56,579.75\,\text{V})$$

The sequence voltages calculated above, as shown in Figure 2-43, complete the analysis of the single-phase-to-ground fault at bus D in the system of Figure 2-39. All the distributed current and voltage values for the system are displayed in Figure 2-43.

5.10 Phase Shifts Through Transformer Banks

In these fault calculations, the phase shifts through the wye-delta transformer banks were not considered. In this example, only a 110-kV system fault, with its currents and voltages, was involved. The effect of the phase shift through the transformer banks could not, however, have been neglected if currents and voltages were required for the opposite side of the power transformers.

If the transformer bank is wye-connected on the high-voltage side, as shown in Figure 2-44, the general equations for one phase are

$$I_A = n(I_a - I_c) \tag{2-30}$$
$$V_{an} = n(V_{An} - V_{Bn}) = nV_{AB} \tag{2-31}$$

The lowercase subscripts represent high-side quantities and the capital letter subscripts low-side quantities. In

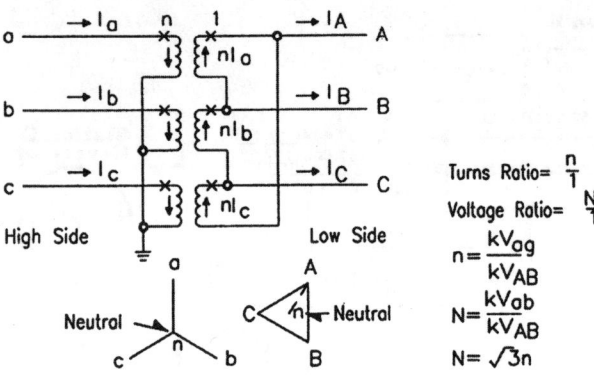

Figure 2-44 Connections and phasors for an ANSI standard power transformer bank with the wye connection on the high side (V_{an} leads V_{AN} by 30°).

the balanced or symmetrical transformer bank, the sequences are independent.

Consequently, positive sequence only is first applied to Eqs. (2-30) and (2-31):

$$I_{A1} = n(I_{a1} - I_{c1})$$
$$= n(I_{a1} - aI_{a1})$$
$$= n(1 - a)I_{a1}$$
$$= n\sqrt{3}I_{a1}\angle -30°$$
$$I_{A1} = NI_{a1}\angle -30° \tag{2-32}$$
$$I_{a1} = \frac{I_{A1}}{N}\angle 30°$$
$$V_{a1} = n(V_{A1} - V_{B1})$$
$$= n(V_{A1} - a^2V_{A1}) \tag{2-33}$$
$$= n(1 - a^2)V_{A1}$$
$$V_{a1} = n\sqrt{3}V_{A1}\angle 30°$$
$$V_{a1} = NV_{A1}\angle 30° \tag{2-34}$$
$$V_{A1} = \frac{V_{a1}}{N}\angle -30° \tag{2-35}$$

Next, only negative sequence quantities are applied to Eqs. (2-30) and (2-31):

$$I_{A2} = n(I_{a2} - I_{c2})$$
$$= n(I_{a2} - a^2I_{a2})$$
$$= n(1 - a^2)I_{a2}$$
$$= n\sqrt{3}I_{a2}\angle 30°$$
$$I_{A2} = NI_{a2}\angle 30° \tag{2-36}$$
$$I_{a2} = \frac{I_{A2}}{N}\angle -30° \tag{2-37}$$

$$V_{a2} = n(V_{A2} - V_{B2})$$
$$= n(V_{A2} - aV_{A2})$$
$$= n(1 - a)V_{A2}$$
$$= n\sqrt{3}V_{A2}\angle -30°$$
$$V_{a2} = NV_{A2}\angle -30°$$
$$V_{A2} = \frac{V_{a2}}{N}\angle 30°$$

If a power transformer bank is connected *delta* on the *high-voltage* side, as shown in Figure 2-45, the general equations for one phase are

$$I_a = \frac{1}{n}(I_A - I_B) \tag{2-38}$$

$$V_A = \frac{1}{n}(V_a - V_c) \tag{2-39}$$

Applying only positive sequence quantities to Eqs. (2-38) and (2-39),

$$I_{a1} = \frac{1}{n}(I_{A1} - I_{B1})$$
$$= \frac{1}{n}(I_{A1} - a^2I_{A1})$$
$$= \frac{1}{n}(1 - a^2)I_{A1}$$
$$= \frac{\sqrt{3}I_{A1}}{n}\angle 30°$$
$$I_{a1} = \frac{I_{A1}}{N}\angle 30° \tag{2-40}$$
$$I_{A1} = NI_{a1}\angle -30° \tag{2-41}$$

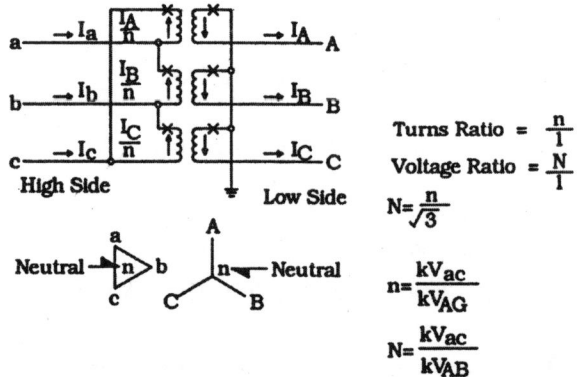

Figure 2-45 Connections and phasors for an ANSI standard power transformer bank with the delta connection on the high side (V_{an} leads V_{AN} by 30°).

$$V_{A1} = \frac{1}{n}(V_{a1} - V_{c1})$$

$$= \frac{1}{n}(V_{a1} - aV_{a1})$$

$$= \frac{1}{n}(1 - a)V_{a1}$$

$$= \frac{\sqrt{3}V_{a1}}{n} \angle -30°$$

$$V_{A1} = \frac{V_{a1}}{N} \angle -30° \qquad (2\text{-}42)$$

$$V_{a1} = NV_{A1} \angle 30° \qquad (2\text{-}43)$$

Then, applying only negative sequence quantities to Eqs. (2-38) and (2-39), we obtain

$$I_{a2} = \frac{1}{n}(I_{A2} - I_{B2})$$

$$= \frac{1}{n}(I_{A2} - aI_{A2})$$

$$= \frac{1}{n}(1 - a)I_{A2}$$

$$= \frac{\sqrt{3}I_{A2}}{n} \angle -30°$$

$$I_{a2} = \frac{I_{A2}}{N} \angle -30° \qquad (2\text{-}44)$$

$$I_{A2} = NI_{a2} \angle 30° \qquad (2\text{-}45)$$

$$V_{A2} = \frac{1}{n}(V_{a2} - V_{c2})$$

$$= \frac{1}{n}(V_{a2} - a^2 V_{a2})$$

$$= \frac{1}{n}(1 - a^2)V_{a2}$$

$$= \frac{\sqrt{3}V_{a2}}{n} \angle 30°$$

$$V_{A2} = \frac{V_{a2}}{N} \angle 30° \qquad (2\text{-}46)$$

$$V_{a2} = NV_{A2} \angle -30° \qquad (2\text{-}47)$$

If the bank is connected according to ANSI standards, the formulas are the same and not dependent on whether the wye or the delta is on the high side. In either case, the positive sequence quantities are shifted 30° in one direction, while the negative sequence quantities are shifted 30° in the opposite direction. These relations for ANSI standard connections are summarized in Table 2-2. Zero sequence quantities are not affected by phase shift. These either pass directly through the bank or, more commonly, are blocked by the connections. Thus, in a wye-delta bank, zero sequence current and voltage on one side cannot pass through the bank to the other side.

Table 2-2 Phase Shift Relations for Power Transformer Banks

High side in terms of low side[a]	Low side in terms of high side[a]
$I_{a1} = \frac{I_{A1}}{N} \angle 30°$	$I_{A1} = NI_{a1} \angle -30°$
$V_{a1} = NV_{A1} \angle 30°$	$V_{A1} = \frac{V_{a1}}{N} \angle -30°$
$I_{a2} = \frac{I_{A2}}{N} \angle -30°$	$I_{A2} = NI_{a2} \angle 30°$
$V_{a2} = NV_{A2} \angle -30°$	$V_{A2} = \frac{V_{A2}}{N} \angle 30°$

[a]The lowercase subscripts represent high-side quantities, and the capital letter subscripts low-side quantities.

5.11 Fault Evaluations

The sample calculation of a phase-to-ground fault on a loop system (see Sec. 5.9) was made at no load; that is, before the fault all currents throughout the system were zero.

With a ground fault, current flows in not only the faulted phase "a," but also the unfaulted "b" and "c" phases. The positive and zero sequence distribution factors on any loop system will be different. Consequently, the positive, negative, and zero sequence currents will not add up to zero in the unfaulted phases. On a radial system (one with a source at one end only for both the positive and zero sequences), the three network distribution factors will all be equal to 1. For a phase-a-to-ground fault on these circuits, I_b equals I_c, which equals 0.

In practice, only $3I_0$ and related $3V_0, V_2$, and I_2 values would be recorded for a phase-to-ground fault. The phase currents and voltages shown in Figure 2-43 were provided for academic purposes.

The reason for showing $3I_0$, rather than the faulted phase current, can be seen from Figure 2-43. In most circuits, there is a significant difference between the I_a and $3I_0$ currents in any loop network. In a radial system, however, I_a is equal to $3I_0$ and ground relays operate on $3I_0$.

On phase-to-ground faults, the phase relays will receive current and may start to operate. Coordination between ground and phase relays is usually not necessary. The principal reason there are so few coordination problems is that phase relays must be set above load (5 A secondary), whereas ground relays are conventionally set at 0.5 to 1.0 A secondary. Since the ground relays are more sensitive, they will generally not miscoordinate with the phase relays. If higher ground settings are used, the likelihood of miscoordination is increased.

Under any fault condition, the total current flowing into the ground must equal the total current flowing up the neutrals. With an autotransformer, however, current can flow down the neutral. In this case, the fault current plus the autotransformer neutral current equals the current up the other transformer neutrals.

The convention that current flows up the neutral when current is flowing down into the earth at the fault has given rise to the idea that the grounded wye-delta transformer bank is a ground source, a source of zero sequence current. This long-established idea is not, in fact, correct. The fault is the true source. It is a

converter of positive sequence into negative sequence and, for ground faults, into zero sequence current.

This is illustrated by a voltage plot for various faults on a simple system (Fig. 2-46). For simplicity, assume Z_1 equals Z_2 equals Z_0. During faults, the voltage inside the generators does not change unless the fault persists long enough for the internal flux to change. No appreciable voltage change should occur in high- or medium-speed relaying.

For a solid three-phase fault, the voltage at the fault is zero. Therefore, high positive sequence-phase currents flow to produce the gradient shown in the plot of Figure 2-46. For a phase-to-phase fault, negative sequence voltage is produced by the fault itself. Negative sequence current, then, flows through-

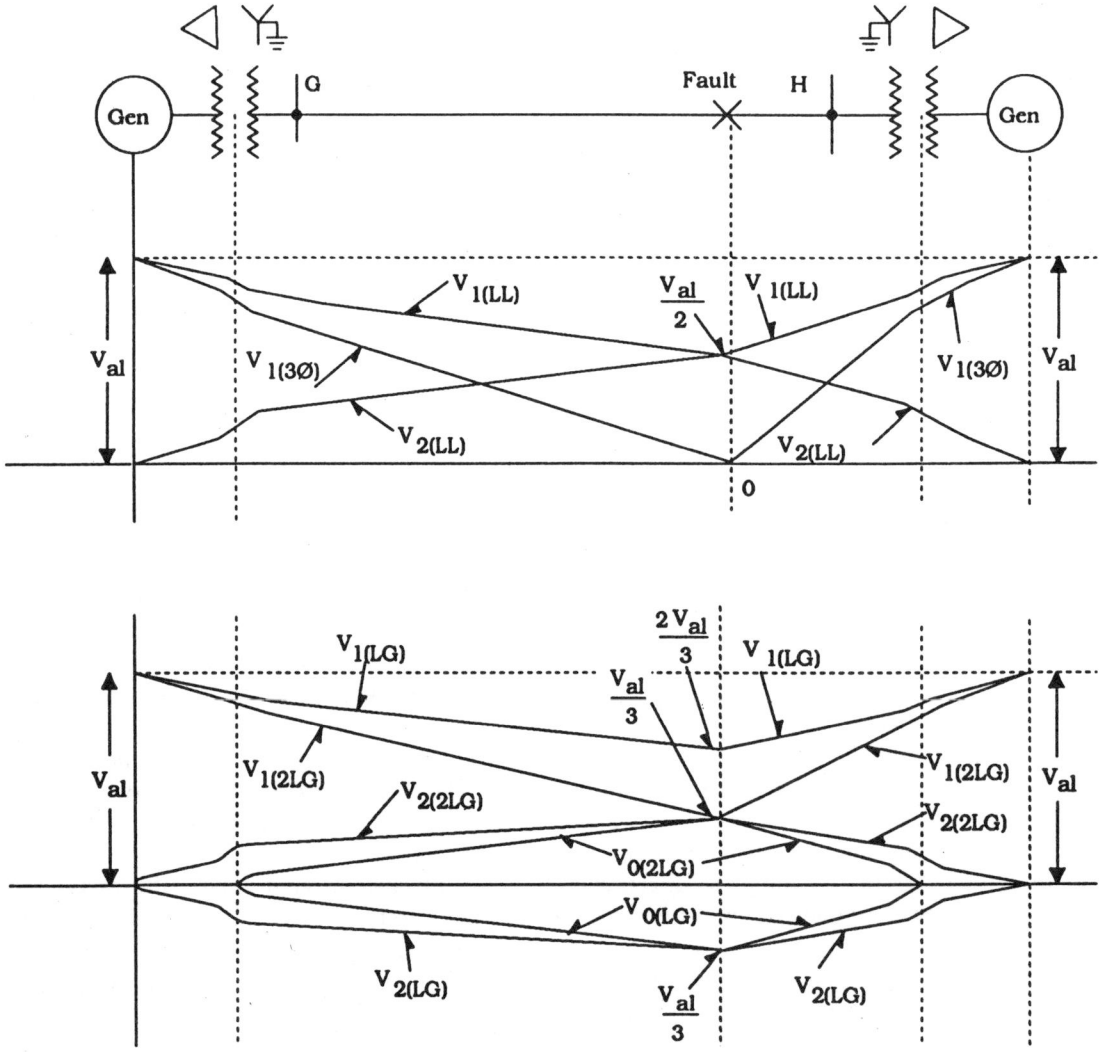

Figure 2-46 Voltage gradient for various types of faults.

out the system. The same general conditions also apply to phase-to-ground faults, except that since V_a is zero, V_2 and V_0 are negative.

In summary, the positive sequence voltage is always highest at the generators or sources and lowest at a fault. In contrast, negative and zero sequence voltages are always highest at the fault and lowest at the "sources."

The phasor diagrams of Figure 2-20 illustrate the same phenomena, from a different viewpoint. In a three-phase fault, the voltages collapse symmetrically, except inside the generator. The three currents have a large symmetrical increase and lagging shift of angle.

Other phase faults shown in Figure 2-20 are characterized by the relative collapse of two of the phase-to-neutral voltages, compared to the relatively

normal third phase-to-neutral voltage. Two of the phase currents have a large lagging increase.

For a single-phase-to-ground fault, on the other hand, one phase-to-neutral voltage is collapsed relative to the other two phases. Similarly, one phase current has a large value and lags the line-to-ground voltage.

With wye-delta transformers between the fault and measurement point, the positive sequence quantities shift 30° in one direction, and the negative sequence quantities shift 30° in the opposite direction. As a result, a phase-to-ground fault on the wye side of a bank has the appearance of a phase-to-phase fault on the delta side.

Figures 2-47 and 2-48 offer a final look at sequence currents and voltages for faults. Note that the positive sequence currents and voltages, shown in the left-hand columns, have approximately the same phase relations

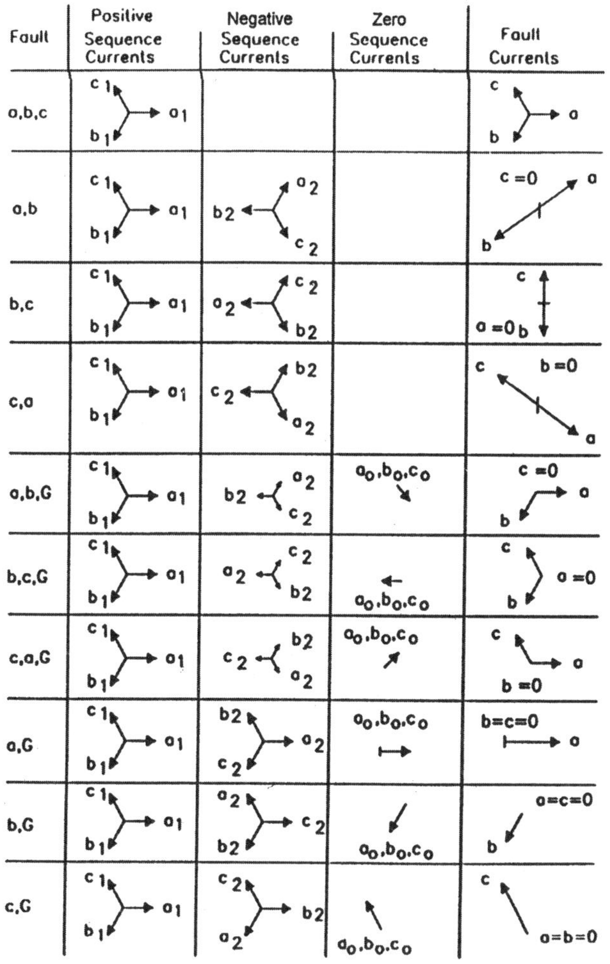

Fault	Positive Sequence Currents	Negative Sequence Currents	Zero Sequence Currents	Fault Currents
a,b,c				
a,b				
b,c				
c,a				
a,b,G				
b,c,G				
c,a,G				
a,G				
b,G				
c,G				

Figure 2-47 Sequence currents for various faults. Assumes $Z_1 = Z_2 = Z_0$.

Fault	Positive Sequence Voltages	Negative Sequence Voltages	Zero Sequence Voltages	Voltages at Fault
a,b,c				
a,b				
b,c				
c,a				
a,b,G				
b,c,G				
c,a,G				
a,G				
b,G				
c,G				

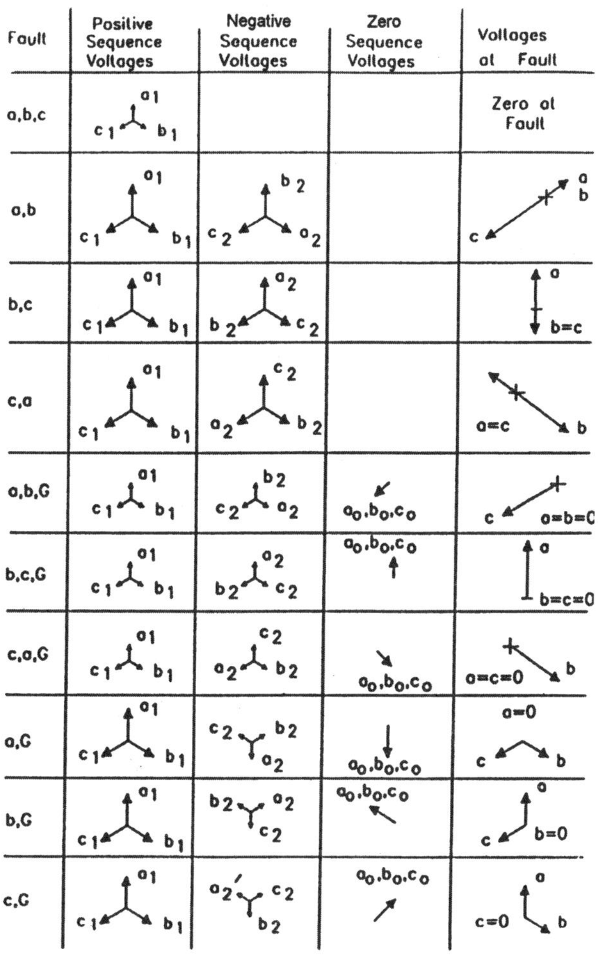

Figure 2-48 Sequence voltages for various faults. Assumes $Z_1 = Z_2 = Z_0$.

for all types of faults. At the fault are various nonsymmetrical currents and voltages, as shown in the far right-hand column. The negative and, sometimes, the zero sequence quantities provide the transition between the symmetrical left-hand column and nonsymmetrical right-hand column. These quantities rotate and change to produce the nonsymmetrical, or unbalanced, quantity when added to the positive sequence.

These phasors can be constructed easily by remembering which fault quantity should be minimum or maximum. In a phase c-a fault, for example, phase-b current will be small. Thus, I_{b2} will tend to be opposite I_{b1}. Since phase-b voltage will be relatively uncollapsed, V_{b1} and V_{b2} will tend to be in phase. After one sequence phasor is established, the others can be derived from Eq. (2-12) and Figure 2-21.

6   SYMMETRICAL COMPONENTS AND RELAYING

Since ground relays operate from zero sequence quantities, all ground relay types use symmetrical components. A number of other protective relays use combinations of the sequence quantities, as summarized in Table 2-3.

A zero sequence ($3I_0$) current filter is obtained by connecting three current transformers in parallel. A zero sequence ($3V_0$) voltage filter is provided by the wye-grounded-broken-delta connection for a voltage transformer or an auxiliary. Positive and negative sequence current and voltage filters are described in Chapter 3.

Table 2-3 Protective Relays Using Symmetrical Component Quantities for Their Operation

Device no.	Application	Sequence quantities used
50N, 51N	Ground overcurrent	I_0
59N	Ground voltage	V_0
67N	Ground directional overcurrent	I_0 with I_0 or V_0I_0 or V_2I_2
32N	Ground product overcurrent	I_0^2 or I_0, V_0
21N	Ground distance	I_0, V_0
		I_0, V_0, $V_1 + V_2$
87	Phase and ground pilot	$K_1I_1 + K_2I_2 + K_0I_0$
46	Phase unbalance voltage	V_2
46	Phase unbalance current	I_2
	Blown fuse detection	V_0 and not I_0

3

Basic Relay Units

Revised by: **W. A. ELMORE**

1 INTRODUCTION

Protective relays for power systems are made up of one or more fault-detecting or decision units, along with any necessary logic networks and auxiliary units. Because a number of these fault-detecting or decision units are used in a variety of relays, they are called basic units. Basic units fall into several categories: electromechanical units, sequence networks, solid-state units, integrated circuits, and microprocessor architecture. Combinations of units are then used to form basic logic circuits applicable to protective relays.

2 ELECTROMECHANICAL UNITS

Four types of electromechanical units are widely used: magnetic attraction, magnetic induction, D'Arsonval, and thermal units.

2.1 Magnetic Attraction Units

Three types of magnetic attraction units are in common use: plunger (solenoid), clapper, and polar. The plunger unit, shown in Figure 3-1, is typically used in SC, SV, and ITH relays; the clapper-type unit (Fig. 3-2) in SG, AR, ICS, IIT, and MG relays; and the polar-type unit (Fig. 3-3) in HCB, HU, and PM-type relays.

2.1.1 Plunger Units

Plunger units have cylindrical coils with an external magnetic structure and a center plunger. When the current or voltage applied to the coil exceeds the pickup value, the plunger moves upward to operate a set of contacts. The force F which moves the plunger is proportional to the square of the current in the coil.

The plunger unit's operating characteristics are largely determined by the plunger shape, internal core, magnetic structure, coil design, and magnetic shunts. Plunger units are instantaneous in that no delay is purposely introduced. Typical operating times are 5 to 50 msec, with the longer times occurring near the threshold values of pickup.

The unit shown in Figure 3-1a is used as a high-dropout instantaneous overcurrent unit. The steel plunger floats in an air gap provided by a nonmagnetic ring in the center of the magnetic core. When the coil is energized, the plunger assembly moves upward, carrying a silver disk that bridges three stationary contacts (only two are shown). A helical spring absorbs the ac plunger vibrations, producing good contact action. The air gap provides a ratio of dropout to pickup of 90% or greater over a two-to-one pickup range. The pickup range can be varied from a two-to-one to a four-to-one range by the adjusting core screw. When the pickup range is increased to four to one, the dropout ratio will decrease to approximately 45%.

The more complex plunger unit shown in Figure 3-1b is used as an instantaneous overcurrent or voltage unit. An adjustable flux shunt permits more precise settings over the nominal four-to-one pickup range. This unit is relatively independent of frequency, operating on dc, 25-Hz, or nominal 60-Hz frequency. It is available in high- and low-dropout versions.

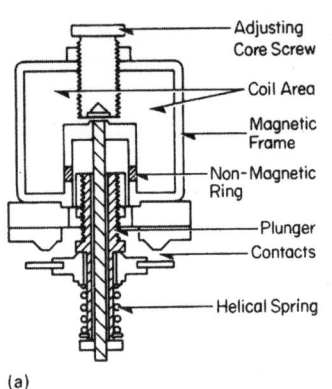
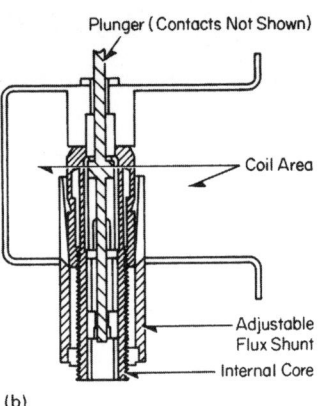

Figure 3-1 Plunger-type units.

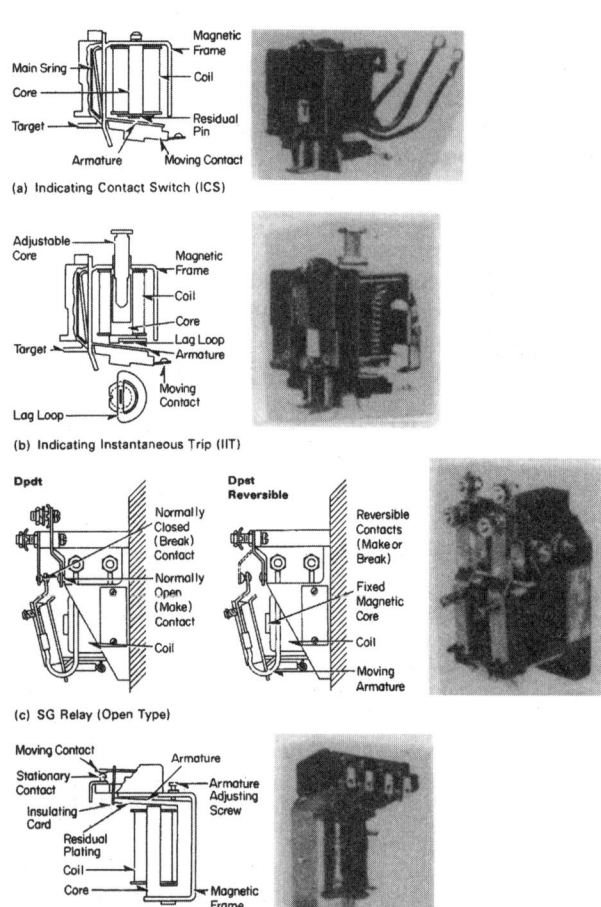

Figure 3-2 Four clapper units.

2.1.2 Clapper Units

Clapper units have a U-shaped magnetic frame with a movable armature across the open end. The armature is hinged at one side and spring-restrained at the other. When the associated electrical coil is energized, the armature moves toward the magnetic core, opening or closing a set of contacts with a torque proportional to the square of the coil current. The pickup and dropout values of clapper units are less accurate than those of plunger units. Clapper units are primarily applied as auxiliary or go/no-go units.

Four clapper units are shown in Figure 3-2. Those illustrated in Figures 3-2a and 3-2b have the same general design, but the first is for dc service and the second for ac operation. In both units, upward movement of the armature releases a target, which drops to provide a visual indication of operation (the target must be reset manually). The dc ICS unit (Fig. 3-2a) is commonly used to provide a seal-in around the main protective relay contacts. The ac IIT unit (Fig. 3-2b) operates as an instantaneous overcurrent or instantaneous trip unit. It is equipped with a lag-loop to smooth the force variations due to the alternating current input. Its adjustable core provides pickup adjustment over a nominal four-to-one range.

The SG (Fig. 3-2c) and MG clapper units provide a wide range of contact multiplier auxiliaries: The SG has provisions for four contacts (two make and two break), and the MG will accept six. The AR clapper unit (Fig. 3-2d) operates in 2 to 4 msec, with four contacts suitable for breaker tripping.

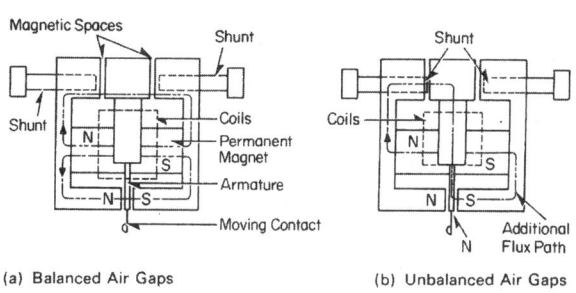

(a) Balanced Air Gaps   (b) Unbalanced Air Gaps

Figure 3-3 Polar-type unit

2.1.3 Polar Units

Polar units (Fig. 3-3) operate from direct current applied to a coil wound around the hinged armature in the center of the magnetic structure. A permanent magnet across the structure polarizes the armature-gap poles, as shown. The nonmagnetic spacers, located at the rear of the magnetic frame, are bridged by two adjustable magnetic shunts. This arrangement enables the magnetic flux paths to be adjusted for pickup and contact action. With balanced air gaps (Fig. 3-3a), the flux paths are as shown and the armature will float in the center with the coil deenergized. With the gaps unbalanced (Fig. 3-3b), some of the flux is shunted through the armature. The resulting polarization holds the armature against one pole with the coil deenergized. The coil is arranged so that its magnetic axis is in line with the armature and at a right angle to the permanent magnet axis. Current in the coil magnetizes the armature either north or south, increasing or decreasing any prior polarization of the armature. If, as shown in Figure 3-3b, the magnetic shunt adjustment normally makes the armature a north pole, it will move to the right. Direct current in the operating coil, which tends to make the contact end a south pole, will

overcome this tendency and the contact will move to the left. Depending on design and adjustments, this polarizing action can be gradual or quick. The left-gap adjustment (Fig. 3-3b) controls the pickup value, the right-gap adjustment the reset value. Some units use both an operating and a restraining coil on the armature. The polarity of the restraint coil tends to maintain the contacts in their initial position. Current of sufficient magnitude applied to the operating coil will provide a force to overcome the restraint, causing the contacts to change position. A combination of normally open or normally closed contacts is available. These polar units operate on alternating current through a full-wave rectifier and provide very sensitive, high-speed operation on very low energy levels.

The operating equation of the polar unit is

$$K_1 I_{op} - K_2 I_r = \frac{K_3}{\phi} \qquad (3\text{-}1)$$

where K_1 and K_2 are adjusted by the magnetic shunts; K_3 is a design constant; ϕ is the permanent magnetic flux; I_{op} is the operating current; and I_r is the restraint current in milliamperes.

2.2 Magnetic Induction Units

There are two general types of magnetic induction units: induction disc and cylinder units. The induction disc unit (Fig. 3-4) is typically used in CO, CV, CR, IRV, IRD, CW, CA, and CM relays. The cylinder unit (see Fig. 3-6) is most commonly used in KD-line, KC, KDXG, KF, KRD, KRC, and KRP relays.

2.2.1 Induction Disc Units

Originally, induction disc units were based on the watthour meter design. Modern units, however, although using the same operating principles are quite different. All operate by torque derived from the interaction of fluxes produced by an electromagnet with those from induced currents in the plane of a rotatable aluminum disc. The E unit in Figure 3-4a has three poles on one side of the disc and a common magnetic member or "keeper" on the opposite side. The main coil is on the center leg. Current I in the main coil produces flux, which passes through the air gap and disc to the keeper. (A small portion of the flux is shunted off through the side air gap.) The flux ϕ_T returns as ϕ_L through the left-hand leg and ϕ_R through the right-hand leg, where $\phi_T = \phi_L + \phi_R$. A short-circuited lagging coil on the left leg causes ϕ_L to lag both ϕ_R and ϕ_T, producing a split-phase motor action. (The phasors are shown in Figure 3-5.)

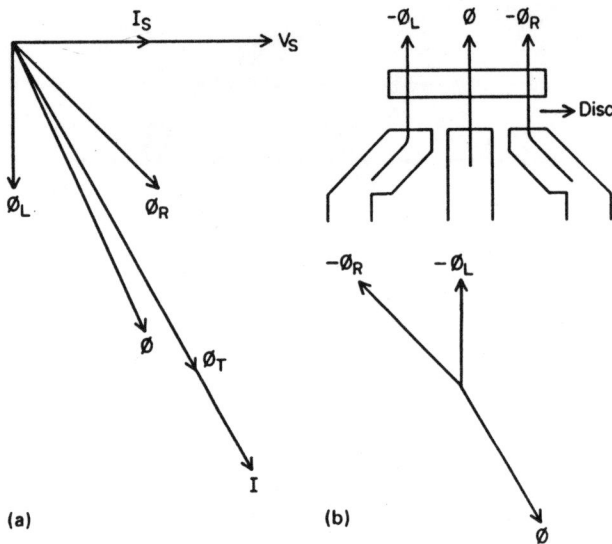

Figure 3-5 Phasors and operations of the "E" unit induction disc.

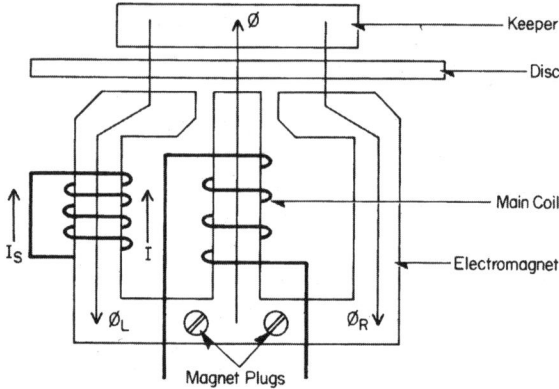

(a) The "E" Unit

(b) The "OA" Unit

Figure 3-4 Induction disc unit.

Flux ϕ_L induces voltage V_s, and current I_s flows, essentially in phase, in the shorted lag coil. Flux ϕ_T is

the total flux produced by main coil current I. The three fluxes cross the disc air gap and induce eddy currents in the disc. These eddy currents react with the pole fluxes and produce the torque that rotates the disc. With the same reference direction for the three fluxes as shown in Figure 3-5b, the flux shifts from left to right and rotates the disc clockwise, as viewed from the top.

There are many alternative versions of the induction disc unit. The unit shown in Figure 3-4, for example, may have a single current or voltage input. The disc always moves in the same direction, regardless of the direction of the input. If the lag coil is open, no torque will exist. Other units can thus control torque in the induction disc unit. Most commonly, a directional unit is connected in the lag coil circuit. When the directional unit's contact is closed, the induction disc unit has torque; when the contact is open, the unit has no torque.

Induction disc units are used in power or directional applications by substituting an additional input coil for the lag coil in the E unit. The phase relation between the two inputs determines the direction of the operating torque.

A spiral spring on the disc shaft conducts current to the moving contact. This spring, together with the shape of the disc (an Archimedes spiral) and design of the electromagnet, provides a constant minimum operating current over the contact travel range. A permanent magnet with adjustable keeper (shunt) dampens the disc, and magnetic plugs in the electro-magnet control the degree of saturation. The spring tension, damping magnet, and magnetic plugs allow separate and relatively independent adjustment of the unit's inverse-time current characteristics.

2.2.2   Cylinder Units

The operation of a cylinder unit is similar to that of an induction motor with salient poles for the stator windings. Shown in Figure 3-6 , the basic unit used for relays has an inner steel core at the center of the square electromagnet, with a thin-walled aluminum cylinder rotating in the air gap. Cylinder travel is limited to a few degrees by the moving contact

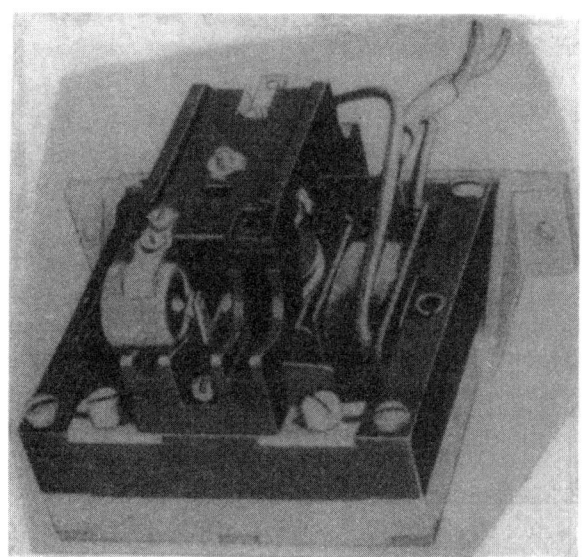

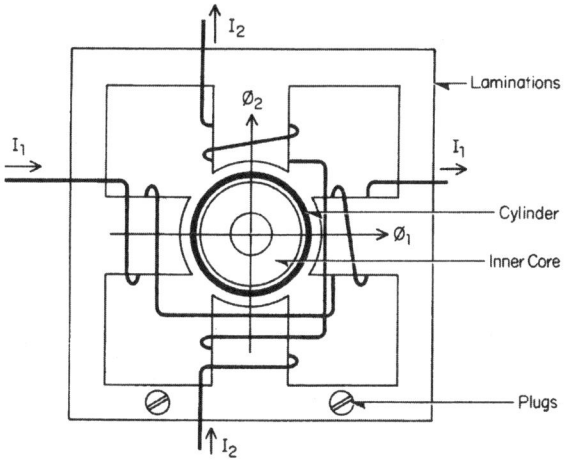

Figure 3-6   Cylinder unit.

attached to the top of the cylinder and the stationary contacts. A spiral spring provides reset torque.

Operating torque is a function of the product of the two operating quantities applied to the coils wound on the four poles of the electromagnet and the sine of the angle between them. The torque equation is

$$T = KI_1I_2 \sin \phi_{12} - K_s \qquad (3\text{-}2)$$

where K is a design constant; I_1 and I_2 are the currents through the two coils; ϕ_{12} is the angle between I_1 and I_2; and K_s is the restraining spring torque. Different combinations of input quantities can be used for different applications, system voltages or currents, or network voltages.

2.3.   D'Arsonval Units

In the D'Arsonval unit, shown in Figure 3-7, a magnetic structure and an inner permanent magnet form a two-pole cylindrical core. A moving coil loop in the air gap is energized by direct current, which reacts with the air gap flux to create rotational torque. The D'Arsonval unit operates on very low energy input, such as that available from dc shunts, bridge networks, or rectified ac. The unit can also be used as a dc contact-making milliammeter or millivoltmeter.

2.4   Thermal Units

Thermal units consist of bimetallic strips or coils that have one end fixed and the other end free. As the temperature changes, the different coefficients of thermal expansion of the two metals cause the free end of the coil or strip to move. A contact attached to the free end will then operate based on temperature change.

3   SEQUENCE NETWORKS

Static networks with three-phase current or voltage inputs can provide a single-phase output proportional to positive, negative, or zero sequence quantities. These networks, also known as sequence filters, are widely used.

3.1   Zero Sequence Networks

In zero sequence networks, three current transformer secondaries, connected in parallel, provide $3I_0$ from I_a,

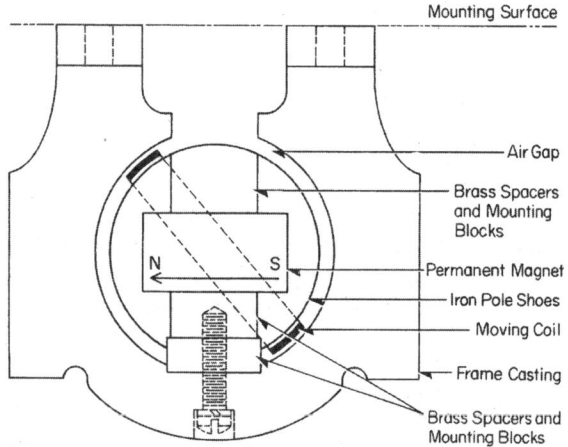

Figure 3-7 D'Arsonval-type unit.

I_b, and I_c inputs. Similarly, the secondaries of three-phase voltage transformers, connected in series with the primary in grounded wye, provide $3V_0$.

3.2 Composite Sequence Current Networks

The network shown in Figure 3-8 can be adapted for a variety of single-phase outputs. Output filter voltage

V_F is obtained from input currents I_a, I_b, and I_c, with neutral ($3I_0$) return. By using Thevenin's theorem, these three-phase networks can be reduced to a simple equivalent circuit, as shown in Figure 3-8b. V_F is the open circuit voltage at the output, and Z the impedance looking back into the three-phase network. Z_s is the self-impedance of the three-winding reactor's secondary with mutual impedance X_m.

The open circuit voltage (Fig. 3-8a) with switch r open and switch s closed is the drop from $V_F(+)$ to $V_F(-)$.

$$V_F = j(I_c - I_b)X_m + I_aR_1 + 3I_0R_0 \qquad (3\text{-}3)$$

From the basic symmetrical component equations [Eq. (2-24) and (2-25) in Chapter 2], we have

$$I_c - I_b = j\sqrt{3}I_1 - j\sqrt{3}I_2 \qquad (3\text{-}4)$$

Substituting this and the sequence equation (2-23) for I_a in Eq. (3-3),

$$\begin{aligned}
V_F &= -\sqrt{3}X_mI_1 + \sqrt{3}X_mI_2 + R_1I_1 \\
&\quad + R_1I_2 + R_1I_0 + 3I_0R_0 \\
&= (R_1 - \sqrt{3}X_m)I_1 + (R_1 + \sqrt{3}X_m)I_2 \\
&\quad + (R_1 + 3R_0)I_0
\end{aligned} \qquad (3\text{-}5)$$

Varying X_m, R_1, R_0, and the connections produces different output characteristics. In some applications, the currents I_b and I_c are interchanged, changing

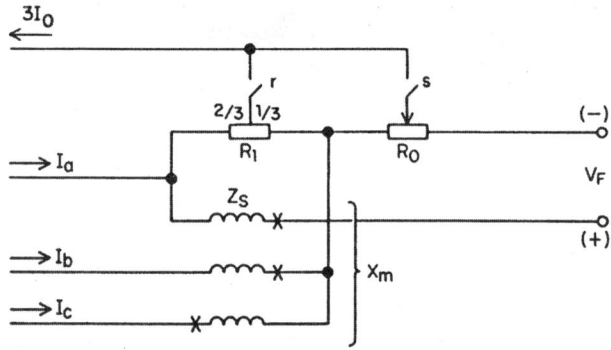

(a) Sequence Network

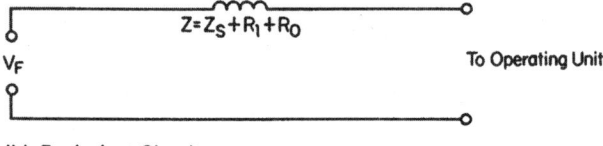

(b) Equivalent Circuit

Figure 3-8 Composite sequence current network.

Table 3-1 Typical Sequence Network Combinations

Network type	Switch r	Switch s	$X_m =$	Figure 3-8 notes	Equation	V_F reduces from: To equal
Positive sequence	closed	open	$R_1/\sqrt{3}$	Interchange I_b and I_c	(3-6)	$2R_1I_1$
Negative sequence	closed	open	$R_1/\sqrt{3}$	As shown	(3-5)	$2R_1I_2$
HCB composite	open	closed	$R_1/\sqrt{3}$	Interchange I_b and I_c	(3-6)	$2R_1I_1 + (R_1 + 3R_0)I_0$
HCB-1 and SKB composites	open	closed	$1.46R_1$ or 0.191 ohms	As shown	(3-5)	$-0.2I_1 + 0.462I_2 + (R_1 + 3R_0)I_0$

Data for Tap C of three taps available.

Eq. (3-5) to

$$V_F = (R_1 + \sqrt{3}X_m)I_1$$
$$+ (R_1 - \sqrt{3}X_m)I_2 + (R_1 + 3R_0)I_0 \qquad (3\text{-}6)$$

Note that the choice of design constant $X_m = R_1/\sqrt{3}$ causes the I_2 term in Eq. (3-6) to become 0. With switch r closed and switch s open, the zero sequence response of Eq. (3-5) and (3-6) is eliminated. The zero sequence drop across R_1 is $2/3R_1I_{ao} - 1/3R_1$ $(I_{b0} + I_{c0}) = 0$. The switches r and s are used in Figure 3-8 as a convenience for description only. Several typical sequence network combinations are given in Table 3-1.

3.3 Sequence Voltage Networks

Sequence voltage networks may be constructed to provide a single-phase output proportional to either positive or negative sequence voltage. A network in common use is shown in Figure 3-9. Since this network is connected phase to phase, there is no zero sequence voltage effect.

The network is best explained through the phasor diagram (Fig. 3-10). By design, the phase angle of $Z + R$ is 60° lagging. For convenience, consider switches s to be closed and switches r open. Impedance $Z + R$ is thus connected across voltage V_{ab}, and the autotransformer across voltage V_{bc}. With only positive sequence voltages (Fig. 3-10a), the current I_{ab1} through $Z + R$ lags V_{ab1} by 60°. The drop V_{by1} across the autotransformer to the tap is in phase with voltage V_{bc} across the entire transformer. The tap is chosen so that $|V_{xb1}| = |V_{by1}|$. The filter output $V_{xy} = V_F$ is the phasor sum of these two voltages.

With only negative voltages applied (Fig. 3-10b), V_{xb2} is equal and opposite to V_{by2}, that is,

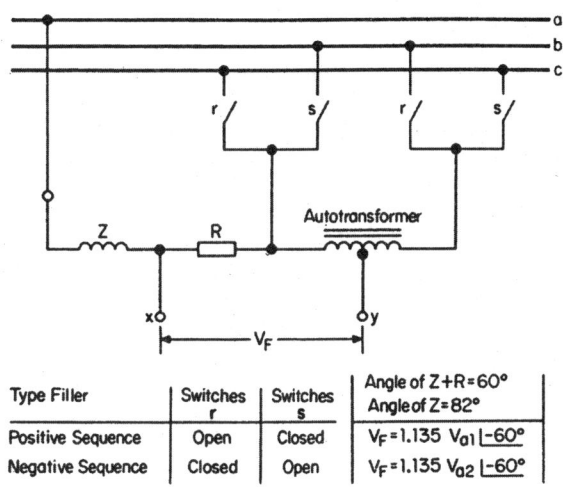

Type Filler	Switches r	Switches s	Angle of Z+R=60° Angle of Z=82°
Positive Sequence	Open	Closed	$V_F = 1.135\ V_{a1}\ \underline{/-60°}$
Negative Sequence	Closed	Open	$V_F = 1.135\ V_{a2}\ \underline{/-60°}$

Figure 3-9 Sequence voltage network.

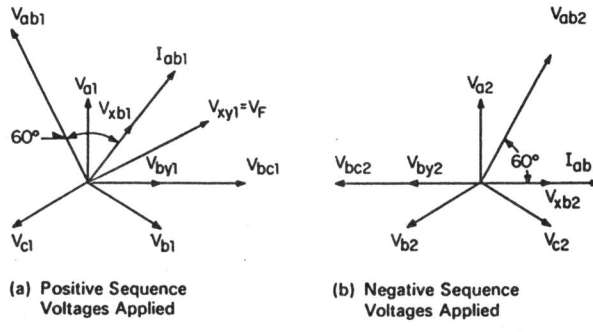

(a) Positive Sequence Voltages Applied

(b) Negative Sequence Voltages Applied

Figure 3-10 Phasor diagrams for the sequence voltage network of Figure 3-9 with "s" closed and "r" open.

$V_{xy} = V_F = 0$. Thus, this is a positive sequence network.

A negative sequence network can be made by reversing the b and c leads or, in Figure 3-9, by opening s and closing r. Then Figure 3-10a conditions apply to a negative sequence, giving an output V_F; Figure 3-10b conditions apply to a positive sequence with $V_F = 0$.

This interchange of b and c leads to either the current or voltage networks offers a very convenient technique for checking the networks. For example, the negative sequence current network should have no output on a balanced power-system load but by interchanging the b and c leads it should produce full output on test.

4 SOLID-STATE UNITS

4.1 Semiconductor Components

Solid-state relays use various low-power components: diodes, transistors, thyristors, and associated resistors and capacitors. These components have been designed into logic units used in many relays. Before these logic units are described in detail, the semiconductor components and their characteristics will be reviewed (Fig. 3-11). Relays use silicon-type components almost exclusively because of their stability over a wide temperature range.

4.1.1 Diode

The diode (Fig. 3-11a) is a two-terminal device that conducts in one direction but does not conduct in the other. The device manifests a voltage drop for conduction in the forward direction of approximately 0.7 V. The limit of voltage to be applied in the reverse direction is defined by the rating of the diode. Failure of the diode is expected if a voltage in excess of the rating is applied in the reverse direction.

These devices are used in dc circuits to block interaction between circuits, for ac test circuits to generate a half-wave rectified current wave shape, or as a protective device around a coil to minimize the voltage associated with coil current interruption.

4.1.2 Zener Diode

The zener diode (Fig. 3-11b) differs from the diode described above in having a sharp and reproducible reverse breakdown voltage, called the zener voltage. If the current is limited to within rated values, the diode

recovers its nonconducting characteristics when the reverse voltage falls below the zener value. They are used for surge protection, voltage-regulating functions, and other applications in which a distinct conduction level is desired.

Where conduction is desired in both directions with a threshold at a level at which conduction occurs, the back-to-back zener (Fig. 3-11c), commonly known as a volt trap or zener clipper, is used. The characteristics of these devices are essentially the same in both the forward and reverse direction.

4.1.3 Varistor, Thermistor

The characteristics of the varistor are shown in Fig. 3-11d. It has a voltage-dependent nonlinear characteristic. The thermistor depicted in Figure 3-11e is a nonlinear device whose resistance varies with temperature.

4.1.4 Transistor

In relaying, the transistor is used primarily as a switch. For this function, it is shifted from a nonconducting to conducting state by the base current I_b. The transistor

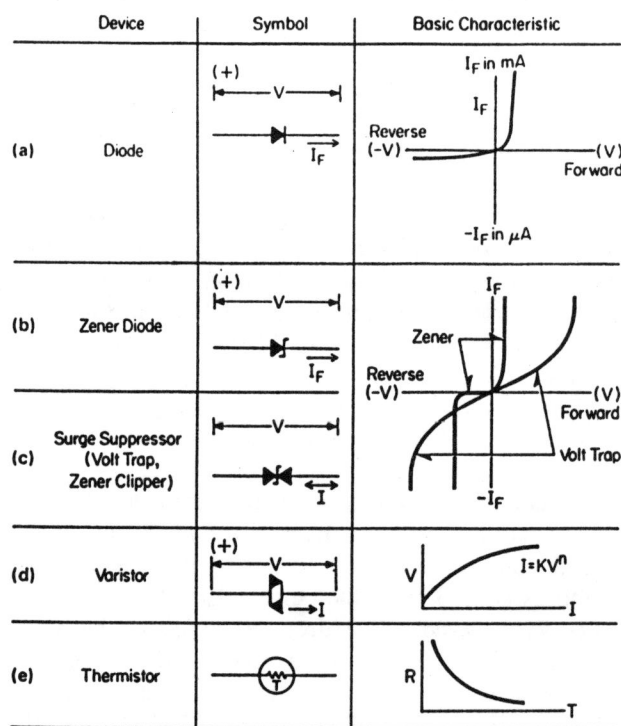

Figure 3-11 Semiconductor components and their characteristics.

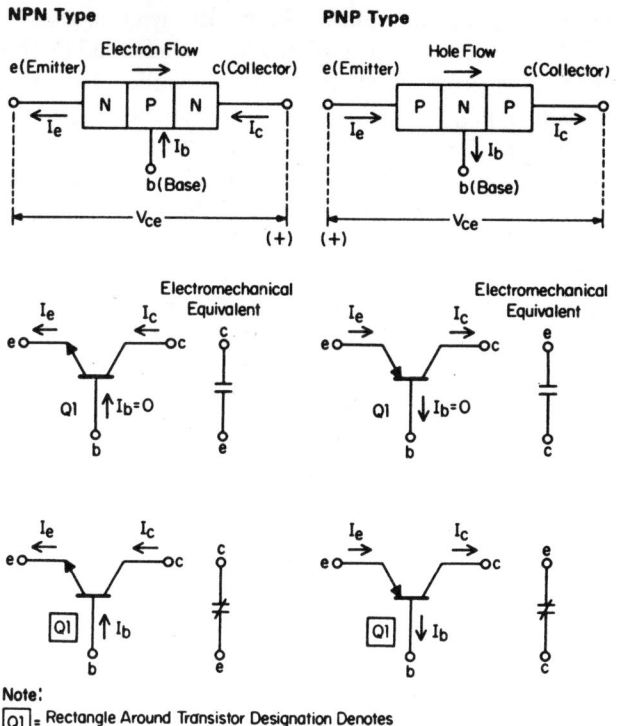

Figure 3-12 The transistor and equivalent electrical symbols.

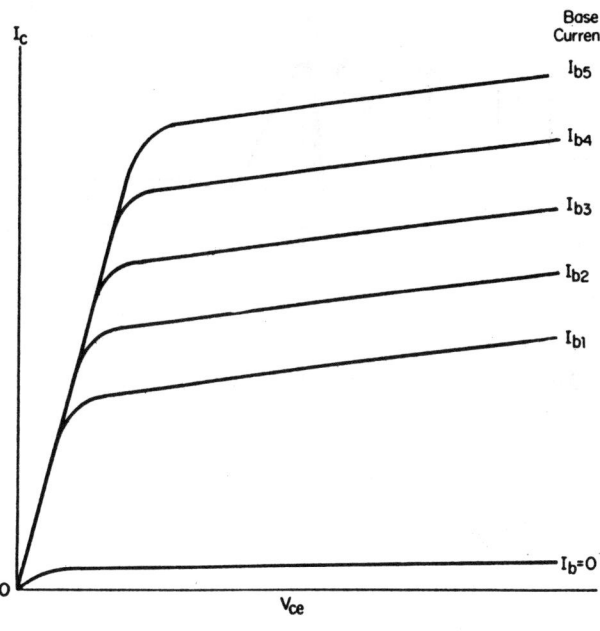

Base Current I_b is Very Much Smaller Than The Collector Current I_c

Figure 3-13 Typical characteristic curves of transistor.

is nonconducting until I_b is increased to a value at which the transistor conducts, and a collector current I_c and emitter current I_e flow (Fig. 3-12). The emitter current I_e is the sum of I_b and I_c. Very small values of I_b are able to control much larger values of I_c and I_e (Fig. 3-13).

4.1.6 Unijunction Transistor

The unijunction transistor (Fig. 3-15) has two bases, b_1 and b_2, and one emitter e. When V_e reaches the peak value V_p, the device conducts and passes current I_e. Current will continue to flow as long as V_e does not fall below the minimum value V_v. The unijunction transistor is used for oscillator and timing circuits.

4.1.5 Thyristor

The thyristor (Fig. 3-14) is a diode with a third electrode (the gate). The thyristor is also known as a silicon-controlled rectifier (SCR). With forward voltage applied, the thyristor will not conduct until gate current I_g is applied to trigger conduction. The higher the gate current, the lower the anode-to-cathode voltage (V_F) required to start anode conduction. After conduction is established and the gate current is removed, the anode current I_F continues to flow. The minimum anode current required to sustain conduction is called the holding current I_H.

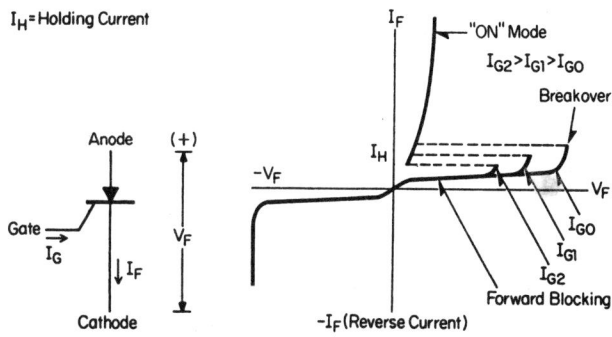

Figure 3-14 The thyristor and its characteristics.

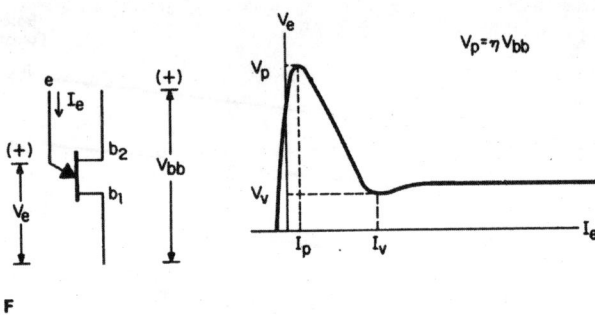

Figure 3-15 The unijunction transistor and its characteristics.

4.2 Solid-State Logic Units

4.2.1 Basic Principles

Solid-state logic units are combinations of solid-state components designed to use dc voltage signals to perform logic functions. A logic unit has only two states: *no output*, represented by 0 (zero), and *output*, represented by 1 (one). Two logic conventions are used to indicate the voltages associated with the 0 and 1 states. In normal logic, 0 is equivalent to zero voltage and 1 to normal voltage. In reverse logic, the corresponding voltage equivalents are reversed; 0 is equivalent to normal voltage and 1 to zero voltage.

In positive logic, inputs and outputs are positive; in negative logic, both inputs and outputs are negative. Relay systems normally use positive logic, although some elements may use negative signal inputs and outputs.

Logic units are shown diagrammatically in their quiescent state, that is, the normal or "at-rest" state. The quiescent state corresponds to the normally deenergized representation in electromechanical relay circuitry.

4.2.2 Logic Unit Representation

Logic units are represented by characteristic function symbols (Fig. 3-16). Two sets of symbols are in common use in the United States. In the commercial/military system, the type of function is indicated by the distinctive geometrical shape of the symbol. In solid-state relaying, the name of the logic function is simply written in a rectangle or block, or a distinctive symbol such as "&" is used inside a block. The European practice is similar to this. Convention dictates that inputs are shown on the left-hand side and outputs on the right-hand side. The symbols and terminology used comply with IEEE Standard 91-1973 (ANSI Standard U32 14-1973), "Graphic Symbols for Logic Diagrams."

When a logic function has only two inputs, its output is usually simple to determine. For three or more inputs, particularly with combination logic functions, a logic or truth table offers a convenient method of determining the output. A logic table for a function with three inputs and one output is shown in Figure 3-17. The table lists all possible combinations of zeros and ones for the inputs. Each output could be 0 or 1, depending on the function.

4.3 Principal Logic Units

In this section, the major units used in relaying will be described. Detailed circuit descriptions will be kept to a minimum. For simplicity, the diagrams will show only two inputs per function and include electromechanical contact equivalents.

4.3.1 AND Unit

The AND logic element is shown in Figure 3-18. The simplest type consists of forward-biased diodes and resistors (Fig. 3-18a). The symbolic representation and electromechanical equivalents for this unit are given in Figure 3-18b, the logic table in Figure 3-18c. The forward-biased diodes shunt the output terminal, and

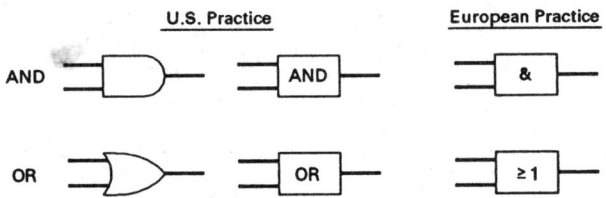

Figure 3-16 Examples of logic symbols.

Inputs			AND Output	OR Output
A	B	C		
O	O	O	O	O
O	O	1	O	1
O	1	O	O	1
O	1	1	O	1
1	O	O	O	1
1	O	1	O	1
1	1	O	O	1
1	1	1	1	1

Figure 3-17 Example of logic table.

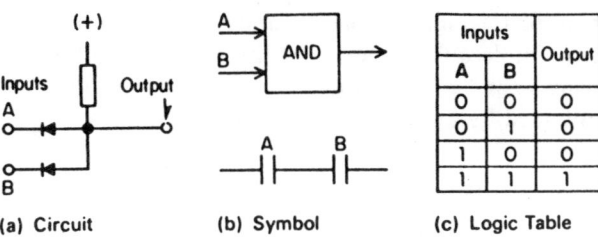

| (a) Circuit | (b) Symbol | (c) Logic Table |

Figure 3-18 AND logic.

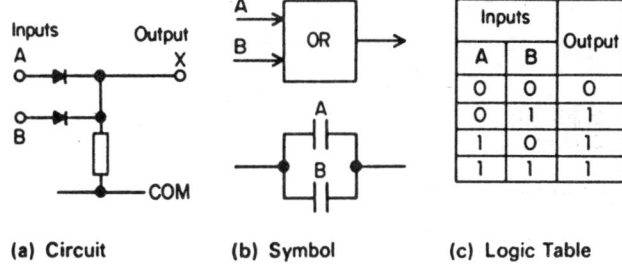

| (a) Circuit | (b) Symbol | (c) Logic Table |

Figure 3-20 OR logic.

no output voltage can appear unless all input diodes have a reverse bias that equals or exceeds the forward bias. Since inputs are either 0 or 1, there is no in-between state that would allow partial output voltage. Thus, the output is either 0 or 1, as shown in the logic table. Three variations of the AND element are provided in Figure 3-19.

4.3.2 OR Unit

The OR unit is shown in Figure 3-20. Again, the simplest type of unit consists of resistors and diodes. The symbolic representation and electromechanical equivalents for the unit are illustrated in Figure 3-20b, the logic table in Figure 3-20c. Since the diodes are not biased, an input voltage applied to any input will produce an output voltage at X.

Three variations of the OR unit, comparable to those of the AND element, are shown together with their electromechanical equivalents in Figure 3-21a, b, and c. By comparing Figure 3-21a with Figure 3-19b, it is clear that the inverse OR unit is equivalent to the negation AND. Similarly, the negation OR of Figure 3-21b is equivalent to the inverse AND of Figure 3-19a.

4.3.3 NOT Unit

The negation, or NOT, unit (Fig. 3-22) is a frequently used logic element. This unit changes the state of the

input from 0 to 1 or vice versa. For convenience, the symbol depicted in Figure 3-22a is replaced by that shown in Figure 3-22b for negated inputs and that shown in Figure 3-22c for negated outputs.

The NOT unit, together with its electromechanical equivalent and logic table, is illustrated in Figure 3-23. Although the NOT unit can be included in logic diagrams as a separate unit, it is usually combined with other units using the symbols shown in Figure 3-22.

4.3.4 Time-Delay Units

Time-delay units are used in the normal manner to provide ON and/or OFF delays. The U.S. and European symbolic representations are presented in Figure 3-24. The X value in Figure 3-24 is the pickup time, that is, the time that elapses between an input signal being received and an output signal appearing. The Y value is the dropout time, that is, the time after the input signal is removed until the output signal goes to 0. In Figure 3-24, W-X is the range of the pickup time and Y-Z that of the dropout time. Any of the values can be 0. Time values are always in milliseconds unless otherwise indicated.

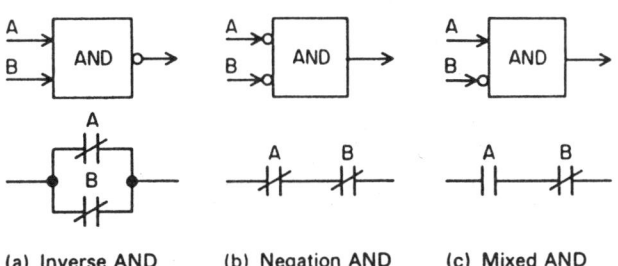

| (a) Inverse AND | (b) Negation AND | (c) Mixed AND |

Figure 3-19 Variations of AND logic.

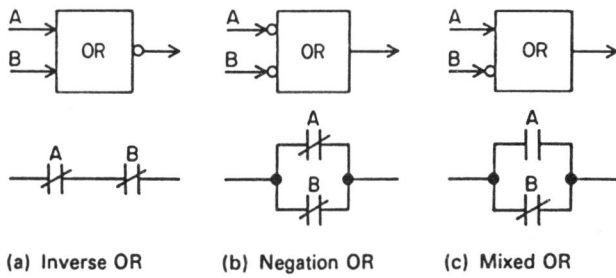

| (a) Inverse OR | (b) Negation OR | (c) Mixed OR |

Figure 3-21 Variations of OR logic.

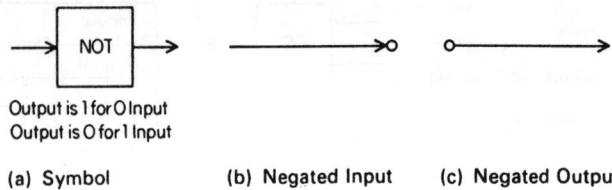

Output is 1 for 0 Input
Output is 0 for 1 Input

(a) Symbol   (b) Negated Input   (c) Negated Output

Figure 3-22 Negation symbol convention.

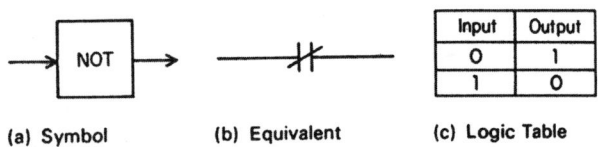

(a) Symbol   (b) Equivalent   (c) Logic Table

Figure 3-23 NOT logic.

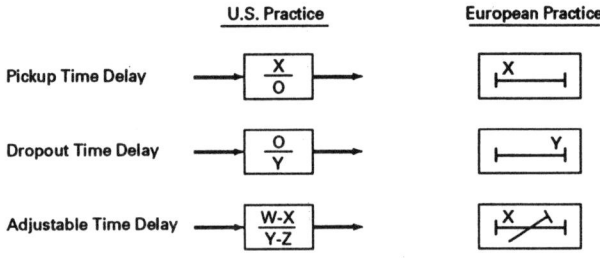

Figure 3-24 Examples of time delay units.

5 BASIC LOGIC CIRCUITS

In describing basic logic circuits, two types of diagrams are used: the logic block diagram and logic circuit schematic diagram. In the logic block diagram, the units are represented by their logic symbols, and the logic symbol blocks are interconnected to provide a complete functional representation of the system. In the logic circuit schematic diagram, the elements are shown schematically. Unit interconnections are depicted in the same way as in normal schematic diagrams. The logic block diagram is useful in showing the complete system in functional form; the logic schematic circuit diagram indicates how the logic units operate. In the following discussion, logic schematic diagrams will be used.

5.1 Fault-Sensing Data Processing Units

In solid-state relays, fault-sensing and data-processing logic circuits use power-system inputs (voltage, cur-

rent, phase angle, frequency, and so on) to determine if any intolerable system conditions exist within the relay's zone of protection. The conventional functions obtained by logic circuits are listed in Table 3-2.

5.1.1 Magnitude Comparison

There are two basic types of magnitude comparison logic units: fixed-reference and variable-reference.

Fixed-Reference

The logic circuit used for an instantaneous overcurrent unit (Fig. 3-25) is basically a dc-level detector. Input current from the current transformer secondary is transformed to a current-derived voltage on the secondary of the input transformer. This voltage is limited by zener clipper Z1 and resistor R2. For low input currents, the voltage is proportional to the current, as determined by R1 and R3. The minimum pickup is adjusted via the setting of R1. A low R1 setting diverts more current through R1 and R3, and less to the phase splitter.

The phase splitter consists of a resistor-capacitor network, transformer, and bridge rectifier. The output voltage of the phase splitter is shown in the upper part of Figure 3-25. When this voltage equals the zener voltage of Z2, Z2 will conduct, providing a base current that turns on Q1. Q1 then turns on Q2, providing an output current through D2 and R9. Q2 provides positive feedback through R7 and D1, compounding the effect on the level detector.

Table 3-2 Conventional Functions Obtained by Fault Sensing and Data Processing Logic Circuits

Conventional function	Logic circuits	Typical relay types
Instantaneous overcurrent	Magnitude comparison with fixed reference	SI-T 50B
Time overcurrent	Magnitude comparison with fixed reference and time	50D 51
Ground distance	Magnitude comparison with variable reference	SDG-T
Phase distance	Block-block comparison	SKD-T
Directional	Coincident-time (ring modulator)	SRGU

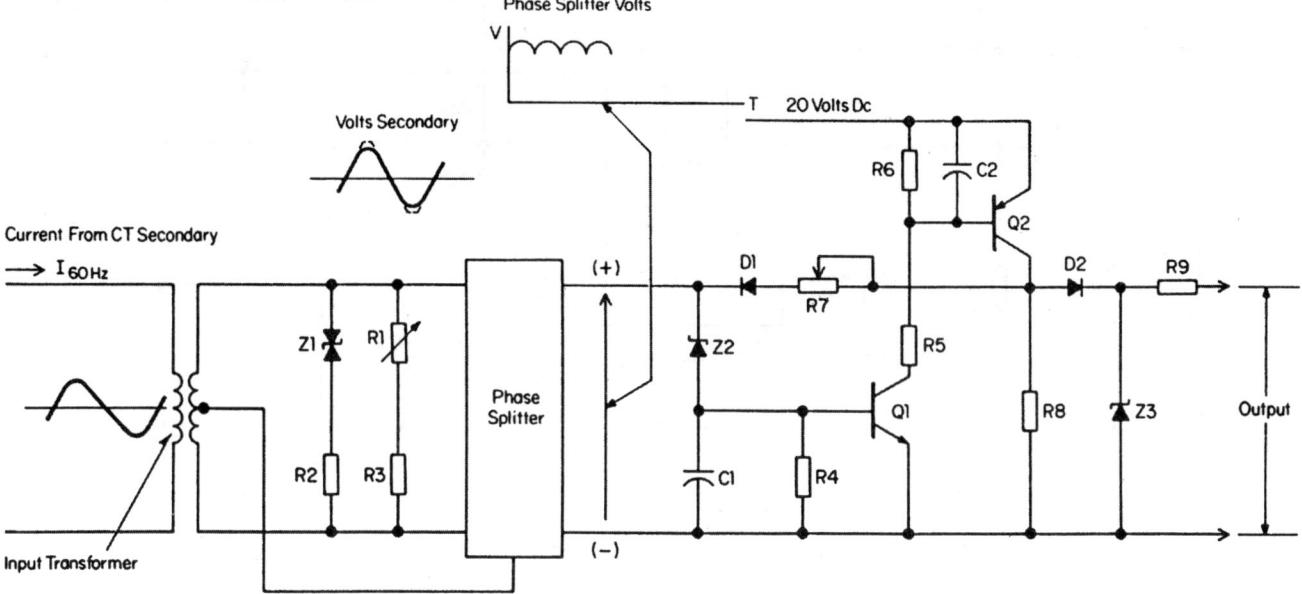

Figure 3-25   Magnitude comparison dc level detector as an instantaneous overcurrent unit.

The dropout current can be adjusted by resistor R7, normally set for a dropout/pickup ratio of about 0.97. Positive feedback provides the equivalent of snap action, and the 3% bandwidth prevents the equivalent of chattering for current values close to minimum pickup. This type of circuit could also be used for overvoltage.

Variable-Reference

This logic unit discriminates between the value of an operate voltage and the smallest of three restraint voltages. Shown in Figure 3-26, this type of circuit forms the decision logic element for the SDG ground distance relay described in Chapter 12. The restraint voltages (V_x, V_y, and V_z) and operating voltage are connected in opposition through tunnel diode TD1 and diodes D25, D26, and D27. When the operating voltage is larger than any of the three restraint voltages, current will flow through TD1. A small current through TD1 drives it to a high voltage, turning on Q1 and producing a voltage across the output terminal. The tunnel diode characteristic provides a sharp turn-on point, which serves as an effective triggering action.

Since double phase-to-ground faults may cause over-reach of the ground distance relay, a desensitizer circuit is included. This circuit consists of three minimum voltage networks. A portion k of each of

the restraint voltages is input to the desensitizer circuit. When any combination of two restraint voltages is smaller than the third restraint voltage, an output produces a blocking action through D86, preventing Q1 from turning on. When the operating voltage becomes larger than the largest restraint voltage, reverse bias is applied to D86 through D38, turning on Q1.

5.1.2   Phase-Angle Comparison

Phase-angle comparator logic circuitry produces an output when the phase angle between two quantities is within certain critical limits. Either of these two quantities, the polarizing (or reference quantity) and operating quantity, may be current or voltage.

Two types of phase-angle comparator logic circuitry are in common use: block-block and coincident-time comparison.

Block-Block Comparison

The block-block type of phase-angle comparator uses the zero-crossing detector principle to generate square waves. Additional logic circuitry provides an output if the operating quantity leads the polarizing quantity. Phase relations for the operating condition are shown in Figure 3-27. An output is obtained if the operating

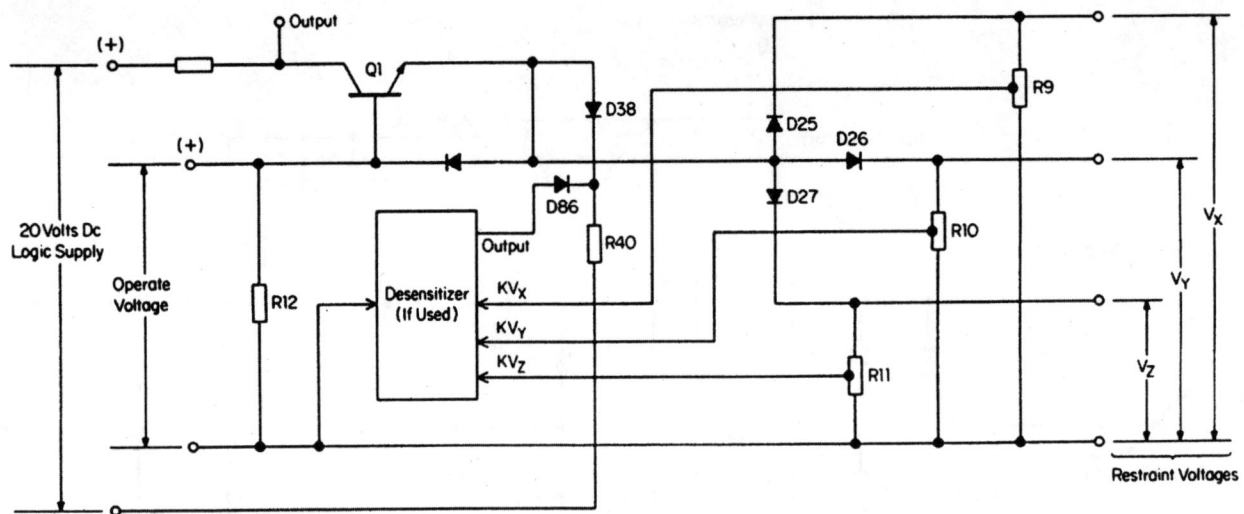

Figure 3-26 Magnitude comparator circuit.

input leads the polarizing input by 0 to 180°. Conversely, no output (restraint) occurs if the operating input lags the polarizing input by 0 to 180°. The phase relation for this restraint condition is given in Figure 3-28.

Half-cycle square waves are generated at each zero crossing of the respective input quantities. The polarity of the square waves is the same as that of the generating quantity during corresponding half-cycles.

One half of the circuit of a block-block type of comparator (Fig. 3-29) makes the comparison during the positive half-cycles. The input diodes, arrays DA and DB, limit the input voltage to 1.5 V and the output of transformers T1 and T2 to about 12 V.

For the operating condition shown in Figure 3-27, the leading operating input makes the base of Q1 positive before the polarizing input can make the base of Q3 positive. Thus, Q1 turns on first, which then

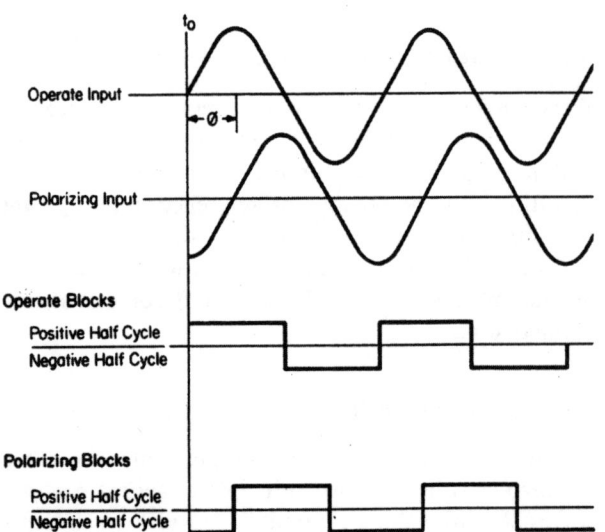

Figure 3-27 Phase relationship of block-block circuit for operate condition.

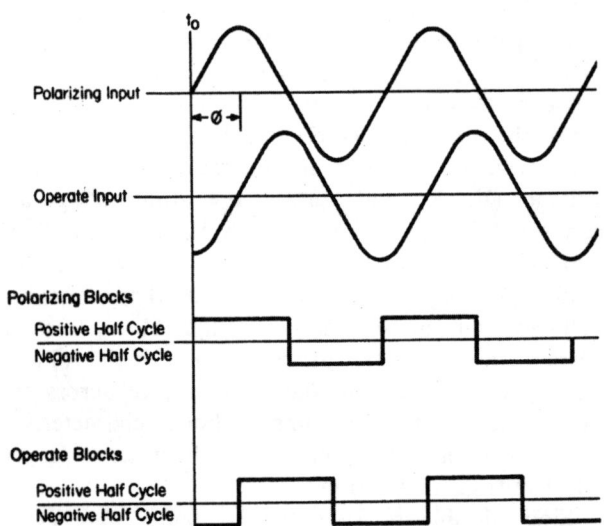

Figure 3-28 Phase relationship of block-block circuit for restraint condition.

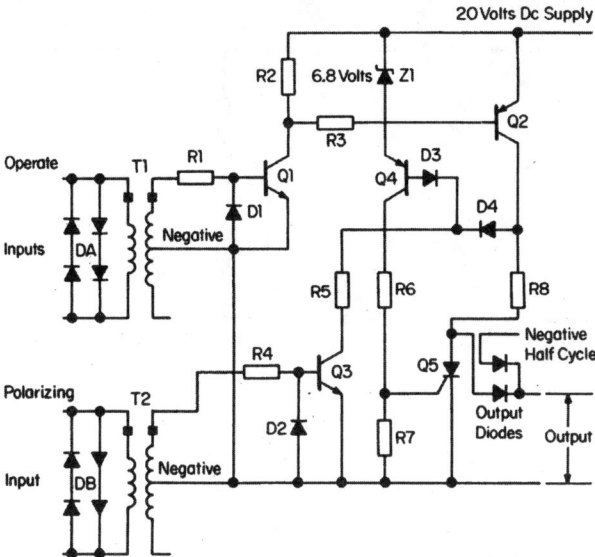

Figure 3-29 Block-block type phase-angle comparator circuit.

turns on Q2. Since Q5 has not been gated, it is in the block state, permitting an output through Q2, R8, and the output diodes. When Q2 turns on, D3 is reverse-biased through D4 from the 20-V supply. This prevents the flow of base current from turning on Q4 as otherwise would occur as the lagging polarizing input becomes positive and turns on Q3. Since Q4 cannot turn on, a half-cycle of output occurs. Similarly, during the negative half-cycle the leading operating input provides an output in the other half of the circuit, which connects through its negative half-cycle diode to the output.

If, however, the polarizing input leads the operating input (Fig. 3-28), the base of Q3 becomes positive before the base of Q1. Q3 turns on first and then turns on Q4. The current flowing through Z1, Q4, R6, and R7 produces a voltage drop across R7. This voltage drop gates thyristor Q5, causing it to conduct and short the output to negative. As the lagging operating input becomes positive, it turns on Q1 and Q2 and (since Q5 is conducting) the current through Q2 and R8 is shunted to negative. The operating input that remains when the polarizing half-cycle is completed cannot produce an output, because Q5 continues to conduct. Recovery is determined by the anode-to-cathode current, and R8 is set to allow sufficient holding current from the 20-V supply to maintain Q5 in a conducting state until the operating quantity is practically 0.

Coincident-Time Comparison (Ring Modulator)

Functioning like a biased bridge rectifier, the ring modulator type of phase-angle comparator produces an output when the operating quantity leads or lags the polarizing quantity by 90° or less. This characteristic makes the ring modulator applicable as a directional sensing unit.

Figure 3-30 shows the operating principles of the bridge under several input conditions. Current inputs are depicted, but combinations of current and voltage can also be used. Solid arrows indicate the input operating quantities, open arrows the input polarizing quantities. Actual current is the phasor sum of the currents shown. The in-phase conditions are illustrated in Figure 3-30.

In the bridge rectifier, two diodes are forward-biased by the larger current, and the magnitude of the output is determined by the smaller current. When the operating current is larger (top half of Fig. 3-30a), D1 and D3 are forward-biased, with the return through R1 and R2 blocked by D4. The polarizing current is shown in two parts, each one half of I_{POL}. The half going down through the transformer from point A flows backward through D3 and R1 to the polarizing terminal. However, since the operating current is larger, net current in D3 is forward. The output voltage $I_{POL}R1$ is proportional to the smaller current.

If the operating current reverses and is still larger than the polarizing current, D4 and D2 are forward-biased, with the return through R2 and R1 blocked by D1. Polarizing current going up from point A flows backward through D2 and up through R2 (net current in D2 is forward). The output voltage $-I_{POL}R2$ is reversed.

With reversed but still smaller polarizing current, part of the polarizing current would flow up through R1, around through D1 (which is forward-biased by the operating current) and back. The other part would flow up through R1, down through D3, and back. Again, the output voltage $-I_{POL}R1$ is reversed.

If the polarizing current is larger, as in the bottom half of Figure 3-30a, D1 is forward-biased through R1, and D4 is forward-biased through R2. If R1 equals R2, the net output from the polarizing current is 0. The smaller operating current flows through D1, R1, R2, and back through D4. Net current in D4 is forward because of the larger value of polarizing current. The net output then is $I_{OP}(R1 + R2)$, or $2I_{OP}R1$. Reversing either the polarizing or operating current will reverse the output voltage.

The output will not be 0 as long as the smaller current is above a threshold or pickup value. When the

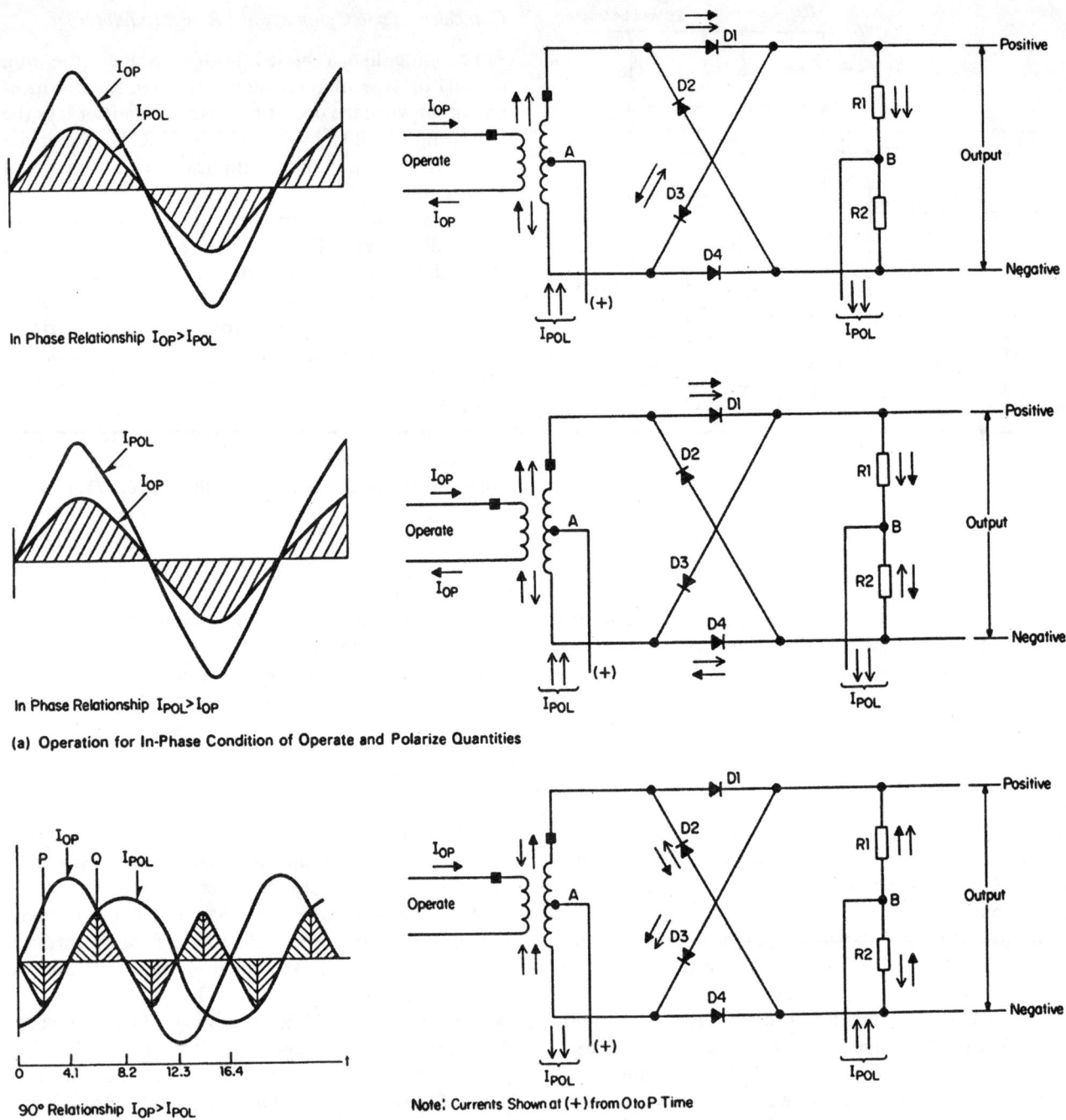

In Phase Relationship $I_{OP} > I_{POL}$

In Phase Relationship $I_{POL} > I_{OP}$

(a) Operation for In-Phase Condition of Operate and Polarize Quantities

90° Relationship $I_{OP} > I_{POL}$

Note: Currents Shown at (+) from O to P Time

(b) Operation for 90° Condition Between Operate and Polarize Quantities

Figure 3-30 Principle of operation of ring-modulator type phase-angle comparator.

sum of the currents through a diode is 0, as through D3 in Figure 3-30a, the output is still $I_{OP}R1$. Any tendency of either current to flow in another path, because D3 is not conducting, will result in one of the two components becoming larger, that is, the zero condition no longer exists.

Figure 3-30a shows the ring modulator operation when the operating current is larger than the polarizing current and leads it by 90°. At time 0, half of I_{POL} flows up through R1, down through D3, and returns up the lower half of the transformer to A. The other half flows down through R2, up through D2, and down the upper half of the transformer to A. The net output is 0 since I_{OP} is 0. As I_{OP} increases from 0 to equal I_{POL} (point P), I_{OP} flows through D2, R2, R1, and D3, producing an output of $-2I_{OP}R1$, where R1 and R2 are equal. When I_{OP} equals I_{POL}, the current in D2 goes to 0. As I_{OP} becomes larger than I_{POL}, D1 conducts. The negative I_{POL} all flows through R1; half passes through D1, and the other half continues through D3. I_{OP} flows through D1 and D3, producing an output of $-2I_{POL}R1$.

At time Q, when I_{OP} again equals I_{POL}, the polarizing current is about to become the larger current and forward-bias D1 and D4. The output changes to $+2I_{OP}R1$ and decreases. When I_{OP} crosses the zero axis, the output is 0. As the operating current becomes negative, so does the output, which reaches a maximum of $-2I_{OP}R$.

Further analysis shows that there is a maximum positive or negative output each time $I_{OP} = I_{POL}$ and alternate one-half-cycle periods (4.17 msec) of positive and negative outputs. These outputs are crosshatched in Figure 3-30b. Similar results are obtained if the polarizing quantity is greater than the operating quantity, but with I_{OP} leading I_{POL} by 90°.

5.2 Amplification Units

5.2.1 Breaker Trip Coil Initiator

The breaker trip coil initiator circuit both provides power amplification for a trip coil and isolates the control circuitry from the tripping energy source (the station battery). A typical circuit is shown in Figure 3-31.

Q1 turns on when the input voltage from the fault-sensing and data-processing circuit exceeds 2 V. Q1 then turns on Q2, allowing C2 to charge through R6. When the voltage across C2 reaches the "firing voltage" of the unijunction transistor Q3, the capacitor energy discharges through T1. This discharge reduces the voltage across the capacitor, turning off Q3 until the charge on C2 builds up again.

In this way, a repetitive train of pulses is generated as long as the input signal exists. These pulses are transformed through LA, Q4, T2 primary, LB, and Z4 to trip the circuit breaker. The time delay of this circuit is approximately 1 msec. T1 has two secondaries, the second of which is connected to a similar Q4 circuitry for double trip.

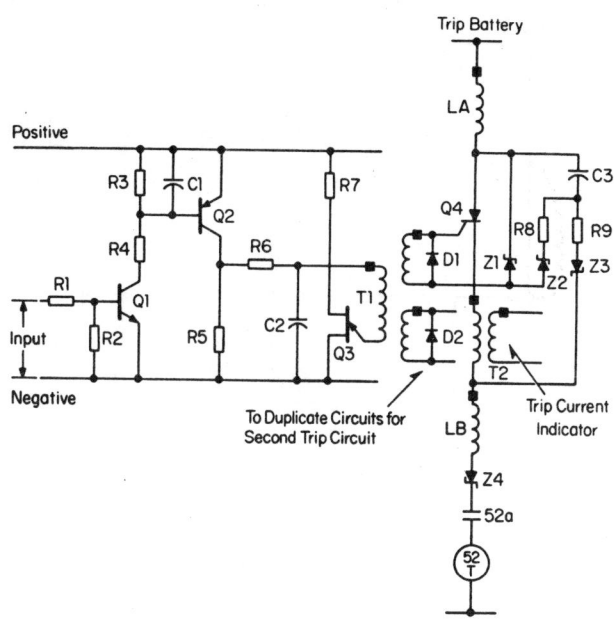

Figure 3-31 Breaker trip coil initiation circuit.

Except for the transformer T2, the devices associated with Q4 provide security. Zener Z1 clips high-voltage transients on the battery leads to a level of one-third of the Q4 rating. This voltage clipping prevents the false operation of Q4 from surges and overvoltage. The two-winding reactor LA-LB suppresses any transients that could be transmitted through the interwinding capacitance of T1 or between the trip circuit and other logic circuit wiring. Zener Z4 prevents shock excitation from setting up high-frequency oscillation, which might reverse the current through Q4 and return it to a blocking state.

Capacitor C3 is initially charged through R9 and Z3 when the breaker or switch is closed, bypassing T2 to avoid a false indication. When Q4 fires, C3 discharges through Q4, Z2, and R8. This discharge provides a holding current for Q4 of about 1 msec, long enough for the current through the inductive trip coil to reach the required holding current for Q4.

5.3 Auxiliary Units

5.3.1 Annunciator Circuits

Two types of circuits are used to provide light and alarm indications: One is for circuit-breaker-trip operations and the other for general use.

Typical breaker-trip indicator and alarm logic are shown in Figure 3-32. Transformer T2 is in the trip

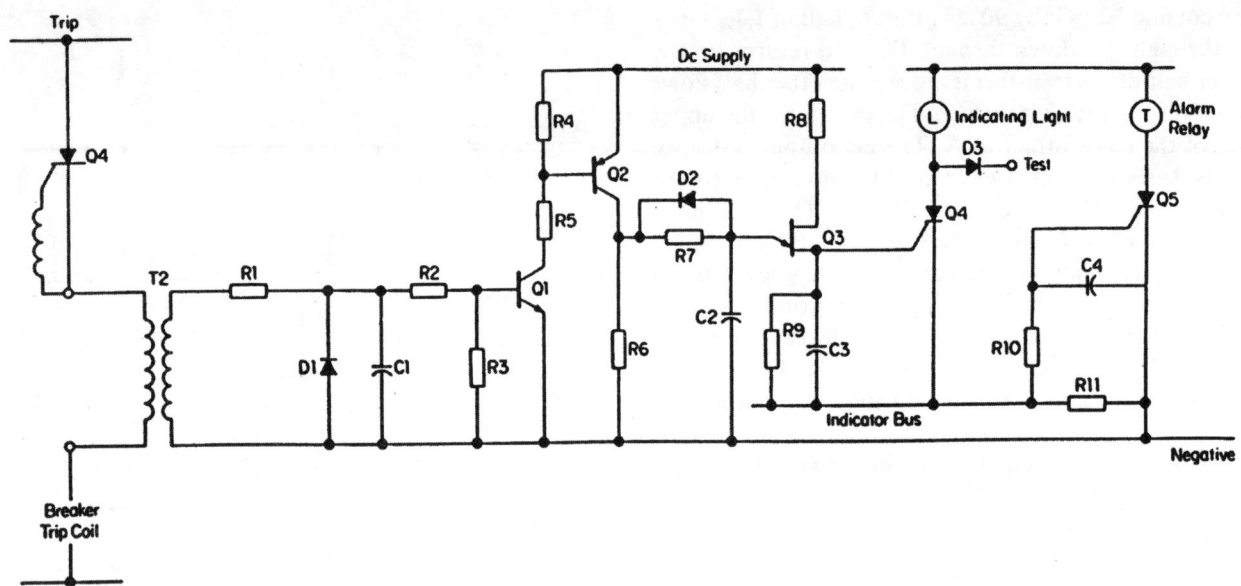

Figure 3-32 Breaker trip indicator and alarm circuit.

circuit, as in Figure 3-31. The transformer core uses square-hystersis loop material to produce a very small exciting current and negligible inductive reactance when saturated. When trip current flows (after Q4 fires), the circuit of R1, C1, R2, and R3 stretches a 2-msec pulse at the secondary of T2 into 6 msec, at 20 V, at the output of Q2. The input signal turns on both Q1 and Q2 to charge capacitor C2. When the voltage builds up to the "intrinsic standoff ratio" of the unijunction transistor (V_P of Fig. 3-15), Q3 fires and gates Q4, energizing the indicating light. The conduction of Q4 also gates Q5 through R10 from the drop across R11. Q5 energizes the alarm relay. Even if the indicating light circuit is open, Q5 will still be gated.

A general indicator circuit is shown in Figure 3-33. The normal condition is a 1 input, which makes Q1 conducting. For indication, the 1 is removed, turning off Q1. Then C1 charges through R3 and R7. When the voltage across C1 reaches the firing point of Q2, Q2 is turned on, gating Q4 and Q5 to energize the indicating light and alarm relay.

The indicating lights are the solid-state equivalent of mechanical indicating targets. Red lights are used to indicate tripping or which sensing unit signaled a trip, amber lights general alarms, blue lights testing. Sixty-volt lamps operated at 48 V or 120-V lamps at 97 V provide a filament life of more than 30,000 hr.

5.3.2 Coordinating and Loop Logic Timers

Fixed time-delay timers are used extensively in logic circuitry. A typical circuit of this type is shown in Figure 3-34. With an input, Q1 is normally conducting and shorts C1 through R4. Removing the input turns off Q1 and permits C1 to charge through R3 and R4. When the voltage across C1 reaches the zener voltage of Z1 plus the potential hill of D1 and Q2, base current will flow, turning on Q2. Turning on Q2 removes voltage from the output. The fixed time interval is between removal of input to removal of output. Although normally used for short delays, the judicious

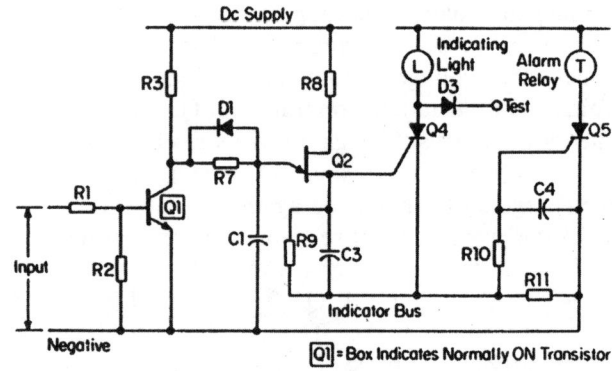

Figure 3-33 Indicator and alarm circuit.

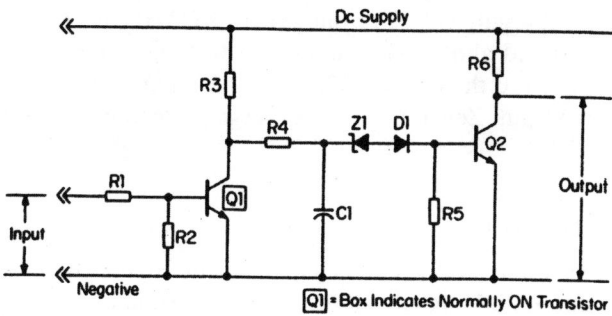

Figure 3-34 Typical logic timer circuit.

selection of values for R3, R4, C1, and Z1 provides a wide range of available time delays. Similar circuitry can provide a delay between an ON input and ON output, or other variations. Also, timers can be made adjustable by making elements such as R3 adjustable.

5.3.3 Toggle or Latching Circuits

Toggle or latching circuits, known as *flip-flops*, are bistable units similar to a latched-in or toggle-type relay. An operating signal will make the unit change state; removal of the signal will leave the unit in the new state. A momentary reset signal will restore the unit to its original state. Normally, a momentary operating signal will change the output from 0 to 1 and a momentary reset signal will change the output from 1 back to 0. The typical circuit shown in Figure 3-35 is simplified to aid in the explanation of its operation.

The circuit depicted is in the reset state with a "clear" output and no "set" output. The voltage dividers R3, R5, and R8 provide base voltage to Q2. Since Q2 is conducting, the set output is shunted to

negative. Q1 is not conducting. Its base, supplied through R6, R4, and R2, is a negative. There is, however, a clear output from the R3-R5-R8 voltage divider.

Closing the set input switch S1 momentarily reverses this condition. Voltage divider R1-R2 provides base drive to turn on Q1. The base drive for Q2 is then shunted through Q1 to negative, and Q2 is turned off. When S1 is opened, Q1 will remain on, through voltage divider R2-R4-R6. Q2 will remain off, since R5-R8 ties the base of Q2 to negative. Thus, with Q1 on and Q2 off, there is a set output but no clear output.

When a momentary signal is applied to the reset input, voltage divider R7-R8 provides base drive to turn on Q2 again. This ties the base of Q1 to negative, turning it off. Q2 then remains on, even after the reset signal is removed. The unit is now back to its reset or normal state.

Figure 3-36a is a symbolic representation of a normal flip-flop. A modification to the normal flip-flop is to desensitize it by holding Q2 in a saturated condition. When saturated, Q2 keeps conducting even when Q1 turns on. This prevents a spurious set signal from producing a set output. The modified flip-flop must first be "armed" by introducing an input arm signal (Fig. 3-36b). This signal removes the desensitizing bias from Q2 and allows it to turn off when the normal set input signal is applied.

The flip-flop can also be combined with AND logic, so that two or more separate set input signals must be received simultaneously to produce an output. This modification may also be provided with desensitizing, again requiring an arming signal. These modified flip-flops are commonly used for the final trip logic unit in solid-state relaying systems.

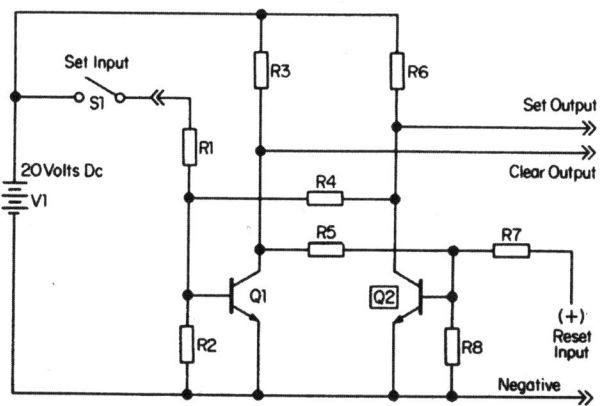

Figure 3-35 Flip-flop circuit.

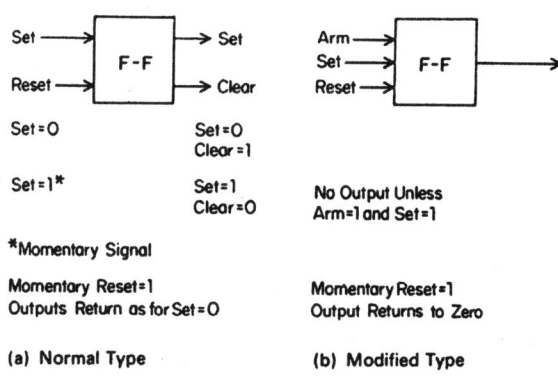

Figure 3-36 Flip-flop logic symbols.

5.3.4 Isolator and Buffer Circuits

Output and input isolators separate and electrically isolate dc circuits between logic units. Used on the input and output of each separately packaged relay, buffers protect the logic circuit from transients and surge on interconnecting leads and circuitry. Both isolator and buffer circuits protect solid-state relays against undesirable operation on spurious signals.

Input Isolator

A typical input isolator circuit is shown in Figure 3-37. A 20-V input to the pulsing circuit of R1-R3-C1-D2 charges the capacitor C1. When the capacitor voltage reaches the breakdown voltage of the four-layer diode D2, a pulse is transmitted through T1. The discharge of C1 turns off D2 until the voltage across C1 builds up again. Thus, a series of pulses continues as long as the input signal exists. Zener Z1 provides surge protection clipping at 20 V. The pulses are rectified and accumulated on C2. C2-R4-R5 provide a steady dc input to Q1 until the input is removed. Q1 conducts, turning on Q2, providing a 20-V output.

Output Isolator

The input section of the output isolator circuit (Figure 3-38) is similar to the breaker-trip coil initiation circuit shown in Figure 3-31. The output isolator circuit differs in that a four-layer diode D1, rather than a unijunction transistor, provides a pulse chain through T1. An input voltage turns on Q1 and

Q2, charging C2. The voltage across C2 triggers D1, as described above. The pulses, rectified and filtered, are applied to the base of Q3, turning it on and producing an output. Zener Z1 provides surge protection clipping at 20 V.

Input Buffer

The input buffer circuit is shown in Figure 3-39. A normal 20-V signal will result in approximately 90% of the voltage appearing across R3 and capacitor C1. When the voltage on C1 builds up to around 5 to 7 V, current flows through Z2, D1, and R4. Q1 then turns on, which produces an output. R5 is required when the output drives a PNP stage, but is omitted for an NPN stage. For internal logic circuitry, Q1 can be turned on by an unbuffered input.

There are three types of buffering: (1) A high-frequency, high-voltage surge on the input, such as the 1.0- to 1.5-MHz, 2500-V standard test surge, is dropped across R1 and clipped to 20 V by Z1; (2) all signals of 150 to 200 µsec are delayed by means of R1-R2-C1; (3) a minimum threshold voltage of 6 V is required to turn on Q1. Thus the maximum "0 level" voltage is 6 V. A bona fide signal must exist for at least 150 µsec. This buffer is described further in Chapter 4 (see Fig. 4-15).

Output Buffer

The output buffer circuit is shown in Figure 3-40. An input greater than 2 V turns on Q1 and Q2 to provide

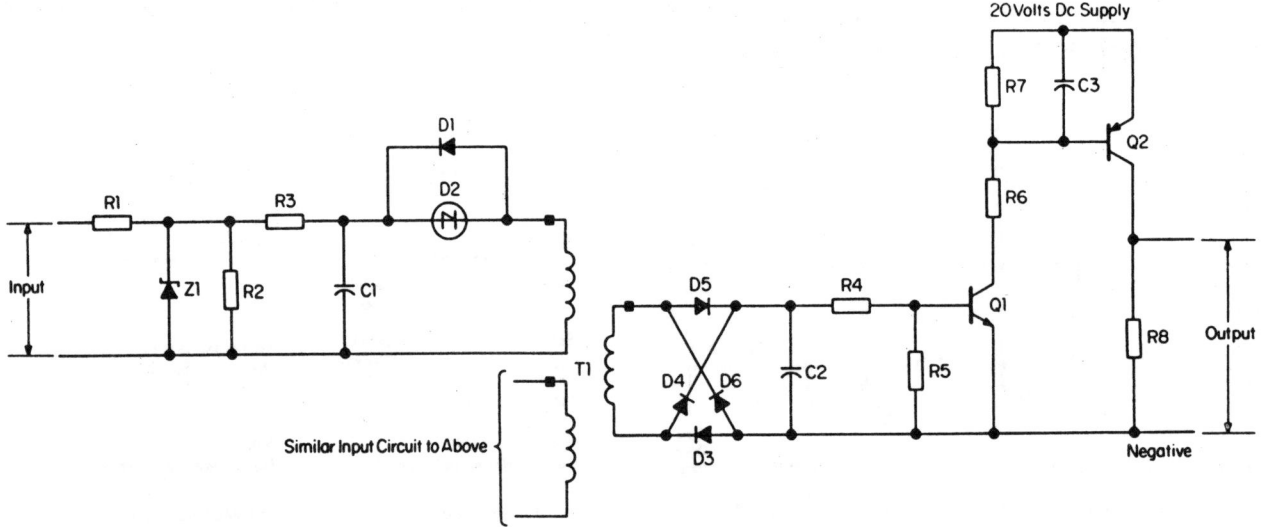

Figure 3-37 Input isolator circuit.

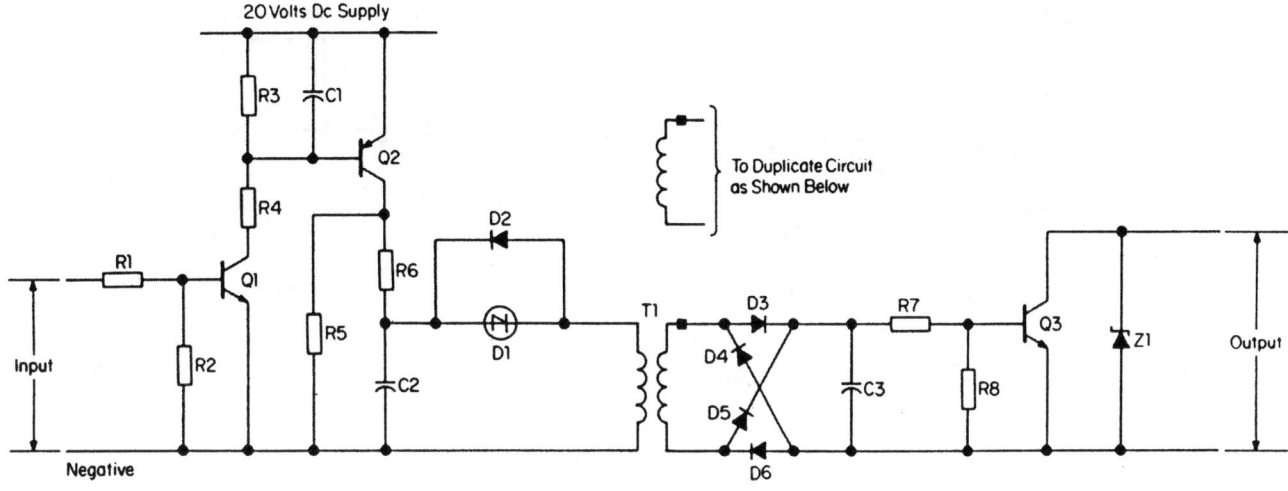

Figure 3-38   Output isolator circuit.

approximately 18 V output. C1 provides a 75-μsec delay through the unit. High-voltage, high-frequency transients on the output are limited and clipped by R6 and the 24-V zener Z1. Should the output be shorted, Q2 is protected by the current-limiting action of R6.

6   INTEGRATED CIRCUITS

The next trend in solid-state relaying was toward the use of linear and digital integrated circuits to replace the discrete transistor circuits described previously. An overview of the linear integrated circuit operational amplifier and its application to basic relay units follows.

Optical Isolator

The isolation of solid-state circuits and components from input and output signals is also accomplished with the use of optocoupler devices. This integrated circuit component uses an internal light-emitting diode (LED) and a photon detector to transmit signals, providing optical isolation between inputs and outputs.

6.1   Operational Amplifier

Figure 3-41 shows the equivalent circuit of a basic operational amplifier. The triangle symbol is used for this device. The supply voltages ± Vcc (generally ± 15 Vdc) with a "common" of 0 V are not shown. The

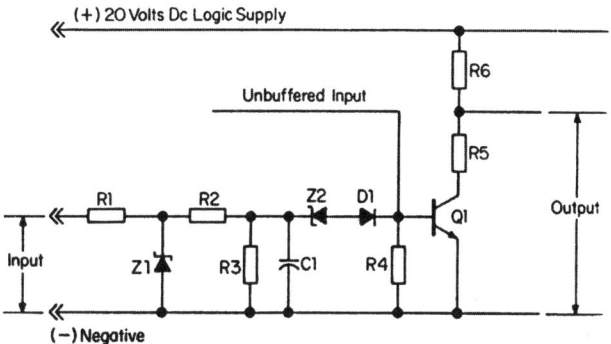

Figure 3-39   Input buffer circuit.

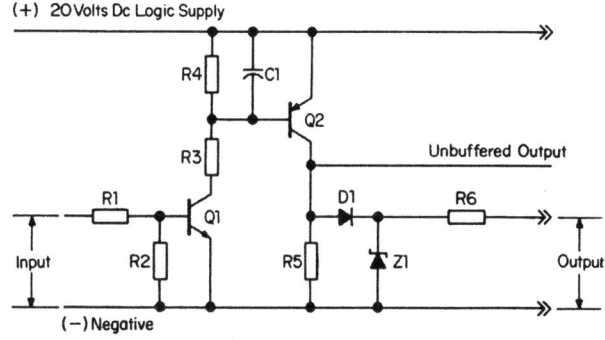

Figure 3-40   Output buffer circuit.

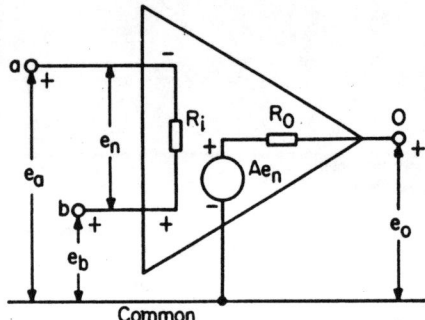

Figure 3-41 The equivalent circuit of an operational amplifier.

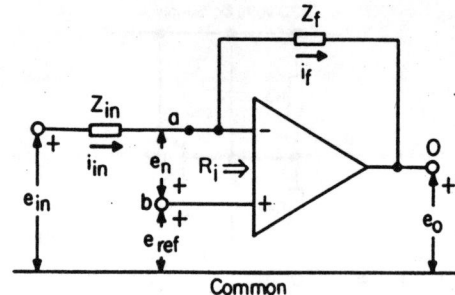

Figure 3-43 An operational amplifier with negative feedback.

input terminals are a and b; b is the noninverting input since a positive voltage produces a positive output. A positive voltage on a, the inverting terminal, will yield a negative output.

The output e_0 is amplified by the open-loop gain A so that

$$e_0 = Ae_n = A(e_b - e_a) \qquad (3\text{-}7)$$

The plot in Figure 3-42 shows that a small differential change drives the amplifier into saturation since the open-loop gain A is very large.

Most applications use negative feedback. In Figure 3-43, where Z_f is connected from the output to the inverting input a, I_{in} can be determined from the drops around the input loop,

$$-e_{in} + I_{in}Z_{in} - e_n + e_{ref} = 0$$
$$I_{in} = \frac{e_{in} + e_n - e_{ref}}{Z_{in}} \qquad (3\text{-}8)$$

If Z_f is much smaller than R_i, the input resistance, the assumption is that no current flows in the a or b terminals, so that

$$I_{in} = I_f \qquad (3\text{-}9)$$

The drops around the Z_f feedback loop are

$$-e_{ref} + e_n + I_fZ_f + e_0 = 0$$
$$I_f = \frac{e_{ref} - e_n - e_0}{Z_f} \qquad (3\text{-}10)$$

Equating Eqs. (3-8) and (3-10) with the assumption of Eq. (3-9) provides

$$e_n = e_{ref} - \left(\frac{Z_f}{Z_f + Z_{in}}\right)e_{in} - \left(\frac{Z_{in}}{Z_f + Z_{in}}\right)e_0 \qquad (3\text{-}11)$$

Substituting Eq. (3-7) and solving for e_0, we obtain

$$e_0\left(\frac{1}{A} + \frac{Z_{in}}{Z_{in} + Z_f}\right) = e_{ref} - \left(\frac{Z_f}{Z_f + Z_{in}}\right)e_{in} \qquad (3\text{-}12)$$

If we assume that A is very large, $1/A$ approaches 0 and is much less than

$$\frac{Z_{in}}{Z_{in} + Z_f}$$

Thus,

$$e_0 = \left(1 + \frac{Z_f}{Z_{in}}\right)e_{ref} - \left(\frac{Z_f}{Z_{in}}\right)e_{in} \qquad (3\text{-}13)$$

Equation (3-13) is the general operational amplifier equation with negative feedback. If it is substituted into Eq. (3-11), the solution for e_n will equal 0. Thus, a and b terminals are of the same relative potential. The inverting input terminal (a) is referred to as *virtual ground*.

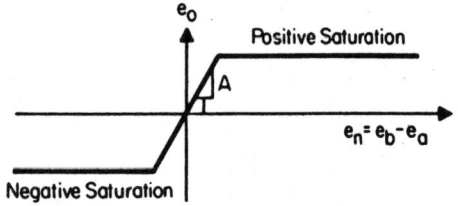

Typical Parameters (Type 741)

Input Resistance = R_i = 2.0 Meg Ohms

Output Resistance = R_0 = 75 Ohms

Open Loop Gain = A = 200,000 Volt/Volt

Figure 3-42 Saturation characteristics for circuit in Figure 3-41.

6.2 Basic Operational Amplifier Units

A number of basic units are derived from a single operational amplifier to use in relay circuits. These are described without the additional components required for accuracy, stability, or compensation.

6.2.1 Inverting Amplifiers

The inverting amplifier of Figure 3-44 is the circuit of Figure 3-43 with terminal b connected directly to common (0 V). From Eq. (3-13), we get

$$e_0 = -\left(\frac{Z_f}{Z_{in}}\right)e_{in} \qquad (3\text{-}14)$$

If resistors are used as shown in Figure 3-44, the output e_0 is the opposite of the input modified by the scale factor

$$\frac{R_f}{R_{in}}$$

6.2.2 Noninverting Amplifiers

If the a input is reduced to 0 through R_{in} (Fig. 3-45a) and the input applied to terminal b instead of e_{ref}, Eq. (3-13) reduces to

$$e_0 = \left(1 + \frac{R_f}{R_{in}}\right)e_{in} \qquad (3\text{-}15)$$

The input and output are in phase with a scale factor of

$$1 + \frac{R_f}{R_{in}}$$

If R_f is made very large compared to R_{in}, then for a sine wave input, the output essentially will be an inphase square wave to provide a squaring circuit.

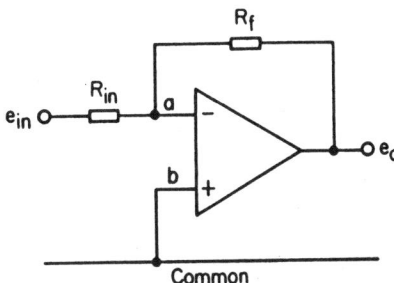

Figure 3-44 An inverting amplifier unit.

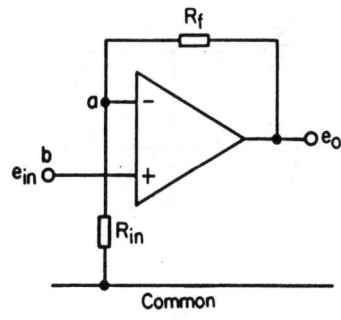

(a) Non-Inverting Amplifier

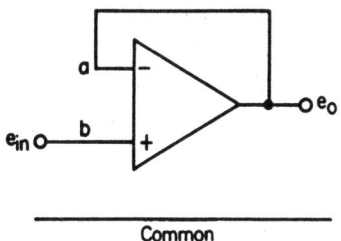

(b) Voltage Follower

Figure 3-45 A noninverting amplifier and voltage follower unit.

Another version is the voltage follower shown in Figure 3-45b. R_f approaches 0, a short circuit, and R_{in} approaches infinity, an open circuit. The gain factor

$$\frac{R_f}{R_{in}}$$

approaches 0 so that the scale factor

$$1 + \frac{R_f}{R_{in}}$$

approaches unity. Thus, the output voltage e_0 equals or follows e_{in}. In this circuit, the input impedance seen by e_{in} essentially is infinite and no current flows into the b terminal.

6.2.3 Adders

An adder unit (Fig. 3-46) has two separate inputs through R_{a1} and R_{a2} to the negative terminal a with terminal b at 0. Equation (3-13) reduces to

$$e_0 = -\frac{R_f}{R_{a1}}e_{a1} - \frac{R_f}{R_{a2}}e_{a2} \qquad (3\text{-}16)$$

If $R_{a1} = R_{a2} = R_f$, then the output equals the negative of $e_{a1} + e_{a2}$.

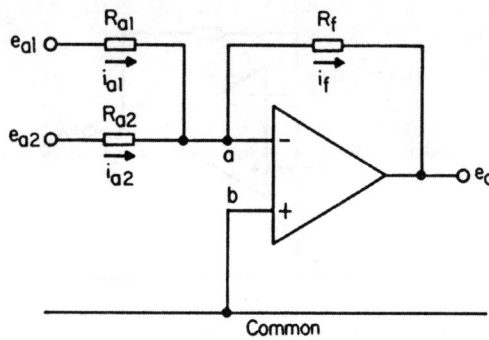

Figure 3-46   An adder unit.

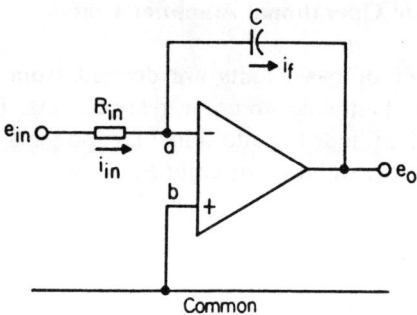

Figure 3-48   An integrator and low-pass filter unit.

6.2.4   Subtractors

The basic circuit is shown in Figure 3-47. The voltage at the plus terminal of the operational amplifier will be

$$\frac{R_f}{R_f + R_{in}} e_b$$

Substituting this in Equation (3-13), we obtain

$$e_0 = \frac{R_f}{R_{in}}(e_b - e_a) \qquad (3\text{-}17)$$

If $R_f = R_{in}$, then $e_0 = e_b - e_a$.

6.2.5   Integrator and Simple Low-Pass Filter

With a capacitor as the feedback component, the inverting amplifier of Figure 3-44 becomes an integrator (Fig. 3-48):

$$e_0 = -\frac{1}{C} \int i_f dt \qquad (3\text{-}18)$$

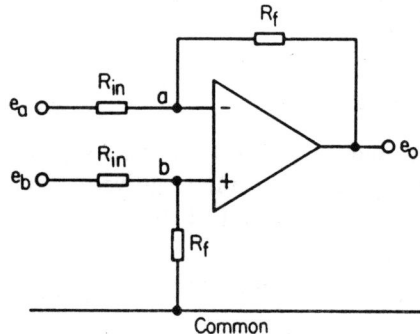

Figure 3-47   A subtractor unit.

and since

$$i_f = \frac{e_{in}}{R_{in}} = i_{in}$$

$$e_0 = -\frac{1}{R_{in}C} \int e_{in} dt \qquad (3\text{-}19)$$

this circuit is a simple low-pass filter. Considering magnitudes only,

$$Z_f = \frac{1}{2\pi fC}$$

so that Eq. (3-14) becomes

$$|e_0| = -\frac{1}{2\pi fC R_{in}} |e_{in}| \qquad (3\text{-}20)$$

Thus, as frequency increases, the magnitude of e_0 decreases.

6.2.6   Differentiator and Simple High-Pass Filter Unit

This circuit is shown in Figure 3-49 and is the inverted amplifier circuit with a capacitor in the input circuit

$$i_{in} = i_f = C\frac{de_{in}}{dt} \qquad (3\text{-}21)$$

so that

$$e_0 = -R_f C\frac{de_{in}}{dt} \qquad (3\text{-}22)$$

with magnitudes from Eq. (3-14), where

$$Z_{in} = \frac{1}{2\pi fC} \qquad \text{and} \qquad Z_f = R_f.$$

Thus,

$$|e_0| = -2\pi fC R_f |e_{in}| \qquad (3\text{-}23)$$

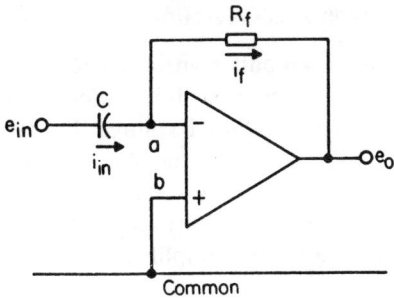

Figure 3-49 A differentiator and high-pass filter unit.

This is a simple high-pass filter since as f decreases, $|e_0|$ decreases.

6.2.7 Phase-Shift Units

A variety of phase-shift units are obtained using capacitor and variable resistor combinations. These are illustrated in Figure 3-50. A phase-angle range of 90 to 180° is obtained with Z_f adjustable from 0 to $-90°$ (Fig. 3-50a), 180° to 270° with Z_{in} adjustable from 0° to $-90°$ (Fig. 3-50b). Inverting operational amplifiers are used in both of these circuits.

Noninverting amplifiers with the RC network connected as a voltage divider (Figs. 3-50c and d) provide a phase-angle range of 0 to +90° or 0 to $-90°$, depending on the position of R and C.

6.2.8 Level Detectors

Figure 3-51a shows a level detector using the operational amplifier in the differential mode. From Eq. (3-7), we have

$$e_0 = A(e_{ref} - e_{in}) \tag{3-24}$$

and

$$e_{ref} = \left(\frac{R_2}{R_1 + R_2}\right) V_{cc} \tag{3-25}$$

As illustrated in Figure 3-42, a change in the level of $e_{in}(e_a)$ slightly before or below $e_{ref}(e_b)$ will cause the amplifier to go into either negative or positive saturations, respectively. With e_{ref} formed by the R1-R2 voltage divider, e_0 becomes low with e_{in} above e_{ref}, and high when e_{in} is less than e_{ref}.

Hysteresis is obtained with positive feedback through resistor R3 (see Fig. 3-51b). With e_0 large, e_{ref} is determined by the voltage divider consisting of R2 in series with the parallel combination of R1 and

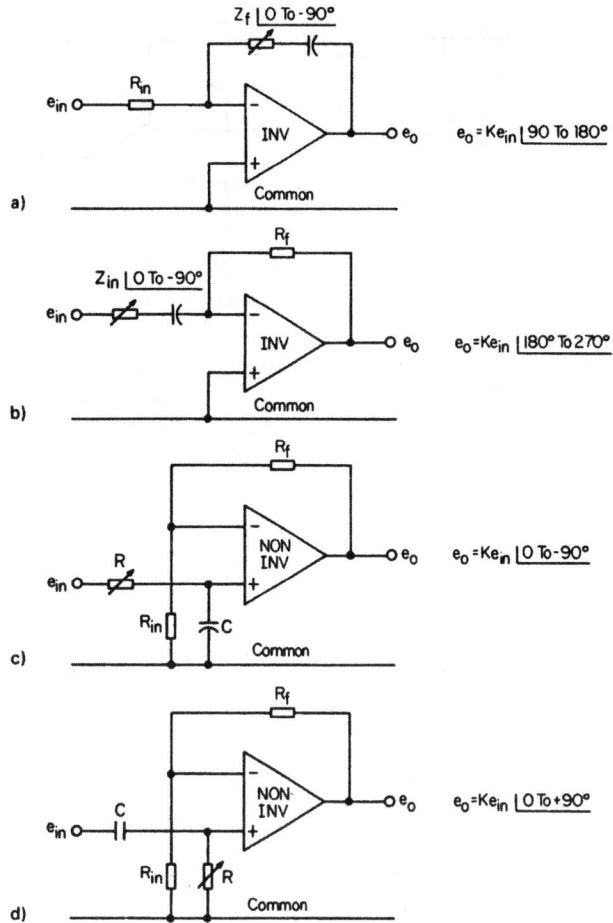

Figure 3-50 Phase shift units.

R3. This voltage is higher than with just R1 and R2: e_0 approaches 0 when e_{in} exceeds the e_{ref}. This causes e_{ref} to be lowered to a potential determined by the divider relationship of R1 in series with the parallel combination of R2 and R3. Thus, the voltage at which e_0 switches from high to low is greater than that when it switches from low to high. This is illustrated in the example in the lower half of Figure 3-51b.

6.2.9 Active Filters

A typical active filter unit is shown in Figure 3-52. High "Q" circuits, different gains, and resonant frequencies are easy to obtain. Inductance is not used in these circuits. The filters can be cascaded and are unaffected by loading.

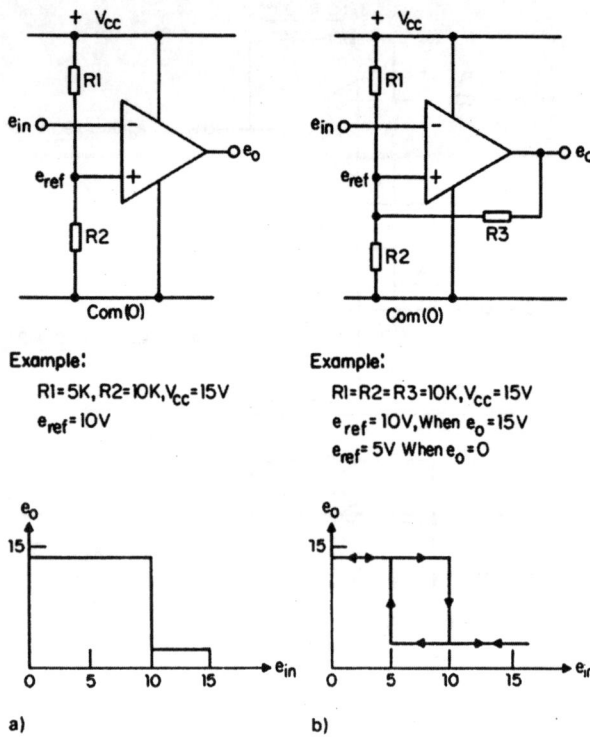

Figure 3-51 Level detector units.

6.3 Relay Applications of Operational Amplifier

Three protective relay applications illustrate the use of the basic operational amplifier units described. The relaying inputs from current and voltage transformers are converted to low-level signals by shunts or auxiliary transformers.

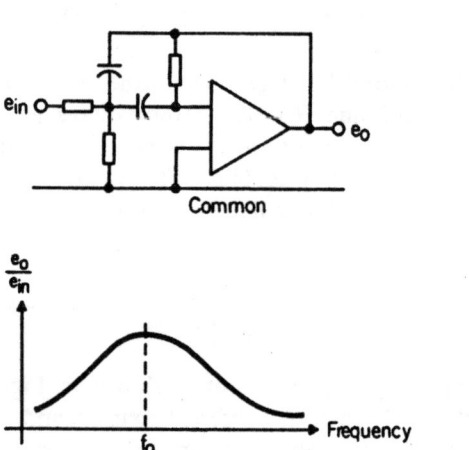

Figure 3-52 A multiple-feedback band-pass filter unit.

6.3.1 Instantaneous Overcurrent Unit

An operational amplifier instantaneous overcurrent unit is shown in Figure 3-53. Input current i is converted to a proportional voltage through shunt R and filtered by an active bandpass filter (OA1). The pickup at other than the system frequency is significantly higher to minimize harmonic effects. OA2 is an adjustable gain inverting amplifier with a gain of

$$-K = \frac{P1}{R4}$$

The amplified signal $-K_i$ is rectified by OA3 and OA4. When the signal from OA2 is negative, OA3 forces its output positive, while the input to the (+) terminal of OA4 will be negative through R7. This back-biases diode D1 to disconnect the OA3 output to OA4. OA4 acts as a voltage follower with its output negative and following the (+) terminal input. When OA2 output goes positive, the output of OA3 goes negative and applies a negative input to the (+) terminal of OA4 through D1. With R5 equal to R6, OA3 is a unity gain inverter with the input to OA4 negative when the OA3 input is positive.

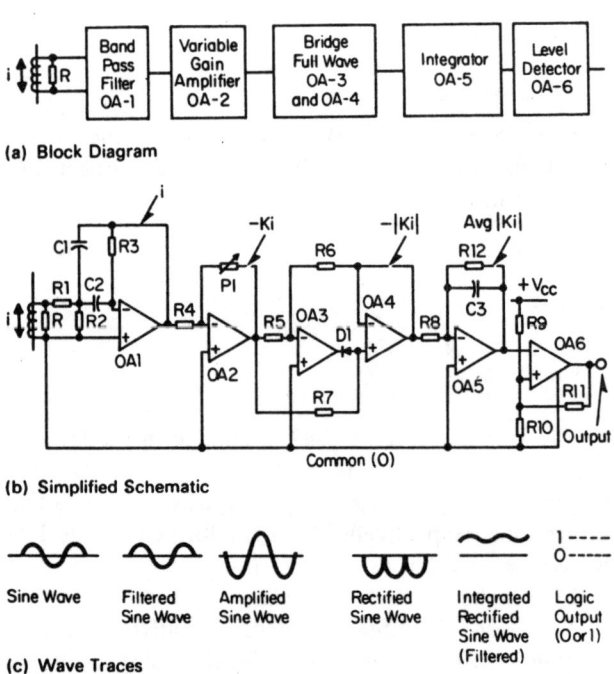

Figure 3-53 An instantaneous overcurrent unit.

6.3.2 Sequence Networks

Sequence networks can be designed by using operational amplifiers. A negative sequence circuit is shown in Figure 3-54. From Chapter 2, we have

$$I_2 = \frac{1}{3}(I_a + a^2 I_b + a I_c)$$

With phase-shift units (Fig. 3-50), I_b is shifted 240° and I_c 120°. With adder units, the final output is

$$e_0 = \frac{R}{3}(I_a + I_b \angle 240° + I_c \angle 120°)$$

and with $R = 1\,\Omega$,

$$e_2 = e_0 = \frac{1}{3}(I_a + a^2 I_b + a 1_c)$$

Positive sequence and composite filters can be designed following the same techniques.

6.3.3 Threshold Squarers and Square-Wave Detectors

These basic units are used in phase-comparison pilot systems. A typical circuit is shown in Figure 3-55. The outputs provide square waves at a low level for keying to a remote terminal, and at a high level for local comparison.

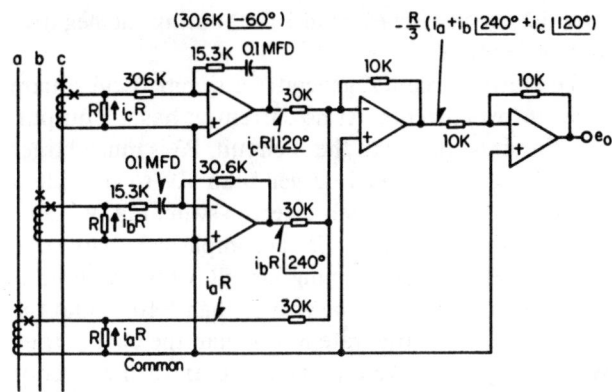

Figure 3-54 An operational amplifier negative sequence network.

The top circuit X has an adjustable noninverting amplifier and a level detector. If the sine wave is of sufficient magnitude to exceed the level detector setting (R4 and R5), the level detector output switches to 0 during the positive half-cycle as shown in the wave traces. P1 determines the magnitude at which the output switches. At low currents, the output remains 1, whereas at high currents, the output is a square wave.

The middle circuit Y is similar to X except that an inverting amplifier is used to provide a positive output from a negative input current. At high currents, the

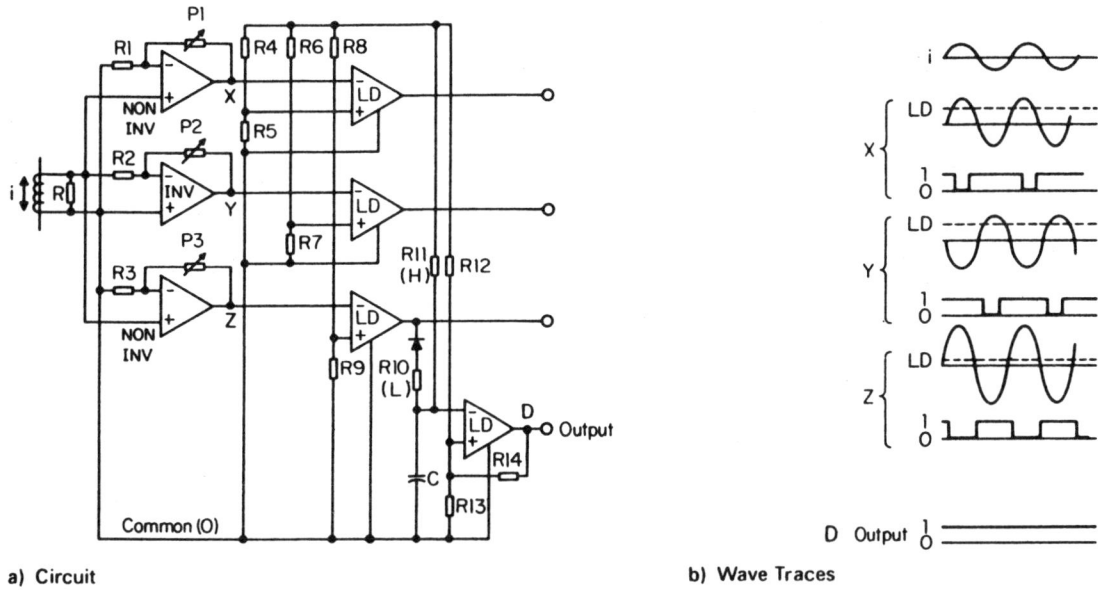

a) Circuit   **b) Wave Traces**

Figure 3-55 A threshold squarer and square-wave detector.

level detector switches from 1 to 0 during the negative half-cycle input.

The lower circuit Z generates a symmetrical square wave at low currents. At no current, it has a 1 output. The circuit is similar to the X circuit. P3 is much larger than R3, and R8 much larger than R9 to provide a high-gain and low-level detector-switching voltage. Square-wave detection is accomplished with the operational amplifier timing circuit whose output is D. Resistor R11 should be greater than R10 so that the capacitor C discharge rate is less than the charge rate. With no square wave (i.e., no current), then capacitor C will remain fully charged, causing the level detector following it to remain switched in the low state. However, if a square wave exists, then the zero transitions allow capacitor C to discharge below the level detector threshold, causing output D to become 1. When the square wave becomes 1 during the opposite half-cycle, the charge rate of R11, C, is high enough to prevent further switching of the level detector.

7 MICROPROCESSOR ARCHITECTURE

Relay design has evolved from electromechanical and solid-state to microprocessor. The functions of electromechanical sensing units, sequence networks, and solid-state logic units are performed through the processing of digital signals. The microprocessor component, integrated with RAM and ROM devices, and software programs make up the basic unit in microprocessor relay design.

4

Protection Against Transients and Surges

W. A. ELMORE

1 INTRODUCTION

The sporadic damped phenomena that occur in electrical systems are generally described as transients and surges. In this book, the two terms are considered synonymous and will be used interchangeably. In some references, however, transients refer to those phenomena related to lumped system parameters; surges refer to those phenomena related to distributed parameters. For any disturbance in an electrical circuit, such as the opening or closing of a switch or breaker, the associated damped transients may be either oscillatory or unidirectional. Surges also appear as traveling waves with a distinct propagation velocity. In such cases, wave reflections may produce voltages substantially greater than the forcing voltage that initiated the phenomenon. Lightning surges must be considered as well. With rare exception, however, experience indicates that only high-voltage systems need be protected against lightning.

From a relaying standpoint, the effect of transients and surges on secondary control circuits is of principal importance. Primary transients affect secondary circuits through common electrical connections, such as "ground" circuits and electrostatic or electromagnetic induction, as well as current and voltage transformers.

1.1 Electrostatic Induction

A simplified version of electrostatic pickup is shown in Figure 4-1. An error signal is introduced into the "signal lead" via the mutual coupling capacitance C_M.

The magnitude of the coupled voltage V_L is $C_M/(C_M + C_G)$ per unit of V_n, as long as R_L and R_S are very high. V_n is the effective noise voltage and R_R the effective load resistance of the noisy lead. The lower R_L and R_S are, the lower the transient voltage. If R_L and R_S are so low that their effect predominates, the voltage on the signal lead becomes approximately $R_T C_M (dV_n/dt)$, where R_T is the parallel equivalent of R_L and R_S, and dV_n/dt the rate of change of the noise voltage. The voltage on the signal lead cannot however exceed $C_M/(C_M + C_G)$ per unit, regardless of the rate of change of the noise voltage.

In some systems such as those used in solid-state relaying, where negative—rather than ground—is the "common," the equivalent circuit is that shown in Figure 4-2a. The basic circuit is rearranged in Figure 4-2b.

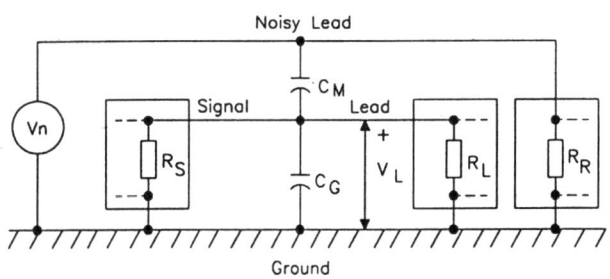

Figure 4-1 Equivalent circuit for electrostatic induction with common ground return.

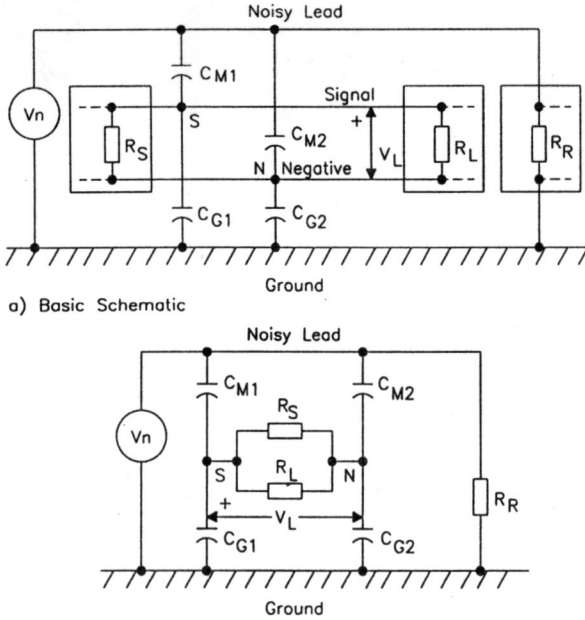

a) Basic Schematic

b) Rearrangement of The Basic Circuit of a).

Figure 4-2 Equivalent circuit for electrostatic induction without common ground return.

1.2 Electromagnetic Induction

Figure 4-3 illustrates electromagnetic pickup. Flux-linking of the signal pair, resulting from current flow in an adjacent circuit, induces a false signal voltage. The total induced loop voltage is MdI/dt, where M is the effective mutual impedance between the two circuits and dI/dt the rate of change of current I. Transposing the signal circuit will reduce the induced voltage, as shown in Figure 4-4.

1.3 Differential- and Common-Mode Classifications

Surges can be classified into two modes: differential (also known as normal or transverse) and common (also known as longitudinal).

Differential-mode surges produce voltage on a pair of conductors in the same way as a legitimate signal. Differential-mode signals are illustrated in Figures 4-1, 4-2, 4-3, 4-8, and 4-9.

Common-mode surges produce equal voltages on a pair of conductors, with respect to some common references. Common-mode surges are generated as shown in Figure 4-7. Common-mode voltage is also produced by the circuit shown in Figure 4-2 if C_{M1} equals C_{M2} and C_{G1} equals C_{G2}.

Differential-mode surges are more likely to produce misoperation of equipment, whereas common-mode surges are more likely to produce dielectric failure. (Note also that purely common-mode surges, when applied to unbalanced circuits, will produce a differential-mode component and vice versa.)

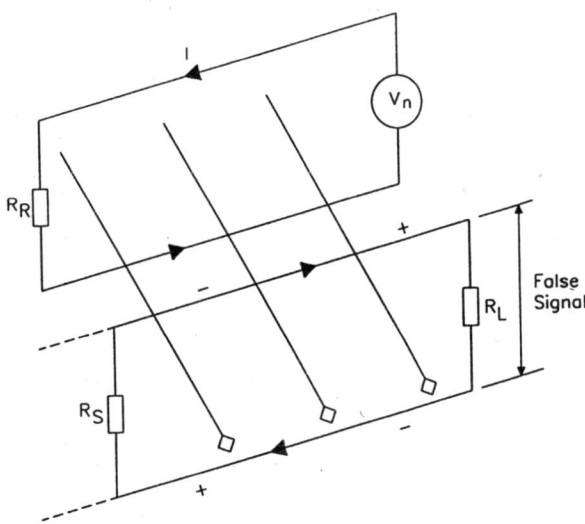

Figure 4-3 Electromagnetic induction.

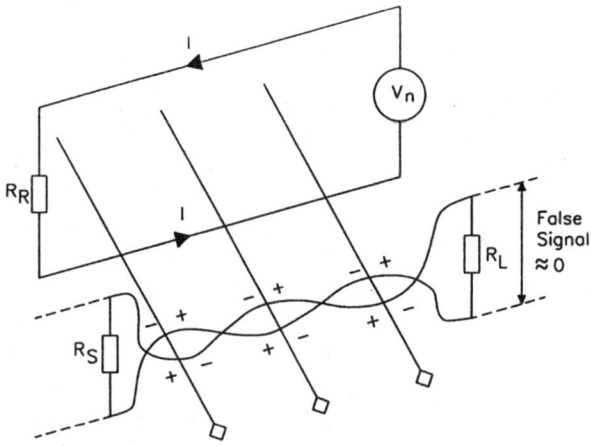

Figure 4-4 Transposing the signal circuit to minimize electromagnetic induction.

2 TRANSIENTS ORIGINATING IN THE HIGH-VOLTAGE SYSTEM

2.1 Capacitor Switching

Primary circuit transients are frequently generated by capacitor switching and are substantially more severe when interruption is accompanied by restriking.

2.1.1 Single-Bank Capacitor Switching

Figure 4-5 demonstrates what happens when a capacitor bank is energized by closing a switch. A high-frequency, high-magnitude current I flows. The capacitance of the bus and connected apparatus causes the same phenomenon to occur when a bus section is energized. Unless precautions are taken to avoid transients, such switching can cause 5- to 6-kV peaks in secondary circuits. Figure 4-6 illustrates what happens if restriking occurs when a capacitive current is interrupted.

At the instant of interruption (current zero), full voltage V_C is trapped on the capacitor bank. This voltage cannot change unless further current flows. The source voltage V_B, however, continues to vary sinusoidally. If the interruption cannot support the recovery voltage, a restrike occurs and current flows again. (The recovery voltage is $V_C - V_B$.) The most unfavorable instant of restriking is shown in Figure 4-6. With continuity reestablished, V_C equals V_B. In the

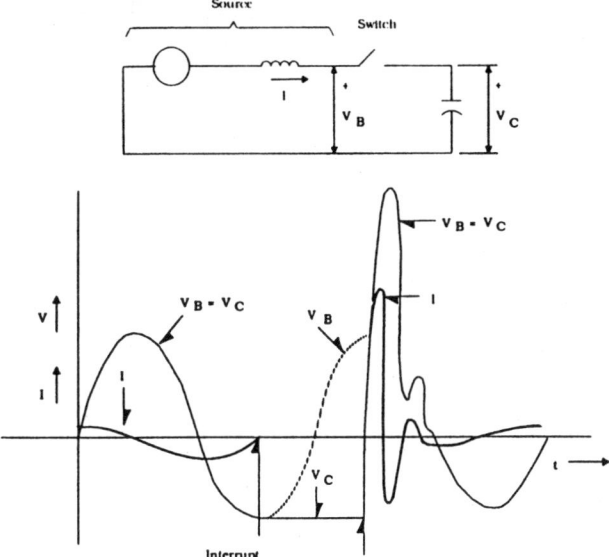

Figure 4-6 Transients generated by opening a capacitive circuit.

process of equalization, considerable overshoot occurs, and both V_B and V_C approach three times normal line-to-neutral peak voltage. The current flow, immediately following the restrike, is also very high. The current oscillates at the natural frequency of the circuit and decays with time, as governed by the circuit time constant.

2.1.2 Back-to-Back Capacitor Switching

Back-to-back capacitor switching consists of the energization of one bank of capacitors adjacent to a previously energized bank. Back-to-back capacitor switching is much like the energization of a single bank, except that the effective inductance is generally very much lower. The capacitance, on the other hand, is only somewhat lower, since it is the series combination of the bank capacitance and capacitance of the unit or units that were energized before the switch was closed. For these reasons, the magnitude and frequency of the current are generally much higher for back-to-back energization than single-bank energization.

2.2 Bus Deenergization

Bus dropping is similar to capacitor bank deenergization, except the capacitance C is very much smaller.

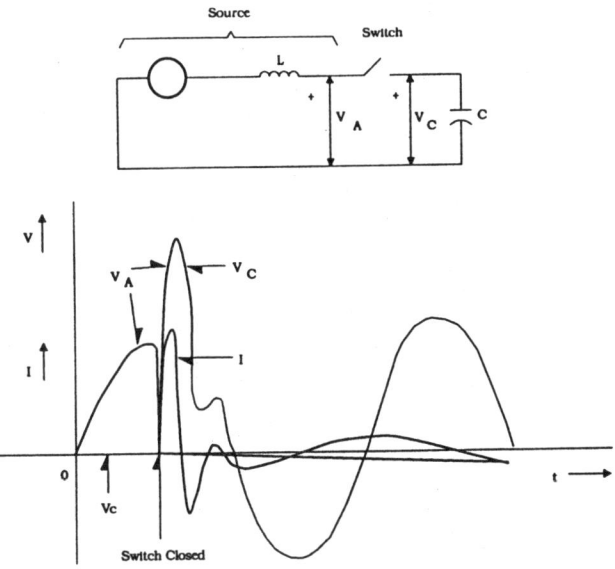

Figure 4-5 Transients generated by energizing a capacitive circuit.

Current magnitude is also generally smaller, and the
frequency is higher. When a simple disconnect is used
to drop the bus, the nonlinearity and prolonged
existence of the restriking arc cause significant
electrical noise. These characteristics together with
the large voltages and currents that accompany
restrikes produce one of the most severe surge
influences in a substation. Surges of up to 8 kV have
been measured in secondary circuits during disconnect
arcing.

2.3 Transmission Line Switching

Transmission line switching is also similar to capacitor
bank switching, except for the distributed nature of the
inductance and capacitance of the line. The inrush
current tends to be substantially less than that for
capacitor bank switching. Frequency is inversely
proportional to the length of the transmission line.

2.4 Coupling Capacitor Voltage Transformer (CCVT) Switching

These transformers contain capacitance voltage-divid-
ing networks. After energization, deenergization, and
restriking, they are subjected to the same high-
frequency, high-current phenomenon experienced in
the other cases of lumped capacitance switching. Even
in a well-designed capacitance voltage transformer,
there is perceptible capacitance between the high-
voltage and low-voltage windings (Fig. 4-7). At the
high frequencies associated with capacitor device
switching, the impedance of this capacitance will be
small. A surge voltage is developed during disconnect
restriking around the path g-g'-p-q-x or y and is
roughly equivalent to $L\, di/dt - M\, di/dt + Ri$. (L and
R are the inductance and resistance of the ground lead
of the voltage device, and M is the mutual impedance
between the ground lead and voltage leads.) If M
equals L, the total surge voltage reduces to Ri. In
practice, M can never equal L, but it will approach it if
the potential leads are placed as close as possible to
the ground lead. This arrangement will lessen the
transient voltage between the voltage leads and
ground.

   Since voltage transformers are inductive devices,
they are not subject to this phenomenon.

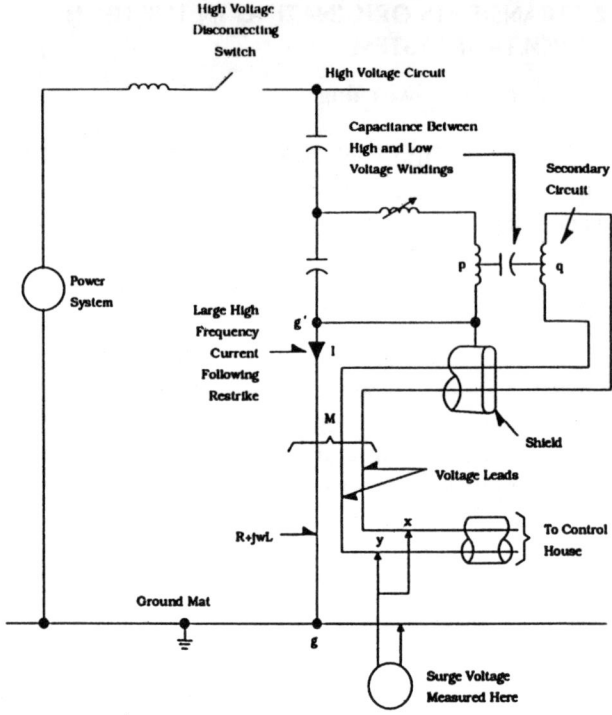

Figure 4-7 Surge in secondary leads during disconnect
switch restriking on a capacitance voltage transformer.

2.5 Other Transient Sources

Many other switching-type operations generate transients: unequal pole-closing of a circuit breaker, fault
occurrence, fault clearing, load-tap changing, line
reactor deenergization, series capacitor gap flashing
and reinsertion, and so forth. In general, the peak
magnitude of such transients is substantially less than
for the phenomenon described above.

3 TRANSIENTS ORIGINATING IN THE LOW-VOLTAGE SYSTEM

3.1 Direct Current Coil Interruption

During interruption of an inductive circuit, such as a
relay coil, the $L\, di/dt$ effect may produce a large
voltage across the coil (Fig. 4-8). In general, the
voltage will be greatest at the instant of interruption.
Voltage magnitude will generally be independent of the
supply circuit characteristics and equal to the difference between the extinction voltage of the interrupting
contact and the battery voltage. The surge voltage

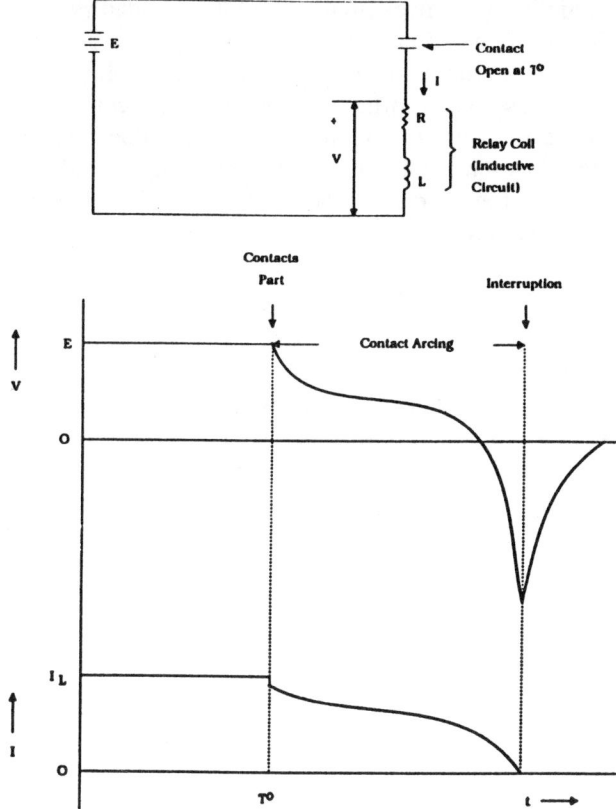

Figure 4-8 Transients produced by interruption of an induction circuit.

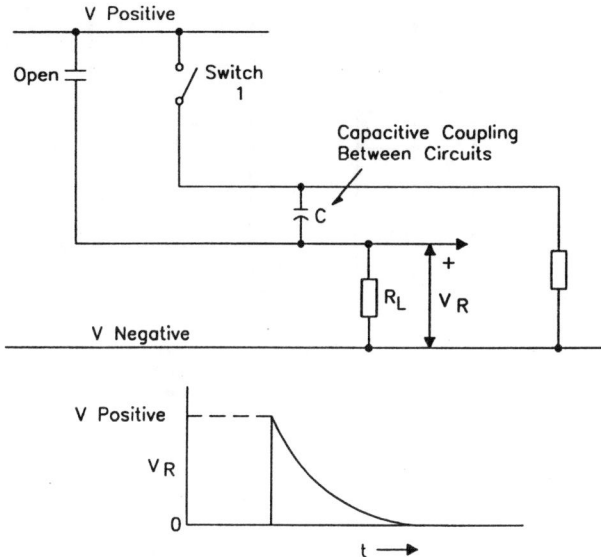

Figure 4-9 Transients produced in adjacent circuits by dc circuit energization.

increases as a function of the speed with which the interruptor forces current zero. Although voltages in excess of 10 kV have been generated across 125-V coils in laboratory tests. 2.5 kV is a more typical value.

3.2 Direct Current Circuit Energization

Energizing a circuit that is capacitively coupled to adjacent or nearby circuits can produce a transient in the latter circuits (Fig. 4-9). When switch 1 is closed, V_R appears as a false signal across the effective resistance of the adjacent circuit. Initially, full battery voltage appears across the coupled circuit. This voltage the decays exponentially, in accordance with the RC tim constant.

3.3 Current Transformer Saturation

Current transformer saturation, which may produce ver high secondary voltage, is caused by high primary current, poor current transformer quality, or excessive burden. The surge repeats during each transition from saturation in one direction to saturation in the other. The voltage appearing at the secondary consists high-magnitude (possibly several kV) spikes with alternating polarity that persist for a few milliseconds eve half-cycle.

3.4 Grounding of Battery Circuit

When a ground occurs on the dc system, the distributed and lumped capacitance of a system may cause sensitive devices to operate. Figures 4-10 and 4-11 illustrate trip circuit behavior in the event of an accidental ground. Comparable phenomena can cause sensitive close circuits and tripping relays to malfunction.

4 PROTECTIVE MEASURES

4.1 Separation

4.1.1 Physical Separation

Noise in critical circuits can be controlled effectively by physically separating quiet and noisy circuits. Since mutual capacitance and mutual inductance are inverse logarithmic functions of distance, small increases in distance produce substantial decreases in circuit interaction.

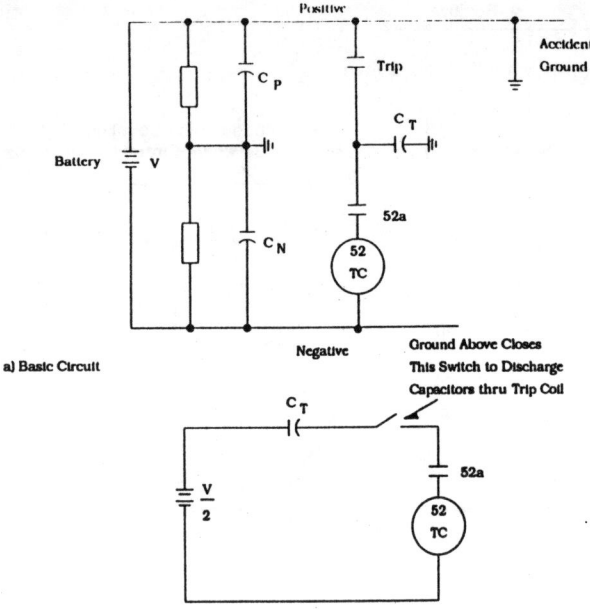

Figure 4-10 Accidental ground on battery positive.

Similarly, control circuits should be routed perpendicular to noisy circuits. For example, a cable duct should be run perpendicular to a high-voltage bus when possible. Another way of effectively controlling surges is to group circuits with comparable sensitivities. Low-energy-level circuits, especially, should be grouped together and placed as far as possible from power circuits.

4.1.2 Electrical Separation

Circuits can, of course, also be separated electrically. For example, surges can be controlled by the discriminate application of inductance to block conduction of high-frequency transients into protected regions. This principle is illustrated by the filter circuit shown in Figure 4-12. High-frequency transients are diverted harmlessly to ground.

Transformer isolation (Fig. 4-13) puts an effective common-mode barrier between segments of a system. High capacitance from each winding to ground and low capacitance from winding to winding further reduce common-mode interaction between windings.

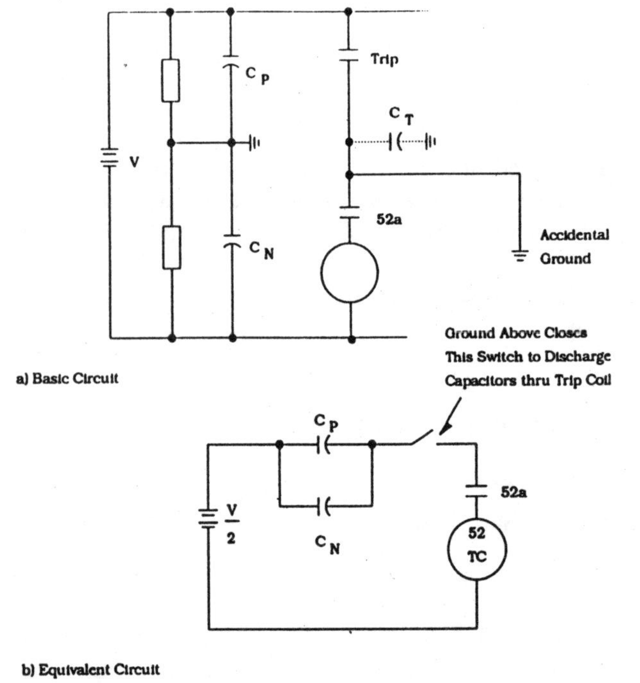

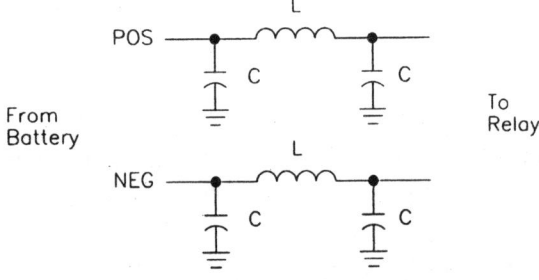

Figure 4-12 Choke coil isolation.

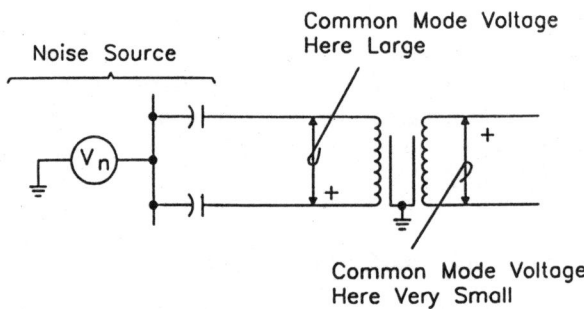

Figure 4-13 Transformer isolation for common mode voltage.

Figure 4-11 Accidental ground on trip lead.

4.2 Suppression at the Source

4.2.1 Resistor Switching

Transient voltages can be kept comparatively low by equipping disconnects and circuit breakers with resistors that are inserted during operation of the device. For reasons of economy, this arrangement is occasionally used to restrict the surge level in substations.

4.2.2 Parallel Clamp

The surge associated with coil interruption can be virtually eliminated by paralleling the coil with a zener diode. Where an extended dropout time is undesirable, a varistor may be substituted for the zener diode arrangement. Although the varistor allows a higher surge than the zener diode, its limiting action is satisfactory.

The zener diode Z_1 in Figure 4-14 performs a dual surge function. First, it minimizes the inductive "kick" produced by the deenergization of the auxiliary coil. Also, when one of the contacts (Trip A in Figure 4-14) closes, the voltage induced in TC BKR A resulting from the interruption of current flow when 52a A opens cannot cause the auxiliary relay to pick up undesirably. Current path I will have no detrimental effect. Zener Z1 will allow forward voltage of only approximately 0.7 V, which is insufficient to operate the auxiliary relay. This scheme prevents undesirable tripping of breaker B.

The transient associated with extreme ac saturation of a current transformer can also be squelched by introducing a voltage-limiting device across the secondary. Silicon carbide devices can be used in this protective function.

4.2.3 Suppression by Termination

Figures 4-1, 4-2, 4-3, and 4-9 illustrate the value of reduced input impedance R_L in restricting the magnitude and/or duration of transients. However, if R_L is reduced, the energy requirement for operation is increased, and more heat is generated when legitimate inputs are applied.

A small capacitor offers another method of reducing input impedance at high frequency, with little effect at 50 or 60 Hz or on dc. This device neither requires a higher input energy for operation nor generates heat. One such widely used capacitor is a 0.01-μf ceramic capacitor. It limits a 2500-V, 1-MHz surge, with a 150-Ω source to 350 V peak to peak. Short leads to the capacitor are imperative.

When sensitive relays, trip circuits, and close circuits exist in a substation, the capacitance on the dc must be restricted if false operations are to be avoided when a ground occurs.

4.3 Suppression by Shielding

A signal lead that is shielded and has one or more grounds will have increased capacitance to ground C_G (Fig. 4-1). For high R_S and R_L values, this increase in capacitance to ground reduces the "false" signal voltage V_L that results from the presence of an adjacent noisy lead. If a shield were used in Figure 4-2, it would surround the signal lead and the common negative and would be grounded in one or more locations. This arrangement tends to force C_{M1}/C_{G1} to equal C_{M2}/C_{G2}, and any capacitively induced signal voltage across R_L to be 0.

Grounding a shield at both ends allows shield current to flow. Shield current resulting from magnetic induction will tend to cancel the flux that created it. The net effect of the shield on the signal lead is to reduce the noise level. Both ends should not be grounded if the shield is the signal return path.

4.4 Suppression by Twisting

Measures that cause the signal and return leads to occupy essentially the same space minimize the effect of differential-mode coupling (Fig. 4-4). As shown by the polarity marks, twisting a pair of leads cancels the effect of adjacent circuit flux. Also, twisting the signal

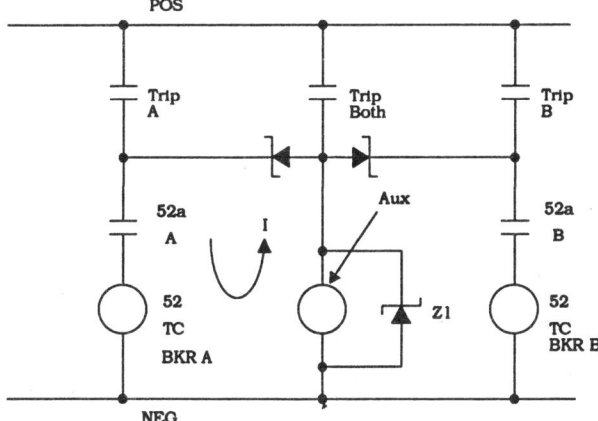

Figure 4-14 Zener Z1 applied for surge suppression.

lead and negative causes C_{M1} to equal C_{M2}, and C_{G1} to equal C_{G2} (Fig. 4-2). This technique substantially reduces the influence of the adjacent noisy lead.

A combination of shielding and twisting effectively minimizes the influence of surges in adjacent circuits. For circuits properly treated with SPP capacitors at the terminal blocks, shielding is not required for static relaying circuits inside a panel or switchboard. These SPP capacitors are 750-V dc oil-filled capacitors. Shielded twisted pair conductors are required for low-energy-level circuits routed outside a panel.

One lead of the shielded twisted pair is normally the signal lead. The other lead (except where it is sensing contact status) connects the negatives of the two devices. Within a panel, electrostatic coupling is the only significant intercircuit transient influence. A single ground on the shield, therefore, is sufficient. For consistency, ground should be at the input end.

4.5   Radial Routing of Control Cables

Circuits routed into the switchyard from the control house should not be looped from one piece of switchyard apparatus to another with the return conductor in another cable. Rather, all supply and return conductors should be in a common cable. This arrangement avoids the large EMI (electromagnetic induction) associated with the large flux loop that would otherwise be produced.

4.6   Buffers

Another effective method of delaying and desensitizing a circuit is to use a buffer (Fig. 4-15). Without causing the transistor to conduct or damaging any element, this buffer can accommodate a test source operating at 1 to 1.5 MHz with a 150-Ω source impedance directly across the input (differential mode) and a 2500-V

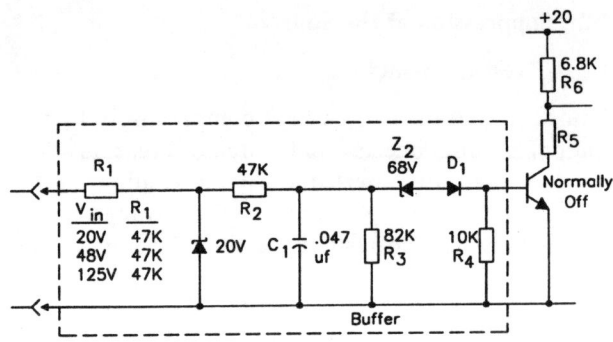

Figure 4-15   A standard input buffer circuit for solid-state relays.

(open circuit) first peak, which decays to 1250 V in 6 or more μsec. The buffer can also withstand a sustained 7-Vdc input, or a high dc input voltage of sufficient duration to produce a minimum 4000-μsec-V product (for example, 20 V for 200 μs).

Buffering low-energy-level circuits greatly decreases the susceptibility of static relays to surge damage or malfunction and, in general, eliminates the need for shielding circuits inside a relaying panel.

4.7   Optical Isolators

Optical isolators can provide excellent electrical separation between two circuits. Figure 4-16 describes an input isolator. When the LED is conducting, as a result of a voltage being applied at the input, base drive is provided for the phototransistor and it conducts, supplying a "logical" input to the protected equipment.

Similar isolation is accomplished for an output by a circuit such as that of Figure 4-17.

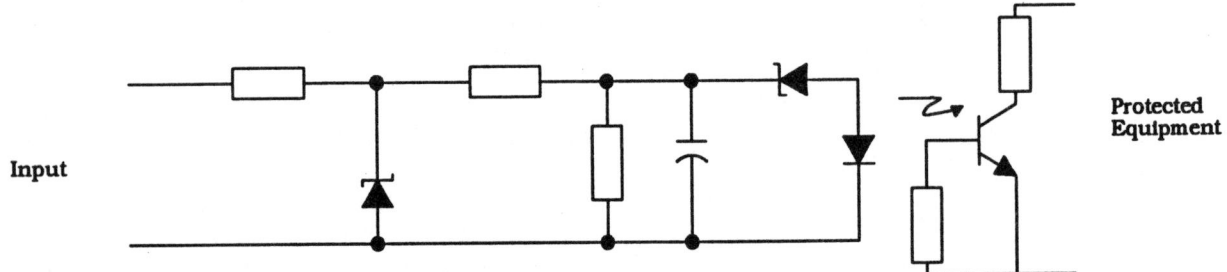

Figure 4-16   Optically isolated input.

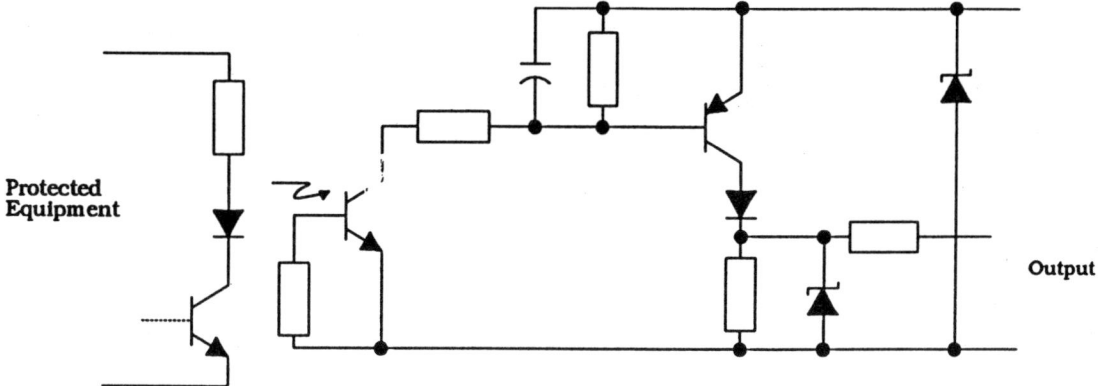

Figure 4-17 Optically isolated output.

4.8   Increased Energy Requirement

Surges can also be endured by raising the threshold voltage or energy level at which operation occurs. The equivalent circuits of Figures 4-10 and 4-11 show that half-maximum battery voltage, applied through the appropriate capacitance, must not be allowed to trip the breaker or operate any other devices. An auxiliary relay designed to pick up at 71 V or more will not respond to a single ground on a dc circuit with a maximum operating voltage of 140 V, regardless of the magnitude of the capacitances on the system. (Note that the higher-voltage design will not solve the problem shown in Figure 4-9; the higher-energy-level restriction will.)

5

Instrument Transformers for Relaying

W. A. ELMORE

1 INTRODUCTION

Instrument transformers are used both to protect personnel and apparatus from high voltage and to allow reasonable insulation levels and current-carrying capacity in relays, meters, and instruments. Instrument transformer performance is critical in protective relaying, since the relays can only be as accurate as the information supplied them by the instrument transformers. Standard instrument transformers and relays are normally rated at 5 or 1 A; 100, 110, or 120 V; and 50 or 60 Hz.

Where the relays operate only on current or voltage magnitude, the relative direction of current flow in the transformer windings is not important. Relative direction (and, therefore, polarity) must be known, however, where the relays compare the sum or difference of two currents or the interactions of several currents or voltages. The polarity is usually marked on the instrument transformer but can be determined if necessary.

2 CURRENT TRANSFORMERS

One major criterion for selecting a current transformer ratio is the continuous current ratings of the connected equipment (relays, auxiliary current transformers, instruments, etc.) and of the secondary winding of the current transformer itself. In practice, with load current normally flowing through the phase relays or devices, the ratio is selected so that the secondary output is around 5 A (or 1 A) at maximum primary load current. When delta-connected current transformers are used, the $\sqrt{3}$ factor must be considered.

Although the performance required of current transformers varies with the relay application, high-quality transformers should always be used. The better-quality transformers reduce application problems, present fewer hazards, and generally provide better relaying. The quality of the current transformers is most critical for differential schemes, where the performance of all the transformers must match. In these schemes, relay performance is a function of the accuracy of reproduction—not only at load currents, but also at all fault currents as well.

Some differences in performance can be accommodated in the relays. In general, the performance of current transformers is not so critical for transmission line protection. The current transformers should reproduce reasonably faithfully for faults near the remote terminal, or at the balance point for coordination or measurement.

2.1 Saturation

For large-magnitude, close-in faults, the current transformer may saturate; however, the magnitude of fault current is not critical to many relays. For example, an induction overcurrent relay may be operating on the flat part of the curve for a large-magnitude, close-in fault. Here it is relatively unimportant whether the current transformer current is accurate, since the timing is essentially identical. The same is true for instantaneous or distance-type relaying for a heavy internal fault well inside the cut-off or balance point. In all cases, however, the current transformer should provide sufficient current to operate the relay positively.

2.2 Effect of dc Component

The presence of dc in the primary current can be particularly detrimental to ct performance. This phenomenon is described in Section 6, "Direct Current Saturation." Having a dc component in fault current on an ac power system is a decaying phenomenon. If a ct is going to saturate due to the dc component of fault current, it will do so in the first few cycles. Until this effect takes place, the fidelity of transformation is reasonably good, and instantaneous overcurrent and distance relays may perform their task before the ct performance collapses.

Following the disappearance of the dc component of fault current, the behavior of the ct will again improve. The error of transformation may then be predicted by using one of the methods described in Section 4.

3 EQUIVALENT CIRCUIT

An approximate equivalent circuit for a current transformer is shown in Figure 5-1. Current is stepped down in magnitude through the perfect (no-loss) transformation provided by windings ab and cd. The primary leakage impedance (Z_H) is modified by n^2 to refer it to the secondary. The secondary impedance is Z_L; Rm and Xm represent the core loss and exciting components.

This generalized circuit can be reduced, as shown in Figure 5-1b. Z_H can be neglected, since it influences neither the perfectly transformed current I_H/n nor the voltage across Xm. The current through Xm, the magnetizing branch, is I_e, the exciting current. The Rm branch produces a negligible influence.

The phasor diagram, with exaggerated voltage drops, is shown in Figure 5-1c. In general, Z_L is resistive and Z_B is resistive or has a lagging angle. I_e lags V_{cd} by 90° and is the prime source of error. Note that the net effect of I_e is to cause I_L to lead and be smaller than the perfectly transformed current I_H/n.

Any simple equivalent diagram for a current transformer is, at best, crude. Exciting current is accompanied by harmonics that, in turn, produce harmonic relay currents. An analysis for application

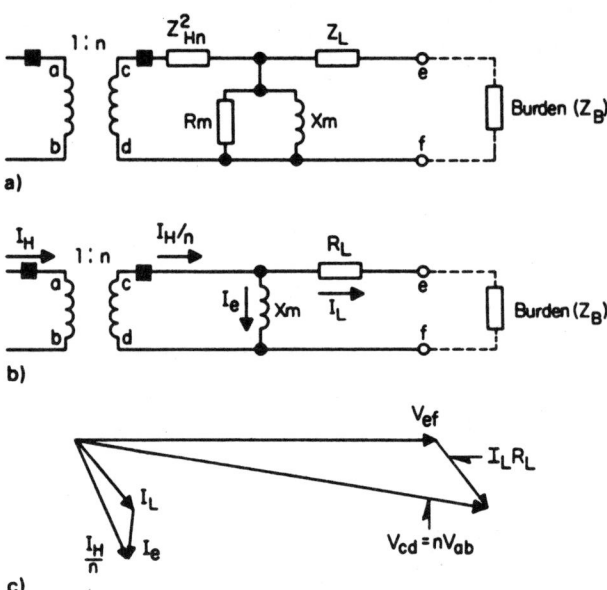

Figure 5-1 The equivalent circuit and phasor diagram of a current transformer.

purposes is usually made on the basis of sinusoidal fundamental quantities. Although this approach is highly simplified, the equivalent diagram is an excellent tool for picturing the phenomenon and estimating the approximate performance to be expected.

4 ESTIMATION OF CURRENT TRANSFORMER PERFORMANCE

A current transformer's performance is measured by its ability to reproduce the primary current in terms of the secondary; in particular, by the highest secondary voltage the transformer can produce without saturation and, consequently, large errors. Current transformer performance with symmetrical (no dc) primary current can be estimated by

> Formula
> The current transformer excitation curves
> The ANSI transformer relaying accuracy classes

The first two methods provide accurate data for analysis; the latter gives only a qualitative appraisal. All three methods require determining the secondary voltage V_{cd} that must be generated

$$V_{cd} = V_S = I_L(Z_L + Z_{lead} + Z_B) \qquad (5-1)$$

where

V_S = rms symmetrical secondary induced voltage Fig. 5-1)

I_L = maximum secondary current in amperes (symmetrical)

Z_B = connected burden impedance in ohms

Z_L = secondary winding impedance in ohms

Z_{lead} = connecting lead impedance in ohms

4.1 Formula Method

The formula method uses the fundamental transformer equation

$$V_S = 4.44\,fANB_{max}10^{-8}(volts) \qquad (5-2)$$

where

f = frequency in hertz

A = cross-sectional area of the iron core in square inches

N = number of turns

B_{max} = flux density in lines per square inch

Both the cross-sectional area of the iron and its saturation density are sometimes difficult to obtain. Current transformers generally use silicon steels, which saturate from 77,500 to 125,000 lines/in.[2]. The lower figure is typical for current transformers built before 1947; a value of 100,000 is typical of most transformers.

The formula method consists of determining V_S using Eq. (5-1), then calculating B_{max} using Eq. (5-2). If B_{max} exceeds the saturation density, there will be appreciable error in the secondary current.

Assume, for example, that a 2000:5, high-permeability silicon steel transformer has 3.1 in.[2] of iron and a secondary winding resistance of 0.31 Ω. The maximum current for which the current transformer must operate is 40,000 A at 60 Hz. The relay burden, including the secondary leads, is 2.0 Ω. Will this current transformer saturate?

If the current transformer does not saturate, the secondary current I_S would be 40,000 divided by 400, or 100 A, since N equals 400. Thus, the current transformer should be able to produce a secondary

voltage V_S of 100(2.0 + 0.31), or 231 V. Equation (5-2), solved for B_{max}, will determine whether the current transformer can reproduce this current

$$B_{max} = \frac{231 \times 10^8}{4.44 \times 60 \times 3.1 \times 400} = 70,000\,lines/in.^2$$

Therefore, the current transformer should have iron that will not saturate below 70,000 lines/in.[2]. Since the current transformer in this example uses high-permeability silicon steel, it will not saturate with symmetrical primary current.

4.2 Excitation Curve Method

A typical excitation curve for a current transformer is shown in Figure 5-2. These data represent rms currents obtained by applying rms voltage to the current transformer secondary, with the primary open-circuited. The curve gives the approximate exciting current requirements for a given secondary voltage.

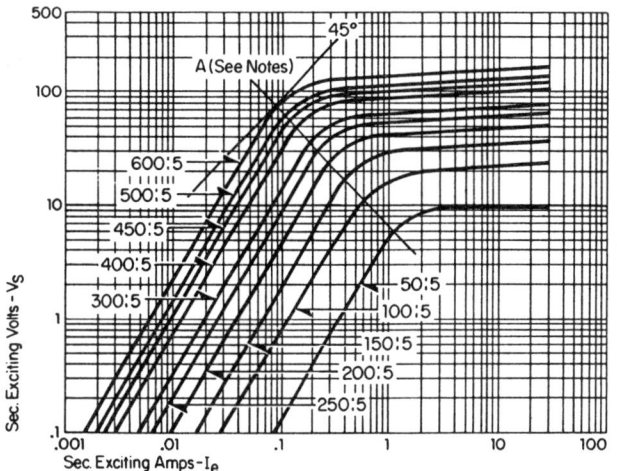

Current Ratio	Turn Ratio	Sec. Res. 1
50:5	10:1	.061
100:5	20:1	.082
150:5	30:1	.104
200:5	40:1	.125
250:5	50:1	.146
300:5	60:1	.168
400:5	80:1	.211
450:5	90:1	.230
500:5	100:1	.242
600:5	120:1	.296

Notes:

1) Above The Line, The Voltage for a Given Exciting Current Will Not be Less Than 95% of The Curve Value.

2) Below The Line, The Exciting Current for a Given Voltage Will Not Exceed The Curve Value by More Than 25%.

Figure 5-2 Exitation curves for a multiratio bushing current transformer with an ANSI accuracy classification of C100.

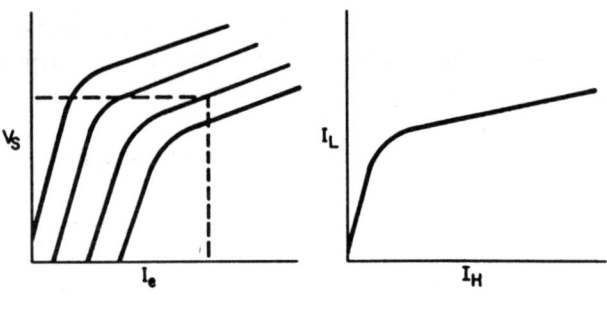

Notes:

a) Assume I_L

b) $V_S = I_L Z_T$ Where $Z_T =$
   $Z_L + Z$ Leads $+ Z$ Burden

c) Find I_e from Curve

d) $(I_L + I_e)n = I_H$

e) Repeat and Plot Curve

Figure 5-3 Excitation curve method.

With this method, a curve relating primary current to secondary current can be developed for the tap, lead length, and burden being used (Fig. 5-3). Any value of primary current can then be entered on the curve to determine the expected value of secondary current.

The following examples will illustrate some of the problems encountered in estimating current transformer performance using the excitation curve method.

Example 1   Phase Relays

The breaker has a multiratio 600:5 bushing current transformer and the feeder is protected with overcurrent relays. The relays should operate for approximately 60 A rms symmetrical primary current. The total burden on the current transformer, including the current transformer secondary resistance, is $1.6\,\Omega$ per phase when the relays are on the 6-A tap, and $3\,\Omega$ per phase on the 3-A tap. The excitation curve for the transformer is shown in Figure 5-2.

One approach would be to use a current transformer ratio of 60:6, or 10 (the 50:5 tap), to take advantage of the lower burden on relay tap 6:

$$N = 10 \text{ turns}$$
$$I_L = 6 \text{ A to operate the relay}$$
$$V_S = I_L Z \text{ total} = 6 \times 1.6 = 9.6 \text{ V}$$

From the excitation curve for V_S of 9.6 V, I_e would be 6 A, and NI_e equals 60. Therefore, the primary pickup current is

$$I_H = NI_S + NI_e = 60 + 60 = 120 \text{ A}$$

This value is considerably higher than the 60 A desired.

In theory, when making the phasor addition, the angle of the burden and the exciting branches should be taken into account. This refinement is not necessary, however, since it is obvious from the curve of Figure 5-2 whether or not the current transformer would be operating in the saturated region.

An alternative approach would be to use a ratio of 60:3, or 20 (the 100:5 tap), with the higher burden of the 3-A relay tap. If we use this ratio, N equals 20, and I_S equals 3 A to operate this relay:

$$V_S = I_L Z \text{ total} = 3 \times 3 = 9 \text{ V}$$

From the excitation curve (Fig. 5-2), I_e equals 0.5 A, and NI_e is 10. The primary pickup current I_H would be

$$I_H = 60 + 10 = 70.0 \text{ A}$$

This value is closer, but still too high.

Now suppose that the breaker has two sets of current transformers, with the secondaries connected in series. Then each current transformer carries one-half the burden, or $1.54\,\Omega$ on the 3-A tap. This value is slightly more than one-half of $3.0\,\Omega$ because of the secondary resistance of the added transformer. Using the 100:5 tap, we obtain

$$N = 20 \text{ turns}$$
$$I_L = 3 \text{ A}$$
$$V_S = 3 \times 1.54 = 4.62 \text{ V per transformer}$$

Then, from Figure 5-2, we have

$$I_e = 0.33$$
$$NI_e = 6.6$$
$$I_H = 3 \times 20 + 6.6 = 66.6 \text{ A}$$

Although this alternative offers some improvement, I_H is not as close to the desired 60 A as might have been expected. In both cases, the current transformer is operating on the straight-line part of the characteristic, making significant improvement difficult. On the other hand, two 50:5 current transformers in series would show a marked improvement over the 50:5 ratio. Here I_H, calculated by the above methods, is 71 A. While much better than 120 A, this value is still not as good as the 66.6-A pickup obtained using the two current transformers with a 100:5 ratio.

Similar evaluations can be made for other configurations.

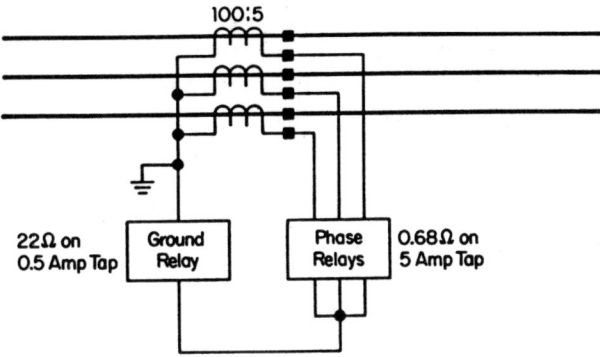

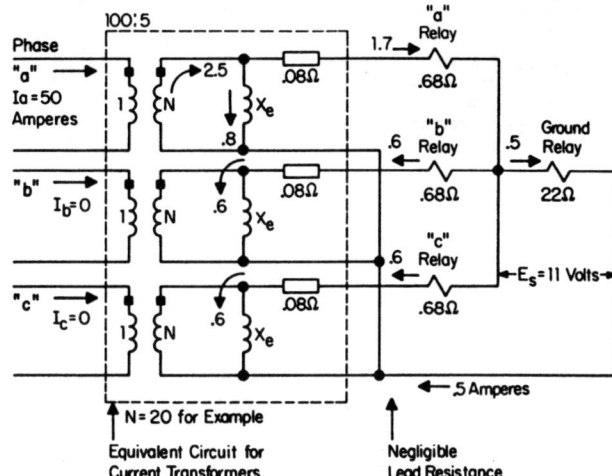

Figure 5-4 Connections for Example 2 illustrating calculation of current transformer performance.

Figure 5-5 Equivalent circuits and distribution of currents for a ground fault with the connections of Figure 5-4 in Example 2.

Example 2   Phase and Ground Relays

The following example, shown in Figure 5-4, will determine the minimum primary current that will operate the phase and ground relays.

PHASE RELAYS.   For the phase relays, the total phase burden Z equals 0.68 plus 0.08, or 0.76 Ω, where 0.08 is the current transformer secondary resistance on the 100:5 tap ($N = 20$):

$$I_L = 5 \text{ A (to operate the relay on the 5-A tap)}$$
$$V_S = 0.76 \times 5 = 3.8 \text{ V}$$

From Figure 5-2, we get

$$I_e = 0.28 \text{ A}$$
$$I_H = N(I_L + I_e) = 20(5 + 0.28) = 105.6 \text{ A primary}$$

If we neglect the exciting current (I_e), this value would become 20 times 5, or 100 A primary, when using the 100:5 current transformer ratio.

GROUND RELAYS.   If we assume the ground current flows only in phase a, the equivalent circuit is shown in Figure 5-5. To obtain 0.5 Ω through the ground relay, with its impedance assumed here of 22 Ω, 11 V must be produced across the ground relay. If we neglect the small unknown voltage across the phase relays, this ground relay voltage will appear across the phase-b and phase-c current transformers to excite them from the secondary side. From Figure 5-2, an I_e of 0.6 A develops 11 V across these current transformers. The accuracy required, generally, does not warrant correction for the small phase relay drop. However, such a correction could be made on a trial-and-error basis.

Thus, the phase-a relay circuit must supply 0.6 plus 0.6 plus 0.5, or 1.7 A. Given the phase relay impedance of 0.68 Ω and current transformer impedance of 0.08 Ω, or 0.76 Ω total, the phase-a current transformer must supply

$$V_S = 11 + (1.7 \times 0.76) = 12.3 \text{ V}$$
$$I_e = 0.8 \text{ A}   \text{(from Fig. 5-2)}$$
$$I_L = 1.7 + 0.8 = 2.5 \text{ A}$$
$$I_H = 2.5 \times 20 = 50 \text{ A primary}$$

Thus, 50 A is required to operate the ground relay. If the exciting requirements of the three current transformers had been ignored, the current required to operate the ground relay would have been estimated to be 0.5 times 20, or 10 A primary. From this, it is apparent that such may be a significant factor.

Using the 200:5 tap on the current transformer could improve sensitivity here. Dramatic improvement would also be possible if a modern, low-impedance ground relay were substituted.

4.3   ANSI Standard: Current Transformer Accuracy Classes

The ANSI relaying accuracy class (ANSI C57-13) is described by two symbols—letter designation and voltage rating—that define the capability of the transformer.

The letter designation code is as follows:

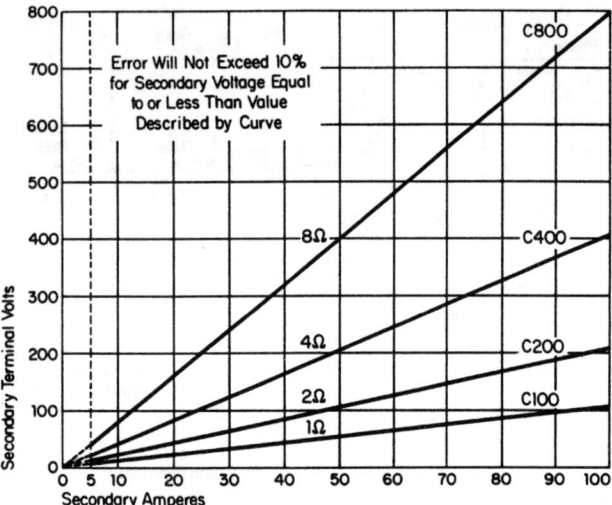

Figure 5-6 ANSI accuracy standard chart for class C current transformers.

C: The transformer ratio can be calculated.

T: The transformer ratio must be determined by test.

The C classification covers bushing current transformers with uniformly distributed windings, and any other transformers whose core leakage flux has a negligible effect on the ratio within the defined limits.

The T classification covers most wound-type transformers and any others whose core leakage flux affects the ratio appreciably.

The secondary terminal voltage rating (Fig. 5-6) is the voltage that the transformer will deliver to a standard burden at 20 times normal secondary current, without exceeding a 10% ratio error.

Figure 5-6 shows the secondary voltage capability for various C-class current transformers, plotted against secondary current. With a transformer in the C100 accuracy class, for example, the transformer ratio can be calculated, and the ratio error will not exceed 10% between 1 and 20 times normal secondary current if the burden does not exceed $1.0\,\Omega$ ($1.0\,\Omega \times 5\,\mathrm{A} \times 20 = 100\,\mathrm{V}$).

ANSI accuracy class ratings apply only to the full winding. When there is a tapped secondary, a proportionately lower-voltage rating exists on the taps.

4.3.1 Current Transformer Data

The following current transformer data, required for relaying service application, should be supplied by the manufacturer:

Relaying accuracy classification.

Mechanical and thermal short-time (1-sec) ratings. Both ratings define rms values that the transformer is capable of withstanding. For mechanical short-time ratings, the rms value is that of the ac component of a completely displaced primary current wave. The thermal 1-sec rating is the rms value of the primary current that the transformer will withstand with the secondary winding short-circuited, without exceeding the limiting temperature of 250 °C for 55 °C-rise transformers, or 350 °C for 80 °C-rise transformers. The short-time thermal current rating for any period of 1 to 5 sec is determined by dividing the 1-sec current rating by the square root of the required number of seconds.

Resistance of the secondary winding between the winding terminals. Data should be presented in a form that allows the value for each published ratio to be determined.

For T-class transformers, the manufacturer should supply typical overcurrent ratio curves on rectangular coordinate paper. The plot should be between primary and secondary current, over the range from 1 to 22 times normal current, for all standard burdens up to the one that causes a ratio error of 50% (Fig. 5-7).

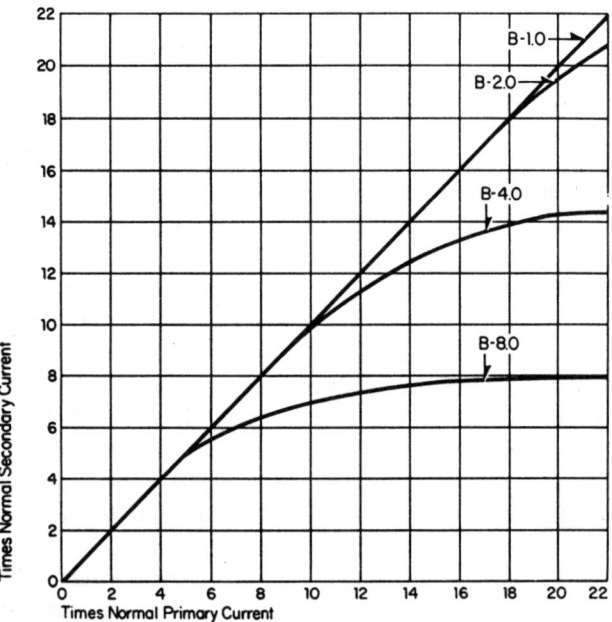

Figure 5-7 Typical overcurrent ratio curves for a T class current transformer.

For C-class transformers, the manufacturer should also supply typical excitation curves on log-log coordinate paper. The plot should show excitation current and secondary terminal voltage for each published ratio from 1% of the accuracy class secondary terminal voltage to a voltage (not to exceed 1600 V) that will cause an excitation current of five times normal secondary current (Fig. 5-2).

The ANSI standard burden is defined with a 50% power factor. These standard ohmic burden values are identified in Figure 5-6. When fewer than the total number of turns are in use on the C-class current transformer, only a portion of that burden can be supplied without exceeding the 10% error. Maximum permissible burden is defined mathematically by

$$Z_B = \frac{N_P V_{cl}}{100} \qquad (5\text{-}3)$$

where

Z_B = permissible burden on the current transformer
N_P = turns in use divided by total turns
V_{cl} = current transformer voltage class

Standard relaying burdens are listed in Table 5-1. B-0.1, B-0.2, B-0.5, B-0.9, and B-1.8 are standard metering burdens. These burdens are defined with a 0.9 power factor.

The following example shows current transformer calculations using ANSI classifications: the maximum calculated fault current for a particular line is 12,000 A. The current transformer is rated at 1200:5 and is to be used on the 800:5 tap. Its relaying accuracy class is C200 (full-rated winding); secondary resistance is 0.2 Ω. The total secondary circuit burden is 2.4 Ω at a 60% power factor. Excluding the effects of residual magnetism and dc offset, will the error exceed 10%? If

so, what corrective action can be taken to reduce the error to 10% or less?

The current transformer secondary winding resistance may be ignored because the C200 relaying accuracy class designation indicates that the current transformer can support 200 V plus the voltage drop caused by ct internal secondary resistance at 20 times rated current, for a 50% power-factor burden. The ct secondary voltage drop may be ignored then if the secondary current does not exceed 100 A:

$$N = \frac{800}{5} = 160$$
$$I_L = \frac{12{,}000}{160} = 75 \text{ A}$$

The permissible burden is given by (from Eq. 5-3)

$$Z_B = \frac{N_P V_{cl}}{100}$$
$$N_P = \frac{800}{1200} = 0.667 \quad \text{(proportion of total turns in use)}$$

Thus

$$Z_B = \frac{0.667(200)}{100} = 1.334 \,\Omega$$

Since the circuit burden, 2.4 Ω, is greater than the calculated permissible burden, 1.334 Ω, the error will exceed 10% at the maximum fault current level (75 A). Consequently, it is necessary to reduce the burden, use a higher current transformer ratio, or use a current transformer with a higher relaying accuracy class voltage.

5 EUROPEAN PRACTICE

In Europe, current transformers are described in terms of protection and measurement classes. The protection classes are those of most interest in relaying, and they carry the designation P, a maximum error of 5 or 10%, a corresponding volt-ampere burden, a rated current, and an accuracy limit factor. For example, a 30-VA class, 5P10, 5-A ct is compatible with a 30-VA continuous burden at 5 A. This corresponds to a 6-V output. It produces no more than 5% error at $10 \times 6 = 60$ V. The permissible burden is $30/(5 \times 5) = 1.2 \,\Omega$.

Three types of ct's are defined by the TPX, TPY, and TPZ designations.

Table 5-1 Standard Relay Burden Designations

Characteristics for 60-Hz and 5-A secondary circuit			
Standard burden designation	Impedance (r)	VA	Power factor
B-1	1.0	25	0.5
B-2	2.0	50	0.5
B-4	4.0	100	0.5
B-8	8.0	200	0.5

5.1 TPX

The TPX is a nongapped core current transformer with a 0.5% ratio error and secondary time constant of 5 sec or more. It may be used with other TPX or TPY ct's in all types of protection applications.

5.2 TPY

This ct has a gapped core and secondary time constant of 0 to 10 sec. It has a ratio error of $\pm 1\%$ and larger cost than the TPX. Its transformation of the dc component of fault current is not as accurate as the TPX. It may be combined with other TPX or TPY ct's in any relaying application. Its advantage is that its remanent flux is quite small compared to that of a nongapped core ct.

5.3 TPZ

TPZ ct's have a linear core with a secondary time constant of 60 ± 6 msec for 50-Hz and 50 ± 5 msec for 60-Hz applications. This provides a very short dc collapse time, making the ct suitable for breaker failure applications in which the overcurrent supervision is susceptible to dc influence. When used in combinations, it should be used only with other TPZ ct's. It has a $\pm 1\%$ ratio error at the rated primary current.

6 DIRECT CURRENT SATURATION

To this point, current transformer performance has been discussed in terms of steady-state behavior only, without considering the dc component of the fault current. Actually, the dc component has far more influence in producing severe saturation than the ac fault current. The dc component arises because (1) the current in an inductance cannot change instantaneously and (2) the steady-state current, before and after a change, must lag (or lead) the voltage by the proper power-factor angle.

Figure 5-8 shows the current immediately following fault inception for two cases: fully offset and with no offset. In the fully offset case, the fault is assumed to occur at the instant that produces the maximum dc component. In the second case, the fault occurs at a time that produces no dc offset.

Figure 5-9 shows an example of the distortion and reduction in the secondary current that occurs as a

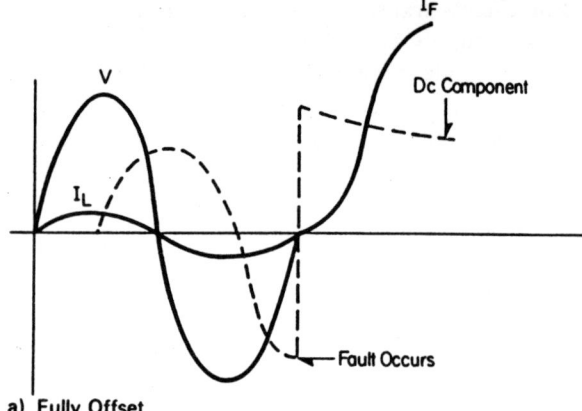

a) Fully Offset

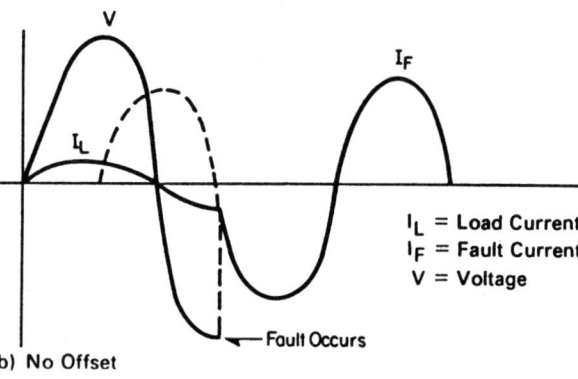

I_L = Load Current
I_F = Fault Current
V = Voltage

b) No Offset

Figure 5-8 Current immediately after fault inception.

result of dc saturation. Note the improvement in performance as the dc diminishes.

If $V_K \geq 6.28$ IRT, the dc component of a fault current will not produce current transformer saturation. In this expression,

V_K = voltage at the knee of the saturation curve, determined by extending the straight-line portions of the curve to find their intersection
I = symmetrical secondary current in amperes rms

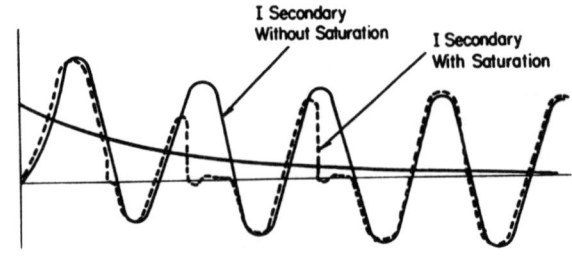

Figure 5-9 Direct current saturation of current transformer.

R = total secondary resistance in ohms
T = dc time constant of the primary circuit in cycles

Here, $T = (L_P/R_P)f$, where

L_P = primary circuit inductance in henries
R_P = primary circuit resistance in ohms
f = frequency

Direct current saturation is particularly significant in bus differential relaying systems, where highly differing currents flow to an external fault through the current transformers of the various circuits. Dissimilar saturation in any differential scheme will produce operating current.

Figure 5-10 shows how current transformer saturation relates to time. Severe current transformer saturation will occur if the primary circuit dc time constant is sufficiently long and the dc component sufficiently high. Curves d, e, and f of Figure 5-10 show that the dc component requires substantially greater flux than that needed to satisfy the ac component.

Time is required to reach saturation flux density. This time can be estimated from Figure 5-11, as follows:

> From the current transformer excitation curve for the tap in use, determine V_K from the intersection formed by extending the two straight-line segments of the curve. Note that both axes must have the same logarithmic scales, as is illustrated in Figure 5-2.
> Calculate V_K/IRT.
> Obtain t/T from Figure 5-11.
> Calculate t, the time to saturate.

V_K must be modified if residual flux is to be considered. For example, with a residual flux of 90%, the saturation voltage value must be multiplied by 1 minus 0.9, or 0.1, to determine the earliest time to saturation. This will give a conservative value for time to saturation.

7 RESIDUAL FLUX

Any iron-core device will retain a flux level even after the exciting current falls to 0. Superimposed on this residual flux are variations in core flux, dictated by the current transformer secondary current and secondary burden.

Figure 5-12 shows the importance of previous loading history on current transformer residual flux level. Suppose a current transformer has a residual flux

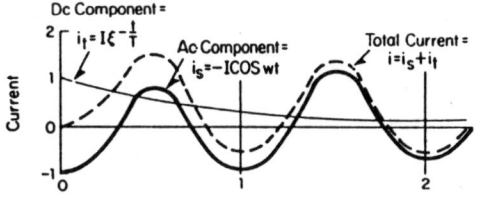

a) Current Wave With Maximum Assymmetry.

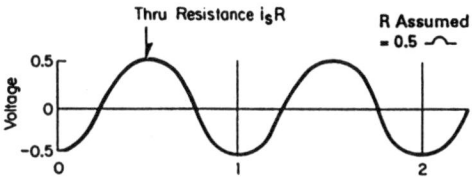

b) Voltage Wave Required to Force Ac Component Through Secondary Resistance.

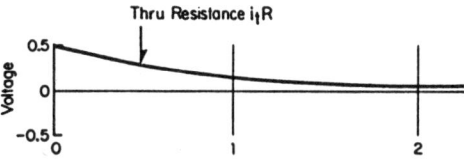

c) Voltage Wave Required to Force Dc Component Through Secondary Resistance.

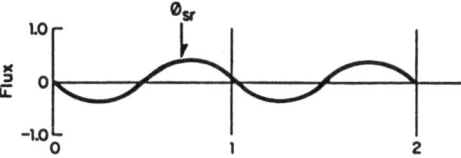

d) Flux Variation to Induce Voltage Wave (b).

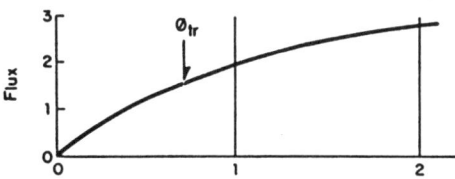

e) Flux Variation to Induce Voltage Wave (c)

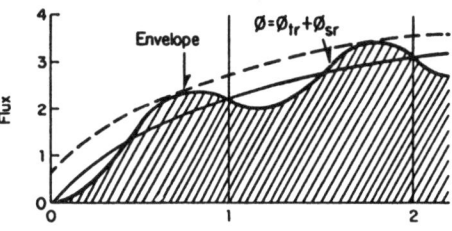

f) Total Flux Variation.

Figure 5-10 Current transformer flux during assymetrical fault (one cycle time constant).

defined by point a. If a symmetrical sinusoidal primary current starts to flow, requiring a flux variation as shown, the pattern between a and A will be traced out.

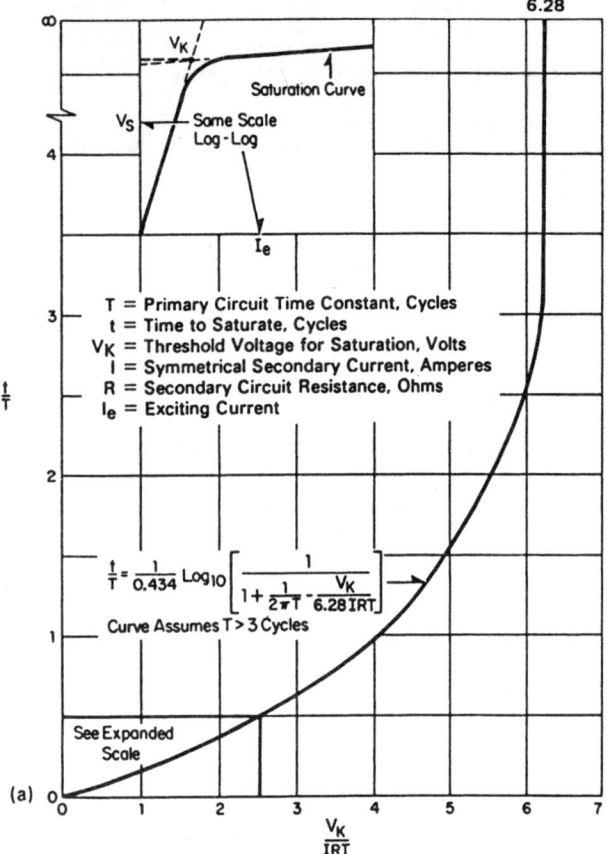

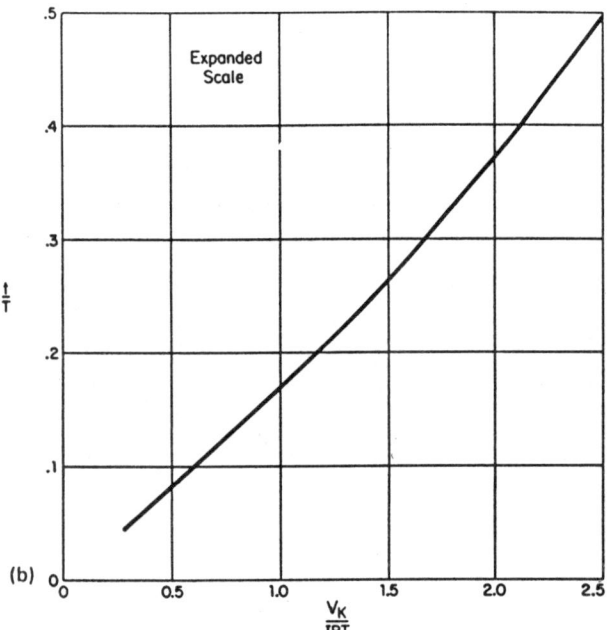

Figure 5-11 (a) Current transformer time to saturate. (b) Expanded scale of (a).

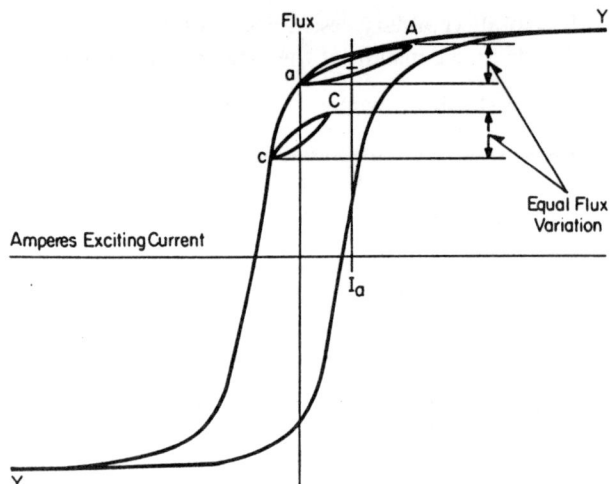

Figure 5-12   Residual flux in current transformer.

The average value of the dc exciting current with this pattern is I_a. This current flows in the secondary and has no counterpart in the primary. It decays with the time constant associated with the secondary circuit. At the completion of this transient, the pattern has moved to cC with an equal flux variation that is symmetrical around the vertical axis. The pattern continues to be traced out. If the circuit were now interrupted, the residual flux would have the value existing at the moment of interruption, which is quite different from the initial value assumed. In fact, any value of flux between 0 and the saturation level may be retained in the core, depending on previous events.

Reducing the residual flux to 0 requires the application of a secondary voltage high enough to produce saturation, followed by a gradual reduction of the voltage to 0.

A current transformer with an air gap in the core has a fairly low residual flux, approximately 10% of saturation density. Residual flux for current transformers with no intentional air gap is approximately 90% of saturation density (max).

Air-gap current transformers do not saturate as rapidly as devices without air gaps subjected to equal current and burden. However, the magnetizing current is higher, resulting in greater ratio and phase-angle errors.

After interruption of the primary current, residual flux decays very slowly (taking approximately 1 sec), and the secondary current collapses slowly. Also, air-gap units are more costly to manufacture, since the small air gap must be both accurate and maintainable.

Although in theory residual flux can cause relaying problems, there have been very few documented cases in which the residual flux has caused a relay misoperation.

8 MOCT

The MOCT (magneto-optic current transducer) relieves many of the problems associated with current transformation. This device utilizes the Faraday effect to produce a high-accuracy analog output that is not influenced by iron saturation. The Faraday effect is the rotation of the plane of polarization when plane-polarized light is sent through glass in a direction parallel to an applied magnetic field. This is illustrated in Figure 5-13. The angle of rotation is directly proportional to the strength of the magnetic field.

In the application of this principle to current measurement, the transmission line current is the source of the magnetic field (see Fig. 5-14). The strength of the field is directly proportional to the instantaneous current magnitude. By placing the rotator (Faraday-effect sensor) in proximity to the transmission line conductor and comparing the angle of rotation of a light beam, a voltage is generated that is directly proportional to current.

This voltage is then used as an input to protective relays at a level and in a way that is virtually identical to that which is used when current transformers supply the relay through a current-to-voltage transformation.

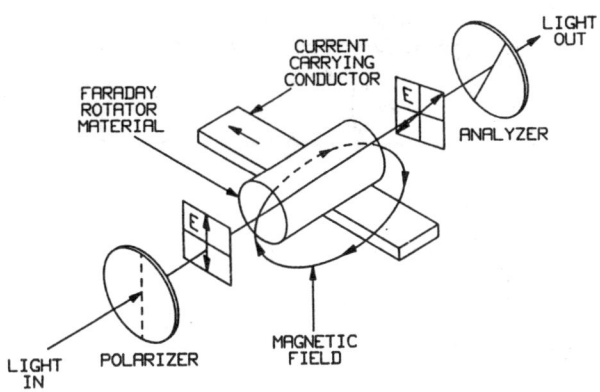

Figure 5-14 Faraday effect current sensing.

9 VOLTAGE TRANSFORMERS AND COUPLING CAPACITANCE VOLTAGE TRANSFORMERS

Voltage transformers (formerly called potential transformers) and coupling capacitance voltage transformers are selected according to two criteria: the system voltage level and basic impulse insulation level required by the system on which they are to be used. Under ANSI, two nominal secondary voltages, 115 and 120 V, are allowed for voltage transformers; the corresponding line-to-neutral values are $115/\sqrt{3}$ and $120/\sqrt{3}$. The applicable voltage depends on the primary voltage rating, as given in ANSI C57.13. The nominal secondary voltages for coupling capacitance voltage transformers are 115 and 66.4 V.

Most protective relays applied in the United States have standard voltage ratings of 120 or 69 V, depending on whether they are to be connected line to line or line to neutral.

9.1 Equivalent Circuit of a Voltage Transformer

The equivalent circuit of a voltage transformer (vt) is shown in Figure 5-15. Since regulation is critical to accuracy, the circuit may be reduced to that shown in Figure 5-15b. The phasor diagram of Figure 5-15c has greatly exaggerated voltage drops to emphasize that, for typical transformers and burdens, the secondary voltage usually lags the "perfectly transformed" primary voltage and is deficient in magnitude. Typical rated maximum errors for these devices are 0.3, 0.6, and 1.2%. Voltage transformers have excellent tran-

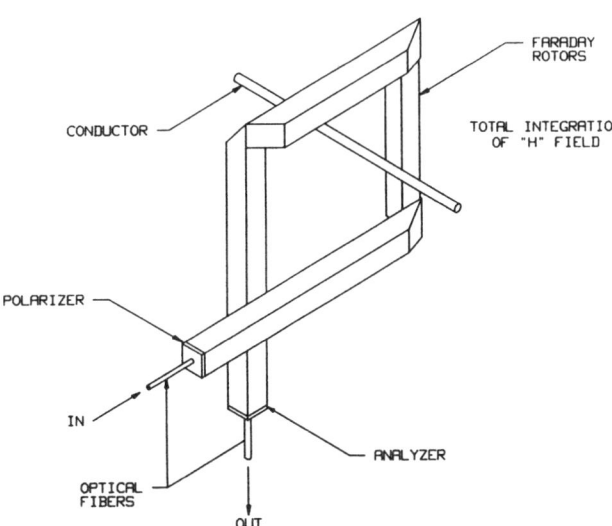

Figure 5-13 Faraday rotator concept.

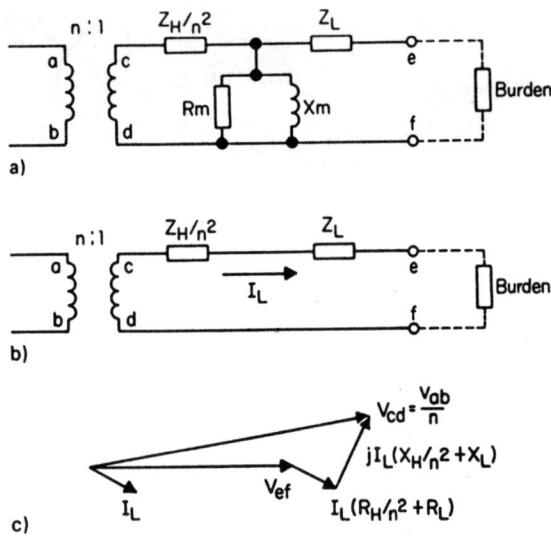

a)

b)

c)

Figure 5-15 The equivalent circuit and phasor diagram of a voltage transformer.

sient performance, faithfully reproducing abrupt changes in the primary voltage.

9.2 Coupling Capacitor Voltage Transformers

Coupling capacitor voltage transformers (ccvt's) and bushing capacitor voltage transformers are less expensive than voltage transformers at the higher voltage ratings, but may be inferior in transient performance. With these voltage devices, a subsidence transient accompanies a sudden reduction of voltage on the primary. This voltage may be oscillatory at 60 Hz or some other frequency, or it may be unidirectional. A representative severe secondary transient is shown in Figure 5-16.

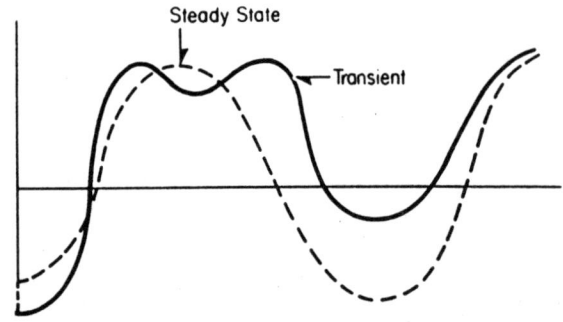

Figure 5-16 A typical subsidence transient of older type coupling capacitor voltage transformers.

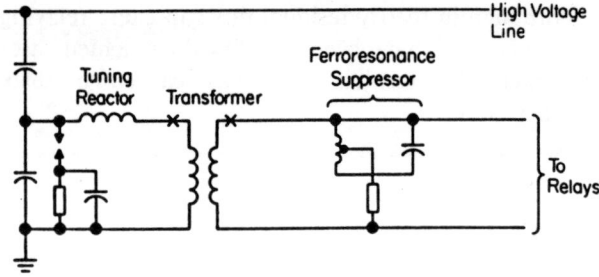

Figure 5-17 Simplified schematic of a coupling capacitor voltage transformer.

Figures 5-17 and 5-18 illustrate the source of the subsidence transient in the ccvt. In Figure 5-18, elements L and C generally contain stored energy when a disturbance, such as a fault, occurs on the primary. Because of the "ringing" tendency inherent in the RLC circuit, a sudden short circuit on the primary does not produce an instantaneous collapse of the voltage applied to the relays. The extent and duration of the deviation from the perfectly transformed voltage depend on the values of R, L, and C. Other transients are introduced by the presence of ferroresonant suppression circuits and the relays themselves.

A voltage transformer is not significantly affected by comparable transients and will reproduce primary transients with excellent fidelity. Modern ccvt's, such as the PCA-7, have capabilities approaching those of the voltage transformer.

The subsidence transient of the ccvt may influence the behavior of some relays. Solid-state phase and ground distance relays, used in a zone 1 direct trip function, may be seriously affected by the temporary excessive reduction of voltage during the decay period. These relays either must be designed with a special provision that allows the subsidence transient to be ignored, be time-delayed to override the transient period or they must have their reach shortened sufficiently to avoid false tripping.

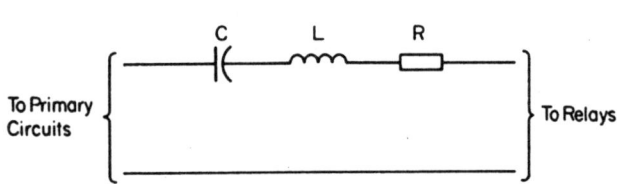

Figure 5-18 Equivalent diagram of a coupling capacitor voltage transformer.

Table 5-2 ABB ccvt's

Type	Cost p.u.	Steady-state accuracy (%)	Max. burden (VA)	Capacitance p.u. (1 p.u. = 0.006 at 115 kV)	Transient response (% voltage)	
					8 msec	16 msec
PCA-5	—	1.2	200	1.0	23.5	12
PCA-7	1.57	1.2	200	4.1	7	6.2
PCA-8	1.0	1.2	200	1.0	23.5	12
PCA-9	1.94	0.3	400	4.1	15	10.5
PCA-10	1.26	1.2	200	4.1	17.5	11
PCM-X	2.51	0.3	400	4.1	12	9.5

Table 5-2 describes a group of ABB ccvt's with differing costs and performance. The transient response column indicates, in general, the degree of compatibility with different relaying systems. High-speed, direct trip, restricted reach relays require either the use of the fast response ccvt's or provision in the relaying system to accommodate or eliminate the effect of the error. The transient response data are based on the percentage of voltage remaining at the ccvt output terminals at the times indicated following sudden reduction of primary voltage from rated voltage to 0, with initiation at zero crossing. The burden is as defined in ANSI C93.1.

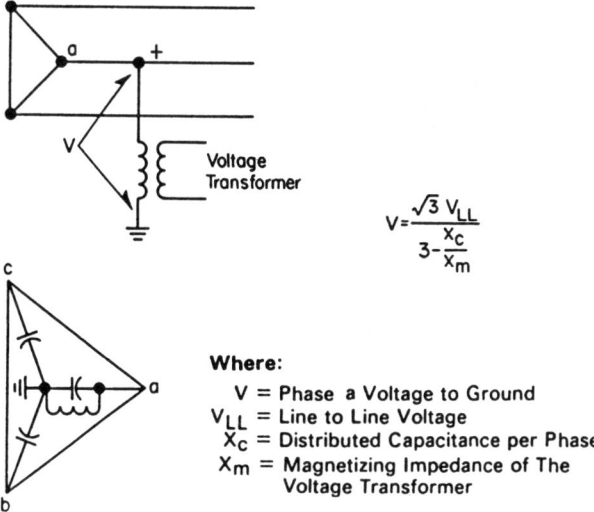

$$V = \frac{\sqrt{3}\, V_{LL}}{3 - \frac{X_c}{X_m}}$$

Where:

V = Phase a Voltage to Ground
V_{LL} = Line to Line Voltage
X_c = Distributed Capacitance per Phase
X_m = Magnetizing Impedance of The Voltage Transformer

Figure 5-19 Neutral inversion on an ungrounded power system.

The impedance of capacitance voltage transformers should not be high enough to produce erroneous behavior in the static compensator distance relays. Excessive impedance may cause false tripping for a reverse fault. For this reason, bushing voltage devices rated below 230 kV should not be used with solid-state distance relays. Bushing voltage devices are, in general, seldom used. Their burden capability is limited and transient performance poor.

9.3 MOVT/EOVT

Voltage sensing also may be accomplished using fiber-optic technology. The MOVT uses the Faraday effect described above, sensing the current flowing through a capacitor stack connected from line to ground. Another voltage-sensing device, the EOVT, uses a Pockel cell rather than the Faraday rotator. Its principle uses light from an optical fiber, which is passed through a special crystal that produces equal components in the X and Y directions. An electric field causes one of these components to be retarded, and this results in a phase difference between the two components. This, in turn, changes the light intensity at the sensor fiber in proportion to the electric field. With additional refinements, it produces an analog output proportional to the electric field present, and this is, in turn, proportional to the instantaneous magnitude of the voltage at the point of measurement.

10 NEUTRAL INVERSION

Neutral "inversion," in which ground becomes external to the system voltage triangle, can occur on ungrounded systems with a single potential transformer connected line to ground. Figure 5-19 shows the possible voltage across an unloaded voltage transformer. Note that an X_c/X_m ratio of 3 would theoretically cause an infinite voltage across the voltage transformer. Such a situation never occurs, of course, because X_m reduces as saturation occurs.

By loading the transformer carefully, this large, sustained overvoltage phenomenon can be avoided. Caution should always be exercised when the secondary voltage of the transformer is used for synchronism check, since the loading will cause a phase shift.

6

Microprocessor Relaying Fundamentals

W. A. ELMORE

1 INTRODUCTION

From the power-system viewpoint, microprocessor or numerical relays are not unlike electromechanical, solid-state, or digital relays. Currents and voltages must be measured and compared with set points, or with each other, and action must be deferred or initiated. Other inputs such as received carrier, 52b switch position, choice of pilot system, etc. temper the action. Figure 6-1 shows a simplified block diagram of a typical microprocessor relay.

Numerical relays must work within the framework of data sampled moment by moment. Currents, for example, are not treated on a continuous basis, but they, like all of the input quantities, are sampled one at a time at as fast a speed as the data handling and storage hardware can accommodate.

The microprocessor has afforded us in protective relaying the remarkable capability of sampling voltages and currents at very high speed, manipulating the data to accomplish a distance or overcurrent measurement, retaining fault information and performing self-checking functions. The utilization of this new technology has also presented us with new challenges regarding the manner in which information is handled and manipulated.

With multiple electromechanical or solid-state devices operating concurrently, time coincidence is no problem. However, a microprocessor literally can handle only one task at a time. Multiplexors can sample only one quantity at a time so voltages and currents are not time-coincident. The awesome task of the programmer is to accommodate these peculiarities and devise ways for the microprocessor to accomplish the tasks in the right order and to cause comparisons to be made based on the correct voltages and currents without the error associated with data skew. Data skew is introduced by the comparison of quantities taken nonsimultaneously.

To allow the sampling of a fixed quantity rather than a rapidly changing quantity, the sample-and-hold (S/H) circuit is usually used in numerical relays. An example of this is shown in Figure 6-2. Since even the simplest relaying function requires that multiple inputs be read, a multiplexer is used. This is a device that allows all the input quantities to be sampled (read) one at a time.

The microprocessor requires that the information be presented to it in digital form, usually an 8- or 16-bit word. The conversion process from the analog signal (which is simply a scaled dc quantity that is representative of the sampled quantity) to the digital signal is accomplished with an A/D converter. Many varieties of these devices have been used over the years. The range and sampling rate required dictate the choice for a particular design for a protective relay.

The microprocessor accepts the sampled data and stores it for future use in RAM (random access memory). The data are acted on by algorithms or comparisons defined by the program memory, which is stored in ROM (read-only memory) or more widely in EPROM (erasable programmable read-only memory). The program stored in ROM or EPROM is nonvolatile.

Another vital element in the architecture required for microprocessor relaying applications is the

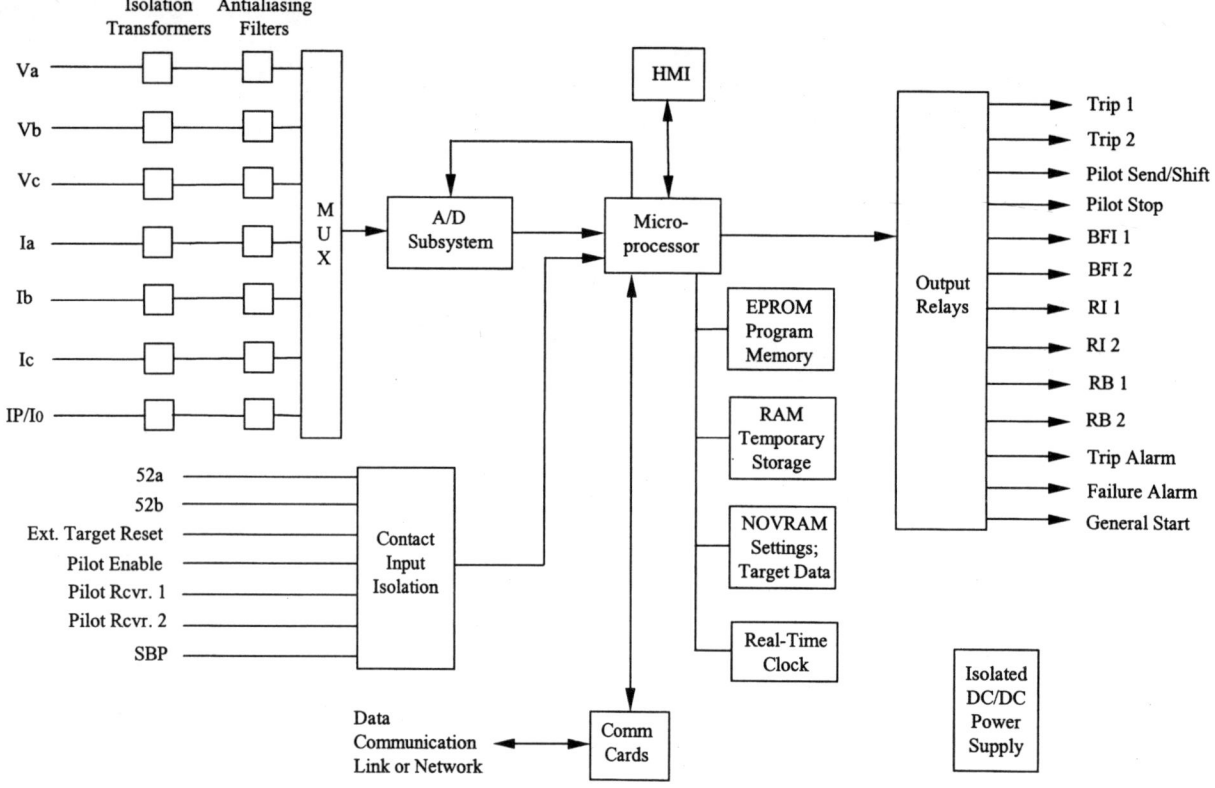

Figure 6-1   Simplified diagram of a typical microprocessor relay.

NOVRAM (nonvolatile RAM) or EEPROM (electrically erasable programmable read-only memory). Data that are stored in this type of memory are not lost when power is removed from the relay. Settings and target data are usually stored here.

Microprocessor-based algorithms typically require time-coincident sampling of the input quantities. Considerable ingenuity is used to address this in real-time processes, particularly in relaying applications where so much is dependent on the time relationship between quantities. There are two basic methods of sampling data, both of which use a sample-and-hold circuit. An example of one method is shown in Figure 6-3. Using separate sample-and-hold circuits for each input, the microprocessor directs a "freeze" to occur at each sampling point. The S/H circuit holds

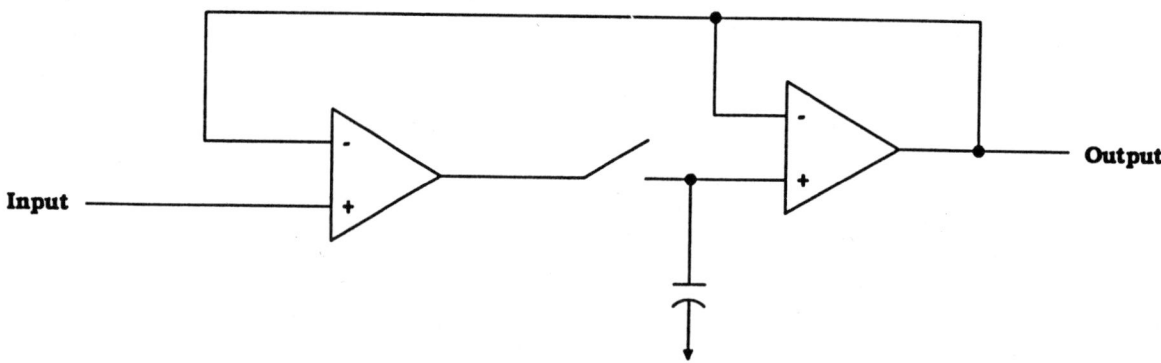

Figure 6-2   Typical sample and hold.

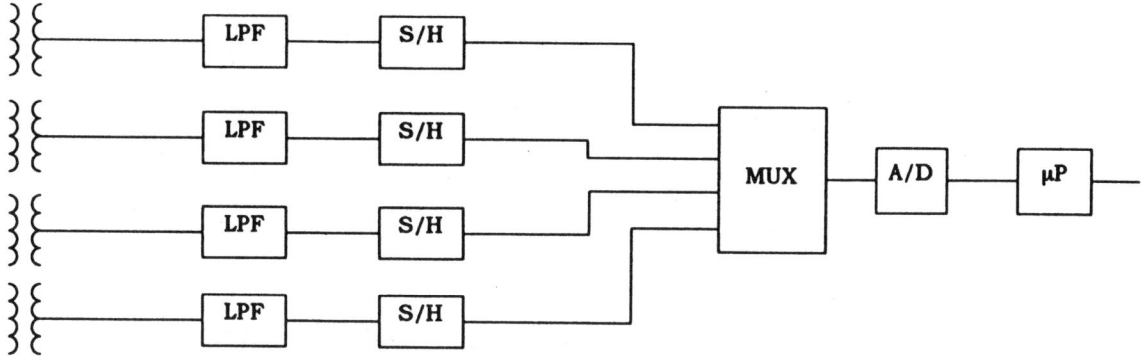

Figure 6-3   Microprocessor relay with individual sample/hold.

this sampled value until the microprocessor can read in each value through the multiplexor and A/D circuit. The microprocessor then directs the S/H circuit to resume the sampling process until the next freeze signal.

An alternative sampling method that is less expensive is to use a single S/H circuit for all inputs. This is shown in Figure 6-4. A time correction factor is applied to each sample after the first in a group.

It is known precisely what difference in sampling times exist and therefore all the samples in a time-sequence-sampled (multiplexed) group can be converted to coincident samples by applying an angle correction. Of course, other methods can be used.

2   SAMPLING PROBLEMS

Because of the practical limitation of sampling rates in a numerical relay, a varying input such as an ac current or voltage will be perceived by the relay considerably differently from its actual continuous waveform. High frequencies in the waveform cannot only fail to be identified due to inadequacies in the sampling process, but may indeed present themselves as a lower-frequency component. Once this error intrudes into the process, it cannot be reconstituted and removed. Either the error must not be allowed to occur in the first place by filtering out the offending frequencies or a process such as asynchronous sampling must be used. The mechanism of a high-frequency component in an input waveform manifesting itself as a low-frequency signal is called *aliasing*. It will now be described using a phasor technique that may clarify the concept for the reader.

3   ALIASING

Figure 6-5 depicts the representation of a simple sinewave by a phasor as it rotates. The projection of

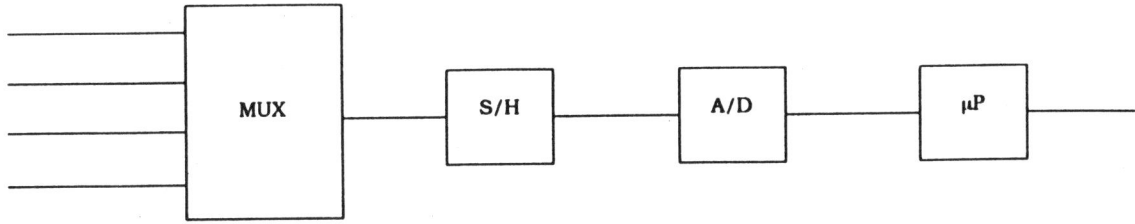

MUX -   Multiplexer

S/H -   Sample and Hold

A/D -   Analog to Digital Converter

μP -   Microprocessor

Figure 6-4   Basic hardware for microprocessor relay.

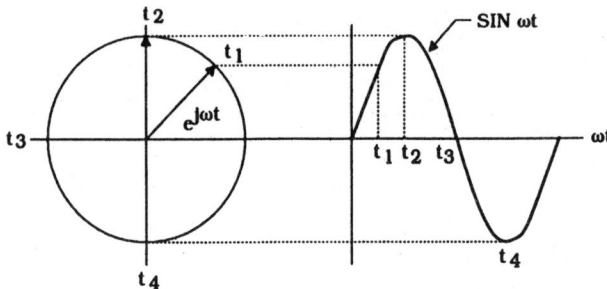

Figure 6-5 Generation of a sine wave by a phasor.

the phasor on the vertical axis, at a given time, represents the magnitude of the sine wave at that time. Note that to do this the phasor must be represented by its peak value, not its rms value. Phasors are generally manipulated, however, using rms values. They are generally shown for a single frequency. Figure 6-6 extends this to show a fifth harmonic phasor superimposed on the fundamental and the distorted sinusoid that is generated by the combination. The fifth harmonic phasor, of course, rotates through 450° in the time required for the 60-Hz fundamental phasor to rotate through 90°.

Now the effect of high-frequency distortion of the waveform on the sampling process can be examined. Figure 6-7 uses an example of a seventh harmonic and an 8-per-cycle sampling rate.

It can be seen that the normal circle for the fundamental is distorted into an ellipse by the presence of the harmonic. Thus, when the total waveform is constructed by the projection on the vertical, it can be seen to be deficient in magnitude. This is the phenomenon known as, aliasing. It is the appearance of a high-frequency signal as a lower-frequency signal that distorts the desired signal.

4 HOW TO OVERCOME ALIASING

4.1 Antialiasing Filters

This effect may be removed by filtering the high-frequency components from the input. The element that accomplishes this function is called an antialiasing filter. The "Nyquist criterion" states that in order to avoid the aliasing error, frequencies above one-half the sampling rate must be removed. Figure 6-8 shows a typical antialiasing filter.

4.2 Nonsynchronous Sampling

When the luxury of time exists in the relay response, an alternative to the antialiasing filter can be used. Nothing is lost in shifting the sampling points for the fundamental frequency component of the quantity being measured. For example, in Figure 6-9, it is not

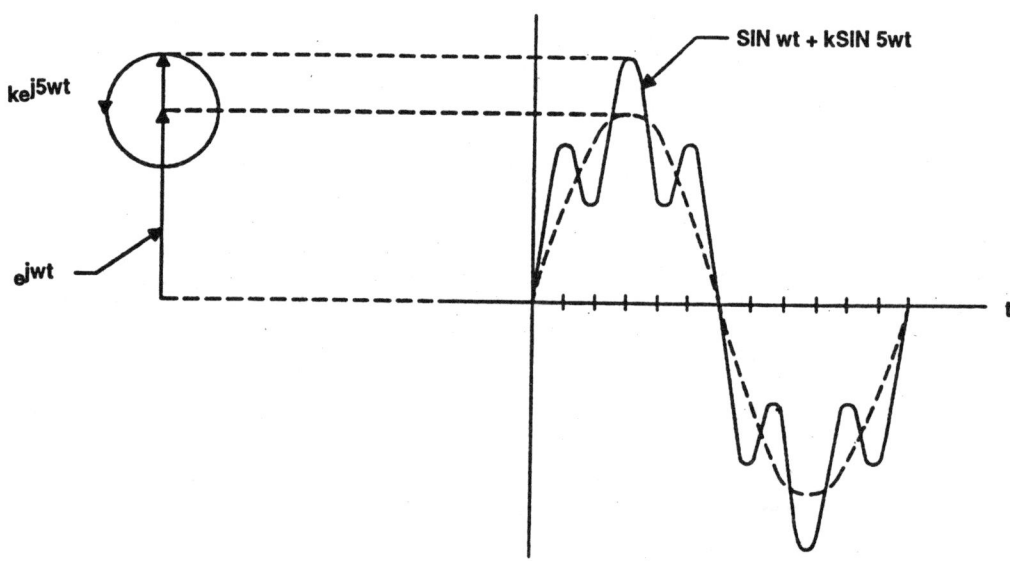

Figure 6-6 Phasor representation for fundamental and fifth harmonic.

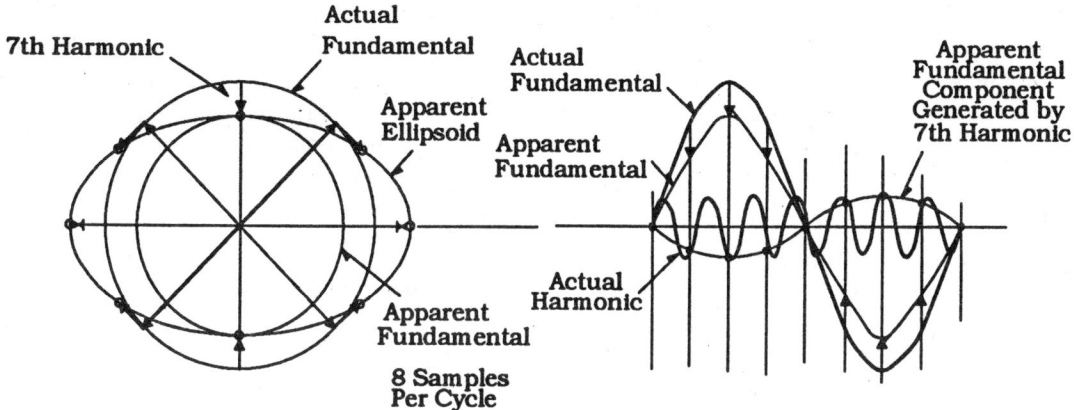

Figure 6-7   Aliasing effect of seventh harmonic.

critical that the first sample be taken at the zero crossing. It may be taken at any arbitrary point with the following samples being equally spaced at the sampling rate. If then, after collecting eight samples in this case, a jump is introduced to delay the beginning of the collection of the next eight samples, by a time corresponding to 180° of a particular high-frequency quantity (25.71° on a 60-Hz basis for the seventh harmonic), an interesting effect takes place. If the measured quantity appears to be too low, as a result of

the presence of the harmonic, for the first fundamental cycle (see Fig. 6-7), it will for the second cycle (following the jump) appear to be too high by the same amount. Thus, a comparison of information in the adjacent cycles allows the effect of the seventh harmonic to be removed.

If the speed requirement of the device allows, then it is possible to eliminate the error associated with a particular harmonic without an antialiasing filter. Note, however, while other harmonic frequency components in the input signal are attenuated by this asynchronous sampling procedure, the effect of only one frequency is eliminated. This concept has been used successfully in the MCO and MMCO relays.

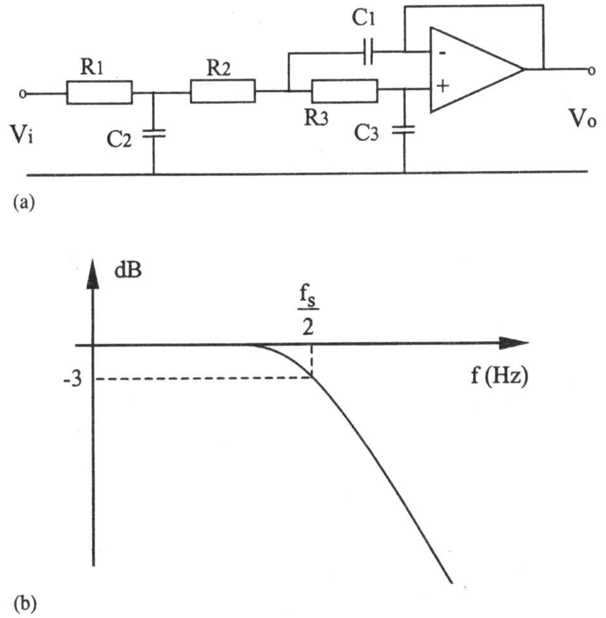

5   CHOICE OF MEASUREMENT PRINCIPLE

With electromechanical relays, the designer has little choice with regard to whether electrical quantities are

Figure 6-8   Antialiasing filter. (a) Diagram; (b) frequency response.

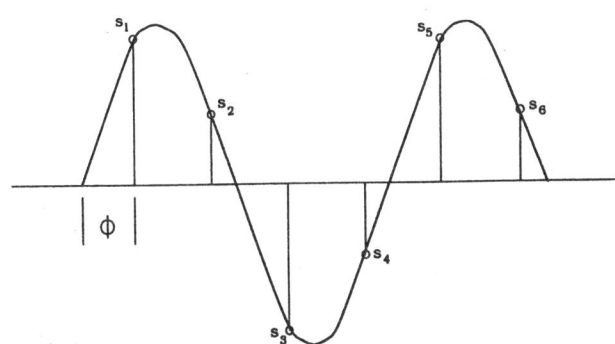

Figure 6-9   Waveform samples taken from a sinusoid.

to be interpreted in terms of peak value, average value, rms value, or fundamental frequency value. With the power of the microprocessor, the designer can apply any of these measurement techniques.

5.1 rms Calculation

The digital determination of root-mean-square values of waveforms is quite similar to the conventional analytical methods. Equations (6-1) and (6-2) illustrate this. By squaring the magnitude of each sample (I_n) over a cycle and summing that with other squares, dividing the sum by the number of samples, and taking the square root, an rms value can be extracted from a complex waveform.

$$\text{Analog rms} = \sqrt{\frac{1}{2\pi} \int_0^{2\pi} I_m^2 \sin^2 \omega t \, dt} \qquad (6\text{-}1)$$

$$\text{Digital rms} = \sqrt{\frac{1}{8} \sum_{n=1}^{8} I_n^2} \qquad (6\text{-}2)$$

Time overcurrent devices that are to be coordinated with other apparatus which experience I^2R heating effects (such as fuses, conductors, and transformers) have been developed with an rms response. For applications in which harmonic effects are generated by apparatus such as six-pulse rectifiers, it may be required that the harmonics be ignored. Relays have been designed using microprocessor techniques that are responsive only to the fundamental frequency component of the input waveform.

5.2 Digital Filters

Any periodic waveform can be represented by a fundamental and series of harmonic frequencies. Any particular frequency can be extracted by utilizing Eqs. (6-3) through (6-5):

$$a_n = \frac{2}{T_0} \int_{-T_0/2}^{T_0/2} f(t) \cos n\omega t \, dt \qquad (6\text{-}3)$$

$$b_n = \frac{2}{T_0} \int_{-T_0/2}^{T_0/2} f(t) \sin n\omega t \, dt \qquad (6\text{-}4)$$

$$f_n(t) = a_n \cos \omega_n t + b_n \sin \omega_n t \qquad (6\text{-}5)$$

The sum of the product of the function and the sine of the frequency that is to be extracted, taken over the period of the fundamental, produces a total that contains only the desired frequency. This is the fundamental premise of Fourier analysis. This calcula-

tion coupled with a similar one using a cosine function instead of a sine allows the complete magnitude and phase position of the desired frequency to be obtained.

With this integration, any desired frequency, such as a 60-Hz fundamental, can be extracted from a distorted periodic waveform, and all other frequencies will be excluded.

5.3 Fourier-Notch Filter

The comparable digital process involves the multiplication of individual samples by stored values from a reference sine wave and summing the products over a full cycle

$$A_C = \sum_{K=0}^{N-1} f_K(t) C_{AK}$$

$$A_S = \sum_{K=0}^{N-1} f_K(t) C_{BK}$$

where

$$C_{AK} = \frac{2}{N} \cos\left(2\pi \frac{K}{N}\right)$$

$$C_{BK} = \frac{2}{N} \sin\left(2\pi \frac{K}{N}\right)$$

K = number of sample
N = samples per cycle

where

$f(t)$ = original function
T_0 = period of the waveform
n = order of the harmonic

Consider, for example, an application in which there are eight samples per (fundamental 60-Hz) cycle. The corresponding samples of a sine wave may be chosen as 0, 0.707, 1.0, 0.707, 0, -0.707, -1.0, and -0.707. These are fixed values, being $\sin(K2\pi/8)$, where K are the samples 0 to 7 and $2\pi/8$ corresponds to 45°. These values are then multiplied by $2/N$ to obtain the constants that are used.

If then the sampled values of the measured quantity over a full cycle are multiplied by these constants in the proper order and summed, the process of Eqs. (6-3) and (6-4) is duplicated. This provides information that excludes all frequencies except the fundamental. By this process, any frequency component can be isolated and utilized to perform a desired function. From this process results a function $A_s = K \sin \omega t$.

Similarly, by using a set of cosine function constants, the sample multiplication and summation can generate a function $A_c = K\cos\omega t$. Since $\sin^2\omega t + \cos^2\omega t = 1, A_s^2 + A_c^2 = K^2$. Thus, the peak value of a particular frequency component can be found by taking the square root of $A_s^2 + A_c^2$. Also since $\sin\omega t/\cos\omega t = \tan\omega t, A_s/A_c = (K\sin\omega t)/(K\cos\omega t) = \tan\omega t$.

The angle of the function can be found by

$$\theta = \tan^{-1}\left(\frac{A_s}{A_c}\right)$$

This algorithm is called a *Fourier-notch* filter.

5.4 Another Digital Filter

Other forms of digital filtering are used for specific applications. In the IMPRS series of relays, four samples per cycle are used. These samples may begin at any point in the cycle, such as at angle ϕ in Figure 6-9. The values of the individual samples can be described as

$$S_1 = \sin(\omega t + \phi)$$
$$S_2 = \sin(\omega t + \phi + 90°)$$
$$S_3 = \sin(\omega t + \phi + 180°)$$
$$S_4 = \sin(\omega t + \phi + 270°)$$
$$S_5 = \sin(\omega t + \phi)$$

Digital filtering can be accomplished with the following procedure:

$$S_s = S_1 - S_2 - S_3 + S_4 = 2\sqrt{2}\sin(\omega t + \phi - 45°)$$

Similarly,

$$S_C = S_2 - S_3 - S_4 + S_5 = 2\sqrt{2}\cos(\omega t + \phi - 45°)$$

These values were related to the simple sine wave of Figure 6-9 with a peak value of 1.0 for a sine wave, $I_m\sin(\omega t + \phi)$. The value of I_m, the peak value, can be obtained by

$$I_m = \frac{\sqrt{S_s^2 + S_C^2}}{2\sqrt{2}}$$

This digital filter also has other useful qualities. If we consider a dc current, it is obvious that $S_1 - S_2 - S_3 + S_4 = 0$ because each sample is the same magnitude. Similarly, a linearly decaying ramp wave shape will also produce a sum equal to 0. Since the dc component of fault current has an exponential decay which is between the constant dc input and linearly decaying

input, it also is severely attenuated by this process of summing.

Using four samples per cycle, this digital filter removes all of the even harmonics. With 60-Hz waveform samples taken at 90° intervals, the samples of 120 Hz would occur at 180° intervals. Every other sample will be equal, so $S_1 - S_3 = 0$ and $S_2 - S_4 = 0$. The second harmonic is eliminated by the summation $S_s = S_1 - S_2 - S_3 + S_4$. With the fourth harmonic, the samples are taken at 360° intervals and being equal also produce $S_s = 0$. All even harmonics are eliminated.

5.5 dc Offset Compensation

Direct current offset in the fault current occurs as a result of two natural laws: (1) Current cannot change instantaneously in an inductance and (2) current must lag the applied voltage by the natural power-factor angle of the system. dc offset produces no desirable effects in overcurrent or distance relays. To make these devices responsive only to the ac component of fault current, it is necessary to remove the dc by some expedient.

The maximum dc component of the fault current is $I_m(1 - e^{-t/T})$, where I_m is the peak value of the symmetrical ac fault current, t the time in cycles, and T the dc time constant of the circuit that limits the fault current. The dc removal algorithm can be exact if T is known. Unfortunately, for a given system it is likely to vary considerably.

Many algorithms have been used. One uses the concept that a sample of the fundamental component of current has the same magnitude as, and the opposite sign to, a sample taken 180° later. The dc components for each of these samples are the same (if we assume this component is truly dc). Thus, for a 480-Hz sampling rate

$$\text{Offset} = \frac{I_K + I_{K-4}}{2}$$

where I_K is the value of a sample of current and I_{K-4} the value taken four samples previously. With eight samples per cycle, these samples would be 180° apart and the effect of the sinusoidal component would be nullified in the summation. The offset may then be used as a correction factor for the samples taken in this interval.

With a decaying dc as opposed to an unvarying value, some error is introduced in this process, depending on the dc time constant.

5.6 Symmetrical Component Filter

Another interesting digital filter, utilizing three samples per cycle, is embodied in the MPR relay. In most applications, time-coincident quantities are necessary, but as the following symmetrical component definitions suggest, quantities that are 120 or 240° displaced in time are useful

$$I_{A1} = \frac{1}{3}(I_A + aI_B + a^2I_C)$$

$$I_{A2} = \frac{1}{3}(I_A + a^2I_B + aI_C)$$

$$I_{A0} = \frac{1}{3}(I_A + I_B + I_C)$$

where

$$a = 1\angle 120°$$

The normal analog process for extracting I_{A2}, for example, from the three-phase currents is to rotate I_C by 120° and I_B by 240°, and add both to I_A. With digital techniques, an alternative procedure can be used. A sample of I_A is added to a sample of I_C that is taken 120° later, then the sum is added to I_B taken 240° later, giving the instantaneous value of $3I_{A2}$ that existed at the time of the I_A sample. Figure 6-10 illustrates this process. A similar procedure is used to extract $3I_{A1}$ from the individual samples of I_A, I_B and I_C that are taken at 5.55-ms (120°) intervals. Note the

50° shift in the sampling interval at S_1 and S_4. This assures over several cycles that a reasonable distribution of samples is obtained and an accurate measurement of negative sequence current is achieved. This algorithm is useful for long-term effects such as motor heating, but is unsatisfactory for fault detection. The accuracy of this method is dependent on the nature of the harmonics.

5.7 Leading-Phase Identification

An interesting task for a microprocessor is to take a collection of nonsimultaneous samples of voltages and currents, determine the related phasors, modify them as the algorithm dictates, and compare the resulting time-coincident phasors to establish which leads the other

$$
\begin{aligned}
A &= a_x + ja_y \\
B &= b_x + jb_y \\
\sin\gamma &= \frac{a_yb_x}{|A||B|} - \frac{a_xb_y}{|A||B|}
\end{aligned}
\tag{6-6}
$$

Equation (6-6) shows the concept that has been developed for digital relays following many years of experience with various analog devices with similar functions. It states simply and remarkably that phasor A leads phasor B if the product a_yb_x is more positive than a_xb_y.

Only the difference in products, $a_yb_x - a_xb_y$, is required to determine which phasor leads the other. No divide function, no sine or tangent calculation, and no table lookup are required, thereby providing very efficient use of the microprocessor. If γ, the angle by which phasor A leads phasor B, is between 0 and 180°, the sine of γ is positive. With $|A||B|$ being always positive, $\sin\gamma$ is positive if a_yb_x is more positive than a_xb_y.

5.8 Fault Detectors

Fault detectors provide an additional level of security in a relaying application. They have been traditionally of the simple-phase overcurrent variety with the frequent addition of ground overcurrent. When the fault current/load current ratio is small, compromises must be made in the settings. Modern technology allows more refined "fault detection."

Phase current change, ΔI, and phase-to-ground voltage change, ΔV, can be implemented simply by

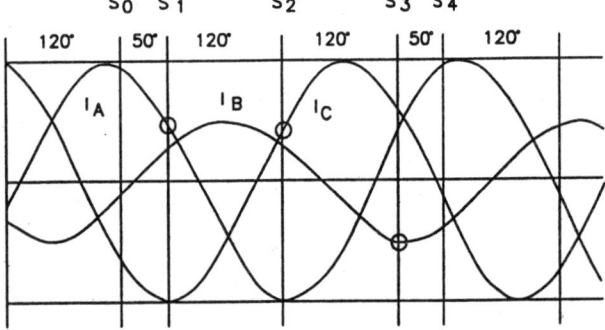

Figure 6-10 Sum of samples taken 120° apart equivalent to $I_A + a\,I_C + a^2I_B$ taken at S_1.

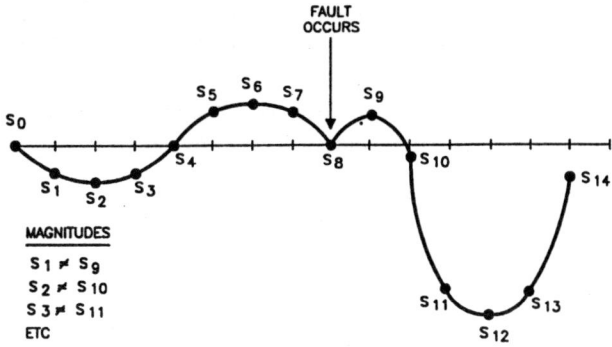

Figure 6-11 ΔI fault detector.

using a comparison of samples at similar points in adjacent cycles as shown in Figure 6-11. These changes, along with a change in zero sequence current, provide a clear indication of the need for a distance or directional unit to make a decision, and therefore they are used to shift some relaying systems (MDAR, REL-301) to "fault mode."

By using ΔI (phase or I_0 current change), a relaying system can discriminate between loss of voltage caused by a fault and that caused by potential circuit problems. ΔI is present for faults and not loss of potential.

6 SELF-TESTING

The ability to monitor much of the hardware is an inherent part of any relaying system equipped with a microprocessor. Some contain a provision for identifying specific types of failure, whereas others indicate only the general condition of check-failure, requiring a more detailed examination to pinpoint the nature of the failure.

6.1 Dead-Man Timer

As part of the housekeeping tasks that are generally performed, a dead-man timer (also called watchdog timer) supervises the fact that the microprocessor is cycling. If the microprocessor fails to perform a given function within a predetermined band of time limits, an alarm output is produced.

6.2 Analog Test

Periodically, a known value of voltage is substituted for the normal inputs to the multiplexer. The output of the A/D converter is then checked for agreement with the known input. If there is disagreement, an alarm output results. If there is agreement, the multiplexer, sample-and-hold, and A/D converter are proven to be in good working order.

6.3 Check-Sum

Any memory segment that is unchanging such as the ROM can be checked through the process of adding up the contents of the memory and periodically verifying that the sum is fixed. Any change in the contents of the ROM following power-up constitutes a failure and will produce an alarm.

6.4 RAM Test

Random access memory is completely checked during the initializing process when power is applied to the relay. Word patterns are written and read. Any inconsistencies are identified.

6.5 Nonvolatile Memory Test

Some relays utilize nonvolatile memory for storing details that are pertinent to the operation of the relay, but will be changed by the user from time to time. An example of this is the settings. By storing the settings in three locations when they are first entered and comparing these three periodically, assurance is obtained that they are correct. Inconsistency produces an alarm.

The ability to test themselves is one of the principal advantages of microprocessor relays. It relieves the need to apply external quantities to them periodically to verify their capability to perform their intended function. At the same time, it should be recognized that no relay is able to completely test itself in all respects. Backup relays are still required even though those failures that do occur in microprocessor relays have a high probability of being identified immediately.

7 CONCLUSIONS

The introduction of microprocessor technology into protective relaying has afforded us the ability to achieve new functions and self-checking provisions not previously possible. At the same time, it has caused a reevaluation of long-established practices, resulting in new approaches to old techniques, as well as encouraging new innovative methods of solving persistent protection problems.

The microprocessor has established its place in protective relaying and will occupy a position of prominence in future designs.

7

System Grounding and Protective Relaying

Revised by: **W. A. ELMORE**

1 INTRODUCTION

Ground fault protection is dependent on the power-system grounding, which can vary from solidly grounded (no intentional impedance from the system neutrals to ground) to "ungrounded" (system grounded only through the capacitance of the system). Ground relaying for effectively grounded systems is discussed in Chapter 12. In these systems, the X_0/X_1 ratio is 3.0 or less, and the R_0/X_1 ratio is 1.0 or less at all points and under all operating conditions. With effective grounding, the line-to-ground fault current is equal to or greater than 0.6 times the three-phase fault current.

Solid grounding is necessary to meet these standard criteria, particularly with overhead lines where the X_0/X_1 ratio averages between 1.6 to 3.5. In solidly grounded systems, the neutrals of the wye-delta power transformers are directly connected to earth through the station ground mat. Considerable design effort is expended to keep the resistance in this connection to a minimum: Typical values of ground mat resistance to earth are on the order of $0.1\,\Omega$ or less in areas of low ground resistivity. Typical values are higher in high ground resistivity areas, resulting in a large station ground mat rise (voltage gradient) between the station area and remote grounds during ground faults.

Earth, *remote ground*, and *true earth* are difficult terms to define precisely, since the earth is a very heterogeneous mass. The terms represent a mathematical fiction needed to identify the zero potential earth plane. In practice, they are considered to exist within the earth at any point remote from the influence of the power system or where current can reasonably flow in the earth structure.

This chapter will cover protective relaying schemes for noneffectively grounded systems. These systems fall into one of three categories:

Ungrounded
Reactance-grounded
Resistance-grounded

In addition, this chapter will discuss the special problems of sensitive ground relaying on distribution circuits and ground fault protection for both ungrounded and multigrounded three-phase, four-wire systems.

2 UNGROUNDED SYSTEMS

2.1 Ground Faults on Ungrounded Systems

The term *ungrounded* is strictly one of definition, indicating no physical connection of any kind between the system and ground. Since, however, there is always distributed capacitance between the three phases of the system and ground, the system is grounded through this capacitance. On such systems, current flows between each conductor and ground under normal conditions. In the event of a single line-ground fault, the corresponding line-to-ground capacitance is shunted out.

Using symmetrical components, Figure 7-1 shows the networks and fault representation. Here, X_{1C}, X_{2C}, and X_{0C} are the total distributed capacitances of each

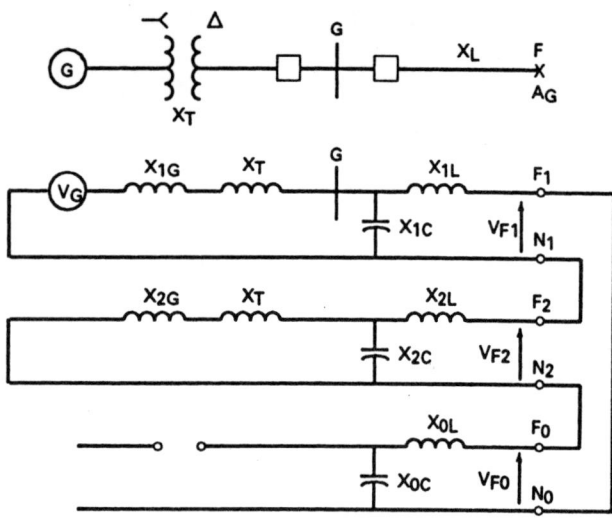

Figure 7-1 Sequence network interconnection for "a" phase-to-ground fault on an ungrounded system.

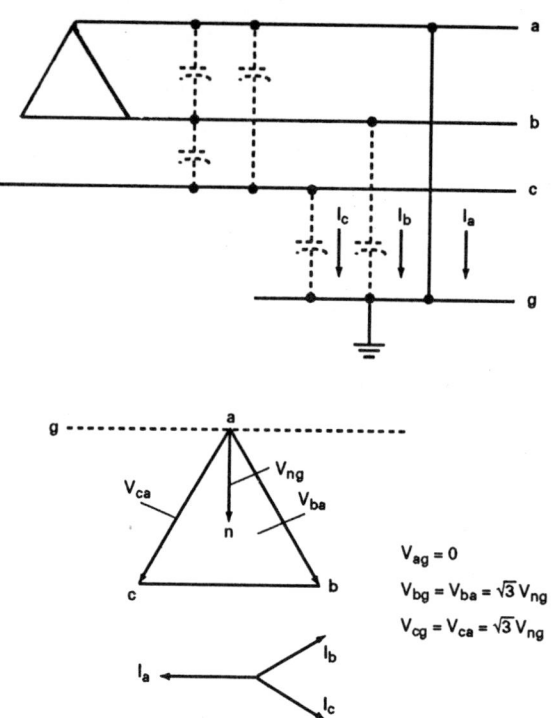

Figure 7-2 Ground fault on ungrounded system.

system. Although they are shown here as a lumped quantity, they are actually distributed parameters. X_{1C} and X_{2C} can be neglected because their effect is insignificant compared to that of X_{0C}. X_{0C} predominates so that approximately

$$I_a = 3I_0 = \frac{3V_G}{X_{0C}} \qquad (7\text{-}1)$$

Since the ground fault current returns through the shunt capacitance, the unfaulted phase currents are not 0 (Fig. 7-2). The phase-b and -c voltages are shown as the prefault line-to-line voltages or $\sqrt{3}V_{LG}$. This relation holds true only for the steady-state condition with zero fault resistance; transient voltages can be considerably higher as shown in Figure 7-3.

When the circuit breaker opens and extinguishes the arc at or near current zero, the voltage is near its peak value. This voltage, shown in Figure 7-3 as 1.0 per unit ($\sqrt{2}$ times the rms value), remains on the line (or right-hand) side of the breaker, while the generator voltage goes to the maximum negative value one half-cycle later. At that time, the voltage across the breaker contacts is essentially 2.0 per unit, the crest value. A voltage of this value can *cause* the arc to restrike across the breaker contacts, sending the line voltage from $+1$ per unit to -1 per unit.

The result is a high-frequency transient voltage, whose first peak overshoots the -1.0 value by -2.0 (the difference between -1 and $+1$), giving a peak voltage value by -3.0. If the arc is again extinguished

at a high-frequency current zero, the trapped charge on the line produces a voltage of -3.0 per unit. If a second restrike occurs at the next voltage positive maximum, the peak voltage will overshoot to $+5$ per unit as it

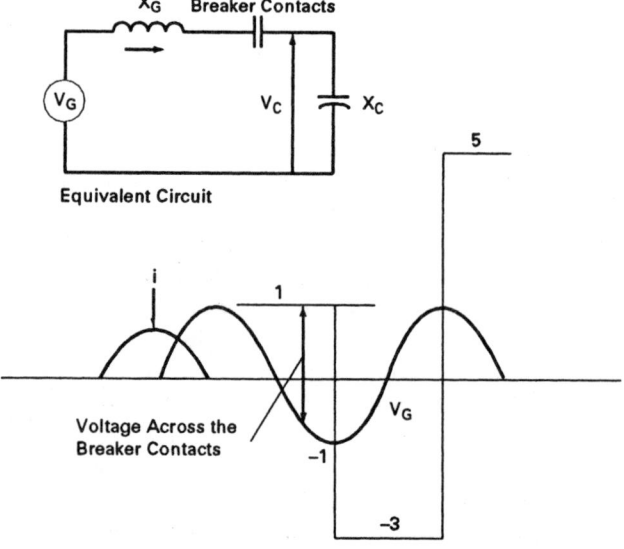

Figure 7-3 Overvoltage due to reignitions and restrikes.

goes from -3 to $+1$ per unit. Theoretically, further cycles of extinguishing and restriking of the arc would build higher and higher voltage values. In practice, however, flash-overs usually occur before these high values are reached.

The peak voltage values shown in Figure 7-3 are maximum theoretical values based on arc extinction at zero current, no damping, and arc restrike at the crest value of the source voltage. In fact, circuit resistance will introduce damping of the transient, reducing the peak value of the first half-cycle overshoot. Further, restrikes may occur before the voltage reaches crest value voltage, which will reduce the value of peak overvoltages. Nevertheless, overvoltages can be very high and represent the major disadvantage of ungrounded systems. An initial fault can cause a second ground fault to occur on a different phase possibly on a different feeder, producing a phase-to-phase-to-ground fault with its associated high current and damage.

2.2 Ground Fault Detection on Ungrounded Systems

Since the fault current for a single line-ground fault on an ungrounded system is very small, overcurrent relays cannot be used for fault detection. Voltage relays will detect the presence of the voltage unbalance produced by the fault, but will not selectively determine its location in the system. The unbalanced phase and zero sequence voltages that occur during ground faults are essentially the same throughout the system. Since selective isolation of the fault is not possible, relay schemes are only useful for providing an alarm.

Figure 7-4a shows the preferred ground fault detection system. The voltage transformers must have a primary voltage rating equal to the line-line voltage, since this is the voltage that will be impressed on the two unfaulted phases during a line-ground fault. Under normal conditions, the voltage across the relay is approximately 0. When a single line-ground fault occurs, the voltage becomes $3V_0$, or approximately 200 V with 69-V secondary windings. The electro-mechanical type CV-8 relay or solid-state type 59G relay shown, with their 200-V continuous rating, will detect fairly high-resistance faults.

Wye-connected transformers with grounded neutral on an ungrounded system may be subject to ferroresonance during switching or arcing ground faults. To avoid this, care must be exercised in planning the relationship between the magnetizing impedance of the transformer, its knee point voltage, and its load.

Users have successfully applied resistors across the break in the broken delta configuration (as in Fig. 7-4) having a value that will limit current in the delta loop to the rated current of the voltage transformer secondaries. Much higher values of resistance have also been used successfully. Karlicek and Taylor in their important paper, "Ferroresonance of Grounded Potential Transformers on Ungrounded Power Systems" (*AIEE Power Apparatus & Systems*, August 1959, pp. 607–618), concluded that the appropriate value of the resistor (called 3R in Fig. 7-4) is 100 La/N^2, where La is the voltage transformer primary inductance in millihenries and N the transformer turns ratio.

Typical resistor values in use are as follows:

Voltage transformer ratio	Resistor (Ω)
2400–120	250
4200–120	125
7200–120	90
14,400–120	60

Although the primary fault current may be low, high secondary currents can flow. This should be checked with the short time or continuous rating of the voltage transformer and resistor.

Applying a grounded-wye-broken-delta transformer with only a relay connected across the break and no shunt resistor is equivalent to very-high-impedance grounding. Any shunt resistor, even as high as $20X_c$, is better than none. It will damp any high transient voltage oscillations and probably hold the peak values to less than twice the normal crest voltage to ground.

The alternative ground fault detection scheme (Fig. 7-4b) is not recommended and should only be applied after careful study. The CVD electromechanical relay or the type 27/59 solid-state relay in this system has separate contacts for operation on either over- or undervoltage. With the vt connected to phase c, line-to-ground faults on phases a and b produce an overvoltage on the relay; faults on phase c produce an undervoltage. For the scheme to work, the capacitance to ground of the lines must be fairly closely balanced and high enough to keep the neutral of the system at close to ground potential.

This scheme can also produce ferroresonance or neutral inversion. When X_C/X_L is 3.0, V_T theoretically would be infinite. Even without faults on the system, the high magnetizing impedance of the voltage transformer can approach resonance with the line

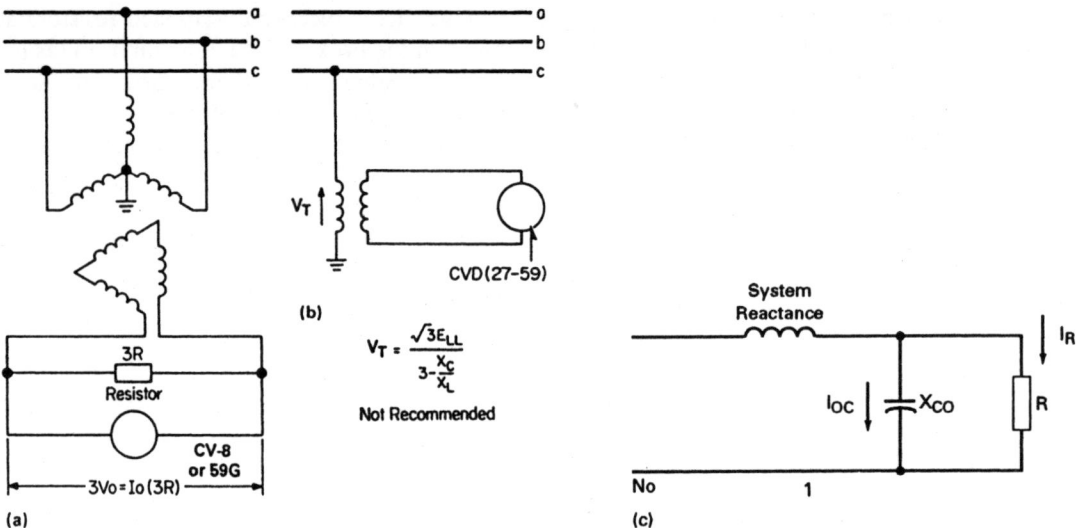

Figure 7-4 Ground fault detection on ungrounded systems.

capacitance to neutral, causing a high overvoltage across the secondary. Neutral inversion can occur during a line-ground fault on a phase other than c. Such a fault produces unbalanced impedances to ground; the resultant current flows can drive the system ground point outside the delta. A loading resistor across the relay or, less desirably, in series with the transformer primary may prevent these problems.

3 REACTANCE GROUNDING

There are three different types of reactance grounding:

High-reactance grounding
Resonant grounding
Low-reactance grounding

3.1 High-Reactance Grounding

Until the early 1940s, some utilities operated their unit-connected generators with the neutral ungrounded. Their purpose was to keep the internal line-to-ground fault current in the generator very low and prevent the iron from being damaged by arcing. Unfortunately, the result was a high insulation failure rate in machine windings.

These failures were caused by high-voltage transients, similar to those discussed earlier. This problem was compounded by an inability to detect single line-

ground faults in the generator. As a result, the faults persisted, causing undue damage.

The initial solution was to connect the generator neutral to ground through the primary of a voltage transformer and put an overvoltage relay across the secondary. In theory, a single line-ground fault would simply cause the generator neutral voltage to shift with respect to ground, activating the relay and tripping the machine or sounding an alarm. In practice, however, this system actually increased the machine failure rate. The cause was arcing grounds—a phenomenon similar to the restrikes that can occur when switching a capacitive reactance.

The arcing ground phenomenon can be explained using Figure 7-5. The equivalent single-line diagram shown in Figure 7-5a is for a generator grounded through a high reactance X_n, with a line-to-ground fault near one terminal. X_c is the distributed capacitive reactance of the windings to ground, connected half-way between the generator reactance X_g. If the arc is extinguished when the small fault current passes through 0, the voltage across the arc path must go from nearly 0 to the normal crest value. In doing so, it must oscillate around the steady-state normal value.

As shown in Figure 7-5b, the resultant voltage transient will reach a peak value of twice the normal crest line-to-neutral voltage, one-half cycle of the high-frequency transient after the arc is extinguished. If the arc restrikes at this point, the fault voltage is driven back to 0. When the arc is initially extinguished, the reactor voltage has to go from the positive maximum

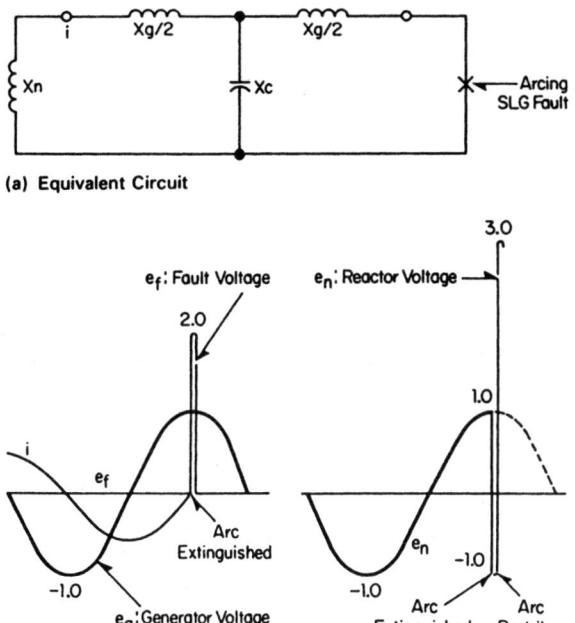

(a) Equivalent Circuit

(b) Fault Voltage   **(c) Reactor Voltage**

Figure 7-5 Overvoltages on reactance grounded system due to arcing fault.

to 0. As a result, it has a transient oscillating period from the positive maximum to negative maximum.

The first half-cycle of this oscillation is shown in Figure 7-5c. If the arc restrikes at the instant when the fault voltage is twice the normal crest value, as was assumed in Figure 7-5b, the reactor voltage has to go from the negative minimum to positive maximum. The result is another transient oscillation, with a peak value of three times the normal maximum line-ground voltage.

Note that in high-reactance grounding, the reactor voltage is applied between the generator neutral and ground. Since the BIL of the reactor is higher than that of the generator windings, insulation failures are more likely to occur in the generator windings.

The switching surges that result from clearing line-to-ground faults for ungrounded systems also occur in high-reactance grounded systems. In the latter case, the resulting transient overvoltages will be even higher. The source voltage for an ungrounded system is the normal line-to-neutral voltage that, theoretically, produces successive line-side voltage peaks of 1.0, 3.0, 5.0, ... of normal crest voltage to neutral. For the high-reactance grounded system shown in Figure 7-5, with the reactor between the neutral and ground, the source voltage is the normal line-line voltage. The

corresponding theoretical transient peaks are $\sqrt{3}$, $3\sqrt{3}$, $5\sqrt{3}$, and so on. For these reasons, high-reactance grounding was discontinued many years ago.

3.2 Resonant Grounding (Ground Fault Neutralizer)

In certain sections of the United States, resonant grounding has been applied successfully in unit-connected generator grounding applications. It is not applied in transmission line applications in the United States, but other countries use it. In this scheme, the total system capacitance to ground is compensated for or cancelled by an inductance in the grounded neutral of the power transformers. The grounding reactor, equipped with taps that permit it to be tuned to system capacitance, was first called a *Petersen coil*. It is now more commonly designated a *ground fault neutralizer*.

Theoretically, if the reactor perfectly matches the system capacitance, a line-to-ground fault will produce zero current, the transient fault arc will be extinguished, and the arc path deionized, without the need for deenergizing the circuit.

In this system, approximately 75% of line-ground faults are self-extinguishing. The remaining faults must be cleared by a line breaker.

In theory, resonant grounding should reduce line outages considerably. This system does, however, have a number of disadvantages:

Transformers connected to the system must have full line-line insulation even when wye-connected.

The entire system must be fully insulated for line-line voltage.

The ground-fault neutralizer must be retuned to accommodate any changes in system configuration: additions, extensions, line removals, or switching.

System effectiveness will be reduced considerably if a substantial number of lines are of wood pole construction. The high insulation to ground will result in a larger portion of line-line faults (conductor swing caused by wind).

A high incidence of faults will occur essentially simultaneously in different parts of the system.

3.3 Low-Reactance Grounding

Low-reactance grounding used to be applied to systems fed at generator voltage. The generator neutral was grounded through a reactor. The reactor was sized to keep the magnitude of a single line-to-ground fault on the machine terminals equal to a three-phase fault. [A reactor value of $(2X_1 - X_0 - X_2)/3$ was used.]

In general, low-reactance grounding was applied to large industrial plant systems with radial distribution feeders, and ground fault protection consisted simply of overcurrent relays. Gradually, this type of generator grounding has been replaced by low-resistance grounding.

Another type of low-reactance grounding provides ground fault current relaying for systems supplied from a delta source. The reactance grounding scheme should

> Supply sufficient ground fault current to operate relays for a fault when the line value $(X_0 + 2X_1)$ is the highest.
> Limit the transient overvoltages attributable to ground faults to a value of 2.5 times normal line-to-neutral crest value, with two restrikes.

In this scheme, either a grounded wye-delta or zig-zag transformer can be used, although the zig-zag (Fig. 7-6) is more common because of its economy. The windings shown parallel are on the same core leg. With this connection, the positive sequence impedance of the bank is very high and equal to the magnetizing impedance. When zero sequence current passes through the bank as shown, the impedance is equal to the leakage reactance.

The rating of the transformer is chosen so that the maximum X_0/X_1 value is 4. When X_0/X_1 equals 4, the line-ground fault current is half the three-phase, short-

circuit value, if we assume that $X_2 = X_1$. Thus, if a ground relay is used in the common neutral connection of the line current transformers, the current level is a maximum of half that of the phase current for a three-phase fault.

4 RESISTANCE GROUNDING

Resistance grounding is applied in systems with three-wire distribution at the generator voltage and for unit-connected generators. The two general types of resistance grounding are low- and high-resistance grounding.

4.1 Low-Resistance Grounding

Whenever low impedance grounding is desired, resistance grounding is generally preferred to the low-reactance systems described above. Specifically, low-resistance grounding is used for systems fed directly at the generator voltage (Fig. 7-7a) or through a delta-wye transformer (Fig. 7-7b). When a line-to-ground fault occurs in the system, the current flowing in the ground resistor results in a sudden change in generator

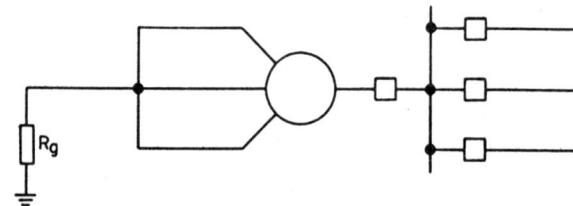

(a) System Fed at Generator Voltage

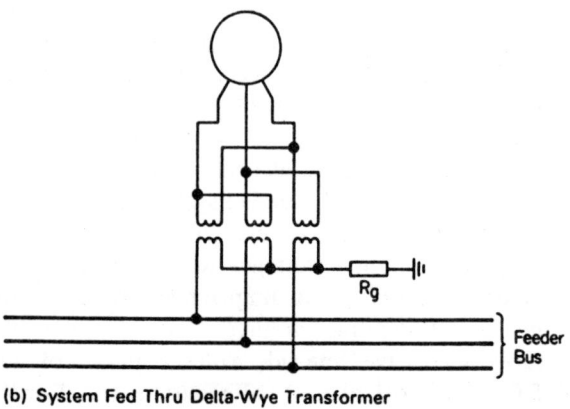

(b) System Fed Thru Delta-Wye Transformer

Figure 7-7 Low-resistance grounding of systems fed through Delta-Wye transformer.

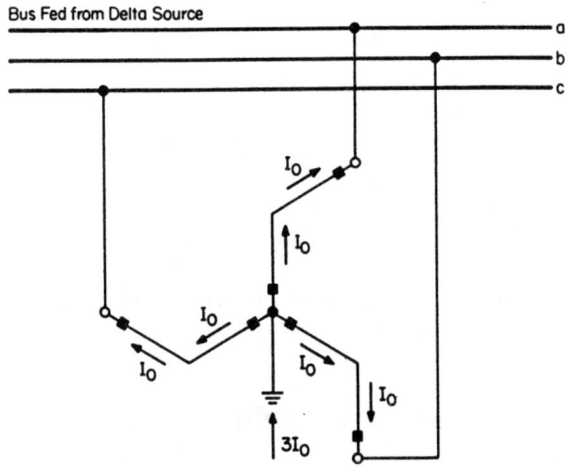

Figure 7-6 Reactance grounding.

load, causing severe generator angular swings and high-peak shaft torques. To keep the ground resistor loss low, the resistor is generally sized to limit the single line-to-ground bus fault to around 100 to 400 A. Ground relaying using residual overcurrent relays may be applied, but zero-sequence-type current transformers will provide greater sensitivity.

For distribution systems, transient overvoltages may be limited to less than 2.5 times the normal crest value to ground, if

$$\frac{R_0}{X_0} \geq 2.0 \quad \text{and} \quad \frac{X_0}{X_1} \leq 20 \quad (7\text{-}2)$$

For a neutral resistor R,

$$R_0 = 3R$$

Assuming that $X_0 = 20X_1$ and $R_0 = 2X_0$, then we have

$$Z_0 = (40 + j\,20)X_1$$

For a line-to-ground fault,

$$
\begin{aligned}
I_g = 3I_0 &= \frac{3}{X_1 + X_2 + Z_0} \\
&= \frac{3.0}{(j\,1 + j\,1 + 40 + j\,20)X_1} \\
&= \frac{3.0}{(40 + j\,22)X_1} = \frac{3.0}{45.65\,X_1} \angle -28.8^\circ \\
&= \frac{0.066}{X_1} \angle -28.0^\circ \text{ per unit}
\end{aligned}
\quad (7\text{-}3)
$$

The three-phase fault current would be

$$I_{3\phi} = \frac{1.0}{X_1} \text{ per unit} \quad (7\text{-}4)$$

Thus, if we use Eq. (7-4), the line-to-ground fault current magnitude is

$$I_g = 0.066\,I_{3\phi} \quad (7\text{-}5)$$

The resistor can be in the neutral of the transformer (Fig. 7-7), or a resistor can be inserted in the neutral of the grounded zig-zag transformer (Fig. 7-6). In either case, the reactance component of the resistor must be considered. Cast-iron grid-type grounding resistors have a power factor of approximately 0.98, stainless steel types one of approximately 0.92. The reactance, while small in itself, is tripled in the zero sequence circuit.

4.2 High-Resistance Grounding

High-resistance grounding is applied to a generator-transformer unit system by connecting a resistor across the secondary of a distribution transformer in the grounded generator neutral (Fig. 7-8). The resistor value is selected so that its KW loss for a solid line-to-ground fault at the machine terminal is equal to or greater than the charging kVA of the low-voltage system. To do this, 3R is chosen to be equal to or less than Xco. Xco is the combined zero sequence capacitive of the generator windings, cable connections to the transformer, low-voltage transformer winding, and station service transformer plus any surge protective capacitors that are applied at the generator terminals and connected phase to ground. This resistor value will limit generator iron burning from ground faults, damp out oscillations, and limit the peak transient voltage to 2.5 times normal line-to-neutral voltage, or less. The rating required for the resistor is

$$\text{Resistor } KW_R = \frac{V^2}{1000R} \quad (7\text{-}6)$$

where V is generator-rated phase-to-neutral voltage (in volts).

The transformer kVA requirement is the same as this resistor KW value, of course, but the kVA rating is higher because of the choice of a higher-voltage rating for the transformer to avoid possible ferroresonance.

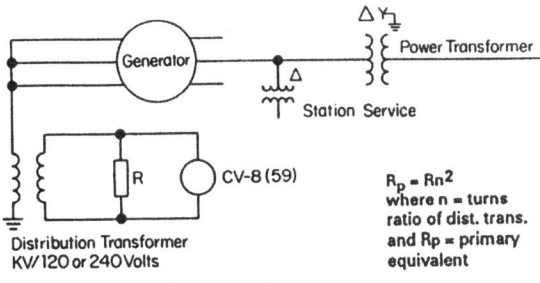

(a) Unit Connected Generator-Transformer

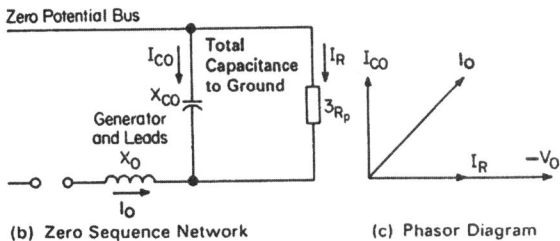

(b) Zero Sequence Network   (c) Phasor Diagram

Figure 7-8 High-resistance grounding of the unit connected generator-transformer.

The magnitude of primary fault currents in these applications is around 8 to 10 A.

Sensitive protection is provided by an overvoltage relay across the resistor. This application is detailed under generator protection (Chap. 8).

5 SENSITIVE GROUND RELAYING

Ground relaying on distribution circuits can be difficult. The range of fault currents can vary from negligible, for a conductor lying on or near the ground with minimum electrical contact, to substantial, for a conductor making good contact with ground. Unfortunately, there is no practical way of distinguishing an intolerable situation from a tolerable one at a breaker or disconnection location.

Some years ago, a utility conducted tests on a 10-ft length of no. 4 bare copper wire energized at 12 kV and laid on a variety of surfaces such as dry grass, green vegetation, dry base soil, and asphalt. Of 128 tests, 7% showed currents of less than 7 A, 7% over 1000 A, and 55% had currents in the range of 150 to 600 A.

Ground fault protection is dictated by the amount of ground fault currents available from the system to operate relays and the ratio of this current to normal system residual unbalance. Load management may help to reduce normal unbalance in some cases. The minimum ground fault current must balance service continuity with equipment protection. That is, it must be low enough to minimize equipment damage, but high enough to be recognizable and allow the faulted area to be selectively isolated without nuisance tripping.

The design of the system grounding should be compatible with the sensitivity of the relaying that is to be used. Three commonly used ground relay schemes, in order of increasing sensitivity, are

Ground relay in the common neutral connection of the line current transformers and/or grounded source (Fig. 7-9).

Ground relay in the common neutral connection of the line current transformers, with a product-type relay to avoid operation on false residual currents (Fig. 7-10). The CWP scheme (Fig. 7-10b) provides increased sensitivity, whereas the CWC scheme will not (see also Fig. 7-11).

Ground relay with a zero sequence (ring) type of current transformer (Fig. 7-12).

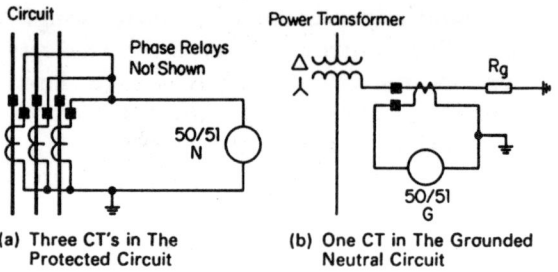

(a) Three CT's in The Protected Circuit

(b) One CT in The Grounded Neutral Circuit

Figure 7-9 Ground protection with conventional current transformers and protective relays.

5.1 Ground Overcurrent Relay with Conventional Current Transformers

In the scheme shown in Figure 7-9, the relays are usually set on the 0.5-A tap. Because of the large burdens of electromechanical ground relays on the minimum tap, the relay pickup current multiplied by the current transformer ratio will not be the primary ampere pickup when using lower-quality current transformers (see Chap. 5). To hold the exciting current to a reasonable minimum, it may be necessary to use a higher tap setting than would otherwise be desired, a solid-state or numerical relay, or a better current transformer.

The unequal performance of current transformers during heavy phase faults or initial asymmetrical motor starting currents may produce false residual currents with the scheme shown in Figure 7-9a. When these currents cause relay operation, an instantaneous relay with a higher pickup should be substituted, or the time overcurrent relay should have a larger time dial and/or pickup setting. Increasing the burden on the current transformers in these cases causes them to saturate more uniformly, reducing the false residual current.

Higher burdens, however, may also decrease the relay sensitivity on light ground faults, depending on the quality of the current transformers. False residual currents do not occur in the scheme shown in Figure 7-9b or 7-12 and do not cause relay operations in Figure 7-10a or b.

With the application of a ground relay set on the 0.5-A tap, the fault current in the relay should not be less than twice pickup, or 1.0 A secondary.

To be able to detect a ground fault that produces a current having a value of 10% of the current produced by a bolted phase-to-ground fault is a reasonable criterion. The maximum load on any circuit off a bus

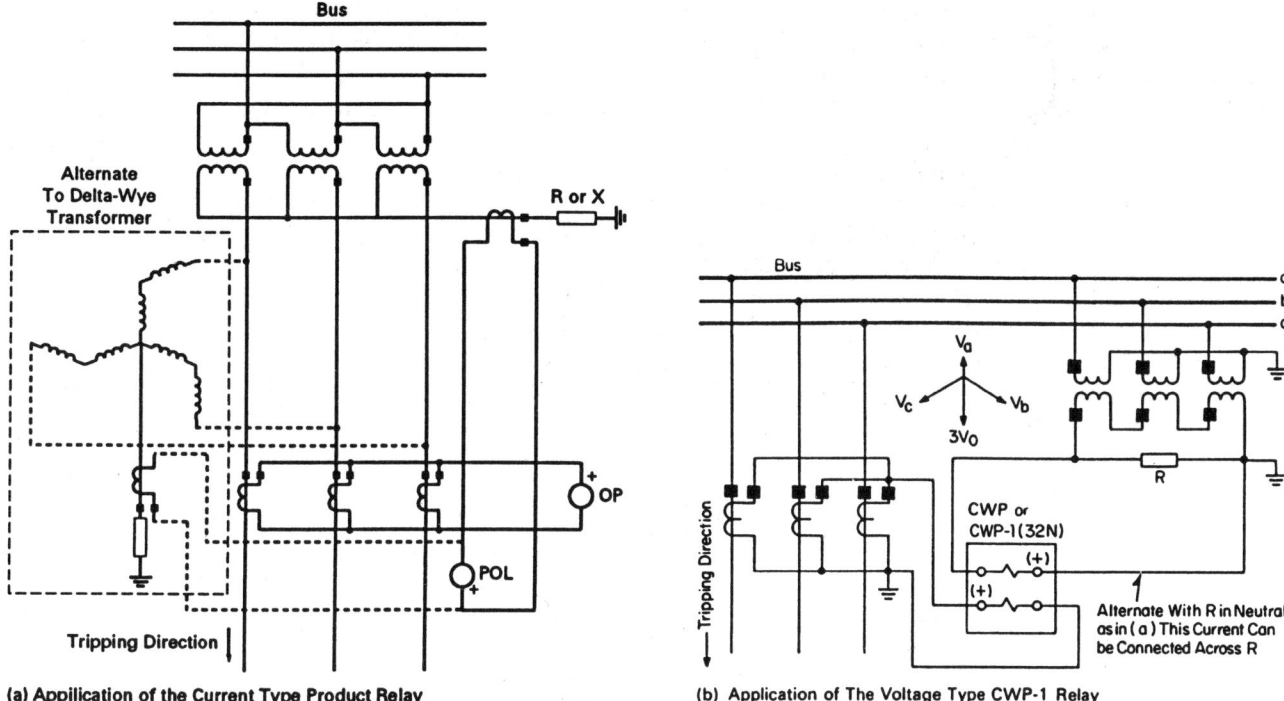

(a) Appilication of the Current Type Product Relay

(b) Application of The Voltage Type CWP-1 Relay

Figure 7-10 Sensitivity ground protection utilizing product-type relays.

dictates the critical ct ratio. With a primary current of $0.10\,I_G$, a ratio of K:5 and minimum secondary current of 1 A produce the following requirement:

$$I_G \geq 2\,K \qquad (7\text{-}7)$$

That is, the current permitted by the grounding device for a bolted phase-to-ground fault should equal or exceed twice the primary rating of the largest ct used on a circuit off of the bus. If the use of this criterion produces an excessively high ground fault current, a lower value of current can be chosen by using a higher resistance value. This will then require a more sensitive

relaying scheme to be used, such as a zero sequence (ring) type of current transformer with its low ratio (typically 50:5) and an instantaneous overcurrent relay, such as an IT or 50D.

5.2 Ground Product Relay with Conventional Current Transformers

Better security is possible using the schemes of Figure 7-10, and increased sensitivity may be provided

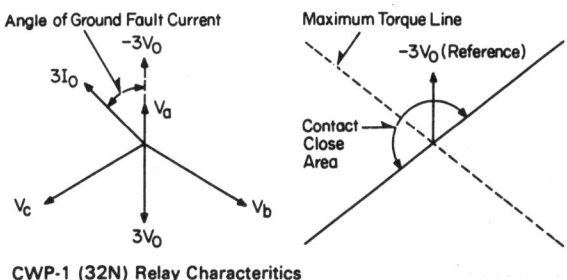

CWP-1 (32N) Relay Characteritics

Figure 7-11 Phasors for Figure 7-10b for a phase "a"-to-ground fault on a high-resistance grounded system.

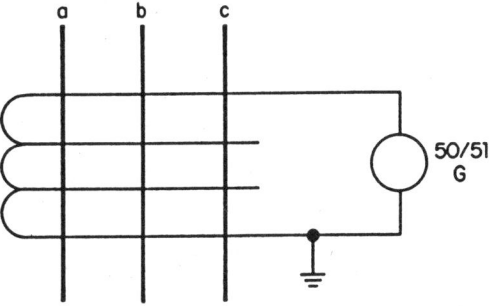

Figure 7-12 Ground protection utilizing the zero sequence (ring) type current transformers.

by the scheme of Figure 7-10b. These schemes will not operate on false residual currents; the relays require current in both windings to operate. No system ground or zero sequence current or voltage will exist for phase faults or motor starting currents.

Two product-type devices are used in the scheme of Figure 7-10b: the CWP and CWP-1. The CWP is applicable for reactance grounded systems and the CWP-1 specifically tailored to high-resistance grounded systems with its 45° *lead* characteristic.

The greatest ground fault sensitivity is provided by the CWP-1 (32N) relay. The relay pickup is adjustable between 5 and 40 mA with 100 V across the potential coil at maximum torque.

The phasors for a high-resistance grounded system where the CWP-1 relay is applicable are shown in Figure 7-11.

5.3 Ground Overcurrent Relay with Zero Sequence Current Transformers

The scheme shown in Figure 7-12 provides maximum sensitivity. There are no false residual currents. The zero sequence type of current transformer has the conductors passed through the center hole, and the ct ratio is *not* dictated by the load current. Secondary current is the transformed system zero sequence current $3I_0$.

The standard ratio for the zero sequence type of transformer (type BYZ) is 50/5; 100/5 ratios were originally used. Various nondirectional relays can be applied, as outlined in Table 7-1.

The IT relay has a large burden when the 0.15-A tap is used (19.6 Ω). Saturation of the 50/5 ct will occur at roughly 5 A primary. Field tests indicate that the secondary pulse width at 1800 A primary is only approximately 30° following each zero crossing, but this relay operates satisfactorily with this extreme

degree of saturation. This relay and ct combination is intended to provide sensitive detection of ground faults and is not expected to perform adequately in the presence of fault currents beyond the moderate (up to 1800 A rms symmetrical primary) range. Where larger fault currents are expected, the system of Figure 7-9a and a larger ct ratio should be used.

When the maximum fault current exceeds the maximum values shown, the output waveform is nonsinusoidal. Relay timing will tend to become variable and longer than indicated in the published literature.

The above schemes are for feeder circuit protection. For the ground protection of equipment, a ground differential scheme can be used with a differential-type relay or product-type (CWC) relay as shown in Figure 7-13. This is also applicable to short-run feeders with three conventional ct's or a zero sequence type ct at each end of the protected zone. The CWC relay is recommended as it provides high sensitivity and is relatively independent of the current transformer performance.

6 GROUND FAULT PROTECTION FOR THREE-PHASE, FOUR-WIRE SYSTEMS

6.1 Unigrounded Four-Wire Systems

Unigrounded, four-wire systems have insulated neutrals; the only ground connection is at the substation. Loads generally are connected phase to neutral, and the net load unbalance returns through the neutral as a residual current. For faults from phase to ground, the current returns through the earth to the substation neutral.

There are three different relay schemes for ground fault protection for unigrounded systems, as shown in Figure 7-14. Figure 7-14a illustrates the conventional

Table 7-1 Relay Settings and Sensitivities Using the 50/5 BYZ Zero Sequence Current Transformers

Relay type	Relay setting	Minimum sensitivity in primary $3I_0$ amperes		Maximum primary $3I_0$ amperes for accurate timing and coordination	
		(4¾ ID)	(7¾ ID)	(4¾ ID)	(7¾ ID)
IT	0.15	5.0	5.0	—	—
CO-8 or 9	0.5	9.0	10.0	25	112
CO-8 or 9	2.5	24.0	24.0	540	1215
CO-11	0.5	6.0	7.0	70	150
CO-11	2.5	24.0	24.0	700	900

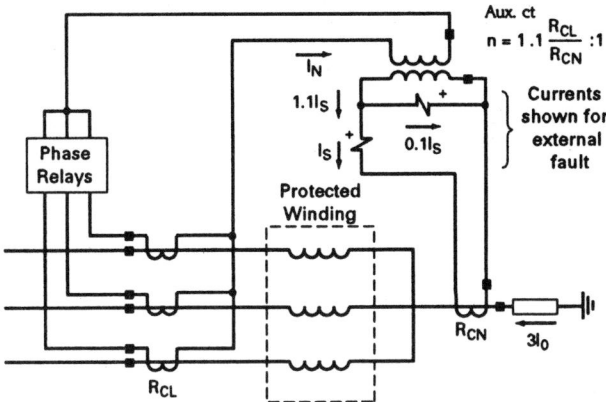

Figure 7-13 Ground differential for wye winding using CWC (product-type relay).

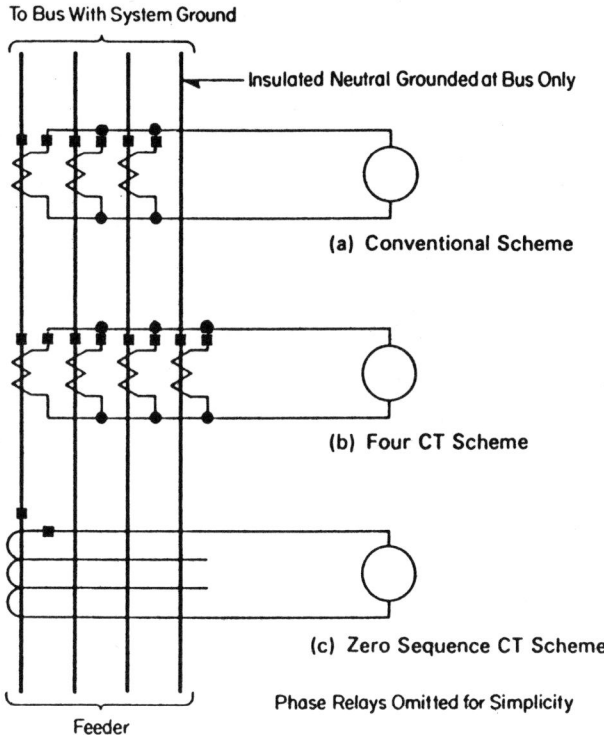

Figure 7-14 Methods of ground protection on unigrounded systems.

scheme used on three-phase, three-wire systems. For a four-wire system, the load unbalance current would flow through the ground relay, requiring a setting to avoid operation on the maximum load unbalance. This scheme is generally not recommended for the uni-grounded system. The four-current transformer scheme shown in Figure 7-14b provides much higher sensitivity, since it does not measure the load unbalance residual current. Even greater sensitivity is provided by the zero sequence type of current transformer depicted in Figure 7-14c. Comparative sensitivities for various relays in this scheme are listed in Table 7-1.

If a line-to-neutral fault occurs on the system, only the conventional scheme (Fig. 7-14a) will respond. The connection of the current transformers in the other two schemes results in cancellation of the fault current, unless it involves ground. The phase relays will provide protection, however, since phase-to-neutral fault current in one phase will be of the same order of magnitude as a three-phase fault.

A fault between neutral and ground is possible even though the neutral is nearly at ground potential, probably as the result of a broken neutral conductor. The schemes of Figures 7-14b and c will measure any current returning through the earth. Because the

ground return may be a high impedance path, causing low voltage at the load points, the more sensitive window-type current transformer scheme is recommended.

6.2 Multigrounded Four-Wire Systems

Many three-phase, four-wire distribution systems are solidly grounded at the substation, with the neutral wire also grounded at each distribution transformer location. Such systems are difficult to protect against ground faults. The scheme of Figure 7-9 is used most often with the sacrifice of sensitivity dictated by maximum load unbalance.

8

Generator Protection

Revised by: **C. L. DOWNS**

1 INTRODUCTION

The frequency of failure in rotating machines is low with modern design practices and improved materials, yet failures will occur and delayed tripping or insensitivity of protection may result in severe damage and long outages for repairs. For these reasons, abnormal conditions must be recognized promptly and quickly isolated to avoid extending the damage or compounding the problem.

Abnormal conditions that may occur with rotating equipment include the following:

Faults in the windings
Overload
Overheating of windings or bearings
Overspeed
Loss of excitation
Motoring
Inadvertent energization
Single-phase or unbalanced current operation
Out of step
Subsynchronous oscillations

Some of these conditions do not require that the unit be tripped automatically, since in a properly attended station, they can be corrected while the machine remains in service. These conditions are signaled by alarms. However, most require prompt removal of the machine from service.

For any particular hazard, the initial, operating, and maintenance costs of protective schemes and the degree of protection they afford must be carefully weighed against the risk encountered if no protection

were applied. The amount of protection that should be applied will, of course, vary according to the size and importance of the machine.

2 CHOICE OF TECHNOLOGY

In the choice of relays to be applied for the various functions described here, it will be recognized that they are available as discrete functions in their individual housing or as a complete complement containing all the pertinent protection plus data acquisition.

Electromechanical, solid-state, and microprocessor-based devices are used depending on personal choice and whether or not a new installation, an upgrade, or a functional addition is involved.

Although electromechanical and single-function solid-state relays have proven their reliability, flexibility, and effectiveness, the trend is toward microprocessor-based integrated packages. Many of these provide event recording, oscillography, self-monitoring, communications, adaptive characteristics, and other features that only a microprocessor-oriented system can provide.

3 PHASE FAULT DETECTION

Internal faults in equipment generally start as a ground in one of the stator windings and may occasionally develop into a fault involving more than one phase. Differential protection is the most effective scheme against multiple-phase faults. In differential protection, the currents in each phase, on each side of the machine, are compared in a differential circuit. Any "difference" current is used to operate a relay.

Figure 8-1 shows the relay circuits for one phase only. For normal operation or a fault outside the two sets of current transformers, I_p entering the machine equals I_p leaving the machine in all phases, neglecting the small internal leakage current. The secondary current of each of the ct's is the perfectly transformed primary current minus the magnetizing current I_e.

The relay current $I_{e1} - I_{e2}$ is the difference in the exciting or magnetizing currents. With the same type of current transformers, this current will be small at normal load. If a fault occurs between the two sets of current transformers, one or more of the left-hand currents will suddenly increase, whereas currents on the right side will either decrease or increase and flow in the opposite direction. Either way, the total fault current will now flow through the relay, causing it to operate.

If perfect current transformers were available, an overcurrent relay in the "difference" circuit could be set to respond very sensitively and quickly. In practice, however, no two current transformers will give exactly the same secondary current for the same primary current. Discrepancies can be traced to manufacturing variations and differences in secondary loading caused by unequal length of relay leads and unequal burdens of meters or instruments connected in one or both secondaries. The differential current produced flows through the relay. Although normally small, the differential current can become appreciable when short-circuit current flows to an external fault. An overcurrent relay would have to be set above the maximum error current that could be expected during an external fault. On a symmetrical basis (no dc offset in the primary current), this would not exceed 10 A if a C class ct were used within rated burden and the ratio were chosen such that the secondary current did not exceed 100 A for the maximum "through" phase fault. To avoid operation on an asymmetrical fault, the trip time would have to exceed three dc time constants.

3.1 Percentage Differential Relays (Device 87)

The percentage differential relay (Fig. 8-2) solves the problems of poor sensitivity and slow operation. The inputs from the two sets of current transformers are used to generate a restraint quantity. This is then compared to the difference of these two currents. Operation (or restraint) is produced as a result of the comparison of the difference to the restraint. This desensitizes the relay for high external fault currents.

The current required for relay operation increases with the magnitude of the through fault current. The percentage of increase may be constant, as in the CA (87) generator percentage differential relay. Alternatively, the percentage of increase may vary with the external fault current, as in the high-speed SA-1 (87) generator relay. The effect of the restraint on internal faults is negligible, because the operating quantity is weighted and responds to the total secondary fault current.

Generator differential relays are available with various percentage differential characteristics. They are typically 10%, 25%, and variable-percentage differential types. The percentage indicates the differ-

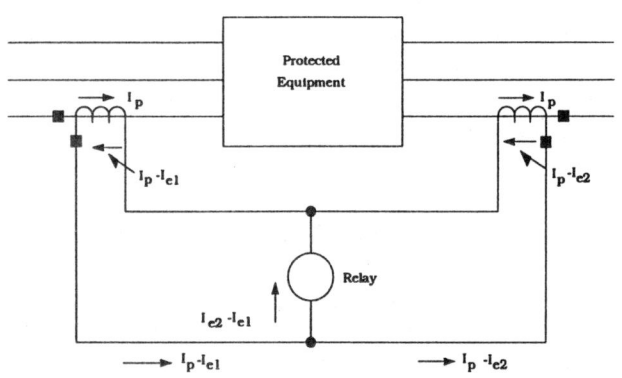

Figure 8-1 The basic differential connection.

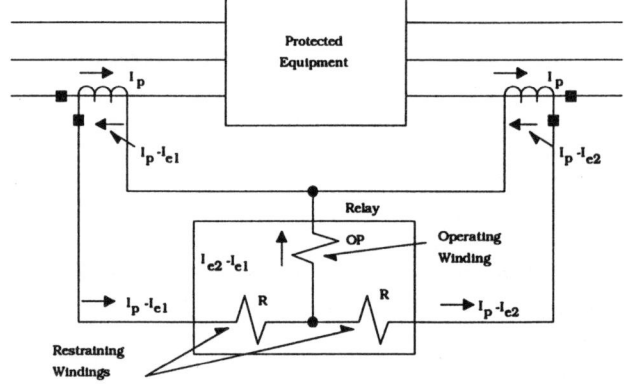

Figure 8-2 Schematic connections of the percentage differential type relay. (Only one-phase connections are shown.)

ence current as a percentage of the smallest restraint current required to operate the relay. The pickup (the value of current into one restraint winding and out the operating winding) is the current required to barely make the relay operate. Its value tends to be smaller for the lower-percentage differential relays and is as low as 0.10 A for some. Operating time, in general, is smaller for solid-state relays, being 25 msec for the solid-state SA-1 compared to 80 to 165 msec for the electromechanical CA relay.

Multifunction microprocessor relays do not have a physical operating winding, the difference of the restraint currents being computed mathematically by the protection algorithm.

In all differential schemes, it is good practice to use current transformers with the same characteristics whenever possible and avoid connecting any other equipment in these circuits.

3.2 High Impedance Differential Relays (Device 87)

High impedance differential relaying is based on the conservative premise that the ct's on one side of the generator perform perfectly for an external fault and the other set of ct's saturate completely. It takes advantage of the fact that the voltage appearing across the relay is limited for an external fault to the voltage drop produced by the maximum secondary current flowing through the leads from the relay to the saturated ct and through its internal resistance. For an internal fault, the voltage will approach the open-circuited ct voltage (usually limited by a varistor internal to the relay). In general, this scheme is not as sensitive as the percentage differential scheme but is more secure.

3.3 Machine Connections

Most generators have wye-connected windings. As shown in Figure 8-2, three relays connected to wye-connected current transformers provide phase and, in some cases, (depending on the type of neutral and system grounding) ground fault protection. Figure 8-3 illustrates a similar protective scheme for delta generators. In this scheme, the delta windings must be brought out so current transformers can be installed inside the delta.

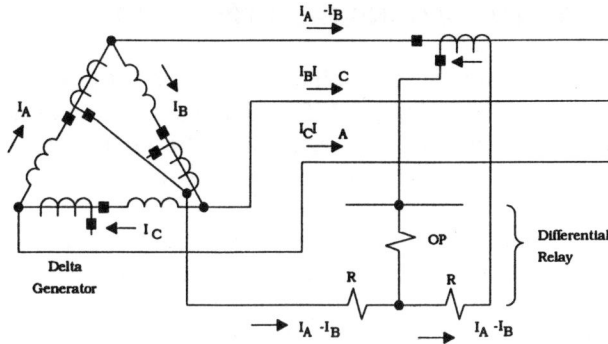

Figure 8-3 Percentage differential relay schematic for a delta-connected machine. (Only one-phase connections are shown.)

3.4 Split-Phase

Generators with split-phase windings can be protected by two sets of differential relays: one connected as in Figure 8-2 and the other as in Figure 8-4. This arrangement protects against all types of internal phase faults, including short-circuited turns or open-circuited windings. This scheme may be extended to accommodate other winding arrangements involving more than two equal windings per phase. Unless the ratios of the current transformers produce an exact match, the scheme of Figure 8-4 must be equipped with auxiliary transformers to provide a balance during normal operation.

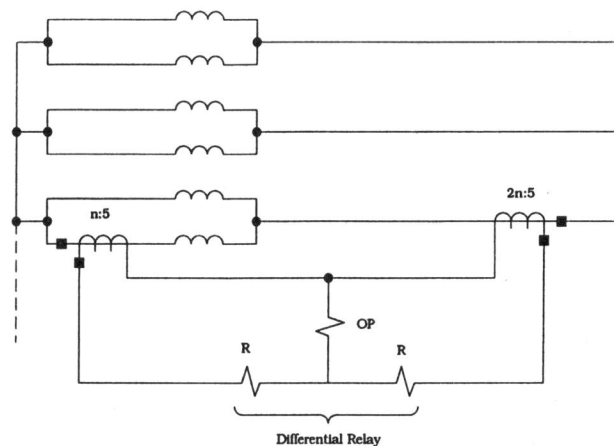

Figure 8-4 Schematic connections for one-phase only for the protection of a machine with split phase windings.

4 STATOR GROUND FAULT PROTECTION

The method of grounding affects the degree of protection afforded by differential relays. The higher the grounding impedance, the less the fault current magnitude and the more difficult it is to detect high impedance faults. With high impedance grounding, the differential relays will not respond to single-phase-to-ground faults. A separate relay in the grounded neutral will provide sensitive protection, since it can be set without regard to load current.

The ground relay may also operate for ground faults beyond the generator. For this reason, a time delay may be necessary to coordinate with any overlapped relays. A typical case is a generator connected directly to a bus with other circuits. A fault on one of these circuits should not trip the machine; the relays in the faulted circuit will clear such faults. A wye-delta transformer bank will block the flow of ground current, preventing faults on the opposite side of the banks from operating ground relays. In the unit-connected scheme, the transformer bank limits the ground relay operation to faults in the generator, the leads up to the transformer bank, and the delta winding.

4.1 Unit-Connected Schemes

The unit-connected system is the most common arrangement for all but small generators. For unit systems high-resistance grounding is used, and the machine is generally grounded through a distribution transformer and resistor combination, as shown in Figure 8-5. Since the secondary is rated at 120 or 240 V, the physical size of the resistor can be somewhat smaller than if it were connected to the primary.

4.1.1 "95% Scheme"

The unit system responds to the voltage shift of the generator neutral with respect to ground that occurs for a ground fault in the machine, bus, or low-voltage winding of the transformer. The relay used must be insensitive to the substantial normal third harmonic voltage that may be present between neutral and ground, and yet sensitive to the fundamental frequency voltage that accompanies a fault.

Since the magnitude of the neutral shift is dependent on the location in the winding of the ground fault (neutral-to-ground fault produces no neutral shift) and the usual choice of relay sensitivity and distribution transformer voltage ratio provides roughly 95% cover-

age of the winding, this relay is often referred to as a 95% relay.

4.1.2 Neutral Third Harmonic Undervoltage

Other schemes take advantage of the *presence* of the third harmonic voltage between neutral and ground and respond to undervoltage for a neutral-to-ground fault.

4.1.3 100% Winding Protection

Other ground relaying schemes provide complete protection of the generator stator by injecting a signal into the stator and monitoring it for change. This concept allows 100% coverage even though the machine is at standstill, whereas the 95% and neutral third harmonic schemes depend on the machine operating at rated speed and voltage.

4.2 95% Ground Relays

The CV-8 or solid-state 59G low-pickup overvoltage relay can be used for unit-generator applications as shown in Figure 8-5. Provided that a full-rated primary winding is used, the maximum voltage for a solid ground fault is $120/\sqrt{3}$ (69.3 V with a 120-V distribution transformer secondary), or $240/\sqrt{3}$ (138.6 V with a 240-V secondary).

The scheme has good sensitivity for internal ground faults while being very insensitive to third harmonic voltages. Various provisions are used to make the relay insensitive to the third harmonic. The third harmonic pickup of the CV-8 relay, for example, is approximately 8 times the pickup at rated frequency.

The voltage appearing from the neutral of the generator to ground is dependent on the location of the ground fault. The more sensitive the ground relay, the greater the percentage of the winding protected. Obviously, a neutral-to-ground fault goes undetected by this relay and other devices must be considered. In some relays, the sensitivity is proportional to its rating, and the best protection is obtained by using the relay having the lowest voltage rating. For a ground fault at the line terminal of the machine, full line-to-neutral voltage will exist from neutral to ground. The voltage on the relay is, of course, dependent on the ratio chosen for the distribution transformer. If this voltage exceeds the rating of the relay and is not removed by tripping the field circuit within the short-time capability of the relay, the SV scheme shown in Figure 8-5 may be used to protect the relay for those fault cases

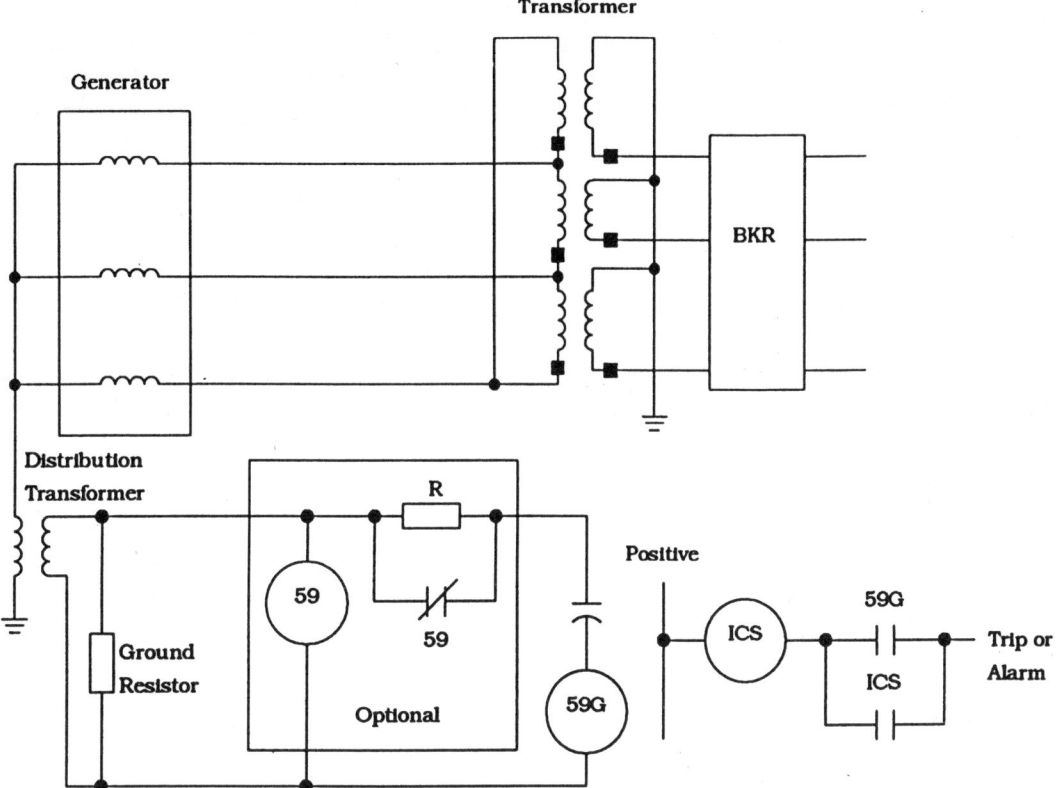

Figure 8-5 Schematic connections for ground fault protection of a unit type machine resistance grounded through a distribution transformer.

producing this high voltage. The SV relay is set to open its contacts at a value somewhat lower than the continuous rating of the protected relay, inserting R to limit the voltage. In general, the 95% type relay is allowed to trip immediately to remove voltage prior to the occurrence of damage to the relay, making the SV unnecessary. Other solid-state relays such as the type 59G have a 208-V continuous rating and 60-Hz sensitivity as low as 1 V.

Time-delay settings of 25 msec to 4 sec are used for this function. These longer delays allow for coordination with voltage transformer fuses, if required.

Operation of the ground relay can be avoided for faults on the main voltage transformer secondary by grounding one phase of the secondary rather than the neutral. Then a ground fault on the voltage transformer secondary will not produce a machine neutral voltage shift and the ground relay will not operate. This is the recommended grounding practice for generator voltage transformers.

Another scheme that is often used in industrial applications uses a product-type relay for ground detection. This arrangement is described in Chapter 7, Figure 7-13. Good sensitivity is achievable because of the ability to use a low-ratio ct in the neutral grounding resistor path that is not related to the machine full-load current.

4.3 Neutral-to-Ground Fault Detection (Device 87N3)

Figure 8-6 describes a scheme for detecting a neutral-to-ground fault on the generator. This fault is, in itself, not hazardous. A second ground fault at the machine terminal, however, causes a line-to-ground fault that is not limited by any neutral impedance. This fault current magnitude will quite likely exceed the current magnitude for which the machine is designed. Machine

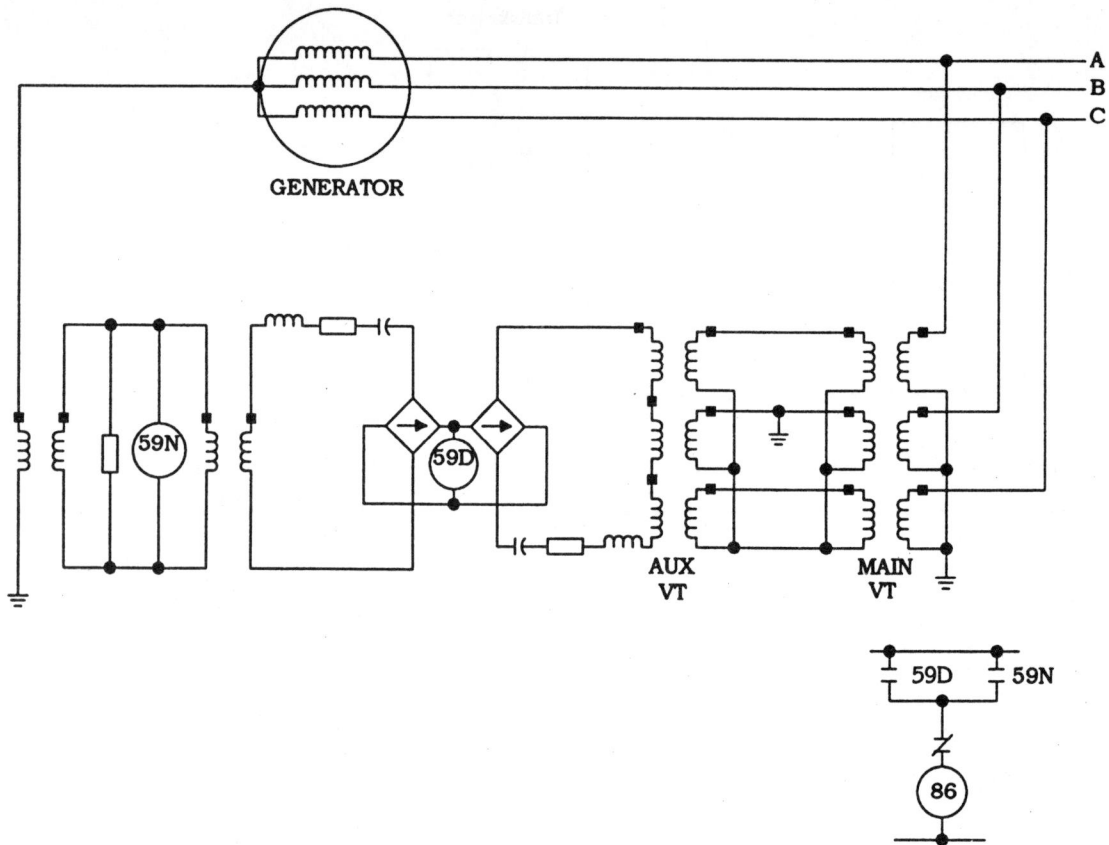

Figure 8-6   180-Hz voltage comparator.

destruction may result. Early detection, then, is imperative.

The scheme of Figure 8-6 compares the third harmonic voltage present between the machine neutral and ground with that at the line terminals. The relative values of these voltages are established by the distributed capacitances of the generator, phase leads, and transformer low-voltage winding plus the grounding system. Though the third harmonic voltage changes with machine loading, the ratio of these values remains relatively constant.

When a ground fault occurs, this relationship changes and therefore allows detection of faults at all points not covered by the "neutral shift" relay. These two relays, 59N and 59D, provide "100% ground fault coverage." Very high impedance faults in the protected zone, of course, cannot be detected. Also for the unit-connected system neither relay will provide any backup for faults on the transmission system.

4.4   100% Winding Protection

Another important ground fault detection scheme (GIX-104) utilizes an injected current that is monitored for magnitude. In the configuration shown in Figure 8-7, the small injection voltage is applied across the lower part of the grounding resistor. As a result of the distributed capacitance of the generator and connected apparatus, a current flows and produces a voltage drop across the measuring transformer. A ground fault occurring anywhere in the protected winding will cause the current to rise and the voltage to increase across the grounding resistor. To differentiate from other conditions that could produce a similar voltage, the injection voltage is commutated, producing a coded character that is readily identifiable.

The GIX-104 may also be equipped to detect, through the injection principle, a ground fault in the generator field.

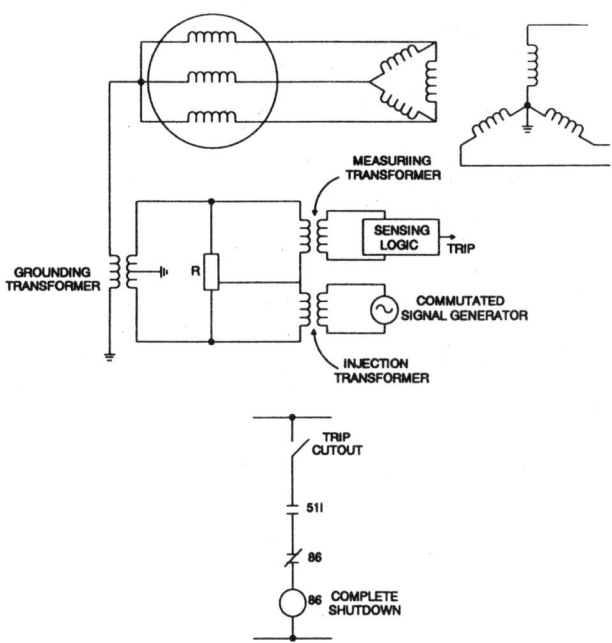

Figure 8-7 Neutral voltage injection.

This scheme has the important advantage that it is able to detect a stator or field circuit ground irrespective of the status of the generator. It may be dead, spinning without field, near synchronizing, on-line, loaded, or unloaded.

5 BACKUP PROTECTION

5.1 Unbalanced Faults

Unsymmetrical faults produce more severe heating in machines than symmetrical faults. The negative sequence currents that flow during these unbalanced faults induce currents in the rotor having twice system frequency. These currents tend to flow in the surface of the solid rotor forging and the nonmagnetic rotor wedges and retaining rings. The resulting I_2^2R loss quickly raises the temperature. If the fault persists, the metal will melt, damaging the rotor structure.

Such faults result from failure of a protective scheme or equipment external to the machine. The relative magnitudes of negative sequence currents for line-to-line faults on a typical turbine generator under different operating conditions are shown in Figure 8-8. The effect of the higher excitation during the fault is included for short circuits with load on the system.

According to ANSI standards, the permissible integrated product I_2^2t, (where I_2 is negative sequence current in per unit of machine rated current and t is seconds duration) that "indirectly cooled" turbine generators, synchronous condensers, and frequency-changer sets can tolerate is 30. The standard for hydraulic turbines or engine-driven generators is 40. Standard "directly cooled" machines up to 800 MVA are capable of withstanding a permissible integrated product of 10, whereas some very large machines (1600 MVA) can only tolerate 5. Early inspection to detect damage is recommended for machines subject to faults between the above limits and 200% of the limit. Serious damage can be expected for faults above 200%.

With so many influences on the I_2^2t as shown in Figure 8-8, it is evident that a relay designed to respond in a way similar to the manner in which heat is generated in the machine is mandatory. Many variations are available in generator negative sequence overcurrent relays (see Fig. 8-9), but each is tailored to match the $I_2^2t = K$ characteristic for K between 10 and 40, with some spanning an even greater range. The ANSI standard explicitly restricts this limit to I_2 values above the full-load level. Earlier relays covered this, but provided no I_2 protection below the full-load value. Later versions of the ANSI standard introduced an additional limit, unrelated to time. More recently designed negative sequence overcurrent relays provide an alarm level and trip capability for currents between a pickup setting and full load. This is pertinent to increased heating effects caused by such factors as unbalanced load or faulty circuit breakers.

The negative sequence protective function is recommended for all machines rated 1000 kVA or larger, though it can be justified for important smaller machines. Examples of schematic connections for unbalanced fault protection are shown in Figure 8-10.

The filter output that is applied across the operating coil of the COQ relay is

$$V_F \propto 2K_2I_{a2} \tag{8-1}$$

when the connection shown in Figure 8-10a is used. If, however, the auxiliary current transformer is not used,

$$V_F \propto K_2(2I_{a2} + I_{a0}) \tag{8-2}$$

where K_2 is the filter constant.

If I_{a0} is small, its effect can be ignored. Otherwise, it will be necessary to use either the auxiliary current transformer to remove it or the relay with neutral made up inside, where its effect can be nullified. The auxiliary current transformer is not normally required

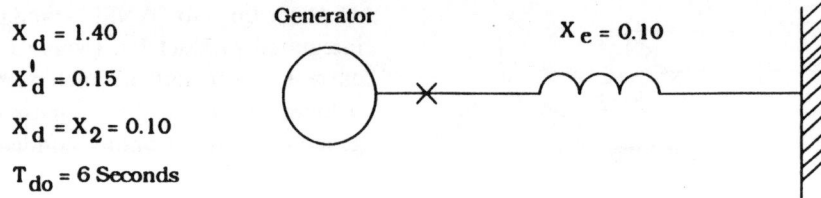

$X_d = 1.40$

$X'_d = 0.15$

$X_d = X_2 = 0.10$

$T_{do} = 6$ Seconds

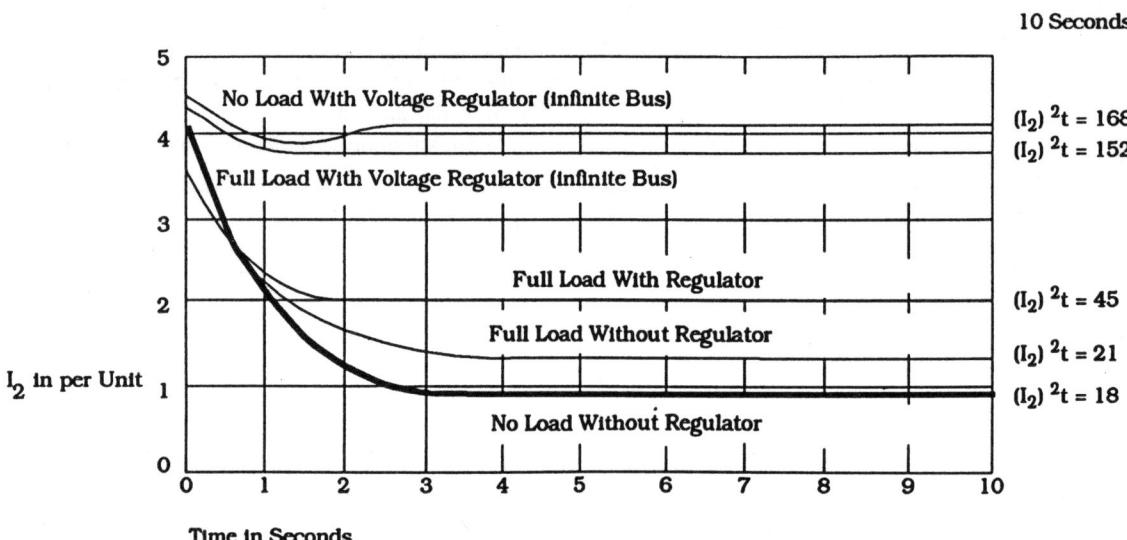

Figure 8-8   Relative magnitudes or negative sequence currents for line-to-line faults on a typical machine under different operating conditions. (From AIEE Transactions, Volume 72, 1953, Part iii, Page 283, Figure 1.)

in unit-connected applications. For relays such as the SOQ and 46Q that use the equivalent of delta currents, any zero sequence current is ignored by the relay.

When a continuous load unbalance in excess of the 5 or 10% of the capability of the particular machine may occur, the SOQ relay, or an additional relay 46Q, set for the desired alarm level may be used to alert an operator. With the SOQ, instrumentation can identify the level of negative sequence current to permit a decision between tripping or decreasing the machine loading.

An excellent backup for ground fault detecting relays for a unit-connected generator utilizes a relay supplied by a current transformer (rated 100:5 or thereabouts) in the secondary leads of the neutral distribution transformer. This is used by some utilities as the primary ground relaying.

Integrated protection relays such as the REG-100 or REG-216 include the negative sequence protective and alarm functions.

5.2   Balanced Faults

5.2.1   Distance Relay (21)

A generator should be protected also against damage that will result from prolonged contribution to a balanced fault. A distance relay such as a KD-11, fed from current transformers in the neutral of the generator and a voltage supply connected at the generator voltage level, provides such protection. A single relay of this type complements the COQ, 46Q, or SOQ in recognizing balanced faults internal and external to the generator. It also supplements these

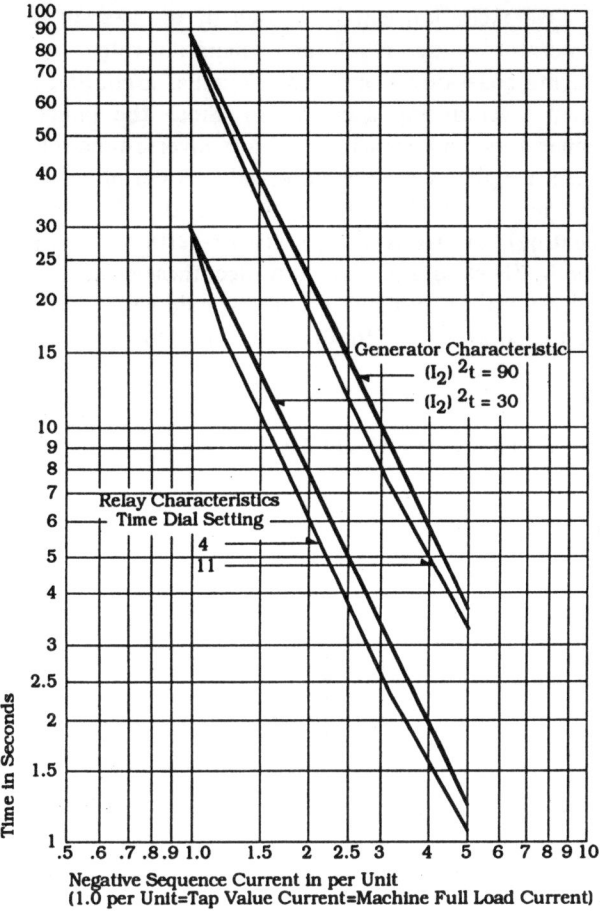

Figure 8-9 Comparison of relay and generator characteristics.

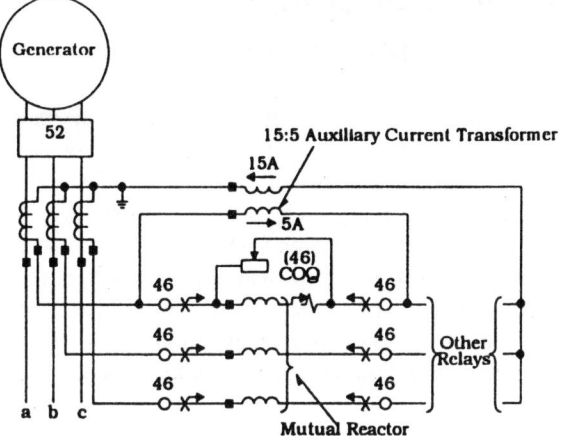

a) Connections When Neutral is Made Externally (Omit 15:5 Current Transformer When Io<0.1 of Machine Rated Current).

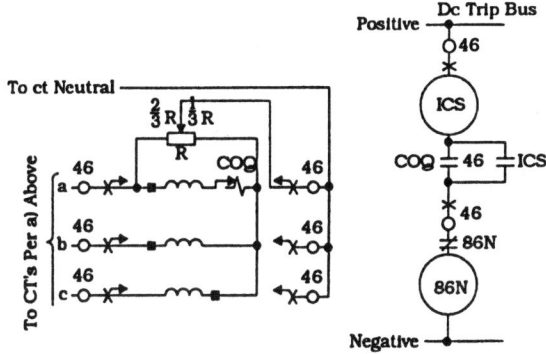

b) Connections When Neutral is Made Internally   C) Trip Circuit Schematic

Figure 8-10 Schematic connections of the COQ relay for unbalanced fault protection.

relays by sensitively recognizing unbalanced faults. The connection described above makes the relay directional from the neutral, but gives it a reach in both directions from the voltage transformer location. As a result, it will sense some generator as well as transformer faults.

The distance relay is usually set to reach through the unit transformer. Unlike single-phase distance relays, the reach of the KD-type relays is not affected by the phase shift through the bank. When set for an impedance greater than the transformer impedance, the KD-11 relay will operate for both generator and line-side phase faults that manifest an impedance within this reach. A timer must be used to ensure only the minimum equipment outage necessary to clear a fault. The timer must be set to coordinate with the high-voltage transmission line relays and all other relays it overreaches.

5.2.2 Voltage-Controlled Overcurrent Relay (51 V)

If a negative sequence overcurrent relay is used, one 2- to 6-A 51-V relay also may be used to provide the balanced fault backup function. A simple overcurrent unit is unsuitable for preventing a sustained machine contribution to a fault because, with a regulator out of service, the bolted sustained (or synchronous) three-phase fault contribution is less than machine full-load current. The 51-V relay, on the other hand, can be set well below full-load current and not operate on load. Its overcurrent unit is supervised or torque-controlled by an undervoltage unit, and therefore voltage must be below the voltage setting to permit the overcurrent unit to function. Both the voltage and current units are independently adjustable, making coordination with other overcurrent devices simpler than if the current unit response were a function of voltage level.

6 OVERLOAD PROTECTION

6.1 RTD Schemes (Device 49)

Most large generators are equipped with resistance temperature detectors (RTDs), which may be used in a bridge circuit to provide sensing intelligence to an indicator or a relay such as the DT-3 or 49 T.

This relay is restrained when the resistance is low, indicating low machine temperature. When the temperature of the machine exceeds some preset level such as 120 °C for class B insulated machines, the bridge becomes unbalanced and the contacts close.

6.2 Thermal Replicas (Device 49)

A thermal replica relay utilizes stator current to approximate the heating effects in the generator. The machine thermal time constants on heating and cooling are represented to take cognizance of previous and present loading effects. When the replica indicates that temperature in excess of the allowable value for the machine insulation has been reached, tripping takes place.

7 VOLTS PER HERTZ PROTECTION

From the fundamental expression for induced voltage in a coil,

$$E = 4.44\ fANB_m\ 10^{-8}$$

where

  E = induced rms voltage in volts

  f = frequency in hertz

  A = cross-sectional area of the core in square inches

  N = number of turns

  B_m = flux density in maxwells per square inch

Since all the elements of the equation are constants except E and f, it can be seen that

$$B_m \propto \frac{E}{f}$$

Flux density is an excellent indicator of no load heating effect. Hysteresis and eddy current losses are each proportional to a power of the flux density. Therefore, impending overheating can be recognized by measuring volts per hertz.

Overexcitation can be caused by an attempt by a generator voltage regulator to maintain rated voltage during coast-down or holding manual excitation at a fixed level during acceleration. Since the limits on generators and transformers are inverse-time-related (i.e., a higher volts per hertz value is permitted for a shorter time to stay within the bounds of acceptable heating), an inverse-time, volts-per-hertz relay such as the MVH should be used to protect these devices when overexcitation is likely. Figure 8-11 shows an example of the coordination that can be achieved.

8 OVERSPEED PROTECTION

A generator accelerates when it becomes separated from its load. The acceleration depends on the inertia

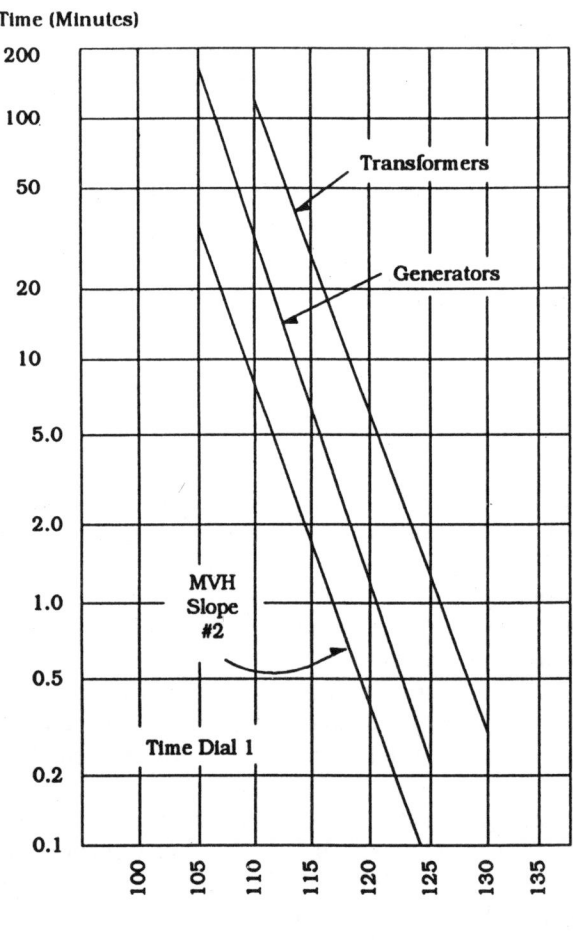

Figure 8-11 Example of volts/hertz protection using MVH relay.

(WR^2), load loss, and governor response. To recognize overspeed, a permanent magnet generator is often connected to the machine shaft to supply a voltage to the governor that is proportional to speed. The governor may also be equipped with a speed-responsive flyball mechanism. Either the permanent magnet generator or flyball mechanism can initiate prime-mover control to remove power input and alleviate overspeed. An overfrequency relay, such as an MDF (device 81), can be used to supplement this overspeed equipment.

9 LOSS-OF-EXCITATION PROTECTION

9.1 Causes of Machine Loss of Field

Loss of excitation can occur as a result of

Loss of field to the main exciter
Accidental tripping of the field breaker
Short circuits in the field circuits
Poor brush contact in the exciter
Field circuit-breaker latch failure
Loss of ac supply to the excitation system
Reduced-frequency operation when the regulator is
    out of service

Some relays such as the KLF and KLF-1 contain multiple operating units. The KLF and KLF-1 contain (1) a directional unit, (2) an offset mho unit, and (3) an instantaneous undervoltage unit.

Loss of excitation produces voltage and current variations as shown in Figure 8-12. This causes a trajectory on the R-X diagram such as that depicted in Figure 8-13b. The initial operating point is dependent on load level and power factor angle prior to the loss of excitation. When reactive power begins to flow into the machine, the locus moves into the −X region. As the transient continues, with the field collapsing, the locus moves into the characteristic circle of the loss-of-field relay. If the relay is equipped with an undervoltage unit and at this time the voltage is sufficiently below normal to operate it, tripping takes place after a short delay (X dropout). (See Fig. 8-13a.)

If reduction in terminal voltage is not appreciable, the undervoltage unit does not drop out. This is indicative of a loss of excitation condition that is not likely to affect system stability nor influence adjacent machines significantly. For this series of conditions, it will suffice to sound an alarm to alert the operator to allow him or her to take action to restore the field or anticipate shutdown. A timer is often run by Z and D

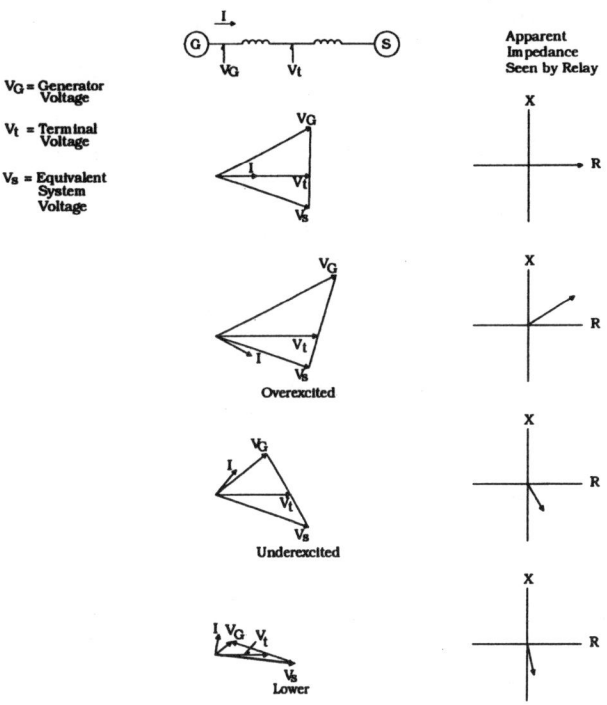

Figure 8-12 Underexcited operation of generator.

and V to initiate partial shutdown if the operator is unable to correct the problem quickly.

These relays can also be used to detect loss of field in a synchronous condenser or motor. External faults will cause the D and/or Z functions to restrain to avoid undesirable tripping of the machine.

Other relays such as the solid-state type 40 relay perform the function with only the reduced-diameter impedance measurement.

9.2 Hazard

The generator must be kept on line, supplying power as long as possible, particularly when the machine represents a sizable portion of the system capacity. To this end, an early warning of low excitation would give the operator an opportunity to restore the field if possible and avoid tripping. Unnecessary tripping and the resultant loss of kW output can precipitate system breakup and a major outage.

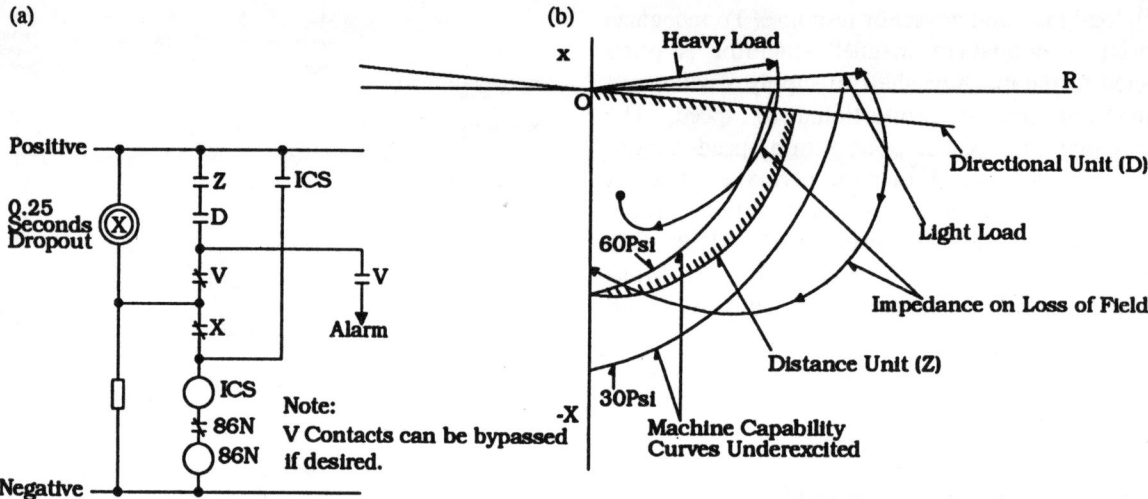

Figure 8-13 Trip circuits and R-X diagram showing operation of the KLF (40) loss-of-field relay.

9.3 Loss-of-Field Relays

A relay designed to detect low excitation should perform the following functions:

Alert the operator to any low excitation that could damage the machine or result in instability

Alert the operator to a loss-of-field condition as early as possible, giving him or her time to correct the condition

Trip·the machine automatically in the case of impending system instability

The KLF-1 differs from the KLF relay in that it has a separate phase voltage supply for each of three different measuring elements. As a result, loss of any one phase voltage to the KLF-1 relay cannot cause incorrect tripping. With the KLF, some combinations of load and phase voltage loss can operate the relay. The KLF-1 must have a wye-wye voltage supply, with the neutral brought to the relay. The KLF may be used with a wye-wye, delta-delta, or open-delta-open-delta supply.

When partial or complete loss of excitation occurs on a synchronous machine, reactive power flows from the system into the machine and the apparent impedance as viewed from the machine terminals (V_t/I) goes into the negative X region (Fig. 8-12). The kW output is controlled by the prime-mover input, whereas kV Ar output is controlled by the field excitation. If the system is large enough to supply the deficiency in excitation through the armature, the synchronous machine will operate as an induction generator, supplying essentially the same kW to the system as before the loss of excitation.

Loss of synchronism does not require immediate tripping unless there is an accompanying decrease in the terminal voltage that threatens the stability of nearby machines. Generally, it takes at least 2 to 6 sec to lose synchronism. Many instances have been reported in which machines have run out of synchronism for several minutes because of loss of excitation without damage to the machines.

9.4 KLF and KLF-1 Curves

Figure 8-14 shows how a generator capability curve can be transformed into an R-X diagram. For each point on the curve, an angle β can be measured from the horizontal and the value of three-phase MVA read. Knowing the line-to-line voltage at which the capability curve applies, a value of Z can be calculated using

$$Z_p = \frac{kV^2}{MVA} \quad \text{primary ohms} \qquad (8\text{-}3)$$

$$Z = \frac{kV^2(R_c)}{MVA(R_v)} \quad \text{secondary ohms} \qquad (8\text{-}4)$$

where R_c and R_v are the current and voltage transformer ratios, respectively.

Point Z_p or Z can then be plotted, at angle β, on the R-X diagram. Other key points on the circle arcs can

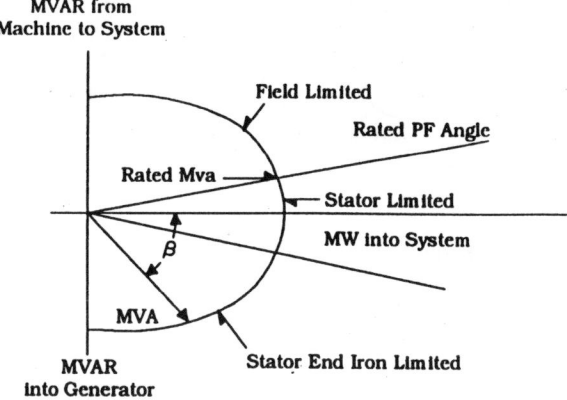

a) Machine Capability Curve

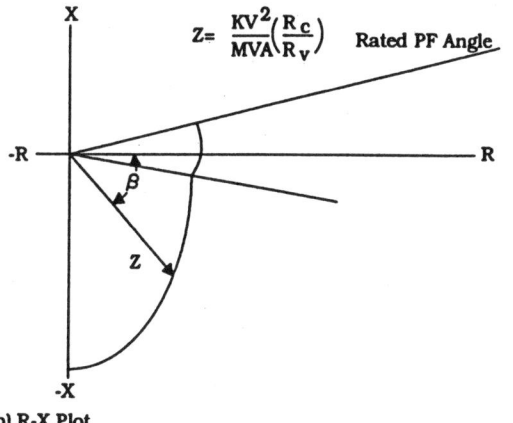

$$Z = \frac{KV^2}{MVA}\left(\frac{R_c}{R_v}\right)$$

b) R-X Plot

Figure 8-14 Transformation from KW-KVAR plot to R-X plot.

be obtained in the same way until the entire capability curve is transformed.

The steady-state stability curve is another significant limit that can be related to a loss-of-field relay with impedance-measuring qualities. The MW-MVAR curve can be developed as shown in Figure 8-15a. In this figure, V is the per unit terminal voltage, X_s the equivalent per unit system impedance as viewed from the generator terminals, and X_d the per unit unsaturated synchronous reactance of the generator. Both X_s and X_d are measured on the machine MVA base.

Figure 8-15b converts the machine's steady-state stability curve to an R-X diagram. Note that the curve of Figure 8-15b can be plotted directly from a knowledge of X_s and X_d without the intermediate

step of Figure 8-15a. To be useful in setting a loss-of-field relay, these per unit values must be converted to secondary ohms.

Figure 8-16 relates KLF (or KLF-1) setting to capability and minimum excitation limiter (MEL) curves. Assume a given kW load on the machine and that the vars into the machine are gradually being increased by decreasing machine field current, producing a trajectory, as in curve A of Figure 8-16. If the regulator is in service, the MEL prevents operation at a level that would jeopardize the machine thermally. If the regulator is out of service, Z continues to decrease until the KLF impedance unit operates. An alarm indicates a hazardous operating condition if the voltage is high. A low voltage, which may seriously jeopardize system stability, trips the machine after 0.25 sec (Fig. 8-13a). The loss-of-field relay must reach into the plus X area if its locus is to follow closely the machine characteristic. A directional unit is included in the relay to avoid tripping for close-in faults beyond the unit transformer.

9.5 Two-Zone KLF Scheme

Like all other elements in generator protection, the loss-of-field relay should be supported by backup to prevent catastrophic failure if a device is out of service or a component should fail.

Two loss-of-field relays provide better protection than one. The first, or zone 1, relay is set to be restrictive (Fig. 8-17) and typically trips through a 0.25-sec timer. It provides fast clearing on loss of field, yet is secure against swings such as that shown passing through points CDEF in Figure 8-17. The zone 2 relay is set wider and typically drives a 1-sec timer to detect partial loss of field, provide an alarm function, and back up the zone 1 relay. A total functional characteristic similar to the KLF type is required to do this. Other setting data are given in Tables 8-1 and 8-2.

Figure 8-18 shows the dc schematic for the KLF or KLF-1 zone 2 relay. Zone 1 may be used without a TD-A timer unless extreme emphasis on security is made. No TD-2 relay is required if the undervoltage contacts of the zone 1 relay are shorted or the type 40, solid-state relay is used. Without TD-2 quick tripping occurs when Z and D operate irrespective of the voltage level.

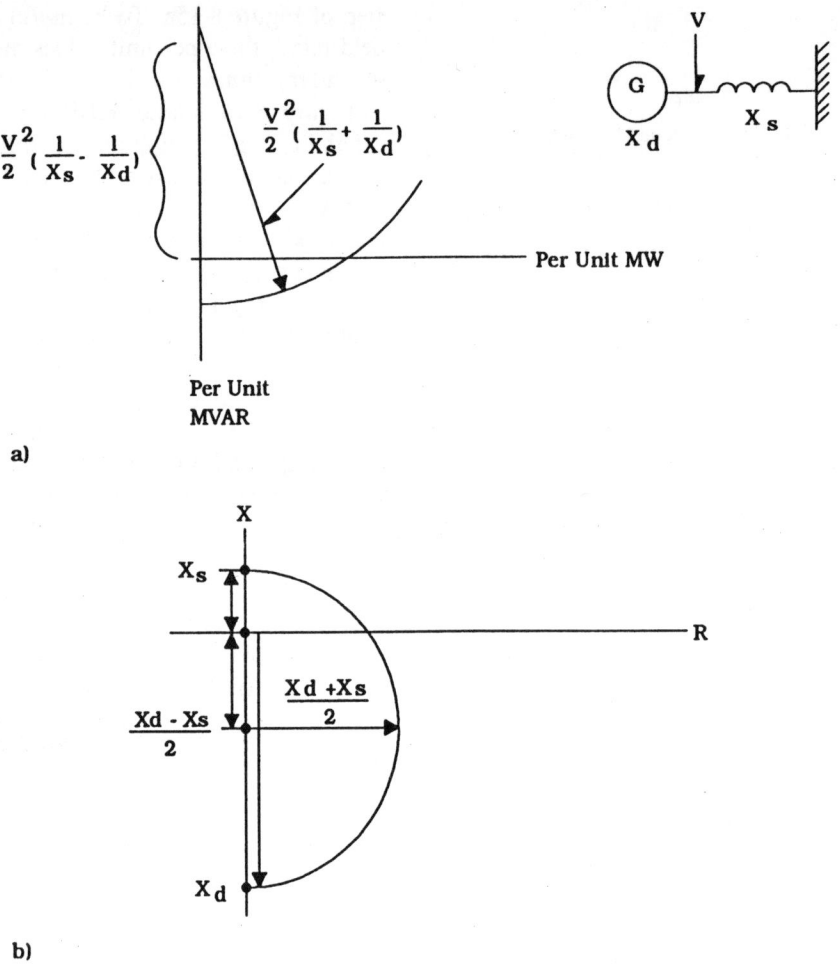

Figure 8-15   Conversion of steady-state stability curve to R-X.

10   PROTECTION AGAINST GENERATOR MOTORING

Generator motoring (sometimes referred to as reverse active power) protection is designed for the prime mover rather than the generator. With steam turbines, for example, turbines will overheat on low steam flow, but will be protected by steam temperature devices. With hydroturbines, hydraulic flow indicators protect against blade cavitation on low water flow. Similar devices are used to protect gas turbines.

Generator motoring protection can be provided by devices such as limit switches or exhaust-hood temperature detectors. However, a reverse-power protective relay is recommended for added safety.

Due to the extreme nature of this hazard, two motoring relays are recommended to operate in an OR mode and respond to different current and voltage. The reverse-power relay is commonly used with diesel engine generating units, particularly when the danger of explosion and fire from unburned fuel exists.

Motoring results from a low prime-mover input to the ac generator. When this input cannot satisfy all the losses, the deficiency is supplied by the generator absorbing real power from the system. Since field excitation should remain the same, the same reactive power would flow as before motoring. Thus, on motoring, the real power will flow into the machine, whereas the reactive power may be either flowing out of or into the machine. Usually, the reactive power will

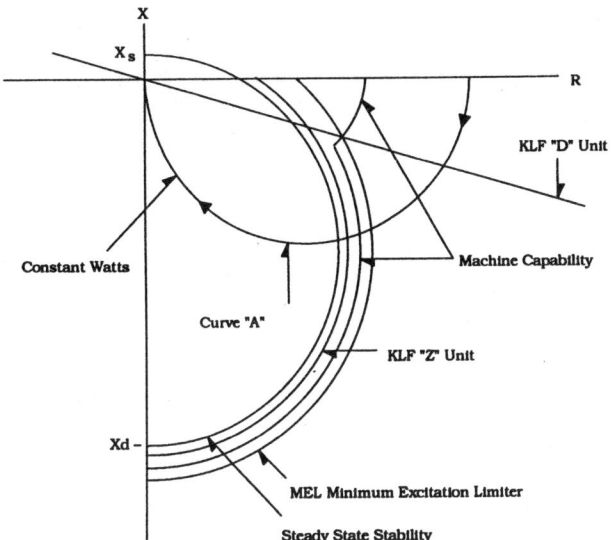

Figure 8-16 KLF setting related to capability and MEL curves.

be supplied to the system as machines are not generally operated underexcited.

A relay designed to detect motoring must be extremely sensitive and even then cannot detect all conditions of reverse power. For example, suppose a

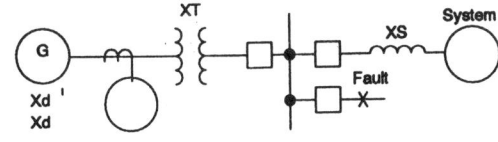

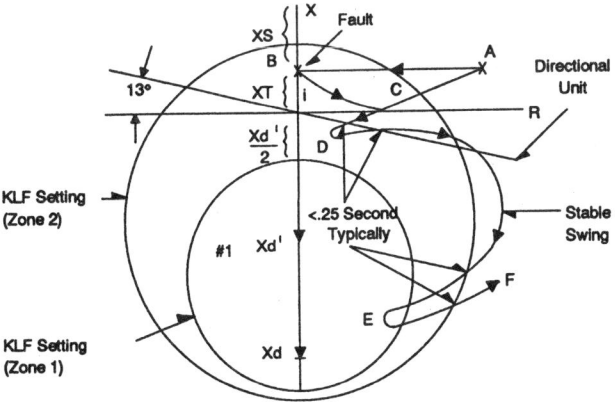

Figure 8-17 KLF relay stable swing following clearing of nearby three-phase fault.

turbine had its valves closed to slightly less than the no-load steam requirements. The turbine would supply, say, 99% of the losses, and the generator (as a motor) would supply 1%. If the total losses were 3.0% of the kW rating, the kW drawn by the generator (as a motor) from the power system would be only 1.0% of 3.0%, or 0.03%, of the nameplate rating. This is a challenge beyond the capabilities of most motoring detection relays. Solid-state and microprocessor technology allows such low sensitivity while, at the other end of the scale, providing full-load current capability. The SRW relay may be set as low as 1 mA.

When the prime mover is spun at synchronous speed with no power input, the approximate reverse power required to motor a generator, as a percentage of the nameplate rating in kW, is as follows:

Condensing steam turbine	1 to 3%
Noncondensing steam turbine	3+%
Diesel engine	25%
Hydraulic turbine	0.2 to 2+%

10.1 Steam Turbines

When operating under full vacuum and zero steam input, condensing turbines require about 3.0% of the kW rating to motor. Noncondensing turbines require 3.0% or more of the rated kW to motor when operating against atmospheric or higher exhaust pressures at zero steam flow.

10.2 Diesel Engines

If no cylinders are firing, diesel engines require about 25% of the rated kW figure. If one or more cylinders are firing at no load, the reverse power will be lower than this depending on the governor action and effect on the system frequency.

10.3 Gas Turbines

The large compressor load of gas turbines represents a substantial power requirement from the system when motoring. Consequently, the sensitivity of the anti-motoring device is not critical.

10.4 Hydraulic Turbines

When the blades are under the tail-race water level, the percent of kW rating required for motoring is

Table 8-1   Recommended Settings for KLF Relay

Setting	Zone 1 (alone)	Zone 2 (alone)	Both zone 1 and zone 2
Impedance setting	See Figure 8-17.	See Figure 8-17.	See Figure 8-17.
Voltage setting	(a) Undervoltage contact shorted or (b) set at 80% for security.	80%	Zone 1 voltage contact shorted. Zone 2 dropout voltage set at 80%.
TD-1 (see Fig. 8-18)	1/4 to 1 sec (1/4 see adequate).	1/4 to 1 sec (1 sec preferred).	Zone 1 timer = 1/4 sec. Zone 2 timer = 1 sec.
TD-2 (see Fig. 8-18)	Not required for (a) above. For (b) above use 1 min.	1 min.	None for zone 1. zone 2 timer = 1 min.
Advantages	Less sensitive to stable system swings.	(1) More sensitive to LOF condition. (2) Can operate on partial LOF. (3) Provide alarm features for manual operation.	(1) Same as (1), (2), and (3) at left. (2) Provides backup protection.

probably well over 2.0%. From 0.2 to 2.0% kW is required for the turbine to motor when the blades are above the tail-race level. For turbines using a Kaplan adjustable-blade propeller, the flat-blade condition probably requires less than 0.2% kW to motor.

Most commonly, a single-phase relay is used for motoring detection, with sensitivities ranging from 1 mA to 5 A. All have adjustable operating times. When the more sensitive relays are used, care should be used in selecting a current transformer for them. Indeed, a metering accuracy-class ct is preferred to a relaying accuracy class. A typical schematic is shown in Figure 8-19.

11   INADVERTENT ENERGIZATION

Many instances of inadvertent energization have occurred over the years. A few of the causes were

- Closing the generator breaker with the machine at standstill
- Closing a station service breaker with the machine at standstill
- High-voltage breaker flashover near synchronism
- Closing of generator disconnect with unit breaker closed

Although careful operating procedures and the judicious use of interlocks can usually prevent these

Table 8-2   Special Settings for Multi-machines Bussed at Machine Terminals

Setting	Zone 1 (alone)	Zone 2 (alone)	Both zone 1 and zone 2
Impedance setting	See Figure 8-17.	See Figure 8-17.	See Figure 8-17.
Voltage setting	(a) Undervoltage contact shorted or (b) set at 87% for security.	87%.	Zone 1 voltage contact shorted with zone 2 set at 87%.
TD-1 (see Fig. 8-18)	1/4 to 1 sec (1/4 sec adequate).	1/4 to 1 sec (1 sec preferred).	Zone 1 timer = 1/4 sec. Zone 2 timer = 1 sec.
TD-2 (see Fig. 8-18)	Not required for (a) above. For (b) above use 10 sec for directly cooled, 25 see for indirectly cooled.	10 sec for directly cooled. 25 sec for indirectly cooled.	None for zone 1. Zone 2 timers: 10 sec for directly cooled, 25 sec for indirectly cooled.

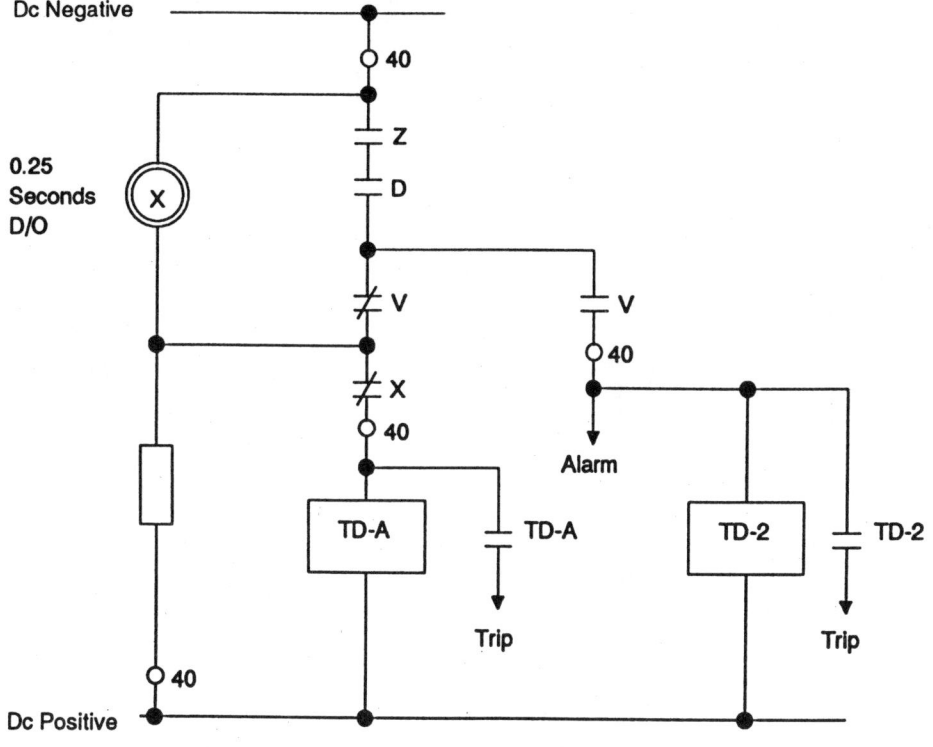

Figure 8-18 Type KLF or KLF-1 dc schematic for zone 2 loss of excitation protection. (Timer settings are given in Tables 8.1 and 8.2.)

occurrences, the ingenuity of humans to circumvent these procedures and interlocks is legendary.

Full-voltage energization of a machine at standstill does not produce an enormous magnitude of current, but it does supply an extreme impact of torque, and mechanical damage to the shaft or bearings may occur. The resulting current is of sufficient magnitude that fast removal is necessary if thermal damage to the generator is to be avoided. Instantaneous separation offers no guarantee that no damage will occur.

Various relays applied for other functions may detect inadvertent energization. Loss-of-field relays may respond, but they usually use single-phase current, and therefore, complete sensing is not afforded for all possible combinations of this phenomenon. Relays applied to detect motoring will operate, but their operating time is set to other criteria and the long time customarily used is unsuitable for this additional function. Distance- and voltage-controlled overcurrent relays applied for generator phase backup protection

may operate, but again the time delay for tripping may be unsuitable.

To further compound the difficulties associated with detecting inadvertent energization is the fact that generator potential circuits are often disconnected in the interests of safety when a machine is shut down. Any of the "normal" relays that are dependent on this voltage supply will be unable to respond at the very time when they are needed. Also, it must be remembered that the flashover of an open breaker cannot be cleared by energizing its trip coil.

Several protective schemes have been used successfully to detect three- or single-phase inadvertent energization. Among them are

Directional overcurrent relays
Pole disagreement relays
Relays containing logic to detect overcurrent for a
    short time following 0 V
Frequency-supervised overcurrent relays

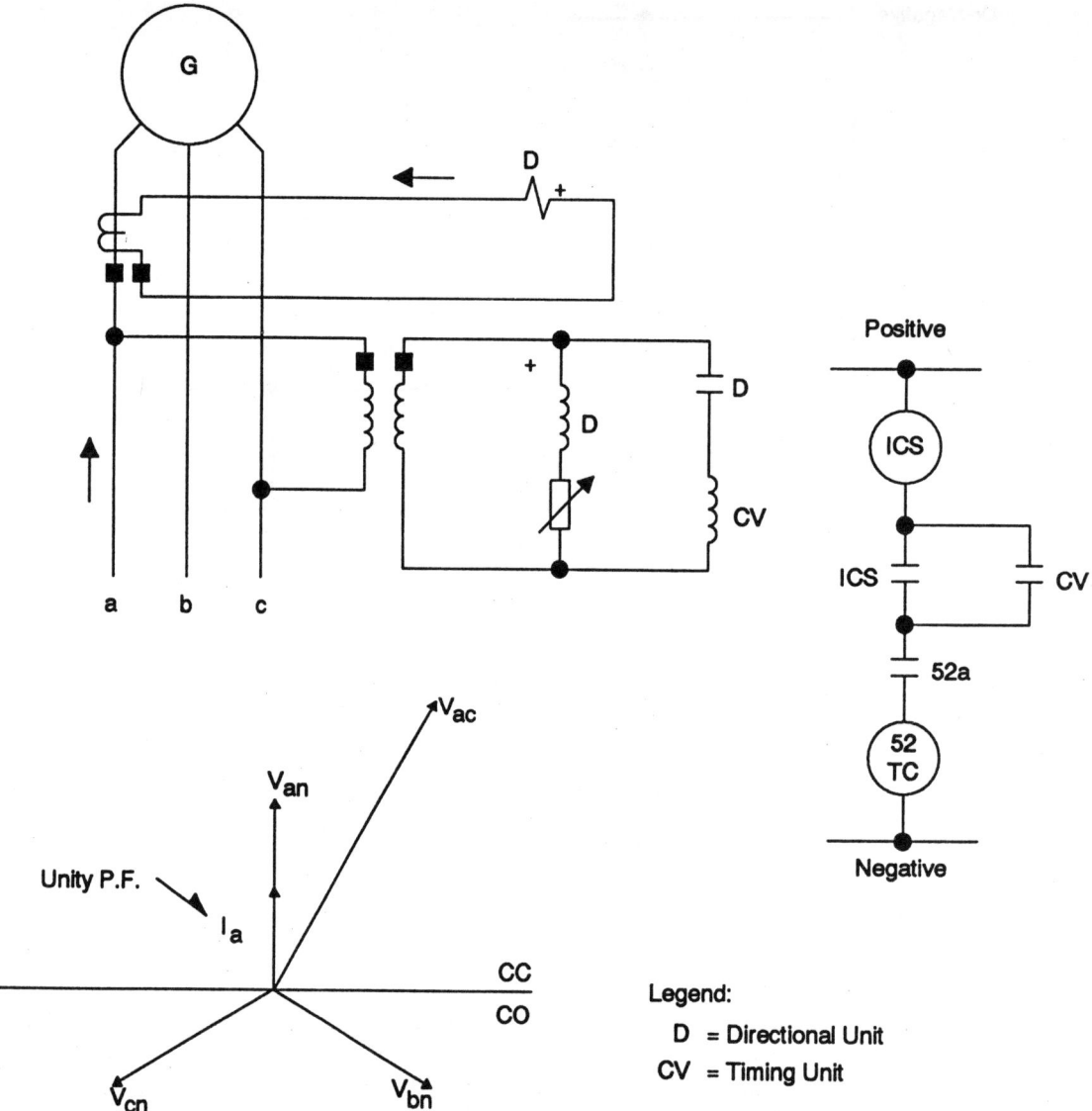

Figure 8-19 Typical schematic for antimonitoring protection using CRN-1 (32) relay.

Voltage-supervised overcurrent relays

Distance relays

Unit transformer neutral overcurrent in special breaker-failure relaying scheme.

Care must be exercised to assure that either a voltage supply is available when operation is required or undervoltage allows operation of the scheme. Further, the circumstances associated with inadvertent energization itself must not be able to circumvent a necessary part of the logic for tripping, such as the requirement for the presence of reduced frequency.

12 FIELD GROUND DETECTION

A single ground on the field of a synchronous machine produces no immediate damaging effect. It must be detected and removed because of the possibility of a second ground that could short part of the field winding and cause damaging vibration. Care should be used in establishing any field ground detecting scheme to assure that any bearing current that is allowed will not cause bearing deterioration. One scheme that has been used is shown in Figure 8-20. A small leakage current flows through the field-to-ground capacitance, which, on a large turbogenerator, can be between 0.3 and 0.5 μf. The relay detects an increase or decrease in the magnitude of this current.

12.1 Brush-Type Machine

One recommended field ground protection scheme for a generator with brushes (i.e., stationary field leads accessible) is illustrated in Figure 8-21. This scheme, which does not require any external source, uses the very sensitive d'Arsonval dc relay, type DGF.

The DGF relay uses a voltage divider circuit, consisting of two linear resistors (R_1 and R_2) and a nonlinear resistor whose resistance varies with the applied voltage. If the field becomes grounded, a voltage will develop between point "M" and ground.

The magnitude of this voltage will vary according to the exciter voltage and point at which the field is grounded. The voltage will be at maximum if the field is grounded at either end of the winding.

A null point will exist in the field winding where a ground will produce no voltage between M and ground. This null point will be located at a point on the field winding from which there is balance between the two field winding resistances and two relay resistances to positive and negative. A ground at the null point will go unrecognized until the field voltage is changed as a result of daily reactive (or voltage-level) scheduling. Faults having impedance of up to 300,000 Ω can be recognized with this scheme.

A pushbutton, connected across a portion of the R_2 resistor, permits a manual check for possible ground faults at the center of the winding. This provision is

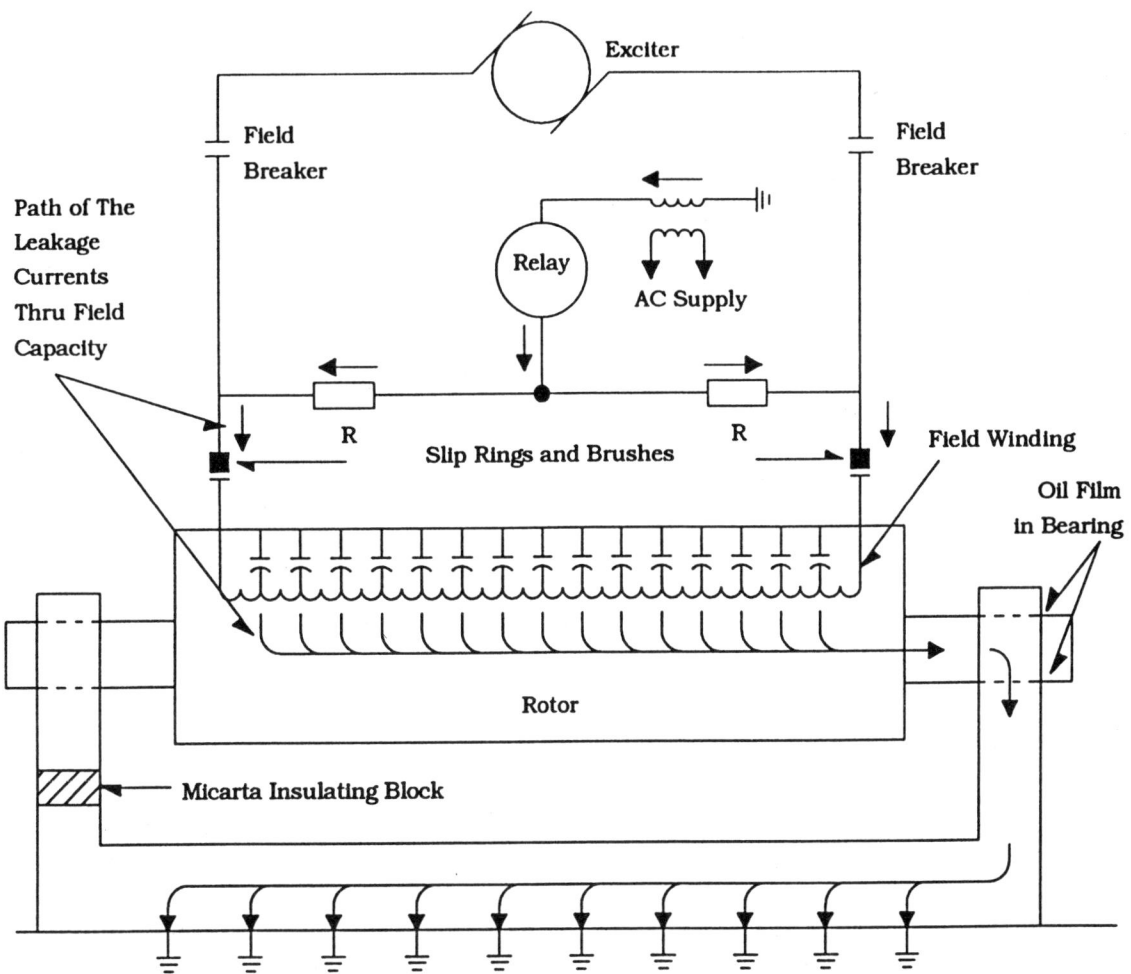

Figure 8-20 Path of the currents in a machine when using an ac field ground relay.

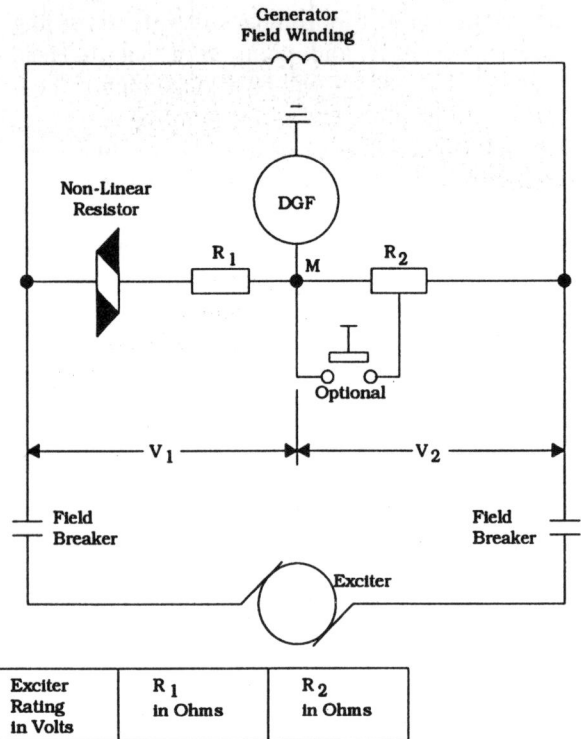

Exciter Rating in Volts	R₁ in Ohms	R₂ in Ohms
125	0	45,000
250	5,000	23,000
375	10,000	23,000

Figure 8-21 Field ground protection scheme for a generator.

desirable when the generator is to be "base-loaded" and will not experience periodic excitation variations.

Another scheme that has been used successfully utilizes a Wheatstone bridge with the field circuit to ground forming one leg of the bridge. A solid ground fault anywhere in the field circuit can be detected immediately through the recognition of the resulting unbalance.

12.2 Brushless Machines

For a "brushless" type of machine, no normal access is available to a stationary part of the generator field circuit, and no continuous monitoring to detect field grounds is possible. One widely employed scheme uses a 60-Hz tuned overvoltage relay connected between the neutral of the three-phase ac exciter and ground. A ground on the exciter, in the three-phase rectifier bridge, in the field, or on the dc leads will be detected.

This requires a pilot brush connected to the neutral of the exciter that is periodically dropped.

Another variation of this form of detection is used by the solid-state type 64F relay. It impresses a dc voltage from the negative dc lead to ground and continuously monitors current flow. An increase in current accompanies a field ground.

The pilot brush arrangement can also be used with the Wheatstone bridge scheme (YWX111) described in Section 12.1.

12.3 Injection Scheme for Field Ground Detection

The scheme described in Section 4.4, GIX-104, is equally applicable to field ground detection, irrespective of the type of excitation system in use. This is shown in Figure 8-22 for a "rotating rectifier" excitation system. As with the stator winding ground detection application, this system is able to detect field circuit grounds even though the machine is at standstill or running, excited or not.

13 ALTERNATING-CURRENT OVERVOLTAGE PROTECTION FOR HYDROELECTRIC GENERATORS

Alternating-current overvoltage protection is recommended for hydroelectric generators subject to overspeed and consequent overvoltage on loss of load. Some hydroelectric generators can go up to 140% or more of rated speed when full load is dropped. The voltage may reach 200% or more.

The ac overvoltage protective scheme is shown in Figure 8-23. The relay, which changes the excitation to reduce the output voltage, can also provide backup protection for the voltage regulator.

14 GENERATOR PROTECTION AT REDUCED FREQUENCIES

Many turbine generators are started on turning gear, which rotates the shaft at about 3 rpm. For a cross-compound machine, the field must be applied before the machine is removed from the turning gear. At this point, excitation should be limited to rated volts per hertz to avoid overexciting the unit or station service transformer. A tandem unit need not have field applied until it is up to speed and ready to synchronize.

Cross-compound generators may be operated for several hours during warmup at frequencies well below their rating. Current transformer and relay performance must be considered at these reduced frequencies because fault magnitudes are approximately the same as at rated frequency. Current transformer performance can be expected to deteriorate badly at low frequency. There may be a small compensating effect, however, in the reduction of burden impedance.

The performances of some electromechanical relays associated with the generator or a generator-transformer unit at 15 and 30 Hz are summarized in Table 8-3 in terms of 60-Hz performance.

For cross-compound turbine generators, the low- and high-pressure units should have their fields applied and be synchronized while on turning gear, so that they are brought together up to rated speed. Synchronizing surges may occur, at these low speeds, that will operate the loss of field relays. These synchronizing surges can reach 60% of the full-load current, since the impedance in the generators is very low. Loss-of-field relays applied to a cross-compound unit should therefore be disabled during startup.

The SC relay (and/or the SV) is recommended as supplementary protection when reduced frequency protection is required. The SC current relay has a flat characteristic and increases slightly in sensitivity as the operating frequency drops. When an SC relay is operated on dc, it picks up at approximately 15% below its normal 60-Hz pickup. The pickup of the SV voltage relay is almost directly proportional to frequency; its sensitivity at 15 Hz is thus 4 times the sensitivity at 60 Hz. For this reason, the SV relay provides excellent reduced-frequency protection.

In this application, an SC relay is frequently located in the differential circuit of each phase of the generator differential relay. The SV relay is connected across the secondary resistor in the generator neutral circuit.

None of the relays listed in Table 8-3 will overheat nor operate incorrectly if left in the circuits when the generator is operated at reduced frequencies. The KLF and KLF-1 relays, when used in a cross-compound configuration, must have their trip incapacitated during startup. The SC and SV relays used for low-frequency protection must be removed from service for normal operation. This may be done by a frequency relay set for approximately 55 Hz (for a 60-Hz system) and an auxiliary relay.

The microprocessor-based multifunction relay Type GPU2000R has frequency-tracking algorithms. These insure that its protective elements which are most

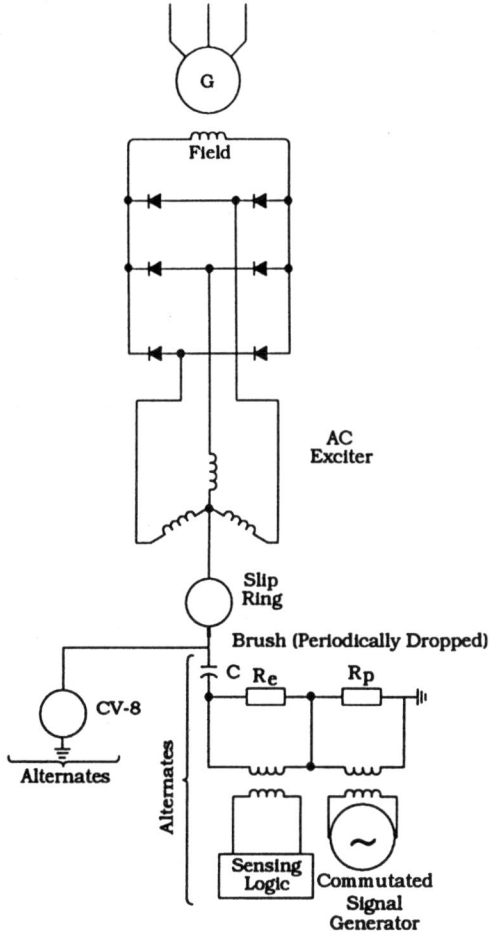

Figure 8-22 Rotating rectifier excitation system.

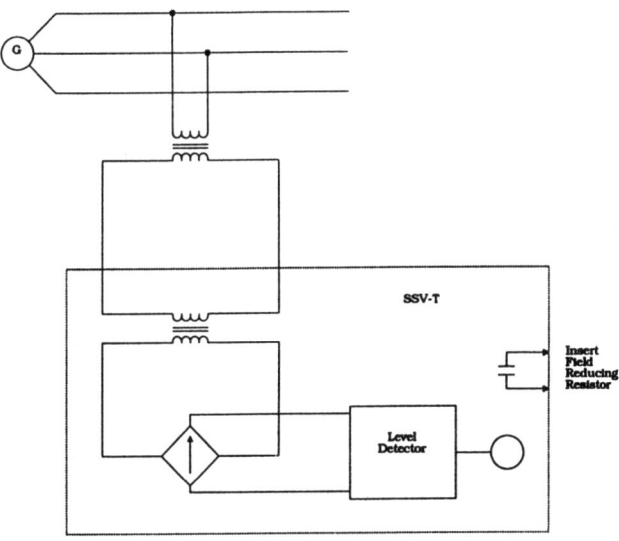

Figure 8-23 Overvoltage protection for generator.

Table 8-3 Performance of ABB Protective Relays at Reduced Frequencies

| | | Pickup in percent of 60-Hz pickup | | |
		15 Hertz	30 Hertz	Classification[a]
Overcurrent (51)	CO-2	165	115	A
	CO-5	[b]	150	B
	CO-6	[b]	143	B
	CO-7	[b]	140	B
	CO-8	262	138	A
	CO-9	260	140	A
	SC	85	93	A
	COV	Performance same as CO Unit used in relay		
Voltage (59)	SV	26	50	A
	CV contact-making voltmeter	122	120	A
	CV-8	[b]	[b]	C
Differential (87)	CA generator	255	123	A
	CA transformer	[b]	149	B
	SA-1	370	175	A
	HU-1	250	130	A
Negative sequence (46)	COQ			O
	SOQ			
Loss of field (40)	KLF (or KLF-1)			Δ

[a](A) Protection available at both 15 and 30 cycles. (B) Protection available at 30 cycles only. (C) Additional protective relays required for start-up or low-frequency operation. (O) The sensitivity of the COQ & SOQ relays to negative sequence currents is a direct function of frequency, while its sensitivity to positive sequence currents is an inverse function of frequency. This relay will operate for heavy three-phase and phase-to-phase faults at reduced frequencies, but should not be relied upon for primary protection during warm-up. (Δ) Since the KLF or KLF-1 relays operate on lagging reactive power into the machine, the relay will neither operate falsely nor provide loss-of-field protection during the warm-up period.
[b]Very insensitive or nonoperable at the frequency indicated.

needed during low-frequency operation maintain their characteristics.

15 OFF-FREQUENCY OPERATION

Turbine blades are carefully designed to have no mechanical resonant conditions when rotating at rated speed. If this were not the case, mechanical deterioration would occur as the blades flex under the stress of loaded operation. At elevated or reduced speed, there are resonant points where prolonged operation produces blade fatigue damage and ultimate failure.

If operating frequency (speed) deviates from the rated value, corrective action must be initiated or tripping must result. Since mechanical fatigue is a cumulative phenomenon, the time of loaded operation at reduced frequency must be monitored and accumulated over the life of the turbine. Figure 8-24 shows the limits that one manufacturer imposes for machines in

two categories: (1) those having the longest stage blading of length 18 to 25 in. and (2) those having the longest stage blading of length $28\frac{1}{2}$ to 44 in. These curves are not a national or international standard. The specific manufacturer of the turbine should be contacted to ascertain specific recommendations for limits.

Under- (or over-) frequency relays as described in the chapter on "load-shedding" may be used to detect frequency excursions, but this must be monitored by watt level because off-frequency operation at a light load (above the level that produces adequate steam flow to remove turbine friction and windage losses) is not hazardous. The combination of these two sensing elements is used to drive an accumulation timer to allow an estimate of the extent of life remaining to be made. The need for blade examination and possible replacement can then be evaluated.

The microprocessor-based system REG-216 has a provision for integrating low-frequency overtime and

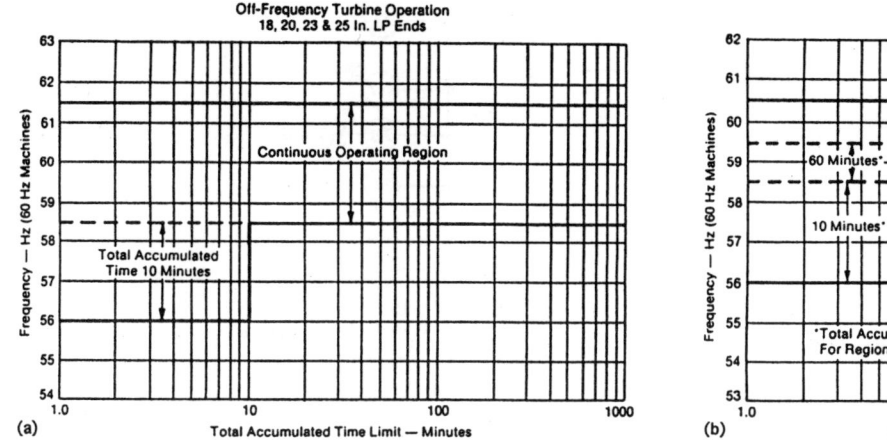
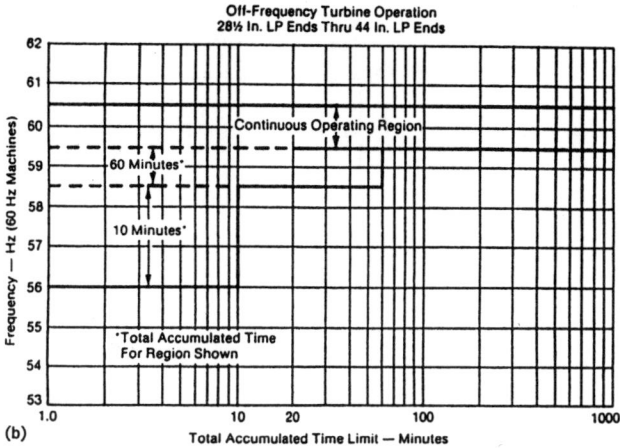

Figure 8-24   One manufacturer's limits for off-frequency operation of combustion turbines.

alarming and tripping to prevent damage to the turbine. Also, to avoid any detrimental effects of the front-end filters in detecting overcurrent at off-frequency levels, the 50, instantaneous-trip function is sensed ahead of these filters. Overcurrent protection, then, is provided at a frequency as low as 2 Hz.

16   RECOMMENDED PROTECTION

Figure 8-25 shows the recommended protection for large tandem-compound unit-connected turbine generators. Figures 8-26 and 8-27 illustrate the recommended protection for machines that are not unit-connected. Generally, such generators are used in industrial applications.

A wide variety of implementations of the numbered functions are in use and some are described in Table 8-4. The REG-100 and REG-216 are complete multi-function microprocessor-based packages.

17   OUT-OF-STEP PROTECTION

As generator impedances become larger in proportion to the system impedance, the electrical center will be closer to the generator. This condition intensifies the need for out-of-step detection as part of the generator relaying complement. Such relaying schemes are

described in Chapter 14, "System Stability and Out-of-Step Relaying."

18   BUS TRANSFER SYSTEMS FOR STATION AUXILIARIES

The automatic transfer of highly essential station auxiliary loads such as reactor coolant pumps, boiler feed pumps, and induced draft fans is common practice. Paralleling the normal and emergency sources on a continuous basis is not generally recommended because the higher breaker-interrupting duties involved can cause problems, as can circulating currents between systems. Transfers should not be made if voltage of the alternate supply is not satisfactory or the load circuits are faulted. Also, supply breaker tripping should be delayed long enough to permit fault sectionalizing in the load circuits.

18.1   Fast Transfer

This is a term applied to the connection of a bus to a second power source with little or no time delay. This process is accomplished with an "open" transfer or "closed" transfer. The open transfer disconnects the normal source before connecting the second. The closed transfer allows the two systems to operate tied together momentarily, and then to have the original source breaker tripped. For all the fast transfer

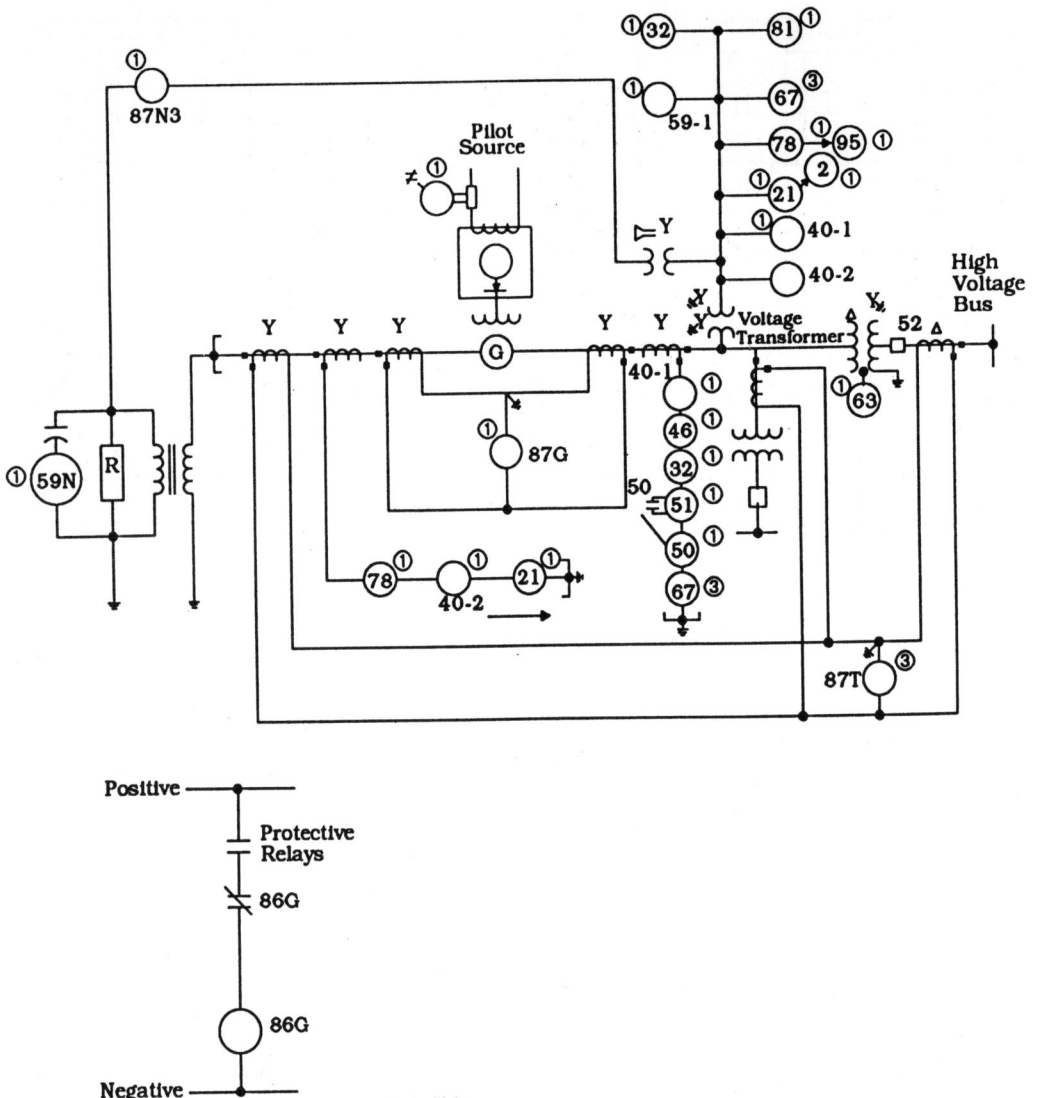

Figure 8-25 Overall protection for a tandem-compound unit-connected generator.

schemes, it is recommended, if a synchronism check relay is required, that it be preenergized so that it may determine the existence (or lack of) synchronism prior to the need to transfer.

Figure 8-28a shows an example of the fast open-transfer scheme. 43 is a switch having automatic and manual positions. With 43A closed, CVX synchronism check relay contact closed due to the two buses being in synchronism prior to the need for transfer, breaker A having been tripped for any reason (as evidenced by its 52b being closed), and no fault having occurred on the bus to produce the opening of 86B, then the close circuit of breaker C becomes energized.

The simple process of tripping breaker A, when in the automatic mode in the absence of a bus fault, produces the closure of breaker C.

18.2 Choice of Fast Transfer Scheme

Open transfer would be selected in a system in which the supportive WR^2 (moment of inertia) is sufficient to keep the switched bus close enough in frequency during the open period to allow reconnection without an excessive shock to rotating machinery connected to that bus.

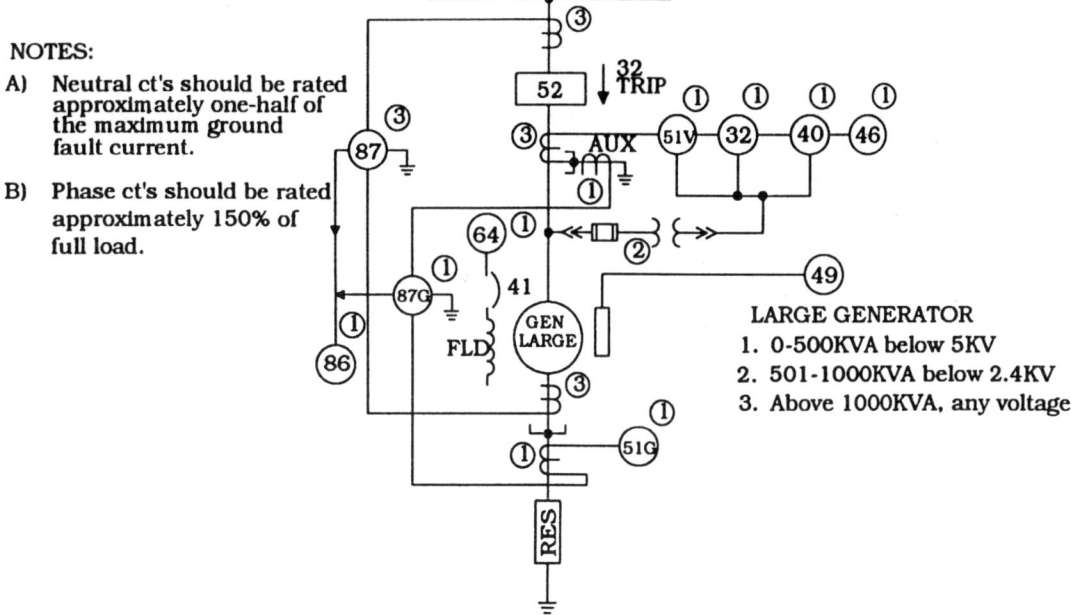

Figure 8-26   Recommended protection for large generators as used in industrial plants.

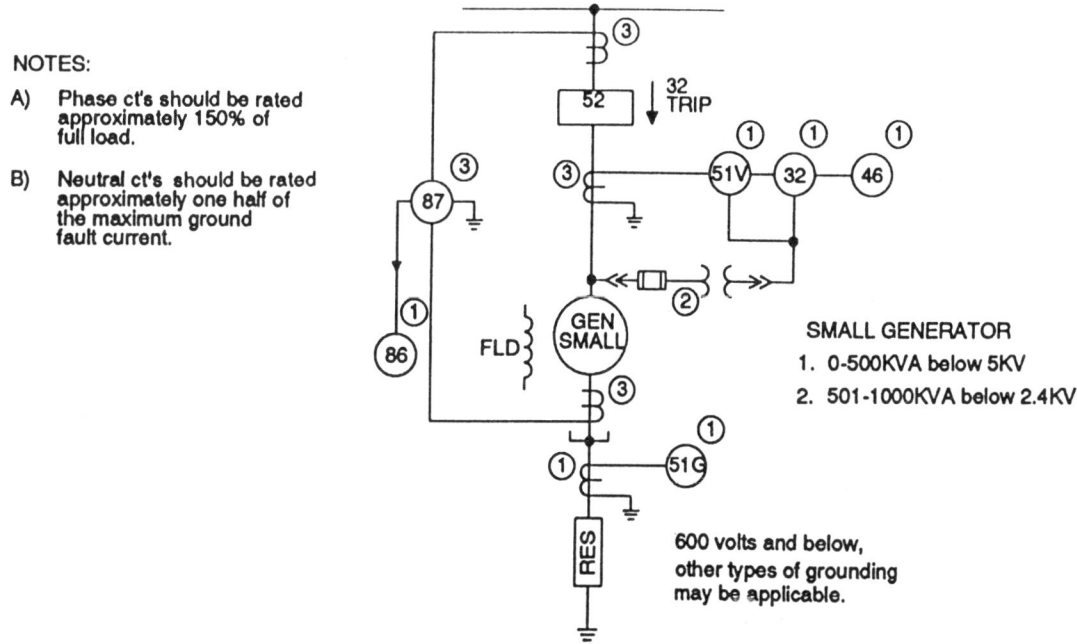

Figure 8-27   Recommended protection for small generators as used in industrial plants.

Table 8-4   Generator Relaying

ANSI device no.	Description	Traditional	Solid-state or numerical	GPU-2000R	REG 216	REG 316
2	Timer	—	62T TD-5	X	X	X
21	Phase backup	KD-11	a	X	X	X
24	Volts/hertz	—	59F MVH	X	X	X
32	Motoring	CRN-1	32R SRW	X	X	X
40	Loss of field	KLF	40	X	X	X
46	Negative sequence	COQ	46Q SOQ	X	X	X
49	Thermal	DT-3	49T	—	X	X
50/51	Stator overcurrent	CO-ITH	Micro 51	X	X	X
59	Overvoltage	CV-5	59	X	X	X
59F	Field ground	CV-8	59G 64F	—	X	X
59N	Stator ground (95%)	CV-8	59G	X	X	X
64S 64R	Current injection (100% ground stator/rotor)	—	GIX-104		—	X
67	Inadvertent-energization	CRG-9	32D	X	X	X
76	Field dc overcurrent	D-3	76H	—	—	—
78	Out-of-step	KST-KD-3	GZX-104 MDAR	—	X	X
81	Underfrequency	—	81 MDF	X	X	X
86G	Generator lockout	LOR	LOR	LOR	LOR	LOR
87G	Generator differential	CA	87M SA-1	X	X	X
87N3	Stator-neutral-ground	DGSH	27G	X	—	X
87T	Overall differential	HU-1	87T	—	X	X

aAlternate 51L + 47H or 51V function.

The "simultaneous" scheme is a variation of the open transfer scheme. With it, the trip-coil of the normal supply breaker and the close-coil of the incoming supply breaker are energized simultaneously. Since the close function is somewhat slower than the trip function of a given type of breaker, the opening breaker wins the race. Figure 8-28c shows the contact array for this sequence.

Closed transfer would be the proper scheme to use when WR^2 associated with the rotating equipment is inadequate to allow separation for even a short time. The impact of reenergization would be too great. Figure 8-28c shows the addition of a "52a" contact of breaker C to delay the tripping of breaker A on a manual transfer. It is not considered wise to delay tripping for a fault that operates 86S, the lockout relay associated with a source circuit fault.

Synchronism check is a satisfactory means for supervising the closing of the bus tie breaker. It is not suitable, however, for a fault-forced transfer unless the system is supervised by an 86B lockout relay that prevents transferring to a bus fault. Otherwise, it would be possible for both sources to be connected to a fault, compounding the damage and producing no useful result.

18.3 Slow Transfer

Slow transfer carries with it the connotation of delaying the reenergization of a bus until the voltage on motors connected to the bus has decayed to the

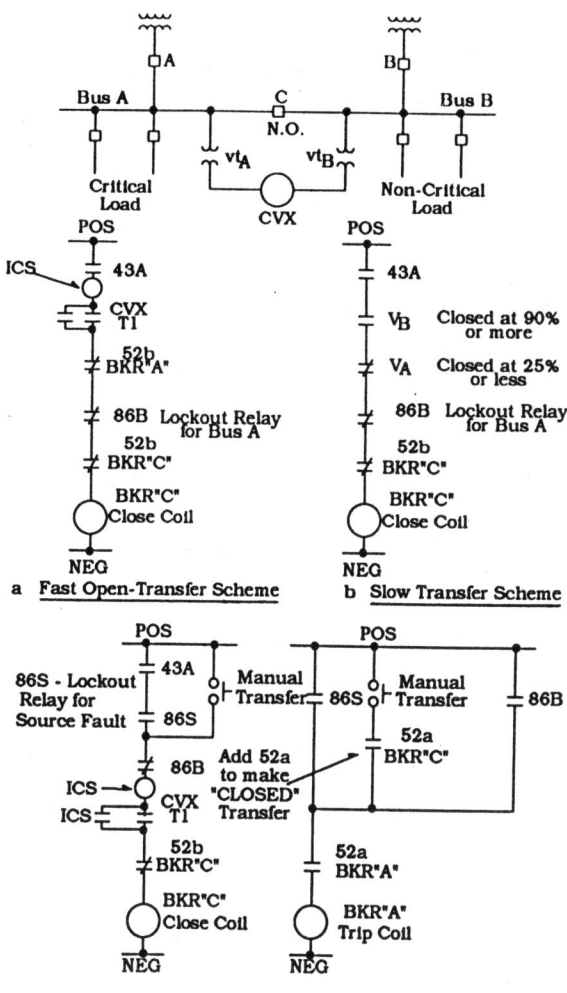

Figure 8-28 Variations of transfer schemes.

point where no damage can be expected with out-of-phase energization. This level is generally considered to be 25% of rated. Figure 8-28b describes this arrangement.

19 MICROPROCESSOR-BASED GENERATOR PROTECTION

Important features related to generator protection can only be achieved by a coordinated generator protection package, utilizing microprocessors. Self-monitoring, communications, oscillography, and adaptive settings are accomplished straightforwardly and reliably. Flexibility in the application is important to allow individual user selections. This is accomplished by a man-machine interface that allows ease of settings and in some cases the choice of software modules.

In the use of multifunction coordinated packages, sight should not be lost of the need for adequate redundancy to provide backup and cover the failure of any element. The REG 216 and REG 100 series address these needs and provide extensive generator protection utilizing proven concepts.

9

Motor Protection

Revised by: **C. L. DOWNS**

1 INTRODUCTION

1.1 General Requirements

Motor protection is far less standardized than generator protection. Although the National Electric Code and NEMA standards specify basic protection requirements, they do not fully cover the many different types and sizes of motors and their varied applications. There are many other schemes, all of which offer different degrees of protection. As with generator protection, the cost and extent of the protective system must be weighed against the potential hazards. The size of the motor and type of service will also influence the type of protection required. Electromechanical, solid-state, or microprocessor-based relays can be used stand-alone or in combination with one another to achieve the desired degree of security and dependability.

Motor protection should involve the detection of the following hazards:

1. Faults in the windings or associated feeder circuits, including both phase and ground fault detection.
2. Excessive overloads. Overloads result in thermal damage to the insulation and can be caused by continuous or intermittent overload, or a locked rotor condition (failure to start or a jam condition).
3. Reduction or loss of supply voltage. Any reduction of supply voltage directly affects the applied torque to the connected mechanical load.

4. Phase reversal. Starting a motor in reverse can be hazardous to the load.
5. Phase unbalance. A small amount of unbalance can result in a significant increase in the motor temperature.
6. Out-of-step operation for synchronous motors.
7. Loss of excitation for synchronous motors.

Protective relays applied for one hazard may operate for others. For example, a relay designed to operate on an excessive overload could also protect against a fault in the windings.

Protective devices may be installed on the motor controllers or directly on the motors. The protection is usually included as part of the controller, except for very small motors, which have various types of built-in thermal protection.

Motors rated 600 V or below are generally switched by contactors and protected by fuses or low-voltage circuit breakers equipped with magnetic trips. Motors rated from 600 to 4800 V are usually switched by a power circuit breaker or contactor (often supplemented by current-limiting fuses to accommodate higher interrupting requirements. Motors rated from 2400 to 13,800 V are switched by power circuit breakers.

Although protective relays may be applied to a motor of any size or voltage rating, in practice they are usually applied only to the larger or higher-voltage motors.

1.2 Induction Motor Equivalent Circuit

Excessive heat in the motor can be caused during starting, a locked rotor condition, load requirements, voltage unbalance, or an open-phase condition and can cause degradation of the mechanical and dielectric strength of the insulation. This thermal deterioration of the insulation sets up the possibility of future faults.

The equivalent circuit of the motor helps in visualizing what occurs in the motor during the above conditions. The motor impedance during these conditions is directly influenced by the slip of the motor:

$$\text{slip } s = \frac{n_s - n_r}{n_s} \text{ per unit}$$

where n_s and n_r are the stator field and rotor speeds. At start, the slip is 1.00, or 100%. During a running condition, the slip is approximately 0.01 to 0.08, or 1 to 8%. The equivalent circuit during a starting condition is as shown in Figure 9-1. The input impedance at start can be approximated by

$$Z_{\text{start}} = R_s + R_r + jX_s + jX_r$$

where R_s, R_r, jX_s, jX_r are the stator and rotor resistances and reactances. During a start condition, the impedance is predominately reactive. As the motor gains speed, the impedance becomes more resistive, and the power factor increases.

The negative sequence impedance of a motor, excluding wound rotor motors, is very nearly equal to Z_{start} above, and from Figure 9-2, can be approximated by

$$Z_2 = R_s + \frac{R_R}{2 - s} + jX_s + jX_R$$

With Z_2 and Z_{start} approximately equal, the negative sequence current can be calculated for a particular negative sequence voltage unbalance. If $V_2 = 5\%$ and

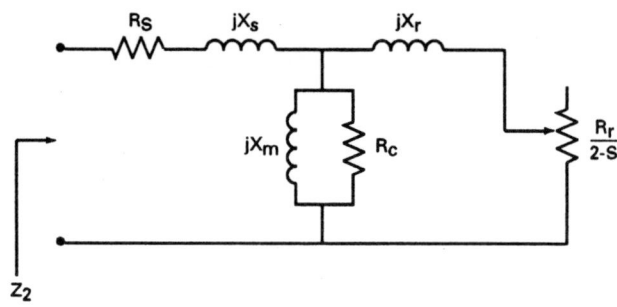

Figure 9-2   Induction motor negative sequence network.

the starting current I_s of the motor is $8I_{FL}$, then I_2 would be 0.40, or 40%. This will result in an increase in stator heating and, in particular, rotor heating. The rotor heating results from the combined effect of the counterrotating flux that causes large currents to be induced at $f_s(2 - s)$ Hz, where f_s is the system frequency, and the increased skin effect in the rotor that can cause its resistance to be 5 to 10 times normal.

1.3 Motor Thermal Capability Curves

Protecting a motor for a variety of hazards requires the protection engineer to know full-load current, permissible continuous allowable temperature rise, locked-rotor current and permissible maximum time at that current, and accelerating time, which is a function of the load characteristics and starting voltage. A typical motor thermal capability curve, which is shown in Figure 9-3, is helpful in determining the temperature endurance of the insulation. The lower part of the curve is usually rotor-limited. The limit arises because of the I^2R heating effect during a locked rotor

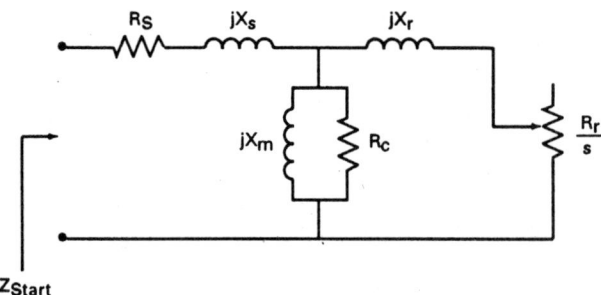

Figure 9-1   Induction motor equivalent circuit at start.

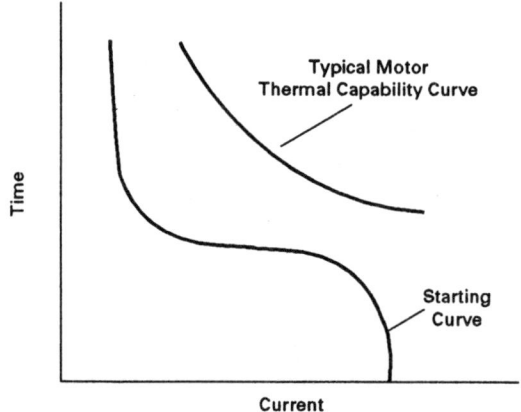

Figure 9-3   Motor thermal capability curve.

condition. This represents the time a motor can remain stalled after being energized before thermal damage occurs in the rotor. This is an I^2t limit, which can also be expressed as a $(V^2/R)t$ limit. The middle portion is the acceleration thermal limit part of the curve. This is from the locked-rotor current to the motor breakdown torque current portion of the curve. The upper section of the curve is the running or operating thermal limit portion. This represents the motor overload capacity.

2 PHASE-FAULT PROTECTION

The phase-fault current at the terminals of a motor usually is considerably larger than any normal current, such as starting current or the motor contribution to a fault. For this reason, a high-set instantaneous-trip unit is recommended for fast, reliable, inexpensive, simple protection. When the starting current value approaches the fault current, however, some form of differential relaying becomes necessary. The sensitivity of the differential relay is independent of starting current, whereas instantaneous-trip units, which respond to phase current, must be set above the starting current (including any dc offset due to asymmetrical transients that may be caused by voltage switching). This difference is shown in Figure 9-4.

To allow for fault resistance and different types of faults and to assure twice pickup on the unit for minimum fault, the instantaneous phase-relay pickup should be set at less than one-third of I_{3ph}, where I_{3ph} is the system contribution, excluding the motor contribution, to a symmetrical three-phase fault on the motor feeder. Also, pickup should be set at 1.6 times I_{LR} or more, where I_{LR} is the actual symmetrical starting current, as limited by source impedance. The ratio I_{3ph}/I_{LR} should thus be greater than approximately 5.0.

In general, then, instantaneous-trip units can be used for phase protection if the motor KVA (or approximately the horsepower) is less than one-half the supply transformer KVA. If not, differential protection, such as that obtained by the CA or 87M (Fig. 9-4), is required for sensitive fault detection.

The logic for this criterion comes from the following. Assume a motor is connected to a supply transformer with 8% impedance. The maximum fault current at the transformer secondary with an infinite source is

$$I_{3ph} = 1/0.08$$
$$= 12.5 \text{ per unit on the transformer base}$$

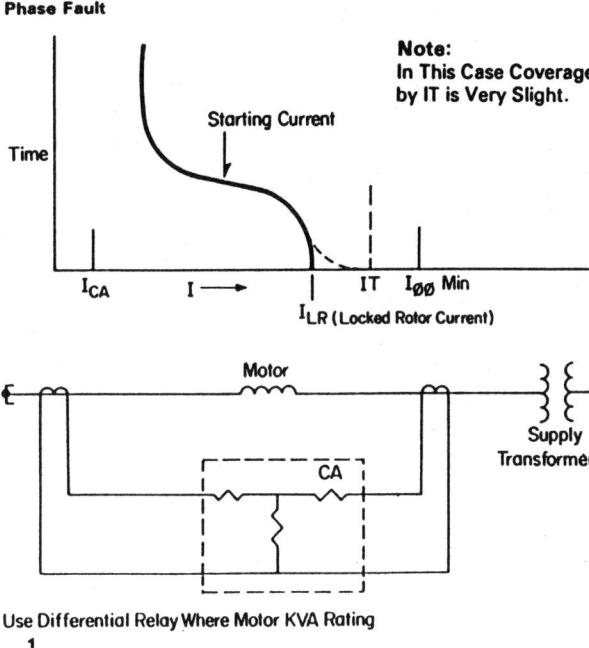

Figure 9-4 Comparison of sensitivities of type CA differential relay and IT instantaneous trip unit.

The maximum motor starting current in this case is

$$I_{LR} = \frac{1}{(0.08 + X_M)}$$

where X_M is the motor impedance. In order that $I_{3ph}/I_{LR} > 5$, X_M must be greater than 0.32 per unit on the transformer-rated KVA base.

If the motor has a full-voltage starting current of six times full load, then $X_M = 1/6 = 0.167$ on the motor-rated KVA base. With a motor KVA of one-half the transformer KVA, an X_M of 0.167 would be 0.333 on the transformer base, and greater than 0.32. Clearly, this rule of thumb should only be applied when there is no appreciable deviation from the parameters assumed above.

3 GROUND-FAULT PROTECTION

A solidly grounded system may be protected by an inverse, very inverse, or short-time induction or microprocessor-based relay connected in the current transformer residual circuit. For a solid fault at the machine terminals, a typical setting is one-fifth of the minimum fault current. Time dial settings of around 1 give operations of four to five cycles at 500% pickup when the CO-2 relay is used.

During across-the-line starting of large motors, care must be taken to prevent the high in-rush current from operating the ground relays. Unequal saturation of the current transformers produces a false residual current in the secondary or relay circuits. Using two- rather than three-phase relays or three-phase relays with different impedances will tend to increase the effects of false residual currents.

False relay operation is unlikely if the phase burdens are limited so that the voltage developed by the current transformer during starting is less than 75% of the relaying accuracy voltage rating of the current transformer for the particular CT tap being used. If false relay operation is a problem, the ground relay burden of an electromechanical relay should be increased by using a lower tap. All three transformers will then be forced to saturate more uniformly, effectively reducing the false residual current. This increased saturation may reduce the sensitivity to legitimate ground faults and this should be checked. Alternatively, a resistor or reactor can be connected in series with the ground relay.

The common practice in 2400- to 14,400-V station service, and industrial power systems, is to use low-resistance grounding. By using the "doughnut current transformer" scheme, such systems offer all the advantages of instantaneous trip units—speed, reliability, simplicity, low cost—without any concern for starting current, fault contributions by the motor, false residual current, or high sensitivity.

Figure 9-5 shows how the BYZ zero-sequence-type current transformer can be used as a supply for the 50 instantaneous-trip (IT) unit or 51 time overcurrent (CO) relay. Typical sensitivities obtainable with these ground-fault protection systems are shown in Table 9-1. A voltage is generated in the secondary

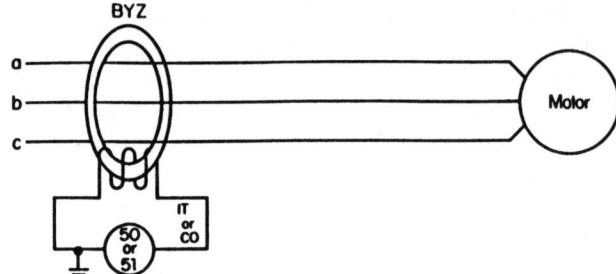

Figure 9-5 BYZ ground relaying scheme.

winding only when zero sequence current is flowing in the primary leads. Since virtually all motors have their neutrals ungrounded, no zero sequence current can flow in the motor leads unless there is a ground fault on the load side of the BYZ. If surge-protective equipment is connected at the motor terminals, however, current may be conducted to earth by this equipment. To date, there has been no reported case of an instantaneous relay connected to a BYZ current transformer tripping because of surge-protective equipment. The presence of such equipment may safely be ignored in choosing a relay.

Solid-state relays such as the type 50D when used with the BYZ give good performance in this application due to a lower burden characteristic. Also, specialized solid-state ground-fault relay systems such as the Ground-Shield® series provide a variety of doughnut CT window sizes, both toroidal and rectangular. As a system, the relay and CT characteristics are properly prematched by design, and thus need not be further considered by the user. These specialized systems usually have relay pickup settings marked in terms of primary amperes.

The BYZ current transformer is also used in the flux balancing differential scheme, in which each phase is

Table 9-1 Relay Settings and Sensitivities Using the 50/5 BYZ Zero Sequence Current Transformers

Relay type	Relay setting	Minimum sensitivity in primary $3I_0$ amperes		Maximum primary $3I_0$ amperes for accurate timing and coordination	
		$4\frac{3}{4}$ ID[a]	$7\frac{3}{4}$ ID[a]	$4\frac{3}{4}$ ID[a]	$7\frac{3}{4}$ ID[a]
IT	0.15	5.0	5.0	—	—
CO-8 or 9	0.5	9.0	10.0	25	112
CO-8 or 9	2.5	24.0	24.0	540	1215
CO-11	0.5	6.0	7.0	70	150
CO-11	2.5	24.0	24.0	700	900

[a] $4\frac{3}{4}$ ID and $7\frac{3}{4}$ ID are the inside diameter of the window in inches.

equipped as shown in Figure 9-6. This scheme combines excellent phase- and ground-fault sensitivity with freedom from load current and starting current problems.

For high-resistance grounded systems, where very high sensitivity is required, the CWP-1 directional ground relay should be considered. The voltage across the transformer grounding resistor may be used as a voltage polarizing source (Fig. 9-7). The relay has a sensitivity of 7 mA at 69 V.

The maximum torque angle occurs when the current leads the polarizing voltage by 45°. It is interesting to note that the maximum sensitivity angle is leading the reference voltage $-3V_0$ by 45°. In high-resistance grounded systems, the predominant impedance for ground faults is the zero sequence network containing the grounding resistor as 3R and the zero sequence distributed capacitance. The resulting fault current calculated is influenced by this RC circuit and thus leads the applied voltage instead of lagging as would normally occur for a ground fault on a low-resistance grounded system. The CWP-1 directional ground relay is intended for use only on high-resistance grounded systems.

4   LOCKED-ROTOR PROTECTION

A rotating motor dissipates far more heat than a motor at standstill, since the cooling medium flows more efficiently. During a failure to start or accelerate after being energized, a motor is subject to extreme heating (approximately 10 to 50 times more than for rated

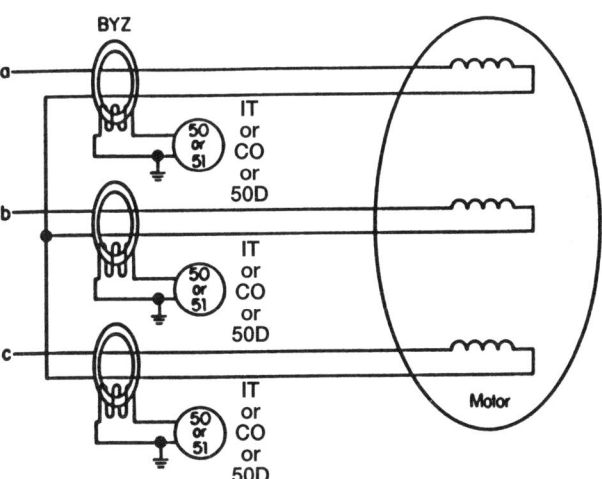

Figure 9-6 Flux balancing differential scheme.

conditions) in both the stator windings and rotor. The equivalent circuit during a locked-rotor condition is similar to a transformer-equivalent circuit with a resistively loaded secondary. The heat distribution between the stator and rotor is contingent on the relative stator resistance and 60-Hz rotor resistance. Unlike an overload condition, in which heat can be absorbed over time by the conductors, core, and structural members, a locked-rotor condition produces significant heat in the conductors, which has little time to be transferred to other sections of the motor. Extreme heating takes place and can be tolerated by the motor for a very limited time. The time that a motor can remain at standstill after being energized varies with the applied voltage and is an I^2t limit. A relay with an I^2t characteristic that could be set for any permissible locked-rotor times and locked-rotor currents would naturally be the best choice for protecting the motor.

The heat generated within the motor can be approximated by

$$I_H^2 = I_1^2 + KI_2^2$$

where

I_1 = per unit stator positive sequence current
K = weighting factor to describe the increased rotor resistance due to skin effect in the rotor bars at $(2f_s - f_{slip})$
I_2 = per unit stator negative sequence current
f_s = system frequency
f_{slip} = slip frequency

Utilizing both the positive and negative sequence currents in an equation relating to I_H allows the motor to be protected throughout the full range of stator current with and without unbalance. The time-current characteristic is as shown in Figure 9-8 for the MPR relay. To insure adequate locked-rotor protection, the curve position can be set slightly below the full-voltage locked-rotor time. Depending on the availability of RTDs, the cutoff can be set to protect the motor during an overload condition. A typical cutoff setting without RTDs would be 115 to 125% of full-load current.

If the time required for the motor to accelerate the load is significantly less than the permissible locked-motor time, the motor can be effectively protected using conventional time-overcurrent relays (Fig. 9-10) or microprocessor relays (Fig. 9-8). If, however, there is little difference in the two time periods, or the

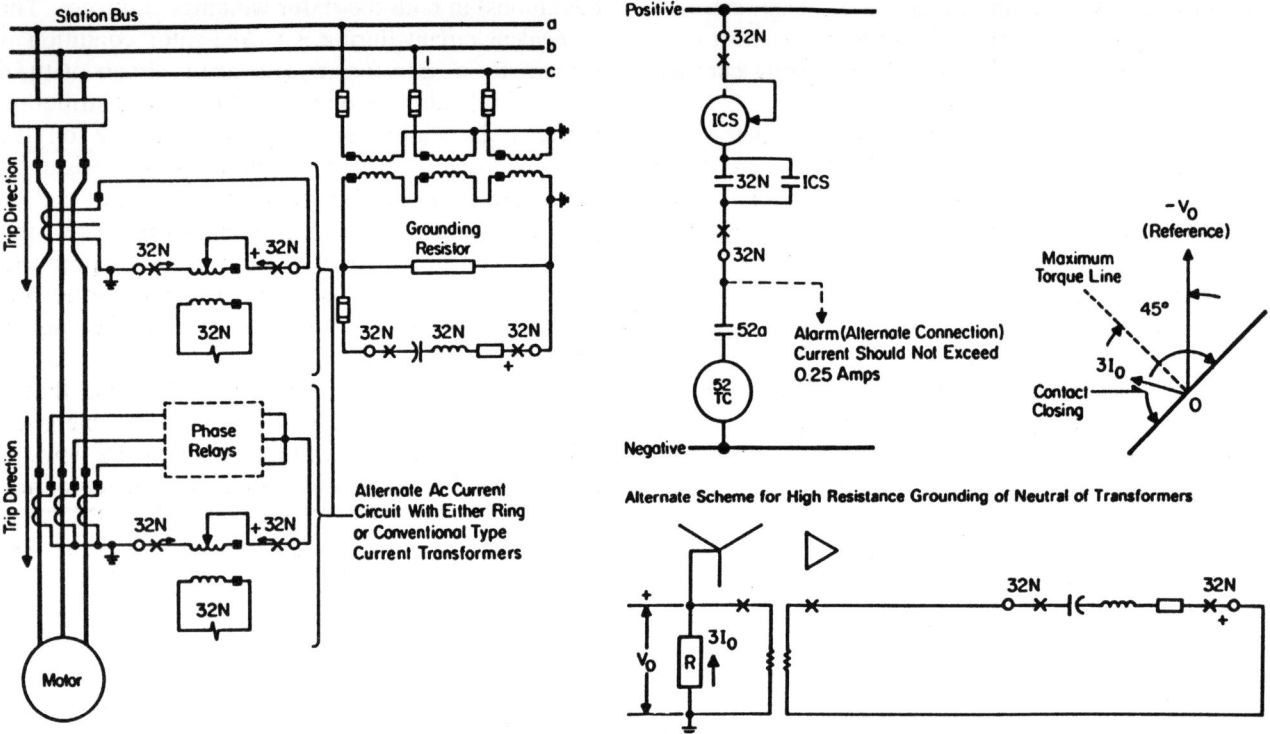

Figure 9-7 Typical connections of the product type CWP-1 (32N) for high-resistance grounded systems.

starting time exceeds the locked-rotor time (Fig. 9-9), other considerations must be taken into account.

For the case shown in Figure 9-9, it is tempting to try to fit an overcurrent relay characteristic between the two curves. It should be remembered, however, that the conventional overcurrent relay characteristic is a *plot* of operating time against sustained current (Fig. 9-10), whereas the starting characteristic is a *trace* of current against time (Fig. 9-9). If I_{LR} is applied to the CO relay for time t_a, the contacts are very nearly closed. Current does not drop below the CO pickup value until time t_b. Contact closure occurs at some point t_c even though the CO relay characteristic is *always* above the current trace.

Over a narrow range, such as that between two and three times pickup, a CO relay can be assumed to operate if the integral of $(I - 1)^n \, dt$ exceeds K, where I is the multiple of pickup and n and K are constants that depend on the relay type and time dial setting. If a linear-linear plot of $(I - 1)^n$ and t is used, a varying current and time can be compared with the relay characteristic on an area basis. In Figure 9-11, for example, the CO relay contact will not close if the

current drops below the CO pickup before area A equals area B.

When both voltage and current are available, an alternative solution to locked-rotor problems for large motors is to use a distance relay and timer. A motor during start behaves like a three-phase balanced fault so a distance relay that responds to three-phase balanced faults should be used. The impedance of the motor will remain fixed (largely reactive at a low power factor) if the motor does not accelerate. If the motor accelerates, both the impedance and power factor will increase (Fig. 9-12). The impedance of a motor with a locked rotor is essentially independent of terminal voltage and, as the motor accelerates, its impedance changes as indicated. This change of impedance with motor acceleration makes the distance relay particularly well-suited to this application. It also affords backup protection for phase faults and some ground faults. The timer can be obtained from one of two types of relays: a time-overvoltage relay (59) or time-over-current relay (50), each supervised by the 52a breaker contact.

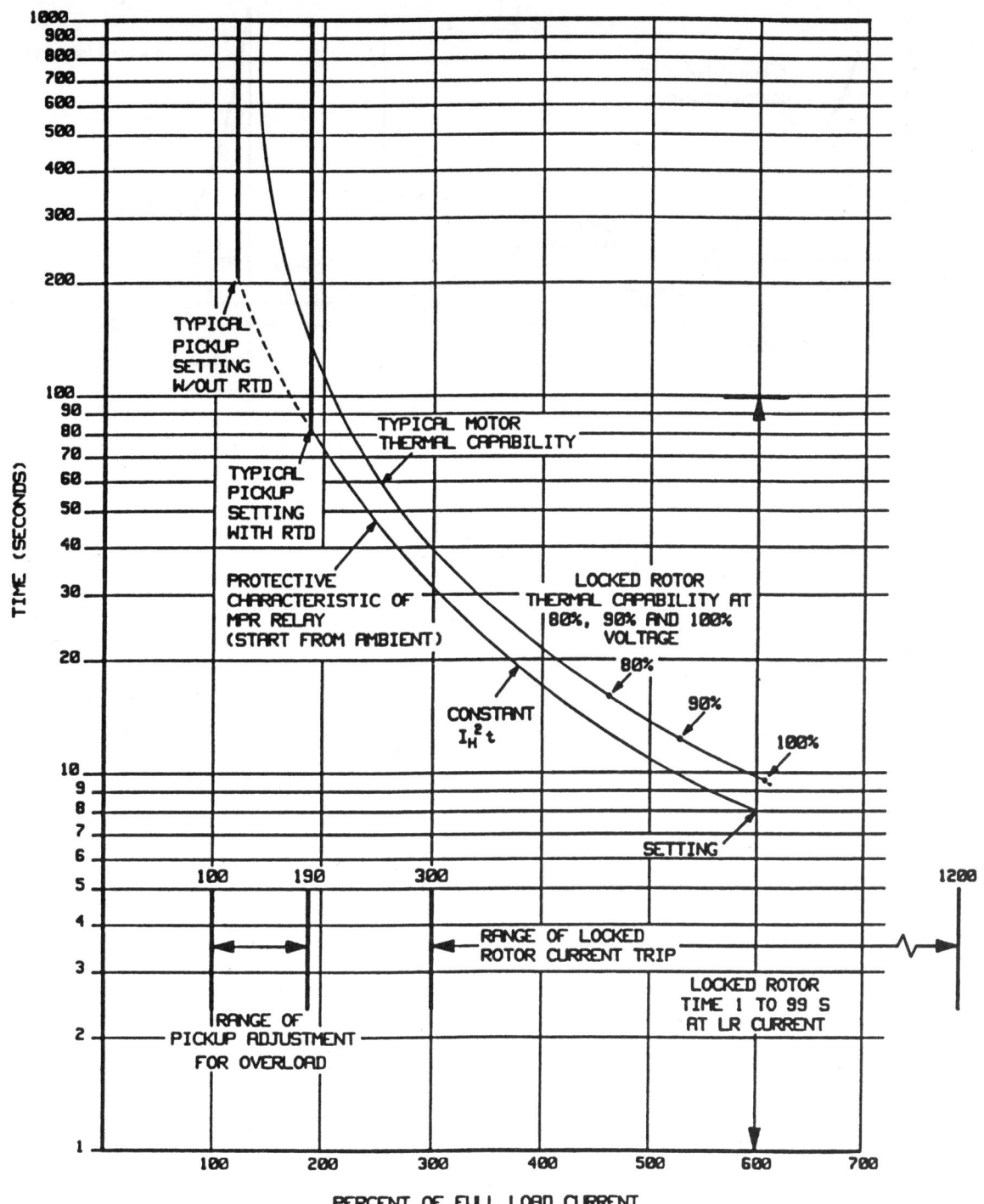

Figure 9-8 Typical MPR characteristics.

An alternative to the distance relay technique is the PRO*STAR® motor protection relay, which determines the speed-dependent heating in the rotor during a start. The relay uses an impedance measurement to estimate the speed of the motor, and the rotor thermal model then accounts for the declining rotor resistance during acceleration. Thus a high-inertia motor where the allowable locked-rotor time is less

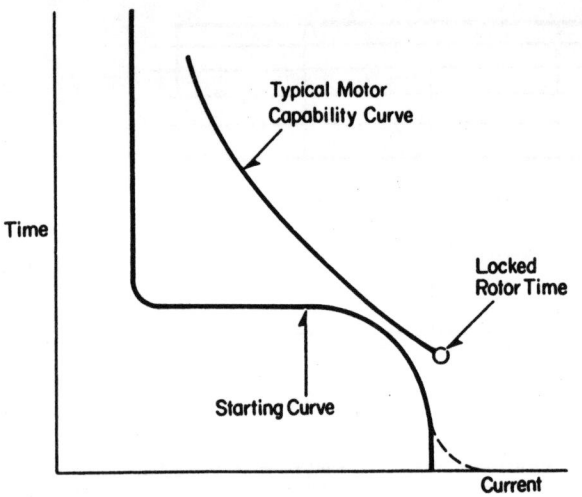

Figure 9-9 Motor starting time exceeding permissible locked rotor time.

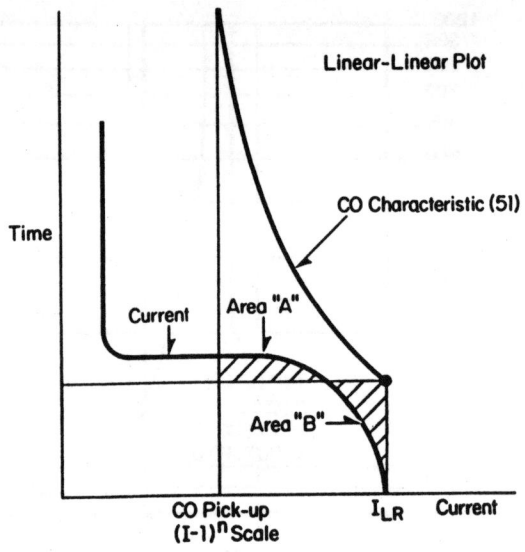

Figure 9-11 Area comparison.

than the normal starting time can be properly protected by this relay.

Some applications use a mechanical zero-speed switch to supervise an overcurrent unit, preventing operation of a timer once motor rotation is detected. This scheme will not detect a failure to accelerate to full speed nor pullout with continued rotation, as the above schemes will.

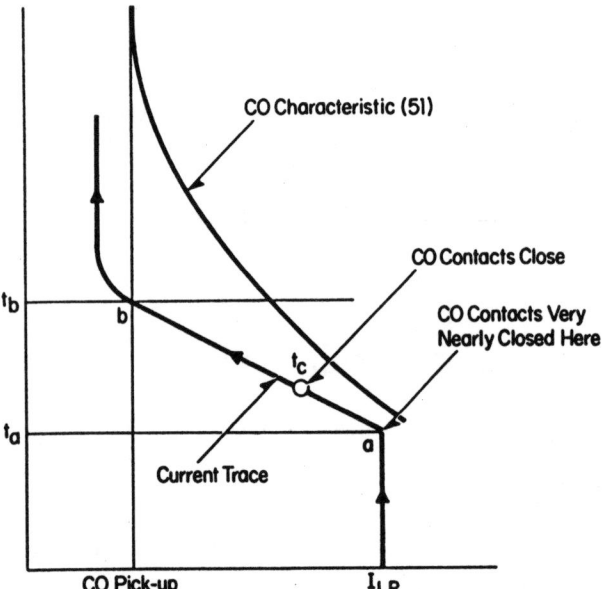

Figure 9-10 CO characteristic compared to current trace.

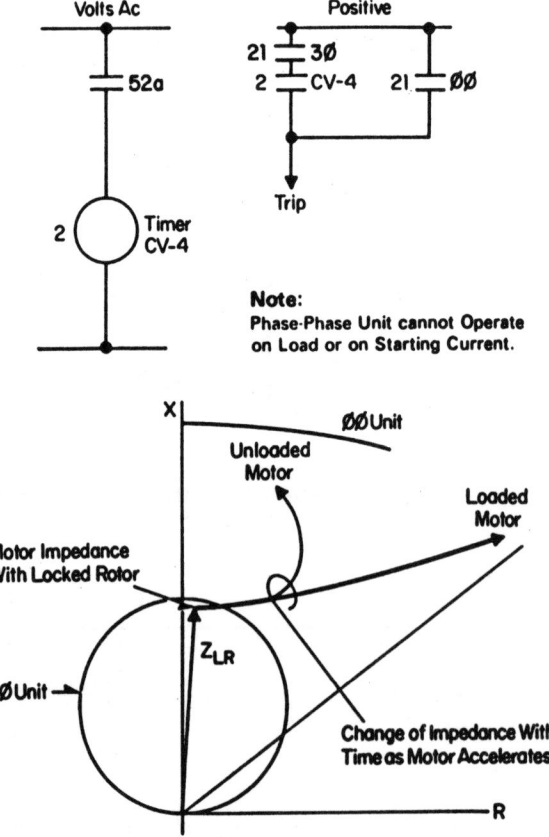

Figure 9-12 KD-10 distance relay (21) used for locked rotor and backup protection for large motor.

5 OVERLOAD PROTECTION

Heating curves are difficult to obtain and vary considerably with motor size and design. Further, these curves are an approximate average of an imprecise thermal zone, in which varying degrees of damage or shortened insulation life may occur. It is difficult, then, for any relay design to approximate these variable curves adequately over the range from light sustained overloads to severe locked-rotor overload.

Thermal overload relays offer good protection for light and medium (long-duration) overloads, but may not for heavy overloads (Fig. 9-13a). The long-time induction overcurrent relay offers good protection for the heavy overloads, but overprotects for light and medium overloads (Fig. 9-13b). A combination of the two devices provides complete thermal protection (Fig. 9-13c).

The National Electric Code requires that an overload device be used in each phase of a motor "unless protected by other approved means." This requirement is necessary because single phasing (opening one supply lead) in the primary of a delta-wye transformer that supplies a motor will produce three-phase motor currents in a 2:1:1 relationship. If the two units of current appeared in a phase with no overload device, the motor would be unprotected. Thus, the NEC requires three overload devices, or two overload devices and another to detect unbalance such as a CM or 46D relay.

6 THERMAL RELAYS

There are two types of thermal relays. The DT-3 and type 49T operate from a resistance temperature detector (RTD) that monitors the temperature in the machine windings, motor or load bearings, or load case. They are typically applied only to large motors, usually 1500 hp and above where RTDs are available. The RTD is an excellent indicator of average winding temperature. It is influenced by the effects of ambient temperature, ventilation variations, and recent loading history.

The second type comprises replica relays such as the BL-1, type 49, and IMPRS, which use an overcurrent element with very slow reset characteristics to replicate the thermal condition of the motor and are applied where RTDs are not available.

Multifunction motor protection relays such as the PRO*STAR, MPR, and REM-543 include multiple RTD inputs. When no RTDs are available, these relays

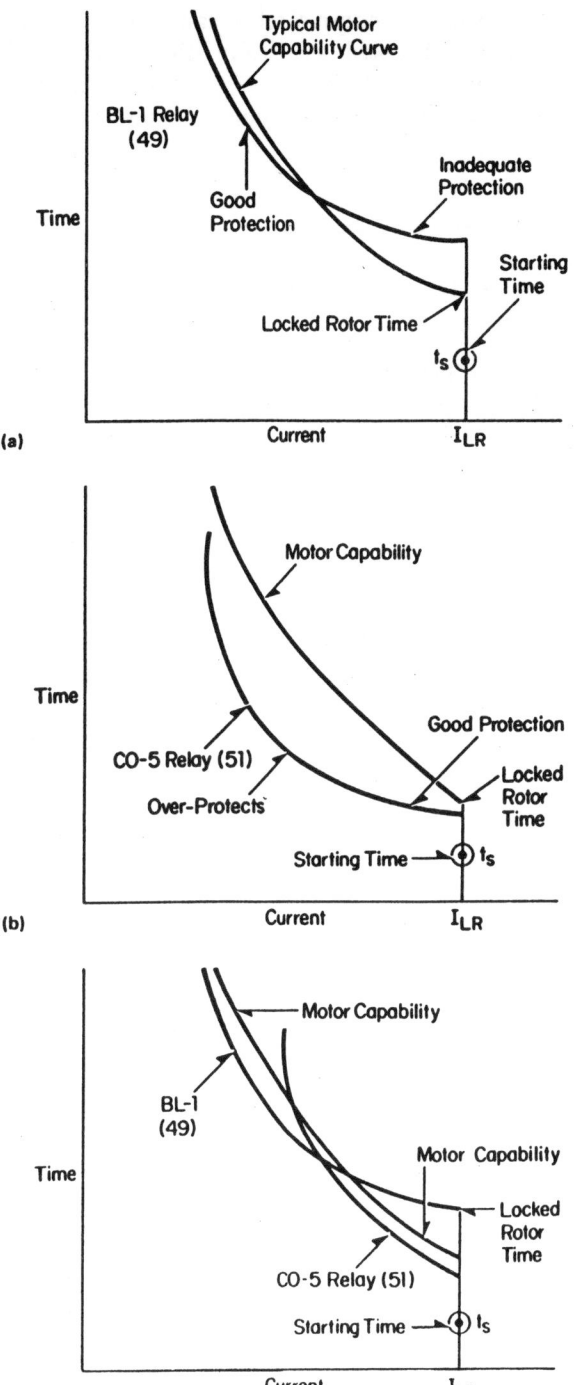

Figure 9-13 Typical motor and relay time current characteristics.

use thermal replica algorithms. When RTDs are available, the direct temperature measurement is also used, and alarm and trip set points are established for each of the RTDs.

No current-responsive relay can protect a motor subjected to blocked ventilation. Relays using RTD inputs for thermal protection overcome this shortcoming by responding to temperature alone.

6.1 RTD-Input-Type Relays

Several RTD types are available for use in temperature monitoring:

$10\,\Omega$ copper
$100\,\Omega$ nickel
$120\,\Omega$ nickel
$100\,\Omega$ platinum

Microprocessor-based motor protection relays typically use three-wire input RTDs. The RTD has a well-defined ohmic characteristic vs. temperature. Accurate detection of the resistance of an RTD requires that the lead resistance be subtracted from the total resistance measured by the relay. One scheme used by the MPR circulates a precision current out the X terminal, through the RTD, and returns to terminal Z via the return lead (Fig. 9-14). The following equation results:

$$V_{XZ} = R_{LEAD}I_{CC} + R_{RTD}I_{CC} + R_{LEAD}I_{CC}$$

where

R_{LEAD} = resistance of the leads to the RTD
I_{CC} = constant current source
R_{RTD} = resistance of the RTD

It next circulates this same current through the Y terminal and return path to Z to obtain

$$V_{YZ} = R_{LEAD}I_{CC} + R_{LEAD}I_{CC}$$

By subtracting V_{YZ} from V_{XZ}, the relay obtains

$$V_{DELTA} = V_{XZ} - V_{YZ} = R_{RTD}I_{CC}$$
$$R_{RTD} = V_{DELTA}/I_{CC}$$

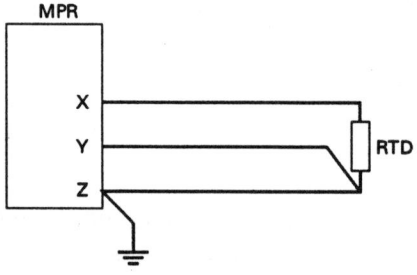

Figure 9-14   MPR three-wire RTD input.

The error associated with the resistance of the leads is removed and translations from resistance to temperature can be performed by the microprocessor relay using setting data stored in nonvolatile RAM or EEPROM that relates to the physical properties of the particular material used by the RTD.

The DT-3 is a bridge-type relay. The exploring coils form part of a Wheatstone bridge circuit, which is balanced at a given temperature. As the motor temperature increases above the balance temperature, operating torque is produced (Fig. 9-15). With the DT-3 relay, only one resistance temperature detector (10, 100, or $120\,\Omega$) or exploring coil is required.

The DT-3 relay is a d'Arsonval-type dc contact-making milliammeter that is connected across the bridge. The bridge is energized by either 125 or 250 Vdc or supplied with 120 Vac through a transformer and full-wave bridge rectifier in the relay. The relay scale is calibrated from either 50 to 190 °C (or 100 to 160°). The right- or left-hand contacts close when the temperature rises or falls to the preset value between 50 and 190 °C (or 100 to 160 °C). The normal setting for class B machines is 120 °C.

6.2 Thermal Replica Relays

Replica-type relays (BL-1 and, additionally, the IMPRS, MPR, and PRO*STAR) are designed to replicate, within the relay operating unit, the heating characteristics of the machine. Thus, when current from the current transformer secondary passes through the relay, its time-overcurrent characteristic approxi-

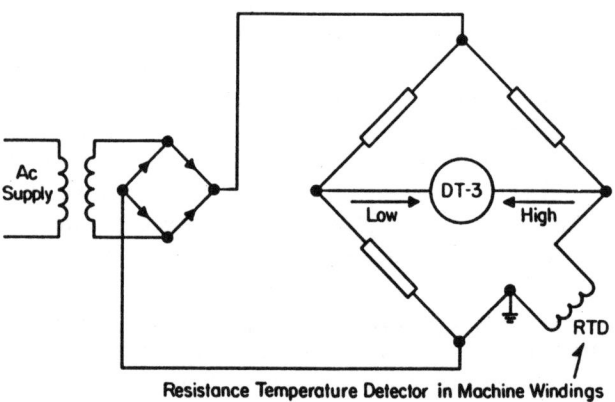

Resistance Temperature Detector in Machine Windings

Figure 9-15   Typical schematic of the type DT-3 relay (49) for motor overload protection. (Its advantages are good protection for overload, blocked ventilation, and high ambient temperature operation.)

mately parallels that of the machine capability curve at moderate overload.

Extreme variations in load, such as jogging, produce a difficult relaying problem. In general, electromechanical thermal replica relays cool at a different rate from the motor they protect. Variations in load may produce a ratcheting effect on the relay and cause premature tripping. Microprocessor motor protection relays typically acknowledge previous loading history through the use of an RTD input, which establishes a starting level.

The thermal replica relay is recommended when embedded temperature detectors are not available, although RTD-input-type relays are recommended when they are. Replica-type relays are typically temperature-compensated and operate in a fixed time at a given current, regardless of relay ambient variations. Although this characteristic is desirable for the stated conditions, it produces underprotection for high motor ambient and overprotection for low motor ambient.

7 LOW-VOLTAGE PROTECTION

Low voltage prevents motors from reaching rated speed on starting, or causes them to lose speed and draw heavy overloads. An equation of the average accelerating motor torque is directly related to the voltage present:

$$T_A = \left(\frac{E}{E_R}\right)^2 T_m - T_L$$

where

  E_R = rated motor voltage
  E = voltage available at motor bus
  T_L = load torque
  T_m = rated voltage motor torque
  T_A = average accelerating torque

It can be seen that the voltage available to the motor significantly affects the accelerating torque of the motor.

Motors should be disconnected when severe low-voltage conditions persist for more than a few seconds. Use of ac contractors, which generally release at 50 to 70% of rated voltage, provides some low-voltage protection. However, time-delayed undervoltage protection is preferred, since it delays contactor release on momentary voltage dips. For switchgear applications, the electromechanical CV (27), CP (27/47), and CVQ

(27/47), or solid-state Types 27/27D and Types 47/47D relays will accurately detect undervoltage and initiate a trip or alarm as required.

8 PHASE-ROTATION PROTECTION

When starting in reverse can be a serious hazard, a reverse-phase relay should be applied. The relay, such as types CP, CVQ, 47, or 60Q, monitors the bus voltage and is wired to supervise motor starting. Another technique, used by the MPR multifunction relay, is to trip instantaneously when reverse-phase currents are detected. The voltage relay method prevents motor energization, whereas the current method requires that the motor be energized and then tripped when reverse-phase sequence exists.

9 NEGATIVE SEQUENCE VOLTAGE PROTECTION

The CVQ (27/47) relay contains a negative sequence voltage unit that operates as shown in Figure 9-16. A negative sequence voltage network, as described in Chapter 3, energizes an induction-disk voltage unit V_2. If a three-phase voltage applied to the relay contains 5% (adjustable to 10%) negative sequence content or more, the negative sequence unit (V_2) operates. A back

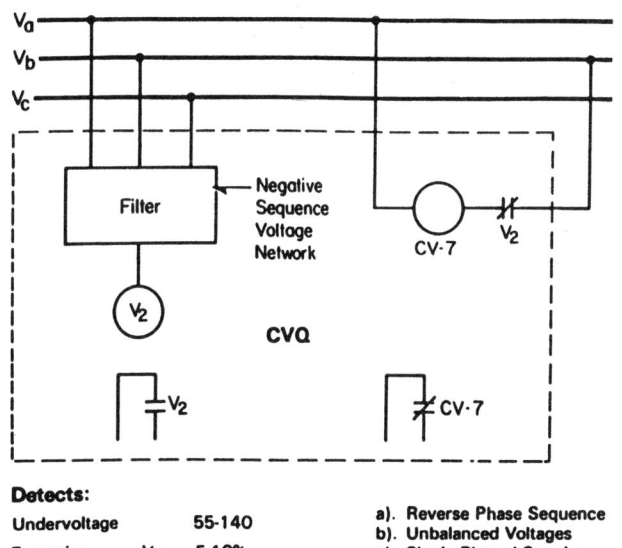

Detects:		
Undervoltage	55-140	a). Reverse Phase Sequence
Excessive V_2	5-10%	b). Unbalanced Voltages
		c). Single Phased Supply

Figure 9-16 Simplified schematic diagram of the CVQ (27/47) negative sequence voltage relay.

contact of the negative sequence unit opens a CV-7 undervoltage unit coil circuit, and after a time delay, the contacts of the undervoltage unit initiate tripping or sound an alarm. This relay operates for

> Reverse-phase rotation (100% negative sequence)
> Unbalanced voltage (partial negative sequence)
> Undervoltage (no negative sequence)

The CVQ relay is recommended for all important buses supplying motor loads.

Although the CVQ relay can detect single phasing of the supply to even a single, lightly loaded large motor if its magnetizing impedance is low enough, it does not respond to single phasing between the point of application of the CVQ and the motor. Figure 9-17 displays two cases of an open phasing condition. The first case is an open phase at A. The resulting sequence network interconnections are shown.

In this first case, the negative sequence voltage relay measures the voltage across the negative sequence impedance of the motor or motors. In the second case, the open phase occurs at B. Figure 9-17 shows the sequence network interconnections. When the open is at location B, the relay now measures the negative sequence voltage across the source.

Very low negative sequence voltage is produced on the source side of the open phase, which makes it extremely difficult for the negative sequence voltage

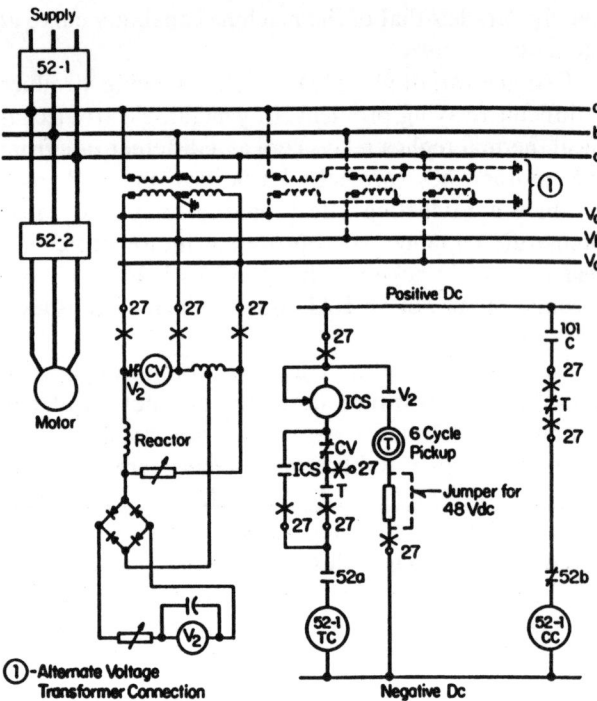

Figure 9-18 External schematic of the CVQ relay used for tripping on negative sequence voltage only.

relay to detect. For practical purposes, the voltage unbalance occurs only on the load side of an open phase. In general, a phase-unbalance current relay is preferred for detecting a feeder circuit open phase.

Figure 9-18 illustrates one type of CVQ relay that responds *only* to negative sequence voltage (not undervoltage). The six-cycle timer prevents operation for non-simultaneous pole closure of the supply breaker, 52–1.

The solid-state relay types 47 and 47D are the functional equivalents of the CVQ providing reverse-phase unbalanced voltage and undervoltage protection. The type 60Q responds only to negative sequence voltage, and will operate only for unbalanced conditions. The 60Q includes an adjustable time delay.

10 PHASE-UNBALANCE PROTECTION

Phase-unbalance protection is applied to a feeder supplying a large motor or group of small motors where there is a possibility of one of the feeder phases opening as a result of a connector failure, fuse failure, or similar cause. The electromechanical type CM relay (device 46) contains two induction-disk units (Fig. 9-19). One unit in the CM relay balances I_a

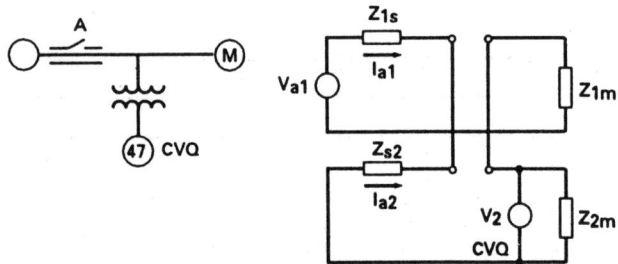

(a) Negative Sequence Voltage Relay Between Open Phase and Motor

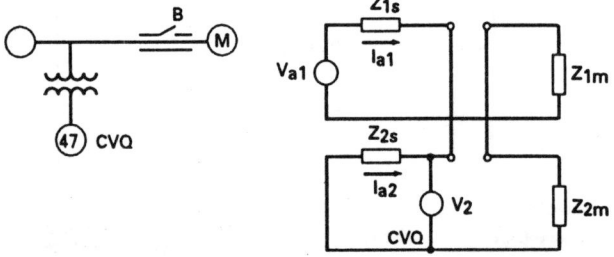

(b) Negative Sequence Voltage Relay Between Source Phase and Open Phase

Figure 9-17 Motor single phasing.

against I_b, and the other balances I_b against I_c. When the currents become sufficiently unbalanced, torque is produced in one or both of the units, closing their contacts (which are connected in parallel in the trip circuit). The solid-state type 46D determines the negative sequence content of the three-phase currents and includes a built-in timer.

One relay can protect many motors subject to collective single phasing. In addition, a phase-unbalance relay may protect up to five motors subject to individual single phasing, depending on how the motors are operated and their relative sizes. For example, the relay will not operate if a motor with a rating of one-fifth of the total feeder load is subject to single phasing while unloaded and while the remaining motors are fully loaded. The CM relay has 7-A continuous capability and operates when the unbalance exceeds approximately 10 to 15% between 2 and 7 A. With no current present in one current circuit and with 1 A in the other, the relay operates (on the 1-A tap).

Multifunction motor protection relays usually include an element for phase unbalance protection. If the unbalance exceeds a set threshold for the set time delay period, tripping is initiated (PRO*STAR, IMPRS, MPR, REM543). In addition, lower levels of unbalance are used to increase the estimated temperature condition of the thermal replica (PRO* STAR, MPR, REM543).

11 NEGATIVE SEQUENCE CURRENT RELAYS

No standards have been established for the $I_2^2 t$ short-time capability for a motor, although $I_2^2 t = 40$ is regarded as a conservative value. (I_2 is the per unit machine negative sequence current, t the time in seconds.)

The electromechanical negative sequence time-over-current relay COQ does not have the sensitivity necessary to properly protect a motor against the overheating caused by a prolonged load current unbalance. The CM and type 46D are the preferred single-function relays. The phase-unbalance protection of the multifunction motor protection relays also provide the required sensitivity. The MPR can be set to pick up when $I_2 = 5$ to 30% of the full load tap setting, and the IMPRS and PRO*STAR when $I_2 = 10$ to 50%. Settings in the range 10–20% would be most typical.

12 JAM PROTECTION

A motor can experience excessive torque and over-current in response to a jam condition that can be caused by a binding action of the motor, bearings, or driven load. To detect a jam condition, the relay has to screen out other possibilities. The motor contribution to a nearby fault, which can last for a few cycles, can be screened out by setting the jam time delay greater than the motor fault contribution time. High current is not recognized as a jam condition unless the motor has been through a start and is in a normal running state. Multifunction relays usually include this element, as they are capable of keeping track of the state of the motor. The advantage of the jam function is much faster tripping than would be provided by the locked rotor protection.

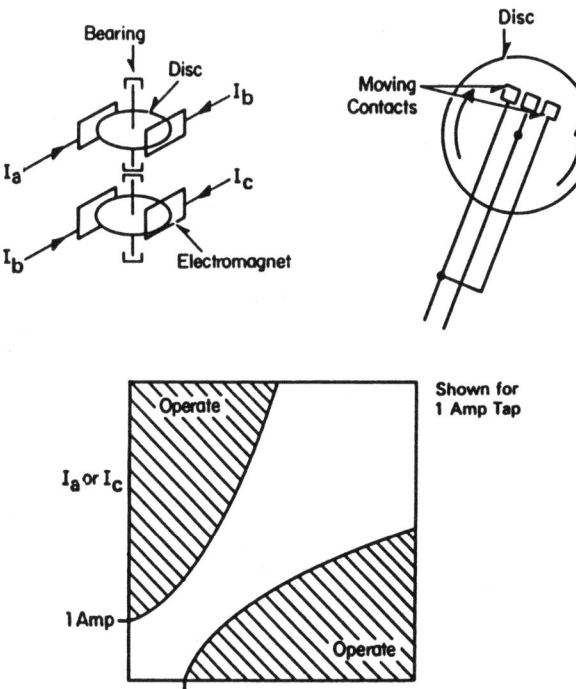

Figure 9-19 The type CM phase unbalance relay (46).

13 LOAD LOSS PROTECTION

The sudden reduction of shaft load is referred to as load loss, which can be caused by a shaft breakage, loss of prime on a pump, or the shearing of a drive pin. To minimize any damage to the driven load, the motor should immediately be taken offline. Detection of load loss requires the recognition of the difference between no load preceding the application of load and no load following the application of load. The load loss element is commonly supplied in multifunction motor protection relays (MPR, PRO*STAR, REM543) that have the ability to track the state of the motor.

14   OUT-OF-STEP PROTECTION

Out-of-step protection is applied to synchronous motors and synchronous condensers to detect pullout resulting from excessive shaft load or too low supply voltage. Other causes of pullout result from a fault occurring on the supply system, whereby the fault type, clearing time, and location are factors relating to the stability of the motor. Underexcitation caused by incorrect field breaker trip or a short or open in the field circuit can also result in loss of synchronism of the motor. For a discussion of the out-of-step protection of large motors, refer to Chapter 8 on generator protection.

Small synchronous motors with brush-type exciters are often protected against out-of-step (or loss-of-excitation) operation by ac voltage detection devices connected in the field. No ac voltage is present when the motor is operating synchronously.

15   LOSS OF EXCITATION

Synchronous motors can be protected against loss of excitation by a low-set undercurrent relay connected in the field. This relay should have a time delay on dropout to trip or alarm the operator. The KLF (or KLF-1) relay (40) (described in Chap. 8) can also be used to protect large motors against loss of field. The under-voltage units of these relays should have their contacts shorted. Loss of excitation of a synchronous motor does not usually depress the voltage enough to operate reliably an undervoltage unit.

Unlike undercurrent relays, the KLF (or KLF-1) relay can detect both partial and complete loss of field, and some out-of-step conditions as well (Fig. 9-20).

Both out-of-step and loss-of-excitation conditions can be detected with a CW watt-type relay (55),

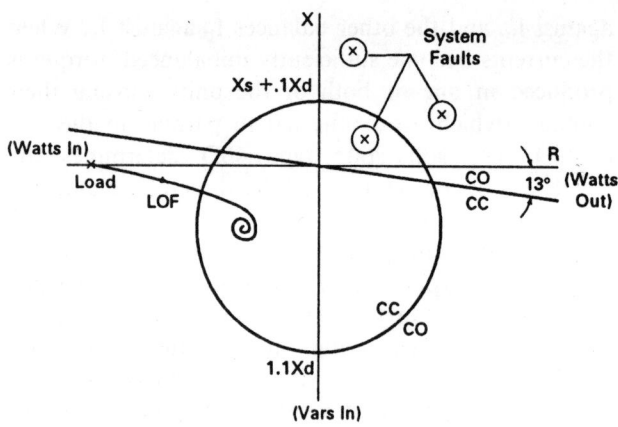

Figure 9-20   KLF used for motor loss of field detection.

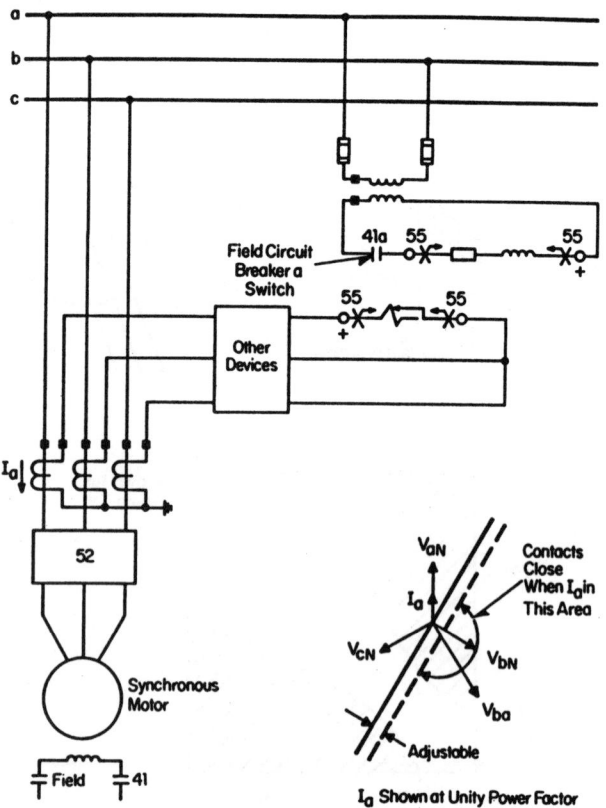

Figure 9-21   CW watt relay used for out-of-step detection.

Table 9-2 Typical Protection for Motors Below 1500 hp (Fig. 9-22)

Device number	Quantity	Description	Typical setting	Remarks
49/50	1	BL-1,2 unit with 2 ITs	Set at full load for motor with 1.15 service factor and 90% of full load for 1.00 service factor motor. IT set 2 times locked rotor.	Good overload protection.
51/50	1	CO-5, 1–12 A; with IIT 10–40 A CO-11, 4–12 A with IIT 10–40 A	Current setting 1/2 locked rotor. Time delay set to give operating time > starting time. IT set 2 times locked rotor.	Locked-rotor protection when starting time 20 to 70 sec. Locked-rotor protection when starting time ≥20 sec.
50G	1	IT, 0.15–0.3 A single unit	0.15 A.	For use with 50/5 BYZ.
47/27	1	CVQ, 5 to 10% V_2 sensitivity, 55- to 140-V range	Low voltage 75 to 80%. $V_2 = 5\%$.	Undervoltage, phase sequence, and unbalanced voltage protection.
51N/50N (alternative to 50G where BYZ cannot be applied)	1	CO-11, 0.5–2.5 A; with IIT 10–40 A	Pickup 0.5 A, time 0.1 sec at IT setting. IT set $4 \times I_{FL}$.	Provides ground protection. Time unit overrides false residual during starting.

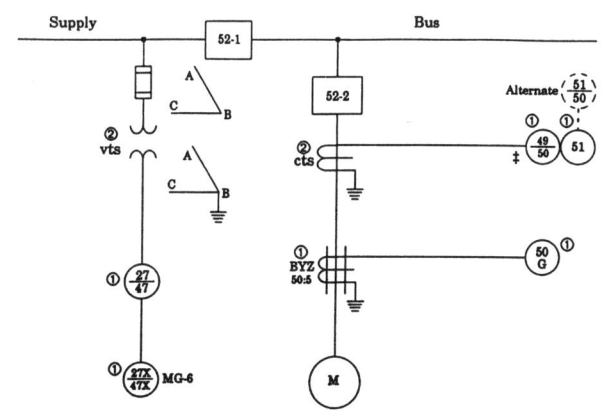

Note: Phase ct's should be approximately 150% of full load.

‡   If device 27/47 is not provided, then three device 49 units are required.

①   Represents the quantity of devices required.

Figure 9-22 Motor protection below 1500 HP.

connected for 0 torque when the current lags the voltage by an appropriate power factor angle, such as 30°. Used in this way, the CW is referred to as a power-factor relay. The connection shown in Figure 9-21 gives maximum contact closing torque when the current lags its unity power-factor position by 120°.

16 TYPICAL APPLICATION COMBINATIONS

Table 9-2 and the associated Figures 9-22 and 9-23 show typical application combinations for motor protection. Table 9-3 lists the combined protection functions that are available by using a microprocessor-based motor protection relay.

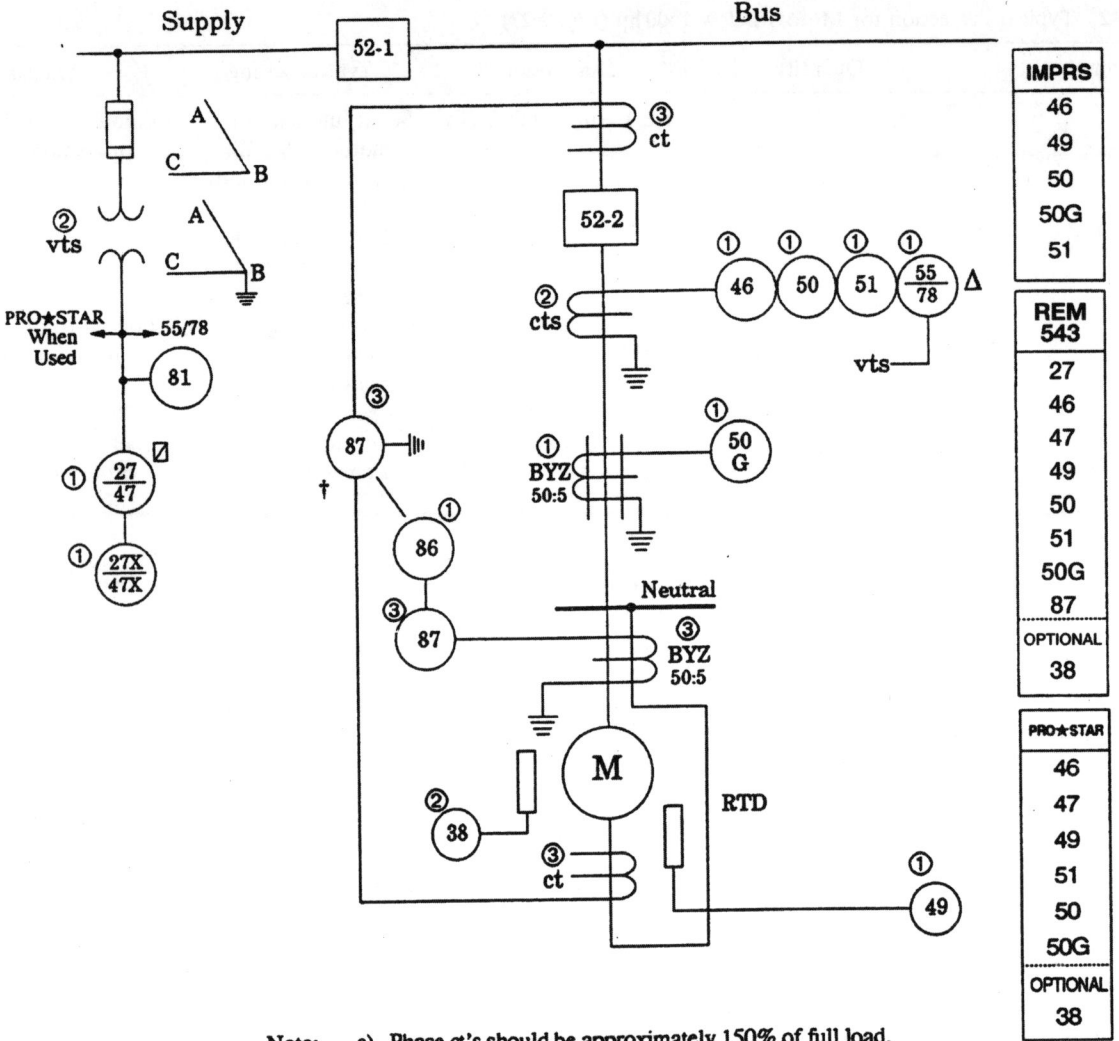

Note:

    a) Phase ct's should be approximately 150% of full load.

    b) The REM 543, PRO★STAR and IMPRS are microprocessor relays containing several functions.

    † Alternate to BYZ scheme. BYZ scheme is preferred.

    ▨ Must apply CVQ or 60Q when Device 46 is omitted.

    Δ Applied on brushless synchronous motors.

    ① Represents the quantity of devices required.

Figure 9-23   Motor protection 1500 HP and above.

Table 9-3 Typical Protection for Motors of ≥1500 hp (Fig. 9-23)

Device number	Quantity	Description	Typical setting	Remarks
49	1	DT-3, 50–190 °C, specify ohms of RTD 120 vac.	Set for motor max. safe operating temperature.	Overload protection: blocked ventilation or high ambient.
51	1	CO-5, 1–12 A	Current setting 1/2 locked rotor.	Locked-rotor protection when starting time is 20 to 70 sec.
		CO-11, 4–12 A	Time delay setting to give operating time > starting time.	Locked-rotor protection when starting time ≤20 sec.
50	1	SC, 2 unit, 20–80 A	Set 2 times locked rotor.	Fault protection.
50G	1	IT, 0.15–0.3 A single unit	0.15 A.	For use with 50/5 BYZ.
46	1	CM, 1–3 A	For $I_{FL} \geq 3$ A: Set 2A. For $I_{FL} < 3$ A: Set 1A.	Unbalanced current protection.
47/27	1	CVQ, 5 to 10% V_2 sensitivity, 55- to 140-V range	Low voltage 75 to 80%, $V_2 = 5\%$.	Undervoltage, phase sequence, and unbalanced voltage protection. Note: CP volt relay can be used in place of CVQ if all three-phase motors on bus are protected by CM relays.
87ϕ	1	IT, 0.15–0.3 A, 3 unit	0.15 A.	Provides phase and ground protection. Use three 50:5 BYZ transformers. 50G still required for cable protection if BYZ at motor.
87 (alternative to 50 and 87ϕ—use where minimum 3ϕ fault current available is less than 5 times motor starting current and 87ϕ cannot be used)	3	CA, 10%	None.	Phase-fault protection.
51N/50N (alternative to 50G where BYZ cannot be applied)	1	CO-11, 0.5–2.5 A, with IIT 10–40 A	Pickup 0.5 A, time 0.1 sec at IIT setting. IIT set 4 times full load.	Provides ground protection.

10

Transformer and Reactor Protection

Revised By: J. J. McGOWAN

1 INTRODUCTION

Differential relays are the principal form of fault protection for transformers rated at 10 MVA and above. These relays, however, cannot be as sensitive as the differential relays used for generator protection.

Transformer differential relays are subject to several factors, not ordinarily present for generators or buses, that can cause misoperation:

Different voltage levels, including taps, that result in different primary currents in the connecting circuits.

Possible mismatch of ratios among different current transformers. For units with ratio-changing taps, mismatch can also occur on the taps. Current transformer performance is different, particularly at high currents.

30° phase-angle shift introduced by transformer wye-delta connections.

Magnetizing inrush currents, which the differential relay sees as internal faults.

Transformer protection is further complicated by a variety of equipment requiring special attention: multiple-winding transformer banks, zig-zag transformers, phase-angle regulators (PAR), voltage regulators, transformers in unit systems, and three-phase transformer banks with single-phase units.

All the above factors can be accommodated by the combination of relay and current transformer design, along with proper application and connections. Magnetizing inrush, the most significant variable in transformer protection, will be discussed first.

2 MAGNETIZING INRUSH

When a transformer is first energized, a transient magnetizing or exciting inrush current may flow. This inrush current, which appears as an internal fault to the differentially connected relays, may reach instantaneous peaks of 8 to 30 times those for full load.

The factors controlling the duration and magnitude of the magnetizing inrush are

Size and location of the transformer bank
Size of the power system
Resistance in the power system from the source to the transformer bank
Type of iron used in the transformer core and its saturation density
Prior history, or residual flux level, of the bank
How the bank is energized

2.1 Initial Inrush

When the excitation of a transformer bank is removed, the magnetizing current goes to 0. The flux, following the hysteresis loop, then falls to some residual value ϕ_R (Fig. 10-1). If the transformer were reenergized at the instant the voltage waveform corresponds to the residual magnetic density within the core, there would be a smooth continuation of the previous operation with no magnetic transient (Fig. 10-1). In practice, however, the instant when switching takes place cannot be controlled and a magnetizing transient is practically unavoidable.

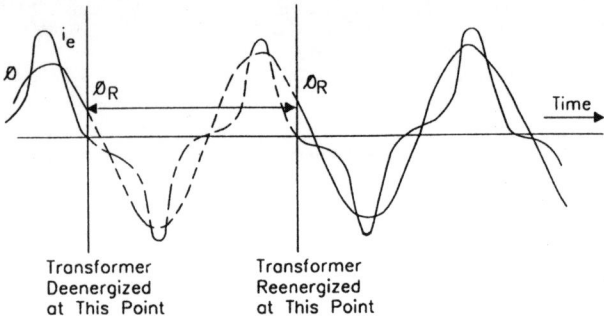

Figure 10-1 Magnetizing current when transformers were reenergized at that instant of the voltage wave corresponding to the residual magnetic density within the core.

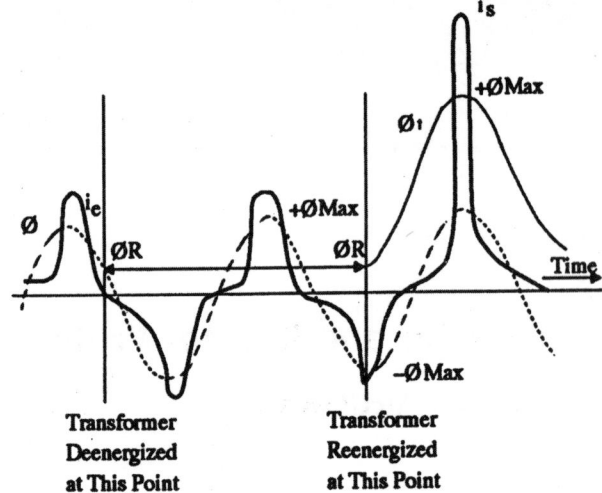

Figure 10-2 Magnetizing current when transformers were reenergized at the instant when the flux would normally be at its negative maximum value.

In Figure 10-2, it is assumed that the circuit is reenergized at the instant the flux would normally be at its negative maximum value ($-\phi_{max}$). At this point, the residual flux would have a positive value. Since magnetic flux can neither be created nor destroyed instantly, the flux wave, instead of starting at its normal value ($-\phi_{max}$) and rising along the dotted line, will start with the residual value (ϕ_R) and trace the curve (ϕ_t).

Curve ϕ_t is a displaced sinusoid, regardless of the magnetic circuit's saturation characteristics. Theoretically, the value of ϕ_{max} is $+(|\phi_R| + 2|\phi_{max}|)$. In transformers designed for some normal, economical saturation density ϕ_s, the crest of ϕ_t will produce supersaturation in the magnetic circuit. The result will be a very large crest value in the magnetizing current (Fig. 10-2).

The residual flux ϕ_R is the flux remaining in the core after the voltage is removed from the transformer bank. The flux will decrease along the hysteresis loop to a value of ϕ_R, where $i = 0$. Because the flux in each of the three phases is 120° apart, one phase will have a positive ϕ_R and the other two a negative ϕ_R, or vice versa. As a result, the residual flux may either add to or subtract from the total flux, increasing or decreasing the inrush current.

A typical inrush current wave is shown in Figure 10-3. For the first few cycles, the inrush current decays rapidly. Then, however, the current subsides very slowly, sometimes taking many seconds if the resistance is low.

The time constant of the circuit (L/R) is not, in fact, a constant: L varies as a result of transformer saturation. During the first few cycles, saturation is high and L is low. As the losses damp the circuit, the saturation drops and L increases. According to a 1951

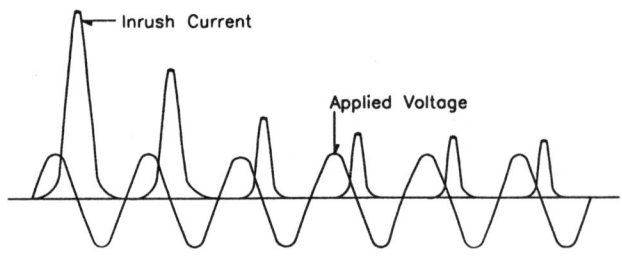

Figure 10-3 A typical magnetizing inrush current wave.

AIEE report, time constants for inrush vary from 10 cycles for small units to as much as 1 min for large units.

The resistance from the source to the bank determines the damping of the current wave. Banks near a generator will have a longer inrush because the resistance is very low. Likewise, large transformer units tend to have a long inrush as they represent a large L relative to the system resistance. At remote substations, the inrush will not be nearly so severe, since the resistance in the connecting line will quickly damp the current.

In addition to the conditions that influence single-phase inrush, the wave shape of the inrush current into a delta winding is influenced by the number of cores affected and the vector sum of the currents from the bank windings. The net wave could, in fact, become oscillatory (Fig. 10-4). The shape of a polyphase or single-phase inrush to a delta winding is affected by the

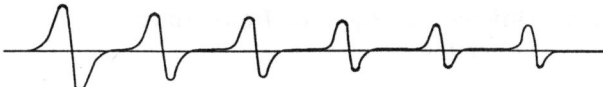

Figure 10-4 Typical magnetizing inrush current wave that can exist in one of the phases to a delta connection or in the secondary of delta connected current transformers.

nature of the line current itself, which is the vector sum of two currents from the bank windings. If we assume that only one core has saturated, the nature of the line current can result in either oscillatory waves or distortion of the single-phase shape.

When there is more than one delta winding on a transformer bank, the inrush will be influenced by the coupling between the different voltage windings. Depending on the core construction, three-phase transformer units may be subject to interphase coupling that could also affect the inrush current.

Similar wave shapes would be encountered when energizing the wye winding of a wye-delta bank or an autotransformer. Here, the single-phase shape would be distorted as a result of the interphase coupling produced by the delta winding (or tertiary).

Maximum inrush will not, of course, occur on every energization. The probability of energizing at the worst condition is relatively low. Energizing at maximum voltage will not produce an inrush with no residual. In a three-phase bank, the inrush in each phase will vary appreciably.

The maximum inrush for a transformer bank can be calculated from the excitation curve if available, and Table 10-1 shows a typical calculation of an inrush current (used phase A voltage as 0° reference).

From these calculated values it can be seen that:

The lower the value of the saturation density flux ϕ_S, the higher the inrush peak value.

The maximum phase-current inrush occurs at the 0° closing angle (i.e., 0 voltage).

The maximum line-current inrush occurs at $\pm 30°$ closing angles.

Because of the delta connection of transformer winding or current transformers, the maximum line-current inrush value should be considered when applying current to the differential relay.

2.2 Recovery Inrush

An inrush can also occur after a fault external to the bank is cleared and the voltage returns to normal (Fig. 10-5). Since the transformer is partially energized, the recovery inrush is always less than the initial inrush.

2.3 Sympathetic Inrush

When a bank is paralleled with a second energized bank, the energized bank can experience a sympathetic inrush. The offset inrush current of the bank being energized will find a parallel path in the energized bank. The dc component may saturate the transformer iron, creating an apparent inrush. The magnitude of this inrush depends on the value of the transformer impedance relative to that of the rest of the system, which forms an additional parallel circuit. Again, the sympathetic inrush will always be less than the initial inrush.

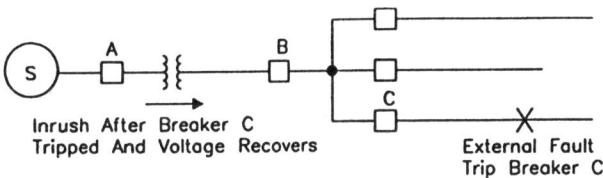

Figure 10-5 Recovery inrush after an external fault is cleared.

Table 10-1 Typical Inrush Current Calculation

		Peak value of inrush current wave (p.u.)					
ϕ_s	Closing angle	I_a	I_b	I_c	$I_a - I_b$	$I_b - I_c$	$I_c - I_a$
1.40	0°	5.60	−3.73	−3.73	8.33	−3.73	−8.33
1.40	30°	5.10	−1.87	−5.10	5.96	5.10	−9.20
1.15	0°	6.53	−4.67	−4.67	10.20	−4.67	−10.20
1.15	30°	6.03	−2.80	−6.03	7.83	6.03	−11.06

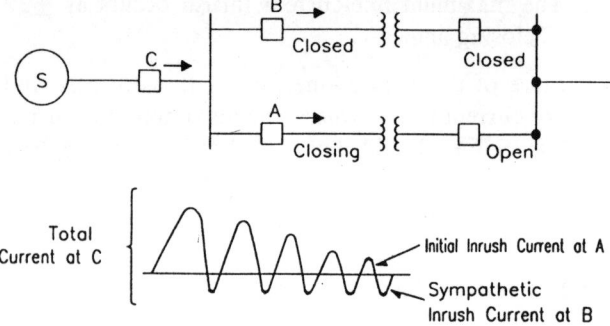

Figure 10-6 Sympathetic inrush when a bank is paralleled with a second energized bank.

As shown in Figure 10-6, the total current at breaker C is the sum of the initial inrush of bank A and the sympathetic inrush of bank B. Since this waveform looks like an offset fault current, it could cause misoperation if a common set of harmonic restraint differential relays were used for both banks.

Unit-type generator and transformer combinations have no initial inrush problem because the unit is brought up to full voltage gradually. Recovery and sympathetic inrush may be a problem, but as indicated above, these conditions are less severe than initial inrush.

3 DIFFERENTIAL RELAYING FOR TRANSFORMER PROTECTION

Since the differential relays see the inrush current as an internal fault, some method of distinguishing between fault and inrush current is necessary. Such methods include

- A differential relay with reduced sensitivity to the inrush wave (such units have a higher pickup for the offset wave, plus time delay to override the high initial peaks), such as types of CA and CA-26 transformer differential relays
- A harmonic restraint or a supervisory unit used in conjunction with the differential relay, such as types of HU, HU-1, HU-4, TPU, and RADSB transformer differential relays
- Desensitization of the differential relay during bank energization

3.1 Differential Relays for Transformer Protection

3.1.1 CA (87) Transformer Differential Relay

The CA transformer differential and the CA generator differential are companion electromechanical relays. Though they have largely been replaced by solid-state and microprocessor relays, their operating principle is still of interest. Since there are thousands of these relays in service, they are included here.

Figure 10-7 shows the basic design of the transformer version of this relay. The generator relay has no taps and is more sensitive than the transformer version. The transformer relay is relatively insensitive to the high percentage of harmonics contained in magnetizing inrush current. Because of this and its relatively slow six-cycle operating time, it has been used successfully for many decades in less critical applications where cost has a significant influence.

Operation occurs when the operating current (which is the differential current) exceeds roughly 50% of the minimum restraint. Putting it in another frame of reference, it operates when the operating current exceeds approximately 20% of the summation restraint. For external faults, note that the restraint ampere-turns are additive; and for an internal fault they are subtractive. For the idealized case of equal contributions from sources on each side of the transformer to an internal fault, the restraint cancels

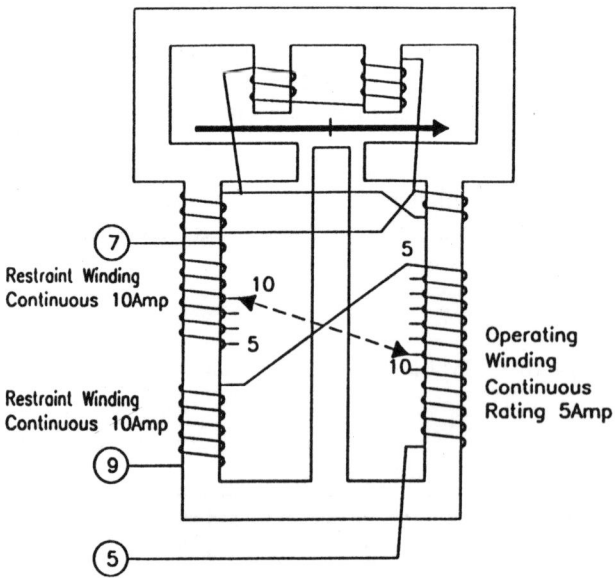

Figure 10-7 Type CA transformer differential relay.

completely, and the currents add together in the operating coil. The transformer differential version of this relay allows currents with as much as a 2:1 ratio to be matched. Figure 10-8 describes the case with 10 A input from one set of current transformers and 5 A to the other set. Balance is accomplished in the operating coil by the autotransformer action. For this case, and

for all such "through" phenomenon that provide this same ratio of currents, cancellation of operating torque occurs. Figure 10-9 shows a typical distribution of currents for a "through" condition with the relay set to balance currents with a 5 to 8 relationship.

3.1.2   CA-26 (87) Transformer Differential Relay

The CA-26 is a similar design to the CA-16 bus differential relay, but with the ability to accommodate all of the necessary input currents for protecting a three-winding transformer. With no taps, the relay requires auxiliary current transformers in two of the input circuits to produce a match. One relay per phase, of course, is required for a three-phase transformer. Figure 10-10 describes the mechanical configuration used with this relay and the principle by which restraint is produced. It is sensitive and reasonably fast, but its lack of harmonic restraint leads one to the selection of more modern relays.

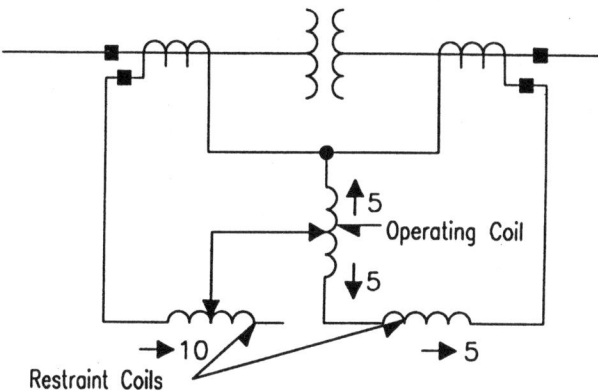

Figure 10-8   Distribution of currents in the type CA relay set on the 5–10 taps.

3.1.3   HU and HU-1 Transformer Differential Relays

Since magnetizing inrush current has a high harmonic content, particularly the second harmonic, this second harmonic can be used to restrain and thus desensitize a relay during energization. The method of harmonic restraint is not without its problems. There must be enough restraint to avoid relay operation on inrush, without making the relay insensitive to internal faults that may also have some harmonic content.

The HU (two restraining winding) and HU-1 (three restraining winding) variable-percentage differential relays have second-harmonic restraint supervision that adequately solves these problems. The connections for these relays are shown schematically in

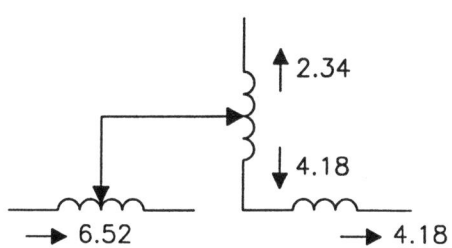

Figure 10-9   Distribution of currents in the type CA relay set on the 5–8 tap for example of Figure 10-14.

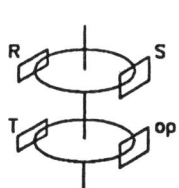

a) 4- Electromagnets Operating on Two Discs Which are Fastened to The Same Shaft.

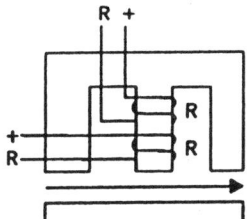

b) Each Restraint Electromagnet With 2 Primary Coils and a Secondary Coil on its Center Leg (Secondary Coil not Shown).

c) Additive or Subtractive Restraining Effect Produced by Two Coils on Same Leg.

Figure 10-10   Type CA-26 transformer differential relay.

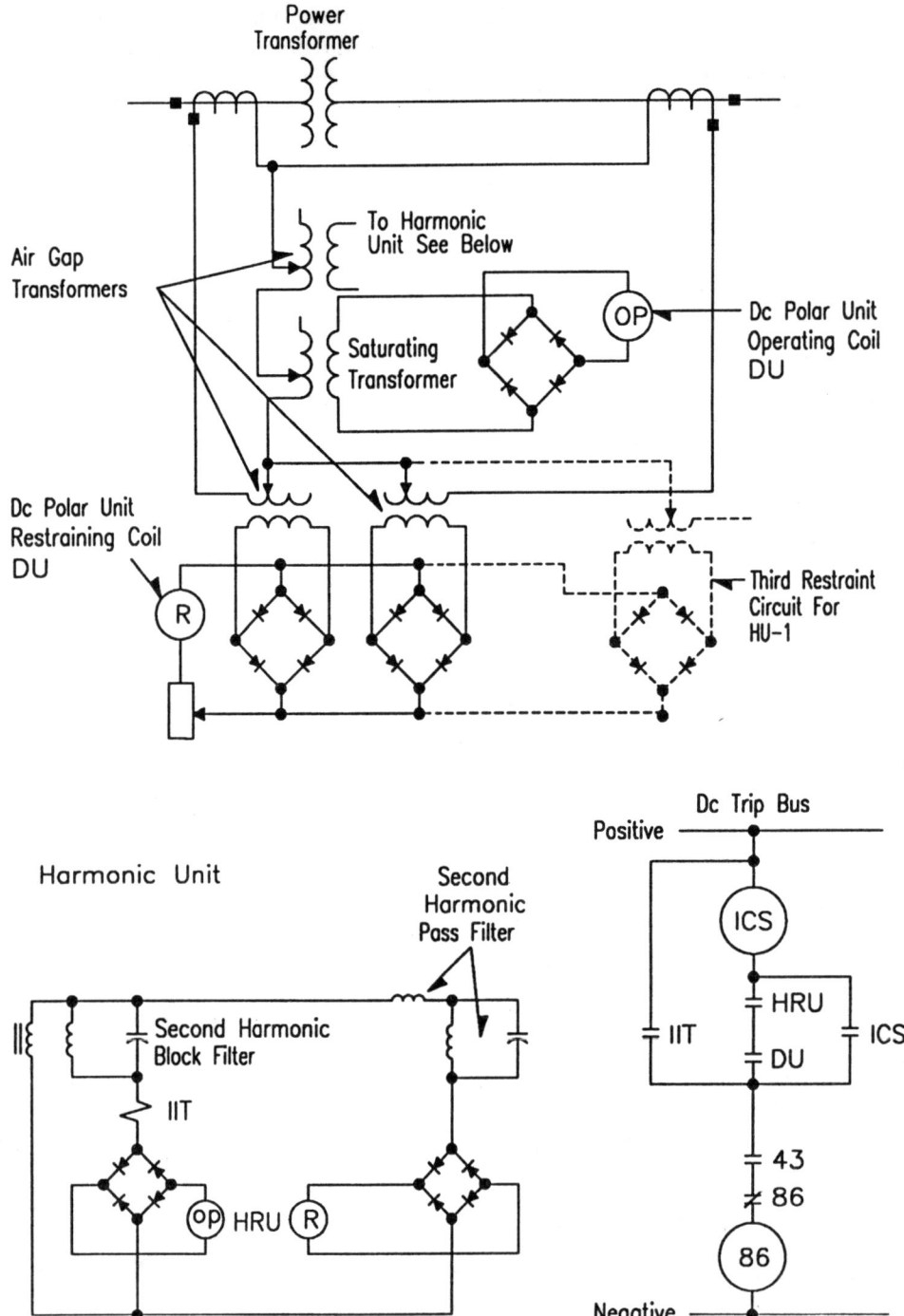

Figure 10-11 Schematic connections of the HU and HU-1 variable percentage differential relays with second harmonic restraint supervision.

Figure 10-11. In the differential unit (DU), air-gap transformers feed the restraint circuits, and a non-air-gap transformer energizes the operating coil circuit. Since the rectified restraint outputs are connected in parallel, the relay restraint is proportional to the maximum restraining current in any restraint circuit.

The percentage characteristic varies from around 20% on light faults, where current transformer performance is good, to approximately 60% on heavy fault, where current transformer saturation may occur. This variable-percentage characteristic is obtained via the saturating transformer in the operating coil circuit. Taps provide a 3:1 difference in the ratio of the main current transformer outputs. These taps are 2.9, 3.2, 3.5, 3.8, 4.2, 4.6, 5.0, and 8.7.

The minimum pickup current is 30% of the tap value for the 30% sensitivity relay and 35% of tap value for the 35% sensitivity relay. The minimum pickup is the current that will just close the differential unit contacts, with the operating coil and one restraint coil energized. The continuous rating of the relay is 10 to 22 A, depending on the relay tap used.

The harmonic restraint unit (HRU) has a second-harmonic blocking filter in the operating coil circuit and a second-harmonic pass filter in the restraint coil circuit. Thus, the predominant second-harmonic characteristic of an inrush current produces ample restraint with minimum operating energy. The circuit is designed to hold open its contacts when the second-harmonic component is higher than 15% of the fundamental. This degree of restraint in the HRU is adequate to prevent relay operation on practically all inrushes, even if the differential unit should operate.

For internal faults, ample operating energy is produced by the fundamental frequency and harmonics other than the second. The second harmonic is at a minimum during a fault. Since the HRU will operate at the same pickup as the DU, the differential unit will operate sensitively on internal faults, as shown in the trip circuit of Figure 10-11. For external faults, the differential unit will restrain.

The relay operating time is one cycle at 20 times tap value. The instantaneous-trip unit (IIT) is included to ensure high-speed operation on heavy internal faults, where current transformer saturation may delay HRU contact closing. The IIT pickup is 10 times the relay tap value. This setting will override the inrush peaks and maximum false differential current on external faults.

3.1.4 HU-4 Transformer Differential Relay

The HU-4 relay is used to protect multiple-winding transformer banks or in the protective zone that includes the bus. The HU-4 relay is similar to the HU and HU-1 relays, but has four restraint windings. Also, the rectified outputs of the restraint transformers are connected in series, and the IIT unit is set at 15 times the tap value. The application of this relay is described in Chapter 11.

3.1.5 Modified HU Relays

In the HU relay, the 15% second-harmonic restraint value is based on a minimum second-harmonic content 15% of fundamental, under an inrush condition of $\phi_S = 1.40$ p.u. and a $0°$ closing angle. In modern transformers, however, saturation density is more often 1.20 to 1.30 p.u. and can even be as low as 1.0 p.u. At these lower saturation densities, the minimum second-harmonic content of the fundamental is significantly lower. Also, service conditions are more severe when the closing angle is $\pm 30°$, rather than $0°$. (See Table 10-1.)

As a result, the second-harmonic content percentage may be as low as 7% of the fundamental, and the percentage of all harmonics may be as low as 7.5% of the fundamental. Under these conditions, the 15% second-harmonic restraint relays may not function properly during the energization of power transformers with low saturation density values.

The modified HU relay is designed to solve this problem. This relay is similar to the 15% second-harmonic HU relay, except that a 33-Ω, 3-W resistor is connected across the HRU operating coil to calibrate the unit for a 7.5% second-harmonic restraint. The characteristics of the modified HU are the same as for the HU relay, and the modification does not affect the characteristics of the IIT unit. The differential unit of the modified relay, however, is about one cycle slower than the unmodified unit.

Ths modified HU relay has been used successfully in inrush current tests at a 138-kV generating station and in several installations where the 15% HU relays experience inrush problems.

3.1.6 Type RADSB Transformer Differential Relay

The type RADSB transformer differential relay is a solid-state three-phase package. Its basic version provides two restraining circuits for two-winding power transformer protection. It can be expanded up to six restraints initially or in the future.

As shown in Figure 10-12, the relay utilizes the second harmonic for inrush current restraint. The relay will restrain if the second harmonic content in any one phase is greater than 17% of its fundamental. This feature is very unique in three-phase package design. It will solve the inrush current problem mentioned in Section 3.1.5, "Modified HU Relays."

The relay utilizes the fifth harmonic for the over-excitation restraint. The relay restrains if the fifth-harmonic content is greater than 38% of its funda-

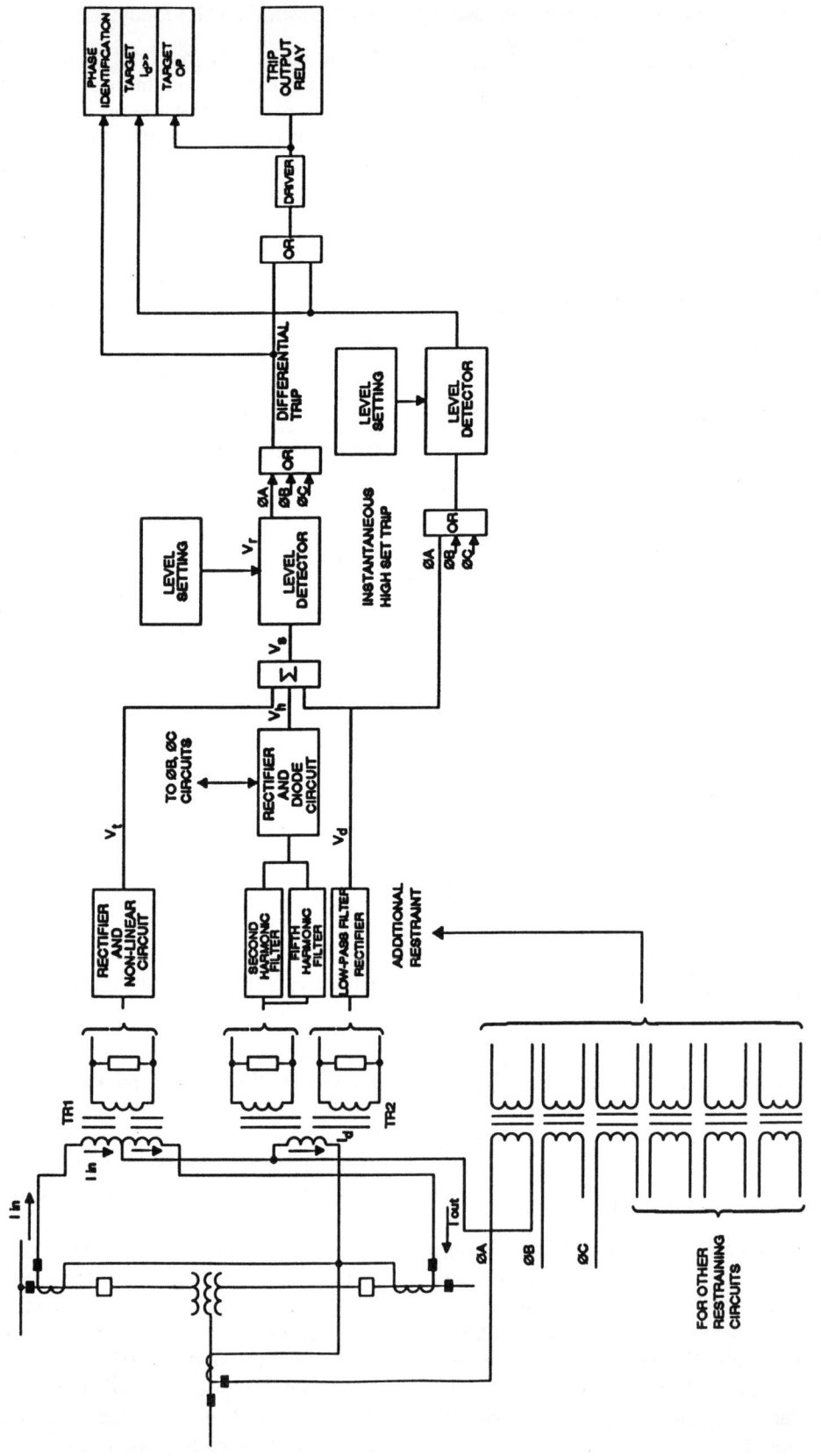

Figure 10-12 Block diagram of RADSB transformer differential relay.

mental. However, refer to Section 5.5, "Overexcitation Protection of a Generator-Transformer Unit," in this chapter. It is necessary to apply the fifth harmonic with caution, or a V/Hz relay, either type MVH or RATUB, should be considered for supervision.

The relay does not provide built-in taps; therefore, it requires external auxiliary ct's for current matching in all applications. Instantaneous high set trip function is also built in the relay.

For applications that required more than two restraining circuits, the RTQTB-061 unit(s) can be added. Each RTQTB-061 unit contains six transformers for two additional restraining circuits, as shown in Figure 10-12.

3.1.7 TPU 2000R Transformer Protection System

This unit takes full advantage of microprocessor technology. It allows the selection of a wide variety of characteristics such as second-harmonic restraint, all-harmonic restraint, and fifth-harmonic restraint. Inputs to it may be from wye- or delta-connected current transformers, irrespective of the transformer winding connections. This allows monitoring of the individual input phase currents rather than a combination of them, as a delta-connected set of ct's would provide. The relay algorithm assures a proper match of the input currents from both (or all three) sides of the transformer. It is compatible (different styles) with a two- or three-winding transformer.

Any conceivable overcurrent unit curve shape can be chosen as a setting for the overcurrent functions of this relay. Similarly, any of the communication protocols that are in popular use can be provided as an inherent part of this device. Operating curve slope and minimum trip level are also obtained by setting. Inrush monitoring is possible, with blocking of tripping being selectable by second harmonic, fifth harmonic, or all harmonics (through the eleventh). Historically, each concept has been used successfully.

Though all of these methods of recognizing inrush as a distinctive phenomenon for which trip blocking is mandatory are useful, each has its own favorable and unfavorable nuances. Second-harmonic blocking is minimally less secure for those cases involving overvoltage. The higher the voltage upon energization, the lower the percentage of second harmonic. In general, the relay is not capable of tripping at an overvoltage level below which the typical transformer can support on a prolonged basis. The relay expresses appropriately the need to trip. However, it would trip far too soon.

Fifth harmonic blocking is favored for those overvoltage cases (long EHV line energization, hydromachine load rejection, etc.) where undesired high-speed tripping on elevated magnetizing current may occur. It should be recognized that while transformers can support short-time overvoltage, they are vulnerable to prolonged overvoltage heating. If differential tripping is to be blocked when inordinate fifth harmonic is observed on overvoltage, the block must be released (or tripping imposed by other means) prior to the occurrence of damage to the transformer.

All-harmonic blocking is moderately less dependable because of the increased harmonics that are present in arcing faults.

Cross-blocking, the feature that blocks all differential tripping when the harmonic restraint setting and the operating current are exceeded in any one or more phases, increases security against an unwarranted operation during inrush. This feature is partially inherent where ct's are traditionally connected in delta for balancing the transformer phase shift and is useful with the harmonic-restraint mode of second or second and fifth. For an internal ground fault, possible elevated voltage on an *unfaulted* phase should be investigated to assure that excessive fifth harmonic in that phase cannot block tripping of the *faulted* phase differential unit. Where this can occur, cross-blocking must *not* be used.

3.2 General Guidelines for Transformer Differential Relaying Application

The following guidelines are designed to assist in selecting and applying relays for transformer protection. When two or more relays appear to be equally suitable, engineering experience and economics will determine the final choice.

1. There is no clearcut answer to the question of which relay or protective method to apply. As a general rule, however, the induction-disk differential relays (CA and CA-26) are used at substations remote from large generating sources where inrush is not a problem and the kVA size of the bank is relatively small. The more complex and more expensive harmonic relays (HU, HU-1, HU-4, TPU and RADSB) are used at generating stations and for large transformer units located close to generating sources, where a severe inrush is highly likely.

2. A current transformer tap that will give approximately 5 A at maximum load is recommended for use with multiratio current transformers. This

arrangement provides good sensitivity without introducing thermal problems in the current transformer, leads, or relay itself. Sensitivity can be improved by using a tap that gives more than 5 A; however, the current transformer, leads, and relay capability must be checked carefully to guard against thermal overload.

3. In general, for all except the TPU, the current transformers on the wye side of a wye-delta bank must be connected in delta, and the current transformers on the delta side connected in wye. This arrangement (1) compensates for the 30° phase-angle shift introduced by the wye-delta bank and (2) blocks the zero sequence current from the differential circuit on external ground faults. As shown in Figure 10-13, zero sequence current will flow in the differential circuit for external ground faults on the wye side of a grounded wye-delta bank; if the current transformers were connected in wye, the relays would misoperate. With the current transformers connected in delta, the zero sequence current circulates inside the current transformers, preventing relay misoperation.

4. Relays should be connected to receive "in" and "out" currents that are in phase for a balanced load condition unless the relay itself is designed or set to accommodate the difference. When there are more than two windings, all combinations must be considered, two at a time.

5. Relay taps or auxiliary current transformer ratios should be as close as possible to the current ratios for a balanced maximum load condition. When there are more than two windings, all combinations must be considered, two at a time, and based on the same kVA capacity.

6. Ground only one point in the differential scheme; never do multiple-point grounding.

7. After the current transformer ratios and relay taps have been selected, the continuous rating of relay windings should be checked for compatibility with the transformer load. If the relay current exceeds its continuous rating, a higher-current transformer ratio or relay tap may be required. The relay's required continuous rating may be determined from the maximum kVA capacity of the transformer bank. If the transformer is allowed to exceed its maximum kVA capacity for a short time, the expected 2-h maximum load should be used. The relay will reach final temperature within 2 h.

8. The percentage of current mismatch should always be checked to ensure that the relay taps selected have an adequate safety margin. When necessary, current mismatch values can be reduced by changing current transformer taps or adding auxiliary current transformers. Percentage mismatch M can be determined from Eq. (10-1):

$$M = \left| \frac{\left| \frac{I_L}{I_H} - \frac{T_L}{T_H} \right|}{S} \right| \times 100\% \tag{10-1}$$

where

I_L, I_H = relay input currents, at the same kVA base, for low- and high-voltage sides, respectively

T_L, T_H = relay tap settings for low- and high-voltage sides, respectively

S = smaller of the two terms, (I_L/I_H) or (T_L/T_H)

When there are more than two windings, all combinations should be calculated, two at a time. When taps are changed under load, the relays should be set on the basis of the middle or neutral tap position. The total mismatch, including the automatic tap change, should not exceed the recommended values shown in Table 10-2.

For example, for a transformer bank with a $\pm 10\%$ on-load tap changer device, the calculated mismatch value should not be greater than $\pm 5\%$ for a 30% HU relay application. However, if the transformer bank does not have an on-load tap changer, then the calculated mismatch value can be tolerated up to the "limit of (M + LTC)" value.

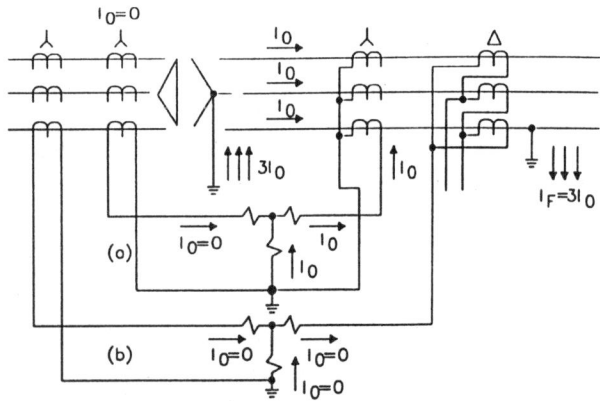

a) I_0 Will Flow in The Differential Circuit if CT's are Connected in Wye Causing Misoperation. DO NOT use This Connection.

b) No Current Will Flow in The Differential Circuits if CT's are Connected in Delta. This is The Recommended Connection for all but the TPU.

Figure 10-13 Reason for delta-connected ct's on wye windings.

Table 10-2 Recommended Mismatch (M) Limitation

Relay	Sensitivity (%)	Limit of (M + LTC) (%)
CA	50	35
HU, HU-1, HU-4, TPU	30	15
HU, HU-1, HU-4, TPU	35	20
CA-26, RADSB	—	10

9. To ensure correct operation of the relaying scheme, the current transformer performance should be checked. A less accurate, but still acceptable, method is to use the ANSI relaying accuracy classification. For units with class C accuracy, performance will be adequate if

$$\frac{N_p V_{CL} - (I_{ext} - 100) R_S}{I_{ext}} > Z_T \qquad (10\text{-}2)$$

where

N_p = proportion of total current transformer turns in use, for example, if 1000/5 tap is used for a 2000/5 MR current transformer, then $N_p = 0.5$

V_{CL} = current transformer accuracy, class C voltage; for example, 200 for a class C200 current transformer

I_{ext} = maximum external fault current in secondary rms A (let $I_{ext} = 100$ if maximum external fault current is less than 100 A)

R_s = current transformer secondary winding resistance in ohms

Z_T = total current transformer secondary circuit burden impedance in ohms, determined by equation:

$$Z_T = 1.13(m \times R_L) + \text{relay burden} + Z_A \qquad (10\text{-}3)$$

where

R_L = one-way lead resistance

1.13 = multiplier used to accommodate temperature rise of the conductors during faults

Z_A = burden impedance of any devices (other than the relay) connected or reflected to the current transformer secondary circuit

m = multiplier, depending on the current transformer connection and type of fault to be considered, as shown in Table 10-3

Table 10-3 Multiplier (m) for Eq. (10-3)

	$3\phi F$	ϕGF
Wye-connected ct	m = 1	m = 2
Delta-connected ct	m = 3	m = 2

4 SAMPLE CHECKS FOR APPLYING TRANSFORMER DIFFERENTIAL RELAYS

The following examples show the importance of the current transformer connections, current ratios, relay ratings, and current transformer performance in applying the differential relay scheme for transformer protection.

4.1 Checks for Two-Winding Banks

A worksheet for connecting differential relays around a two-winding bank is shown in Figure 10-14. This does not apply to the TPU relay.

4.1.1 Phasing Check

It is very important to note that the transformer bank, as shown in Figure 10-14, is connected so that the high side lags the low side by 30°, which is not an ANSI-standard-connected bank. (In a standard connection, the high side leads the low side by 30°.) In this example, the nonstandard connection is used for illustration only. Practically, the actual connection of the bank should be confirmed with the information from its nameplate before proceeding to the next step of the phasing check.

Procedures for phasing check can be simplified as below (refer to Figures 10-14 and 10-15).

Step 1 Assume that $I_a, I_b,$ and I_c on the wye side flow through the bank to an external three-phase fault or maximum load.

Step 2 See Figure 10-14; use the statement on page 12 to define the current in the windings: "the current flowing out at the polarity-marked terminal on the secondary side is substantially in phase with the current flowing in at the polarity-marked terminal on the primary side."

Three-phase transformers do not carry polarity marks as do their single-phase counterparts. However, from the nameplate data for the transformer, it can clearly be seen which of the low-voltage windings is drawn in parallel with

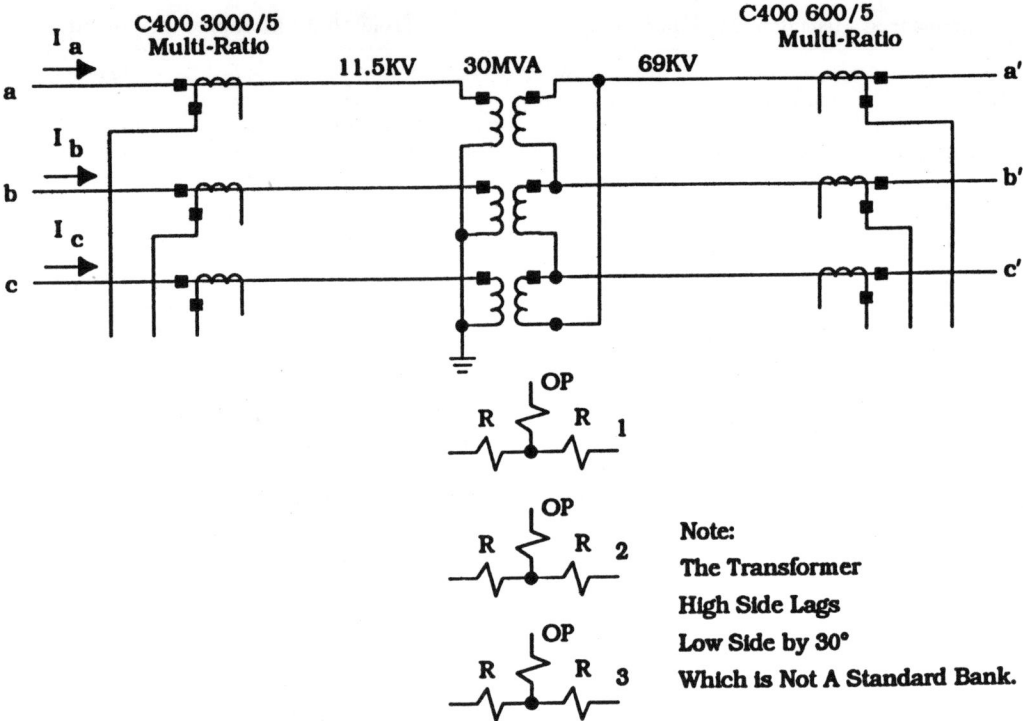

Figure 10-14 Worksheet for connecting differential relays around a two-winding transformer bank.

one of the high-voltage windings. This is a symbolic identification that these two windings are on the same core-leg. If these windings (actually lines) are vertical on the nameplate diagram, it may be assumed that the polarity markings could be placed at the upper end of each of these windings. Similar identification can be made of the other windings. The terminals of a three-phase transformer are identified as H1, H2, H3, and, if there is a neutral, H0 (note this says *terminals*, not *windings*). The low-voltage terminals are called X1, X2, and X3 (and possibly X0). If there is a third set of windings and the terminals are brought out of the case, they will be identified as Y1, Y2, and Y3 (and again Y0 if there is a neutral and it is brought out of the case).

The nameplate will show whether the voltage drop from H1 to H3 is in-phase with X1 to X0 (high leads low by 30°), or H1 to H2 is in-phase X1 to X0 (high lags low by 30°) or some other combination. It is assumed that the system "A-phase" on the high-voltage side will be connected to the H1 terminal and that the

"A-phase" on the low-voltage side will be connected to the X1 terminal, but surprisingly this is seldom the case. This simply adds to the complexity of the relay engineer's task, sorting out all of these factors. In fact, the transformer connection dictates the relationship between the high- and low-voltage systems, and the phasors can be named anything with which the user feels comfortable.

Step 3 Trace these currents through the delta to the delta-side phase wires, then through the wye-connected current transformers to the relays.

Step 4 Repeat the above relay currents to the other-side restraint windings.

Step 5 Trace the currents on the wye-winding phase wires; then determine the secondary currents (phase and direction) on each wye-side current transformer.

Step 6 Match up the information from steps 4 and 5, enabling the wye-side current transformers to be properly connected in delta for correct phasing under all conditions. The completed check for the example of Figure 10-14 is shown in Figure 10-15.

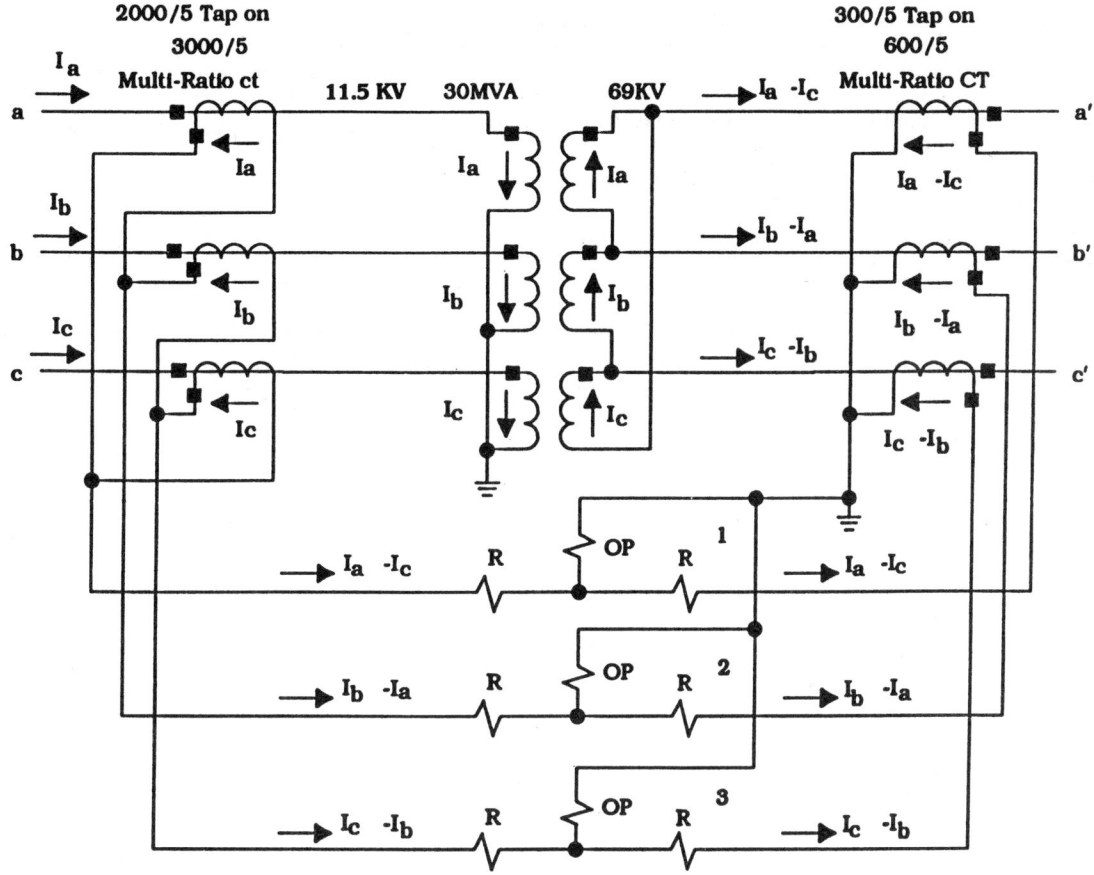

Figure 10-15 Complete phasing check for the example of Figure 10-14.

4.1.2 Ratio Check

For the example in Figure 10-14, the ratio check should be executed as shown in Table 10-4. The steps that follow depend on the type relay being applied: CA (steps 7 to 10); CA-26 and RADSB (steps 11 to 13); or HU (steps 14 to 17):

4.1.3 For Type CA Relays (Steps 7 to 10)

For the application of CA relays, the tap settings, continuous coil ratings, and mismatch must also be checked as described in steps 6 to 9. The taps in CA relays are 5-5, 5-5.5, 5-6.6, 5-7.3, 5-8, 5-9, and 5-10. Tap ratios are 1.00, 1.10, 1.32, 1.46, 1.60, 1.80, and 2.00, respectively.

Step 7 Select relay taps with a ratio as close as possible to the relay current ratio in step 5. In this case, tap 5-8, with a ratio of 1.60, is the closest.

Step 8 Connect relay terminal 9 to the 69-kV side and terminal 7 to the 11.5-kV side. Always set the time dial at position number 1.

Step 9 Check the continuous rating of the relay coils. As shown in calculations, the continuous currents flowing in the restraint coils are less than 10 A, and any through currents flowing in the operating coil are less than 5 A. Therefore, the relay windings will not be subject to a thermal problem.

Step 10 Calculate mismatch; use Eq. (10-1). This mismatch is well within the 35% mismatch limit of Table 10-2.

4.1.4 For Type CA-26 and RADSB Relays (Steps 11 to 13)

To calculate mismatch for CA-26 relays, perform steps 11 to 13 as follows:

Table 10-4 Example of Ratio Check for Two-Winding Transformer

Step	LV (wye)	HV (delta)
1. For the example shown in Figure 10-14, assume that the maximum load is 30,000 kVA. Then the rating of the bank I_{FL} is	$\dfrac{30,000}{\sqrt{3}\times 11.5}=1506\,A$	$\dfrac{30,000}{\sqrt{3}\times 69.0}=251\,A$
2. For increased sensitivity, select current transformer ratios as close to the I_{FL} value as possible. Practically, a calculated value of $(I_{FL}/0.8)$ can be used as the reference for determining the current transformer ratios for this example. Then use	$\dfrac{1506}{0.8}=1882.5$ $n=\dfrac{2000}{5}$ $=400$	$\dfrac{251}{0.8}=313.7$ $n=\dfrac{300}{5}$ $=60$
3. Calculate current transformer secondary currents $I_S=(I_{FL}/n)$	$=\dfrac{1506}{400}$ $=3.77\,A$	$=\dfrac{251}{60}$ $=4.18\,A$
4. Calculate relay current	$I_{RL}=3.77(\sqrt{3})$ $=6.52\,A$	$I_{RH}=4.18\,A$
5. Calculate relay current ratio	$\dfrac{I_{RL}}{I_{RH}}=\dfrac{6.52}{4.18}=1.560$	

Step 11 Calculate mismatch; use Eq. (10-1).

$$M=\frac{\frac{6.52}{4.18}-1}{S}\times 100\%$$
$$=\frac{1.560-1}{1}\times 100\%=56\%$$

Step 12 Since the percent mismatch is higher than the recommended value (Table 10-2), an auxiliary current transformer or a current-balancing autotransformer is required to decrease the current to the relay. (*Note*: Use auxiliary current transformer to decrease the current to the relay.) The turns ratio of the balancing current transformer is $(4.18/6.52)\times 100\%$, or 64.1%. Either a two-winding auxiliary current transformer with a turns ratio of 3/2 (for example, ABB style no. 7881A026G06), or a current-balancing autotransformer (for example, ABB type A auxiliary current transformers) would be satisfactory.
The connections for these two transformers are shown in Figure 10-16. The continuous rating of the auxiliary current transformers should be checked to guard against possible thermal problems.

Step 13 After selecting the auxiliary current transformer ratio, the mismatch should be rechecked.

If a 3/2 ratio auxiliary current transformer is used, refer to Figure 10-16 and apply Eq. (10-1):

$$M=\frac{\frac{4.34}{4.18}-1}{1}\times 100\%=3.80\%$$

If a current-balancing autotransformer is used, refer to Figure 10-16 and apply Eq. (10-1):

$$M=\frac{\frac{4.17}{4.18}-1}{1}\times 100\%=0.24\%$$

Both values are well within the 10% mismatch limit of Table 10-2.

4.1.5 For Type HU Relays (Steps 14 to 17)

To apply HU relays, the tap settings, continuous coil ratings, and mismatch must be checked, as described in steps 13 to 16. The taps in HU relays are 2.9, 3.2, 3.5, 3.8, 4.2, 4.6, 5.0, and 8.7. The tap ratios are given in Table 10-5.

Step 14 Select relay taps that have a ratio as close as possible to the relay current ratio in step 5. In this case, tap 5.0/3.2, with a ratio of 1.563, is the closest.

Step 15 Use tap 3.2 on the 69-kV side and tap 5.0 on the 11.5-kV side. In most HU applications, auxiliary current transformers are not required for current balancing if the current ratio is between 1 and 3.

Step 16 Check the continuous thermal rating of the relay coils. Since the continuous rating of the relay is 12 A for the 3.2 tap, there should be no thermal problem with the relay coils.

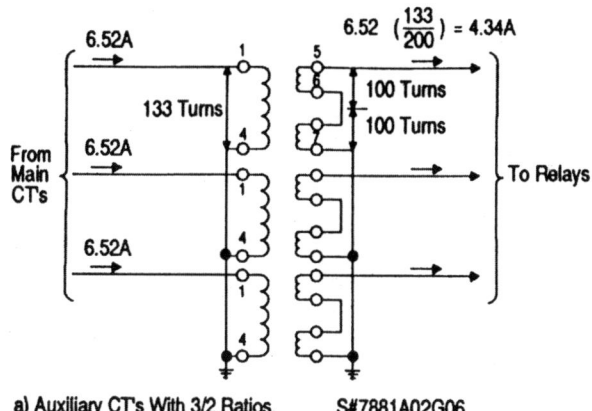

a) Auxiliary CT's With 3/2 Ratios.   S#7881A02G06

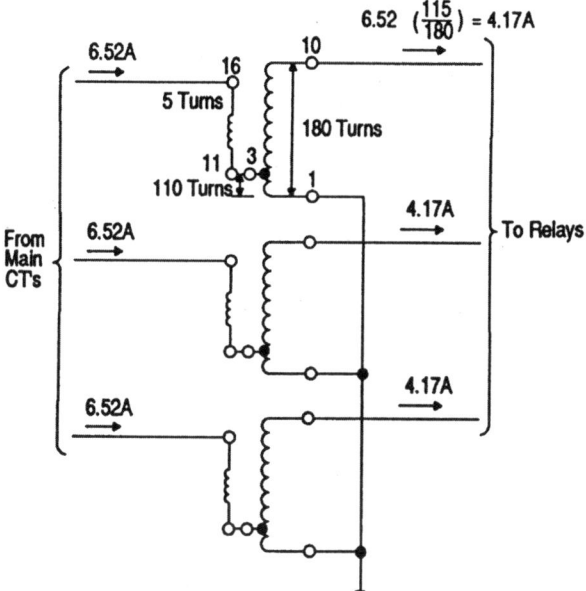

b) Type "A" Current Balancing Transformers.

Figure 10-16 Current balancing transformer connections for Figure 10-15 when type CA-26 or RADSB relay is used.

Step 17 Calculate mismatch:

$$M = \frac{\frac{6.52}{4.18} - \frac{5.0}{3.2}}{S} \times 100\%$$

$$= \frac{1.560 - 1.563}{1.560} = 100\% = 0.2\%$$

This value is well within the 15% mismatch limit of Table 10-2. Also, the 30% sensitivity of the HU relay would be satisfactory for this application.

4.1.6 For Type TPU Relays

The TPU is applied similarly to the HU relays with added benefits. Its wider tap range and finer tap increments yield an even smaller mismatch than that achieved with the HU relay. Also, the ct's can be connected in wye on both sides of the transformer, with compensation for transformer phase shift accomplished internally for phase current metering and overcurrent protection on the wye side of the transformer. For the example in Table 10-4, the tap calculations and selections would be the same as that for the HU relay. The $\sqrt{3}$ factor for I_{RL} would still apply because of the TPU internal compensation and not because of delta-connected ct's.

4.1.7 Current Transformer Performance Check

Assume that the three-phase external fault currents are higher than the single-phase fault currents, with values of 15,000 A on the 11.5-kV side and 2500 A on the 69-kV side. In this case, the current transformer burden limit can be calculated as follows:

	Low voltage	High voltage
Maximum external fault current (primary amperes), I_p	15,000	2500
Current transformer turns ratio, n	400	60
Secondary amperes $\dfrac{I_p}{n}$	37.5	41.7
N_P	$\dfrac{2000}{3000} = 0.67$	$\dfrac{300}{600} = 0.50$

From Eq. (10-2) where $(I_{ext} - 100)R_S = 0$ since I_{ext} is less than 100 A secondary (Figure 10-14 shows the

Table 10-5 HU Relay Tap Ratios

	2.9	3.2	3.5	3.8	4.2	4.6	5.0	8.7
2.9	1.000	1.103	1.207	1.310	1.448	1.586	1.724	3.000
3.2		1.000	1.094	1.188	1.313	1.438	1.563	2.719
3.5			1.000	1.086	1.200	1.314	1.429	2.486
3.8				1.000	1.105	1.211	1.316	2.289
4.2					1.000	1.095	1.190	2.071
4.6						1.000	1.087	1.890
5.0							1.000	1.740
8.7								1.000

C400 current transformers are used in this example),

$$\frac{0.67 \times 400}{100} = 2.67\,\Omega \qquad \frac{0.5 \times 400}{100} = 2.0\,\Omega$$

Current transformer performance will be satisfactory if the total burden impedance values, as calculated from Eq. (10-3), are less than the above values.

4.2 Checks for Multiwinding Banks

The same types of phasing, ratio, continuous rating, and current transformer performance checks are used for multiwinding transformer as for two-winding transformers. To determine the correct direction and phase of the restraint currents, one side of the transformer is considered the primary and the other windings the secondaries. For ratio checks, any two windings must be checked based on a same kVA value, as if the bank were a two-winding unit with no current

in the other winding. Any other pair is then checked in the same manner. This process ensures that all ratios are correct for any distribution of fault or load current.

A worksheet showing the connection of differential relays around a typical three-winding bank is given in Figure 10-17. The completed phasing checks are shown in Figure 10-18.

If the ratios are not correct for the relay, auxiliary current-balancing autotransformers or current transformers are required. In general, one or two sets are required for the three-winding bank, depending on the unbalanced condition and relay type (CA-26, HU-1, or RADSB).

For the three-winding bank shown in Figure 10-18, the ratio check is performed per Table 10-6, steps 1 to 5.

4.2.1 For Type CA-26 and RADSB Relays (Steps 6 to 8)

For CA-26 relay application, mismatch is checked according to steps 6 to 8 below.

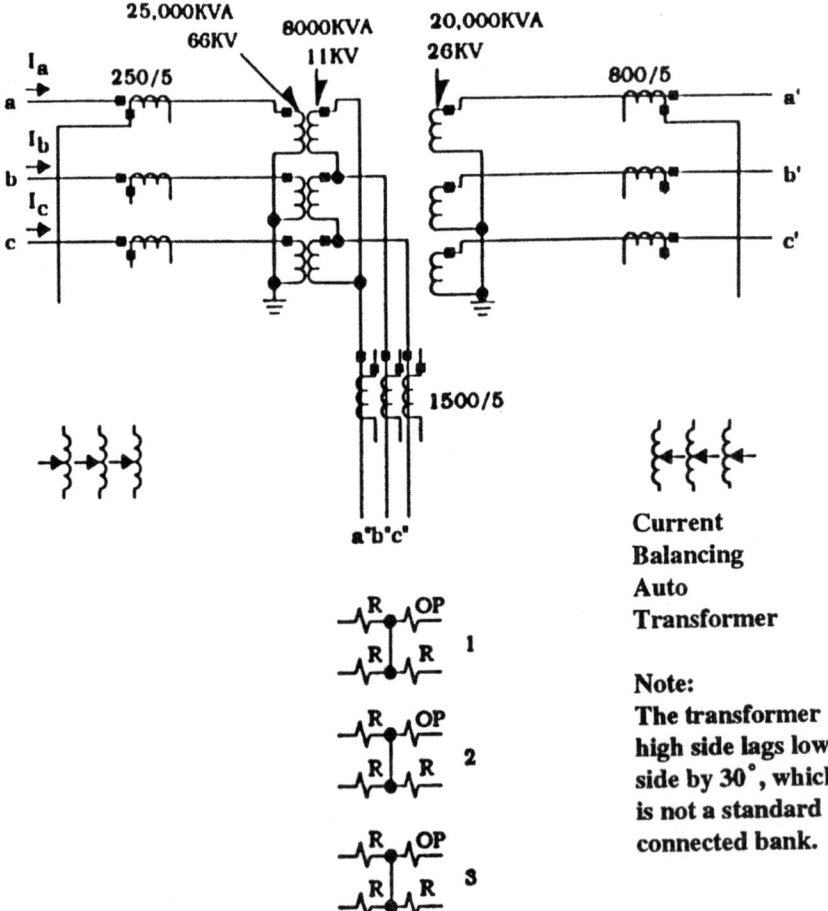

Figure 10-17 Worksheet for connecting differential relays around a three-winding transformer bank.

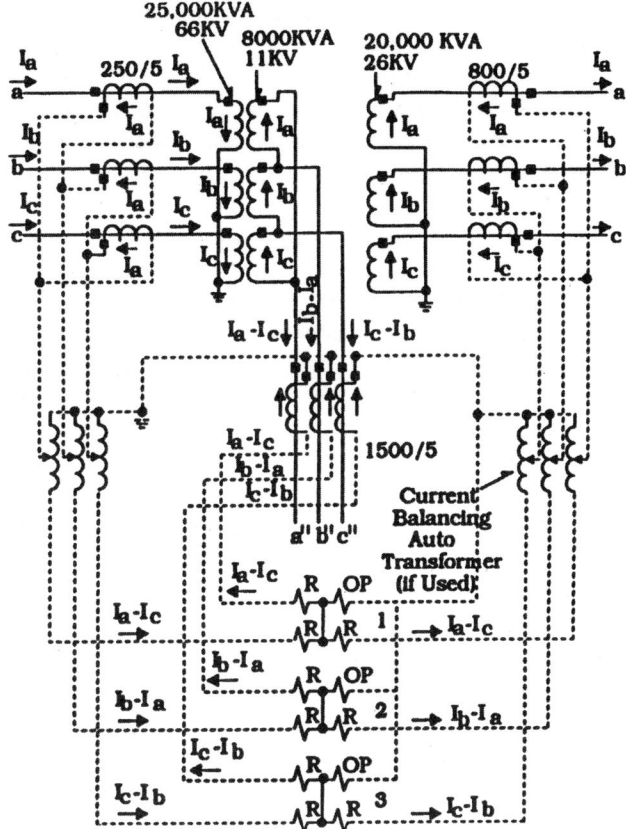

Figure 10-18 Complete phasing check for the example of Figure 10-17. The dotted lines show the connections for a phasing check between the 66 and 11 KV windings. Assuming the 26 KV circuit does not exist. The dashed lines show the connections for a phasing check between the previous connections made for the 66 KV winding, assuming that the 11 KV circuits do not exist. With this method, phasing is correct for any distribution of currents through the three windings.

Step 6 Calculate mismatch; use Eq. (10-1):

$$M_{HM} = \frac{\frac{7.577}{6.012} - 1}{1} \times 100\% = 26\%$$

$$M_{ML} = \frac{\frac{6.012}{4.380} - 1}{1} \times 100\% = 37.3\%$$

$$M_{HL} = \frac{\frac{7.577}{4.380} - 1}{1} \times 100\% = 73\%$$

Step 7 To decrease the currents to the relay, add current-balancing autotransformers at the 66- and 26-kV sides. The turn ratios of these transformers are

$$\frac{4.380}{7.577} \times 100\% = 57.8\%$$

Use 73 turns ratio.

$$\frac{4.380}{6.012} \times 100\% = 72.8\%$$

Use 58 turns ratio.

Then the currents to the relay are

$$I_{RH} = 7.577 \times 58\% = 4.395\,A$$

$$I_{RM} = 6.012 \times 73\% = 4.389\,A$$

$$I_{RL} = 4.380\,A$$

Step 8 Recalculate the mismatch:

$$M_{HM} = \frac{\frac{4.395}{4.389} - 1}{1} \times 100\% = 0.1\%$$

$$M_{ML} = \frac{\frac{4389}{4.380} - 1}{1} \times 100\% = 0.2\%$$

$$M_{HL} = \frac{\frac{4.395}{4.380} - 1}{1} \times 100\% = 4.4\%$$

Table 10-6 Example on Ratio Check for Three-Winding Transformer

Step	66 kV (Y)	26 kV (Y)	11 kV (Δ)
1. If 25,000 kVA flows through the bank, the currents in each winding are I_{FL}	$\frac{25,000}{\sqrt{3} \times 66} = 219$	$\frac{25,000}{\sqrt{3} \times 26} = 556$	$\frac{25,000}{\sqrt{3} \times 11} = 1314$
2. If we assume current transformer turn ratios of	$n = 250/5$ $= 50$	$= 800/5$ $= 160$	$= 1500/5$ $= 300$
3. Then the current transformer secondary currents are I_{FL}/n	$= 219/50$ $= 4.380$	$= 556/160$ $= 3.475$	$= 1314/300$ $= 4.380$
4. Relay currents are	$I_{RH} = \sqrt{3} \times 4.380 = 7.577$	$I_{RM} = \sqrt{3} \times 3.475 = 6.012$	$I_{RL} = 4.380$
5. Relay current ratios are	$\frac{I_{RH}}{I_{RH}} = \frac{7.577}{6.012} = 1.260$	$\frac{I_{RM}}{I_{RL}} = \frac{6.012}{4.380} = 1.373$	$\frac{I_{RH}}{I_{RL}} = \frac{7.577}{4.380} = 1.730$

4.2.2 For Type HU-1 Relays (Steps 9 to 12)

For HU-1 relay application, tap settings, mismatch, and current transformer performance are calculated as follows:

Step 9 To select the relay taps, use Table 10-5 and start from the highest current ratio, $I_{RH}/I_{RL} = 1.730$. The nearest tap ratio is $5/2.9 = 1.724$. Select $T_H = 5$ and $T_L = 2.9$. Next, select the tap ratio for the second higher-current ratio, $I_{RM}/I_{RL} = 1.373$, by using the lower tap from the first tap ratio ($T_L = 2.9$) as a reference. This ratio would be $3.8/2.9 = 1.310$. In other words, $T_M = 3.8$.

Step 10 Calculate the mismatch:

$$M_{HM} = \frac{\frac{7.577}{6.012} - \frac{5}{3.8}}{1.26} \times 100\% = 4.4\%$$

$$M_{ML} = \frac{\frac{6.012}{4.380} - \frac{3.8}{2.9}}{1.31} \times 100\% = 4.8\%$$

$$M_{HL} = \frac{\frac{7.577}{4.380} - \frac{5}{2.9}}{1.72} \times 100\% = 0.3\%$$

Step 11 Check current transformer performance in the same way as for the two-winding bank above.

Step 12 Even though the current ratios are within 3.0 (step 5), auxiliary current transformers may be required for current balancing. For example, if $I_{RH} = 3.75$, $I_{RM} = 8.109$, and $I_{RL} = 6.222$, the current ratios are 2.162, 1.303, and 1.659. Even though these values are within 3, auxiliary current transformers are necessary in this application.

4.2.3 For Type TPU Relays

Again, the TPU would be applied similarly to the HU relay. In the three-winding TPU, all ct's must be connected in wye, resulting in a $\sqrt{3}$ factor being applied to I_{RH} and I_{RM} due to internal compensation as they were applied with the HU due to the ct's being connected in delta.

4.3 Modern Microprocessor Relay

The modern microprocessor relay, such as the TPU relay, is a protection system, rather than a simple transformer differential overcurrent relay. Settings are required for phase and ground (time and instantaneous) as well as negative sequence overcurrent functions. For the differential function, the tap ranges are wider and the steps are smaller than those employed in electromechanical or solid-state relays. Differential comparisons are made between the I/tap values of all currents entering and leaving the relay. The taps are chosen to be proportional to the (actual or adjusted) input currents to the relay for a "through" fault (or load). Contrary to the traditional method of selecting the current transformer connections to deliver currents to the relay with appropriate phase relationships, with this relay flexibility exists in those connections. Factors are applied in setting the relay that compensate for the connections of the transformer and the ct's on the high and low (and third winding where present) sides of the protected transformer. The selection is based on the angle by which the winding-1, typically high voltage, input current leads the comparable winding-2, typically low voltage (and winding-3, typically tertiary voltage, where applicable) outgoing current.

The differential slope (I_{OP}/I_R) of the relay is linear (15 to 60%) with a minimum operating current. Both slope and minimum trip level are selectable. Typically, a slope of 20 or 30% is selected, depending on whether or not the transformer is equipped with a load-tap changer. A minimum operating current of 20 to 40% of tap value may be chosen, with 30% being a traditional and successful setting.

The relay system provides load current metering (including peak demand), menu-driven programming of settings, three groups of setting tables, programmable input and output contacts (with and without time delay), oscillographic data storage, fault records, and communications ports.

5 TYPICAL APPLICATION OF TRANSFORMER PROTECTION

5.1 Differential Scheme with Harmonic Restraint Relay Supervision

By setting this relay with second- and fifth-harmonic blocking, it will refrain from tripping on the disparity between incoming and outgoing currents that result from inrush or exciting current due to overvoltage. Any combination of connections of a transformer can be accommodated. Contrary to conventional practice, the current transformers should be connected in wye on each side, irrespective of the phase shift between the currents on one side of the transformer and those on the other. Figure 10-19 describes a typical external connection of the current transformers for a TPU 2000R relay.

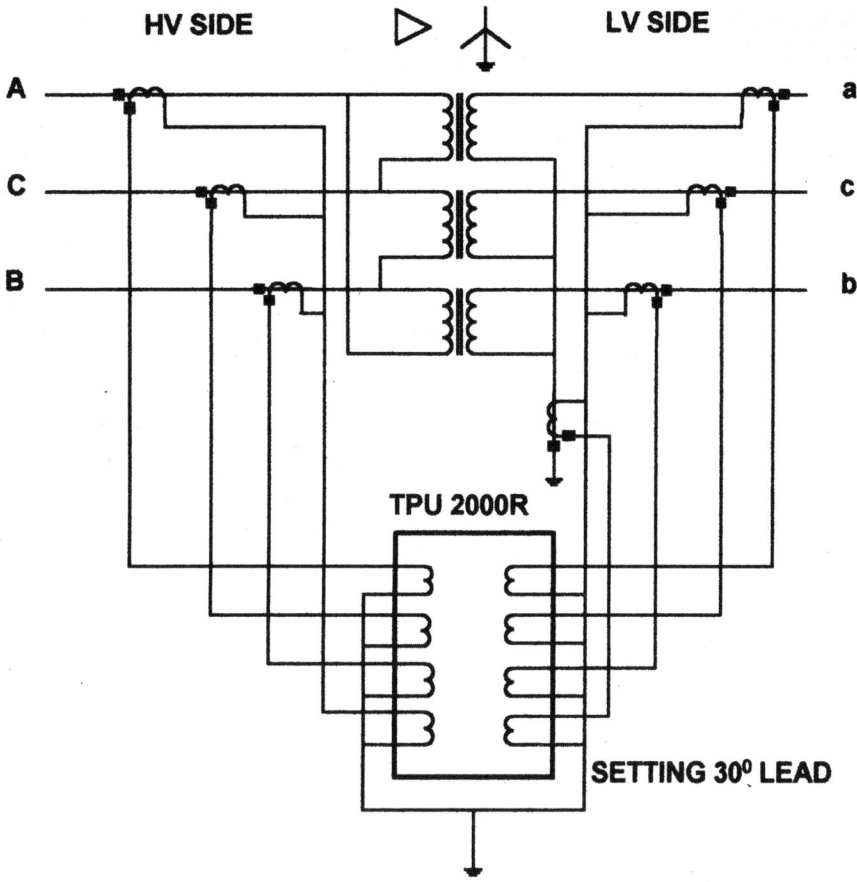

Figure 10-19 Typical external connection for TPU relay.

The appropriate phase shift is taken into consideration internally in the relay. The currents that are delivered to the relay are modified in the setting process to cause them to appear to have the same value they would have had if conventional ct connections had been used. This modifier is called the compensation factor. Figure 10-20 shows the majority of connections that are in use, the appropriate phase-angle settings, and the compensating factors that should be used. Consider, for example, that a delta connection of the ct's were in use in the conventional setup. The current delivered to the relay would be multiplied, by the connection, to be $\sqrt{3}$ times the phase current. Using wye currents only instead of the conventional delta currents, the currents are multiplied, in the setting process, by $\sqrt{3}$ (the compensating factor) to obtain equivalency. This internal compensation is only applied to the phase differential algorithm. With the wye-wye connection, where the conventional practice would dictate the ct's be connected in delta, the compensating factors of $\sqrt{3}$ and $\sqrt{3}$ are recom-

mended. This preserves the proportionality between the input and output currents and is consistent with the other combinations. Similar settings to those applied for two-winding transformers are used for three-winding transformers.

5.2 Ground Source on Delta Side

As shown in Figure 10-21a, the differential relay will operate falsely on external ground faults if the differential zone covers a grounding bank and a conventional wye-connected current transformer set is used. This misoperation can be eliminated by inserting a zero sequence current trap in the circuit (Fig. 10-21b). The zero sequence current trap consists of wye-delta-connected auxiliary current transformers, which can have any ratio. These auxiliary current transformers provide a low-impedance path for the zero sequence current component and high-impedance path for the positive and negative sequence current

TRANSFORMER CONNECTIONS HIGH	LOW	ANGLE	COMPENSATING FACTOR H	L
▷	⅄	30°	1	$\sqrt{3}$
▷	▷	0°	1	1
⅄	◁	30°	$\sqrt{3}$	1
⅄	⅄	0°	$\sqrt{3}$	$\sqrt{3}$
◁	⅄	330°	1	$\sqrt{3}$
⅄	▷	330°	$\sqrt{3}$	1

ANGLE HV LEADS LV

120°
90°
60°
150°
30°
180°
REFERENCE
210°
330°
240°
270°
300°

ALL CT'S ASSUMED WYE CONNECTED

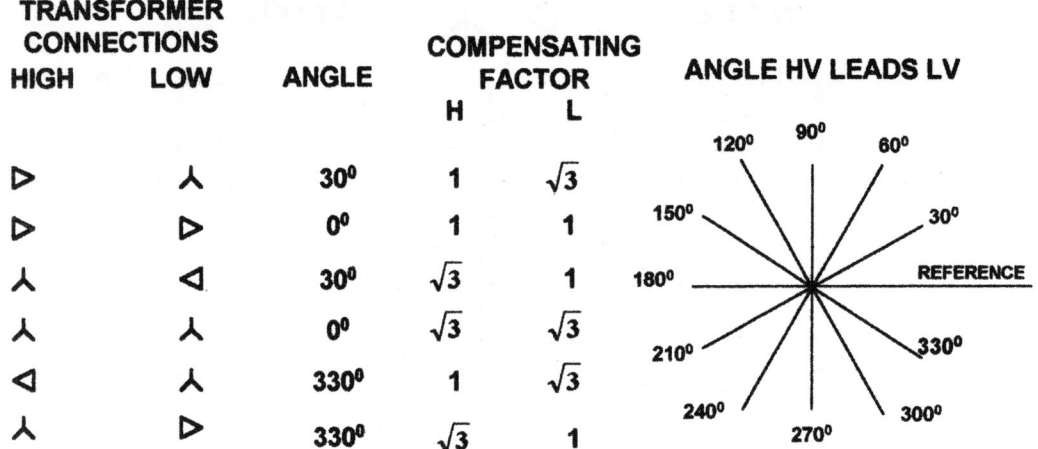

Figure 10-20 TPU settings dictated by transformer connections.

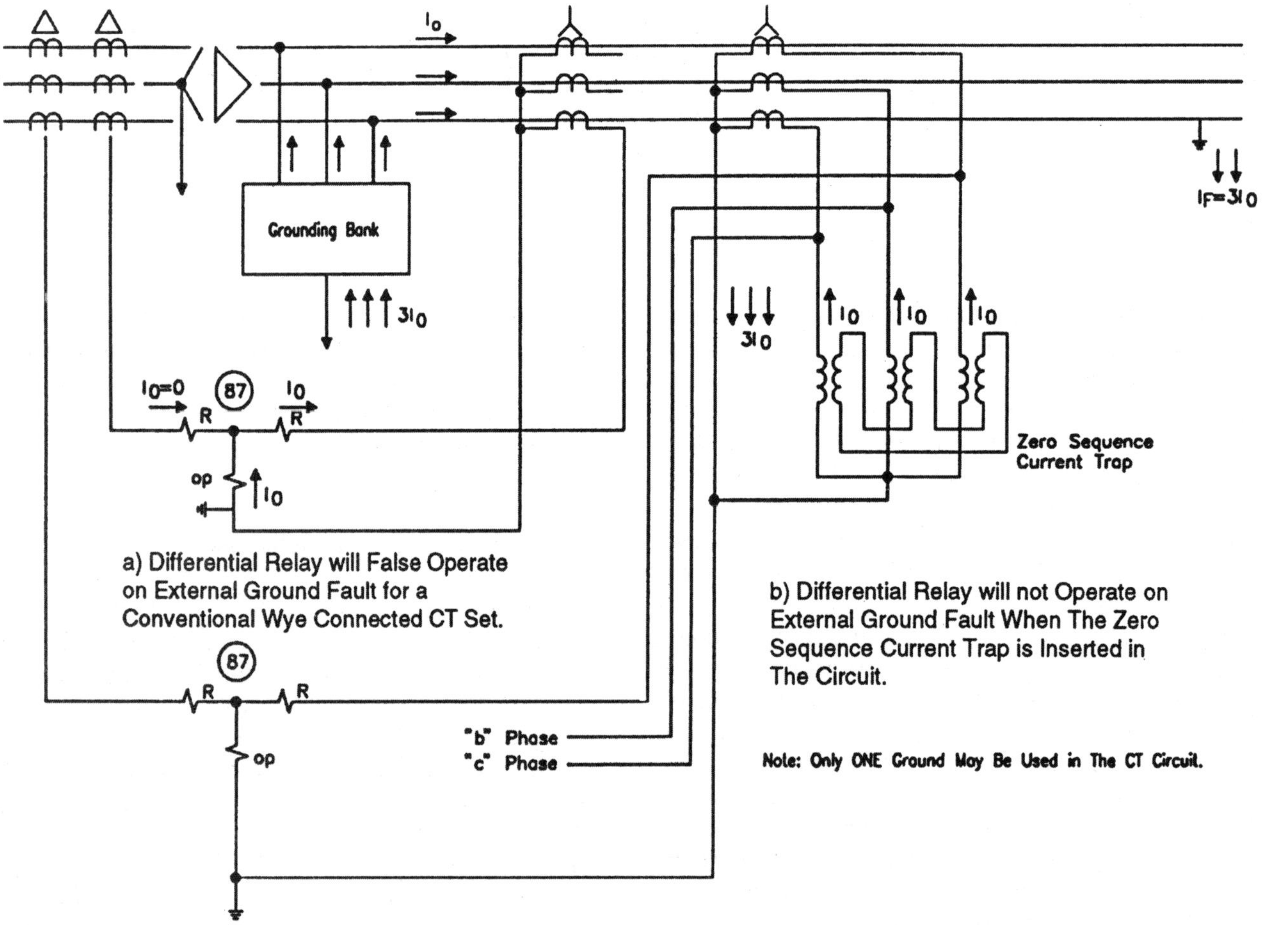

Io
Grounding Bank
3Io
IF=3Io
3Io
Io
Io
Io
Zero Sequence Current Trap
Io=0
87
Io
R
R
op
Io
a) Differential Relay will False Operate on External Ground Fault for a Conventional Wye Connected CT Set.
b) Differential Relay will not Operate on External Ground Fault When The Zero Sequence Current Trap is Inserted in The Circuit.
87
R
R
op
"b" Phase
"c" Phase
Note: Only ONE Ground May Be Used in The CT Circuit.

Figure 10-21 Ground source on delta side.

components of the fault current. The scheme can be grounded at the differential relay or trap, but only one ground point may be used.

5.3   Three-Phase Banks of Single-Phase Units

Figure 10-22 shows one phase of the transformer differential connection for transformer bushing current transformers used in a three-phase bank of single-phase units. In such cases, conventional current transformer connection cannot be used in the three circuits to the delta, such as those shown in Figure 10-15. This differential relaying scheme will not detect internal bushing flashovers if the power system is grounded as illustrated in Figure 10-22a. This differential relaying scheme will not detect internal bushing

flashovers if the power system is ungrounded, as illustrated in Figure 10-22b. The relay scheme also will not detect an internal bushing flashover and external ground fault on another phase.

Protection for these internal faults is obtained by placing current transformers on both bushings of a single-phase transformer winding that forms part of the three-phase delta connection (Fig. 10-22c). All current transformers can be wye-connected.

An unbalance of 2/1 results from connecting the two current transformers in the delta winding in parallel with one ct in the wye winding. When selecting current transformer ratios and/or relay taps, this unbalance must be taken into account.

5.4   Differential Protection of a Generator-Transformer Unit

In the unit-type system shown in Figure 10-23, the transformer differential relay is often connected to include the generator as well as the transformer. This arrangement provides additional and overlapping protection for the rotating machine. Separate current transformers on the generator neutral are recommended to keep the burden low.

Because the station service transformer bank is much smaller than the generator unit, the transformer differential relay may not protect against secondary or internal faults in the station service transformer unless such faults occur near the high-voltage end of the primary winding. The main units require a high current transformer ratio to limit secondary currents under continuous operation and high faults. Since the station service unit is small, it will have a high impedance and light fault currents—frequently below the transformer differential relay sensitivity. In this case, a separate differential relay around the station service bank can be operated with current transformer ratios appropriate to the size of the bank. Overload relays, without a separate differential, may be used to protect this station service bank.

If no appreciable current is fed back through the small station service unit for faults external to the large main unit, faults on the low side of the station service unit may be well below the main differential relay sensitivity. If so, no current transformer connection to the main differential relay circuit is required. Otherwise, a connection is required to prevent tripping on station service bus faults.

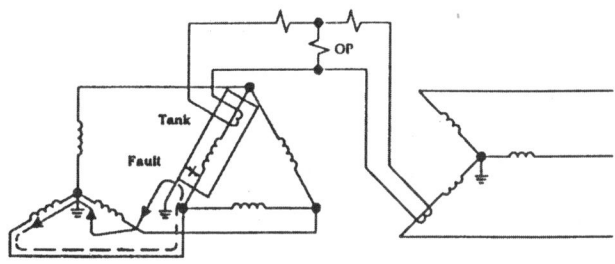

a) Grounded System on Delta Side.

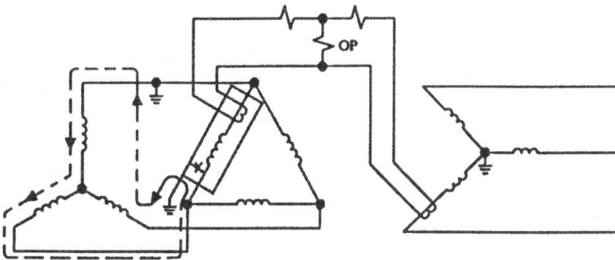

b) Ungrounded System on Delta Side.

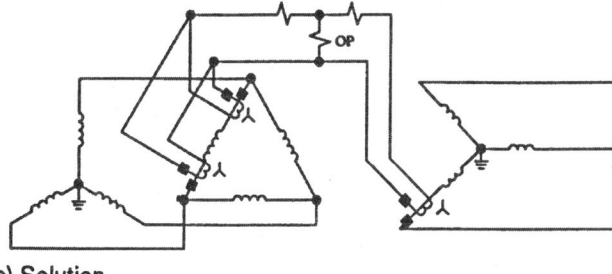

c) Solution

Figure 10-22   Protection problems and solution for internal faults on delta side of a three-phase bank consisting of single-phase units.

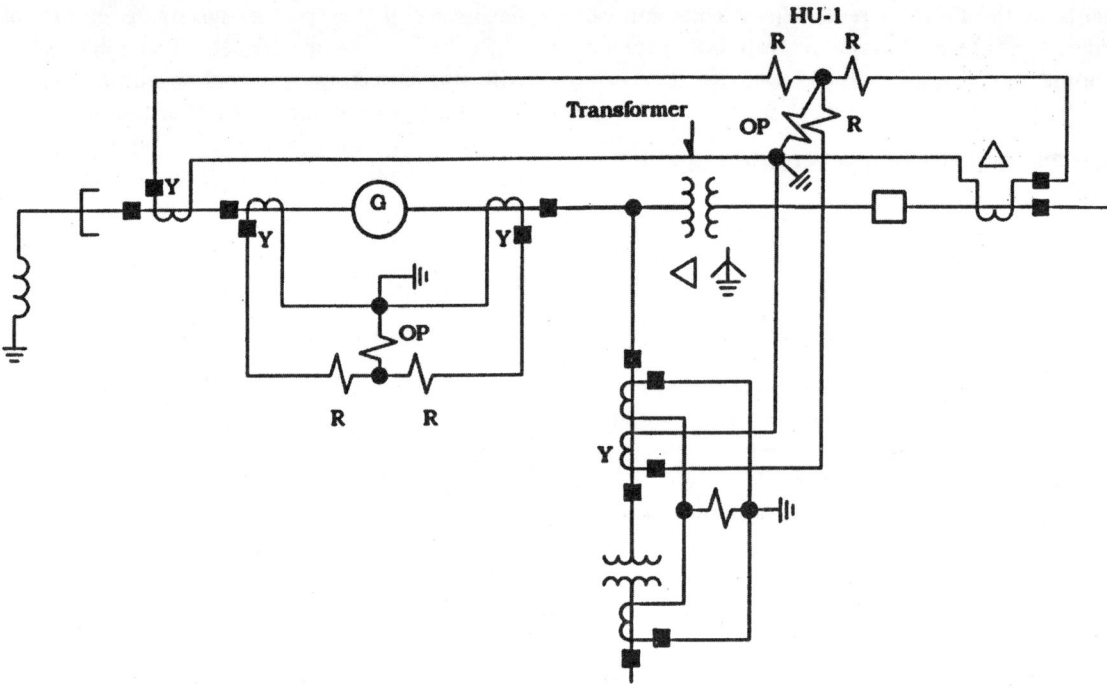

Figure 10-23 Differential protection of the unit-type generator-transformer system with separate differential protection for the station service unit.

5.5 Overexcitation Protection of a Generator-Transformer Unit

Overexcitation of a transformer may result in thermal damage to cores due to excessively high flux in the magnetic circuits. Excess flux saturates the core steel and flows into the adjacent structure, causing high eddy current losses in the core and adjacent conducting materials.

Since flux is directly proportional to voltage and inversely proportional to frequency, the unit of measure for excitation is defined as per unit voltage divided by per unit frequency (volts/hertz). Overexcitation exists whenever the per unit volts/hertz exceeds the design limits of the equipment; for example, a transformer designed for a voltage limit of 1.2 per unit at rated frequency will experience overexcitation whenever the per unit volts/hertz exceeds 1.2. Should the voltage exceed 120% at rated frequency, or the frequency go below 83.3% at rated voltage, overexcitation will exist. Severe overexcitation can cause rapid damage and equipment failure. Figure 10-24 shows the curves of transformer overexcitation limitations for various manufacturers (curves are from Figure 11 of ANSI/IEEE Standard

C37.106—1987 "IEEE Guide for Abnormal-Frequency Protection for Power Generating Plants").

In a generator-transformer unit system, the transformer may be subjected to an overvoltage or overexcitation condition on load rejection or as external faults are cleared by the high-side breaker. During periods of high overexcitation, conventional transformer differential relays and relaying schemes may operate. Some users, however, consider this "misoperation" an advantage, since it protects against transformer damage from overvoltage.

Figure 10-25 shows a scheme for preventing undesired tripping of the differential relay and of protecting a transformer-generator combination against overvoltage. The differential relay must be equipped with some form of restraint that will recognize excessive volts/hertz. Either fifth-harmonic or "all-harmonic" restraint is available for this function.

Figure 10-24 provides some help in establishing permissible limits for transformers, but the individual manufacturer should be contacted for assurances.

In Figure 10-25, device 87 is chosen to provide protection to both the transformer and generator (and provide backup to the generator differential relay, not shown). As voltage increases, due to, for

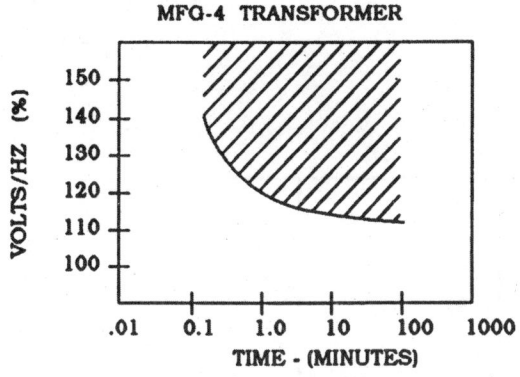

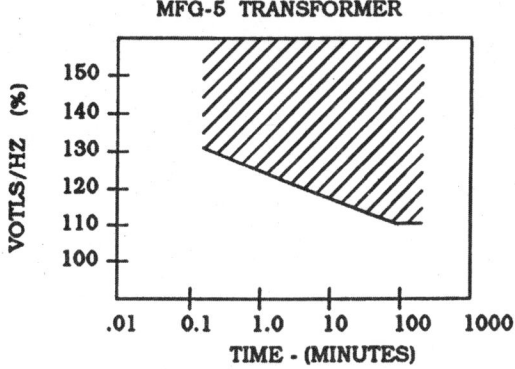

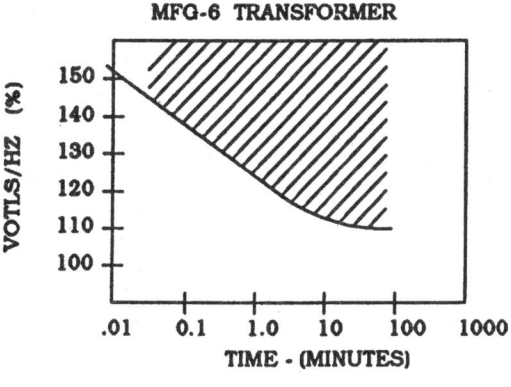

Figure 10-24 Transformer overexcitation limitations for various manufacturers transformer under no-load conditions.

example, load rejection, the tendency of the differential relay to operate is blocked by, say, the detection of increased fifth harmonic. As the voltage increases farther, to the point where damage may occur to the protected apparatus, device 24 (volts/hertz relay) takes over to provide tripping that is time-delayed to coordinate with the capability of the transformer and generator.

Care must be exercised in choosing both the level of fifth harmonic at which blocking takes place as well as the level of volts/hertz and time at which tripping takes place. Information such as that provided in Figure 10-26 should be obtained from the individual manufacturer to identify what may be expected for a particular transformer or generator.

5.6 Sudden-Pressure Relay (SPR)

With the application of a gas-pressure relay, many transformers can be protected by a simple differential relay set insensitively in the inrush current. The sudden-pressure relay (SPR), which operates on a rate of rise of gas in the transformer, can be applied to any transformer with a sealed air or gas chamber above the oil level. The relay is fastened to the tank or manhole cover, above the oil level. It will not operate on static pressure or pressure changes resulting from normal operation of the transformer.

The SPR relay is recommended for all units of 5000 kVA or more. It is extremely sensitive to light faults as it will operate on pressure changes as low as 0.33 lb/in.^{2}/sec. In one case, this represented a fault of 50 A. The SPR relay is far more sensitive to light internal faults than the differential relay. The differential relay, however, is still required for faults in the bushing and other areas outside the tank.

The SPR relay operating time varies from one-half cycle to 37 cycles, depending on the size of the fault.

In the past, large-magnitude through-fault conditions on power transformers have caused rate-of-change-of-pressure relays to occasionally operate falsely. There has been reluctance on the part of some users to connect these rate-of-change-of-pressure relays to trip, and they have therefore used them for alarming only. Schemes have been devised to restrict tripping of the rate-of-change-of-pressure device only to levels of current below which the transformer differential relay cannot operate. One means of doing this is with a high-speed current-blocking type RAICB relay (or a simple overcurrent relay) to supervise the SPR trip circuit. By wiring the SPR output to a TPU 2000R input, recording, supervision, and time delay of an SPR operation can be accomplished.

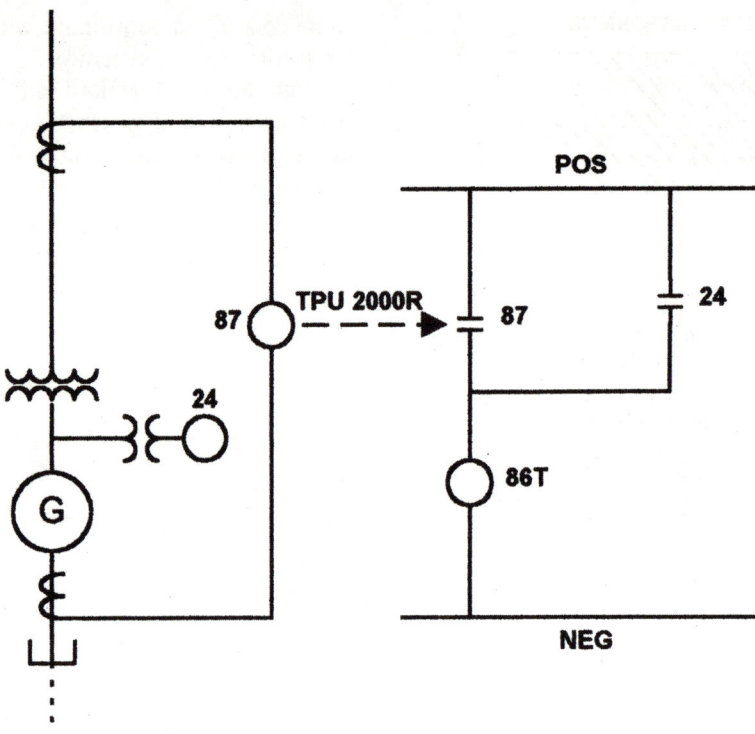

Figure 10-25 Overexcitation protection for generator and transformer.

5.7 Overcurrent and Backup Protection

To allow transformer overloading when necessary, the pickup value of phase overcurrent relays must be set above this overload current. An inverse-time characteristic relay usually provides the best coordination. Settings of 200 to 300% of the transformer's self-cooled rating are common, although higher values are sometimes used. Fast operation is not possible, since the transformer relays must coordinate with all other relays they overreach.

Overcurrent relays cannot be used for primary protection without the risk of internal faults causing extensive damage to the transformer. Fast operation on heavy internal faults is obtained by using instantaneous trip units in the overcurrent relays. These units may be set at 125% of the maximum through fault, which is usually a low-side three-phase fault. The setting should be above the inrush current. Often, instantaneous trip units cannot be used because the fault currents are too small.

An overcurrent relay set to protect the main windings of an autotransformer or three-winding transformer offers almost no protection to the tertiary windings, which have a much smaller kVA. Also, these tertiary windings may carry very heavy currents during ground faults. In such cases, tertiary overcurrent protection must be provided.

A through fault external to a transformer results in an overload that can cause transformer failure if the fault is not cleared promptly. It is widely recognized that damage to transformers from through faults is the result of thermal and mechanical effects. The thermal effect has been well understood for years. The mechanical effect has recently gained increased recognition as a major concern of transformer failure. This results from the cumulative nature of some of the mechanical effects, particularly insulation compression, insulation wear, and friction-induced displacement. The damage that occurs as a result of these cumulative effects is a function of not only the magnitude and duration of through faults, but also the total number of such faults.

The transformer can be isolated from the fault before damage occurs by using fuses or overcurrent relays. The former ANSI C37.91, "Guide for Protective Relay Applications to Power Transformers," was based on the former U.S. Standard C57 for Power Transformer, in which mechanical effect was not considered. The latest published Standard C57.109–

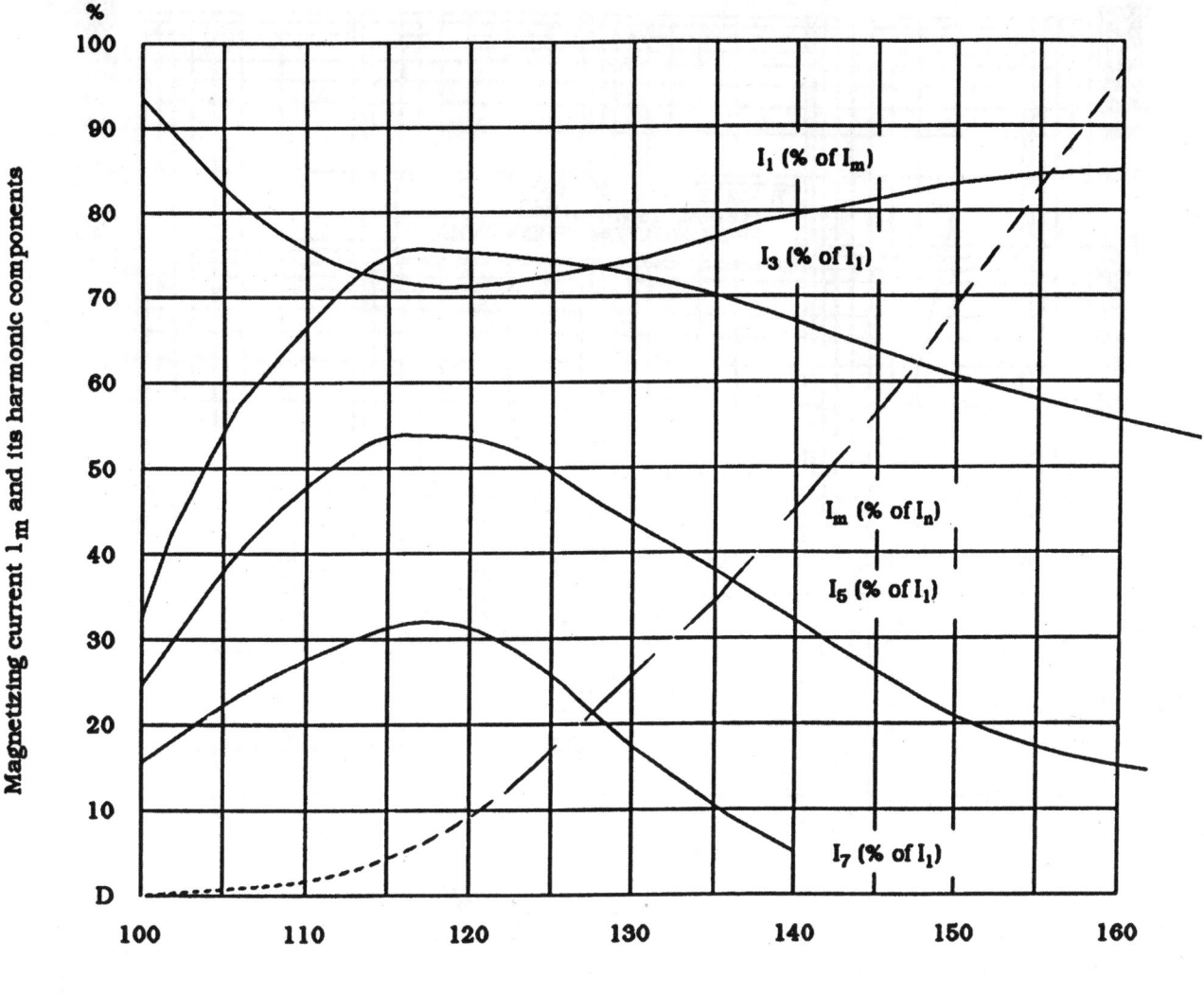

Figure 10-26   Example of harmonic components on transformer overexcitation.

Table 10-7   Transformer Category (ANSI/IEEE Standard C57.109-1985 Curves)

| Category | Minimum nameplate (kVA) | | Reference protective curve |
	Single-phase	Three-phase	
I	5–500	15–500	Fig. 10-27
II	501–1667	501–5000	Fig. 10-28
III	1668–10,000	5001–30,000	Fig. 10-29
IV	above 10,000	above 30,000	Fig. 10-30

1985, "IEEE Guide for Protective Relay Applications to Power Transformers," considers both the thermal and mechanical effects.

For purposes of coordination of overcurrent protective devices, ANSI/IEEE Standard C57.109–1985 presents different curves for different-size transformers as listed in Table 10-7. For applications including the category I transformer, only the thermal effect from the through-fault current is considered in Figure 10-27; on the contrary, for applications including the category IV transformer, the thermal and mechanical effects from the through-fault current should be considered as

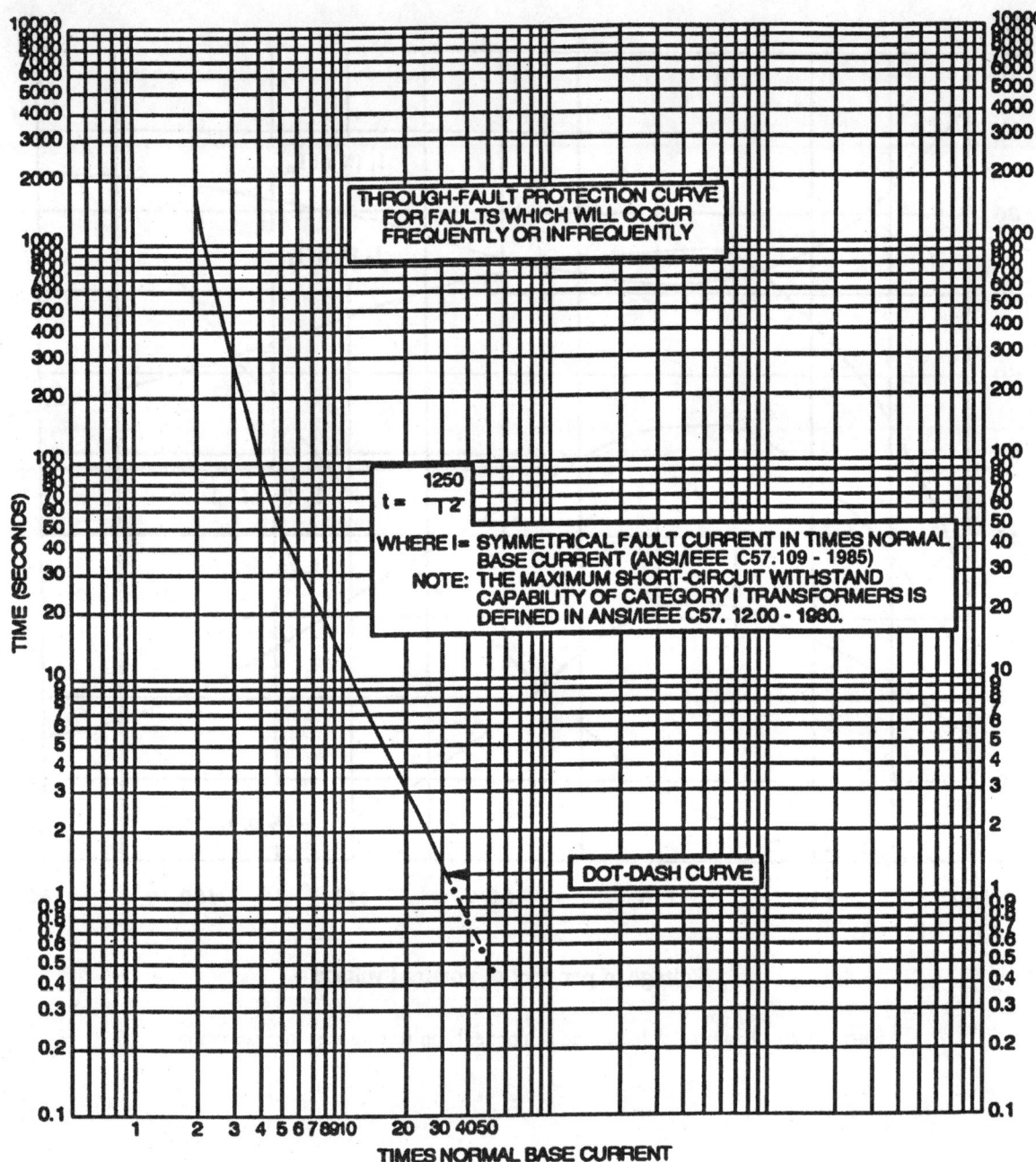

Figure 10-27   Through-fault protection curve for category 1 transformers 5 to 500 kVA single-phase 15 to 500 kVA three-phase.

shown in Figure 10-30. For applications including the category II or III transformer, Figures 10-28 and 10-29, whether or not the mechanical effect from the through-fault current should be considered depends on the frequency of the external fault.

For applications in which external faults occur infrequently, for example, transformers with secondary-side conductors enclosed in conduit or isolated in some other fashion, the through-fault protection curve should reflect primarily thermal damage considerations, since the cumulative mechanical damage effect of through faults will not be a problem.

For applications in which external faults occur frequently, for example, transformers with secondary-side overhead lines, the through-fault protection curve should reflect the fact that the transformer will be

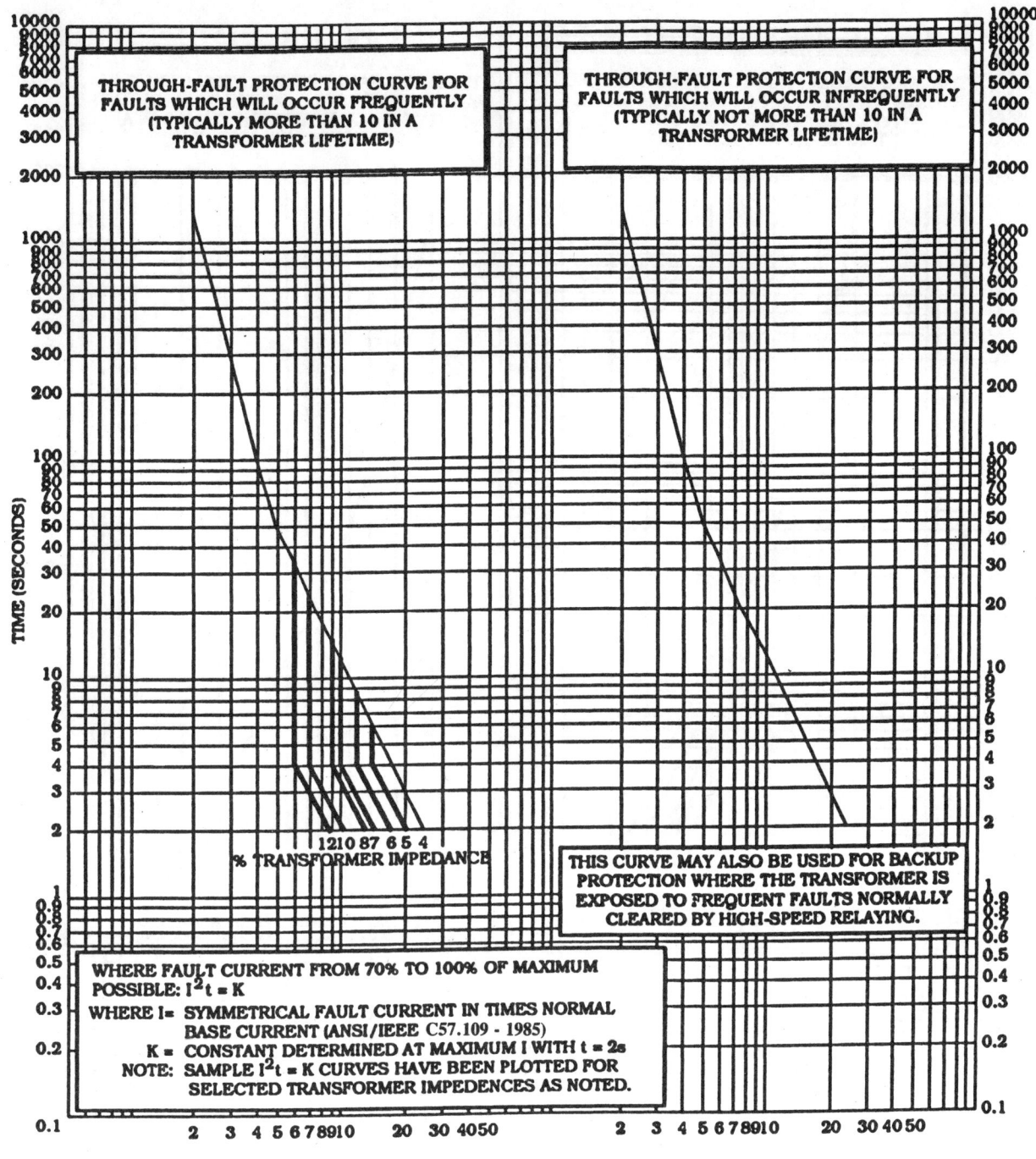

Figure 10-28 Category II transformers 501 to 1667 kVA single-phase 501 to 5000 kVA three-phase.

subjected to the thermal and cumulative mechanical damage effects of through faults.

Figure 10-31 shows the infrequent-frequent fault incidence zones for the determination of curve selection.

The following example describes the procedures for constructing the dog-leg portion of the thermal/mechanical limit curves: a 230/25-kV, 30/50-MVA transformer with an impedance of 10% on a 30-MVA base and with secondary-side overhead lines.

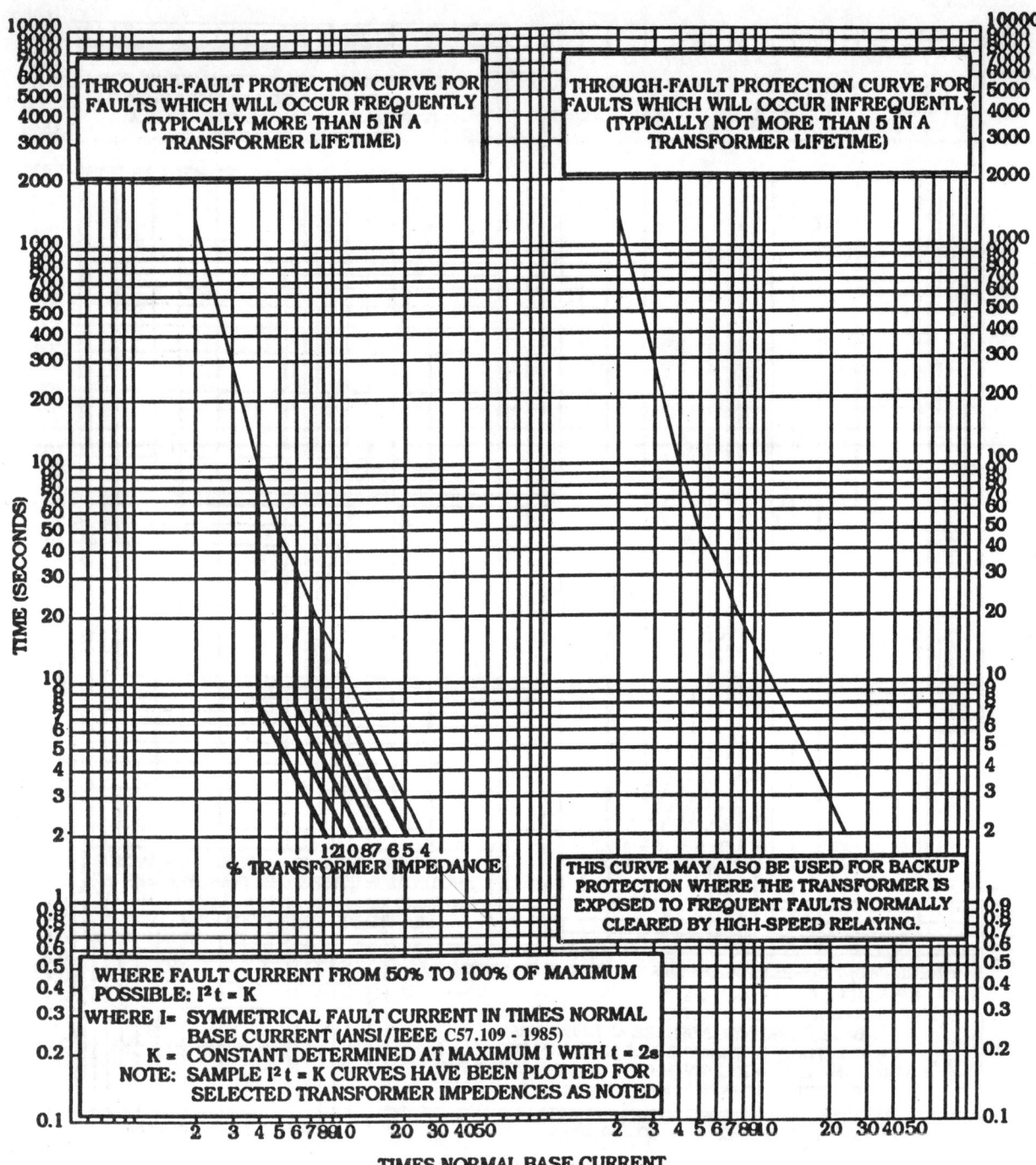

Figure 10-29 Category III transformers 1668 to 10,000 kVA single-phase 5001 to 30,000 kVA three-phase.

Step 1 Select the category from the minimum name-plate rating of the principal winding. For this example, it is a category III transformer, and the Figure 10-29 curve should be used for the coordination.

Step 2 Plot the infrequent through-fault curve of category III (Fig. 10-29), as shown in Figure 10-32, portion A.

Step 3 Determine the dog-leg portion of the curve as follows:

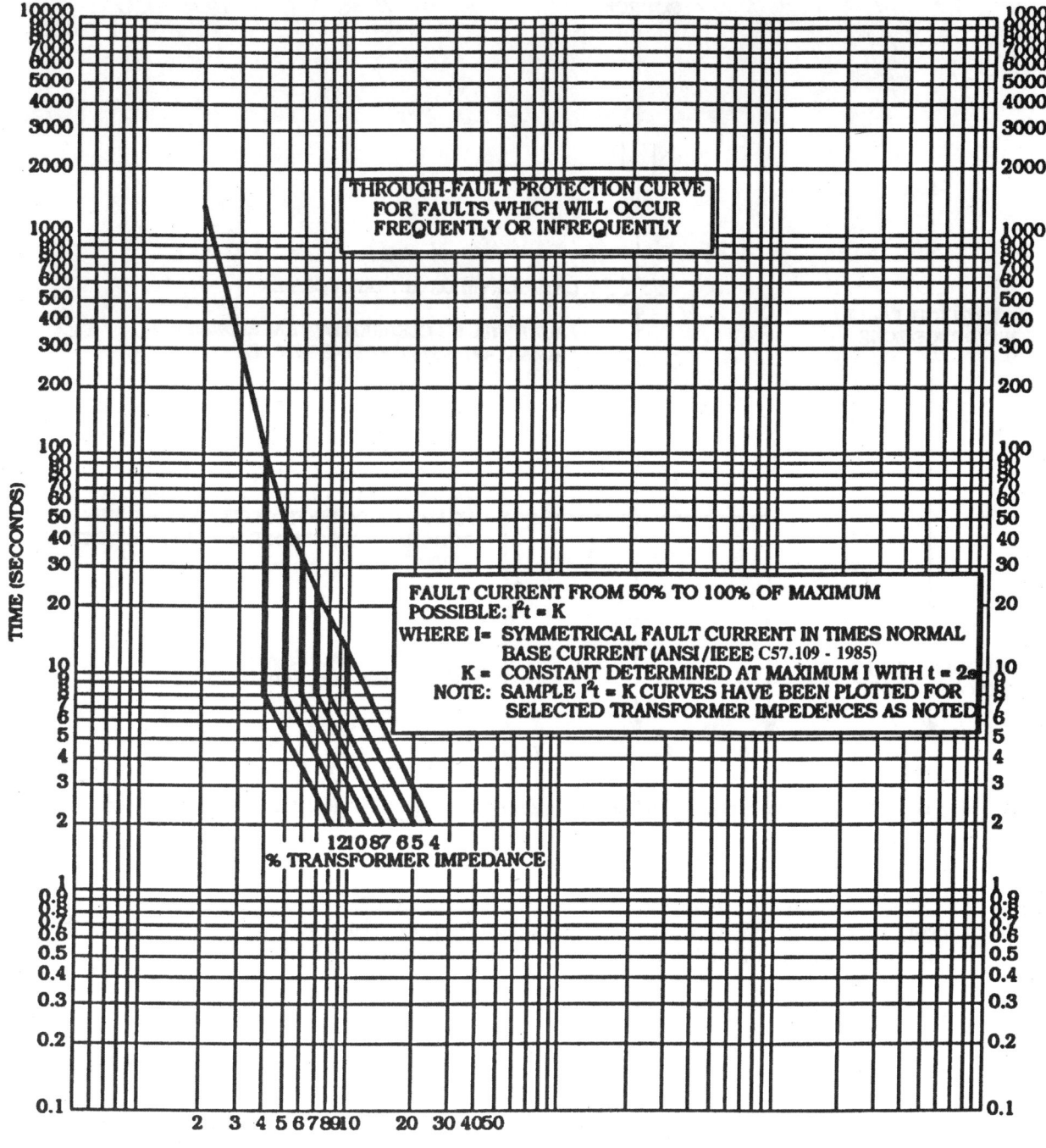

Figure 10-30 Category IV transformers above 10,000 kVA single-phase above 30,000 kVA three-phase.

1. Calculate the maximum per unit through-fault current:

$$I = \left(\frac{1}{0.10}\right) = 10 \times \text{base current at 2 sec}$$

This is point 1 in Figure 10-32.

2. Calculate the constant $K = I^2t$ at $t = 2$ sec:

$$K = \left(\frac{1}{0.10}\right)^2 \times 2 = 200$$

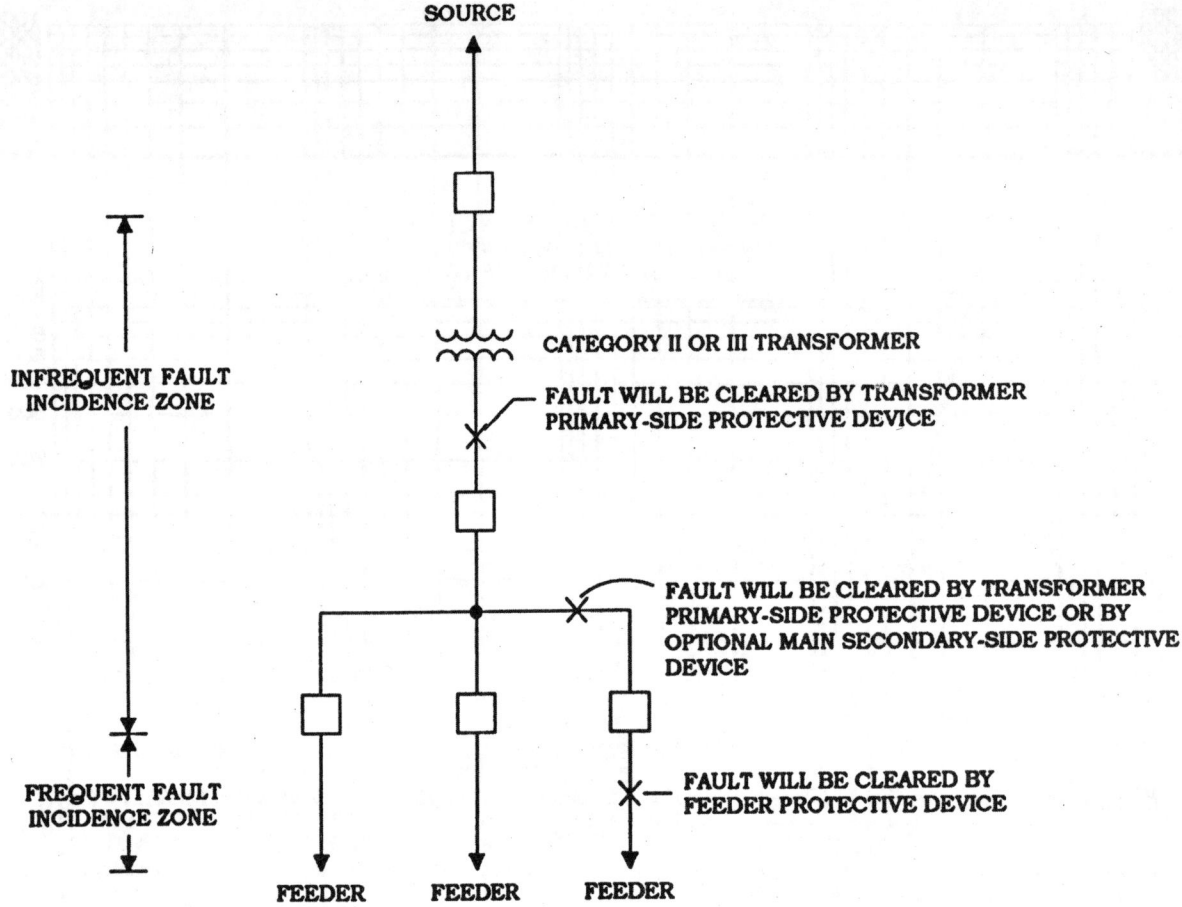

Figure 10-31   Infrequent-frequent fault incidence zones for category II and category III transformers.

3. Calculate the time at 50% (*note*: use 50% for category III and IV, 70% for category II) of the maximum per unit through-fault current:

$$t = \frac{K}{I^2} = \frac{200}{[0.5(10)]^2} = 8 \text{ sec}$$

This is point 2 in Figure 10-32.

4. Connect points 1 and 2 and draw a vertical line from point 2 to the infrequent curve to complete the dog-leg portion of the curve as shown in Figure 10-32.

5.8   Distance Relaying for Backup Protection

Directional distance relaying can be used for transformer backup protection when the setting or coordination of the overcurrent relays is a problem. The directional distance relays are connected to operate when the fault current flows toward the protected transformer. They are set to reach into, but not beyond, the transformer.

5.9   Overcurrent Relay with HRU Supplement

Three single-phase HRU units with instantaneous trip elements can be used to supplement the time-delay overcurrent relays when inrush is a problem and no current transformers are available on the secondary side of the protected bank. As illustrated in Figure 10-33, this arrangement provides high-speed tripping when the transformer is energized on faults. The scheme is not recommended, however, unless transformer loads are supervised by individual local breakers, or load pickup does not occur during transformer energization.

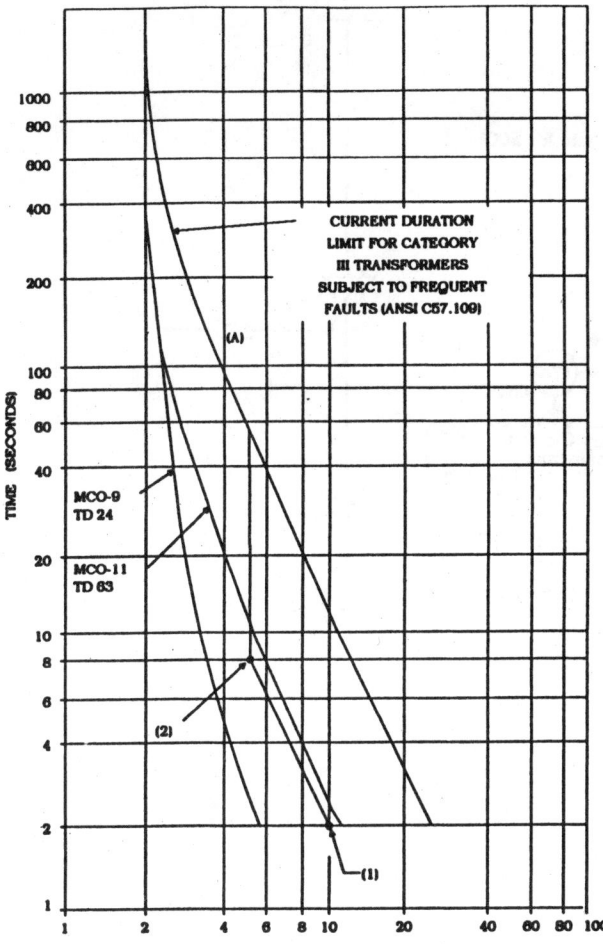

Figure 10-32 Multiple of transformer full load (per unit).

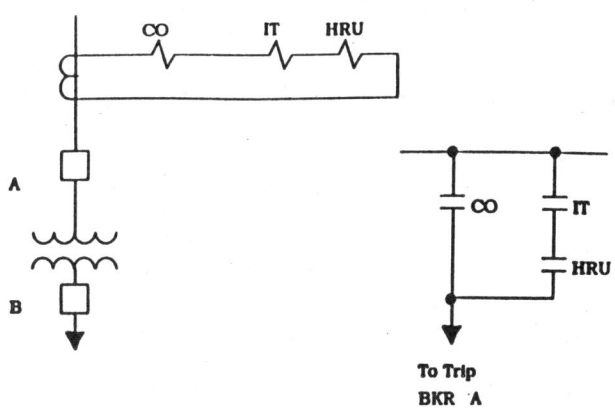

Figure 10-33 Overcurrent relay with single-phase HRU supplement for speed improvement.

6 TYPICAL PROTECTIVE SCHEMES FOR INDUSTRIAL AND COMMERCIAL POWER TRANSFORMERS

The protection of industrial and commercial power transformer banks is somewhat different from the protective schemes used by utilities. The differences in protective schemes are a function of several major factors, including system configuration, method of grounding, speed, coordination, operation, and cost. Some of the more commonly used industrial and commercial protective schemes are shown in Figures 10-34 to 10-39.

Figure 10-34 illustrates how a primary breaker can be used for transformer protection. The basic protection is provided by the 87T transformer differential relays. Either type TPU, CA, or HU relays can be used, depending on the severity of inrush and operating speed requirements. Device 50/51, an inverse-time CO relay with IIT unit, provides trans-

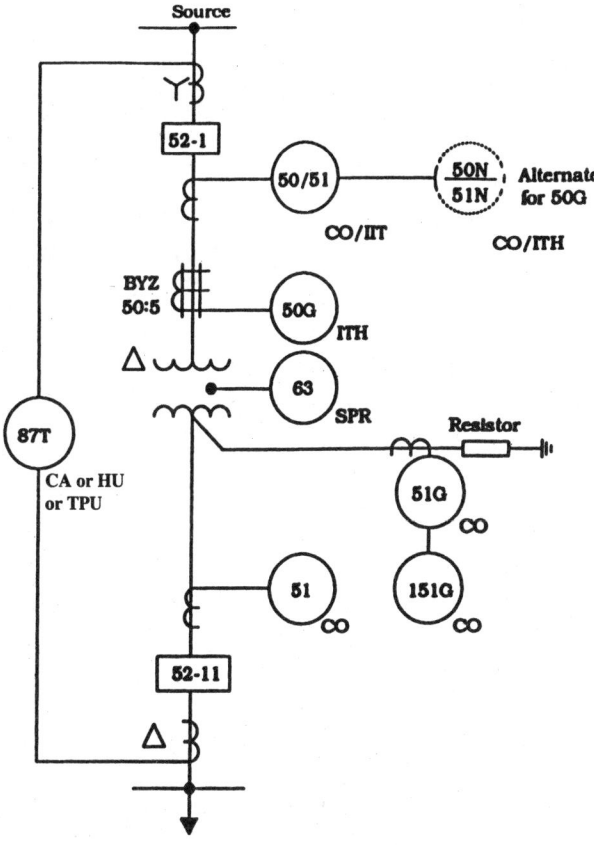

Figure 10-34 Transformer protection with primary breaker. The MSOC relay may be used for the overcurrent functions.

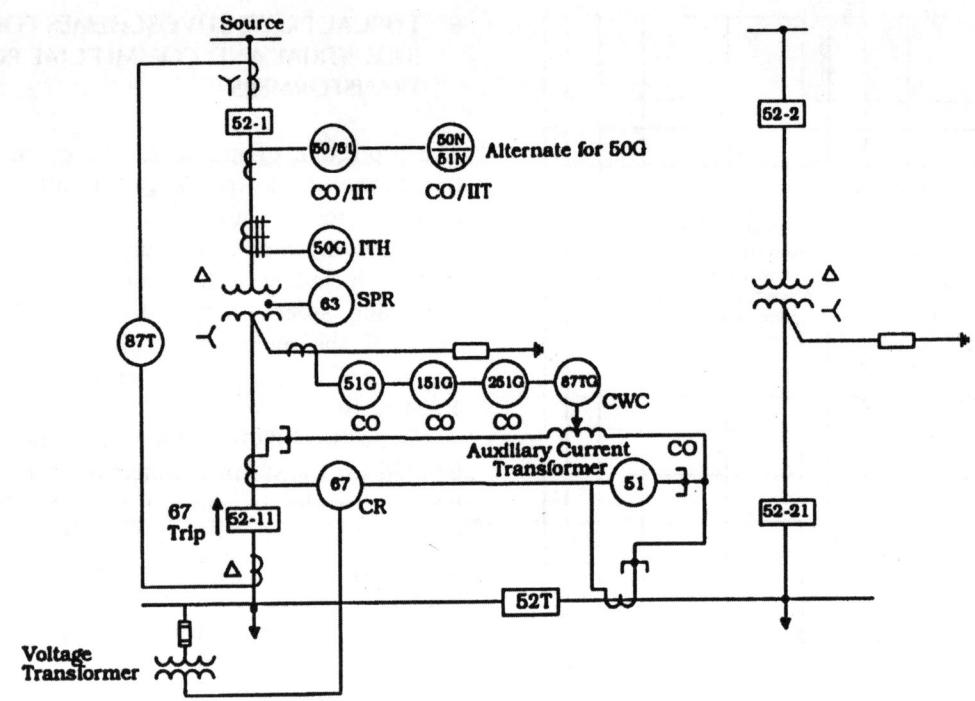

Figure 10-35 Paralleled transformer protection with primary breaker.

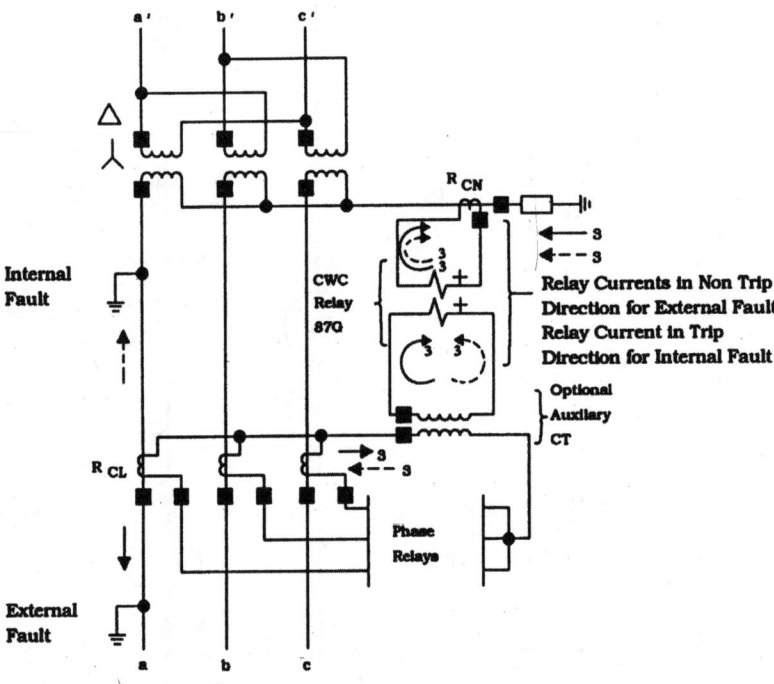

Note:
Solid Arrows Show Zero Sequence Current for External Fault:
Dotted Arrows for Internal Fault.

Figure 10-36 Connections and operation at the CWC (87TG) ground differential relay where the connected system is grounded.

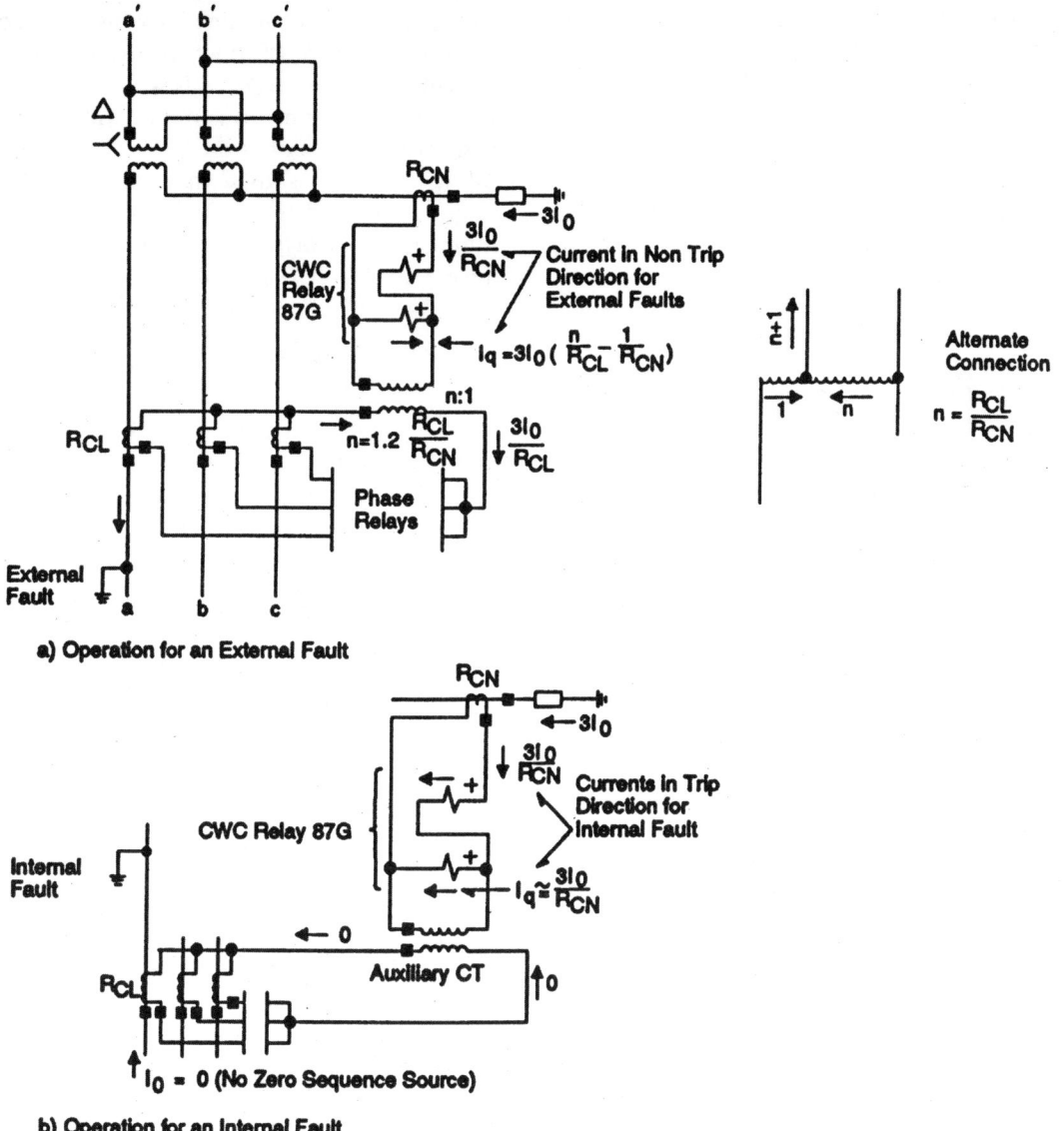

Figure 10-37 Connections and operation of the CWC (87TG) ground differential relay where the connected system is ungrounded or the external ground source is not always available.

former primary winding backup protection for phase faults; either device 50G (type ITH with a zero sequence current transformer) or 50N/51N can be used as transformer primary winding backup for ground faults. Transformer overload, low-voltage bus, and feeder backup protection are provided by device 51 on the transformer secondary side. Since the low-voltage side is medium-resistance-grounded, a ground relay (51G) should be used to trip breaker 52-1 for low-side ground faults and for resistor thermal protection. Device 151G, which trips breaker 52–11,

provides feeder ground backup, whereas device 63, such as a type SPR relay, offers highly sensitive protection for light faults.

The current transformer ratings in this scheme should be compatible with the transformer short-time overload capability: approximately 200% of transformer self-cooled rating for wye-connected current transformers and 350% ($\sqrt{3} \times 200\%$) for delta-connected current transformers. The neutral current transformer rating should be 50% of the maximum resistor current rating.

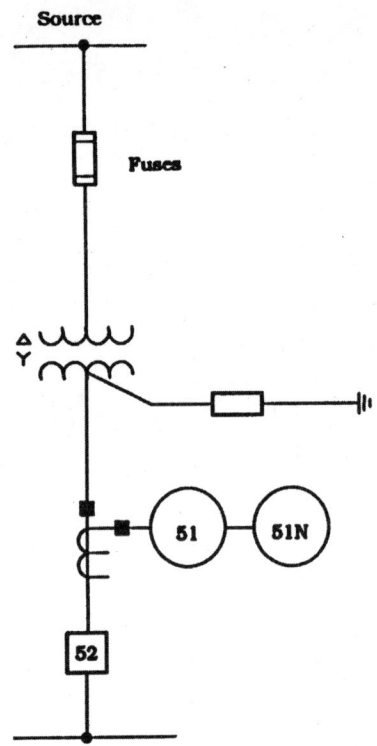

Figure 10-38 Transformer protection with primary fuses.

When a normally closed secondary bus tie breaker is used for paralleled transformer protection (Fig. 10-35), there are several differences, with the primary breaker scheme shown in Figure 10-34. First, a type CWC relay (87TG) provides selective and sensitive protection for ground faults within the secondary circuit of the differential zone. The CWC relay, an induction-disk relay, has two windings that operate on the product of the two currents. The operating torque is proportional to the product times the cosine of the angle between the two currents. Maximum torque occurs when the currents are in phase; the connections and operation of the scheme are shown in Figures 10-36 and 10-37.

For an external fault, the two currents in the relay coils are essentially 180° out of phase. In this case, the CWC relay has no operating torque. For an internal fault, the currents in the relay are essentially in phase, producing operating torque. The relay sensitivity is 0.25 VA. Make sure that the external ground source is always available when applying Figure 10-36; otherwise, the CWC relay will not operate on internal faults.

In these applications, the ratios of the current transformers do not have to be identical. Increased sensitivity can be obtained by using a lower-ratio neutral current transformer. It is desirable to keep the currents in the two relay coils within a 2:4 ratio so that an auxiliary current transformer will be required if a large ground resistor is used.

Figure 10-37 shows the connections when the external system is not grounded or the external ground source is not always available. I_q must be positive for the external fault. This can be done by using

$$n = 1.2 \frac{R_{CL}}{R_{CN}} \text{ or higher}$$

for the auxiliary ct ratio as shown in Figure 10-37a. Current transformer ratios, as well as any effect of saturation of the line current transformer, must be considered.

In addition to devices 51G and 151G shown in Figure 10-35, a 251G relay is used to trip breaker 52T on ground faults. The trip sequence of these three ground relays is as follows: (1) 251G trips 52T, (2) 151G trips 52-11, (3) 51G trips 52-1. The 87TG trips 52-1 and 52-11. Device 67 (type CR relays) provides reverse overcurrent protection.

If the transformer is too small to warrant the basic schemes described above, the scheme shown in Figures 10-38 and 10-39 is recommended. Here, fuses provide the primary fault protection. Solid grounding will assure sufficient primary phase fault current to operate the fuses for most secondary ground faults. The opening of a single primary fuse will result in single phasing of the transformer secondary system. This may be difficult to detect, particularly at light loads, and appropriate precautionary measures should be taken.

If the primary source is grounded and there is a power source on the secondary side, a ground fault on the incoming line will be interrupted by the source breaker; the transformer primary or secondary breaker, however, will not be relayed open because of the delta primary transformer connection. The failure of these breakers to open can result in hazards to personnel, possible damaging transient overvoltages produced by an arcing-type fault, and problems with automatic reclosing of the source breaker. Several schemes can be used to ensure opening of the source and transformer breakers, including pilot protection of the incoming line, transfer-trip, or potential ground detection relaying schemes on the transformer primary. Automatic reclosing is a special problem, requiring that the secondary breaker be opened before the primary source breaker is reclosed.

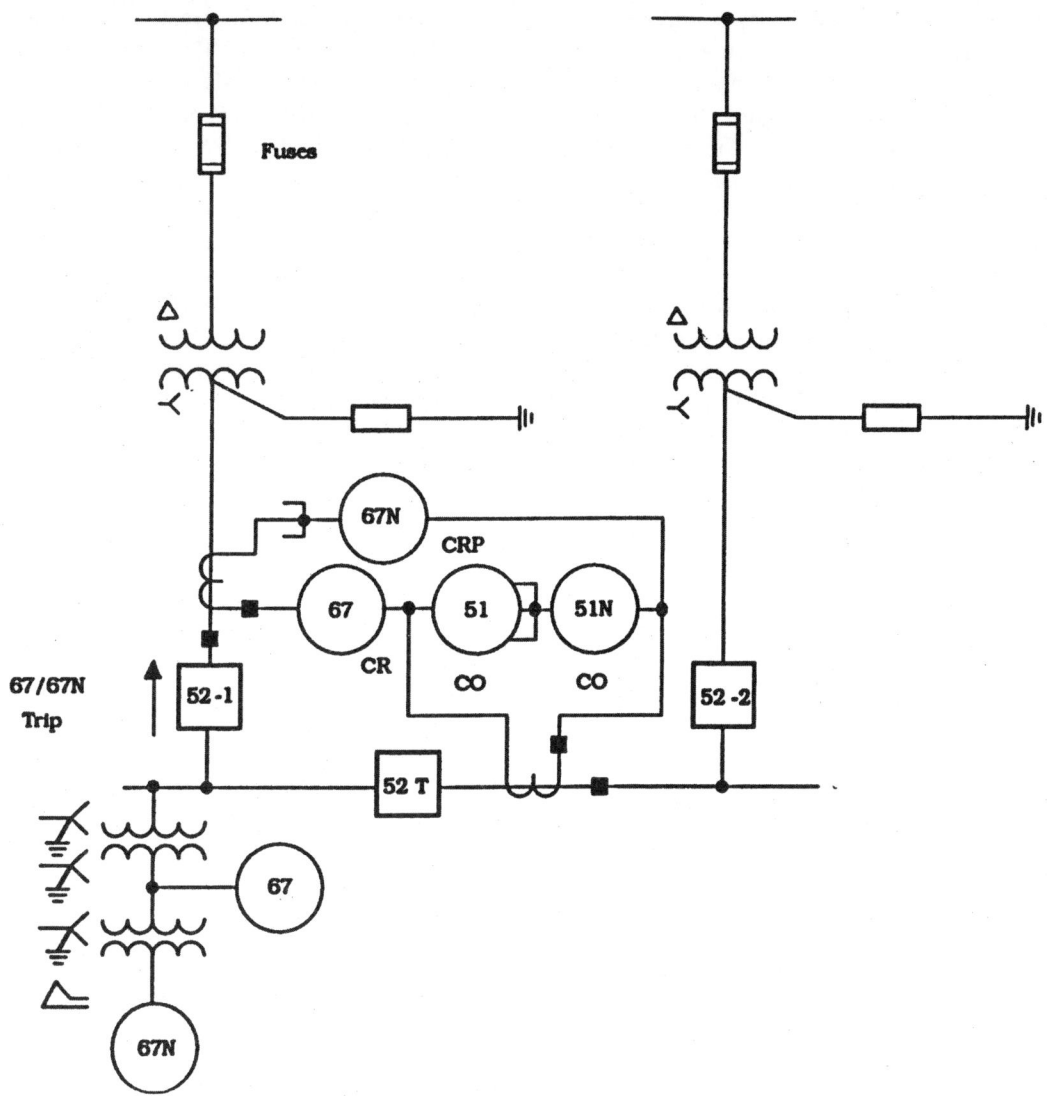

Figure 10-39   Paralleled transformer protection with primary fuses.

When a normally open bus tie breaker is used, as in Figures 10-35 and 10-39, devices 67 and 67N are not required.

7   REMOTE TRIPPING OF TRANSFORMER BANK

Transformer banks are often applied as a part of the line section, with no high-side breakers. The protection problems associated with this combination are described in Chapters 12, "Line and Circuit Protection."

8   PROTECTION OF PHASE-ANGLE REGULATORS AND VOLTAGE REGULATORS

A phase-angle regulating transformer, which inserts or impresses a regulated voltage on a line in quadrature with its line-to-ground voltage, is used to control power flow in the system. A voltage-regulating transformer compensates for drops in IR by inserting or impressing a regulated voltage on a line in phase with its line-to-ground voltage. These two transformers consist basically of a series unit and an exciting unit, located on at least two separate cores and in separate tanks. Depending on the design and size of the bank, the exciting unit may be wye- or delta-connected, and the series unit may be constructed as one unit or split into two identical units. By mixing the control elements in the exciting unit, as shown in Figure 10-40, a single bank can sometimes provide both the phase-angle regulation and voltage control functions. The protective schemes for the phase-angle regulator and voltage regulator are as varied as the ways in which the bank is constructed. Figure 10-41 shows a typical scheme that could be used for the system depicted in Figure 10-40.

Device 87E (HU-1 relays, one per phase; TPU relays, one per bank; or KAB relays, one per phase) provides overall regulator protection. As shown in Figure 10-42, the currents I_S, I_L, and I_e are applied vectorially to the relay. Since all these currents flow in the primary circuit of the series windings, saturation of the series windings from external-fault overvoltages will not affect the 87E relays.

In application, the current transformer ratio and relay tap's selection for 87E would be similar to the conventional differential scheme for three-winding transformer protection. The use of an equal current transformer ratio for the source, load, and primary of the exciting unit is recommended. This allows the use of equal tap settings in the differential relay. For most applications, the current transformer ratio is determined by the full-load current through the series winding. However, for larger-angle shifts, the current transformer ratio is determined by the current of the exciting-unit primary winding.

To set the 87E relays at their minimum tap, current transformer ratios should be identical. For HU-1 relay application, a setting of 2.9 for all restraint elements will provide the best sensitivity. The series-unit primary and exciting-unit primary are electrically

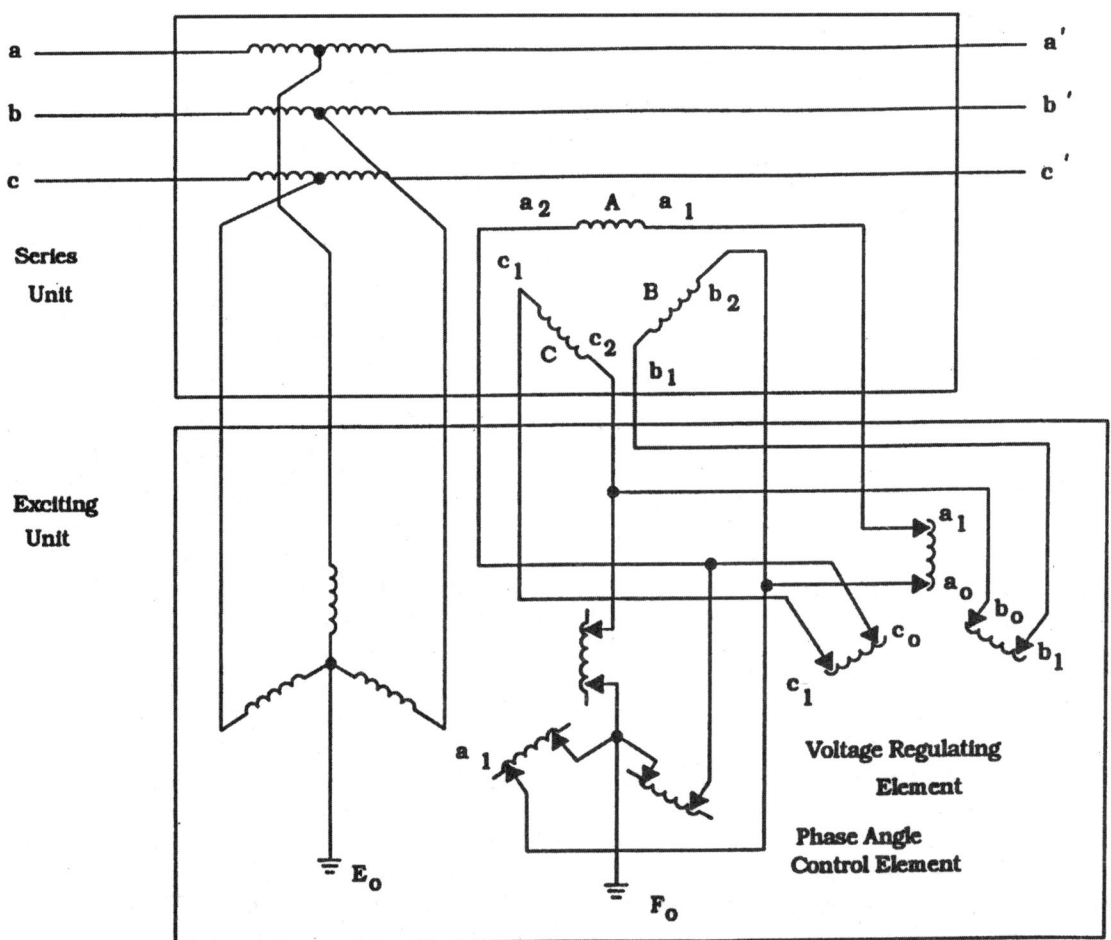

Figure 10-40   Typical 400 MVA 115 KV phase-angle regulator, $\pm 26°$ with voltage control.

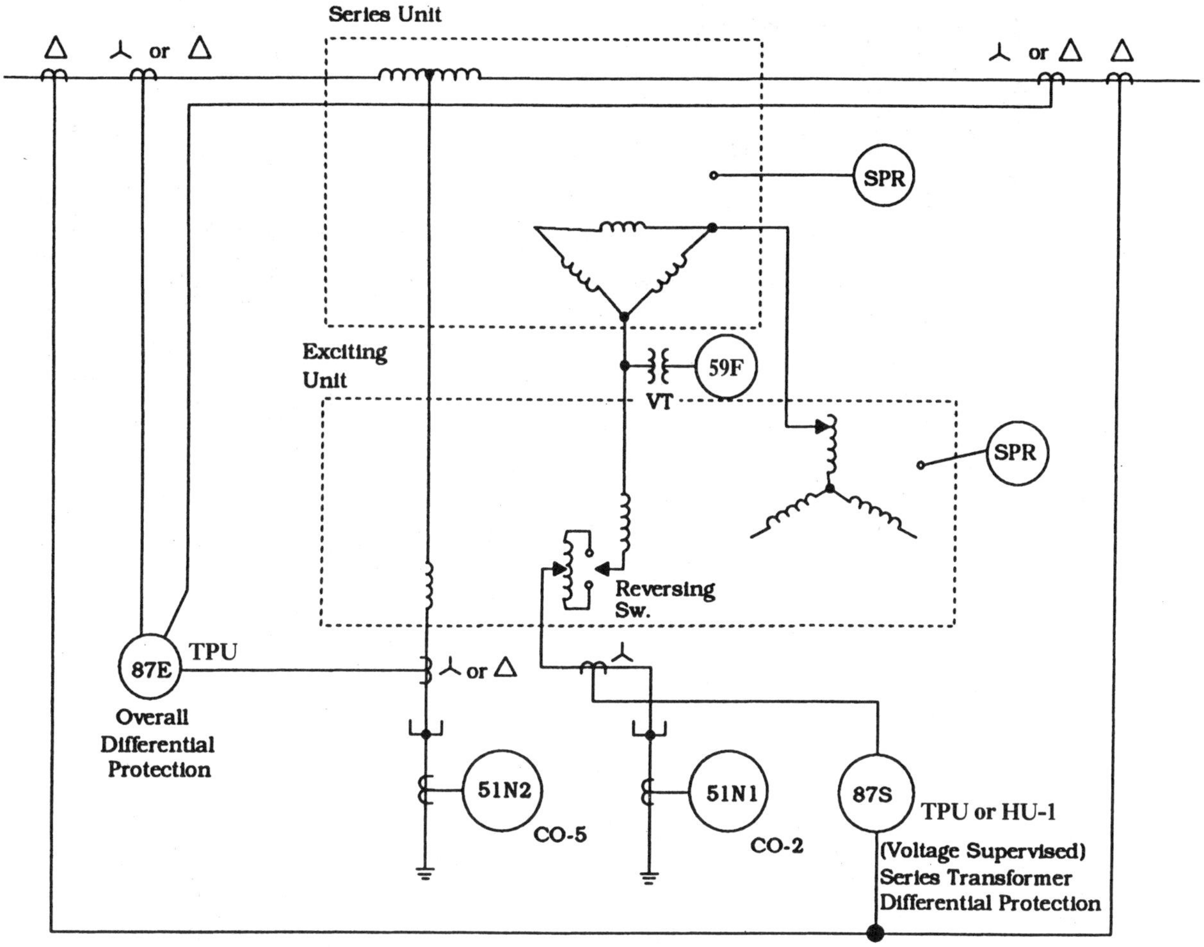

Figure 10-41 Typical scheme for protecting the phase-angle regulator of Figure 10-40.

connected (instead of magnetically), and there is no phase shift. Consequently, the current transformers for 87E relays can be either wye- or delta-connected. Wye connection will permit faulted-phase identification, whereas delta connection will produce more current

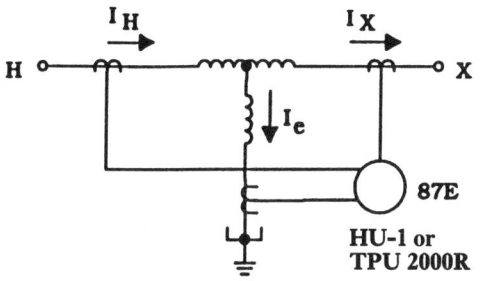

Figure 10-42 Overall differential protection for the phase-angle regulator of Figure 10-40.

($\sqrt{3}$) to operate the relay and two out of the three relays in the scheme will pick up on an internal fault providing redundant backup.

Device 87S (HU relays, one per phase or TPU) provides differential protection for the series unit. As shown in Figure 10-44, the current transformers for current Ie' should be located at the neutral end of the windings to provide some backup protection for the exciting unit.

The series winding has very low impedance and is designed for rated voltage equivalent to the quadrature voltage at maximum phase-angle shift. For the example shown in Figures 10-43 and 10-44, it is approximately [2 sin (26°/2)] or 45% of the line-to-neutral voltage. As a result, the transformer windings are subject to over-voltage or overexcitation conditions on external fault, which may produce saturation of the series windings and cause false operation of the 87S relays.

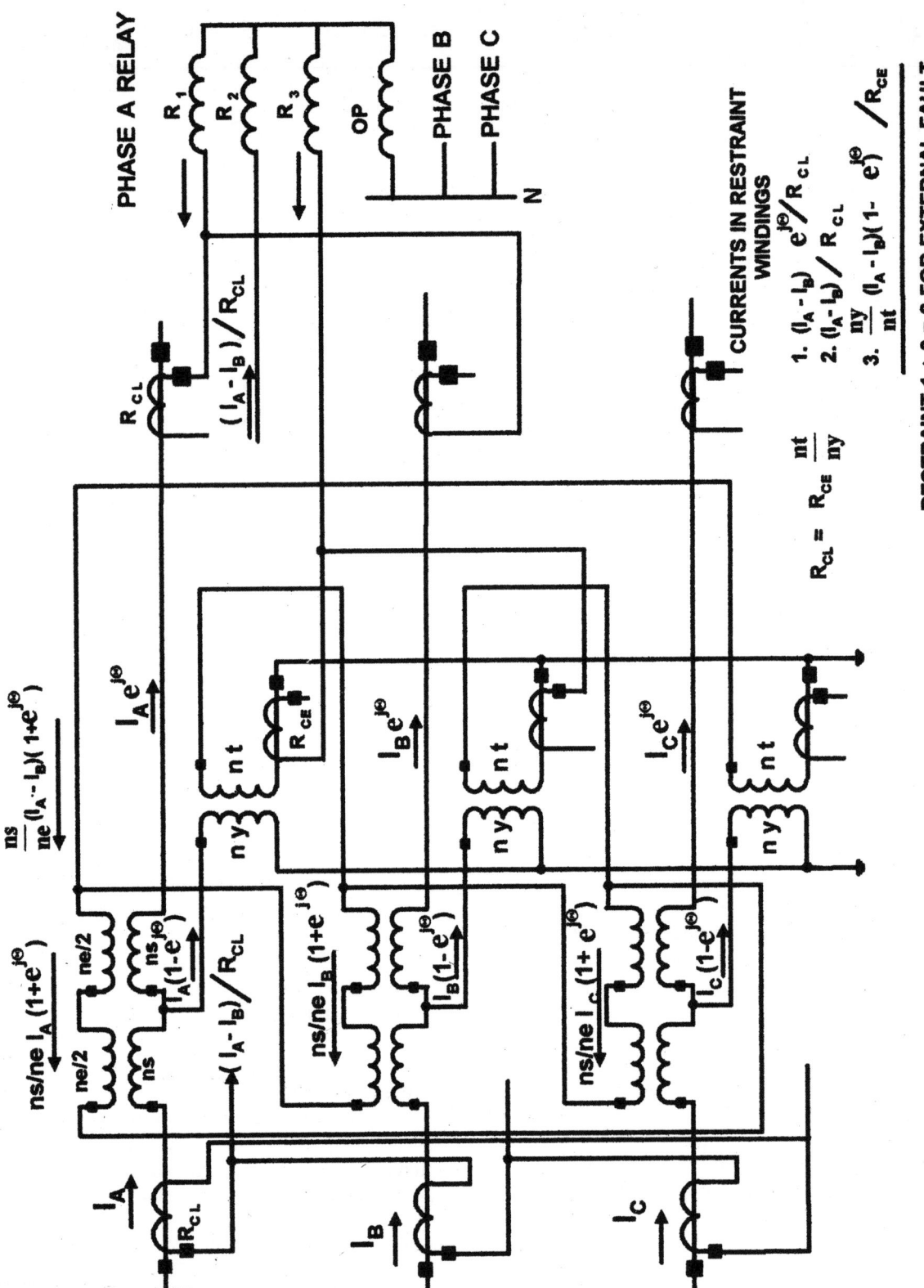

Figure 10-43 Phase-angle regulator (87S).

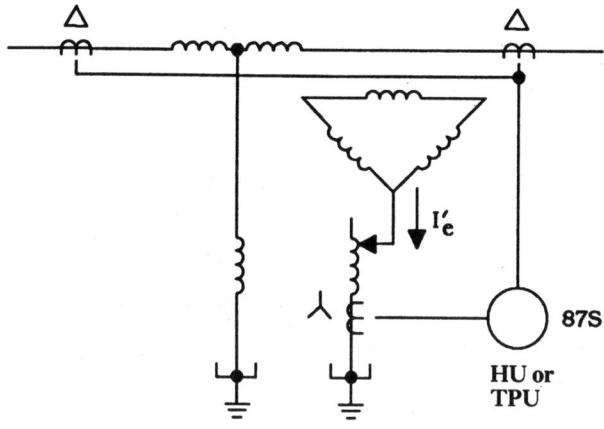

Figure 10-44 Differential protection for series winding of the phase-angle regulator of Figure 10-40.

Whether or not an external fault will cause such overvoltage on the winding and false relay operation depends on several factors, including the characteristic of the series winding (saturation curve, impedance, and tap position), location and type of fault, and power-system condition. A typical analysis shows that faults on one side of the bank produce false operation of the relay, whereas faults on the other side of the bank do not. Other cases indicate no overvoltage problems at all. If overvoltage is a problem, therefore, the 87S relays should be supervised by volts/hertz or excessive fifth harmonic.

Note in Figure 10-43 that for this phase-shifting-regulator protective scheme that the currents are summed in the relay at the right. As shown in this drawing, the current in R_2 (restraint 2) is equal to the sum of the currents in restraints R_1 and R_3. Differences in the tap settings are required only if the R_{CL} does not equal R_{CE} (nt/ny).

For illustration purposes, a typical example of current transformer ratio and relay tap selection for the 87S device is given below.

It should be noted that the ampere turns in the series winding, primary and secondary, are always balanced for any "through" condition.

An example illustrates the 87S relay taps selection for a phase shifter as shown in Figure 10-43. It has the following information:

345 kV, plus/minus 60°	189/252/315 MVA
Series winding primary	2 × 280 turns
Series winding secondary	232 turns
Exciting winding	334 turns

Use the maximum load-current value to determine the source-/load-side ct ratios:

$$\frac{315,000}{1.73 \times 345} = 528 \text{ A}$$

That is, use the 1200:5 ratio for the series-unit primary-side ct's:

Series-unit primary current	$= 528$ A
Series-unit secondary current $= 528\ (2 \times 280/232)$	$= 1274$ A
Exciting-unit secondary current $= 1.73 \times 1274$	$= 2204$ A

Therefore, the ct ratio at this location should not be lower than $2204/0.8 = 2755$ to 5, i.e., use the 3000/5 ct. 87S relay taps selection would be as follows:

Series-unit primary-side ct	$= 1200/5$ delta
Exciting-unit secondary-side ct	$= 3000/5$ wye
Current from series-unit primary to relay $= 1.73(528)(5/1200)$	$= 3.804$
Current from exciting-unit secondary to relay $= 2204\ (5/3000)$	$= 3.673$
Current ratio $= 2 \times 3.804/3.673$	$= 2.071$

Note the following:

1. A factor of 2 is included in the current's ratio calculation for this particular example. This is different from the approach in a conventional differential scheme.
2. The following illustration shows a simpler way for finding the current's ratio:

$$n_e = 232 \qquad R_{ce} = 600$$
$$n_s = 280 \qquad R_{cL} = 240$$
$$\text{Current ratio} = \frac{n_e \times R_{ce}}{n_s \times R_{cL}} = \frac{232 \times 600}{280 \times 240} = 2.071$$

Selected relay taps' ratio $= 8.7/4.2 = 2.070$
Calculated mismatch $= <0.01\%$

Use 8.7 taps for the series-unit ct's and 4.2 tap for the exciting-unit ct.

In the event that ct ratios are to be determined, the following expression may be helpful. Make M% approach 0; then the ct ratios can be determined:

$$M\% = \frac{\frac{n_e \times R_{ce}}{n_s \times R_{cL}} - \frac{T_L}{T_e}}{S} \rightarrow 0.0$$

where

$M\%$ = percent mismatch
T_L = relay line/restraint tap setting
T_e = relay exiting-unit/restraint tap setting
n_e = secondary series-winding number of turns
n_s = half-primary series-winding number of turns
R_{cL} = line/source ct ratio
R_{ce} = secondary of exciting-unit ct ratio
S = smallest of the two ratios

Devices 51N1 (short time) in the neutral circuit of the exciting-unit secondary provides sensitive ground fault protection for single-phase-to-ground faults on the secondary side of the exciting unit. The zero sequence current distribution for a ground fault in this area is shown in Figure 10-45.

Device 51N2 provides backup for devices 51N1 and 87E during single-phase-to-ground faults on the exciting-unit primary. The current flow in neutral depends on the autotransformer action of the faulted winding (Fig. 10-46).

Refer to Figure 10-47. Since there is no delta-connected winding in the exciting unit for this particular phase shift to provide a path for circulating zero sequence current, it will provide no zero sequence current for external ground fault. Therefore, the 51N1 and 51N2 devices do not have a coordination problem on external faults. However; coordination should be considered if there is a zero sequence current path for external ground faults in this area.

The sudden-pressure relays are recommended for these units, especially when there is high potential for an arcing fault in the tap-changing equipment.

9 ZIG-ZAG TRANSFORMER PROTECTION

Since the connected system will be ungrounded in some applications, a zig-zag grounding transformer can be protected against ground faults by the scheme shown in Figure 10-48. The overcurrent relays for the delta-connected current transformers provide phase-fault protection (Fig. 10-48). The time-overcurrent relay (device 51N) in the neutral provides backup ground protection. The ground relay must be set to coordinate with ground relays in the connected system. Rate-of-pressure-rise relays, such as the sudden-pressure relays, are recommended for light internal faults.

The grounding banks are seldom switched by themselves. When they are switched, however, they are subjected to magnetizing inrush—just as for other types

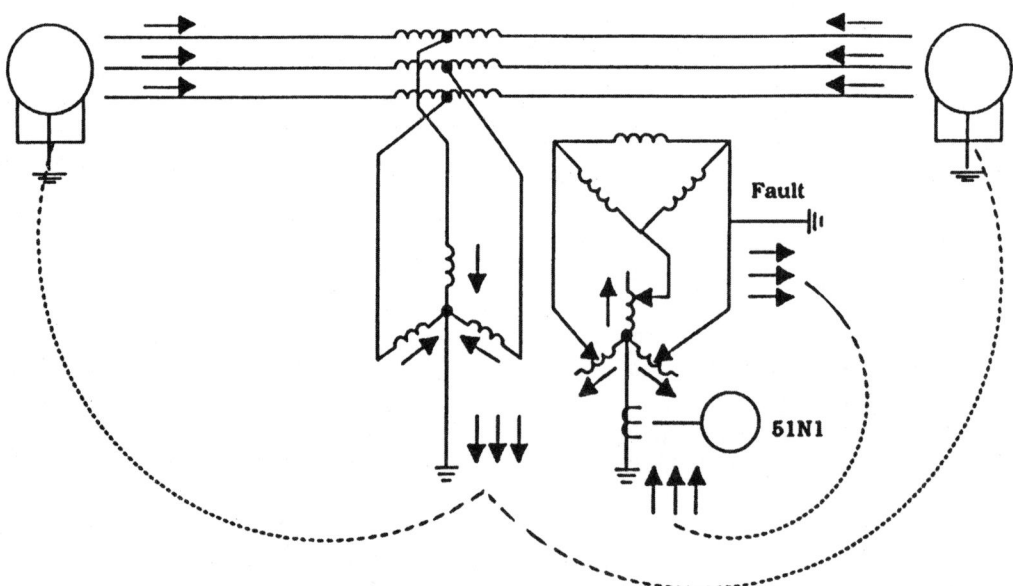

Figure 10-45 Sensitive ground protection zero sequence currents for a secondary ground fault in the exciting unit of the phase-angle regulator of Figure 10-40.

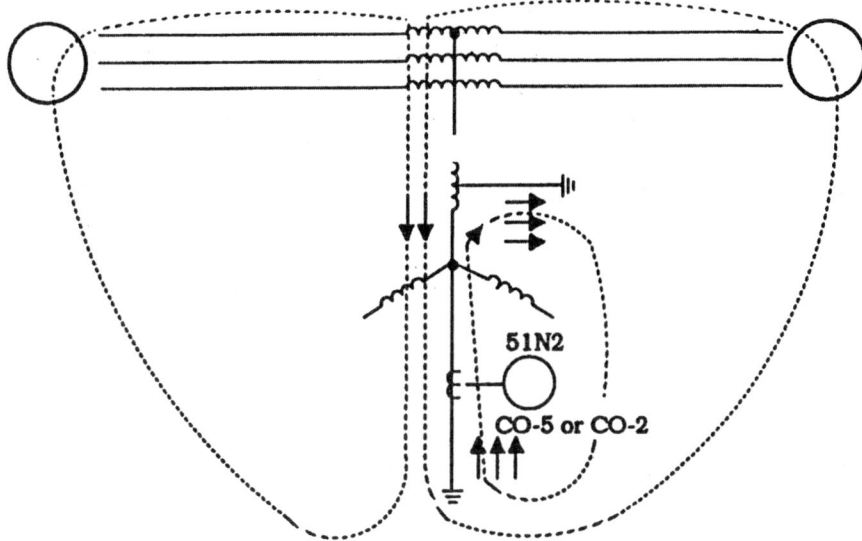

Figure 10-46 Ground backup protection and zero sequence currents for a primary ground fault in the exciting unit of the phase-angle regulator of Figure 10-40.

of transformers. The harmonic restraint relay (single-phase-type HRU), as shown in Figure 10-48, can be used to prevent inadvertent tripping during energization.

Some power transformer banks consist of zig-zag-connected windings for phase correction or system grounding. As shown in Figures 10-49 and 10-50, the phase-angle shift between the primary and secondary sides of the banks depends on their connections. In these examples, windings on one side are delta-connected. As shown in Figure 10-51, however, a wye connection could also be used, introducing a phase-angle shift of either plus or minus 30°. The grounded wye is not a zero sequence current source for ground faults on the wye side, even if both windings are grounded. It is, however, a good zero sequence current source for ground faults at the zig-zag side. Both the phase-angle shift and zero sequence current source should be considered when applying the differential scheme for these transformers.

10 PROTECTION OF SHUNT REACTORS

10.1 Shunt Reactor Applications

Both EHV transmission lines and long HV transmission lines and cables require shunt reactance to compensate for their large line-charging capacitance. This capacitance produces VAR generation that the system generally cannot absorb. This VAR genera-

tion increases as the square of the voltage and is a function of line length and the conductor configuration. In many cases, it is necessary to absorb these VAR and provide voltage control at both terminals during normal operation. High overvoltage on sudden loss of load must be limited as well. System switching and operation may require a different amount of VAR absorption and even, at times, some VAR generation.

Shunt reactance for VAR control is obtained by

Fixed shunt reactors
Switched shunt reactors or capacitors
Synchronous condensers
Static VAR compensators

Fixed shunt reactors are generally used for EHV and long HV lines and for HV cables. Switched shunt reactors or capacitors and synchronous condensers are applied in the underlying system and near load centers.

Shunt reactors vary greatly in size, type, construction, and application. Their capacities range from 3 to 125 MVA, at voltage levels from 4.6 to 765 kV. They can be single- or three-phase, oil- or dry-type, with either air or gapped-iron cores. The connections may be directly (1) to the transmission circuit, (2) to the tertiary winding of a transformer bank that is part of the line, or (3) to the low-voltage bus associated with the line transformer bank. (This third application is not common.)

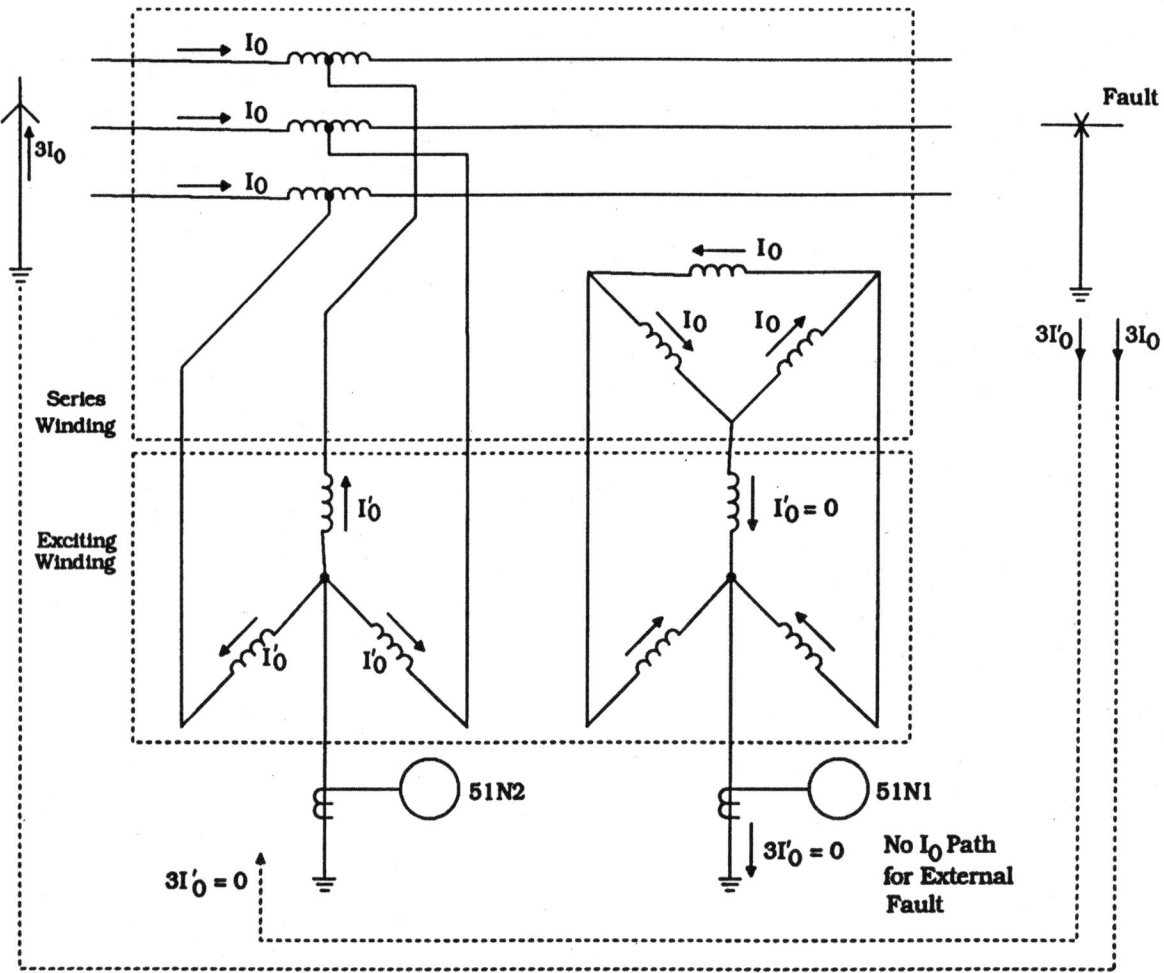

Figure 10-47   The exciting unit provides no zero sequence current path for external ground fault. Devices 51N1 and 51N2 do not have coordination problem on external faults.

Line reactors, which are connected directly or through a disconnect switch, are a part of the transmission circuit. Circuit breakers are seldom used. The neutrals of the reactors are solidly grounded or grounded through a neutral reactor. Reactor faults require that all line terminals be open.

When connected to the tertiary of a transformer bank, circuit breakers are generally used, either in the supply or on the neutral. Opening the neutral breaker does not isolate a reactor fault. Tertiary applications are operated either ungrounded or grounded through impedance.

Line operation without a reactor can result in very high overvoltage when load is lost, such as when one end is opened. This factor encourages the use of direct-connected reactors to avoid the accidental loss of service should load be lost.

Line-connected reactors are generally included within the line protection zone and are often well protected by the line relays adjacent to the units. Separate reactor relays are recommended, however, since the remote terminal may have difficulty detecting a reactor fault. These relays can be applied with current transformers sized to the reactor MVA and should include some way of transfer-tripping the remote line terminals—especially on long lines or when the remote terminal is a relatively weak source. With separate reactor relays, the line relays provide additional backup.

Tertiary-connected reactors can be included in the transformer bank differential zone. Separate reactor-protection relays are recommended. When practical, the transformer protection zone overlap should be used as backup. Line-side reactor breakers allow the

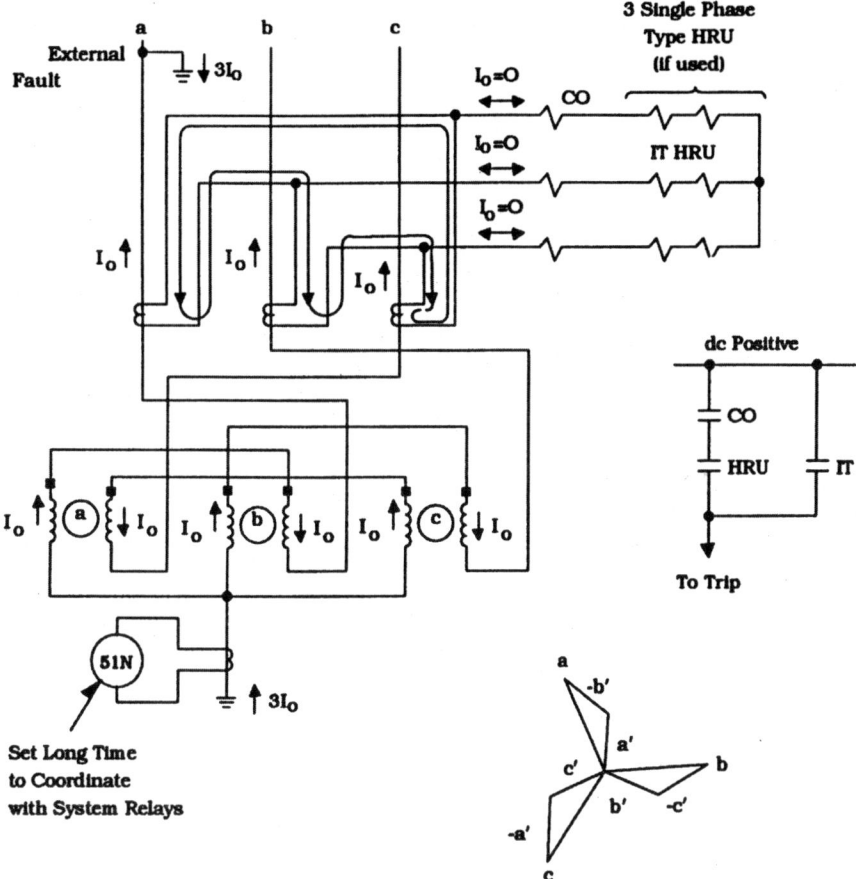

Figure 10-48 Protection of a zig-zag grounding transformer and the zero sequence currents for an external ground fault.

protection to be separated, so that the transformer bank need not be tripped for reactor faults. In such cases, the possibility of high voltage during operation without the reactors should be examined.

The protective techniques commonly used for reactor primary and backup protection are

Rate-of-rise-of-pressure (applicable to oil units with a sealed gas chamber above the oil level)
Overcurrent (three-phase and/or ground)
Differential (three-phase or ground only)

Other protective relaying techniques, such as distance, negative sequence, and current balance, have been used to a limited extent.

10.2 Rate-of-Rise-of-Pressure Protection

Rate-of-rise-of-pressure protection provides the most sensitive protection available for light internal faults.

Tripping is recommended, although such protection is sometimes used for alarm purposes only. An alarm operation should be monitored carefully since there are cases where a fault left no tangible evidence after the first pressure relay operation but later developed into a severe fault. Even on the severe fault, the pressure relay was distrusted because of the initial assumed-false operation.

Rate-of-rise-of-pressure protection can be used as separate primary protection only if line or transformer differential protection is available for faults outside the reactor tank and for backup protection. Rate-of-rise-of-pressure protection is, of course, not applicable to dry-type units.

10.3 Overcurrent Protection

Overcurrent phase and ground protection for reactors are shown in Figure 10-52. To avoid operation on

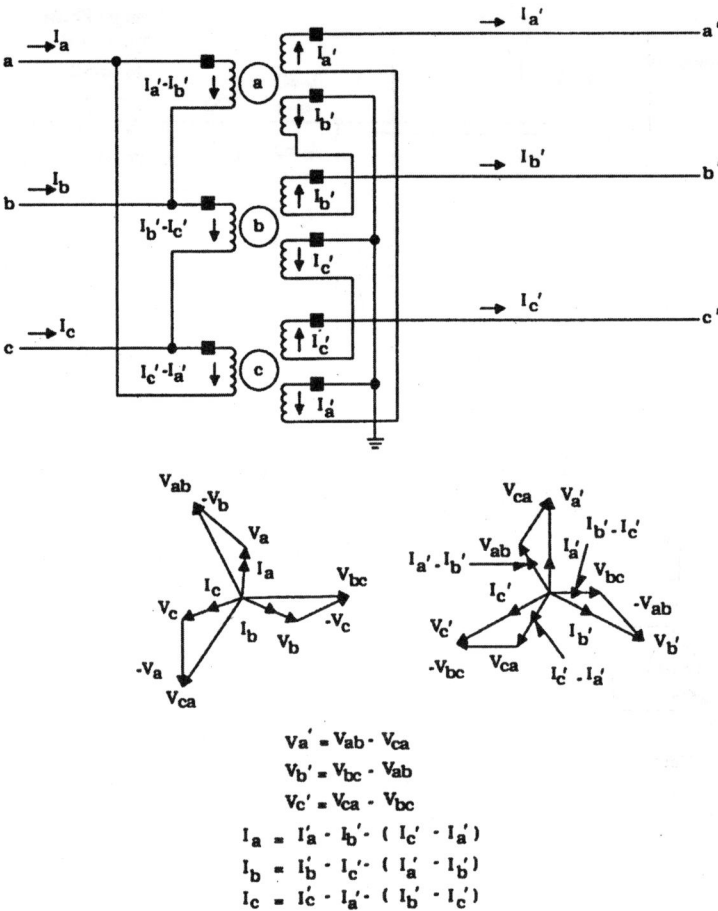

$$V_a' = V_{ab} - V_{ca}$$
$$V_b' = V_{bc} - V_{ab}$$
$$V_c' = V_{ca} - V_{bc}$$
$$I_a = I_a' - I_b' - (I_c' - I_a')$$
$$I_b = I_b' - I_c' - (I_a' - I_b')$$
$$I_c = I_c' - I_a' - (I_b' - I_c')$$

Figure 10-49   Interconnected delta zig-zag transformers with voltages in phase on the two sides.

transients, the phase-type CO time-overcurrent units (51) are set at 1.5 times the rated shunt reactor current: the IIT instantaneous units (50) are set at five times the rated current. The ground relay unit (51N) can be set at 0.5 to 1.0 A and the relay (50N) at five times the 51N setting. Both ground units should be set above the zero sequence current ($3I_0$) contribution of the reactor for faults outside the reactor protection zone. This setting will avoid operation on line-deenergized oscillations. If the reactor is connected to an ungrounded system, 50N and 51N should be omitted. This scheme requires only one set of current transformers.

10.4   Differential Protection

Separate-phase differential relays (87), as shown in Figure 10-53, are applicable for either three- or single-phase reactor units. With single-phase units, the separate differential relays aid in identifying the fault. The relays detect both winding and bushing faults. Since the relays will see magnetizing inrush as a "through" condition, generator-type relays can be used. Either the SA-1 or generator CA-type relays may be applied; both provide sensitive internal fault protection (0.14 A for the SA-1 and 0.18 A for the CA). The ground relay (50N/51N) provides backup protection.

A single CA-16 or HU-4 relay can be used for a ground differential. The four restraints are connected to the three lines and one neutral current transformer, as shown in Figure 10-54. The minimum pickup of the CA-16 is 0.15 A and of the HU-4 is 0.87 A.

The scheme shown in Figure 10-55 provides an excellent combination of phase instantaneous and time overcurrent with ground differential. For single-phase reactors, phase faults that do not involve ground cannot occur—at least within the tank. Therefore, the

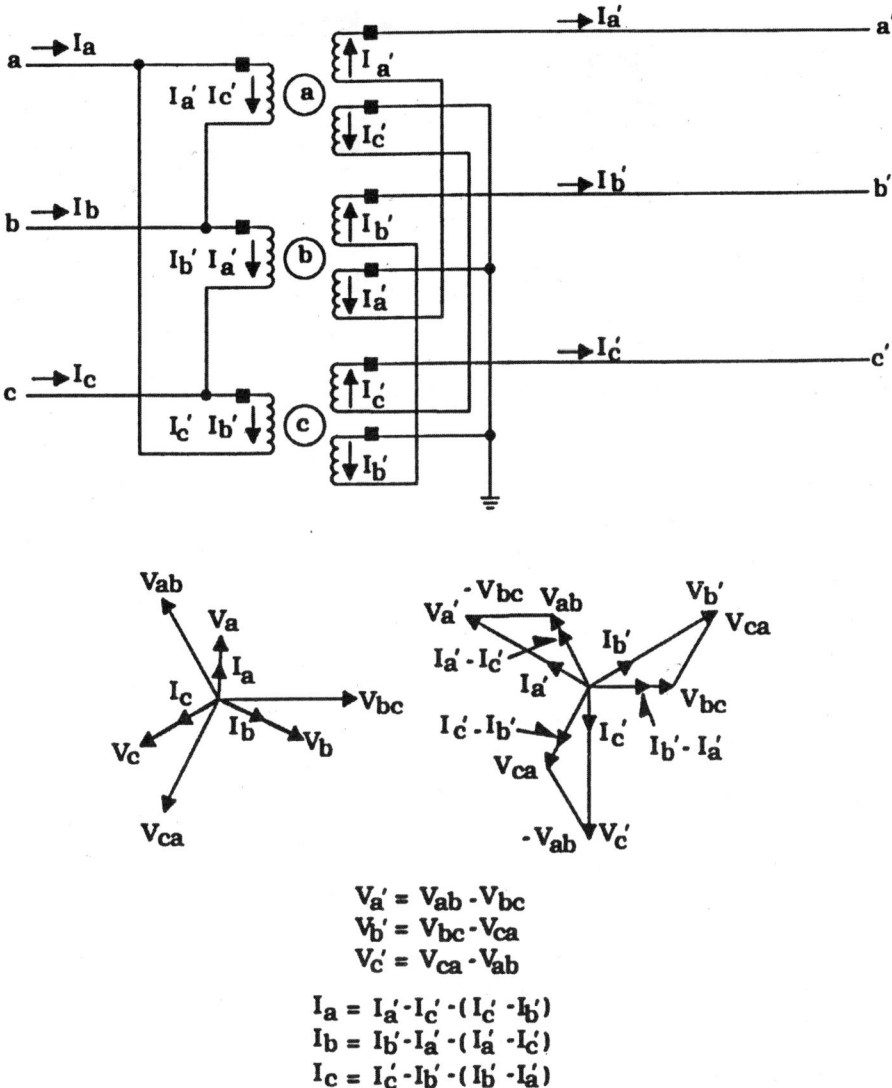

$$V_{a'} = V_{ab} - V_{bc}$$
$$V_{b'} = V_{bc} - V_{ca}$$
$$V_{c'} = V_{ca} - V_{ab}$$

$$I_a = I_{a'} - I_{c'} - (I_{c'} - I_{b'})$$
$$I_b = I_{b'} - I_{a'} - (I_{a'} - I_{c'})$$
$$I_c = I_{c'} - I_{b'} - (I_{b'} - I_{a'})$$

Figure 10-50 Interconnected delta zig-zag transformer with voltages 60° apart on the two sides.

three 50/51 relays represent backup protection, which could be omitted.

The type SA-1 or CA generator differential relays or bus differential type KAB relays can also be used for the ground differential (87N). Additional security can be obtained by using the type CWC relay, particularly if the current transformer performance is inferior. This connection is shown in Figure 10-36. Also, the ratios do not have to be identical. Sensitivity can be increased by using a lower-ratio neutral current transformer, as described above.

When the shunt reactor is grounded and connected to an ungrounded system, a CWC relay can be used, with the connections shown in Figure 10-37.

10.5 Reactors on Delta System

On delta systems, shunt reactors are usually connected to the tertiary of a power transformer associated with the line. Since most faults will involve ground, the units or associated system are grounded through high resistance for detection purposes. Neutral resistance grounding is shown in Figure 10-56, and voltage transformer grounding in Figure 10-57 (see Chap. 7). To limit both transient overvoltage and ground fault current, the resistor is sized so that I_{0R} equals or exceeds I_{0C}. Since the system capacitance to ground is very large, the impedance of the associated system is essentially negligible and is not shown in the

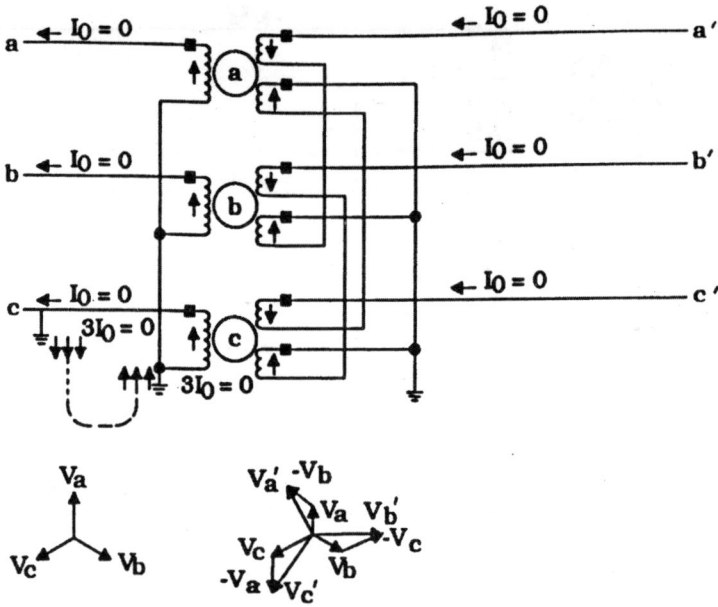

a) The Grounded Wye is Not a Ground Source for Ground Faults on The Wye Side.

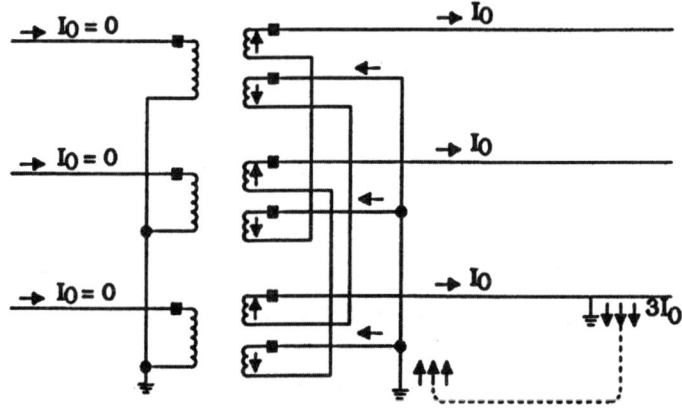

b) Zig-Zag Windings Provide a Good Ground Source Only on that Side.

Figure 10-51   Interconnected wye zig-zag transformer with wye phase voltages lagging 30° from zig-zag phase voltage.

zero sequence diagrams. Whereas the primary current for a ground fault is quite small, the secondary current will be large. If 59N is used for alarm instead of tripping, the secondary current may exceed the continuous thermal rating of the voltage transformers.

The CV-8 relay for 59N provides sensitive protection: Its pickup is 8% of its continuous rating. For alarm applications, the $3E_0$ voltage should not exceed the 69- or 199-V rating unless a series resistor is used to limit the voltage across the relay to its rating. (See Chap. 8.)

Phase protection for three-phase reactors can be obtained by overcurrent or differential relay schemes. Overcurrent protection is the same as for Figure 10-52, without 50N/51N; differential protection is as shown in Figure 10-54 or 10-53, without 50N/51N. The arrangements offer little protection for single-phase reactors unless a second ground fault should develop in another unit.

Although including the reactor within the transformer differential circuit provides some phase-fault protection, it offers no ground-fault protection with high impedance grounding. Even the phase-fault

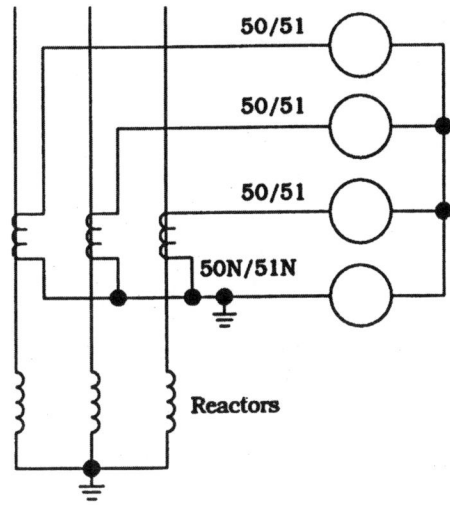

Figure 10-52 Phase and ground instantaneous and time overcurrent protection for shunt reactors.

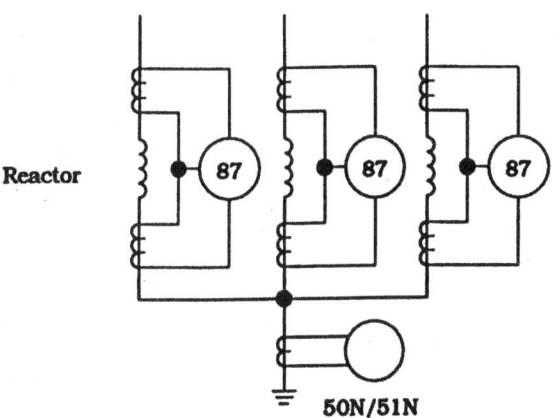

Figure 10-53 Separate phase differential protection with ground time overcurrent backup protection for shunt reactors.

protection is limited, since the current transformers of the transformer differential are sized for transformer capacity and not the smaller-reactor MVA.

Low impedance or solid grounding of the reactors may be used. In this case, either the 50N/51N neutral overcurrent relay (Figs. 10-56 and 10-57) or 87N ground differential of Figure 10-37 should be applied.

10.6 Turn-to-Turn Faults

Light turn-to-turn faults are extremely difficult to detect. Although the rate-of-rise-of-pressure relay

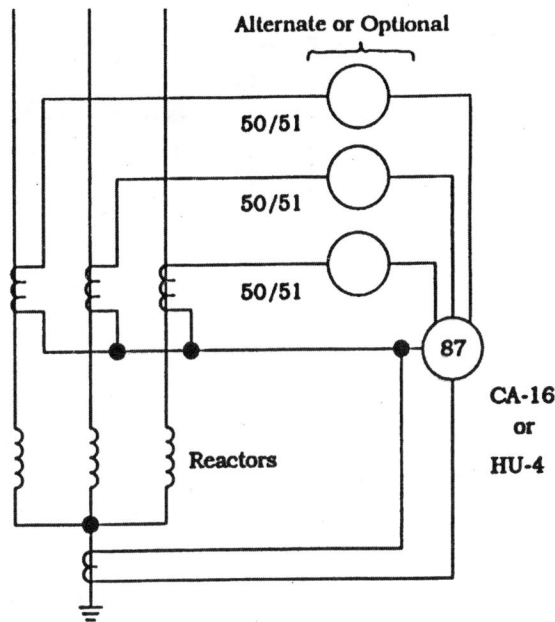

Figure 10-54 Phase instantaneous and time overcurrent with ground differential protection for shunt reactors.

offers the greatest sensitivity, its application is limited. The reactors must be oil-type, and the fault must cause a sufficient pressure change to operate the unit. While transformer action in a turn-to-turn fault can produce a large current within the shorted turn, there is very

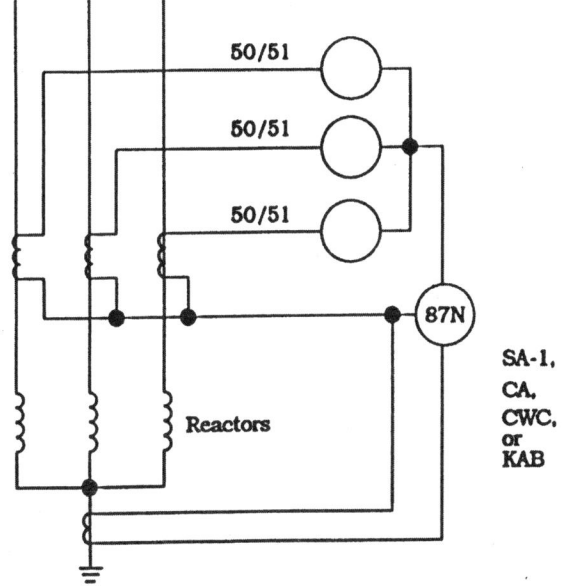

Figure 10-55 Combined phase and ground differential protection for shunt reactors.

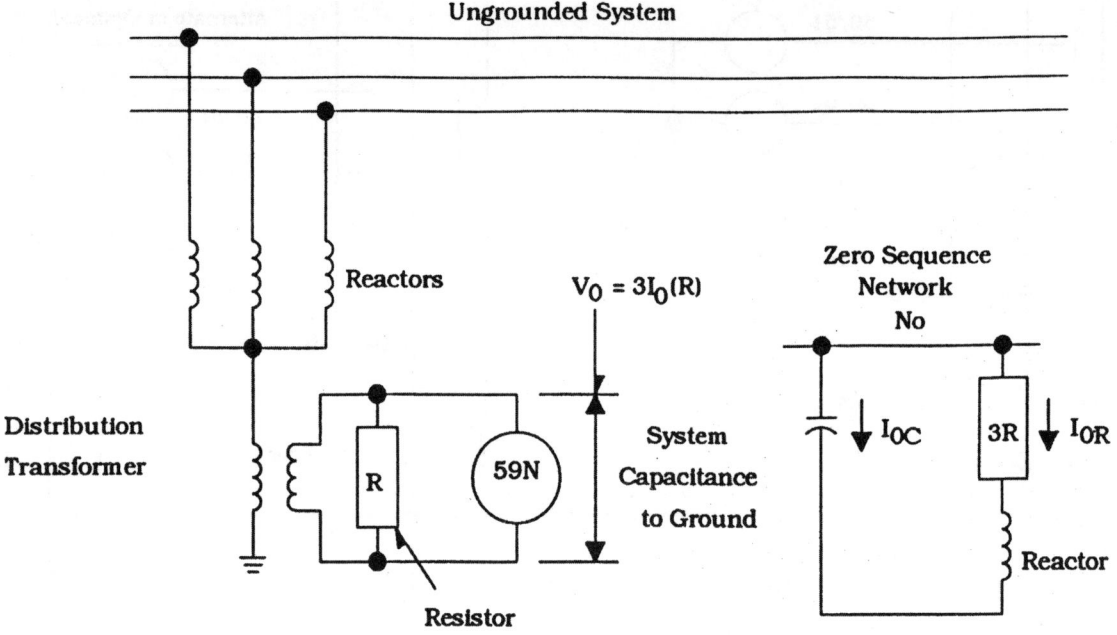

Figure 10-56 Neutral resistance grounding for ground fault detection.

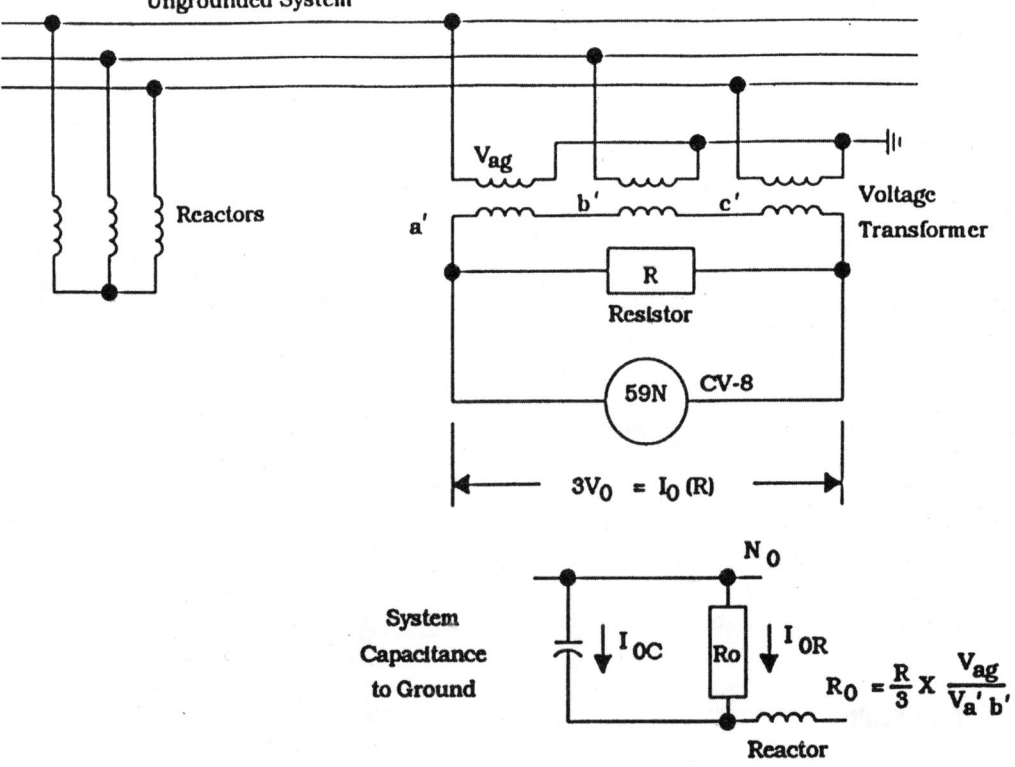

Figure 10-57 Resistance grounding through voltage transformer for ground detection.

little current change at the terminals of the unit. The effect is equivalent to an autotransformer with a shorted secondary. The impedance change that will occur in one phase can be represented by symmetrical components as a shunt unbalance. As shown in Figure 10-58, impedance ZA of phase A is not equal to the other two phases, shown with a total reactor impedance of ZB. For this condition, the sequence networks are connected as shown in Figure 10-58.

Because of the transformer action, the change in impedance of the total phase circuit for a shorted turn is difficult to calculate. As a rough estimate, assume a change of 3% in the phase with the shorted turn. Also assume that the fault has not yet involved ground or other phases. Given these assumptions and if we neglect phase angles, distributed winding capacitance, and transformer action, negative and zero sequence currents will be less than 1%. Removing the ground from the units does not change the positive and negative sequence currents significantly, although it does eliminate the zero sequence. The magnitudes of the currents are largely a function of the total reactor impedance; the source impedance is relatively low compared to the reactor impedance.

The small unbalances and sequence currents associated with turn-to-turn faults generally are no larger than the normal or tolerable unbalances. Consequently, there seems to be no reliable "handle" to distinguish between the intolerable and tolerable

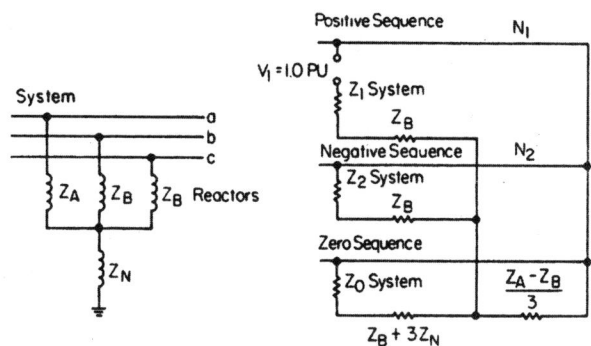

Figure 10-58 Shunt reactor with a shortened turn in phase "a" so that ZA < ZB and the sequence connections for an unbalanced impedance in phase "a."

conditions. Although special schemes or relays have been reported, they will require very careful "customized" applications.

As the turn-to-turn fault spreads to more turns, the current will increase. A negative sequence relay may be set at 0.2 A negative sequence. The relay should be applied with a timer to avoid operation on system transients and external faults and should be disabled when the breaker is opened. This latter safeguard avoids possible operation on low-frequency line oscillations after the line is deenergized. With very little resistance in the line, such oscillations can last an appreciable time.

11

Station-Bus Protection

Revised by: **SOLVEIG WARD**

1 INTRODUCTION

A bus is a critical element of a power system, as it is the point of convergence of many circuits, transmission, generation, or loads. The effect of a single bus fault is equivalent to many simultaneous faults and usually, due to the concentration of supply circuits, involves high-current magnitudes. High-speed bus protection is often required to limit the damaging effects on equipment and system stability or to maintain service to as much load as possible. The bus protection described refers to protection at the bus location, independent of equipment at remote locations.

Differential protection is the most sensitive and reliable method for protecting a station bus. The phasor summation of all the measured current entering and leaving the bus must be 0 unless there is a fault within the protective zone. For a fault not in the protective zone, the instantaneous direction of at least one current is opposite to the others, and the sum of the currents in is identical to the sum out. A fault on the bus provides a path for current flow that is not included in these summations. This is called the differential current. Detection of a difference exceeding the predictable errors in the comparison is one important basis for bus relaying.

In dealing with high-voltage power systems, the relay is dependent on the current transformers in the individual circuits to provide information to it regarding the high-voltage currents. Figure 11-1 shows typical examples of the location of current transformers that are used for this purpose. The arrowheads indicate the reference direction of the currents.

1.1 Current Transformer Saturation Problem and Its Solutions on Bus Protection

Bus differential relaying is complicated by the fact that for an external fault on one circuit, all of the other circuits connected to the bus contribute to that fault. The current through the circuit breaker for the faulted circuit will be substantially higher than that for any of the other circuits. With this very high current flowing through the current transformer and its circuit breaker, there is a very high likelihood that some degree of saturation will occur. A saturated current transformer will not deliver its appropriate current to the bus relay. With the lower currents in the other circuits for this external fault, the degree of saturation is expected to be considerably lower. This may lead to a large differential current that will tend to cause the relay to sense an internal fault rather than the actual external fault that exists. The relay must accommodate this error current without misoperation.

A widely used equivalent diagram for a current transformer appears in Figure 11-3b. It consists of a perfect transformation from the high current side to the low current side (e.g., 600:5). All of the significant imperfections are lumped into R_p, R_s, and X_m. The Rs represents the internal secondary resistance of the ct (current transformer), and the X represents a current path that accommodates the exciting requirements. The ct is assumed to have a uniformly distributed winding and, therefore, to manifest no significant leakage reactance.

When the ct is subjected to excessive flux, the ct is said to "saturate," meaning that the core of the ct has

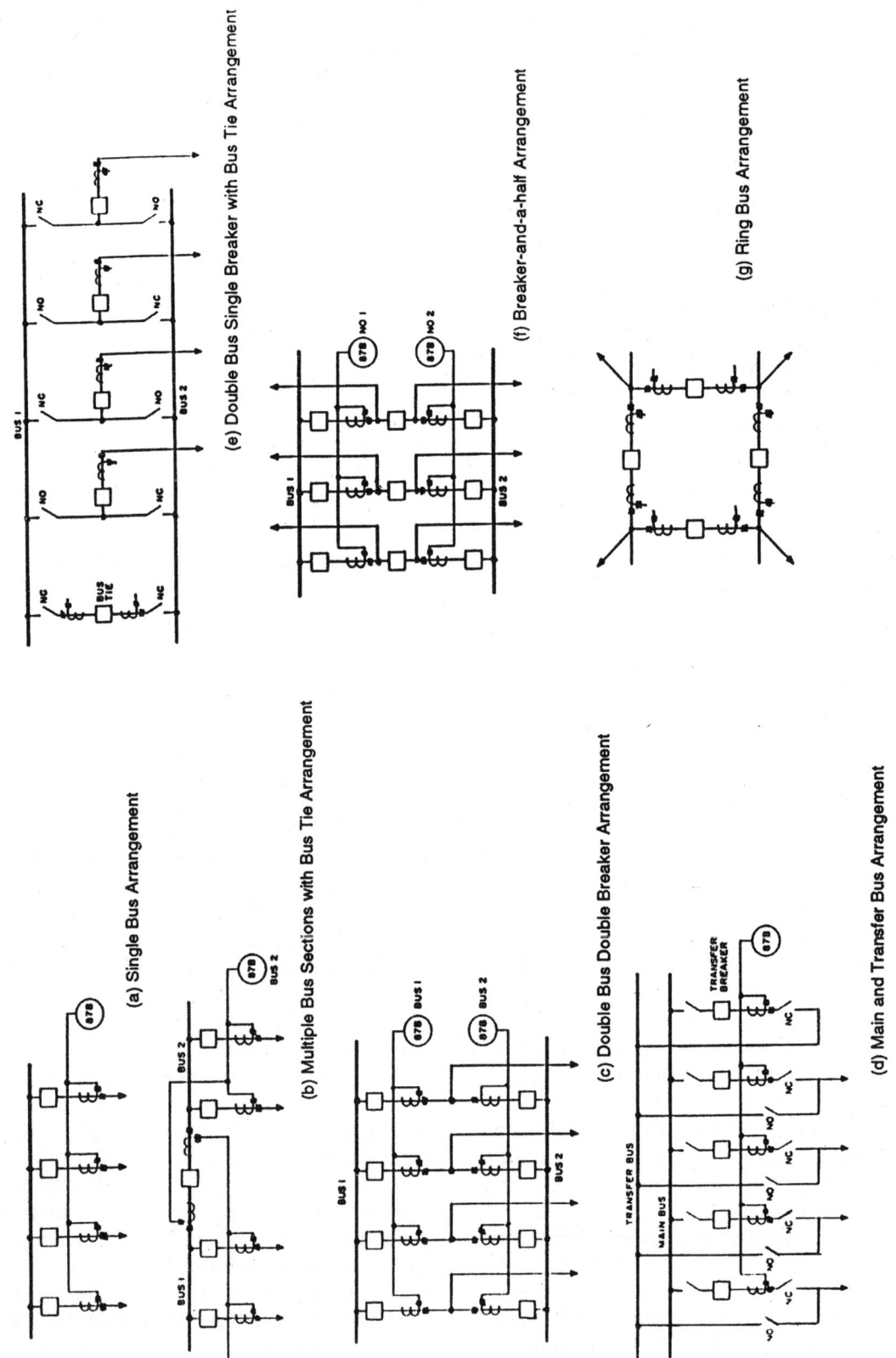

Figure 11-1 Common bus arrangements with relay input sources.

been forced to carry more flux than it can handle. The flux then spills into the area surrounding the core, causing the magnetizing reactance to have a much lower value than normal. It can be seen that any current that flows in X subtracts from the perfectly transformed current, producing a deficiency in the current that is delivered to the devices connected to the ct. The black blocks are the polarity markers. A single polarity marker has no significance. With two, it is acknowledged that, at the instant of time at which current is flowing *into* the polarity marker on the high current side of the ct, current is flowing *out* of the polarity marker on the low current side. Of course, the current reverses every half cycle, but both the high and low reverse together.

Direct current saturation is much more serious than ac saturation because a relatively small amount of dc from an asymmetrical fault wave will saturate the current transformer core and appreciably reduce the secondary output. The L/R ratio of the power-system impedance, which determines the decay of the dc component of fault current, should strongly influence the selection of the bus protective relaying. Typically, the dc time constants for the different circuit elements can vary from 0.01 sec for lines to 0.3 sec or more for generating plants. The nearer a bus location is to a strong source of generation, the greater the L/R ratio and the slower the decay of the resulting dc component of fault current.

Of the several available methods for solving the unequal performance of current transformers, four are in common use:

1. Eliminating the problem by eliminating iron in the current transformer [a linear coupler (LC) system]
2. Using a multirestraint, variable-percentage differential relay which is specifically designed to be insensitive to dc saturation (CA-16 relay system)
3. Using a high impedance differential relay with a series resonant circuit to limit sensitivity to ct saturation (KAB relay system)
4. Using a Differential Comparator relay with moderately high impedance to limit sensitivity to ct saturation (RED-521)

1.2 Information Required for the Preparation of a Bus Protective Scheme

Some bus protection schemes rely on the operation of a remote breaker. It is simple and economic, but slow

(zone-2 trip) and may interrupt unnecessarily a tapped load. When local bus protection is applied, the following information is required for the scheme selection, relay selection, and setting calculations:

1. Information about the bus configuration is required. The common bus arrangements are as shown in Figure 11-1, such as single bus, double bus, main-and-transfer bus, ring bus, breaker and a half, bus tie-breaker, double-bus-single-breaker, etc.
2. Maximum and minimum bus fault currents (single-phase-to-ground fault and three-phase fault)
3. Current transformer information, including

   Current transformer location
   Current transformer ratios
   Current transformer accuracy class
   Current transformer saturation curves

4. Operating speed requirement

1.3 Normal Practices on Bus Protection

The normal practices on bus protection are

1. There is one set of bus relays per bus section.
2. Use a dedicated ct for bus differential protection. If possible, the connection of meters, auxiliary ct's, and other relays in differential-type bus schemes should be avoided since these devices introduce an additional burden into the main circuit.
3. Lead resistance, as well as ct winding resistance, contributes to ct saturation. Therefore, the length of secondary lead runs should be held to a minimum.
4. Usually, the full-ct secondary winding tap should be used. This has two advantages. It minimizes the burden effect of the cable and, second, leads by minimizing the secondary current and makes use of the full-voltage capability of the ct.
5. Normally, there is no bus relay required for the transfer bus on a main-and-transfer bus arrangement. The transfer bus is normally deenergized and will be included in the main bus section when it is energized.
6. No bus relay is required for a ring bus because the bus section between each pair of circuit breakers is protected as a part of the connected circuit.

7. Special arrangements should be considered if there is any other apparatus, such as station service transformers, capacitor banks, grounding transformers, or surge arresters, inside the bus differential zone.

8. There is no simple scheme available for a double-bus-single-breaker arrangement (Fig. 11-1e), because its current transformers are normally located on the line side. These applications greatly benefit from numerical schemes, such as the RED-521. (Refer to Sec. 9 of this chapter for more information.)

2 BUS DIFFERENTIAL RELAYING WITH OVERCURRENT RELAYS

2.1 Overcurrent Differential Protection

This differential scheme requires that a time-overcurrent relay be paralleled with all of the current transformers for a particular phase, as shown in Figure 11-2. It is permissible to use auxiliary ct's to match ratios, but it is preferred that all of the ct's have the same ratio on the tap chosen and that the use of auxiliary ct's be avoided.

In this scheme, the overcurrent relay must be set to override the maximum error current that results from an external fault (phase or ground). It may also be necessary to have sufficient time delay to refrain from tripping during the time that one or more of the current transformers is severely saturated by the dc component of the primary current. To assure this, using a simple overcurrent relay, the current transformers must be chosen to have no more than 20 times rated current flowing in their primary for the worst-case external fault, and each have a burden no more than the rated value (relaying-accuracy-class voltage/100). The operating time of the relay must not be less

than three primary time constants, and its setting must be greater than the exciting current of the current transformer under worst-case conditions. This may require a setting of 10 or more amperes and a time setting of, say, 18 cycles. These values may be acceptable for smaller substation buses, but more sophistication and faster relaying speed are generally mandatory for more extensive and higher-voltage buses.

In these applications a "short time" or "extremely inverse" characteristic overcurrent relay is used in the interests of getting faster tripping speeds at high current. Operating times of 8 to 18 cycles are expected. Although the relay cost is low, the engineering cost may be high because of the usual need for considerable study for the application to assure correct operation.

2.2 Improved Overcurrent Differential Protection

The sensitivity of the overcurrent differential scheme (Fig. 11-2) can be improved by externally connecting a series resistor with each overcurrent relay, as shown in Figure 11-3. These resistors are called stabilizing resistors. If we assume that an external fault causes the ct on the faulted feeder to be saturated completely, the ct excitation reactance will approach 0. As shown in Figure 11-3, the error current I_d that flows through

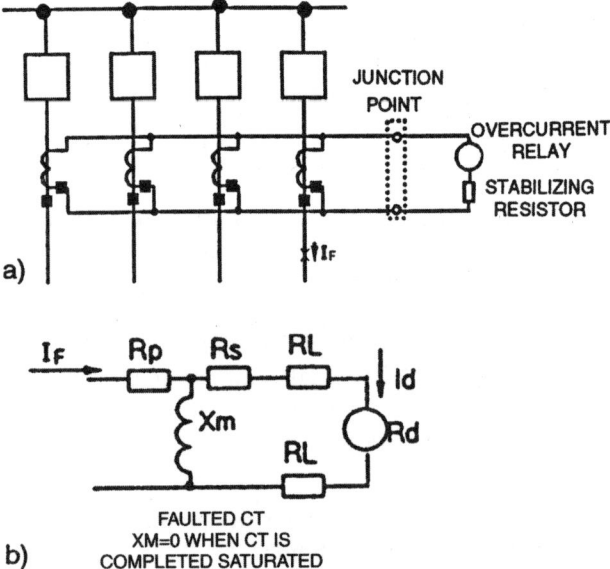

Figure 11-3 The improved overcurrent differential bus protection.

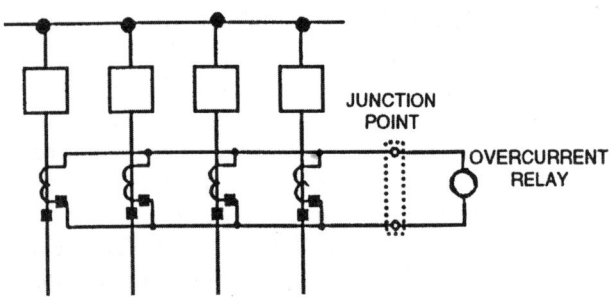

Figure 11-2 The overcurrent differential bus protection.

the overcurrent unit would be

$$I_d = I_F \left(\frac{2R_L + R_S}{2R_L + R_S + R_d} \right) \qquad (11\text{-}1)$$

where R_d is the resistance in the differential path.

In order to reduce the error current I_d in the differential path for improving the sensitivity of the scheme, the most effective way is to increase the value of R_d. The limitations of this additional resistance are determined by (1) the overvoltage to the ct circuit and (2) the minimum available internal fault current. It should be limited to

$$R_d = \frac{V_{CL}}{4 \times I_{min} \text{pickup}} \Omega \qquad (11\text{-}2)$$

Note: The multiplier 4 includes a safety factor of 2.

3 MULTIRESTRAINT DIFFERENTIAL SYSTEM

Multirestraint differential schemes use conventional current transformers, which may saturate on heavy external faults. For this reason, the secondary current output may not represent the primary. In a differential scheme, the current transformers and relay function as a team. When the current transformers do not perform adequately, the relay can within limits make up for the deficiency.

The multirestraint differential scheme uses the CA-16 variable-percentage differential relay, which consists of three induction restraint units and one induction operating unit per phase. Two of the units are placed opposite each other and operate on a common disc. In turn, the two discs are connected to a common shaft with the moving contacts. All four of the units are unidirectional; that is, current flow in either direction through the windings generates contact-opening torque for the restraint units or contact-closing torque for the operating unit. Each restraint unit (called R, S, and T) also has two windings to provide restraint proportional to the sum or difference, depending on the direction of the current flow. If the currents in the two paired windings are equal and opposite, the restraint is cancelled. Thus, the paired restraint windings have a polarity with respect to each other. With this method six restraint windings are available per phase.

In addition to providing multiple restraint, the variable-percentage characteristic helps in overcoming current transformer errors. At light fault currents, current transformer performance is good, and the

percentage is small for maximum sensitivity. For heavy external faults, current transformer performance is likely to be poor, and the percentage is large. The variable-percentage characteristic is obtained by energizing the operating unit through a built-in saturating autotransformer.

The saturating autotransformer also presents a high impedance to the false differential current, which tends to limit the current through the operating coil and to force more equal saturation of the current transformers. On internal faults, in which a desirable high differential current exists, saturation reduces the impedance. A further advantage of the saturating autotransformer is that it provides a very effective shunt for the dc component, appreciably reducing the dc sensitivity of the operating units. At the minimum pickup current of $0.15 \pm 5\%$ A, the restraining coils are ineffective.

When using the CA-16 relay, the current transformers should not saturate when carrying the maximum external symmetrical fault current; that is, the exciting current should not exceed one secondary ampere rms. This requirement is met if the burden impedance does not exceed

$$\frac{[N_P V_{CL} - (I_{EXT} - 100)]R_S}{1.33 \, I_{EXT}} \qquad (11\text{-}3)$$

where

N_P = proportion of total current transformer turns in use

V_{CL} = current transformer accuracy-class voltage

I_{EXT} = maximum external symmetrical fault current in secondary (amperes rms) (use $I_{EXT} = 100$ if $I_{EXT} < 100$)

R_S = current transformer secondary winding resistance of the turns in use (in ohms); for example, if the 400:5 tap of a 600:5 wye-connected class C200 current transformer is used, then $N_P = 400/600 = 0.67$ and $V_{CL} = 200$

If $I_{EXT} = 120$ A and $R_S = 0.5\,\Omega$, then the burden of the ct's secondary circuit, excluding current transformer secondary winding resistance, should not exceed

$$\frac{0.67 \times 200 - (120 - 100)0.5}{1.33 \times 120} = 0.78\,\Omega$$

Settings for the CA-16 relay need not be calculated. Field experience indicates that one CA-16 relay per phase is satisfactory for the vast majority of applications.

External connections are as shown in Figures 11-4 through 11-6. Figure 11-4 may be used if only three circuits are involved. The term circuit refers to a source or feeder group.

When several circuits exist and the bus can be reduced to four circuits, then the scheme of Figure 11-5 may be used. For example, assume a bus consists of two sources and six feeders, and that the feeders are lumped into two groups. The bus now reduces to four circuits.

In paralleling current transformers, each feeder group must have less than 14 A load current (restraint coil continuous rating).

If the bus reduces to more than four circuits, then the scheme of Figure 11-6 should be used. In applying the scheme of Figure 11-6, each primary circuit must be identified as either a source or feeder. As defined here, a feeder contributes only a small portion of the total fault current for a bus fault. All other circuits are sources. Next, a number of feeders are lumped into a feeder group by paralleling feeder current transformers. Each feeder group must have less than 14 A load current and not contribute more than 10% of the total phase- or ground-fault current for a bus fault. Then connect the "source" and "feeder groups" alternately as shown in Figure 11-6.

Note that in Figures 11-4 through 11-6, electromagnets R, S, and T are referred to. Each of these elements has two windings. The polarity markings are extremely significant as related to one another on the same electromagnet, but have no significance with respect to one another on different electromagnets. If the current into a polarity marker is equal to the

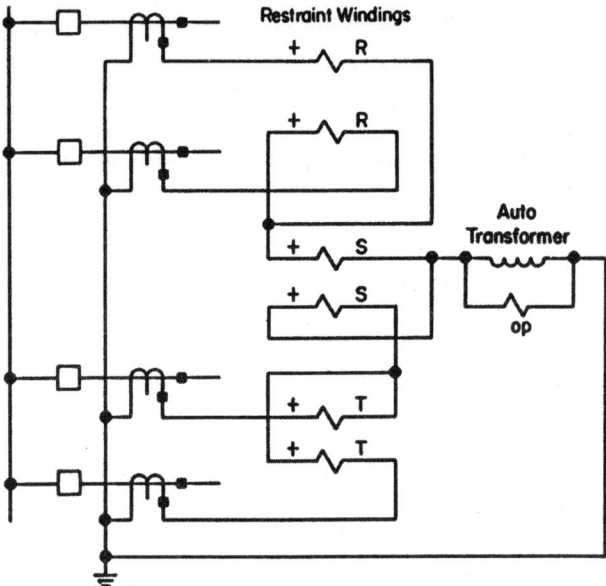

Figure 11-5 Connection of one CA-16 relay per phase to protect a bus with four equivalent circuits. (Connections for one phase only are shown.)

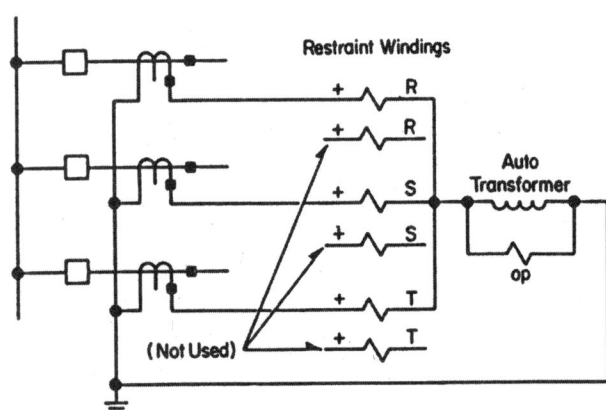

Figure 11-4 Connection of one CA-16 relay per phase to protect a bus with three equivalent circuits. (Connections for one phase only are shown.)

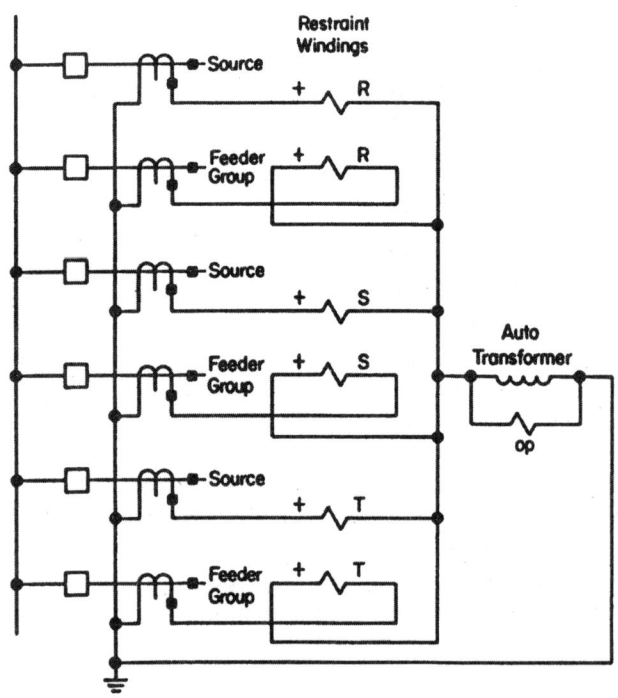

Figure 11-6 Connection of one CA-16 relay per phase to protect a bus with six equivalent circuits. (Connections for one phase only are shown.)

current out of the polarity marker on the same electromagnet, there will be no restraining torque produced by that electromagnet. The sum of all of the restraint torques is compared to that produced by the operating coil. Current into the operating coil circuit produces a much stronger effect than the same current through a single restraint winding. For an external fault, there is no current through the operating coil if the current transformers perform perfectly. There will be substantial restraint for this same condition, even though in some restraint electromagnets some (or even total) cancellation may take place.

Consider a fault on the bus of Figure 11-5 in which all of the high-voltage circuits contribute the same value of current. *All* of the restraint cancels because in each of the electromagnets the current into the polarity marker equals the current out of its paired coil. All of the internal fault current (in secondary terms, of course) flows into the operating coil circuit and fast tripping occurs. Practical cases with widely differing fault contributions produce similar effects even though considerable restraint torque may be present.

Consider, now, an external fault on the upper circuit off of the bus with the equal fault current contributions that were assumed in the previous case. Torque cancellation occurs in electromagnet T, as before. Substantial restraint torque is produced by R and S. The operating coil current cannot exceed the error current in the faulted circuit (which may well be extreme due to the effect of saturation).

This is a very sensitive bus relaying scheme, and it is very secure against operation for external faults even though severe ct saturation may occur for one or more ct's. It is reasonably fast. Another advantage is that it can accept auxiliary ct's in the circuit, which allows different ratios of the main ct's. Two shortcomings are its comparative inflexibility as other circuits are added to the bus and the need to bring all circuits back from the switchyard to the relay location.

4   HIGH IMPEDANCE DIFFERENTIAL SYSTEM

Although the high impedance differential scheme also uses conventional current transformers, it avoids the problem of unequal current transformer performance by loading them with a high impedance relay (Fig. 11-7).

This arrangement tends to force the false differential currents through the current transformers rather than the relay operating coil. Actually, the high impedance differential concept comes from the above "improved

overcurrent differential" approach. It uses a high impedance voltage element instead of "a low impedance overcurrent element plus an external resistor."

The high impedance differential KAB relay consists of an instantaneous overvoltage cylinder unit (V), a voltage-limiting suppressor (varistor), an adjustable tuned circuit, and an instantaneous current unit (IT).

On external faults, the voltage across the relay terminals will be low, essentially 0, unless the current transformers are unequally saturated. On internal faults, the voltage across the relay terminals will be high and will operate the overvoltage unit. Since the impedance of the overvoltage unit is $2600\,\Omega$, this high voltage may approach the open-circuit voltage of the current transformer secondaries. The varistor limits this voltage to a safe level.

Since offset fault current or residual magnetism exists in the current transformer core, there is an appreciable dc component in the secondary current. The dc voltage that appears across the relay will be filtered out by the tuned circuit, preventing relay pickup.

The IT current unit provides faster operation on severe internal faults and also backup to the voltage unit. The range of adjustment is 3 to 48 A.

The KAB relay has successfully performed operations up to external fault currents of 200 A secondary and down to an internal fault current of 0.27 A secondary. Its typical operating speed is 25 msec.

The overvoltage unit is set by calculating the maximum possible voltage for an external fault as follows:

$$V_R = K(R_S + R_L)\frac{I_F}{N} \qquad (11\text{-}4)$$

Where

V_R = pickup setting of the V unit in volts rms
R_S = dc resistance of current transformer secondary winding, including internal leads to bushing terminals
R_L = resistance of lead from junction points to the most distant current transformer (one-way lead for phase faults, two-way lead for phase-to-ground faults)
I_F = maximum external primary fault current, in amperes rms, contributed by the bus
N = current transformer turns ratio
K = margin factor

The maximum voltage occurs for the external fault when the faulted circuit current transformer is

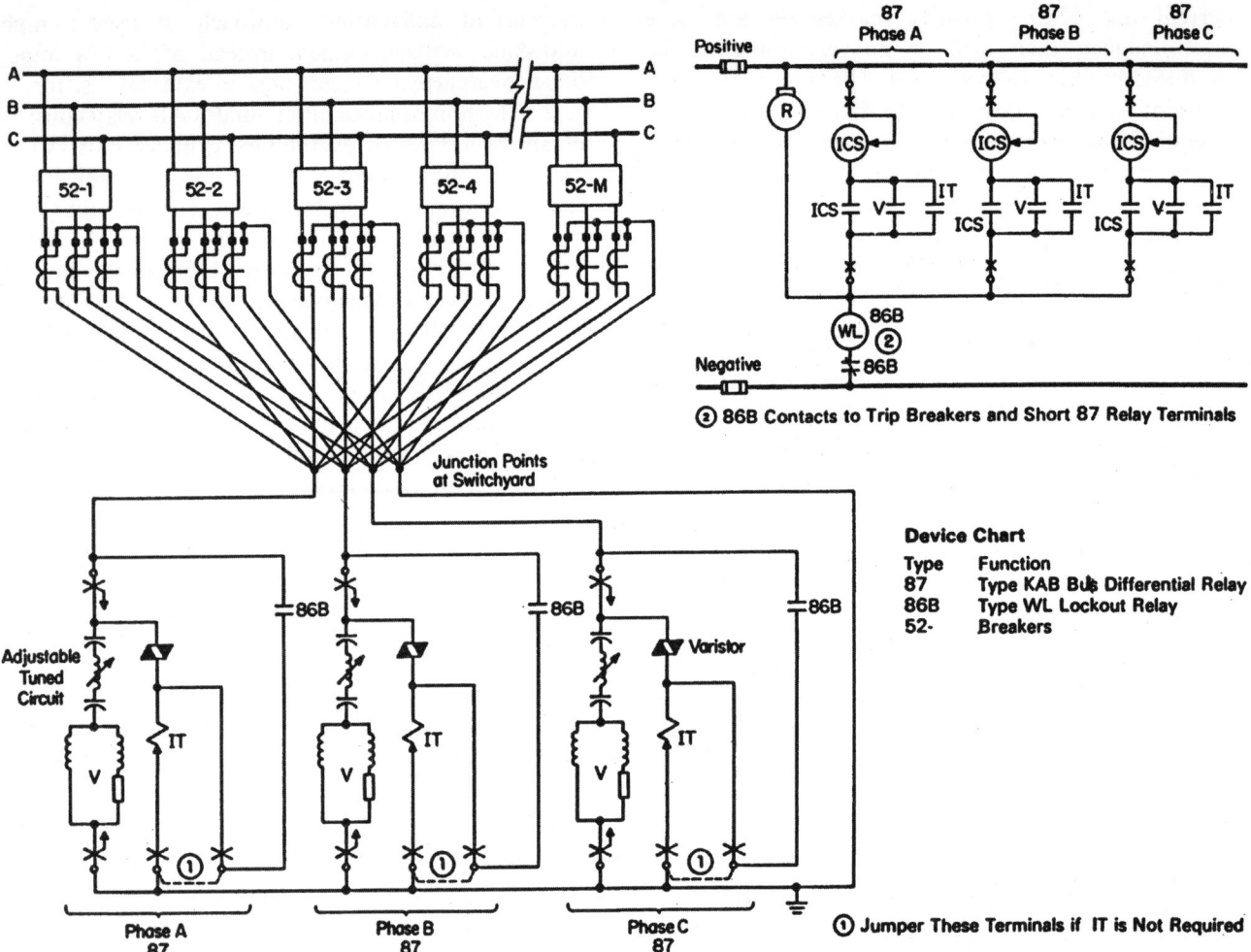

Figure 11-7 External connection of type KAB bus differential relay.

completely saturated, and there is no saturation in the source current transformers. The maximum voltage is equal to the resistance drop produced by the secondary current through the leads and secondary winding of the saturated current transformer. In practice, the faulted current transformer will never completely saturate, and the source current transformers will tend to saturate. As a result, the actual maximum voltage is less than the theoretical value. The margin factor K, which modifies this voltage, varies directly with the current transformer saturation factor SF:

$$\frac{1}{\text{SF}} = \frac{(R_S + R_L)I_F}{NV_k} \tag{11-5}$$

where

V_k = knee voltage value of the poorest current transformer connected to the relay. For type

KAB relay application, the knee voltage is defined as the intersection of the extension of the two straight-line portions of the saturation curve. The ordinate and abscissa must use the same scales.

The margin factor curve, shown in Figure 11-8, is based on tests of the KAB relay in the high-power laboratory. A safety factor of 2 has been included in constructing this curve.

The maximum number of circuits that can be connected to the relay, or the minimum internal fault current required to operate the relay, can be estimated from the following equation:

$$I_{min} = (XI_e + I_R + I_V)N \tag{11-6}$$

where

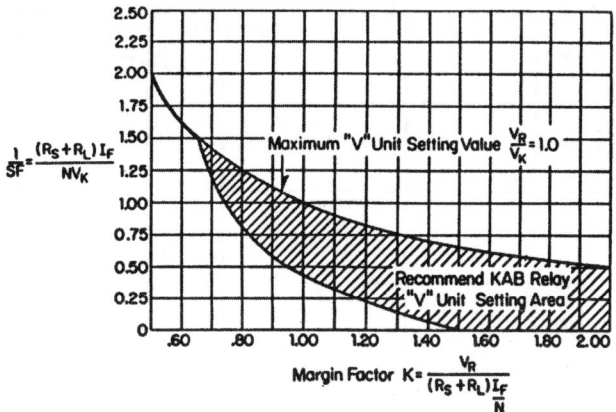

Figure 11-8 Empirical margin factor for setting the V-unit of the KAB relay.

I_{min} = minimum primary fault current in amperes rms

I_e = secondary excitation current of the current transformer at a voltage equal to the setting value of the V unit in amperes

I_R = current in the V unit at setting voltage V_R in amperes, that is, $I_R = V_R/2600$

I_V = current in varistor circuit at a voltage equal to the setting value of the V unit in amperes (generally negligible)

N = current transformer turns ratio

X = number of circuits connected to the bus

In general, the following factors should be considered when applying a high impedance bus differential relay.

4.1 Factors that Relate to the Relay Setting

The V-unit setting of the KAB relay is based on the calculated result of Eq. (11–4), which is determined by the values of K, R_S, R_L, and I_F. In order to keep this setting value within the available relay range of 75 to 400 V, it is necessary to keep the values of $(R_S + R_L)$ and any additional burden in the ct secondary as low as possible. This includes the consideration of the following:

Use fully distributed winding current transformers, such as bushing ct's or current transformers with toroidally wound cores, such as those used in metal-clad switchgear. These ct's provide a negligible leakage reactance and therefore do not contribute to the internal impedance in the equivalent circuit of the ct. Only the R_S resistance is needed in series with R_L in Eq. (11–4).

The use of auxiliary ct's is discouraged, though, with proper consideration of their resistance in series with the lead resistance (raising the effective R_L), they may be used at the sacrifice of some sensitivity of fault recognition. The same comment applies to the introduction of other devices in the current transformer circuits.

The junction point for all of the ct's in the bus differential system should be in such a location as to equalize as much as possible the distance from each ct to this point. This will minimize R_L, the value used in the setting calculation and thus allow better sensitivity to be achieved. Departure from this requirement is permissible in metal-clad switchgear because of the comparatively short distances usually involved.

The lead resistance from the junction point to the relay terminals is not critical.

Note that with this system total saturation of the current transformer on a circuit feeding an external fault is allowed and the relay remains secure.

4.2 Factors that Relate to the High-Voltage Problem

All ct's in the bus differential circuit should be operated on their full-tap position. Refer to Figure 11-9; a high voltage will be induced on the unused portion of the ct circuit due to auto-transformer action.

All current transformers should have the same ratio. If taps must be used, the windings between the

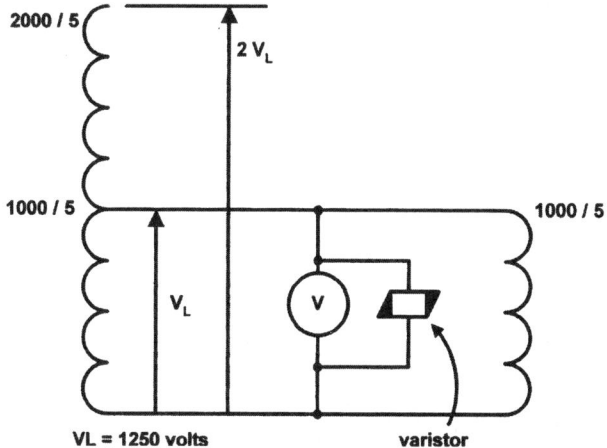

Figure 11-9 High voltage induced by autotransformer action.

taps must be completely distributed, and any high voltage at the full-tap terminal caused by autotransformer action should be checked to avoid insulation breakdown. In general, auxiliary ct's should not be used to match ratios.

4.3 Setting Example for the KAB Bus Protection

Assume a six-circuit bus for which the maximum external three-phase fault current is 60,000 A rms, symmetrical; the maximum external phase-to-ground fault current is 45,000 A, and the minimum internal fault current is 10,000 A. The current transformer ratios are 2000:5, ANSI class C400, V_k is 375 V. The secondary winding resistance R_S is 0.93, and one-way lead resistance to junction point R_L is 1.07 Ω.

4.3.1 Settings for the V Voltage Unit

For the three-phase fault condition [using Eq. (11-5)],

$$\frac{1}{SF} = \frac{(0.93 + 1.07)60,000}{400 \times 375} = 0.8$$

From Figure 11-8, $1.2 > K \geq 0.82$ (use the lower value of 0.82 for sensitivity); therefore, using Eq. (11-4), we get

$$V_R \geq 0.82(0.93 + 1.07)\frac{60,000}{400} = 246\,V$$

For the phase-to-ground fault condition,

$$\frac{1}{SF} = \frac{(0.93 + 2 \times 1.07) \times 45,000}{400 \times 375} = 0.92$$

And from Figure 11-8, $1.1 > K \geq 0.77$; therefore, using Eq. (11-4) yields

$$V_R \geq 0.77(0.93 + 2 \times 1.07)\frac{45,000}{400} = 266\,V$$

The minimum setting of the V unit in the KAB relay, therefore, should be 266 V, the larger value for either the three-phase or phase-to-ground conditions, as calculated.

4.3.2 Setting for the IT Current Unit

The IT setting is determined from Figure 11-10. The higher value is used as the ordinate as determined from the three-phase and phase-to-ground fault. Thus, for

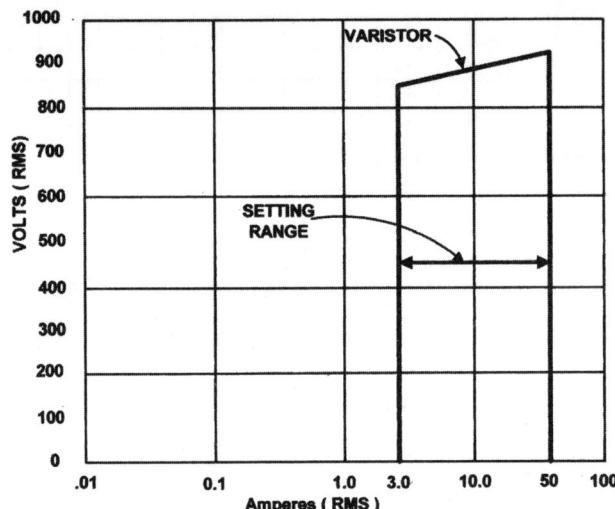

Figure 11-10   Setting of KAB instantaneous unit.

the example, the ordinate value is

$$\text{Three-phase fault} = \frac{(0.93 + 1.07)60,000}{400} = 300$$

Phase-to-ground fault

$$= \frac{(0.93 + 2 \times 1.07)45,000}{400} = 345$$

From these numbers, it is obvious from Figure 11-10 that the IT unit is incapable of operating for an external fault. The lowest available setting of 3 A will usually be adequate because of the high conduction level of present-day varistors. The principal tripping function is accomplished at high speed by the voltage unit, and only in extreme circumstances will the IT unit operate for an internal fault.

5 DIFFERENTIAL COMPARATOR RELAYS

These relays use the fundamental principle described in Figure 11-11. The RADSS is a solid-state version, the REB-103 is similar to this, but the logic is accomplished with a microprocessor, while the RED-521 is entirely a numerical relay. All are very high-speed relays (9- to 16-msec tripping) and are very secure against misoperation for external faults; all reliably and sensitively detect internal faults and are quite flexible in accommodating additional circuits. They may also be used for generator stator protection and for shunt reactor protection though their prime application area is for bus protection.

The RADSS and the REB-103 relays use external auxiliary current transformers which allow substan-

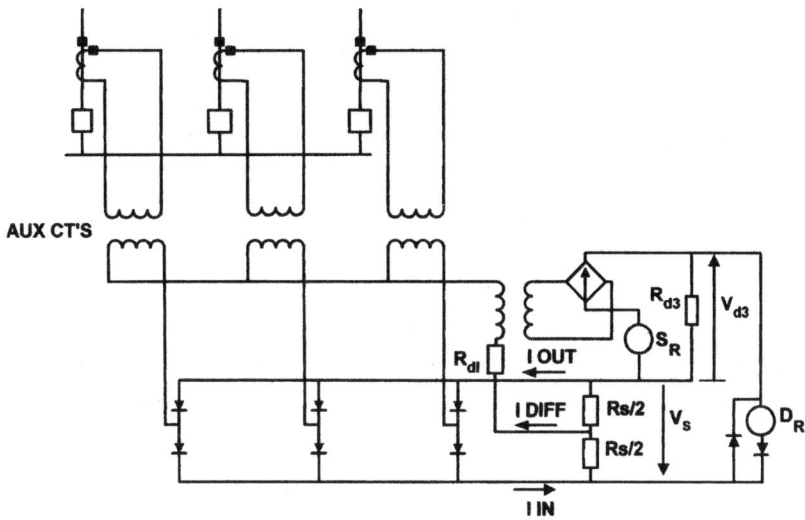

Figure 11-11 Differential comparator relay.

tially different main circuit current transformers to be accommodated and also reduce current to a suitable level for the relay. The RED-521, being a microprocessor relay, is able to accept widely varying inputs from the main current transformers and to provide, internally, the appropriate scaling factors. The RED-521 is therefore very suitable for double-bus-single-breaker arrangements as no external ct switching takes place. The ct is connected to the appropriate protection zone numerically inside the relay.

Taking advantage of Kirchoff's law, the scheme compares the sum of all of the currents entering the bus with the sum of all of the currents leaving the bus. These are instantaneous currents (as opposed to rms or average currents.)

In the circuit of Figure 11-11, the currents are delivered to the relay through the diodes. The sum of the currents through the lower group of diodes is representative of the instantaneous sum of the incoming currents to the bus, and the current flowing to the upper group of diodes is representative of the instantaneous sum of the currents leaving the bus. These two sum currents are always in perfect balance provided the current transformers perform their job faithfully and there is no fault on the bus (or to state it more correctly, provided there are no current paths off of the bus that are unaccounted for).

If an internal fault (phase or ground) were to occur, the currents in and out would no longer match. They would differ by the amount of the fault current. This difference current appears as I DIFF in the relay.

To accommodate the inherent errors in the current transformers for an external fault, particularly in the ct

associated with the circuit on which the external fault occurred, restraint is developed across the resistor Rs.

Any condition that produces I DIFF current will, through the transformer and the full-wave bridge, generate a voltage V_{d3}. For the through fault case, the restraint voltage V_s will exceed the operating voltage V_{d3}, and the relay will refrain from operating. For the internal fault case, I DIFF will be large, V_{d3} will exceed V_s, current will be passed through the diode and the reed relay D_R, and tripping will occur. S_R is a "start" relay whose contact supervises tripping to add to the overall security of the relay. It is obvious that this relay is extremely fast because the decision to trip is based on instantaneous currents.

The RED-521 numerical relay uses this principle, but is not encumbered by need for the auxiliary matching current transformers, the diodes, or any other of the components inherently required in the comparison process. The individual samples of currents are collected and summed appropriately to develop numerically the I IN and I OUT values and the corresponding restraint quantity. This is compared with the difference of these individual sums, I DIFF, and a determination of the need to trip is established.

6 PROTECTING A BUS THAT INCLUDES A TRANSFORMER BANK

Ideally, when the bus includes a power transformer bank, separate protection should be provided for the bus and transformer, even though both protection schemes must trip all breakers around the two units.

Such a system offers maximum continuity of service, since faults are easier to locate and isolate. Also, using a bus differential relay for bus protection and transformer differential relay for transformer protection provides maximum sensitivity and security with minimum application engineering.

However, economics and location of current transformers often dictate that both units be protected in one differential zone. For these applications, either the multirestraint HU-4 or CA-26 relays should be used.

The HU-4 relay is similar to the HU and HU-1 relays, except that it has four restraint windings. Also, the rectified outputs of the restraint transformers are connected in series, providing a higher restraint force when a through fault occurs on the bus. Since the dc saturation of current transformers will allow current to pass into the HRU transformers and possibly pick up the IIT, the IIT unit of the HU-4 relay is set at 15 times the rms tap value to prevent false tripping for external faults.

Similar to the CA-16, the CA-26 relay has a stronger contact spring and higher pickup of $1.25 \pm 5\%$ A to help override inrush. Its variable restraint curve is steeper than the CA-16, and its operating time is approximately three cycles.

Of the two types, the HU-4 relay is preferred, as it is immune from operation on transformer magnetizing inrush. The HU-4 should always be applied for large transformer banks or those associated with HV and EHV buses. A typical application, shown in Figure 11-12, protects a three-winding transformer bus with four circuits. Figure 11-13 illustrates another typical application used in EHV systems.

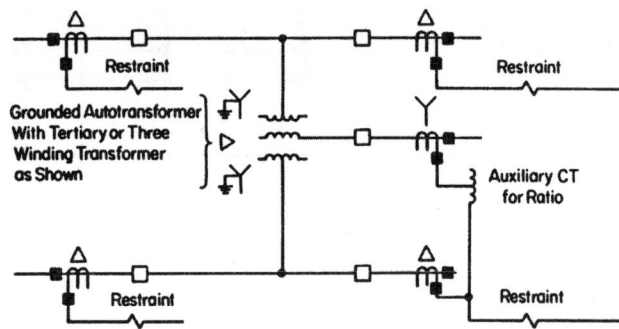

Figure 11-13 Protection of a typical transformer section where the transformer tertiary is brought out for load or connected to an external source.

The CA-26 relay is applicable to relatively small transformers remote from generating stations, HV, and EHV buses. Here, inrush will usually be light and not cause the CA-26 to operate. If, however, complete security against inrush is required, the HU-4 must be applied.

With CA-26 relays, the four-circuit bus connections of Figure 11-5 are not recommended for bus protection, since the relay may have too much restraint for a bus fault.

The bus CA-16 relay should not be used for the transformer differential, since it is too sensitive to override magnetizing inrush.

7 PROTECTING A DOUBLE-BUS SINGLE-BREAKER WITH BUS TIE ARRANGEMENT

The double-bus single-breaker with bus tie (Figure 11-1e) provides economic and operating flexibility comparable to the double-bus double-breaker arrangement (Fig. 11-1c). However, the ct's are normally on the line-side location, which results in increased differential relaying problems. Two different approaches have been used in the bus protection of such arrangements: the fully switched scheme (Fig. 11-14) and the paralleling switch scheme (Fig. 11-15). They are both complicated (inserting switch contacts in the ct circuits) and/or imperfect in protection. These schemes either require switching ct's and/or disabling the bus protection before any switching operation. This is a period when the probability of a bus fault occurring is high and it is most desirable that the bus protection be in service. A third scheme as shown in Figure 11-16 can be considered. It is similar to the paralleling switched scheme except a check-zone relay is added as shown.

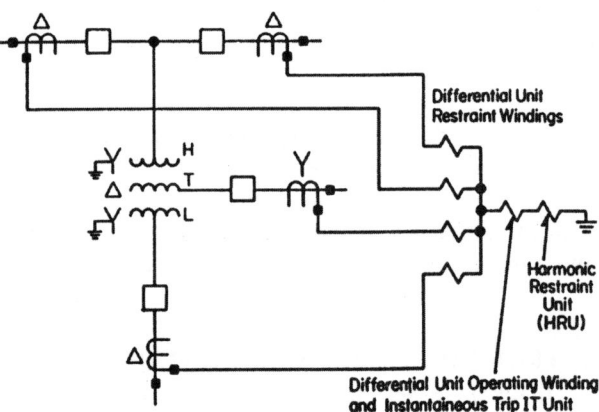

Figure 11-12 Typical application of HU-4 relay for protecting a large transformer bank associated with HV and EHV buses. (Auxiliary current transformers for ratio matching are not shown.)

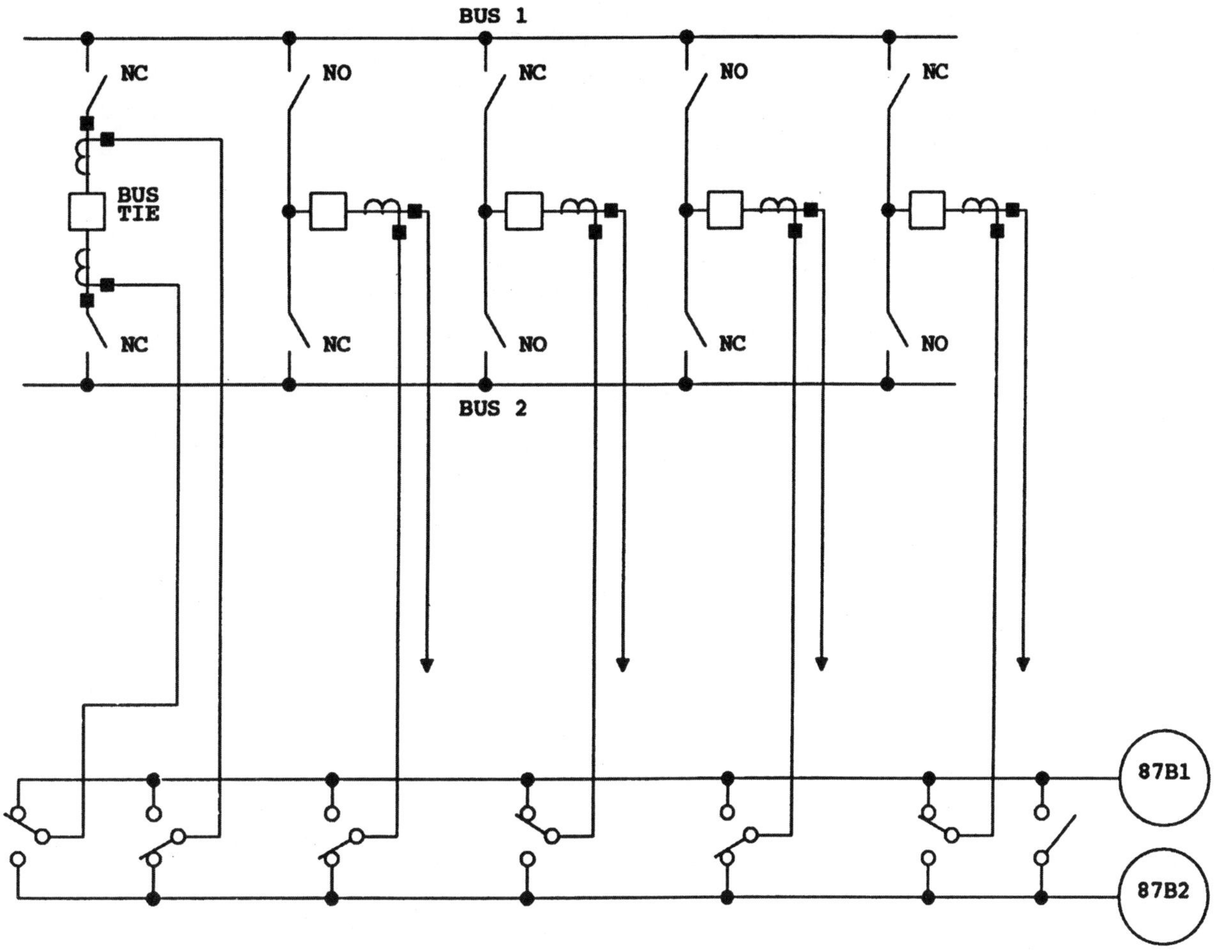

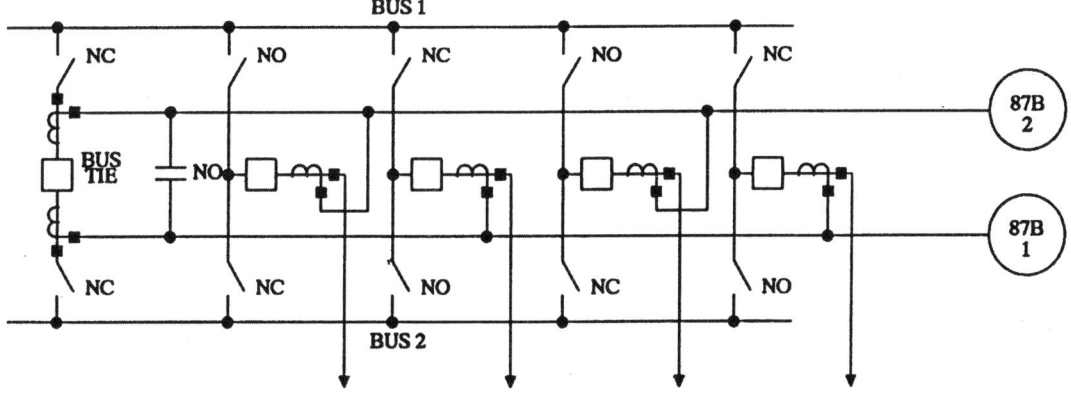

Figure 11-14 Fully switched scheme.

Figure 11-15 Paralleled switched scheme.

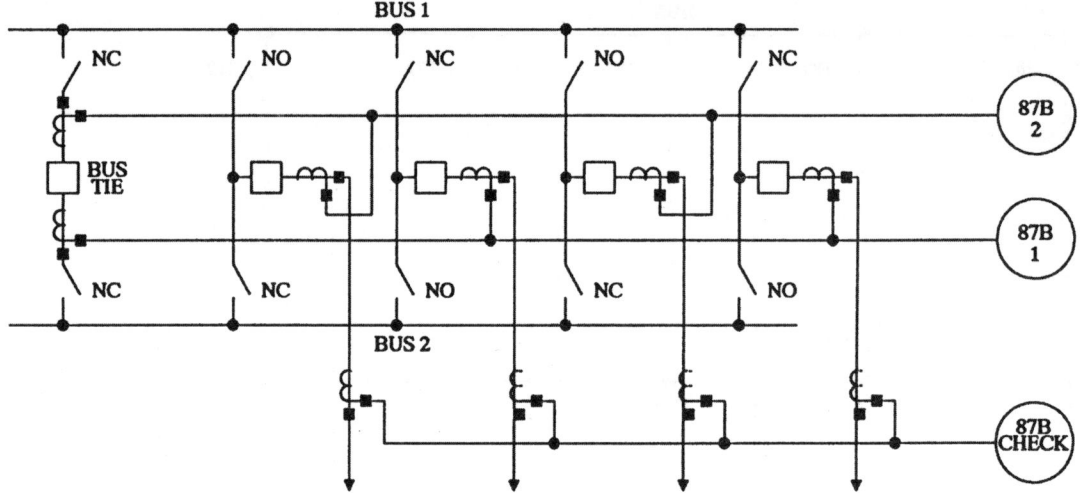

Figure 11-16   Paralleled switch with check zone scheme.

Two bus differential zones are provided, one for each bus, with each one overlapping the bus breaker. Each primary circuit is normally switched to a specific bus, and relay input circuits and breaker control circuits are wired accordingly. The additional check-zone device supervises the trip circuits. If it becomes necessary to clear one of the buses, all the primary circuits may be switched to the opposite bus and it is needless to disable the bus protection before any switching operation. However, this scheme still has two drawbacks when any one or all of the primary circuits is switched to the opposite bus: (1) It will lose its selectivity, and (2) it will reduce its sensitivity since the two relays are paralleled.

A numerical scheme, such as RED-521, overcomes these drawbacks as there is no external ct switching involved. The ct's are connected to the appropriate zone by numerical switching in the relay.

8   OTHER BUS PROTECTIVE SCHEMES

Other methods for protecting buses are in limited use: (1) partial differential schemes, (2) directional comparison relaying, and (3) the fault-bus method. Except for the latter, these schemes are most often applied as economic compromises for the protection of buses.

8.1   Partial Differential Relaying

This type of protection is also referred to as "bus overload" or "selective backup" protection. It is a variation of the differential principle in which currents

in one or more of the circuits are not included in the phasor summation of the current to the relay.

In this scheme, only the source circuits are differentially connected, as shown in Figure 11-17b, using a high-set overcurrent relay with time delay. The ct's protecting the feeders or circuits are not in the differential connection.

Essentially, this arrangement combines time-delay bus protection with feeder backup protection. The sensitivity and speed of this scheme are not as good as with complete differential protection. This method may be used as a backup to a complete differential scheme, as primary protection for a station with loads protected by fuses, or to provide local breaker failure protection for load breakers.

In modern microprocessor systems, provision has been included to allow communications between the feeder breaker relaying and the source breaker relaying. The feeder breakers are each equipped with a nontrip-

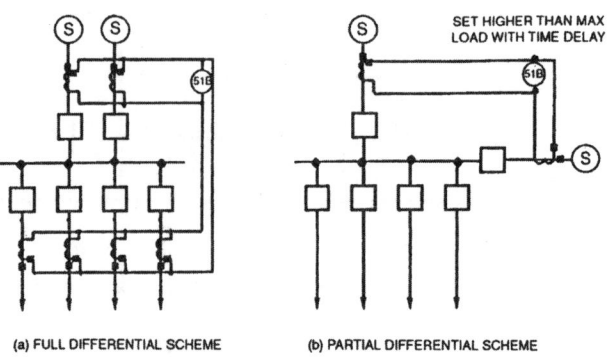

Figure 11-17   Partial differential protection.

ping low-set instantaneous overcurrent function that is set somewhat above their maximum load. The source breakers have an instantaneous overcurrent unit with slight time delay that is set above the maximum total load current for the bus, and they are equipped to receive a status input from the feeder breakers. For a fault on one of the feeder circuits, the low-set instantaneous overcurrent unit operates and applies a block signal to the source relay. The instantaneous unit of the source breaker operates, but is unable to trip because of the block signal. The time-delayed and coordinated tripping of the source breaker is not affected so its backup function stays intact.

For a bus fault, the block signal is absent, and tripping of the source breaker occurs at high speed.

Some partial differential circuits use distance-type relays in the scheme. The use of a distance relay for this scheme produces both faster and more sensitive operation than the overcurrent scheme.

8.2 Directional Comparison Relaying

Occasionally, it is desirable to add bus protection to an older substation where additional ct's and control cable are too costly to install. In this instance, the existing ct circuits used for line relaying can also be used for the directional comparison bus relaying protection.

As shown in Figure 11-18, the directional comparison relaying uses individual directional overcurrent relays on all sources and instantaneous overcurrent relays on all feeders. The directional relays close contacts when fault power flows into the bus section. Back contacts on the overcurrent relays open when the fault is external on the feeder. All contacts are connected in series, and when the fault occurs on the bus, the trip circuit is energized through a timer. A time delay of at least four cycles will allow all the relays to decide correctly the direction of the fault and to permit contact coordination.

In this scheme, the ct's in each circuit do not require the same ratio and can be used for other forms of relaying and metering.

The disadvantage of this scheme is the large number of contacts and complex connections required. There is also the remote possibility of the directional elements not operating on a solid three-phase bus fault as a result of 0 voltage.

8.3 Fault Bus (Ground-Fault Protection Only)

This method requires that all the bus supporting structure and associated equipment be interconnected

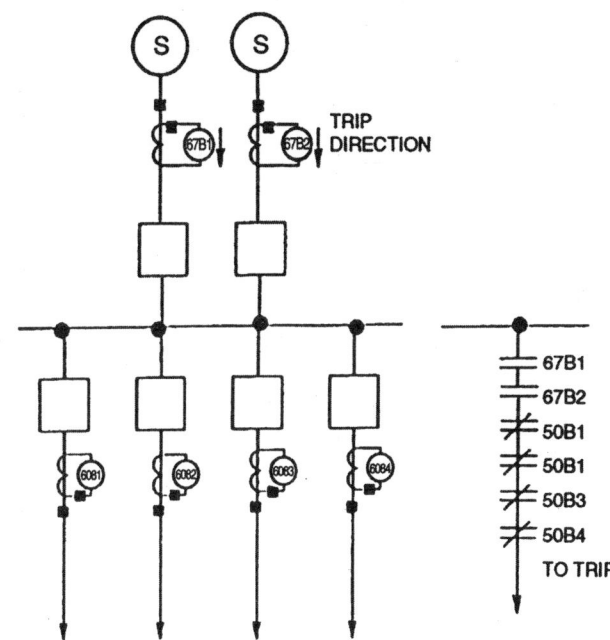

Figure 11-18 Directional comparison bus protection.

and have only one connection to ground. An overcurrent relay is connected in this ground path as shown in Figure 11-19. Any ground fault to the supporting structure will cause fault current to flow through the relay circuit, tripping the bus through the multiple-contact auxiliary tripping relay. A fault detector, energized from the neutral of the grounded transformer or generator, prevents accidental tripping. This scheme requires special construction measures and is expensive.

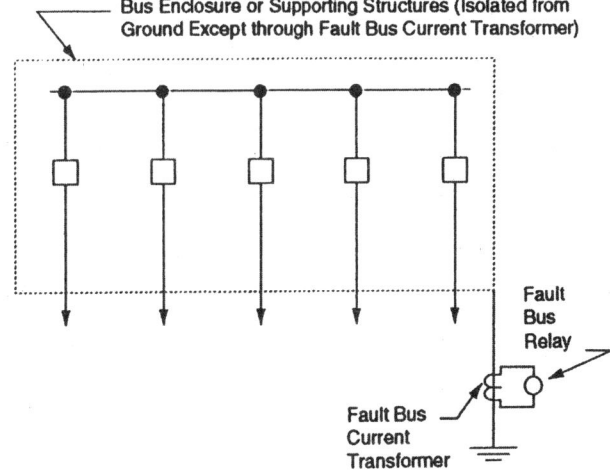

Figure 11-19 Fault bus.

12

Line and Circuit Protection

Revised by: **ELMO PRICE**

1 INTRODUCTION

1.1 Classification of Electric Power Lines

Alternating current lines are commonly classified by function, which is related to voltage level. Although there are no utility-wide standards, typical classifications are as follows:

1. *Distribution (2.4 to 34.5 kV)* Circuits transmitting power to the final users.
2. *Subtransmission (13.8 to 138 kV)* Circuits transmitting power to distribution substations and to bulk loads.
3. *Transmission (69 to 765 kV)* Circuits transmitting power between major substations or interconnecting systems, and to wholesale outlets. Transmission lines are further divided into

   High voltage (HV): 69 to 230 kV
   Extra-high voltage (EHV): 345 to 765 kV
   Ultra-high voltage (UHV): greater than 765 kV

1.2 Techniques for Line Protection

Most faults experienced in a power system occur on the lines connecting generating sources with usage points. Just as these circuits vary widely in their characteristics, configurations, length, and relative importance, so do their protection and techniques.

There are several protective techniques commonly used for line protection:

1. Instantaneous overcurrent
2. Time overcurrent
3. Directional instantaneous and/or time overcurrent
4. Step time overcurrent
5. Inverse time distance
6. Zone distance
7. Pilot relaying

1.3 Selecting a Protective System

Several fundamental factors influence the final choice of the protection applied to a power line:

1. *Type of circuit* Cable, overhead, single line, parallel lines, multiterminals, etc.
2. *Line function and importance* Effect on service continuity, realistic and practical time requirements to isolate the fault from the rest of the system
3. *Coordination and matching requirements* Compatibility with equipment on the associated lines and systems
4. *Influence on power system stability*

To these four considerations must be added economic factors and the relay engineer's preferences based on his or her technical knowledge and experience. Because of these many considerations, it is not possible to establish firm rules for line protection. This chapter, however, focuses on basic application rules and coordination procedures to aid the engineer in the selection of proper protective systems for both phase and ground faults with the techniques as listed in Section 1.2, except for pilot relaying. Also, this chapter covers the basic protective concept of series-compensated transmission lines using distance relay techniques.

Pilot relaying is covered in a companion book, *Pilot Protective Relaying* (Marcel Dekker).

1.4 Relays for Phase- and Ground-Fault Protection

Relay systems for the phase-fault protection of power lines are outlined in Table 12-1, those for ground faults in Table 12-2.

1.5 Multiterminal and Tapped Lines and Weak Feed

The protection of multiterminal and tapped lines and weak feed will also be discussed in this chapter. Multiterminal and tapped lines, although usually economical in their breaker requirements, need complex relaying for adequate protection and operation. In

fact, these lines are the most difficult to protect, particularly with weak feed (limited-fault current) and when high-speed reclosing is desired at one or more terminals. Weak-feed protection may also be required for two terminal lines. Although the weak-feed terminal can maintain the fault arc, the current may not be sufficient to operate conventional protective relays adequately.

The following definitions will be used throughout this chapter:

Multiterminal lines. Transmission lines with more than two terminals, each connected to a major power source. The source will provide positive sequence fault current and, usually, zero sequence as well. A transformer bank may be included as part of the transmission line at one or more of the terminals.

Tapped lines. Transmission lines that are tapped (usually through a transformer bank) primarily to supply loads. Behind the tapped line there may be a positive sequence source, either local generation or an interconnecting tie with another part of the power system. There may also be a zero sequece source.

Table 12-1 Relay Protection Systems for Phase Faults

		Basic relay type	
Type of protection	Device no.	Electromechanical	Static or numerical[a]
Time overcurrent	51	CO	MCO, MMCO, 51 IMPRS, MICRO-51
Instantaneous and time overcurrent	50/51	CO with IIT	MCO, MMCO, 51 IMPRS, MICRO-51
Directional time overcurrent	67	CR	32 + MMCO, 32 + 51 32 + MICRO-51
Directional instantaneous overcurrent	67	KRV	32 + 50D[b]
Step time overcurrent	51	CO-4	51 + 50D
Directional instantaneous and directional time overcurrent	67	IRV	32 + MMCO, 32 + MICRO-51
Inverse time distance system	21/51	Two KD-10, plus two-element CO	
Zone distance system	21	Two KD-10, plus KD-11, plus two TD-5 (or one TD-52)	
Complete zone phase distance system			REL-300 (MDAR), REL-301/REL-302, REL-512
Complete distribution package			DPU 1500R, DPU 2000R, MSOC (nondirectional)

[a] Type numbers refer to ABB circuit-shield types. Certain functions require two relays, with the output of the controlling relay wired to the torque-control input of the second relay.
[b] Select 50D with a 0.01 to 0.03 adjustable range.

Table 12-2 Relay Protection for Ground Faults

| Type of protection | Device no. | Basic relay type | |
		Electromechanical	Static or numerical[a]
Time overcurrent	51N	CO	MSOC
			MICRO-51
Instantaneous and time overcurrent	50N/51N	CO with IIT	MSOC
			MICRO-51
Product overcurrent	67N	CWC or CWP	
Directional time overcurrent	67N	CRC, CRP, CRD, or CRQ	MSOC + 32D or 32Q, MICRO-51 + 32D or 32Q, 51 + 32D or 32Q
Directional instantaneous overcurrent	67N	KRC, KRP, KRD, or KRQ	50D + 32D or 32Q[b]
Directional instantaneous and time delay	67N/50N	IRC, IRP, IRD, or IRQ	MSOC + 32D, 51 + 32D, or 32Q
Complete zone ground distance system			REL-300 (MDAR), REL-301/REL-302, REL-512
Complete distribution package			DPU 1500R, DPU 2000R

[a] Type numbers refer to ABB circuit-shield types. Certain functions require two relays, with the output of the controlling relay wired to the torque-control input of the second relay.
[b] Select 50D with a 0.01 to 0.03 adjustable range.

Weak-feed terminal. A terminal whose source does not supply enough current for faults on the line to operate the line protective relays at that terminal. This situation can occur for either phase (positive sequence), ground (zero sequence), or both. The terminals may only be "weak" during some operating periods, but "strong" or have only load at other times. A tapped terminal is frequently a weak feed source if the tapped load has limited local generation, synchronous motors, and does not have many large induction motors.

2 OVERCURRENT PHASE- AND GROUND-FAULT PROTECTION

2.1 Fault Detection

Most of the faults on power lines can be detected by applying overcurrent relays, since the fault currents are normally higher than the load current.

Radial circuits can be protected by nondirectional overcurrent relays. Figure 12-1 shows several sections of a typical radial circuit. Because the circuit is radial, each section requires only one circuit breaker at the source end. To clear a fault at (1) and other faults to the right, then, only the breaker at R needs to be

tripped. To clear faults at (2) and (3) and in the area between them, the breaker at H must be tripped. Likewise, to clear faults at (4) and (5) and between them, the breaker at G must be tripped.

However, none of the relays at the breaker locations can distinguish whether the remote fault is on the protected line, the remote bus, or an adjacent line. The relays at H, for example, cannot distinguish between faults at (1) and (2), since the current magnitude measured at H will be the same in either case. Opening breaker H for fault (1) is not desirable, since it would interrupt the load at R unnecessarily. Two techniques are available to solve this problem: time delay or pilot relaying. The latter requires a communication channel between the two stations and is covered in the companion book, *Pilot Protective Relaying.*

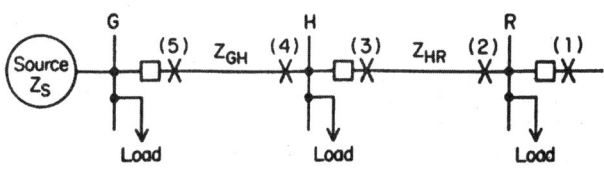

Figure 12-1 A typical radial feeder.

2.2 Time Overcurrent Protection

2.2.1 Time-Delay Relaying

Time relaying delays the operation of the relay for a remote fault, allowing relays and breakers closer to the fault to clear it, if possible. In the example shown in Figure 12-1, relays at H will delay for faults at (1) or (2). If the fault is at (1), this delay will allow the R relays and breaker to operate before H. Thus, although H would not open for a fault at (1) (unless the R relays or associated breaker failed), it would operate for a fault at (2). This technique, called coordination or selectivity, is designed to combine minimum operating time for the close-in faults with a long enough delay for remote faults. In Figure 12-1, e.g., the relays and breaker at R must coordinate or select with those to the right (not shown), H must coordinate with R, and G with H.

2.2.2 Coordination

Relays are coordinated in pairs. If, in Figure 12-1, breaker H relay-tripping characteristics have already been coordinated with whatever protective devices exist at R and beyond, the breaker at G must then be coordinated with those at H.

For the three critical fault points, (5), (3), and (2), the following data are required:

1. *Fault at (5).* Maximum and minimum fault currents
2. *Fault at (3).* Maximum fault current, which determines the required coordination between breakers G and H
3. *Fault at (2).* Minimum fault current, which determines when the G relays must operate to provide backup protection for faults on line HR not cleared by the breaker at H

Relays within a system can be coordinated using graphs or tables, although graphs are generally more useful for radial systems. Semilog (log abscissa for current and linear ordinate for time) or log-log paper can be used. Log-log is preferred when a number of different types of devices, including fuses, are being coordinated on one graph. The current scale can be in primary amperes or per unit. Any difference in current transformer ratios must be taken into consideration when determining actual relay currents at different locations.

The coordination procedure is conducted as follows (Fig. 12-2). First, assume that the desired relay type (tap range and time characteristic) and current

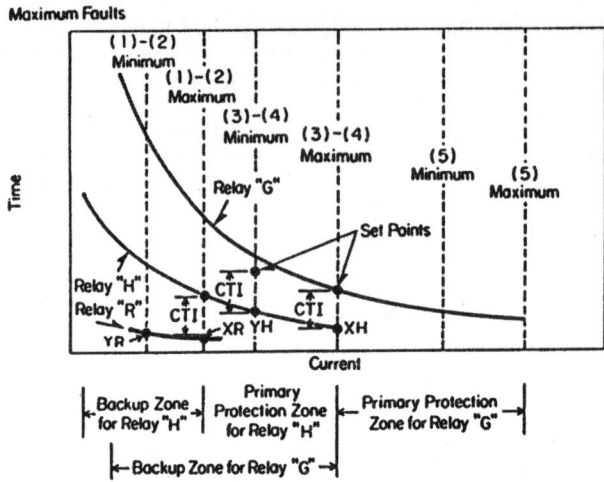

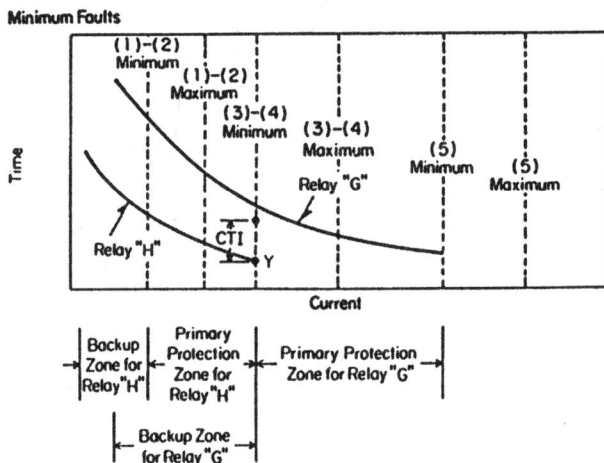

Figure 12-2 Coordination setting procedure for relays at breaker "G" of Figure 12-1.

transformer ratio have been determined. (The selection of these variables will be discussed later in this chapter.) Then perform the following:

1. Determine the critical fault locations and fault current values.
2. Plot these variables on the time-current graph, drawing vertical lines at the various values.
3. Determine the setting for the most downstream relay for the maximum and minimum fault currents. Set the relay as sensitive and fast as possible if there is no other device downstream that has to be coordinated with. For example, consider the relay R in Figure 12-1. If there is some other device, such as a power fuse, to the right of this relay, then relay R should be

coordinated with the power fuse first. If there are no other devices to coordinate with down-stream, set the overcurrent relay equal to or greater than 2.0 maximum load.

4. Plot the operating time of relay R on the time-current graph, shown as XR and YR points in Figure 12-2, respectively.

5. Add a one-step coordinating time interval (CTI); (see Sec. 2.2.3) to points XR and YR. This step gives two set points for the characteristic curve of the relay at H.

6. Select a tap for relay H to operate for fault (1) minimum and, for a phase relay, not to operate on maximum load. The fault (1) minimum should operate the relay on at least twice pickup, although compromises may be necessary (see Sec. 2.2.4). For phase relays, the setting must always be above the maximum load.

7. Select a time lever such that the relay H time-current curve passes through or above one or both of the set points XR and YR and provides the minimum operating time for maximum and minimum fault.

8. Repeat the above steps for each time section "up-stream." For example, add a one-step CTI to XH and YH in Figure 12-2 for relay G, respectively; then select a tap and time lever for relay G, etc.

2.2.3 Coordinating Time Interval

The coordinating time interval is the minimum time between the operating characteristics of two series devices. Factors influencing the CTI are as follows:

1. Breaker fault interruption time
2. Relay-impulse-time overtravel of the induction disk or solid-state relay after the fault current has been interrupted
3. Safety margin to compensate for possible deviations in calculated fault currents, relay tap selection, relay operating time, and current transformer ratio errors

For coordinating at above approximately three times minimum trip current (at least two times the setting value), the CTI should be in the range of 0.2 to 0.5 sec. Larger CTIs should be used on the steep part of the curve to compensate for errors below a multiplier of 3. A CTI of 0.3 sec is commonly used. Lower values should be used only after careful consideration of 1 through 3 above.

2.2.4 Selecting an Overcurrent Relay Tap

As indicated above, phase overcurrent relays must not operate on the maximum load current that can occur on the line. Situations in which temporary overloads may occur, such as the cold loads discussed in the next section, must be factored into the value used for setting the overcurrent relays. Thus, it is important for the relay engineer to cooperate with the operating engineers in determining the maximum possible load for each circuit. This maximum-value STM (short-time maximum) load can differ from the rating of the line and is the value that should be used for setting the relays.

The tap (minimum pickup value of the phase overcurrent relays) should be at least 2 (a safety factor for security) times normal maximum load and never less than 1.5 times. If we assume that the STM is greater than the normal maximum load, the tap can be selected as the next available tap greater than 1.25 STM.

Dependability should be checked once the relay tap is selected. In Figure 12-8, the minimum fault current I_{2min} through the relay for a fault on the remote bus H, divided by 2 (a dependability factor), should be greater than the selected tap value if the relay is set for protecting the line only; or the minimum fault current I_{3min} through the relay for a fault on the remote bus R, divided by 2, should be greater than the selected tap value if the relay is set for protecting the line and remote backup of the line (HR) beyond.

Current transformers are normally selected to provide secondary currents between 4 and 5 A during rated maximum load. As a result, the phase relay pickup will usually be above 5 A.

The above limitation does not apply to ground relays, since load current does not produce current in their operating windings unless it is unbalanced. To avoid operation on possible imbalances in a normally balanced circuit, a good rule of thumb is to set the ground relays for not less than 10% of the maximum load current. Four-wire distribution circuits will, in general, require a much higher setting than this.

2.2.5 Selecting an Overcurrent Relay Time Curve

Five different curve shapes have been established by the vast number of electromechanical relays that are in service on power systems and that dictate coordination requirements. Solid-state and numerical relays implement these shapes while often allowing others. These widely used time-current characteristics are described as:

1. Definite time, CO-6
2. Moderately inverse, CO-7
3. Inverse, CO-8
4. Very inverse, CO-9
5. Extremely inverse, CO-11

These time-current characteristics are compared in Figure 12-3. The time lever settings are selected so that all relays operate in 0.2 sec at 20 times the tap setting.

The microprocessor-based overcurrent relay, type MCO, is a single-phase one, and type MMCO is a three-phase and ground package. All the above time-current curves are built in these relays, and can be selected by settings. The equations for the MCO or MMCO time-current curves are

$$T(\text{sec}) = \left[T_0 + \frac{K}{(M-C)^p} \right] \frac{D}{24,000} \quad \text{for } M \geq 1.5$$

$$= \left[\frac{R}{(M-1)} \right] \frac{D}{24,000} \quad \text{for } M < 1.5$$

where

T = trip time in seconds
D = time dial setting from 1 to 63
M = operating current in terms of multiple of tap setting
T_0 = definite time term
K = scale factor for the basic inverse time
P = an exponent determining inverseness

T_0, K, C, P, and R are constants and are shown as below:

Curve no.	T_0	K	C	P	R
CO-2	111.99	735.00	0.675	1	501
CO-5	8196.67	13,768.94	1.130	1	22,705
CO-6	784.52	671.01	1.190	1	1475
CO-7	524.84	3120.56	0.800	1	2491
CO-8	477.84	4122.08	1.270	1	9200
CO-9	310.01	2756.06	1.350	1	9342
CO-11	110.00	17,640.00	0.500	2	8875

Operating time as shown in Table 12-3 illustrates that the MCO or MMCO provides a fairly good time-current characteristic for coordinating with the conventional type of CO relay. Similar curves and other

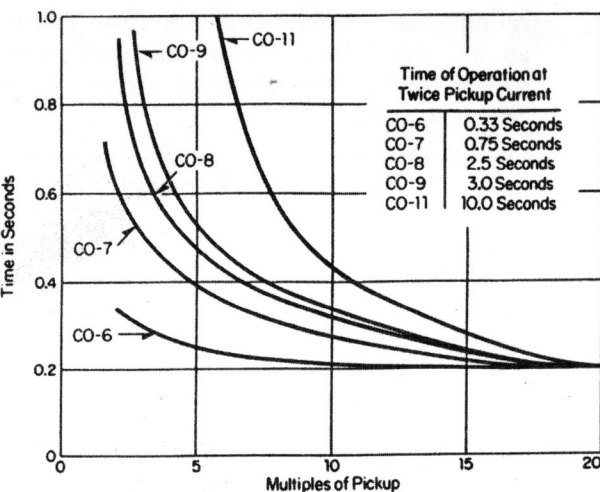

Figure 12-3 Type CO curve shape comparison.

variations are available in the Micro-51, DPU 2000R, DPU 1500, and Microshield relays.

The choice of a relay time-current characteristic is a function of the sources, lines, and loads. Since these factors vary throughout a system, a characteristic that is ideal for one line and one operating condition requires compromises for other conditions and associated lines.

If possible, time curves with the same or approximately the same characteristics should be used. Identical or similar curves applied at different places in the system tend to "track" together as operating conditions change. If different time characteristic curves must be used, all possible operating conditions must be checked carefully to ensure that the CTI is maintained for selective tripping. (Using similar characteristics, in other words, minimizes coordination studies.)

Fixed- and inverse-time characteristics for a system are compared in Figure 12-4. The last feeder supplying one load center can be protected with an instantaneous overcurrent device set into the load. Since no coordination at the load is involved, no time delay is required, as shown with bus R in both time-distance charts.

In the upper chart of Figure 12-4, the fixed-time characteristics approximate the definite minimum time (CO-6). The relay at H is coordinated with R, and G is coordinated with H, as shown. The advantage of this arrangement is that the operating times are relatively constant and independent of changes in fault levels from maximum to minimum generation. On the other hand, the operating times for heavy

Table 12-3 Relay Operation Times

Type of relay	Conventional CO		MCO or MMCO	
	Time dial set at	Time of operation at 4 × pickup current (Fig. 12-3) (sec)	Time dial set at	Time of operation at 4 × pickup current (sec)
CO-6	0.6	0.25	6	0.25
CO-7	1.0	0.40	7	0.44
CO-8	1.2	0.60	7	0.58
CO-9	2.1	0.70	11	0.62
CO-11	5.0	2.00	31	2.00

faults near the source are very long. For this reason, this arrangement is not practical when there are more than one or two radial feeders from the distribution substation.

The lower chart of Figure 12-4 shows inverse-time relay characteristics. For faults near the relay, particularly for the maximum conditions, operating times are very short. Unfortunately, as system conditions change from maximum to minimum, operating times vary considerably. Even though this arrangement can

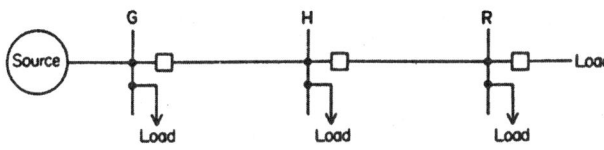

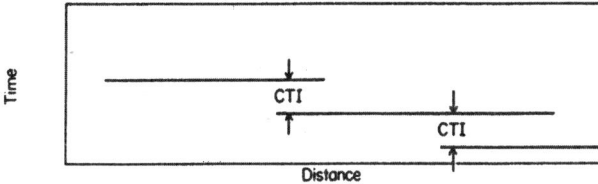

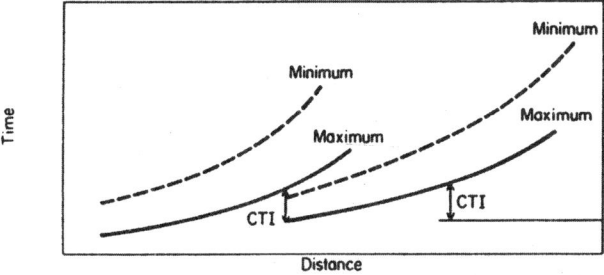

Figure 12-4 Comparison of fixed time vs. inverse time overcurrent relays on radial feeder circuits.

produce very long operating times for minimum faults near the remote bus, it is commonly used.

Line length is also an important factor. For a short line, one whose impedance is low compared to the source impedance, the fault currents for the close-in and far-end faults are essentially the same; that is, the inverse-time characteristic gives a relatively fixed operating time over the line. In such cases, the definite minimum time characteristic is preferred, since the operating time will not vary as much for different generation levels as with inverse relays.

In general, the following apply:

1. The flatter curves (CO-6 and CO-7) are more suitable when:

    (a) There are no coordination requirements with other types of protection devices farther out in the system.

    (b) The variation in current for faults at the near and far ends of the protected circuit is too small to take advantage of the inverse characteristic.

    (c) Instantaneous trip units give good coverage (see Sec. 3.1).

2. The "inverse-time" relay (CO-8) provides faster clearing time than the "more inverse-time" relays for low-current faults. This would be advantageous on long lines where the available fault current is much less at the end of the line than at the local end. It does not provide much margin for cold load pickup.

3. The steeper (more inverse) curves (CO-9 and CO-11) are more suitable when

    (a) Fault currents are significantly different for the close-in and remote faults (for example, when the line impedance is large compared to the source impedance).

(b) There is an appreciable current inrush on service restoration (cold load).

(c) Coordination with other types of devices with very inverse characteristics, such as fuses and reclosers, is required.

General Comments on Curve Shape Selection

There is no known scientific means of determining the ideal curve shape for a specific application, except to make preliminary setting calculations. However, some general comments can be made:

1. Use a CO-6 definite minimum time relay when coordination is not a problem.
2. Use a CO-6 relay for short line application.
3. Use a CO-11, extreme inverse-time relay when fuses are involved.
4. The more inverse shape (CO-8, CO-9, and CO-11) is more suitable in loop systems.
5. Use a comparable shape within a system segment for easier coordination.

One of the advantages of using a microprocessor-based overcurrent relay is that a different time curve can be selected in the same device without changing the unit.

2.2.6 Effect of Extended Load Outage/Cold-Load Inrush

A particularly critical phenomenon for distribution circuits serving residential and commercial loads is the high transient current inrush that may occur when a feeder is energized after a prolonged outage. For these "cold-load" conditions, the diversity of intermittent loads is lost: Consumers tend to leave more than the normal load connected, and thermostatically controlled equipment will start as soon as the voltage is restored. The overall effect is a very high initial current, or cold-load inrush.

In general, the pickup of time-overcurrent relays cannot be set above this transient without severely compromising protection. Setting the relay below the transient will cause it to begin to operate on the cold load; however, the current will decrease below the pickup value before the relay has time to operate.

A current-time curve for the cold-load inrush on a typical feeder is shown in Figure 12-5. Since such curves vary considerably with different feeders, each utility must develop its own system history and probability data. If we assume that the time-overcurrent CO relays are set at twice the normal

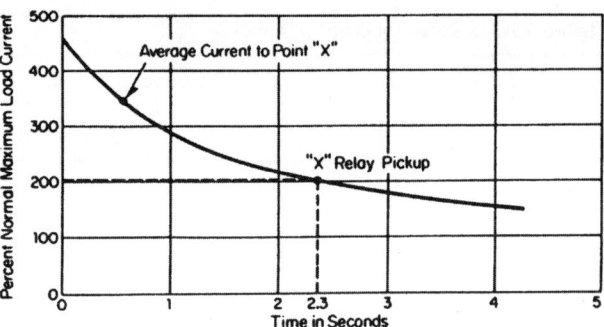

Figure 12-5 Typical example of feeder load current following extended outage (cold load inrush).

maximum load, they will receive operating current for the first 2.3 sec that the feeder is energized (Fig. 12-5). The average current for this period is 3.4 p.u., an equivalent of 1.7 times pickup. To prevent tripping the breaker, then, the relay operating time at 1.7 times the tap value should be slightly more than 2.3 sec (about 2.5 sec). For the CO-9 very inverse relay, this condition requires a time dial setting of 1.25; for the CO-11 extremely inverse relay, a time dial setting of 0.75 is required. Time dial settings of 1.5 for the CO-9 relay and 1.0 for the CO-11 relay are suggested, unless operating experience indicates otherwise.

Although the extremely inverse relays may provide faster fault operations than the less inverse type of relays, they still override the cold-load inrush.

Modern microprocessor relays contain logic that allows sensing of the energization of a feeder circuit. A short time delay in instantaneous tripping can be introduced to prevent operation on cold-load pickup while still permitting sensitive and relatively fast fault recognition.

Often, it is necessary to sectionalize the feeder and to pick up the load in increments in order to reenergize the cold load without undesired tripping.

2.2.7 Fuse and Relay Coordination

Because fuses have a time-current characteristic that is much more inverse than most induction-disc time-overcurrent characteristics, coordinating these relays and fuses can be difficult (Fig. 12-6). A fuse curve and two sets of relay curves, one for the extremely inverse relay and one for the very inverse relay, are plotted on a linear time vs. logarithmic current scale. The right-hand set of relay curves provides a good margin of protection at high levels of fault current, but is unsatisfactorily slow for medium values of fault current, particularly with extremely inverse character-

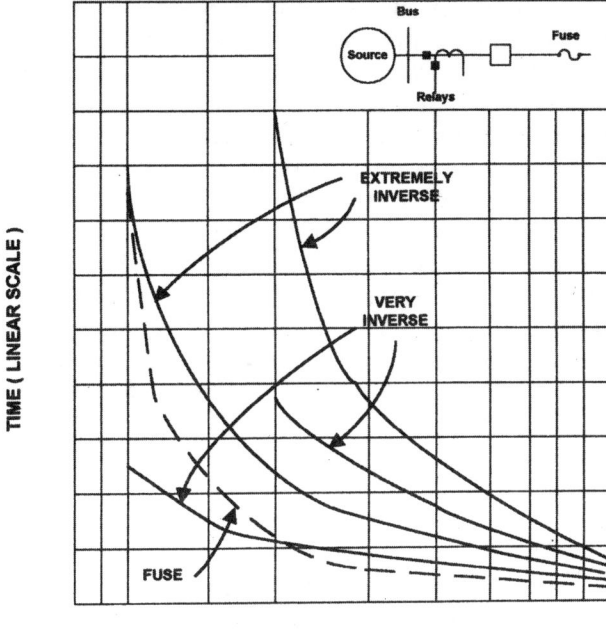

Figure 12-6 Current (logarithmic scale) comparison of fuse and relay curves.

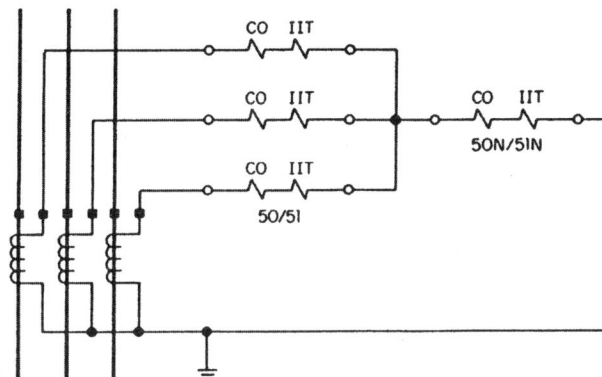

Figure 12-7 Connections for overcurrent ground relay.

2.3 Instantaneous Overcurrent Protection

Adding instantaneous trip units to time-overcurrent relays provides high-speed relay operation for close-in faults and may also permit faster settings on the relays in the adjacent section.

Instantaneous trips may be used on a circuit if the maximum close-in fault current is on the order of 1.1 to 1.3 or more times the maximum fault current on bus H (see Fig. 12-8).

$$I_{1\,max} > (1.1 \text{ to } 1.3) \, I_{2\,max} \qquad (12\text{-}1)$$

The factor of (1.1 to 1.3) in Eq. (12-1) is for preventing the instantaneous unit from overreaching. In other words, the instantaneous unit must operate for as many of the line faults as possible, but to avoid miscoordination, it must not operate for the far-end fault. The greater the ratio of close-in to far-end faults, the more of the line the instantaneous unit will protect. In terms of system constants and setting, the reach or coverage of three-phase faults on a line can be determined as follows:

$$n = \frac{SIR(1 - K_i) + 1}{K_i} \qquad (12\text{-}2)$$

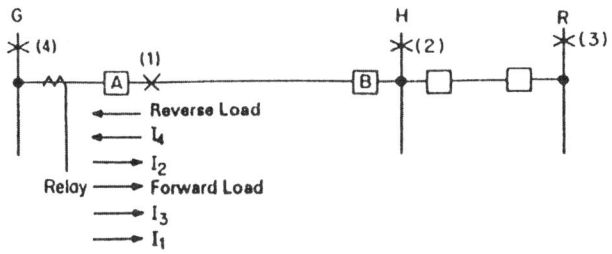

Figure 12-8 Criteria for a directional unit requirement at relay breaker A.

istics. If the right curve were moved to the left, it would coordinate better at higher current values. As shown in Figure 12-6, the curve would then cross the fuse curve at lower values of fault current. Usually, either the very inverse or extremely inverse characteristic can be set to coordinate with fuses. By adjusting the tap and time lever settings, areas of crossing such as those shown in Figure 12-6 may impose impractical or impossible operating conditions on the circuit.

When plotting fuse curves, the following three time characteristics must be considered:

1. Maximum time that the fuse will carry current without suffering damage
2. Melting time for the fuse links
3. Total clearing time for the fuse to clear the circuit

The first two characteristics are used for coordination with protective devices beyond the fuse. Normally, curves based on the melting-time characteristic are provided with a "safety band"; in this case, maximum time curves are not required. The total clearing-time characteristic is used for coordination with other protective devices, including relays, ahead of the fuse.

where

$n =$ per unit of line section length protected by the instantaneous unit

$K_i = \dfrac{\text{instantaneous unit pickup current, } I_{IT}}{\text{maximum far-end fault current, } I_F}$

$SIR =$ source impedance ratio

$= \dfrac{\text{source impedance, } Z_S}{\text{protected line impedance, } Z_L}$

Refer to Appendix A in this chapter for more information about Eq. (12-2).

Recommended values of K_i, are 1.3 for the solenoid or plunger units with transient overreach (IIT, SC, IT units), 1.2 for static or mumerical units (MCO, LI units), and 1.1 for the cylinder units with negligible transient overreach (KC-2, KC-4, KO, KR, and IR types). The value of 1.25 can be used as a general factor.

The minimum value that can justify the use of an instantaneous unit for line protection is a matter of choice. Since the relative cost of adding the instantaneous units is quite low, they are recommended even when the line coverage is low for maximum faults and 0 for minimum faults. The arrangement provides fast protection for the most severe, heavy, close-in faults.

Cold-load inrush may be above the instantaneous unit setting desired for maximum fault protection. To avoid operation when a setting above this inrush is not practical, the instantaneous trip circuit can be manually opened at restoration and left open until the instantaneous trip unit resets. For manual operation, a slip contact on the control switch that is open while the switch is held in the "close" position (or its equivalent) prevents operation until the inrush subsides to drop out the instantaneous unit. If set above cold-load inrush, the instantaneous trip unit setting should be at least three times the overcurrent tap setting, or around six times the normal maximum load.

2.4 Overcurrent Ground-Fault Protection

Ground overcurrent relays are for faults involving zero sequence quantities, primarily single-phase-to-ground faults and sometimes two-phase-to-ground faults. With a few significant differences, the general application rules for phase relays also can be applied to ground relays.

The directional or nondirectional overcurrent types are used widely at most voltage levels. In addition to their lower cost and complete independence of load, the power system provides more rapid attenuation of current with distance and the relatively higher independence of system changes. This makes their application and setting easier than for phase relays.

Ground relays usually can be set and coordinated independently of phase relays, even though the faulted phase current does flow through the one or more phase relays for a single-phase-to-ground fault. The primary reason for this independence is that ground relays are set at one-fifth to one-tenth of the sensitivity of phase relays. The more sensitive settings obtainable with ground overcurrent relays may mean (for electromechanical relays) a higher burden on the current transformers, and their performance should be checked as described in Chapter 5.

A circuit may be protected with a single, nondirectional overcurrent ground relay, as shown in Figure 12-8. Positive and negative sequence currents are balanced out at the current transformer neutral, so only $3I_0$ currents pass through the ground relay (50N/51N). Since, under normal balanced conditions, $3I_0$, is at or approaches 0, a very low pickup current is used, typically 0.5 to 1.0 A. Although ground-fault currents on distribution circuits are generally higher at the substation than phase-fault currents, they decrease at a much greater rate with the distance from the substation because X_0 is considerably larger than X_1 for the feeder circuits. With the exception of fault current values, the application and coordination of the nondirectional overcurrent relays are the same as for phase relays, as given above.

As with phase relays, instantaneous ground trip can be used to improve relaying, particularly for close-in faults. Instantaneous ground-trip units are more applicable in general, with the higher attenuation of the fault currents with distance. Unless care is exercised in the choice of settings, high transient overloads and unequal current transformer performance can give rise to "false residual currents" and, hence, misoperation.

The choice of a relay time characteristic for line protection is usually limited to the inverse or very inverse type. The very inverse type is the more commonly used. However, when coordination with fuses and/or series trip reclosers is required, an extremely inverse characteristic would probably be preferable.

The foregoing descriptions can be summarized as follows:

1. Factors that are favorable to ground-fault protection:

Normally load current does not affect the ground-relay operation. This means that the ground relay can be set more sensitively than the phase relay.

Ground relays are not affected by out-of-step conditions.

Ground relays always have available the unfaulted phase voltages for polarizing. They may not require a memory circuit.

The higher zero sequence line impedance Z_{0L}, as compared with the positive line impedance Z_{1L}, may allow one to use a high-set ground overcurrent unit and make coordination easier than for phase faults.

The zero sequence isolated system may make coordination easier.

2. Factors that are not favorable to ground-fault protection:

Most of the time, ground faults involve higher fault resistance; this may introduce an over- or under-reach problem to the relay.

Zero sequence mutual effect may cause a ground-relay directionality problem.

The zero sequence current distribution factor is not equal to the positive sequence current distribution factor, except for a single end feed condition; this makes the ground relay more complicated than the phase relay in design.

The ground relay faces more problems than the phase relay on reverse fault clearing (e.g., contact bounce, unequal pole clearing, etc.).

3 DIRECTIONAL OVERCURRENT PHASE- AND GROUND-FAULT PROTECTION

3.1 Criteria for Phase Directional Overcurrent Relay Applications

When there is a source at more than one of the line terminals, fault and load current can flow in either direction. Relays protecting the line are therefore subject to fault power and reactive flowing in both directions. If nondirectional relays were used, they would have to be coordinated with not only relays at the remote end of the line, but also the relays behind them. Since directional relays operate only when fault current flows in the specified tripping direction, they avoid both this complex coordination and the possibility of compromising line protection.

Figure 12-8 shows a line, with a source at each end, that could be a section of a loop. The following procedure determines the criteria for a directional unit by comparing the currents flowing through the relay for faults at either bus.

3.1.1 For Phase Directional Time-Overcurrent Relays

A directional time-overcurrent relay should be applied at G if, as in Figure 12-8, the maximum reverse-fault current (I_{4max}) for a fault on bus G or the maximum reverse-load current I_{RLoad} through the relay exceeds 0.25 (this value includes the safety factor 2 and dependability factor 2) times the minimum fault current I_{2min} through the relay for a fault on the remote bus H for the protection of the line only; or the minimum I_{3min} through the relay for a fault on the remote bus R if the relay is set for the protection of the line on the remote bus R and remote backup of the line (HR) beyond. In other words, a directional relay should be used when

$$\frac{I_{4max} \text{ or } I_{RLoad}}{I_{2min} \text{ or } I_{3min}} > = 0.25 \tag{12-3}$$

3.1.2 For Phase Directional Instantaneous-Trip Overcurrent Relays

A directional instantaneous-trip overcurrent unit should be used if the maximum reverse-fault current I_{4max} is greater than the maximum I_{2max} (Fig. 12-8). Also, as mentioned in Section 2.3 before, for instantaneous-trip overcurrent application, the criteria as shown in Eq. (12-1) should be met.

3.2 Criteria for Ground Directional Overcurrent.Relay Applications

With a few significant differences, the general rules of application for directional overcurrent phase relays also apply to ground relays. Normal balanced load current is not a consideration. However, the ground overcurrent unit still should be set above any maximum expected unbalanced load current.

3.3 Directional Ground-Relay Polarization

To determine the direction to a fault, a directional relay requires a reference against which line current can be compared. This reference is known as the polarizing quantity and, in this context, reference and polarizing are synonymous terms. With zero sequence line current, either a zero sequence current or voltage or both must be used. In power systems with mutual induction problems, the trend is toward the use of negative sequence quantities for the ground directional unit.

3.3.1   Voltage (Potential) Polarization

The zero sequence voltage at or near any bus in an inductive power system can be used for polarization. The voltage measured on the bus or just on any line near the station will have the same direction for any fault location. However, the current through the line breaker will change direction according to the fault location.

The polarizing zero sequence voltage is obtained from the broken-delta secondary of grounded-wye voltage transformers (Fig. 12-9). Phase voltages are also required for the phase relays, instrumentation, etc. In such cases, either a double-secondary voltage transformer or device, or a set of auxiliary wye-grounded, broken-delta auxiliary transformers, can be used. The voltage across the broken delta, V_{XY}, always equals $3V_0$, or V_{AG} plus V_{BG} plus V_{CG}.

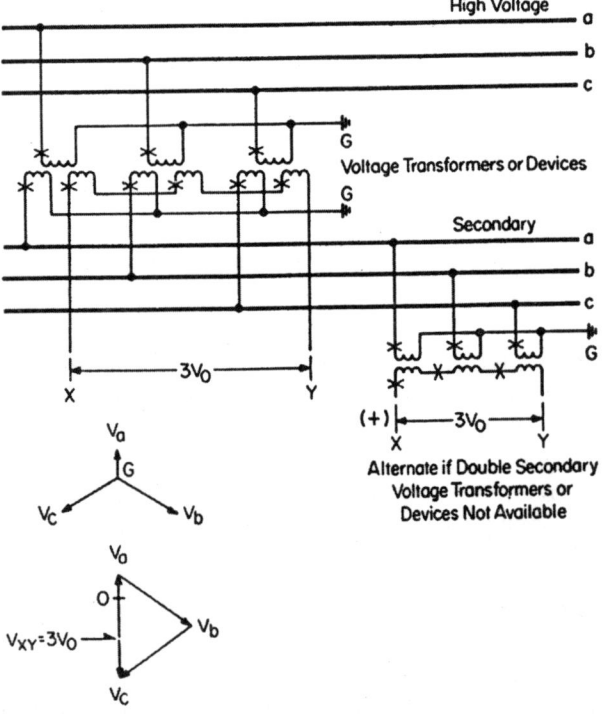

Figure 12-9   Zero sequence polarizing voltage source.

3.3.2   Current Polarization

The current polarization reference depends on the availability and connection of the power transformer at the relay location. Various bank connections are diagrammed in Figure 12-10. The current in the neutral of a wye-grounded delta power transformer can be used for polarizing. For almost any ground fault on the wye-side system, this current flows up the neutral when current is flowing to the fault. (An exception with high mutual induction will be discussed later.) Thus, a current transformer in the neutral measures $3I_0$ for I_P current polarization, the polarizing current shown in Figure 12-10a for a wye-delta bank, and that shown in Figure 12-10d for a zig-zag bank. Wye-wye banks, either grounded or ungrounded, cannot be used for polarizing (Fig. 12-10b and 12-10c). The grounded wye of three-winding wye-wye-delta banks can be used (Fig. 12-10e, f, and g).

The separate neutral currents of the three-winding wye-delta-wye transformers cannot be used for polarizing, but current transformers in each grounded neutral must be parallel with inverse ratios, as shown in Figures 12-10f and 12-10g. If we assume that both the high- and low-voltage sides connect to a ground source, ground faults on the low-voltage side (Fig. 12-10g) result in current flowing up the low-voltage neutral and down the high-voltage neutral. Conversely, for faults on the high-voltage side (Fig. 12-10f), current flows up the high-voltage neutral and down the low-voltage neutral. Hence, the reversal in either neutral does not provide a reference. By paralleling the two neutral current transformers, however, I_P always falls in the same direction for faults on either side since, on a per unit basis, the current flowing down the neutral is always less than the current flowing up the other neutral. The actual current distribution will vary as determined by the zero sequence network.

The tertiary or delta winding can also be used as a polarizing source. If there are no external circuits from the delta, one current transformer connected in any leg of the delta will provide I_0. A current transformer is required in each of the three windings if the delta is connected to external circuits, so that positive and/or negative sequence currents can exist during load or faults. These current transformers must be connected in parallel to cancel out positive and negative sequence and provide $3I_0$ only.

Autotransformers should not be used for current polarizing without careful analysis because they are frequently unreliable as a reference. Autotransformers

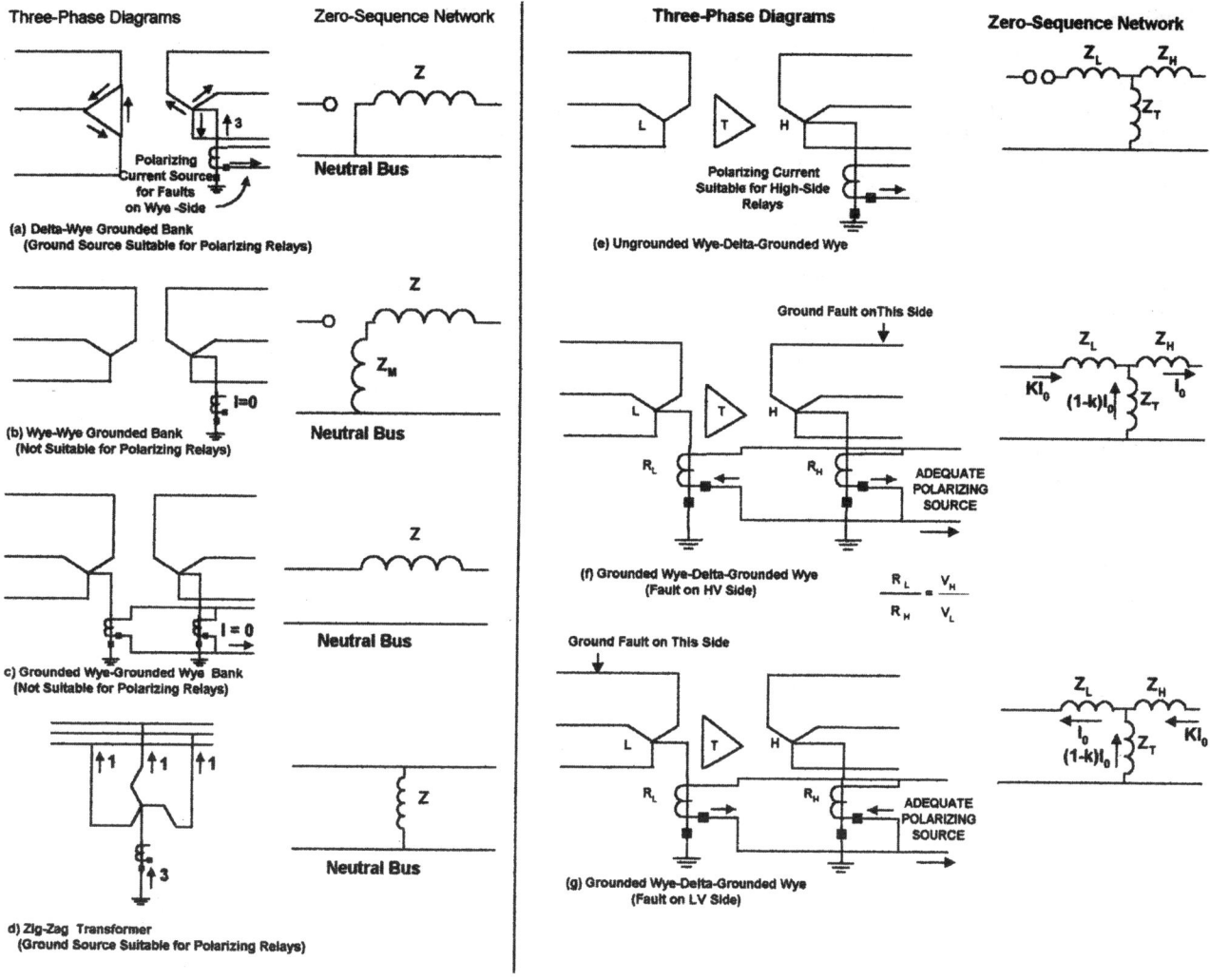

Figure 12-10 Zero sequence polarization from power transformer banks.

that are ungrounded or without a delta tertiary cannot be used. Such units can bypass zero sequence in roughly the same way as the wye-wye-grounded banks. A grounded autotransformer with a tertiary is shown in Figure 12-11.

For Ground Faults on the High-Voltage Side

(Figure 12-11a, if we assume the high-side source is opened for simplicity). The zero sequence current flowing through the high-voltage winding of the autotransformer is thus I_{0H}. The current flowing in the low-voltage circuit will be $P_0 I_{0H}(V_H/V_L)$. Because of the physical connections, the current in the neutral

of the bank is

$$I_{NH} = 3\left(I_{0H} - P_0 \frac{V_H}{V_L} I_{0H}\right)$$
$$= 3I_{0H}\left(1 - P_0 \frac{V_H}{V_L}\right) \qquad (12\text{-}4)$$

From this, it is evident that a positive value of $(1 - P_0 V_H/V_L)$ will assure that the neutral is a reliable polarizing source.

From the zero sequence network, we obtain

$$P_0 = \left(\frac{Z_T}{Z_{0LS} + Z_L + Z_T}\right) \qquad (12\text{-}5)$$

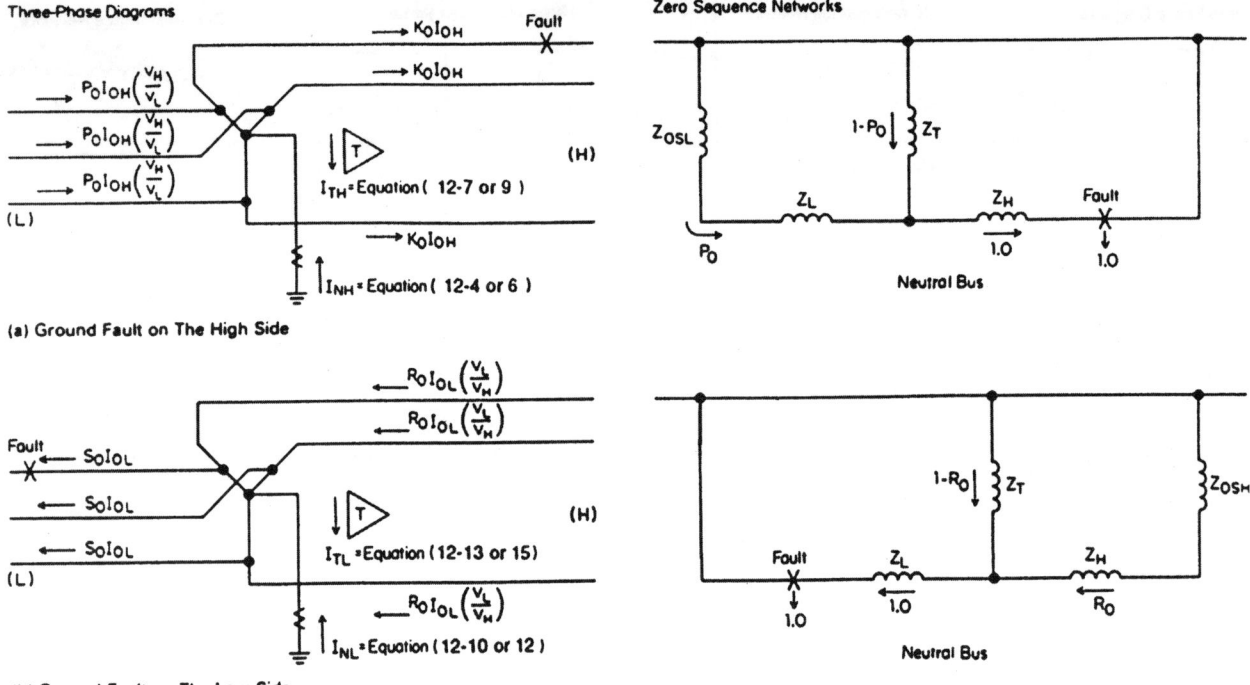

Figure 12-11   Polarizing from grounded autotransformer banks.

Substituting yields

$$I_{NH} = 3I_{0H} \times \left(1 - \frac{Z_T}{Z_{0S}L + Z_L + Z_T} \times \frac{V_H}{V_L} \right) \quad (12\text{-}6)$$

The tertiary current (inside the delta) is

$$I_{TH} = I_{0H}(1 - P_0)\frac{V_H}{\sqrt{3}V_T} \quad \text{at } V_T \qquad (12\text{-}7)$$

and since

$$(1 - P_0) = \left(\frac{Z_{0SL} + Z_L}{Z_{0SL} + Z_L + Z_T} \right) \qquad (12\text{-}8)$$

$$I_{TH} = I_{0H}\left(\frac{Z_{0SL} + Z_L}{Z_{0SL} + Z_L + Z_T} \right)$$
$$\times \frac{V_H}{\sqrt{3}V_T} \quad \text{at } V_T \qquad (12\text{-}9)$$

For Ground Faults on the Low Voltage Side

(Figure 12-11b, if we assume the low-side source is opened for simplicity). I_{0L} is the zero sequence current in amperes from the autotransformer. The current

from the high voltage system will be $R_0I_{0L}(V_L/V_H)$ at V_H. Again, the bank neutral current is

$$I_{NH} = 3\left(I_{0L} - R_0\frac{V_L}{V_H}I_{0L} \right)$$
$$= 3I_{0L}\left(1 - R_0\frac{V_L}{V_H} \right) \qquad (12\text{-}10)$$

From the zero sequence network, we get

$$R_0 = \left(\frac{Z_T}{Z_{0SH} + Z_H + Z_T} \right) \qquad (12\text{-}11)$$

Substituting yields

$$I_{NL} = 3I_{0L} \times \left(1 - \frac{Z_T}{Z_{0SH} + Z_H + Z_T} \times \frac{V_L}{V_H} \right)$$
$$\qquad (12\text{-}12)$$

The tertiary current (inside the delta) is

$$I_{TL} = I_{0L}(1 - R_0)\frac{V_L}{\sqrt{3}V_T} \quad \text{at } V_T \qquad (12\text{-}13)$$

and since

$$(1 - R_0) = \left(\frac{Z_{0SH} + Z_H}{Z_{0SH} + Z_H + Z_T} \right) \qquad (12\text{-}14)$$

$$I_{TL} = I_{0L} \left(\frac{Z_{0SH} + Z_H}{Z_{0SH} + Z_H + Z_T} \right)$$

$$\times \frac{V_L}{\sqrt{3}V_T} \text{ at } V_T \qquad (12\text{-}15)$$

If the neutral or tertiary current is used for polarizing, Eq. (12-6) and (12-12) or (12-9) and (12-15) must give a positive operating value for all possible variations of Z_{0SH} and Z_{0SL}. However, in some transformer designs, especially those for autotransformers, the Z_H or Z_L may be negative, the direction of the I_{NH} or I_{NL} may be reversed, and this current is not a reliable source for polarization, when the combined impedance of $(Z_{0SH} + Z_H)$ or $(Z_{0SL} + Z_L)$ is a negative value.

3.3.3 Dual Polarization

The approach is called dual polarization if the directional ground overcurrent relay uses both zero sequence voltage $3V_0$ and zero sequence current I_p for polarization. It provides more flexibility in applications.

3.3.4 Negative Sequence Polarization

The negative sequence directional ground unit is operated by the quantities V_2 and I_2. One typical design operates when I_2 leads V_2 by 98°. The output of the negative sequence directional unit D_2 can be used for either supervising or the torque control of an I_0 or I_2 unit.

Negative sequence relays can be tested easily and quickly using load current flow. The directional unit is checked by simply interchanging B and C currents and voltages to provide negative sequence from the balanced load quantities.

Negative sequence directional sensing has become increasingly necessary because of a mutual induction problem. Refer to Section 3.5 in this chapter for more details on applying negative directional sensing relays.

3.4 Mutual Induction and Ground-Relay Directional Sensing

Transmission lines on the same tower or parallel along the same right of way are mutually coupled. For a positive and negative sequence, mutual impedances are less than 10% (usually they do not exceed 3 to 7%) of self-impedances and can be considered negligible. For zero sequence, however, mutual impedance can be 50 to 70% of the zero sequence self-impedance Z_{0L} and is, therefore, significant. Mutual impedance affects the magnitude of ground-fault currents and can result in incorrect directional sensing.

Figure 12-12 shows two parallel three-phase lines with zero sequence isolation, except for mutual coupling. The two lines can be completely isolated, or more commonly, tied together to common generation sources. The lines can be at the same or different voltage levels.

A ground fault at or near G on line GH involves no directional sensing problems at either station G or H, as shown by the I_0 current directional arrows. The $3I_0$ current flowing from H to G induces a zero sequence voltage in the parallel line RS, causing current to flow from R to S as shown. The polarizing and operating quantities are properly oriented to operate the zero sequence quantity-polarized directional ground relays at both R and S, so that this current appears as an internal line RS fault to the relays at terminals R and S.

The zero sequence current I_{0M} in line RS induced by uniform mutual coupling (Z_{0M}) for length GH will be

$$I_{0M} = \frac{-K_0 I_0(nZ_{0M}) + (1 - k_0)I_0(1 - n)Z_{0M}}{Z_{0SR} + Z_{0L} + Z_{0SS}}$$

$$= \frac{[1 - (n + K_0)]I_0 Z_{0M}}{Z_{0SR} + Z_{0L} + Z_{0SS}}$$

$$= \frac{\Delta V_{SR}}{Z_{0SR} + Z_{0L} + Z_{0SS}} \qquad (12\text{-}16)$$

where

$n =$ per unit fraction of Z'_{0L} from bus G to the fault

$\Delta V_{SR} =$ induced voltage drop from S to R

As the fault moves from G to H, the induced current I_{0M} in line RS will decrease, reverse, and then increase

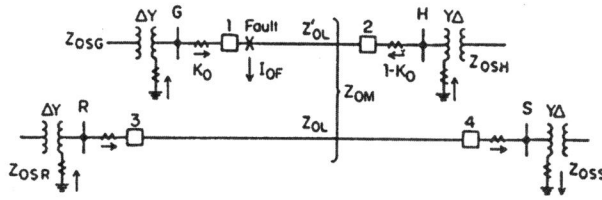

Figure 12-12 Mutual coupled transmission lines with zero sequence isolation.

in the opposite direction. For example, when $(n + K_0)$ equals 1, I_{0M} will be 0. If $(n + K_0)$ is greater than 1, then I_{0M} reverses.

For some conditions and fault locations, it may be difficult to set the overcurrent units to distinguish between faults on line GH and those on line RS. Mutual inductance will also cause incorrect directional sensing on line GH for faults on line RS. A similar analysis applies.

Correct directional sensing can be obtained for Figure 12-12 if the neutrals of the current transformers can be paralleled at G and R, and at H and S as well. A practical example of this situation would be parallel lines terminating in transformer banks at each end with grounded wye on the line side and delta on the bus. *On a per unit basis, the current flowing up the neutral on the faulted line is always greater than the current, resulting from induction, flowing down the neutral. Hence, paralleling the current transformer provides a net current up the neutral for any fault.*

Mutually coupled lines, with only one common bus (Fig. 12-13a), provide no directional sensing problems in most of the conditions, as long as all breakers are closed. The problems are as follows:

1. See Figure 12-13a. For example, zero sequence current flow arrows are for a ground fault near or at bus G. The voltage drop between the zero potential bus along line SG starting from the transformer at S is

$$I_{0S}Z_{0SS} + I_{0H}Z_{0M} + I_{0S}Z_{0L} - I_{0G}Z_{0SG} = 0$$

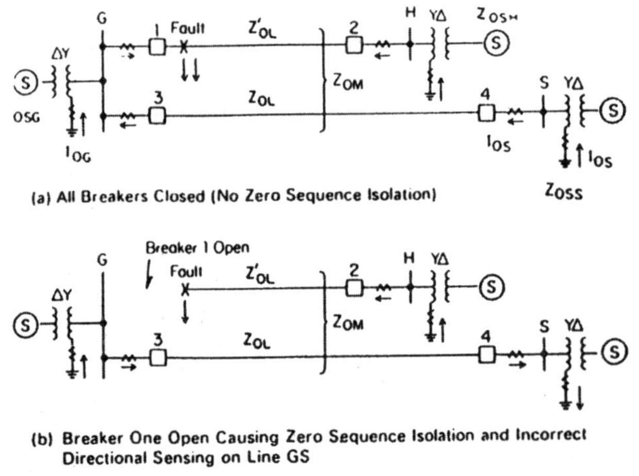

Solve for I_{0S}:

$$I_{0S} = \frac{I_{0G}Z_{0SG} - I_{0H}Z_{0M}}{Z_{0SS} + Z_{0L}} \tag{12-17}$$

I_{0S} will reverse and flow down the transformer bank neutral at station S if the mutually induced voltages are greater than the drop across the bank at G. I_{0S} then will reverse when

$$I_{0H}Z_{0M} > I_{0G}Z_{0SG} \tag{12-18}$$

2. When one breaker is opened, zero sequence isolation and incorrect directional sensing can occur. In Figure 12-13b, for example, a fault near G would be cleared by breaker 1 relays. When breaker 1 opens, line GS current flow reverses, and incorrect directional sensing occurs at S.

3.5 Application of Negative Sequence Directional Units for Ground Relays

3. A negative sequence directional ground unit can be considered in application for any one of the following system configurations, as shown in Figure 12-14:

1. In Figure 12-14a, there are no high-side VT and no I_{0S} source available for the relay polarization.
2. Figure 12-14b cannot get $3V_0$ from high-side VT; it is an open delta connection and/or there is no I_{0S} source available.
3. Figure 12-14c, an autotransformer neutral current, is not a reliable polarizing quantity.
4. In Figure 12-14d, there is mutual induction between parallel lines.
5. In Figure 12-14e, there is a mutually coupled transmission line with one common terminating bus.

3.6 Selection of Directional Overcurrent Phase and Ground Relays

All of the desired functions are available in solid-state versions, usually with curve shape selection inherent in the relay. Multifunction microprocessor relays are so flexible they allow not only curve shape selection, but also the choice of the type of polarization (including dual). Electromechanical relays require a foreknowledge of the requirements of the power system before the equipment can be chosen.

(a) All Breakers Closed (No Zero Sequence Isolation)

(b) Breaker One Open Causing Zero Sequence Isolation and Incorrect Directional Sensing on Line GS

Figure 12-13 Mutual coupled transmission lines with one common terminating bus.

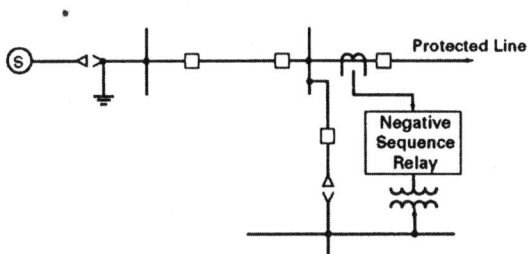

(a) No Bus VT and No Ground Current Source Are Available at Station

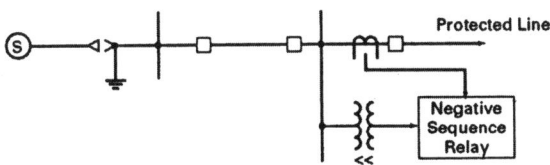

(b) Open Delta Bus VT and/or No Ground Current Source at Station

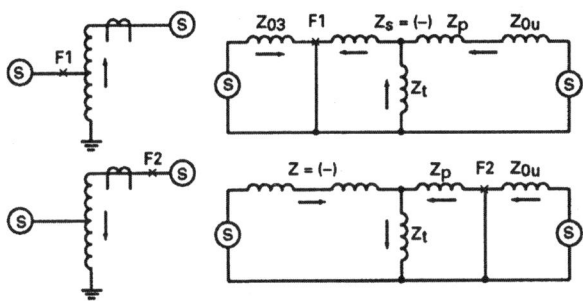

(c) Ground Current Source is Not Suitable for Polarization

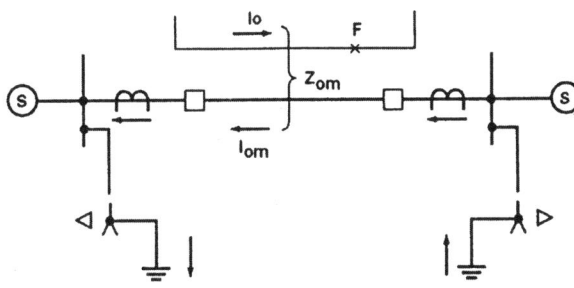

(d) Strong Zero Sequence Mutual Induction

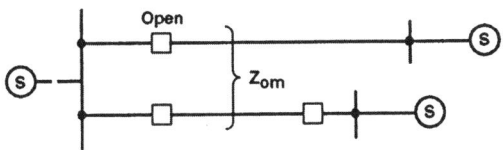

(e) Mutual Coupled Lines with one Common Terminating Bus

Figure 12-14 Typical applications of negative sequence directional ground relay.

3.6.1 Available Phase and Ground Overcurrent Relays

Nondirectional Phase and Ground Overcurrent Relays

Type Hi-Lo CO (single-phase, electromechanical) and type MMCO (three or four elements, microprocessor-based) relays are suitable for this application.

Directional Phase and Ground Overcurrent Relays

The available electromechanical and solid-state directional phase and ground overcurrent relays are as shown in Table 12-4.

3.6.2 Type of Relay Selection

The procedures for selecting a type of relay for either phase or ground protection are as below (refer to Fig. 12-8):

Step 1 A directional relay should be used if

$$4(I_{4\,max} \text{ or } I_{RLoad}) \geqq (I_{2\,min} \text{ or } I_{3\,min})$$

Step 2 An instantaneous relay can be applied and set to underreach. If

$$I_{1\,max} > 2.5\, I_{2\,max}$$

the unit can be set $IT > (1.1 \text{ to } 1.3)\, I_{2max}$ and step 3 should be checked.

Step 3 The instantaneous unit should be directional if

$$I_{4\,max} > I_{2\,max}$$

3.6.3 Evaluation of Ground-Relay Polarizing Methods

Both zero sequence and negative sequence polarizing methods require some evaluation to ensure that sufficient operating quantities are available. Fault studies provide the $3I_0$, $3V_0$, V_2, and I_2 values that exist for various faults.

It should be noted that I_0 is equal to I_2 only for a single-phase-to-ground fault on a radial circuit.

Typical profiles for V_0 and V_2 are shown in Figure 12-15. For remote faults, where nZ_{0L} can be quite large compared to Z_T, $3V_0$ for the relay at bus G can be quite low. Since negative sequence flows through the transformer to the source, V_2 can be larger than $3V_0$. The factor of 3 for both I_0 and V_0 helps, but it is still not uncommon for V_2 to be larger than $3V_0$ near a strong ground source.

Table 12-4 Summary of Available Independent Single-Function Directional Phase and Ground Relays (67/67N)

Relay type	Directional unit polarizing or operating quantities	Fault-sensing unit operating quantities		
		Directional controlled time-overcurrent unit	Nondirectional controlled instantaneous unit	Directional controlled instantaneous unit
CR[a]	L-L voltage	Yes	Option	
CRC	I_0	Yes	Option	
CRP	V_0	Yes	Option	
CRD	I_0 and/or V_0	Yes	Option	
CRQ	I_2 and V_2	Yes	Option	
IRV[a]	L-L voltage	Yes		Yes
IRC	I_0	Yes		Yes
IRP	V_0	Yes		Yes
IRD	I_0 and/or V_0	Yes		Yes
IRD	I_2 and V_2	Yes		Yes
KRV[a]	L-L voltage			Yes
KRC	I_0			Yes
KRP	V_0			Yes
KRD	I_0 and/or V_0			Yes
KRQ	I_2 and V_2			Yes
CWC	I_0	Yes		
CWP	V_0	Yes		
CWP-1	V_0	Yes		
32[b]	L-L voltage			
32D[b]	I_0 or V_0			
32Q[b]	I_2 and V_2			

[a] Phase units.

[b] Solid-state high-speed directional units that can be used to control overcurrent relay units such as MMCO, MICRO-51, 51, or 50D.

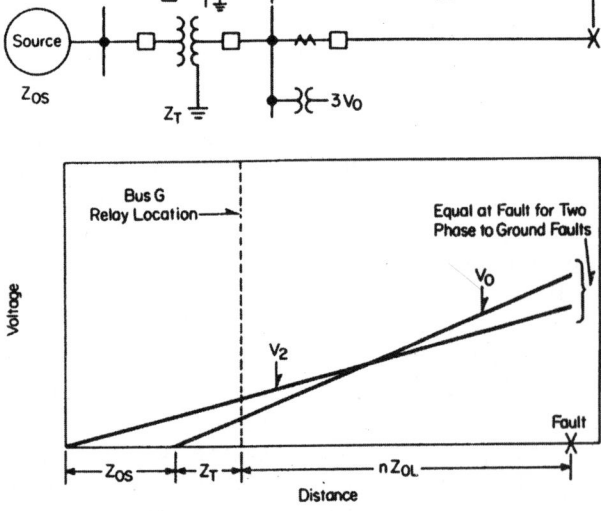

Figure 12-15 Typical voltage profiles for negative and zero sequence voltages for ground faults.

The variation in negative sequence energy is considerably less than that for zero sequence. Also, the negative sequence energy will often be higher than the zero sequence energy for remote phase-to-ground faults. The higher zero sequence impedance of lines, in conjunction with the remote ground source, highly attenuates the zero sequence distribution to a remote fault.

Negative sequence directional ground relays are widely used because they are not subject to polarization reversals or mutual induction. Neither are they subject to switching isolation, which produces mutual reversal; they require neither a transformer bank grounded neutral nor zero sequence voltage transformers.

Zero sequence generally provides higher relay operating quantities for short- and medium-length lines, although values can be quite low for remote faults on long lines. Current polarization is preferred at stations where there are ground sources, since V_0 can be very small. When there are several power transformer banks in a station, the current transformers in all the grounded neutrals should be paralleled. This method will avoid the loss of polarization if one of the banks is removed from service. Zero sequence polarization from autotransformers should be used only after a careful analysis.

Dual-polarized ground relays offer flexibility. One relay type can be used in various applications, as current polarized only, potential polarized only, or both, depending on system conditions.

4 DISTANCE PHASE AND GROUND PROTECTION

Fault levels are usually high for high-voltage transmission circuits and, if faults are not cleared rapidly, they can cause system instability as well as extensive damage and hazards to personnel. For these reasons, phase distance relays are generally used in place of the directional overcurrent relays, except at the lower-voltage levels. Even at the lower-voltage levels, the trend is toward distance relays. For the higher-voltage lines, one or two pilot systems are used in conjunction with or as a supplement to phase distance relays.

The advantages of the application of a distance relay in comparison to that of an overcurrent relay are

1. Greater instantaneous trip coverage
2. Greater sensitivity (overcurrent relays have to be set above twice load current)
3. Easier setting calculation and coordination
4. Fixed zone of protection, relatively independent of system changes, requiring less setting maintenance
5. Higher independence of load

4.1 Fundamentals of Distance Relaying

A distance relay responds to input quantities as a function of the electrical circuit distance between the relay location and point of faults. There are many types of distance relays, including impedance, reactance, offset distance, and mho.

4.1.1 Operation of Distance Relays

Basically, a distance relay compares the current and voltage of the power system to determine whether a fault exists within or outside its operating zone. The

pioneer beam-type distance relay can be used to illustrate the operating principle. Consider a horizontal beam pivoted in the center, with a voltage coil on one end and current coil on the other. In Figure 12-16, the relay coils are connected to a power line through instrument transformers. Suppose a solid fault occurs on the line at a distance of nZ_{1L} Ω from the relay. Since the voltage at the fault is 0, the voltage V_R on the relay will be the $I_R Z_L$ drop from the relay to the fault. This voltage provides a magnetic force or "pull" on one end of the beam. If, for this fault, the current or operating force I_R on the other end of the beam is adjusted to equal the voltage or restraint force V_R, the beam will be balanced. That is,

$$\frac{V_R}{I_R} = \frac{I_R n Z_{1L}}{I_R} = nZ_{1L} \tag{12-19}$$

Should a fault occur between the relay and Z_L distance, say at $(Z_L - \Delta Z_L)$ Ω from the relay, then the restraint force $I_R(Z_L - \Delta Z_L)$ would be less than the operating force at the same current magnitude. As a result, the beam would tilt down at the current end, closing the contacts. If the fault were beyond the Z_L distance, say at $(Z_L + \Delta Z_L)$ Ω from the relay, then the restraint force V_R would be greater than the operating force I_R. The beam would then tilt down at the voltage end, and the contacts would not close. The expression

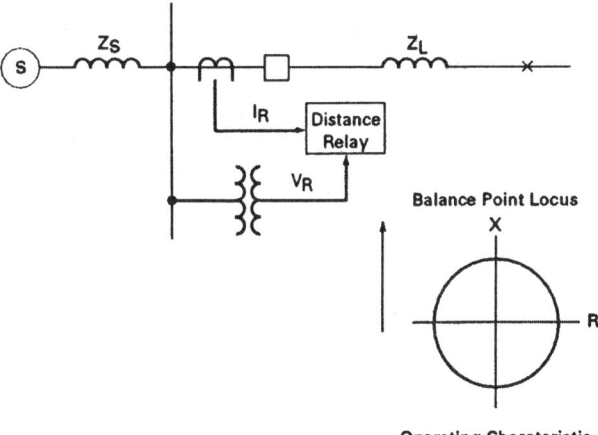

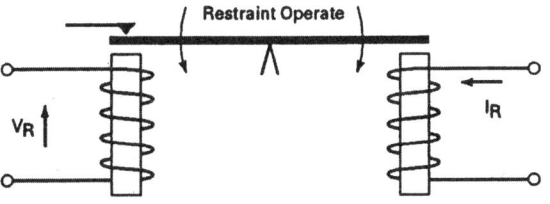

Figure 12-16 Beam type distance relay.

"balance point," which describes this threshold locus of operation for distance relays, is still commonly used, even though modern distance relays operate on quite different principles.

In general, the operating torque T of a cylinder-type distance relay is

$$T = K_1 I_R^2 - K_2 V_R^2 \tag{12-20}$$

At the threshold or balance point, T is 0. Then,

$$\frac{K_1}{K_2} = \frac{V_R^2}{I_R^2} \tag{12-21}$$

and the reach, or ohms to the balance point, is

$$\frac{V_R}{I_R} = Z_L$$
$$= \frac{\sqrt{K_1}}{\sqrt{K_2}} \tag{12-22}$$

If delta voltage and delta currents are used, a set of three relays will provide operation by one or more relays for all types of phase faults within the set-balance point impedance (Z_L).

4.1.2   Application of Distance Relays

The major advantage of distance relays is apparent from Eq. (12-19) or Eq. (12-22). The relay's zone of operation is a function of only the protected line impedance, which is a fixed constant, and is relatively independent of the current and voltage magnitudes. Thus, the distance relay has a fixed reach, as opposed to overcurrent units, for which reach varies as source conditions change.

Any line section in a power system can be represented as shown in Figure 12-17. In this figure, Z_L is the impedance of the line to be protected from bus G to bus H. Z_S is the equivalent source impedance up to bus G, and Z_U the equivalent source impedance up to bus H. Z_E could be a parallel line equal to Z_L, but, more generally, Z_E represents the equivalence of

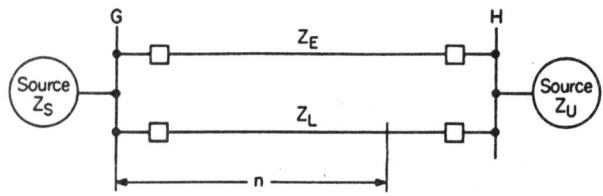

Figure 12-17 General representation of a line section between two buses.

the interconnecting system between buses G and H, except for line Z_L.

Figure 12-18 shows a simplified representation of the protected line for which distance relays are to be applied at the bus G line terminal. The system can be plotted on an R-X diagram as follows. With G as the origin, the phasor impedance Z_L of the line is drawn to scale in the first quadrant. Either per unit or ohms can be used, although secondary or relay ohms are generally preferred. Modern distance relays are normally connected to wye-connected current transformers,

$$Z_{sec} = Z_{relay}$$
$$= \frac{Z_{pri} R_C}{R_V} \tag{12-23}$$

where R_C and R_V are the ratios of current transformer and voltage transformer, respectively. Z_S, the source impedance, can be plotted from G into the third quadrant; at H, the source impedance Z_U can be extended, both impedances at their respective magnitudes and angles. Z_S in this figure is assumed to be very large or infinite relative to the others. In applications involving several line sections, Z_U would be the remote line section beyond bus H; Z_S would be the line section behind the G line relay or to the left of bus G (if we assume there were no other lines or sources at either bus G or H).

4.1.3   General Characteristics of Distance Relays

A number of distance-relay characteristics plotted on the R-X diagram are shown in Figure 12-19. The

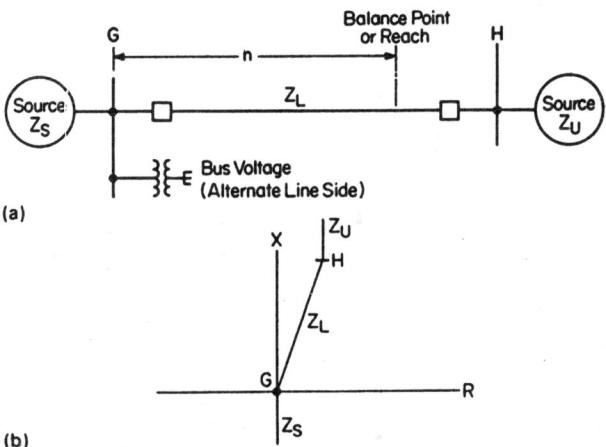

Figure 12-18   (a) Representation of a line section; (b) the R-X diagram.

operating zones are inside the circles for the types labeled a, b, and c. That is, whenever the phasor ratio of V/I falls inside the circle, the distance unit operates. The beam-type distance relay, described earlier, would have the nondirectional impedance characteristic shown in Figure 12-19a. When used for fault protection, a separate directional unit is added to limit the tripping to line faults. Since the beam-type relay is no longer manufactured, this characteristic is obtained by other techniques.

By modifying either the restraint and/or operating quantities, the circle can be shifted as shown in Figures 12-19b and 12-19c. The characteristics given in Figures 12-19d and 12-19e can be obtained in the same general way. There are a number of methods for obtaining these characteristics, the details of which are beyond the scope of this section. Appendix B of this chapter provides more information for the distance-relay characteristics.

Load can be represented on these R-X diagrams as an impedance phasor, generally lying near the R axis (depending on the power factor of the load current on the line). The phasor lies to the right (first quadrant of the R-X diagram) when flowing into the protected line from the bus and to the left (third quadrant of the R-X diagram) when flowing out of the line to the bus. Load is between 0 and 5 A secondary at or near rated voltage; faults generally produce much higher current levels and lower voltages, so that the load phasor usually falls outside the distance operating circles. Since these conditions do not hold for the reactance type shown in Figure 12-19d, this unit cannot be used alone. Also, because the reactance unit is not directional, it would, without supervision, operate for faults behind the relay. The reactance unit then needs very careful supervision and is not a particularly desirable characteristic for most applications. A blinder characteristic (Fig. 12-19e) unit can be used for load restriction on impedance (Fig. 12-19a), modified impedance or offset (Fig. 12-19b), and mho (Fig. 12-19c) distance units.

The blinder characteristic, Figure 12-19e, is essentially two reactance-type units shifted to the line impedance angle. The right unit operates for a wide area to the left, and the left unit operates for a wide area to the right of the X axis. Together, the two provide operation in the band shown.

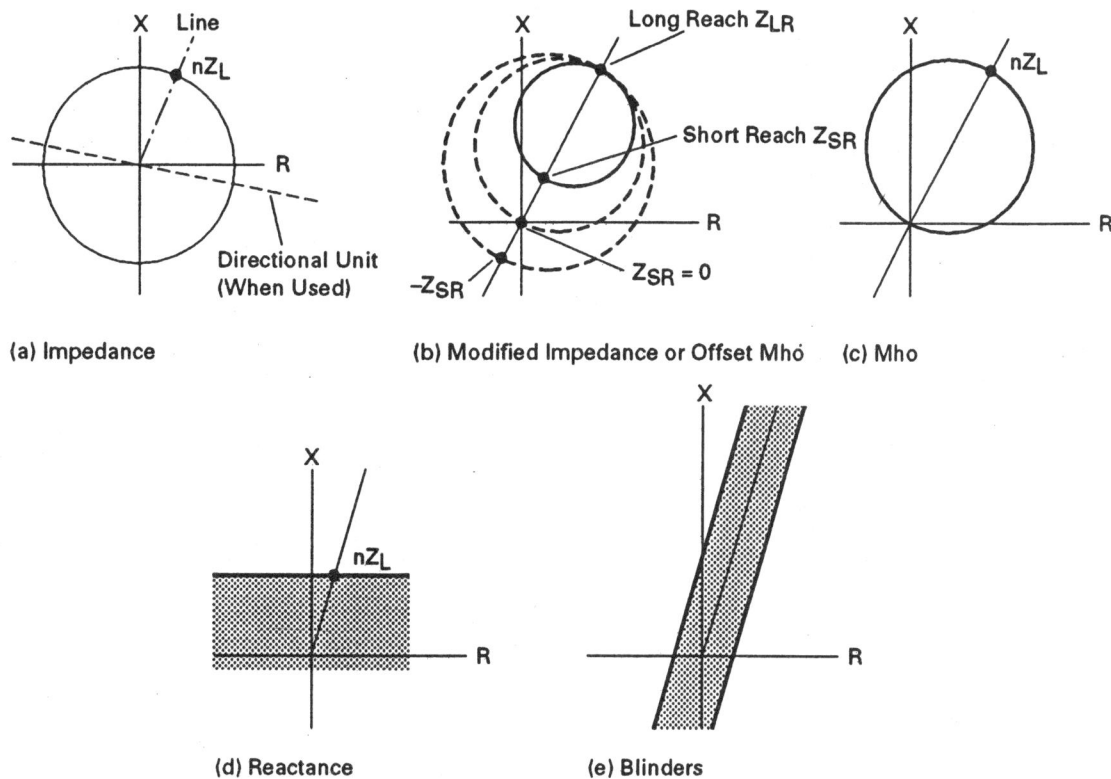

Figure 12-19 Distance relay characteristics.

With single-phase-type relays connected as phase-to-ground elements, those with one distance element connected to each phase, the R-X characteristic of Figure 12-19 applies to only the unit on the faulted phase. For three-phase faults, load, and three-phase power swings, all three single-phase distance relays have a reach of nZ_{1L} along the line (if arc resistance is neglected). All three will operate for any three-phase fault between the relay and nZ_{1L} set point.

Three-phase-type relays respond to all phase faults, regardless of the specific phases involved. For example, the KD-10 relays have two operating units. One unit responds to any three-phase fault between the relay and set reach of nZ_{1L}; the other responds to any phase-to-phase fault (AB, BC, and CA) between the relay and set reach of nZ_{1L}. One or both of the two units will respond to all two-phase-to-ground (ABG, BCG, and CAG) faults from the relay to the set reach of nZ_{1L}.

4.1.4   Characteristics of Mho-Type Distance Relays

The operation of the mho-type distance relays on the R-X diagram is given in Figure 12-20. The three-phase unit is a mho circle, where the locus at any point on the R-X diagram is

$$Z_{reach} = \frac{Z_C - Z_C \angle\theta}{2} \qquad (12\text{-}24)$$

This defines a circle whose center is offset from G at

$Z_C/2$ and radius is $Z_C \angle\theta/2$, where $\angle\theta = 0$ to $360°$. When θ is $0°$, Z_{reach} is 0 at the origin. When θ is $180°$, Z_{reach} equals Z_C, the forward reach or set point.

For the phase-to-phase unit, the reach at any point is

$$Z_{reach} = \frac{(Z_C - Z_S) + (Z_C + Z_S)\angle\theta}{2} \qquad (12\text{-}25)$$

In Eqs. (12-24) and (12-25), the first term in parentheses is the center, and the second the radius of the circle.

For the phase-to-phase unit, the circle does not pass through the origin and varies with source impedance, except at the set point.

If Z_s equals 0, then the characteristic is a mho circle through the origin, as Eq. (12-25) reduces to Eq. (12-24). When θ equals $180°$, a reverse reach is established. This function has no practical significance, since the current reversal that occurs when the fault moves from the line side of the current transformers to the bus side always produces restraint (see Fig. 12-21 and 12-23) faults (1). As a result, the relay is directional, and the operating area exists only in the first quadrant, within the area of the sector of the curve from Z_C to the R axis.

When θ equals 0, the terms containing Z_S cancel, so that Z_{reach} equals Z_C. Thus, the phase-to-phase characteristic is fixed at the set point and variable at all others. This variation poses no disadvantage, since neither load, power swings, nor any type of balanced conditions can produce operating torque. In other words, the operating and setting of the phase-to-phase unit are completely independent of load and swings.

4.2   Phase-Distance Relays

4.2.1   The KDAR Phase-Distance Relays

The KDAR phase-distance relays, types KD-10 and KD-11, are three-phase, single-zone packages in design. These relays provide mho-type characteristics, as shown in Figures 12-19c and 12-20. They consist of a three-phase unit for three-phase fault detection and a phase-to-phase unit for two-phase fault detection. The operating units are a electromechanical cylinder element. They use line-drop compensator theory that produces two phasor voltages, which are in phase when a fault occurs at the balance or reach set point. This condition produces no output. Faults inside the balance point shift the voltages in a direction to provide operation; faults beyond the balance point produce a shift in the opposite direction to provide restraint (Figs. 12-21 and 12-22).

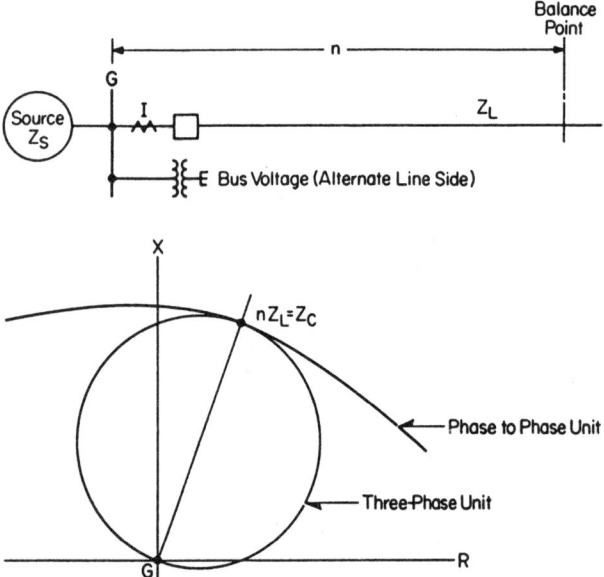

Figure 12-20   The K-DAR relay on the R-X diagram.

Figure 12-21 illustrates the operation of the three-phase unit of the KD-10 relays. With three-phase potential V_A, V_B, and V_C applied with line current $(I_A - 3I_0)$,

$$V_X = 1.5 V_A - 1.5(I_A - 3I_0)Z_C$$
$$V_Y = V_B$$
$$V_Z = V_C \qquad (12\text{-}26)$$

The I_0 component in Eq. (12-26) provides double-phase-to-ground fault coverage for systems with a very low Z_{0s} relative to Z_{2s}. For these faults, the three sequence networks are connected in parallel. If, in the limit, Z_{0s} goes to 0, the negative sequence network is shorted out, leaving only the positive sequence network, as for a three-phase fault. For very low Z_{0s} systems, therefore, double-phase-to-ground faults "look like" three-phase faults, and the three-phase unit responds. The $3I_0$ helps cover these faults but does not enter into the phasors of Figure 12-21, which are for three-phase faults.

The compensator is set so that Z_C equals the positive sequence line impedance from the relay to the balance point (Z_L). For the three-phase fault at the balance (fault 3), V_X terminates on the line between, Y and Z. The result is a zero-area triangle and no operating torque on the cylinder unit connected to X, Y, and Z (see Chap. 3, "Basic Relay Units," for a discussion of the cylinder unit). Faults beyond the balance point (4) and behind the relay (1) provide an XYZ triangle that produces opening torque. Faults (2) inside the balance point produce an XYZ triangle and operating torque proportional to the area of the triangle.

Close-in, solid three-phase faults produce an ABC triangle with a very small area. That is, Y and Z would collapse to the origin with very little or no operating area. To avoid this, memory action is provided by delaying the collapse of Y and Z when B and C voltages approach or equal 0. The memory circuit, as shown in Figure 12-21, consists of a reactor and a capacitor in phase "c" of the voltage circuit and tunes the oscillating voltage drop to the power-system frequency.

This delay is long enough to allow the instantaneous (zone 1) three-phase unit to operate and trip a breaker for 0-V three-phase faults. For backup (zone 3) and other fault applications where continuous torque is

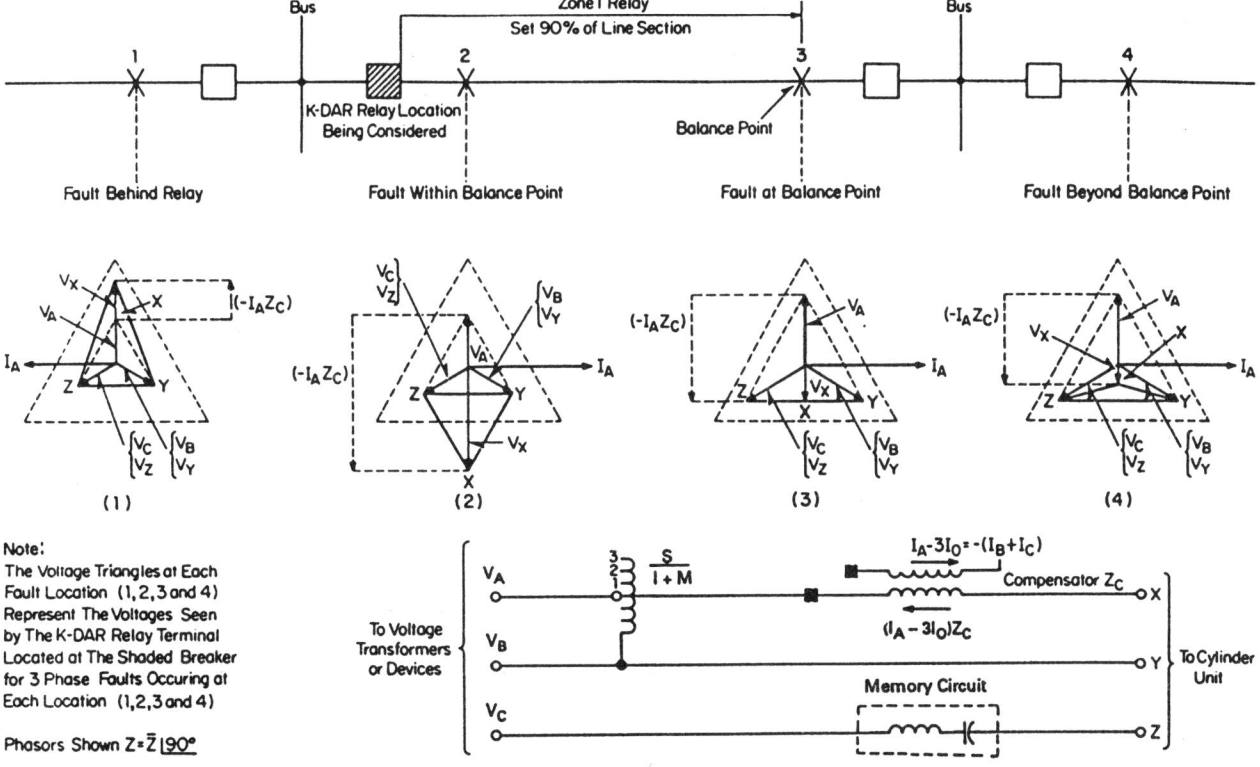

Figure 12-21 Phasor diagrams of KD-10 phase distance relays for faults at various locations (three-phase unit).

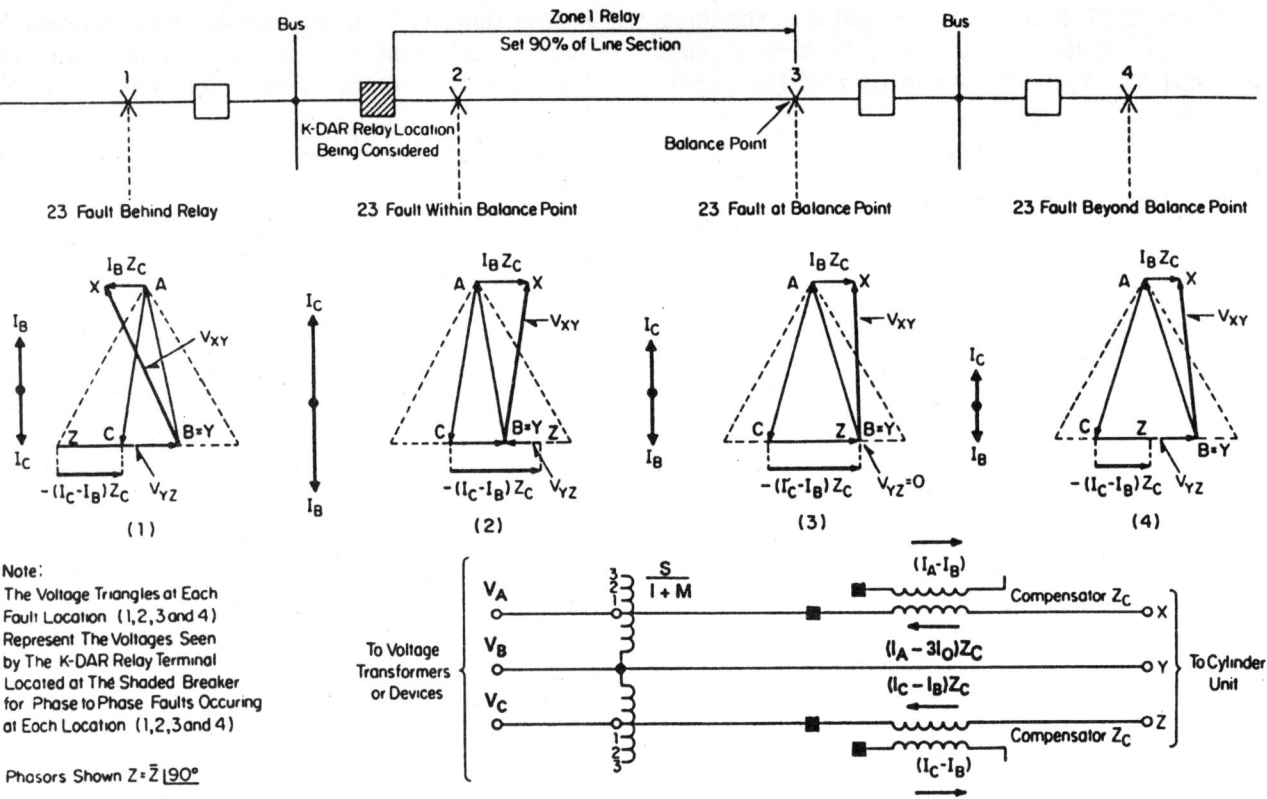

Figure 12-22 Phasor diagrams of KD-10 phase distance relays for faults at various locations (phase-phase unit).

required, a different unit with added current-only torque is applied. This arrangement makes the units nondirectional for the heavy, close-in three-phase faults. This is the KD-11 relay.

The operation of the phase-to-phase unit of the KD-10 phase-distance relay is shown in Figure 12-22. With three-phase applied voltages and current I_A, I_B, and I_C from wye-connected current transformers,

$$V_{XY} = V_{AB} - (I_A - I_B)Z_C$$
$$V_{ZY} = V_{CB} - (I_C - I_B)Z_C \qquad (12\text{-}27)$$

The compensator is set so that Z_C equals the positive sequence line impedance from the relay to the balance point (nZ_{1L}). With a phase-to-phase fault at the balance point, the XYZ triangle has 0 area for an AB, BC, or CA fault. Figure 12-22 shows the phasors for various BC faults. At the balance point, a BC fault (3) results in V_{ZY} being equal to 0. For BC faults beyond (4) or behind (1), a restraint triangle XYZ is produced. An operating triangle XZY is produced for internal BC faults (2).

The KD-10 phase-to-phase unit will also operate for most double-phase-to-ground faults. Together, the two units (three-phase and phase-to-phase) provide complete coverage for all types of double-phase-to-ground faults from the relay to the balance point setting. With the positive, negative, and zero sequence networks in parallel for these faults, the fault tends to "look like" a phase-to-phase fault when Z_{0s} is large compared to Z_{2s}. If Z_{0s} goes to infinity, the double-phase-to-ground network becomes equivalent to the phase-to-phase fault network.

Memory action is not required for this unit, and high torque exists for the solid, 0-V, phase-to-phase fault at the relay. As seen in Figure 12-22, when V_{BC} equals 0, V_{ZY} and V_{XY} are large, providing a large operating XYZ triangle.

4.2.2. The Microprocessor-Based Phase-Distance Relays:

The microprocessor-based relay, type REL-300 (MDAR) or REL-301/REL-302, is a numerical transmission line protection system with three nonpilot zones and one optional pilot zone of distance protection.

These relays use a single processor approach in design. All measurements and logic in REL-300 and REL-301/REL-302 use microprocessor technology. All the distance units provide a mho-type characteristic, as shown in Figure 12-23. The zone 3 units may be chosen to have a forward or reverse application.

Equations (12-28) and (12-29) show the phasors of the operating and reference (restraint) quantities for three-phase and $\phi\phi$ faults, respectively. The unit will produce a trip output when the operating quantity leads the reference quantity:

	Operating	Reference	
For three-phase fault	$V_{XG} - I_X Z_C$	V_Q	(12-28)
For phase-to-phase faults	$V_{AB} - I_{AB} Z_C$	$V_{CB} - I_{CB} Z_C$	(12-29)

where

$$V_{XG} = V_{AG}, V_{BG}, \text{ or } V_{CG}$$
$$I_X = I_A, I_B, \text{ or } I_C$$
Z_C = zone reach settings in terms of positive line impedance
V_Q = quadrature phase voltages, i.e., V_{CB}, V_{AC}, and V_{BA} for ϕ_A, ϕ_B, and ϕ_C, respectively
V_{AB}, V_{CB} = line-to-line voltages
I_{AB}, I_{CB} = delta currents (e.g., $I_A - I_B$)

4.2.3 The Microprocessor-Based Phase-Distance Relay: Type REL-100

The REL-100 uses a multiprocessor design with three processors for the basic distance measuring function and up to three additional signal processors performing the optional functions. The impedance units measure the apparent impedances of the fault loops. The resulting impedance is compared against reactance and resistance limits determined by the relay setting. Either zone of the relay can be selected as reverse-looking. The quadrilateral characteristics with individual settings of reactive and resistive reach are as shown in Figure 12-24.

4.2.4 Microprocessor-Based REL-512

This microprocessor-based relay utilizes the full power of digital techniques to identify whether or not a fault is within the reach of the distance unit. The same fundamental concept is used, in which two developed quantities as described by Eqs. (12-28) and (12-29) are compared. Contrasted with previous methods that developed the equivalent phasor quantities for comparison, REL-512 uses individual samples of voltage and current to determine fault location.

In the various measurements (phase fault, blinder, directional, etc.) quantities S1 and S2 are defined. In Eq. (12-28), for "a-phase," for example, $S1 = V_{AG} - I_A Z_C$, and $S2 = V_{CB}$. Operation occurs when S1 phasor leads S2 phasor, with maximum sensitivity occurring when S1 leads S2 by 90°.

Figure 12-25 shows the manner in which the REL-512 system determines fault location by using two adjacent samples of two waveforms. The two quantities are similar to those previously described such as

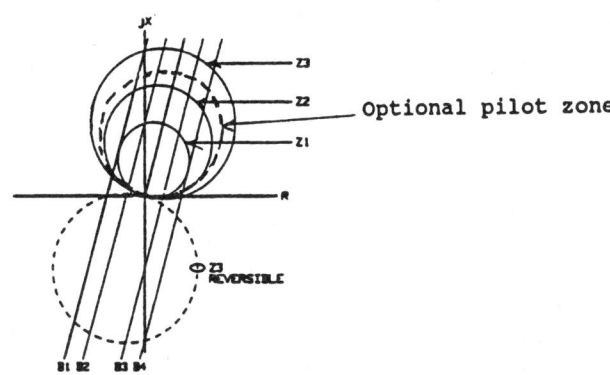

Figure 12-23   REL-300 (MDAR) characteristics.

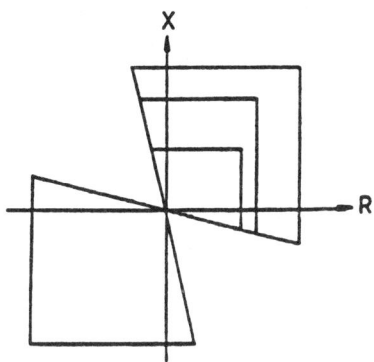

Figure 12-24   REL-100 characteristics.

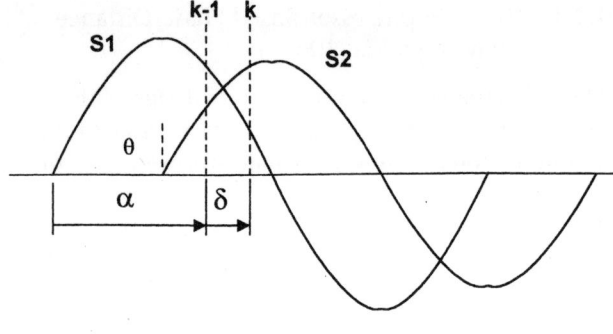

$$S1 = A \sin \omega t$$

$$S2 = B \sin (\omega t - \theta)$$

$$S1k = A \sin (\omega t + \alpha + \delta)$$

$$S1k - 1 = A \sin (\omega t + \alpha)$$

$$S2k = B \sin (\omega t + \alpha + \delta - \theta)$$

$$S2k - 1 = B \sin (\omega t + \alpha - \theta)$$

$$S1k - 1 \; S2k = A \sin (\omega t + \alpha) \; B \sin (\omega t + \alpha + \delta - \theta)$$

But, $\sin \alpha \sin \beta = 1/2 \cos (\alpha - \beta) - 1/2 \cos (\alpha + \beta)$

$$S1k - 1 \; S2k = AB/2 \; [\cos (\omega t + \alpha - \omega t - \alpha - \delta + \theta)$$
$$- \cos (\omega t + \alpha + \omega t + \alpha + \delta - \theta)]$$

$$S1k - 1 \; S2k = AB/2 \; [\cos (\theta - \delta) - \cos (2 \omega t + 2 \alpha + \delta - \theta)]$$

Also, $S1k \; S2k - 1 = A \sin (\omega t + \alpha + \delta) \; B \sin (\omega t + \alpha - \theta)$

$$S1k \; S2k - 1 = AB/2 \; [\cos (\omega t + \alpha + \delta - \omega t - \alpha + \theta)$$
$$- \cos (\omega t + \alpha + \delta + \omega t + \alpha - \theta)]$$

$$S1k \; S2k - 1 = AB/2 \; [\cos (\theta + \delta) - \cos (2 \omega t + 2 \alpha + \delta - \theta)]$$

Now, obtaining the difference of the products:

$$S1k - 1 \; S2k - S1k \; S2k - 1 = AB/2 \; [\cos (\theta - \delta) - \cos (\theta + \delta)]$$

Applying $\cos (\alpha + \beta) = \cos \alpha \cos \beta - \sin \alpha \sin \beta$

and $\cos (\alpha - \beta) = \cos \alpha \cos \beta + \sin \alpha \sin \beta$

Then,

$$(S1k - 1 \; S2k) - (S1k \; S2k - 1) = AB/2 \; [\cos \theta \cos \delta + \sin \theta \sin \delta$$
$$- \cos \theta \cos \delta + \sin \theta \sin \delta]$$

$$= AB \sin \theta \sin \delta$$

$\sin \delta$ is always positive and therefore does not influence the sign of the product.

θ is positive when $\sin \theta$ is positive and less than 180°. Therefore, θ is positive, <u>indicating that S1 leads S2</u>, whenever the difference of the products is positive.

Figure 12-25 Criterion for Lead-Lag Relationship.

in Figure 12-22. In this example,

$$S1 = V_{XY} = V_{AB} - I_{AB}Z_C$$
$$S2 = V_{ZY} = V_{CB} - I_{CB}Z_C$$

Inputs of V_{AG}, V_{BG}, V_{CG}, I_A, I_B, and I_C allow this

digital algorithm to be implemented. Operation is produced when S1 leads S2.

In Figure 12-25, θ represents the angle by which S1 leads S2. This quantity is negative when S2 leads S1. K represents the instant at which a sample is taken of S1 and S2. K-1 is the point at which the previous samples of S1 and S2 were taken. A and B are the peaks of the two waveforms.

S1 and S2 are shown as two arbitrary sine functions. S1k is the instantaneous value of the sample of S1 taken at time k. S1 k − 1 is the value of S1 taken at the previous sample. Similar values are taken of S2.

The purpose of this example is to show the ability to identify which of two waveforms leads the other based upon two adjacent samples of each of the waveforms. In the actual implementation, samples of all of the voltage and current inputs are combined appropriately (along with acknowledgment of the various settings) to produce the proper samples for comparison. This, in turn, allows the location of a fault to be identified as within or outside of the operate zone of the relay.

4.3 Ground-Distance Relays

4.3.1 Fundamentals of Ground-Distance Relaying

The general representation of a line section, given in Figure 12-17, is expanded in Figure 12-26 for a line-to-ground fault at F. The equations of the relay current and voltage at bus G are developed for phase "a." For the phase a relay, consider the following combinations for a ground-distance relay.

1. Using zero sequence quantities $3V_0$ and $3I_0$, we get

$$Z_{relay} = \frac{3V_{0G}}{3I_{0G}}$$
$$= \frac{-I_0 P_0 Z_{0S}}{K_0} \tag{12-30}$$

Unfortunately, this method is not useful. It measures the source impedance and current distribution factors, both of which are variable, rather than the protected line impedance.

2. Using phase voltage and phase current. This method depends on voltage compensation. By using only V_{AG} and I_{AG}, we obtain

$$Z_{relay} = \frac{V_{AG}}{I_{AG}}$$
$$= \frac{(K_1 I_1 + K_2 I_2) n Z_{1L} + K_0 I_0 n Z_{0L}}{K_1 I_1 + K_2 I_2 + K_0 I_0} \tag{12-31}$$

The lines are not Mutually Coupled, P and K are The Sequence Network Distribution Factors. Each Sequence Network to F Reduces to an Equivalent Impedance of:

$$Z_1 = P_1 Z_{1S} + n K_1 Z_{1L}$$
$$Z_2 = P_2 Z_{2S} + n K_2 Z_{2L}$$
$$Z_0 = P_0 Z_{0S} + n K_0 Z_{0L}$$

For a Phase A to Ground Fault at F, at The Relay Location at Bus G:

$$V_{1G} = V - I_1 P_1 Z_{1S}$$
$$V_{2G} = 0 - I_2 P_2 Z_{2S}$$
$$V_{0G} = 0 - I_0 P_0 Z_{0S}$$

$$V_{aG} = V - I_1 P_1 Z_{1S} - I_0 P_0 Z_{0S}$$

From The Sequence Network:

$$V = I_1 P_1 Z_{1S} + I_1 Z_{1L} + I_2 Z_{2S} +$$
$$I_2 n K_2 Z_{1L} + I_0 P_0 Z_{0L} + I_0 n K_0 Z_{0L}$$

$$V = I_1 P_1 Z_{1S} - I_2 P_2 Z_{2S} - I_0 P_0 Z_{0S} =$$
$$(K_1 I_1 + K_2 I_2) \, n Z_{1L} + K_0 I_0 \, n Z_{0L}$$

Substituting:

$$V_{aG} = (K_1 I_1 + K_2 I_2) \, n Z_{1L} + K_0 I_0 \, n Z_{0L}$$
$$I_a = K_1 I_1 + K_2 I_2 + K_0 I_0$$

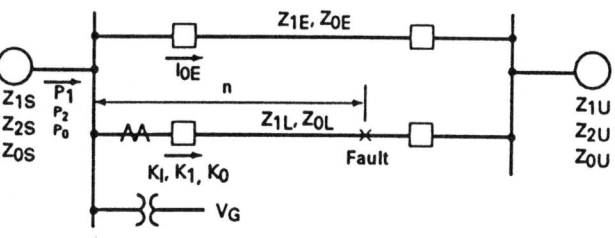

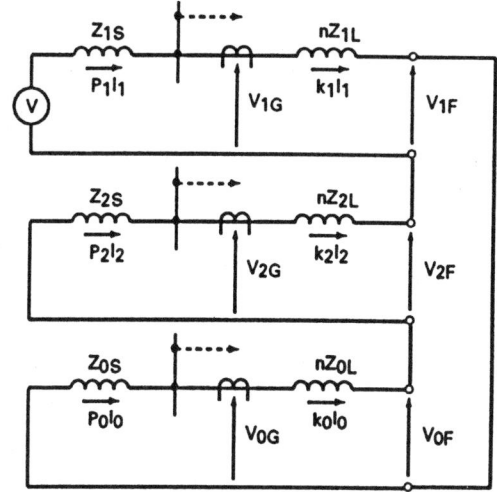

Figure 12-26   A line-to-ground fault on line section "GH".

Again, the method is unsatisfactory.

3.   Using V_{AG} modified by subtracting the positive and negative sequence drop of $(K_1 I_1 + K_2 I_2) n Z_{1L}$, and I_0, we get

$$Z_{relay} = \frac{V_{AG} - (K_1 I_1 + K_2 I_2) n Z_{1L}}{K_0 I_0}$$
$$= n Z_{0L} \qquad (12\text{-}32)$$

This method is satisfactory and has been used.

4.   With phase-to-ground voltage and modified phase current. This is the current compensation method. Assuming that $n Z_{0L}$ equals $p n Z_{1L}$, we obtain

$$Z_{relay} = \frac{V_{AG}}{I_{AG}}$$
$$= \frac{(K_1 I_1 + K_2 I_2) n Z_{1L} + K_0 I_0 p n Z_{1L}}{K_1 I_1 + K_2 I_2 + K_0 I_0}$$
$$= n Z_{1L} \frac{K_1 I_1 + K_2 I_2 + p K_0 I_0}{K_1 I_1 + K_2 I_2 + K_0 I_0} \qquad (12\text{-}33)$$

By modifying the relay current I_R to $(K_1 I_1 + K_2 I_2 +$

$p K_0 I_0)$ instead of I_{AG}, we get

$$Z_{relay} = n Z_{1L}$$
$$I_R = K_1 I_1 + K_2 I_2 + p K_0 I_0$$
$$= K_1 I_1 + K_2 I_2 + K_0 I_0 + (p - 1) K_0 I_0$$
$$= I_{AG} + (p - 1) K_0 I_0 \qquad (12\text{-}34)$$

where

$$p = \frac{Z_{0L}}{Z_{1L}} \qquad \text{and} \qquad (p - 1) = \frac{Z_{0L} - Z_{1L}}{Z_{1L}}$$

Substituting yields

$$I_R = I_{AG} + \left(\frac{Z_{0L} - Z_{1L}}{Z_{1L}} \right) K_0 I_0 \qquad (12\text{-}35)$$

This method is the one used in the KDXG, MDAR and REL-512 ground-distance units. The complete formula, including arc resistance and the mutual effect of the parallel line, can be written as

$$Z_{relay} = \frac{V_{AG}}{I_R} = nZ_{1L} + R_G\frac{3I_0}{I_R} \qquad (12\text{-}36)$$

where

$$I_R = I_{AG} + \left(\frac{Z_{0L} - Z_{1L}}{Z_{1L}}\right)K_0 I_0$$
$$+ \left(\frac{Z_{0M}}{Z_{1L}}\right)I_{0E} \qquad (12\text{-}37)$$

In these formula,

R_G = arc plus tower footing resistance and includes ground wires, when used

Z_{0M} = mutual impedance to a parallel line where I_{0E} current flows

I_0 = total zero sequence current, of which $K_0 I_0$ is the portion through the relay

4.3.2 The KDXG Reactance Ground-Distance Relay

The KDXG is an electromechanical cylinder unit, single-phase multizone relay. It permits three reactance zones of protection.

The operating principle of the reactance unit of the KDXG relay is similar to the description in Section 4.3.1, Step 3, except the compensator setting jX is the reactance to the balance point, which equals the reactance part of nZ_{1L} in Eq. (12-38).

Figure 12-27 shows the phasor diagrams for the reactance unit of the KDXG relay for faults at various locations. The unit will produce contact-closing torque when the operating quantity $V_{op} = V_L - jXI_R$ lags the reference quantity $V_{ref} = I_R$.

Voltage switching in the three single-phase reactance relays changes the reach through all three zones, in order. Switching is initiated by the zones 2 and 3 timers.

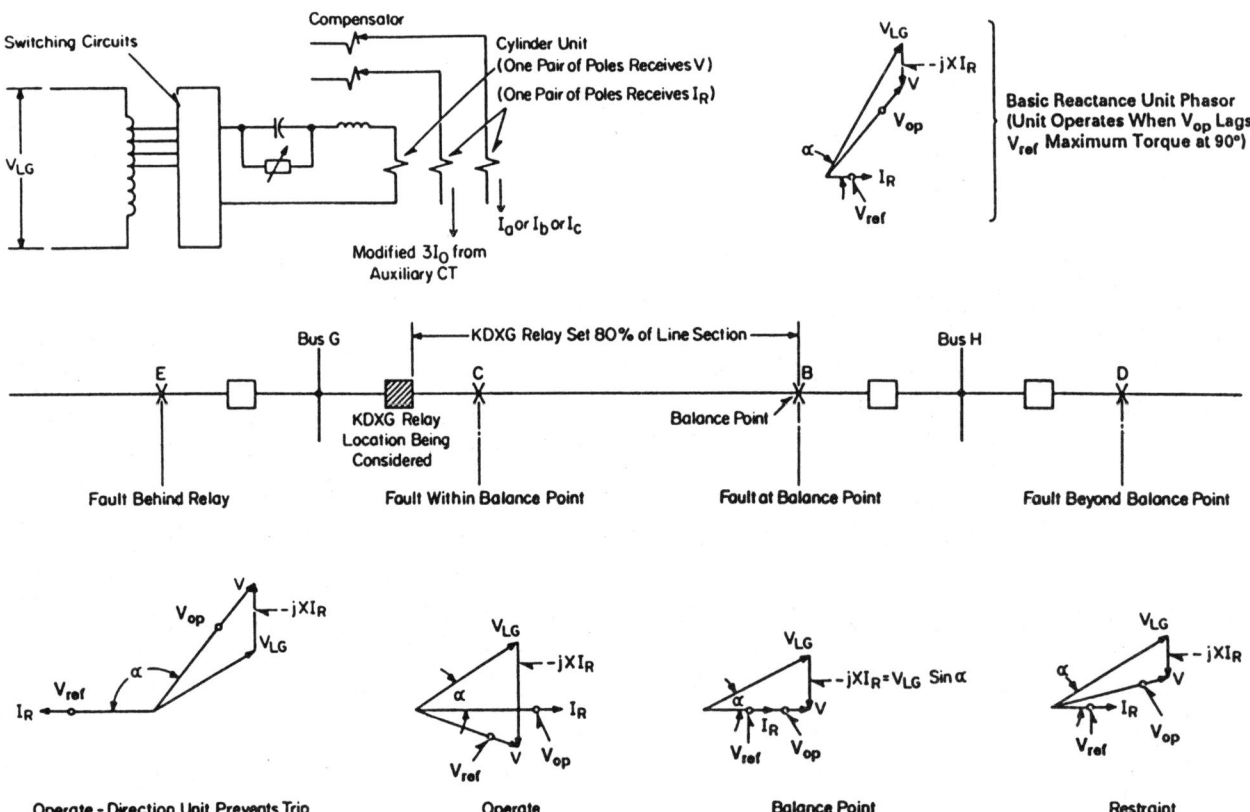

Figure 12-27 Phasor diagrams for KDXG relay for faults at various locations.

The reactance units are nondirectional and must be supervised by two units: (1) an external connected directional (KRT or KDTG), and (2) an internal ratio discriminator, RD unit. The RD unit determines the faulted phase and also provides the phase-to-phase fault blocking feature for the relay.

A terminal consists of three KDXGs (reactance and ratio discriminator units) and one KRT or KDTG (directional unit and two zone timers). The basic trip and control circuits are illustrated in Figure 12-28.

4.4 Effect of Line Length

4.4.1 Zone Application of Distance Relaying

Historically, three zones of protection have been used to protect a line section and provide backup for the remote section (Fig. 12-29). Each of the three zones uses instantaneous operating distance relays. Zone 1 is set for 80 to 90% of the line impedance. Zone 2 is adjusted for 100% of the line, plus approximately 50% of the shortest adjacent line off the remote bus. Zone 3 is set for 100% of both lines, plus approximately 25% of the adjacent line off the remote bus. These classical settings define the protective zones only if there are no infeed effects. In practice, there is almost always an infeed effect at the buses, which reduces the reach as described later.

Since zone 1 (Z1) tripping is instantaneous, the zone must not reach the remote bus, hence the 80 to 90% settings. The 10 to 20% margin provides a safety factor, for security, to accommodate differences or inaccuracies in relays, current, potential transformers, and line impedances. The 10 to 20% end zone is protected by the zone 2 (Z2) relay, which operates through a timer T2, set with one step of CTI (coordination time interval), as for overcurrent relays. Two zones at each terminal are required to protect all of the line section, with 60 to 80% of the line having simultaneous instantaneous protection. This protection is independent of system changes and loading.

The backup zone 3 also operates through a timer T3, set as shown to coordinate with the zone 2 unit of the remote bus. Coordinating distance relays, with their fixed reach and time, is much easier than coordinating overcurrent relays.

For directional comparison blocking pilot relaying, zone 3 is used to start the carrier. It must consequently be set with reverse reach, in the opposite direction to the protected line section. T3 must be coordinated with relays operating on the lines behind, rather than ahead.

An application of distance relays to parallel lines is shown in Figure 12-30. Here, the T2 settings on each parallel line should be the same. Thus, the remote ends of both lines will trip if a fault in any end zone is not cleared by the nearby breaker and relay operation.

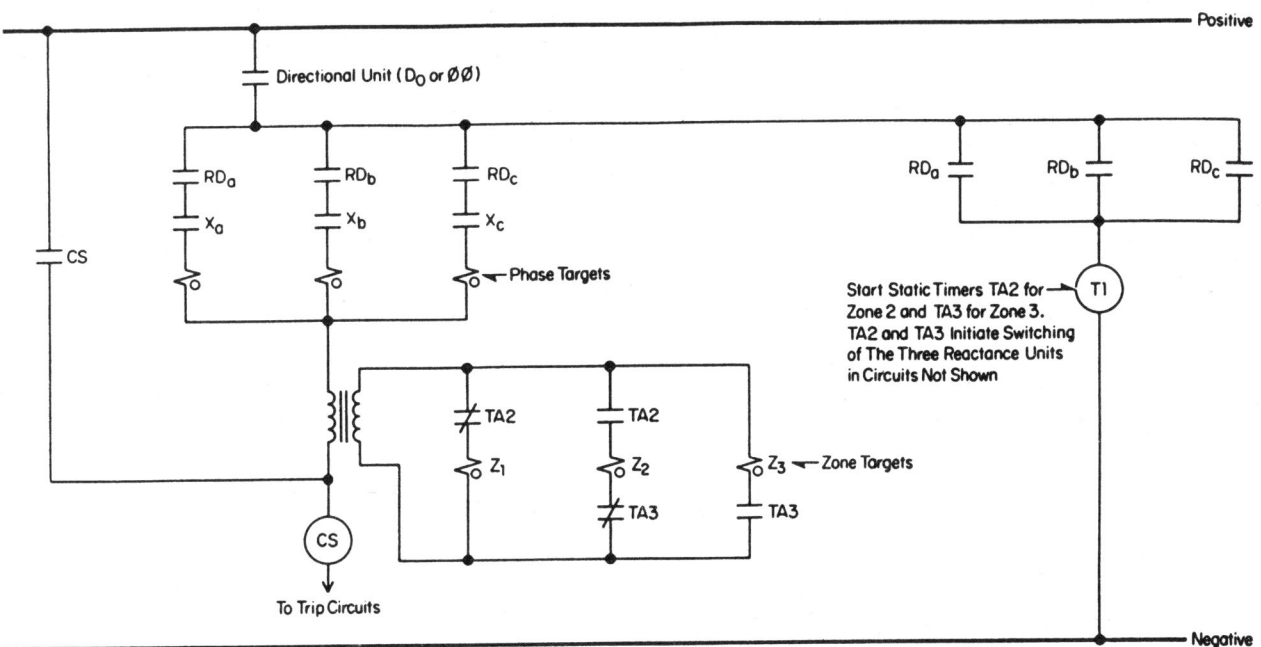

Figure 12-28 Basic trip and control circuit for the KDXG relay system.

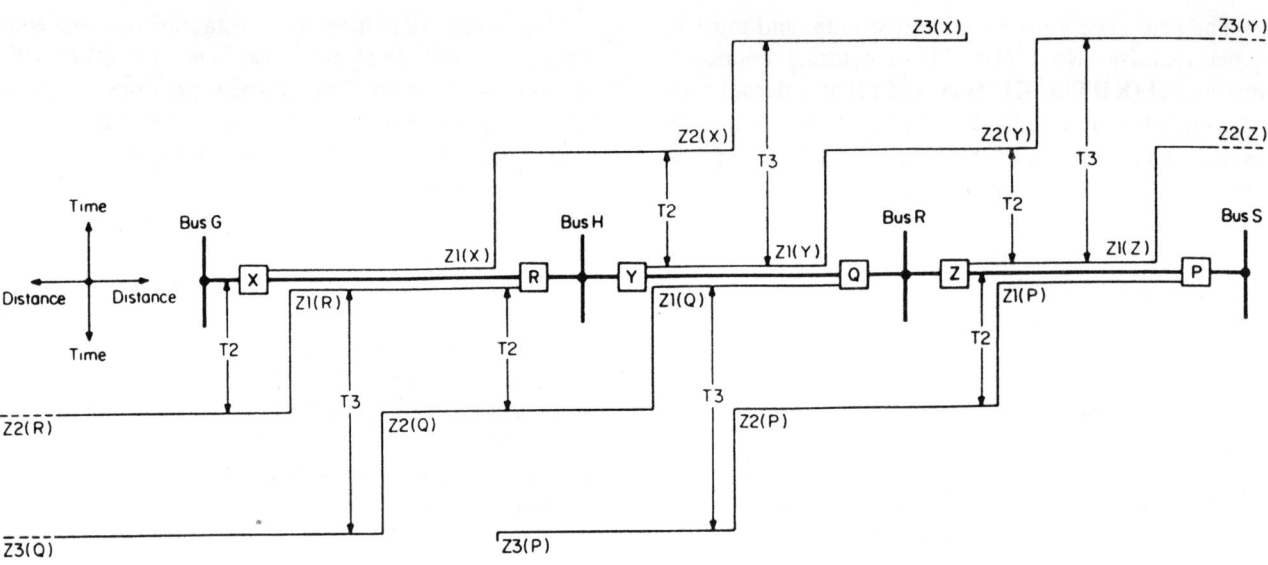

Figure 12-29 Step time zones of distance relay protection.

Suppose that for fault F1 the H-end breaker does not clear as it should on zone 1. At G, both the top- and bottom-line zone 2 relays "see" this fault and operate in time T2. A similar operation would result if the fault occurred at F2.

If both T2 settings were not the same, the unfaulted line may be cleared first. If T2 on the bottom line were greater than T2 on the top line, for example, the top-line zone 2 relay at G would operate to clear the F1 fault first. But if the fault were at F2, the top-line zone

2 relay at G would operate, even though the bottom-line zone 2 relay should clear this fault.

When several remote lines have different lengths, the zone 2 and 3 settings involve compromises (Fig. 12-31). Since line HV is short compared to lines HS and HR, setting zone 2 at G for 50% of line HV provides a maximum of 5.5% coverage for line HR and 8.4% for line HS. This coverage is further reduced by the infeed effect. Additional coverage could be obtained by increasing the G zone 2 setting and the corresponding

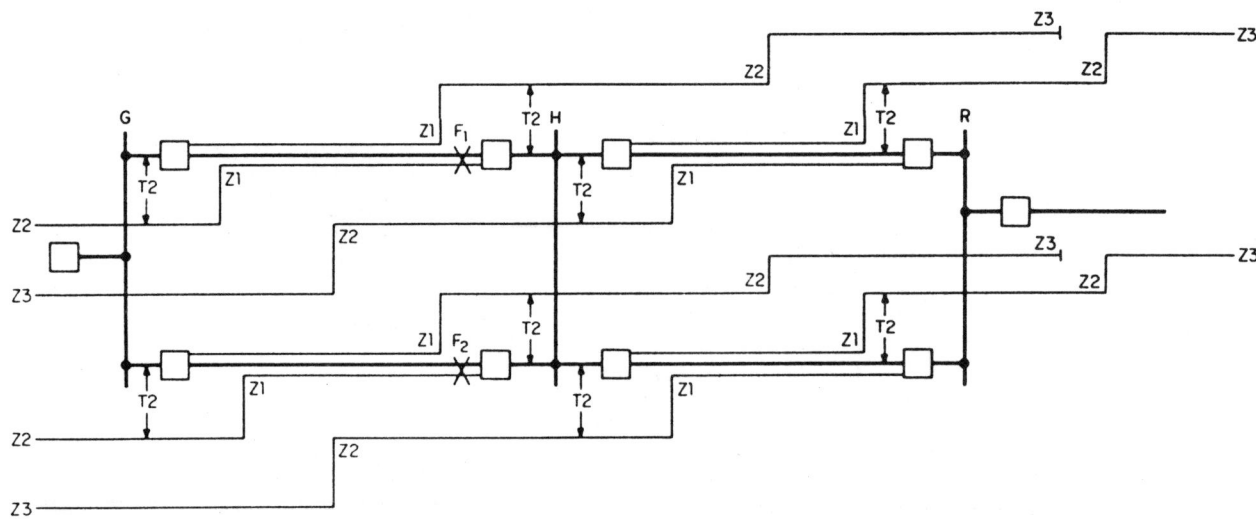

Figure 12-30 Distance relays on parallel lines.

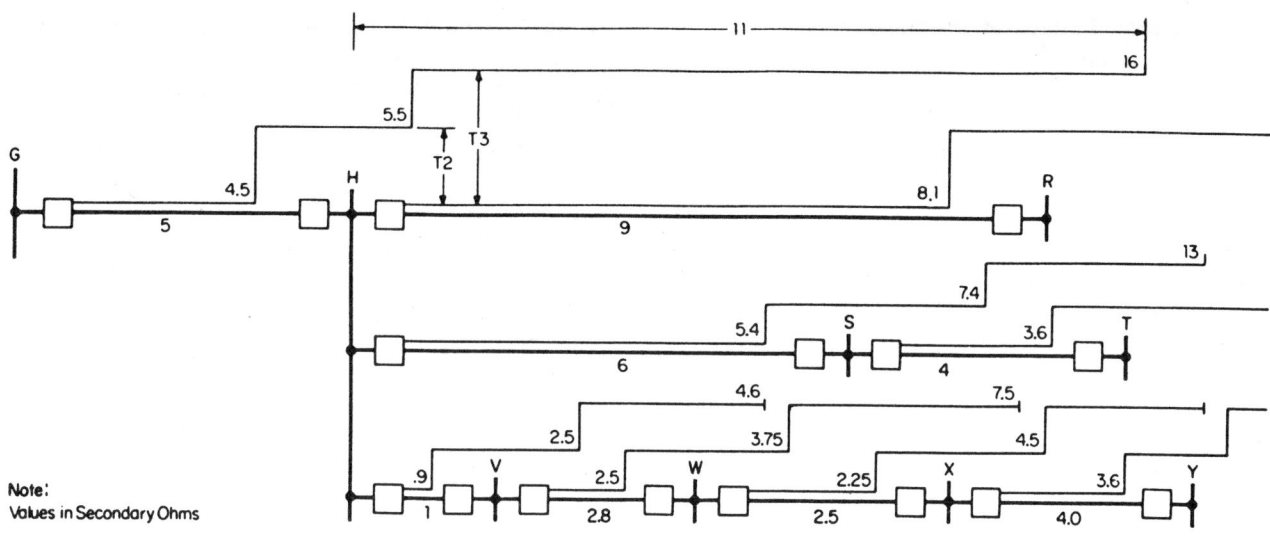

Figure 12-31 Distance relays looking into a bus with various lengths of adjacent line sections.

T2 setting to coordinate with the T2 times on lines HV, VW, and WX. The result would be long end-zone clearing for line G. If pilot relaying is used for primary protection, increased backup with longer T2 times could be employed.

Setting zone 3 to cover line HR would provide coverage through several sections HV, VW, WX, and XY requiring a longer T3 setting. Again, the infeed effect from lines HS and HV probably would not provide T3 coverage for line HR. *This fact reemphasizes the need for local backup in modern power systems.*

4.4.2 Classes of Line Length

Transmission lines are often referred to as short, medium, or long, and relays are recommended on this basis. In reality, the length of the transmission line is not a predominant factor. The significant criterion is SIR (source/line impedance ratio).

The SIR establishes the positive sequence voltage that will be present at the relay location for a three-phase fault at the far end of the protected line. While solid-state and microprocessor relays require little energy to operate, there is a voltage level below which operation is not clearly predictable or below which the operating speed is unsatisfactory. For a distance relay, a fault at the balance point (the point to which the relay is set to reach) is the point at which the operating voltage for the relay is zero. For a fault closer to the relay, a positive operating voltage will exist. The lower the SIR, the greater the magnitude of this operating voltage, and thus the more positively the relay will operate (with the greater speed).

The IEEE/PES/Power System Relaying Committee has chosen guidelines for line length criteria (IEEE Standard C37.113-1999). They have found the following to be reasonable:

Short line	$SIR > 4$
Medium line	$0.05 < SIR < 4$
Long line	$SIR < 0.05$

For short lines, there may be little difference in voltage and current at the relay location for faults that are a considerable distance from one another. The distinction between internal and forward external fault locations may be difficult. This encourages, for short lines, the use of relaying systems that establish fault location on the basis of current transformer location, such as current differential or phase comparison.

Long lines allow excellent use to be made of distance relays alone or in pilot applications (see the companion volume, *Pilot Protective Relaying*).

For medium lines, many choices of relaying types exist, including overcurrent or directional overcurrent. Other factors such as the relaying on adjacent line sections, presence of taps on the line, tripping speed required, channel availability, etc., must be considered.

4.5 The Infeed Effect on Distance-Relay Application

4.5.1 Infeed Effect on Phase-Distance Relay

When there is a source of fault current within the operating zone of the distance relay, its reach will be reduced and variable. This infeed effect can be seen from Figure 12-32, where there are other lines and sources feeding current to a fault at F from bus H. The relays at bus G are set beyond this fault point to F′. With a solid 0-V fault at F, the voltage for the relay at G is the drop along the lines from the fault to the relay, or

$$V_G = I_G Z_L + (I_G + I_H)Z_H \qquad (12\text{-}37)$$

Since relay G receives only current I_G, the impedance appears to be

$$Z_{G\ apparent} = \frac{V_G}{I_G}$$

$$= Z_L + Z_H + \frac{I_H}{I_G}Z_H \qquad (12\text{-}38)$$

$$= Z_L + \frac{Z_H}{K} \qquad (12\text{-}39)$$

where K is the current distribution factor (phasor), which equals $I_G/(I_G + I_H)$. This apparent impedance compares to the actual impedance to fault F of

$$Z_{G\ actual} = Z_L + Z_H \qquad (12\text{-}40)$$

If I_H is 0 (no infeed), Z apparent equals Z actual. As the infeed increases in proportion to I_G, Z apparent increases by the factor $(I_H/I_G)Z_H$. Since this impedance, as "measured" by the distance relay, is larger than the actual, the reach of the relay decreases. That is, the relay protects less of the line as infeed increases. Since the reach can never be less than Z_L in Figure 12-32 zones 2 and 3 provide protection for the line.

However, remote backup for the adjacent line (s) may be limited. Since infeed is very common and can be quite large in modern power systems, the trend is toward local backup (see Chap. 13).

Note that the infeed effect varies with system configuration and changes, and that the apparent impedance may be a maximum under either maximum or minimum system conditions.

For example, for Figure 12-33, assume that the line impedances (relay ohms) are $2\,\Omega$ from G to the tap point, $8\,\Omega$ from the tap point to H, and $2\,\Omega$ from the tap point to R. The zone 1 relay at G is set to reach $3.6\,\Omega$. A three-phase fault occurs at $1.0\,\Omega$ from the tap point toward H. Fault-current contributions (relay amperes) are 10.95 A from bus G and 14.61 A from R. The voltage drop from G to the fault is

$$2 \times 10.95 + 1 \times 25.56 = 47.46\,V\,(\text{relay side})$$

Since the current through the relay at G is 10.95 A,

$$Z_{apparent} = \frac{47.46}{10.95} = 4.33\,\Omega\ (\text{relay side})$$

The relay apparent impedance of $4.33\,\Omega$ is higher than the relay actual setting of $3.6\,\Omega$, i.e., the relay will not operate on this fault, i.e., underreach.

However, if the fault-current contribution is changed to 11.56 A from G and 11.56 A from R, due to the source impedances change, the voltage drop from G to the fault would be

$$2 \times 1.56 + 1 \times 11.56 = 34.69\,V\ (\text{relay side})$$

Since the current through the relay at G is 11.56 A,

$$Z_{apparent} = \frac{34.69}{11.56} = 3.00\,\Omega\ (\text{relay side})$$

the relay will see the fault. Hence, the reach of the

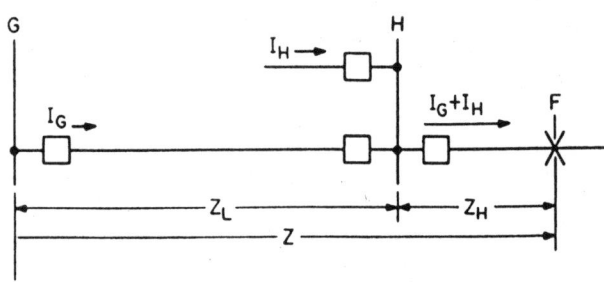

Figure 12-32 Effect of infeed on impedance measured by distance relays.

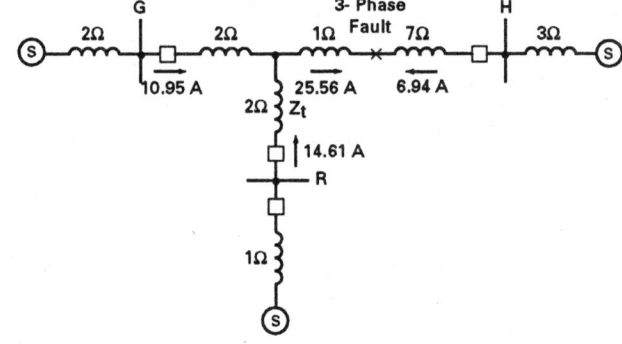

Figure 12-33 Typical infeed effect on a multiterminal line.

distance relay varies as a function of fault-current distribution, as well as fault location.

4.5.2 Infeed Effect on Ground-Distance Relay

The descriptions and results in Section 4.5 cannot be directly used for ground-distance relay application. For ground-distance units, the operating current, e.g., the phase A ground unit, is

$$I_A + \frac{Z_{0L} - Z_{1L}}{Z_{1L}} I_0$$

Therefore, when applying Eq. (12-38) the currents I_H and I_G should be replaced with

$$I_H = I_H + \left(\frac{Z_{0L} - Z_{1L}}{Z_{1L}}\right) I_0' \qquad (12\text{-}41)$$

$$I_G = I_G + \left(\frac{Z_{0L} - Z_{1L}}{Z_{1L}}\right) I_0 \qquad (12\text{-}42)$$

where I_H and I_0' are phase and zero sequence currents in the tap, I_G and I_0 are phase and zero sequence currents at the relay. (For more detail, see Appendix C to this chapter.)

4.6 The Outfeed Effect on Distance-Relay Applications

When a tap has no source except a tie line to a remote bus, fault current can flow out from this tap terminal for an internal fault near the remote bus (Fig. 12-34). Although the fault is shown on bus H, it could be near or at the breaker on the GH line. With no source at R other than line RH, current flows out of R and over RH to the internal fault on line GH, reducing the apparent impedance.

For example (a) (Fig. 12-34a), assume that the line impedances (relay ohms) are $2\,\Omega$ from G to the tap point, $8\,\Omega$ from the tap point to H, and $2\,\Omega$ from tap point to R, and the tie line between RH is $2\,\Omega$. The zone 1 relay at G is set to reach $3.6\,\Omega$ (90% of line GR). A three-phase fault occurs on bus H. Fault-current contributions (relay amperes) are 10.42 A from bus G, 3.47 A from the tap point toward bus H, and 6.95 A from the tap point out to bus R. The voltage drop from G to the fault along line GH is

$$2 \times 10.42 + 8 \times 3.47 = 48.6 \text{ V (relay side)}$$

Since the current through the relay at G is 10.42 A,

$$Z_{\text{apparent}} = \frac{48.6}{10.42} = 4.66\,\Omega \text{ (relay side)}$$

The relay apparent impedance of $4.66\,\Omega$ is higher than the relay actual setting of $3.6\,\Omega$, i.e., the zone 1 relay G is not affected by this outfeed current.

However, in example (b) (Fig. 12-34b), if the line impedance from the tap point to R is $6\,\Omega$, the zone 1 relay at G would have to be set to $7.2\,\Omega$ (90% of line GR). The fault-current contributions (relay amperes) would be 8.67 A from bus G, 4.33 A from the tap point toward bus H, and 4.33 A from the tap point out to bus R. The voltage drop from G to the fault along line GH is

$$2 \times 8.67 + 8 \times 4.33 = 51.78 \text{ V (relay side)}$$

Since the current through the relay at G is 8.67 A,

$$Z_{\text{apparent}} = \frac{51.78}{8.67} = 5.99\,\Omega \text{ (relay side)}$$

The relay apparent impedance of $5.99\,\Omega$ is less than the relay actual setting of $7.2\,\Omega$, i.e., the relay will operate on this fault, overreach. The setting must be lowered.

4.7 Effect of Tapped Transformer Bank on Relay Application

The effect of a transformer bank tapped off a line must be considered, although it often presents no problem. The typical case is shown in Figure 12-35. The infeed

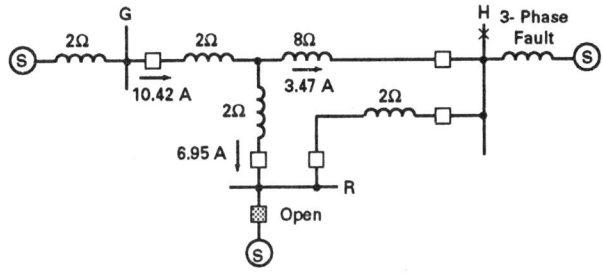

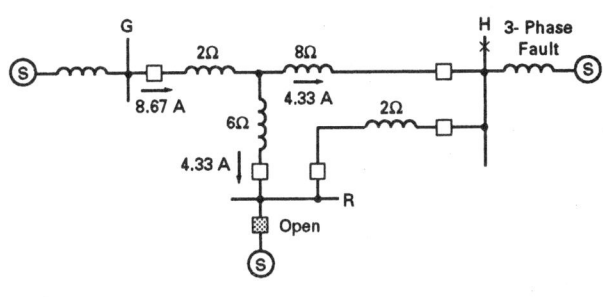

Figure 12-34 Typical outfeed effect on a multiterminal line.

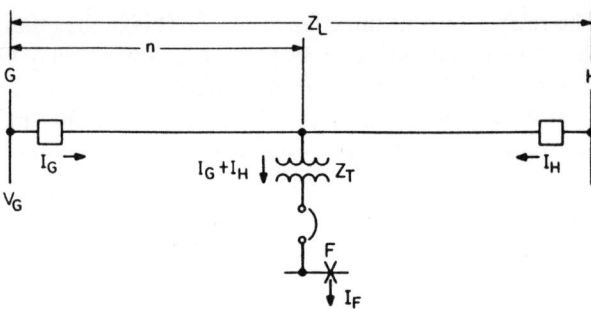

Figure 12-35 Apparent impedance for low-side faults on a tapped transformer bank.

fault current from bus H causes the relay at bus G to "see" an apparent impedance for faults in the transformer and low-side system circuits. As described above, for a fault at F, relay G sees

$$Z_{G\ apparent} = \frac{V_G}{I_G} = \frac{nZ_L I_G + Z_T(I_G + I_H)}{I_G}$$

Since $I_F = (I_G + I_H)$ and $I_G = KI_F$, where K is the current distribution factor for the current through relay G,

$$Z_{G\ apparent} = nZ_L + \frac{Z_T}{K} \qquad (12\text{-}43)$$

Since K is always less than 1 when $I_H > 0$, $Z_{G\ apparent}$ is always greater than the actual impedance to the fault $(nZ_L + Z_T)$.

Similarly, relay H would see a higher apparent impedance of

$$Z_{H\ apparent} = (1 - n)Z_L + \frac{Z_T}{1 - K} \qquad (12\text{-}44)$$

The zone 1 relay at G must be set for 80 to 90% of the smallest value of the actual impedances, Z_L or $(nZ_L + Z_T)$, and not the apparent impedances. Otherwise, zone 1 may overreach either the transformer or bus H, which would result in miscoordination. When Z_T is less than $(1 - n)Z_L$, relay G cannot protect as much of the line as it could without the tap. Usually, the tap bank is relatively small, so that Z_T is large compared to Z_L.

Zone 2 must be set greater than Z_L to protect the line. When Z_T is less than $(1 - n)Z_L$ for the relay at G or the setting for zone 2 is greater than $(nZ_L + Z_T)$, then zone 2 requires coordination with the relays in the low-voltage system. As long as H is in service, the

apparent infeed will shorten the reach. When H is open, however, the infeed effect disappears.

A complex relationship affects the reach of single-phase-type distance relays through star-delta or delta-star banks. To these relays, a phase-to-phase fault on one side of the bank appears as a phase-to-ground fault on the other, and vice versa. Using the principle described in Figure 12-22, however, all phase-to-phase faults on one side of a wye-delta or delta-wye bank appear to the relays on the other side to be a distance Z_T away. In other words, the phase shift does not affect their reach, as it does for single-phase relays.

The above discussion of the transformer tap on the line assumed no source of fault power on the low-voltage side. If such a power source exists, it produces an apparent impedance and relay underreach for line faults: on the $(1 - n)Z_L$ section of the line for the relays at G, and on the nZ_L section for the relays at H. Again, zone 1 relays G and H must be set for the actual, rather than apparent, impedances. Zones 2 and 3 must be set for the maximum apparent impedance to cover the line section. When the tap source is open, of course, the reach will be greater. This situation presents a significant problem for multiterminal lines.

4.8 Distance Relays with Transformer Banks at the Terminal

The reach of a distance relay is measured from the location of the voltage transformers; directional sensing occurs from the location of the current transformers. Voltage and current transformers are usually at approximately the same location for most applications.

4.8.1 Connections Using I_H and V_H Quantities

When the line includes a transformer bank without a breaker on the line side, there are several ways of applying distance relays (Fig. 12-36). Using I_H and V_H for the line-distance relays is preferred, since the reach is a function of nZ_L only.

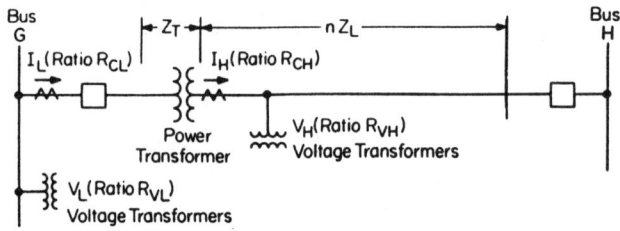

Figure 12-36 A line terminating in a transformer bank.

4.8.2 Connections Using I_L and V_H Quantities

Alternatively, I_L could be used with V_H, which would also make the reach a function of nZ_L only. In setting the relay with high-side primary ohms nZ_L, the current transformer ratio R_C of Eq. (12-23) must include the ratio of the power transformer $[R_C = R_{CL} \times (KV_L/KV_H)]$ and ratio R_{VH} of the high-side voltage transformers. Taps on the power transformer will change this ratio and, therefore, the reach. Unless the relay setting can be adjusted each time the taps are changed, zone 1 must be set at 90% of the minimum secondary ohms using Z_L. Zone 2 must be set at more than 100% of the maximum secondary ohms. The first requirement prevents zone 1 from overreaching the remote bus. The second requirement provides end-zone coverage.

If the power transformer is a wye-delta bank, either the low-side current transformers must be connected in delta or auxiliary current transformers used. In this way, the relay current will be equivalent to the wye current that would be measured on the line or high side. Figure 12-37a shows the connections for the delta on the high side, and Figure 12-37b those for the wye on the high side.

The advantage of this arrangement is that, although distance is measured from V_H and includes only nZ_L, the relay will operate for some faults in the transformer Z_T. These faults fall within the nZ_L setting, since the relay is directional from the CT location. The impedance for transformer faults is apparent: The voltage is a function of the current from the remote end, while the current flows from the near end. For the system shown in Figure 12-36, assume a solid three-phase fault of total value I between the breaker and transformer, with K per unit flowing through the low-side current transformer and $(1 - K)$ per unit flowing from the far bus to the right. Then

$$V_H = (1 - K)IZ_T,$$

and

$$Z_{apparent} = \frac{(1 - K)IZ_T}{KI}$$
$$= \left(\frac{1 - K}{K}\right)Z_T \qquad (12\text{-}45)$$

Normally, Z_T would be larger than nZ_L for zone 1 applications, limiting instantaneous protection. Zones 2 or 3, however, could be on the order of Z_T.

4.8.3 Connections Using I_L and V_L Quantities

A more common method is to use V_L since it may not be economical to provide high-side potential. It is more convenient with V_L and I_L to use primary ohms on the low-side (V_L) base. With *single-phase-type* zone-distance *phase* (not for ground units) relays set through wye-delta banks, it is necessary to shift both the currents and voltages to provide high-side quantities equivalent to those that would be measured at I_H and V_H. Otherwise, for high-side faults, the distance relay would "see" a complex impedance, which would be a function of the line transformer and source.

For the KDAR three-phase-type relays, however, this shift is not necessary. The conventional wye-current and line-to-line voltage connections of I_L and V_L will provide phase-fault protection in the transformer and on the line. The reach is a function of Z_T and nZ_L only, a distinct advantage. Since the impedance of the power transformers is known accurately, the zone 1 relay should be set through the transformer bank as

$$Z_C \text{ for zone } 1 = 0.99Z_T + 0.90Z_L \qquad (12\text{-}46)$$

Again, the power transformer taps must be taken into account, as discussed above.

This method has the disadvantage of limiting line protection when Z_T is large compared to Z_L. For example, if $Z_T = 10\,\Omega$ and $Z_L = 1\,\Omega$ then Z_1 is set for

$$9.9 + 0.9 = 10.8\,\Omega$$

Subtracting the Z_T of $10\,\Omega$ leaves only $0.8\,\Omega$ of the line protected, or 80% rather than 90%, if low-side voltage V_L were used.

Taps on the bank can change the value of Z_T and the reflected value of Z_L such that setting zone 1 to never overreach on any tap can result in very little or no line protection on another tap. This assumes that the settings are not changed with the taps.

Zone 2 should always be set for the maximum apparent ohms so as to protect the line for any tap. This may cause considerable overreach for other taps.

4.8.4 Connections Using I_H and V_L Quantities

The fourth possible arrangement is to use V_L and I_H. There is little advantage in this method, since the ohms are still measured from the low-side bus. Directional sensing would be from the line-current transformer location, rather than the bus-side current transformers.

When the line terminates in a transformer bank at the remote end, zone 1 can be set into the bank to provide 100% high-speed line protection. Only with the

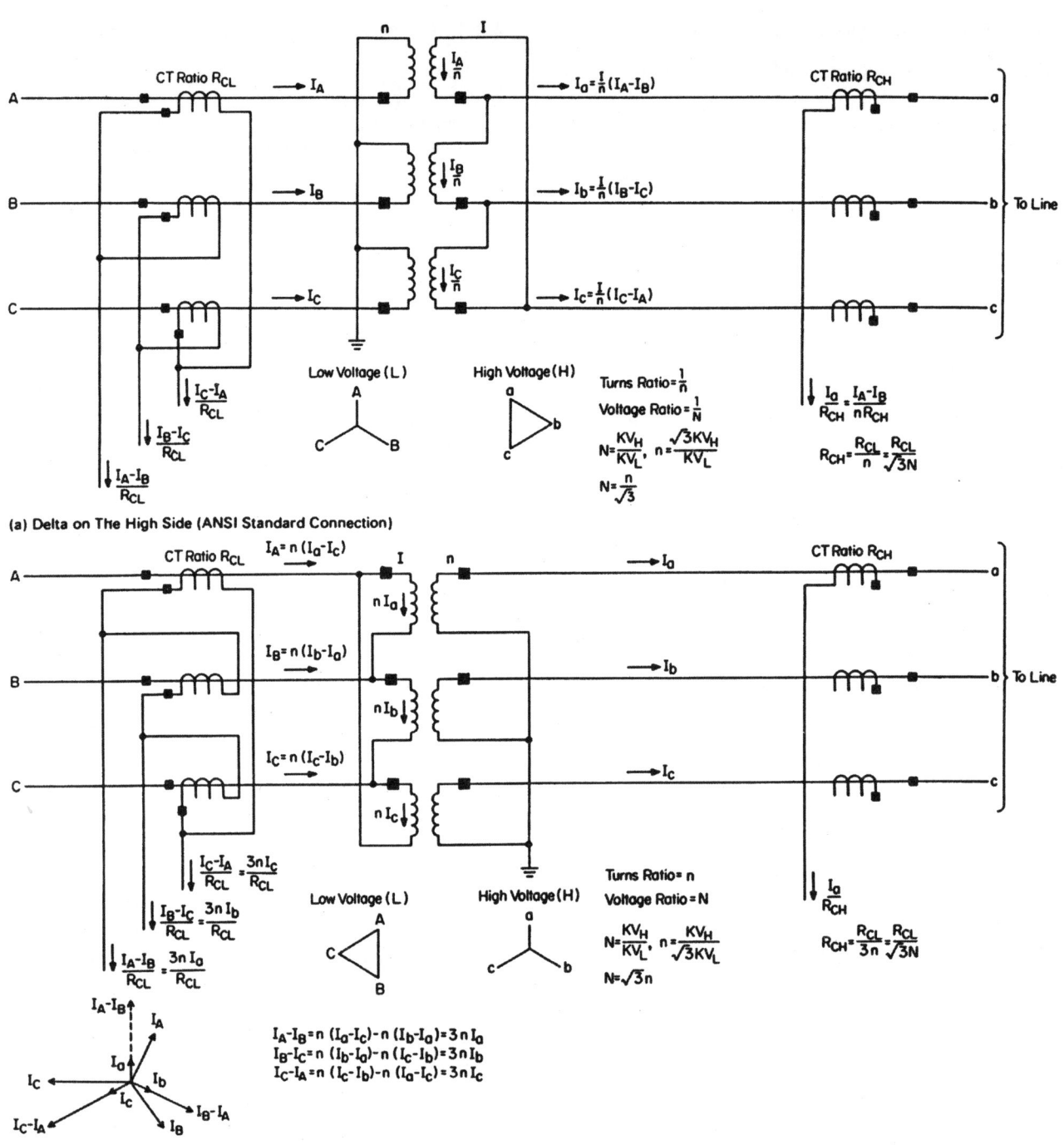

Figure 12-37   Connections for low-side ct's to provide equivalent of high-side CT's

three-phase KDAR type design relays can zones 2 and 3 be set through the bank with an accurate balance point of $Z_L + Z_T + Z_{LV}$ system for all phase faults. With single-phase-type relays, the reach through wye-delta banks is variable.

Light internal faults in the transformers will probably not produce enough variation in current and voltage to operate the remote distance relays. Consequently, transformers should have individual protection. Such protection does present a problem in that remote breakers must be tripped to clear the fault when the local transformer relays operate. A transfer trip system should be used.

4.9 Fault Resistance and Ground-Distance Relays

Two additional factors must be considered for ground faults that are not present with phase faults: tower footing resistance and ground wires. Tower footing resistance can vary from less than $1\,\Omega$ to more than $200\,\Omega$. This resistance term, multiplied by 3, must be added to the relay reach equations. As described before, the infeed to the fault from the remote terminals further magnifies this fault resistance and can cause the ground-distance relays to over- or underreach. Although, in theory, this apparent reactance effect can be quite significant, relatively few problems have been encountered in many years of field experience.

Overhead ground wires substantially reduce the line zero sequence impedance and tower footing resistance component. To the relay, however, the effect is anything but a resistance component. For fault calculations, the ground wire is assumed to be parallel with the earth. In practice, its impedance is parallel with the earth through the tower footing resistance at each tower.

The effect of arc resistance is shown in Figure 12-38 for a 15-mile, 138-kV line with a source at each end. Calculations were made assuming no angle between the two voltage sources and, hence, no angle between the current distribution factors. In practice, there is always an angle between the current distribution factors. The single phase-to-ground fault is assumed to occur at the balance point 90% of the line from bus G. The zero sequence impedances were calculated in the conventional manner, using $\rho = 100\,\text{m-}\Omega$. For the 5000 ft on either side of the fault, however, the mutual and self-impedances of the ground wires were separated for each span, a tower footing resistance of $10\,\Omega$ added, and a modified zero sequence impedance calculated.

The apparent impedance "seen" is given by Eq. (12-47) and plotted in Figure 12-38:

$$Z'_L = nZ_{1L} + \frac{3(R_{SG} + R_{TF})}{K_1 + K_2 + \frac{Z_{0L}}{Z_{1L}}K_0}$$

$$= Z_C + Z'_S + Z'_{TF} \qquad (12\text{-}47)$$

From the study, Z'_{TF} is $2.8\angle 51°\,\Omega$. Although the ground

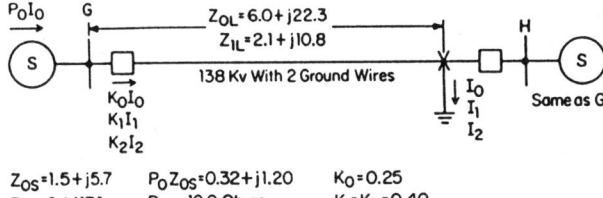

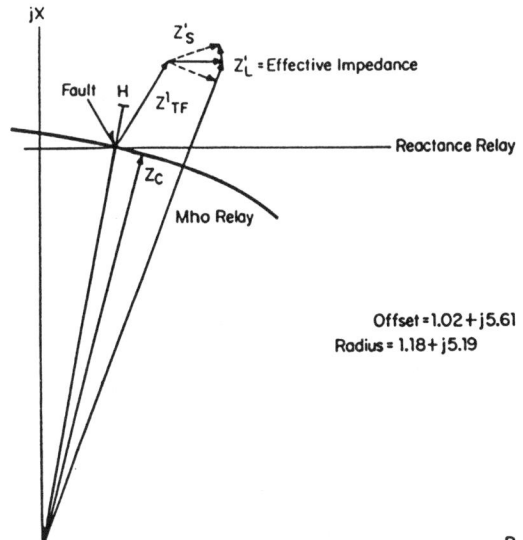

Figure 12-38 Example of effect of ground wires on tower footing and arc resistances and on reach of distance ground relays

wires have reduced the tower footing effect from 10 to $2.8\,\Omega$, there is a significant angle that causes problems with all types of distance-relay characteristics. A significant number of tree faults on EHV and HV lines have shown that, when fault resistance is significant, distance relays are less sensitive than ground overcurrent relays and often will not respond properly.

4.10 Zero Sequence Mutual Impedance and Ground-Distance Relays

Mutual induction from a parallel line will affect the reach of all ground-distance relays, but rarely cause serious problems. The mutual induction effect can be studied from parallel lines bussed at both ends as shown in Figure 12-39. Systems with more coupled lines and/or different terminating stations, although more complex, can be analyzed on the same basis.

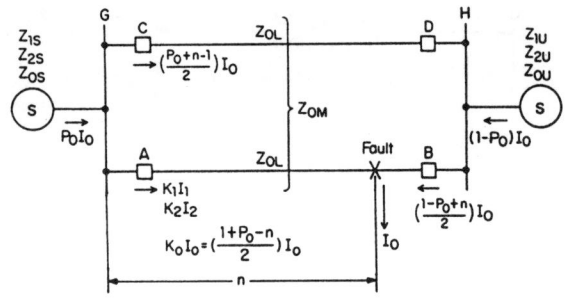

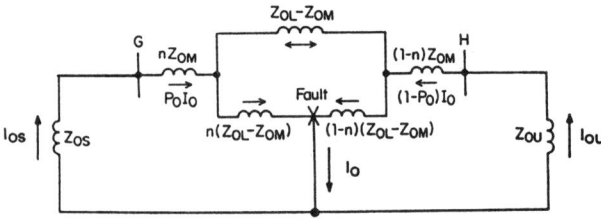

Figure 12-39 A parallel line section and its zero sequence network.

Figure 12-39 also shows the zero sequence network, including mutual effect. An equivalent zero sequence impedance to the fault can be obtained by reducing the delta to an equivalent wye and combining it with the two source impedances. By working back, the various distribution factors can be calculated. These factors are shown in Figure 12-39.

Consider the ground relays at breaker A. When the parallel line current flows in the same direction (C to D) as the fault current for fault F, the mutually induced voltage is added to the faulted line voltage. This phenomenon causes an apparent impedance greater than the line impedance. That is, mutual induction causes the relay to underreach, unless compensated for by using the parallel line current given in Eq. (12-37).

However, if the current in the parallel line flows from D to C, the mutual effect is added to the flow of the faulted line current. The result is a lower apparent than actual impedance, causing the relay to reach farther. The current flowing in line CD is

$$\left(\frac{P_0 + n - 1}{2}\right)I_0 = \left[\frac{Z_{0U}/(Z_{0S} + Z_{0U}) + n - 1}{2}\right]I_0$$

$$(12\text{-}48)$$

When $[Z_{0U}/(Z_{0S} + Z_{0U}) + n]$ is less than 1, the current reverses and flows from D to C, causing overreach. For zone 1, the critical area is that for faults around the balance-point setting. For faults close to the relay at A, the current would flow from D to C. The effect, however, is of no importance. For end-zone faults at B or on bus H, current will always flow from C to D. Consequently, *zone 1 cannot overreach the end of the line because of the mutual effect.*

When breaker B opens for faults at F, the current through the parallel line flows from D to C, causing an overreach of relay A. In this case, overreach is desirable, since it will often cause zone 1 to operate sequentially for 100% of the line.

The response of ground-distance relays under various system conditions is given in Figure 12-40. The three curves for different values of K_0/K_1 show the reach of relay A on line AB. The mutual induction effect is shown as a function of the P_0/K_0 current distribution ratio. Curves A to E are superimposed, showing the system constraints for various values of

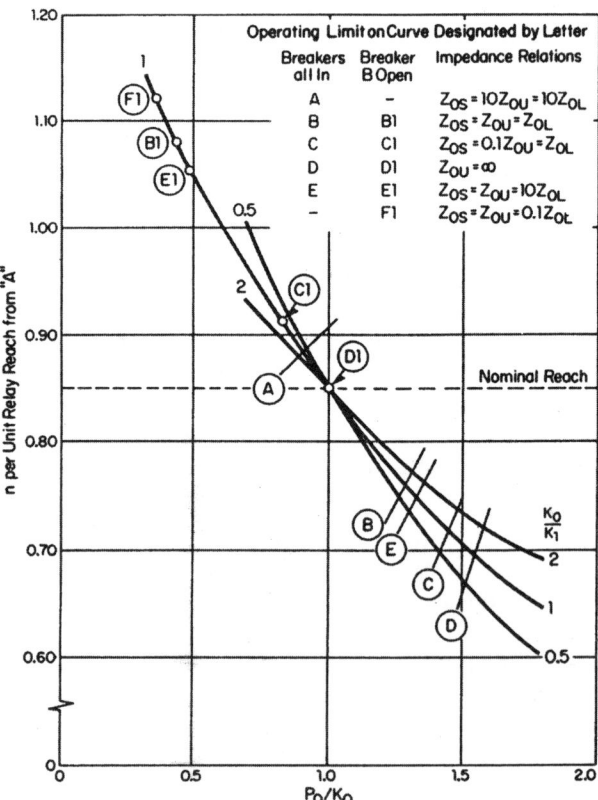

Figure 12-40 Zone 1 reach without mutual compensation for the system of Figure 12-39.

zero sequence source and line impedance relationships.

The circled points in Figure 12-40 show the sequential reach after breaker B opens. The region between A and C represents the practical area of operation corresponding to zone 1 and reaches between 70 and 88%, when nominally set for 85% of the line. If there is no zero sequence source at bus H, then curve D shows a reach of 68% with $K_0/K_1 = 0.5$.

Since mutual induction can reduce the reach of the ground-distance relay, zone 2 should be set to provide a minimum of 100% line protection. Figure 12-41 illustrates the setting required for a variety of system variables. A setting of 150% will provide good protection, including an adequate margin for the majority of systems.

After breaker C opens, mutual induction can extend the reach of zone 2 at A (Fig. 12-39) for faults at or near C on line CD. The fault current flows from A to B and from D to the fault. This same condition can cause zone 1 at D to extend its reach up to 100% of the line. If zone 2 is not set greater than 150% of the line, the reach of zone 2 at A will coordinate with that of zone 1 at D.

The above discussion applies to most relays without mutual compensation. Although ground-distance relays can use the parallel-line current to cancel the mutual induction effect, this method is generally not recommended, particularly for zone 1.

In summary, for the parallel-line cases shown in Figure 12-39:

1. Without mutual compensation, a zone 1 distance relay set for 85% of the line will cover from 70 to 88% if the breakers are all in, and from 85 to 100% of the line after the far breaker opens.

2. For most applications, a zone 2 distance relay without mutual compensation will provide complete end-zone coverage when set for 150% of the line.

3. Compared with an uncompensated relay, mutual compensation usually increases zone 1 coverage with the breakers all in, but decreases sequential coverage.

4. Mutual compensation must be used with caution when "looking into" a weak source. In these cases, $K_0 I_0$ flows from A toward B; and for faults near C, the large mutual compensation current from the parallel line can cause misoperation.

To use mutual compensation, the parallel line must terminate in the same station in order to have its current available. This cross-connection is complex and increases the possibility of an incorrect connection or testing mistake.

If currents flow in the source direction in all the lines, additional parallel lines can cause greater relay under-reach for end-zone faults. In this situation, the relay must be mutually compensated with all parallel currents. However, mutual compensation with three or more lines is inordinately complex. If the compensation of a given line is necessary, it should therefore be limited to the insertion of current from just one parallel line. Although this arrangement minimizes complexity, as long as mutual compensation is employed, over-reach hazards still exist.

5 LOOP-SYSTEM PROTECTION

5.1 Single-Source Loop-Circuit Protection

5.1.1 Using Directional Overcurrent Relays

A loop circuit with a single source is shown in Figure 12-42. For the purposes of the following discussion, all breakers in this circuit will be considered closed during operation, at least for a significant amount of time. (Should a breaker open, the system becomes radial.)

1. Relay application rules are similar to those for radial circuits. Typical uses of the phase overcurrent

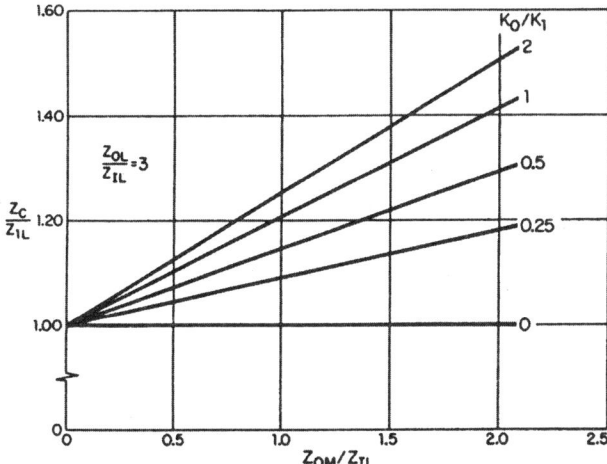

Figure 12-41 Required zone 2 setting to provide complete coverage of the protected line.

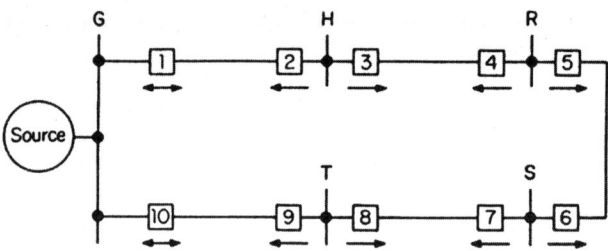

Figure 12-42 A single source loop circuit and its protection.

relays for the single-source loop circuit shown in Figure 12-42 are summarized in Table 12-5.

2. Nondirectional overcurrent relays can be used at 1 and 10, since no current flows through these locations for faults in the source system and on bus G. (This application is indicated by the double-arrow line below the breaker.)

3. At all other locations, fault current can flow in either direction through the relays for faults to the right of bus G. These relays will operate when the pickup current is above the setting, but only if the current (either load or fault) flows in the direction of the arrows, which, in each case, is into the line from the bus.

4. In practice, directional relays are usually applied at all locations to accommodate any future system changes.

5. For this single-source system, lines HG and TG may be protected by directional instantaneous overcurrent relays that can be applied to 2 and 9 and set very sensitively. Since the fault current through relays 2 and 9 goes to 0 as the fault location reaches bus G, the

directional instantaneous overcurrent relays will not overreach. For these same faults, however, the current through relay 1 or 10 will be maximum. Either the inverse-time relays or, if applicable, instantaneous-trip units will operate in minimum time. When breaker 1 opens, for line HG faults, the current increases through relay 2 in the tripping direction. Thus, for line HG faults close to bus G, breaker 2 opens sequentially after breaker 1. Similarly, for line TG faults close to bus G, breaker 9 opens sequentially after breaker 10.

6. In principle, the coordination procedure for these relays is the same as for a radial circuit. Coordination is based on maximum fault current on the remote bus. It must be assumed that the loop is open at one end, because when the loop is opened between the fault and source bus, the current in any branch will increase. Instantaneous-trip units, when applicable, must also be set on the assumption of open-loop conditions that yield maximum relay currents for remote end faults.

5.1.2 Using Inverse-Time and Distance Relays

With single-source loop circuits (Fig. 12-42), the inverse-time distance-relaying scheme offers distinct advantages. The scheme is especially advantageous for long loops with many sections, in which the relays at the source end breaker would require a long time delay for a far-end fault.

Figure 12-43, the inverse-time distance scheme, consists of a zone 1 distance relay (21) along with a similar zone 2 distance relay (21) that torque-controls a two-unit, inverse-time overcurrent relay (51). The reach of the distance relay is independent of source impedance variations; it can be set below and is independent of load current.

Zone 1, set for 90% of the line impedance, protects a much larger portion of the line than an instantaneous unit. Zone 2 is set through the next adjacent section to

Table 12-5 Phase Overcurrent Relays for Single-Source Loop-Circuit Protection

Breaker locations	Relays if instantaneous-trip units are not applicable	Relays if nondirectional instantaneous-trip units are applicable	Relays if directional instantaneous-trip units are applicable
1 and 10	Inverse time	Inverse time with instantaneous trip	—
2 to 9	Directional inverse	Directional inverse with instantaneous trip	Directional inverse with directional instantaneous trip

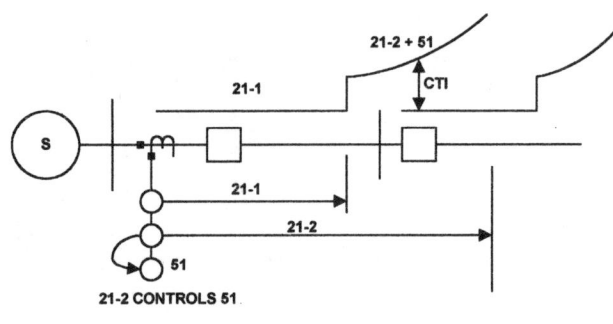

Figure 12-43 Distance controlled overcurrent scheme.

provide end-zone and adjacent line backup protection. Since the two-unit overcurrent relay is torque-controlled, it will not operate unless the zone 2 relay has operated.

The principle of application is shown in Figure 12-43. Ninety percent of the line is tripped at high speed. For the remaining 10% of the line, the operating time of the 51 unit can be made comparatively fast (equal to or less than the coordinating time intervals) since coordination is with the next zone 1 instantaneous relay, rather than the time overcurrent unit.

The 51 unit may be set on a tap whose value is less than the maximum load current, since the controlling zone 2 distance unit will not operate on load current. This arrangement provides faster end-zone faults and backup protection for the adjacent section than directional overcurrent relays with instantaneous-trip units.

5.1.3   Protecting Loop Circuits with Tap

Circuits with load taps present coordination problems if the tap impedance to its bus has the same order of magnitude as the impedance from the tap point to the remote line terminal. In such cases, time overcurrent relays must be coordinated with the protection at and beyond the tap bus, as well as with that at the remote bus and beyond.

With the inverse-time distance scheme, the zone of instantaneous protection can be quite limited if the tap is near the relay terminal, particularly if zone 1 must be set so as not to operate for faults protected by the fuse (Fig. 12-44). Generally, zone 1 would be set into but not through the transformers. This arrangement imposes no limitation as long as Z_T is greater than Z_L. Fast reclosing and the subsequent lockout of zone 1 permit the fuse to clear transformer faults.

The zone 1 setting must be made on the basis of actual ohms, since the infeed effect disappears if the loop is open at some point. Also, the zone 1 reach must remain short of both the remote end and the low-voltage bus.

If zone 2 reaches through the transformer, the 51 relay must coordinate with the fuse and low-side breakers. For faults in and on the low side of the transformer, the current from the remote end tends to make the distance relay underreach. The zone 2 setting may also be based on actual ohms unless the zone-2-controlled 51 relays are acting as backup to the transformer secondary main-breaker relays. In this case, the effect of far-end contribution requires that the

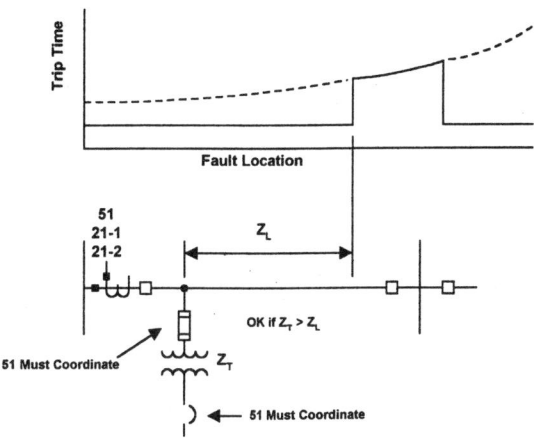

Figure 12-44   Protection of line with fused transformer tap.

zone 2 setting be determined on the basis of total apparent impedance including the transformer.

5.2   Multiple-Source Loop Protection

5.2.1   Using Directional Overcurrent Relays

In general, directional relays are necessary to protect loop circuits with multiple sources. The coordination problems for such systems are very complex and frequently require compromises in protection. Imagine a system that is similar to the one shown in Figure 12-42, except that there are sources at each of the buses (instead of only at bus G). In this case, relays 1 and 10 must be directional, and two coordination loops will exist. Relay 1 must be coordinated with 3, 3 with 5, 5 with 7, 7 with 9, and 9 with 1. Similarly, relay 10 must be coordinated with 8, 8 with 6, 6 with 4, 4 with 2, and 2 with 10. In addition, each relay must be coordinated with any other circuits or loops connected to that relay's remote bus.

These multiple-source loops tend to close on each other, so that there is no specific starting point in the coordination procedure and the last relay inevitably does not properly coordinate with the first relay. A trial-and-error process is necessary to adjust the settings. This laborious procedure must be performed for both maximum and minimum operating conditions, as well as the various lines in or out of service.

A good starting point for the coordination process is at the largest sources, where the close-in fault current will be large and the relay time correspondingly short. Many computer coordination programs have been developed and used. They are invaluable aids in this procedure.

For illustration, an example for coordination on a multiple-loop system is described in Appendix D of this chapter.

5.2.2 Using Inverse-Time Distance Relays

As discussed in Section 5.1.2, inverse-time distance relays can provide high-speed protection for 90% of the line section, as well as improved protection for the remaining 10% of the line section and adjacent lines.

The relays are used in the same way as described in Section 5.1.2. If the protected line is tapped, the relay application and settings should be modified as described in Section 5.1.3.

In multiple-source systems, backup protection may be limited, since the fault-current infeed from the remote bus to adjacent line faults can reduce the reach of the zone 2 distance relays. The infeed also affects the fault levels for overcurrent relays. Nevertheless, inverse-time distance relays are generally easier to apply, permit more sensitive settings of the inverse-time overcurrent units (settings below load), and improve fault clearing times. Selecting an inverse-time characteristic compatible with the characteristics of other similar relays in the system simplifies coordination.

6 SHORT-LINE PROTECTION

6.1 Definition of Short Line

Many problems associated with so-called short lines are actually related to the SIR value. SIR, the source impedance ratio, is the ratio of the source impedance to the line impedance. As the SIR value increases, so do application complexities. A 10-mile line with a low SIR value may be considered a "long" line, whereas a 100-mile line with a high value may face many of the problems associated with "short" lines.

6.2 Problem Associated with Short-Line Protection

In the past, many short lines were protected by a current-only scheme. Today, distance relays are being used more than ever before. The difficulty associated with short-line protection is the zone 1 overreach problem. Overreach may be caused by the following factors:

1. Current and/or voltage transformer inaccuracy
2. Ratio of the source impedance to line impedance
3. Relay sensitivity
4. Voltage transformer transient problem

6.3 Current-Only Scheme for Short-Line Protection

A current-only scheme does not require voltage information, and this greatly reduces the complexity of protection. Also, it provides better coverage on arc and fault resistance than the distance schemes. However, most of the time, the application of current-only schemes on a short line may be a problem in terms of directionality, and also the application is a function of the SIR. Table 12-6 shows that the coverage is limited as the source impedance is increased. Table 12-6's data are based on the following assumptions, which can be calculated from Eq. (12-2):

Source voltage = 69.3 V

Line impedance = 0.5 Ω

I_{F1}, I_{F2} = faults at 0 and 100% of the line section, respectively

Instantaneous – trip unit setting = 1.3 (I_{F2})

6.4 Distance Relay for Short-Line Protection

6.4.1 Sensitivity

In general, as the reach setting of a distance relay is reduced, the fault current required to operate the relay increases and the operating time of the relay also increases. Therefore, for short-line application, more fault current is required for the relay operation.

Table 12-6 Three-Phase Fault Currents

Z_s (Ω)	SIR	I_{F1} (A)	I_{F2} (A)	Instantaneous-trip unit setting	Coverage (Ω)	Coverage (%)
0.5	1	138.6	69.3	90.1	0.27	53.8
1.0	2	69.3	46.2	60.1	0.15	30.6
1.5	3	46.2	34.6	45.0	0.04	8.0
2.0	4	34.6	27.7	36.0	No	No

For many short-line applications, the sources are quite strong and the fault currents high; however, for those systems with high source impedance ratios, the current sensitivity of the relay must be considered.

6.4.2 Effect of Source Impedance Ratio

Table 12-7 shows the per unit values of V_R, the fault voltage at the relay location, and V_R-$I_R Z_C$, the relay operating voltage, for the simple system of Figure 12-45 with a fault applied at 85% of the relay reach. The relay is set for 90% of the line, i.e., $Z_C = 0.9Z_L$.

The relay operating quantity, V_R-$I_R Z_C$, becomes very small when the SIR value increases. For example, on SIR = 30, the voltage that can be applied to the relay V_R and the relay operating quantity are less than 3 and 0.5%, respectively. When the signals are so small, any error in the voltage or current can be substantial relative to the theoretical values.

6.4.3 Current Source

Current transformers used with distance relays should not saturate for faults occurring at the balance point. Limited saturation for faults inside the operating zone presents no problems for the relays, as long as the relay current is not reduced or shifted enough to cause the impedance phasor to fall outside the operating zone. However, since this determination requires a complex calculation, it is desirable to use good-quality current transformers for short-line applications.

The dc offset component of the fault current may cause a transient overreach for a distance relay. This tendency is particularly important for zone 1 applications. If the compensators of the distance relay are air-gap-type transformers, the transient overreach will be negligible.

Table 12-7 Fault and Relay Operating Voltages for System in Figure 12-45

SIR	V_R	V_R-$I_R Z_C$
0.25	0.754	0.133
1.00	0.433	0.076
10.00	0.071	0.012
30.00	0.025	0.004
100.00	0.007	0.001

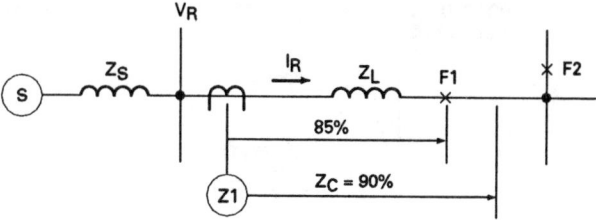

Figure 12-45 Sample system for short line relaying.

6.4.4 Potential Source

Most conventional voltage transformers are adequate for use with distance relays. The subsidence transient of some capacitor voltage devices, CCVT, requires a special setting consideration or an added time delay for static zone 1 relay applications.

As the source impedance ratio increases, the fault voltage at the relay location decreases. The lower accuracy of the potential source at this lower voltages may limit the usefulness of a zone 1 unit in short-line applications.

The transients associated with the capacitive coupling devices used to obtain the line voltage for the relay systems have been known to cause problems in static relay applications. In general, there are two areas, directionality and transient overreach, where the subsidence transient behavior of the CCVT can significantly affect static relay performance.

1. *Directionality* The transient error of a CCVT is greatest for zero-voltage faults because any output is pure error. Therefore, the most severe condition for the directionality of a distance relay would be a zero-voltage reserse fault. Transiently, the voltage seen by the relay is a function of the design of the particular CCVT, the fault initiation angle, the fault location, and the burden connected to it. A typical waveform is shown in Figure 12-46. For this case, the polarity of the voltage from the CCVT is out of phase with the prefault voltage in the second half-cycle after the fault occurs. For a distance relay, a trip output is produced when the operating signal (V_R-$I_R Z_C$) is out of phase with the polarizing signal V_{pol}. For a relay without memory action, this can result in a reversal of the polarizing quantity relative to the operating quantity (V_R-$I_R Z_C$) that will result in a misoperation. The use of memory action in the polarizing circuit will insure proper direc-

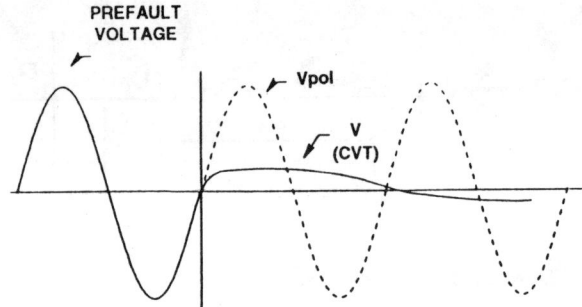

Figure 12-46 Typical CCVT transient voltage for a zero voltage fault.

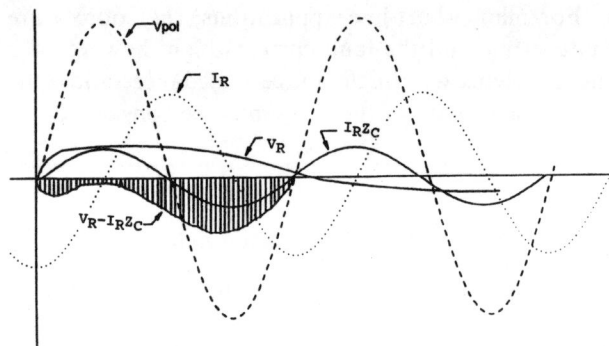

Figure 12-47 Distance relay transient overreach caused by subsidence transient voltage.

tional action for this case if it can ride over the CCVT transient error. However, misoperation may still occur, if the magnitude of the $I_R Z_C$ signal is less than that of the CCVT transient. This misoperation is caused by the phase reversal of the $I_R Z_C$ operating signal, not the polarizing signal; therefore, the use of cross-polarization or memory voltage will not prevent misoperation. This problem is most evident when the magnitude of the $I_R Z_C$ signal in the relay is small. Thus, it can occur when the fault current I_R is low and/or the reach of the relay Z_C is small.

2. *Transient overreach* In general, all relays include memory action for a short time after the fault occurrence. This memory action is important to the performance of a mho distance relay regardless of the transient under- or overreach. As the location approaches the end of the line, the magnitude of the operating signal V_R-$I_R Z_C$ approaches 0. As the magnitude of V_R-$I_R Z_C$ is reduced, the effect of any erroneous voltage, such as the CCVT transient, is increased. If the output voltage is momentarily lower than the true value, the relay may overreach. For example (Fig. 12-45), with an external fault F2 on the remote bus with zone 1 set for 90% of the line, the current and voltage, as well as the V_{pol} and V_R-$I_R Z_C$ relay signals, are as shown in Figure 12-47. Note that in the second half-cycle after the start of the fault, the CCVT transient has caused the V_R-$I_R Z_C$ signal to reverse polarity with respect to the polarizing voltage. This is the operating condition for the unit. The problem is most evident when the magnitude of the $I_R Z_C$ signal in the relay is small. This can occur when the

fault current is low and/or the reach of the relay is small. It can be the situation when the relay is applied to a high source impedance ratio condition.

6.4.5 Arc/Fault Resistance

As the source impedance ratio increases, the voltage drop in the arc becomes a significant percentage of the voltage at the relay location. With a source impedance ratio of 100, the arc voltages are greater than—several times over—the voltage drop in the line. The apparent arc impedance seen by the relay is thus greater than the line impedance, again several times over.

Because the magnitude of the apparent arc impedance is greatest at high source impedance ratios, the performance of both phase- and ground-distance functions will be similarly affected under those conditions.

The infeed from another source will not affect the magnitude of the apparent arc impedance seen by the relay, but the angle of the apparent impedance. The change in angle of the impedance could cause a distance function to overreach, or underreach, or have no effect at all, depending on the design of the relay.

If faults involve ground, the total resistance in the fault is composed of arc resistance plus fault resistance that can be very large depending on the components involved in the fault. For example, a tree fault, or a fault to ground through a fire, can have a very large resistance component relative to arc resistance. The effect of infeed on this component is to magnify the resistance, as well as shift it in phase angle.

7 SERIES-CAPACITOR COMPENSATED-LINE PROTECTION

7.1 A Series-Capacitor Compensated Line

Transmission lines are inherently inductive. The purpose of a series capacitor is to tune out part or all of the transmission-line inductance. In a network without series capacitors, faults are inductive in character and the current will always lag the voltage by some angle. Commonly used types of line protection can detect a fault and by operating circuit breakers clear it fast and selectively. With the series compensation of the transmission line, capacitive elements are introduced, and the network will no longer be inductive under all fault conditions. The degree of this change is dependent on the line and network parameters, extent of series compensation, type of fault, and fault location.

The capacitive or apparent capacitive nature of the fault current may cause the line protection to fail to operate, or to operate incorrectly, unless careful measures are taken to acknowledge this problem. Due to the capacitive nature of the fault loop, a complication with respect to protection may arise both on the compensated line as well as adjacent lines.

Series-capacitor banks are equipped with spark-gaps that bridge the capacitor and often with metal oxide protective devices. The spark-gaps are set to flash over at a voltage two to three times the nominal voltage of the bank. When the spark-gaps flash over, the network is restored to an inductive nature. In spite of this, the protection complications remain. Spark-gaps flash instantaneously when breakdown voltage is reached, but following a fault, time is required to reach this level and some faults will not cause the gaps to fire at all. The time to gap-flashing is often longer than the operating time of high-speed line protection. The effect of capacitive reactances must be evaluated even for faults that flash the spark-gaps. Adding to these complications is the fact that transients are generated due to the presence of the series capacitor at the occurrence of the fault, as well as at the instant of spark-gap flashing.

The use of metal oxide devices in parallel with the series capacitor introduces another element of concern. These units are never removed unless they themselves are jeopardized. Their level of conduction is approximately 1.5 times the rated peak voltage of the series capacitor. When voltage in excess of their conduction level appears across the metal oxide device, their impedance is reduced markedly, causing the series capacitor to be partly bypassed. However, when the voltage decreases to a level below the threshold, the impedance of the device becomes very high, and the capacitor is effectively reinserted. This action provides another level of transient generation, but, in general, it causes a softer impact on protective relaying than simple spark-gaps.

The metal oxide devices are bridged with triggered spark-gaps to limit the energy generated in the device during fault conditions. Therefore, the protective relays then must be able to handle the effect of both the metal oxide device and spark-gaps.

7.2 Relaying Quantities Under Fault Conditions

The effect of series compensation on transmission-line protection depends on the location of the capacitors and degree of compensation. Figure 12-48a is an example of a one-line diagram of a series capacitor and transmission line. Figure 12-48b is the steady-state R-X diagram. Because of the fact that the capacitor bypass protective equipment may be conducting or not, the apparent impedance as viewed from location A for a fault at B may appear vastly different.

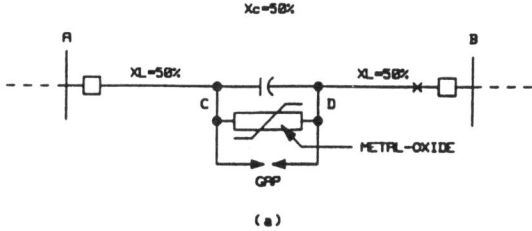

(a)

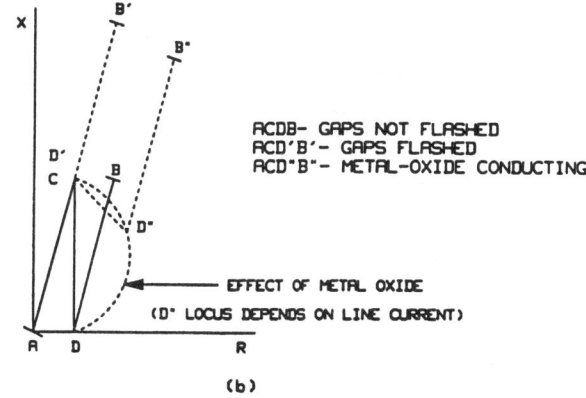

(b)

Figure 12-48 Apparent impedance as viewed from station "A" for a fault at B.

Figure 12-49 illustrates the influence of a nearby series-capacitor bank. Faults nearer the capacitor-line junction as viewed from location A will have a very large negative reactance character. This negative reactance is actually due to a reversal of the voltage at the relaying point or, under certain conditions, reversal of the current through the series capacitor.

Voltage reversal occurs at the bus if the negative reactance of the series capacitor is greater than the positive reactance of the line section to the fault location. Current reversal occurs if the negative reactance of the series capacitor is greater than the sum of the source reactance and line reactance to the fault location. Figure 12-50 depicts this condition.

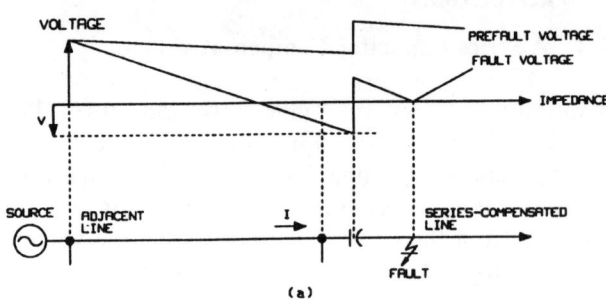

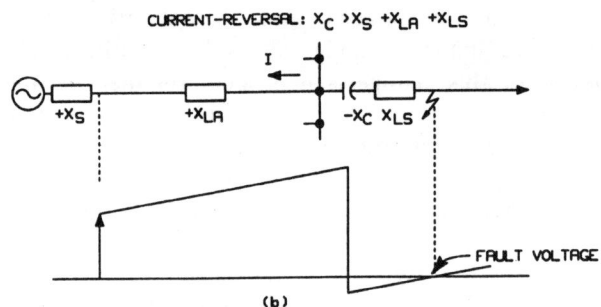

Figure 12-50 Voltage and current reversal.

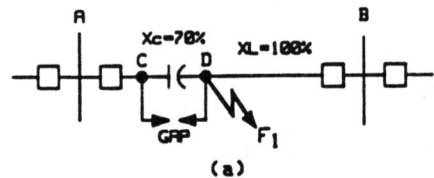

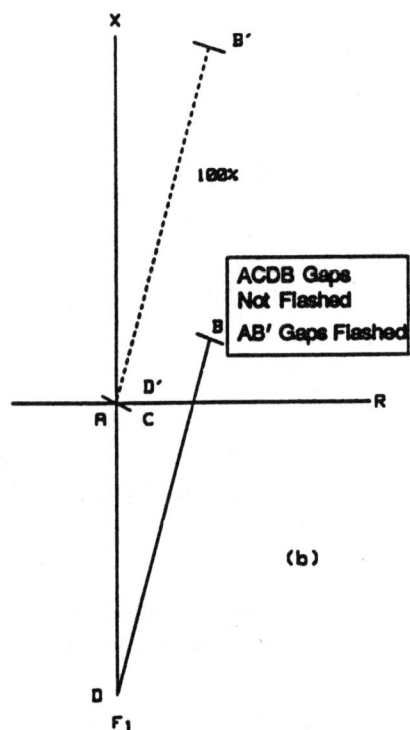

Figure 12-49 Apparent impedance at 60 Hz under fault conditions.

"Current reversal" or "outfeed" can also occur in some applications where a fault is at the capacitor-line junction and a parallel line exists between buses A and B. Figure 12-51 describes for a typical case the variation of voltage to be expected at various locations in the power system. Note that there is no voltage inversion in this case. "Current inversion" occurs at 2. Current at 4 also falls in the direction opposite to that for the same case without series capacitors. Whether line-side or bus potentials are used for the relays makes no difference in establishing the direction for this fault.

As can be seen in Figure 12-51, the zero voltage point in the system can be moved farther back as a result of multiple lines contributing to a fault near the capacitor-line junction. The negative reactance of the capacitor is enlarged compared to the positive reactance of the adjacent lines. This can result in zero voltage occurring on lines that are located far away from the series capacitor. The voltage can be 0 only in a network with negligible resistances. In a real network, the remaining voltage is so small that it can be regarded as 0.

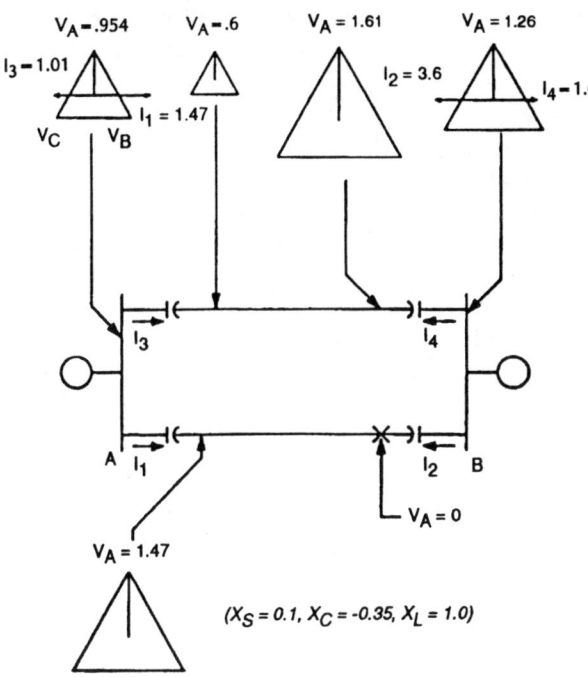

Figure 12-51 Typical voltages and currents for fault at capacitor-line function.

7.3 Distance Protection Behavior

Series compensation of a network will affect the distance protection on both the compensated line and adjacent lines connected to buses where a voltage reversal can occur. Generally, the most severe problems occur with the relaying associated with the adjacent line.

The following problem areas can be identified:

Determination of direction to a fault
Low-frequency oscillation
Transients caused by flashing of bridging gaps
Transfer of capacitor reactance to resistance by a metal oxide element bridging the capacitor
Zone reach measurement
False voltage zeros

There will be difficulties with distance protection in determining the correct direction of a fault in a station where a voltage reversal can occur. When direct polarization (polarization voltage from the faulty phase) is used, the protection on both the faulty and healthy lines may see the fault in an improper direction. This false determination of direction will

take place with both mho relays and plain directional elements.

To overcome this and achieve correct directional measurement, polarization quantities from the healthy phases are utilized. Healthy phase quantities will not be reversed and a correct directional measurement achieved for all unsymmetrical faults for an unlimited time.

Cross-polarized mho relays would under some fault conditions overtrip for faults on adjacent lines because of the use of a single comparator for both the direction and reach measurement, and therefore additional measuring criteria are required.

In the case of the three-phase fault, where all phase voltages reverse, only memorizing the prefault polarizing voltage can correctly determine direction. Normally in distance protection, memory voltage is used only when the voltage is reduced to some percentage of the nominal voltage. These criteria cannot be used when a voltage reversal occurs. The use of memory voltage must be controlled by general nondirectional three-phase fault criteria.

The time the memory voltage can be used must be limited to approximately 100 msec. Today, memory can be made very accurate, but in the case of a three-phase fault, the prefault condition should only be extrapolated for a limited time after the fault. The network is in a changing state and will run out of synchronism with the memory. Therefore, directional measurements have to be sealed in after the time the memory becomes unreliable.

The transient caused by flashing of bridging gaps will jeopardize the security of the relaying system. Also, line-energizing transients are high frequency in character and could cause some relays to operate. To avoid unwanted tripping, low-pass filtering of the measuring quantities is necessary.

The problems above require that bandpass filtering be used on the measuring quantities. The requirement of bandpass filtering exists in all distance protection, but is much more pronounced in applications involving series-compensated networks to avoid unwanted operation.

With an increase in current through the capacitor bank and an increased "conducting angle" of the parallel metal oxide element, the capacitive reactance will start to diminish and the combination will have a resistive component as seen in Figure 12-48. When setting impedance relays on the compensated line, allowance for this apparent resistance is necessary to assure tripping at all fault-current levels.

7.4 Practical Considerations

The diversity of problems associated with the introduction of series capacitors to transmission lines makes the selection of relaying systems and principles difficult. Voltage-related problems are the main reason for which current-only systems, like REL-350, are preferred for protecting lines equipped with series capacitors.

For high-speed relaying of a series-compensated transmission line, the use of a pilot system is unavoidable. If directional comparison systems are used, distance elements provided with an acceptable duration of "memory" and very special logic should be available. A reverse-looking unit with memory action is used to block high-speed tripping.

If phase-comparison systems are used, like REL-350, a relaying channel is required to transmit the information of the currents from side to side such that a phase comparison of the currents could take place. The simplicity of the logic is attractive, and the well-proven concept has an advantage over directional comparison systems with extra logic. Directional and phase-comparison systems are inherently dependent on the channel.

In REL-350, provisions have been made to include as backup both time-delayed zone 2 and zone 3 distance units. Both zones are composed of one phase-to-phase unit and three phase-to-ground units.

The phase-to-phase unit, described before and in the appendix, is inherently directional and will sense faults in the forward direction only. Therefore, the series capacitor(s) will always be in the protective zone regardless of a flashing-gap state.

The phase-to-ground units have their limitations. The units have a forward and reverse reach that is selectable. The unit is described in the appendices. The main purpose of the intentional reverse reach is to include negative reactance in the operating area of the unit.

These zone 2 and zone 3 protective zones being nondirectional, the time delay introduced into the units should coordinate with any step-distance zone outside the transmission line. Generally, zone 2 should not overreach zone 1 of any adjacent line as zone 3 should not overreach zone 2 if possible.

The apparent impedance to the distance relay depends on the state of the surge protective devices in parallel with the series capacitor. It is important for a zone 2 application that the transmission line be totally covered. For this reason, zone 2 and zone 3 settings should be calculated for a totally uncompensated line. This way, the settings will make sure that the line is fully covered under all operating conditions. Figure 12-52 illustrates the coverage of a zone 2 application.

The reverse-reach characteristic of the ground units enables the proper coverage of the negative reactance of the series capacitor as shown in Figure 12-52. It is unfortunate that the unit is not inherently directional since it will operate for reverse faults; however, for a time-delayed backup zone, this is acceptable.

In applying stepped-distance protection to a series-capacitor environment, coordination may be a problem since the reach of the zones depends on the conducting state of the surge protective equipment for the capacitor. Some compromises will result from the application and they may include the time delaying of the compensated line protective zones to also time-coordinate with adjacent lines.

8 DISTRIBUTION FEEDER PROTECTION

For the purpose of relay application, a feeder is considered to be radial if, at a particular relay location, the maximum backfeed (fault current in the

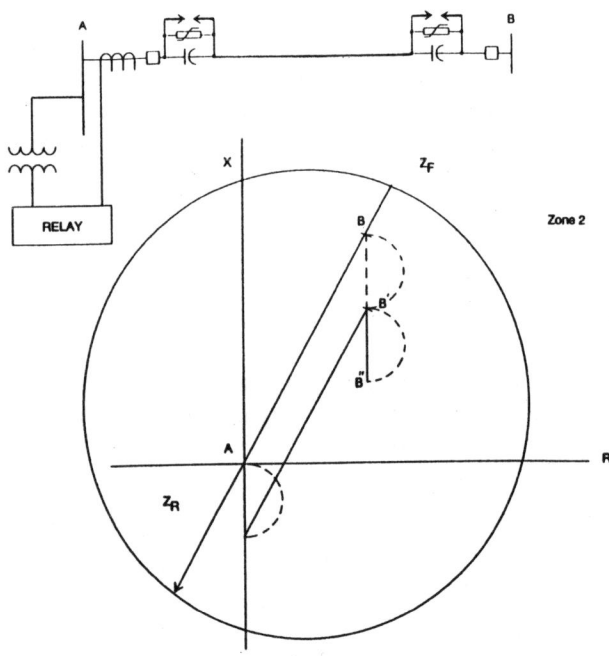

Figure 12-52 Series-capacitor line protection.

nontrip direction) is less than 25% of the minimum fault current for which the protective relay must operate.

8.1 Relay Coordination with Reclosers and Sectionalizers on a Feeder

Figure 12-53 shows a typical feeder circuit using a circuit breaker, recloser, sectionalizers, and fuses. The three reclosures of breaker G should be time-delayed to allow clearing of faults beyond recloser H. The first reclosure can be instantaneous, however, if the instantaneous-trip units of the relays can be set short of H, and the reclosing relay can lock out subsequent instantaneous-trip operations after the first reclosure. The recloser at H can be set for either one or two instantaneous reclosures; the other two or three should be time-delayed.

Reclosures are circuit-interrupting devices, similar to circuit breakers, that include automatic tripping and reclosing facilities. Normally, there are four reclosures before lockout: one instantaneous and three adjustable, time-delayed. Three types of controls are used: series trip, relay trip, and static control. Series-trip and static-control reclosers have adjustable time characteristics over a wide range of minimum pickup and curve shapes. To simplify coordination with protective relays, the relay-trip recloser can be equipped with any of the inverse and instantaneous-time overcurrent relays. Whenever possible, relays with the same types of time characteristic should be used. Adjustment of the time curve shape for the reclosers usually simplifies coordination with relays. There may be problems, however, in cases requiring coordination with relays, reclosers, and fuses, in that order.

Sectionalizers are usually single-pole devices, which do not have fault-current-interrupting capability but can sectionalize a distribution feeder during a permanent fault.

The sectionalizer is opened by an integrator that, in turn, is operated by the fault-current pulses resulting from the initial fault and subsequent opening and reclosing cycle of a recloser, or by the reclosing of circuit breakers ahead of the recloser. The integrator counts the number of current pulses and opens the sectionalizer after the count has reached a preset value and the circuit is dead.

Sectionalizers simplify the coordination of reclosers and fuses, since, for currents above the recloser's minimum trip, the sectionalizer can be set to open for any 0 current point in the reclosing cycle. This sequence ensures that the fuse is not subject to any additional fault current.

8.2 Coordinating with Low-Voltage Breaker and Fuse

Low-voltage breakers, used for circuits of 600 V and below, have built-in solid-state overcurrent trip devices that actuate a solenoid trip mechanism. Tripping energy is obtained from the primary fault current, rather than a separate station battery. The time-current characteristics of these devices may be different from those for the time-overcurrent relays.

As shown in Figure 12-54, there are in some cases four different characteristics: a long delay phase with an inverse characteristic (top right section of the curves), short delay phase (middle section), phase instantaneous with no intentional time delay (lower right section), and ground (left curve). The settings are continuously adjustable and can be easily set and tested in the field with a portable test set. An operating band rather than a curve is used to describe their characteristics.

A typical application of these breakers at a 480-V secondary unit substation and load center is shown in Figure 12-55. A typical application with coordination curves is shown in Figure 12-56. The key system current data are plotted first as an aid in coordinating and setting the various devices. These are (1) motor starting, (2) transformer full load, (3) transformer magnetizing inrush estimated at 8 to 12 times the full load at 0.1 sec, (4) the three-phase fault current (19,600- and 4800-A) backfeed from the motor loads, and (5) the total load center bus three-phase fault current (20,000 A). This example assumes several motor feeders, although only one is shown. Circuit breakers C and E coordinate with B.

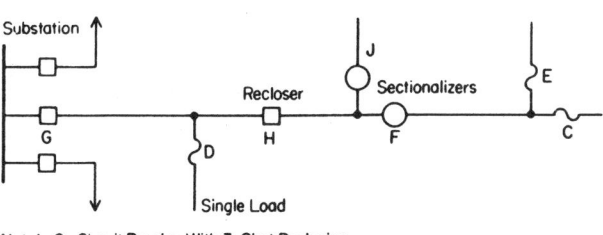

Note: G = Circuit Breaker With 3-Shot Reclosing
      H = Circuit Recloser
      C, D and E = Fuses
      J and F = Sectionalizers

Figure 12-53 Typical distribution feeder protection.

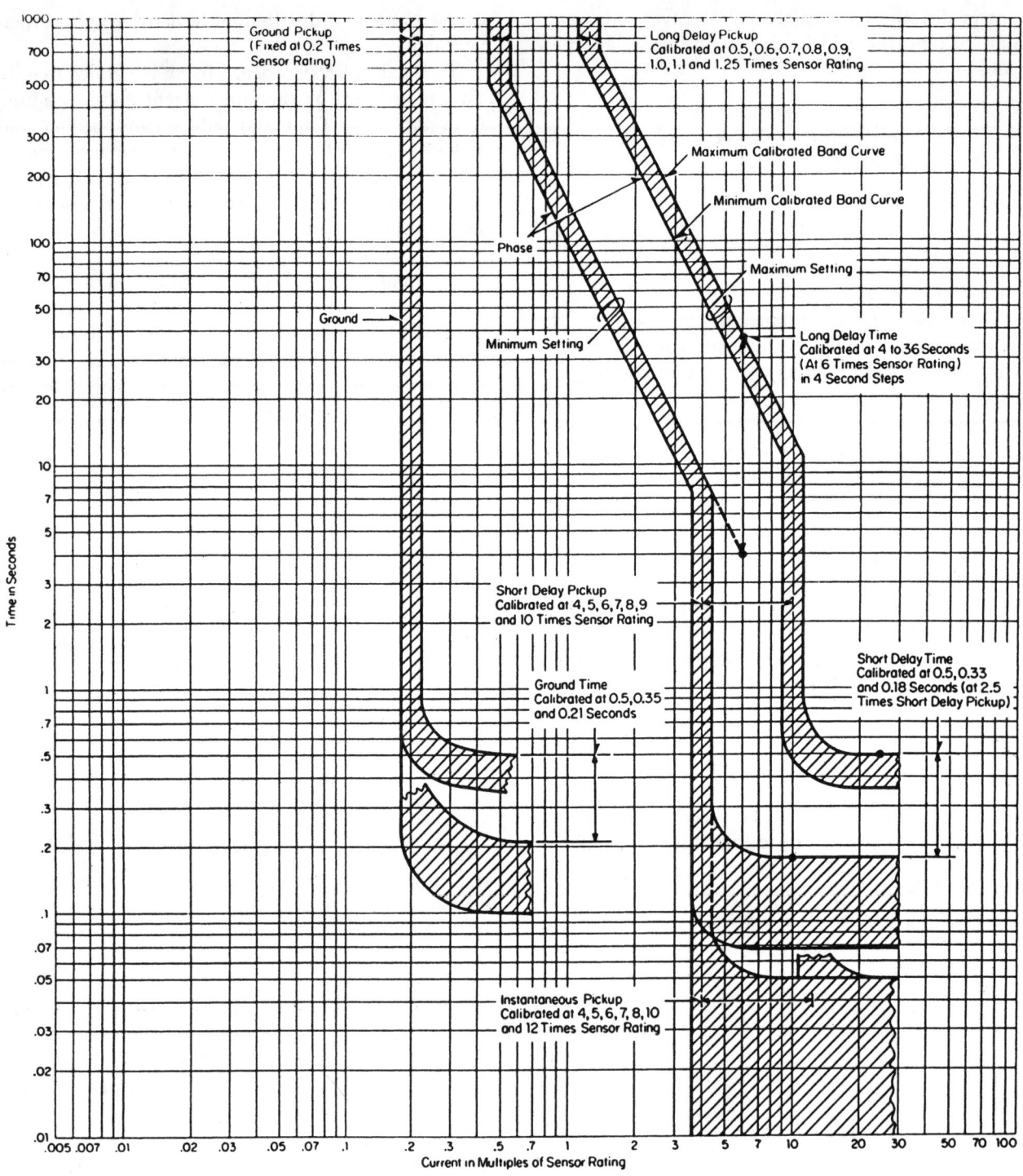

Figure 12-54 Low voltage air breaker time characteristics (type DS).

On double-ended substations where a normally open tie breaker (E) is used, the incoming (C) and tie (E) breakers have duplicate settings, except for the short time delay. The incoming line breaker (C) is set to operate one time interval longer than the tie breaker (E).

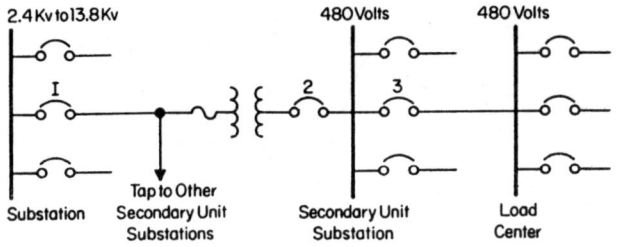

Figure 12-55 Distribution at a typical secondary unit substation and load center.

The primary protection for the high-voltage side of the secondary unit substation normally is furnished by high-side current-limiting fuses. As illustrated in Figure 12-56, with a high capacity source, the current-limiting fuse is current-limiting for faults as low as 33% of a maximum fault. The fuse curves (D) provide very adequate phase-fault protection, but not for ground faults. These are normally restricted and not isolated by a fuse.

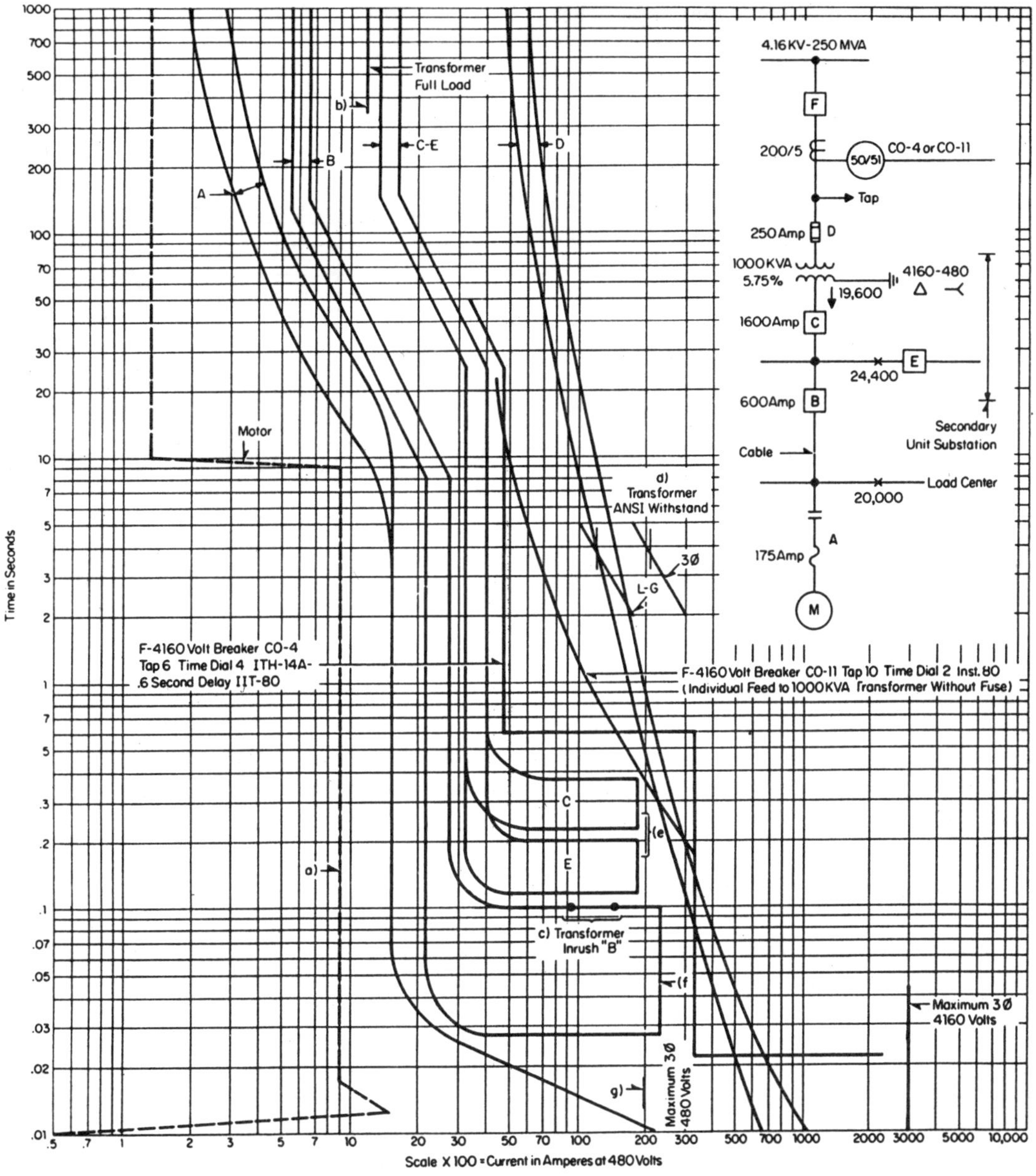

Figure 12-56 Protection and coordination for a typical secondary unit substation and load center.

This fuse protection is very inadequate for transformer secondary faults located between the transformer and main secondary breaker (C). The maximum secondary fault would take approximately 1 sec to blow the fuse. For a more probable 50% fault, the fuse would take between 15 to 20 sec. This illustrates that a high-side fuse is very good for primary maximum phase-to-phase and three-phase faults, but very inadequate for the more probable high-side ground faults and restricted secondary faults.

The best protection and coordination with the secondary-unit substation secondary protection are provided by a step-time characteristic that approaches the low-voltage breaker characteristic very closely (Fig. 12-56). The relay consists of a long time-overcurrent unit and two instantaneous-trip units, one of which operates through a timer that can be adjusted from 0.25 to 3.0 sec. Normally, the IT unit has a range of 10 to 40 A; the second unit (IIT) has a range of 20 to 80 A. As illustrated in Figure 12-56, the relay will recognize a transformer secondary fault down to 25% of the maximum value and operate in 0.6 sec.

Extremely inverse relays could be applied and set as shown as an alternative. Usually, it is very difficult to coordinate this characteristic between the breakers' time interval (CTI). Long operating times will occur for light secondary faults as compared to the step-time protection

In areas of high load density, the trend in secondary-unit substations is toward larger and fewer units to transform the distribution voltage down to a utilization voltage of 600 V or less.

Associated with the larger units are higher interrupting requirements that may necessitate much larger-frame breakers than the load requirements dictate. When larger-frame breakers are required, smaller-frame integrally fused circuit breakers may be economically applied.

A fused breaker is a standard breaker with special current-limiting fuses (limiters) to extend the upper limit of interrupting capability to possibly 200,000 symmetrical rms A. This breaker-fuse combination may consist of integral or separately mounted apparatus.

The current-limiting fuse restricts the peak "let-through" current on the first cycle to a value that is within the air breakers' interrupting capability. This limiting is illustrated in Figure 12-57. Figure 12-58 shows the instantaneous peak currents for various fuse ratings as a function of the available short-circuit rms symmetrical current and system x/r ratio of 6.6. Note

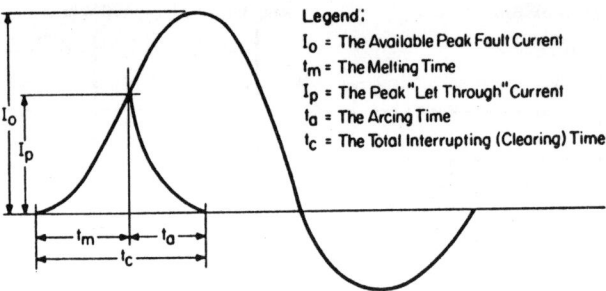

Figure 12-57 Peak let through current of a current limiting fuse.

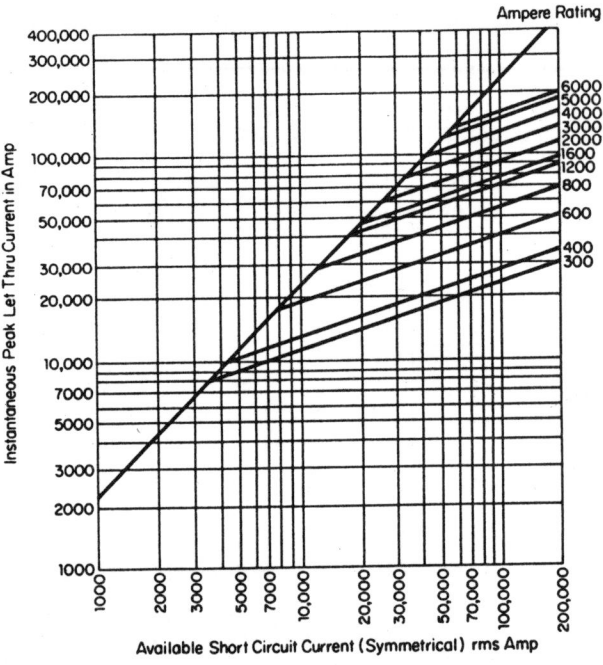

Figure 12-58 Characteristics of current limiting fuses for a system X/R ratio of 6:6.

that the peak let-through values obtained from the curves take into account the 1.414 factor from the peak-to-rms ratio and a factor of 1.62 for the maximum offset effect. The total ratio is, therefore, $1.414 \times 1.62 = 2.29$.

Data on the use of current-limiting breakers and minimum recommended fuse sizes are available from manufacturers. Loads to which these breakers are applied are protected against single phasing by an interlock device that trips the breaker when any one fuse blows.

APPENDIX A: EQUATION (12-2)

Let

$n =$ per unit of line section length protected by the instantaneous unit

$$K_i = \frac{\text{instantaneous unit pickup current, } I_{IT}}{\text{maximum far-end fault current, } I_F}$$

$SIR =$ source impedance ratio

$$= \frac{\text{source impedance, } Z_s}{\text{protected line impedance, } Z_L}$$

Since

$$I_F = \frac{1}{Z_S + Z_L} \qquad I_{IT} = \frac{1}{Z_S + nZ_L}$$

$$K_i = \frac{I_{IT}}{I_F} = \frac{Z_S + Z_L}{Z_S + nZ_L} = \frac{SIR + 1}{SIR + n}$$

Solve for n:

$$n = \frac{SIR(1 - K_i) + 1}{K_i} \qquad (12\text{-}2)$$

APPENDIX B: IMPEDANCE UNIT CHARACTERISTICS

B.1 Introduction

There is no topic in protective relaying more challenging and interesting than high-speed relaying using distance concepts. It is also true that the operation of different impedance units remains a mystery for most engineers and has generated several misconceptions on impedance relay applications. In most cases, as it should for any application, the characteristics of the distance elements are not an issue. Most of us tend to disregard the influence that particular system parameters have on the performance of the units.

The purpose of this appendix is to illustrate the performance and characteristics of the different distance elements found in ABB relays. More than stressing the operating characteristics of the units, the reader will notice that the equations and derivations to be presented are nothing more than academic exercises. Perhaps, as will be described, the simple mathematical models of the units are overshadowed by the superior performance of the distance elements in real life.

This appendix will introduce step by step the various factors influencing the performance of distance elements. Symmetrical components analysis is important and will be used throughout the development of the equations.

For a good understanding, the basic idea of a comparator is introduced first and a general procedure described. Some of the common units in ABB relays will be described as an illustration of the use of comparators. Finally, a brief discussion of derived characteristics will complement the contents of this appendix.

B.1.1 Basic Idea of a Comparator

Phasors are fundamental quantities in the analysis of ac systems. A comparator is a design element used in relays to compare two phasors either in magnitude or phase. A distance relay will always have a phase comparator or magnitude comparator regardless of the technology used, i.e., electromechanical, solid-state, and microprocessor-based relays.

Protective relaying is a binary science; either it is a "go" or a "no go" condition. The diverse units used in any discipline of protective relaying determine that the system is normal or abnormal. A comparator will give the relay system an output when the conditions for operation are satisfied. Since phasors are expressed in magnitude and phase, there are two types of comparators: phase and magnitude.

Phase Comparators

In Figure 12B-1, given two arbitrary phasors, S_1 and S_2, the output of a phase comparator is a logic "1" (the comparator has operated) if

$$\frac{S_1}{S_2} = Me^{\pm j90°} \qquad (B\text{-}1)$$

as a limiting condition, and

$$(-90°) < \angle S_1 - \angle S_2 < (+90°) \qquad (B\text{-}2)$$

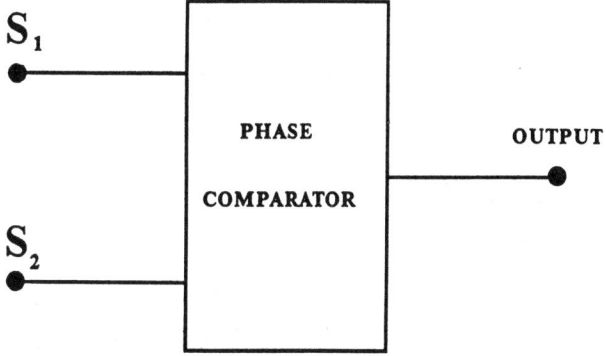

Figure 12B-1 Phase comparator concept.

if we define the operating characteristic of the phase comparator, as shown in Figure 12B-2.

The quantity

$$M = \frac{|S_1|}{|S_2|}$$

is any arbitrary magnitude for the condition to be met and it does not affect the operation of the phase comparator.

The characteristic angle of $\pm 90°$ determines a symmetric characteristic. This does not mean, however, that other limits have not been used, as will be described later.

Magnitude Comparators

Given two arbitrary phasors, S_A and S_B, the output of a magnitude comparator is a logic "1" (the comparator has operated) if

$$\frac{S_A}{S_B} = Ce^{jr} \qquad (B-3)$$

This is a boundary condition, and

$$|S_A| > |S_B| \qquad (B-4)$$

defines the operating characteristic of the magnitude comparator (see Figs. 12B-3 and 12B-4). The quantity C is a constant and *r* any arbitrary angle. Generally, C has a value of 1.

Relationship Between a Phase and Magnitude Comparator

The characteristics of a phase comparator and magnitude comparator are totally different. However, the inputs to a magnitude comparator, S_A and S_B, can

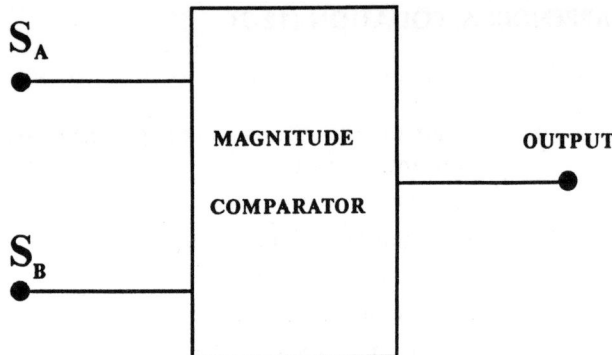

Figure 12B-3   Magnitude comparator concept.

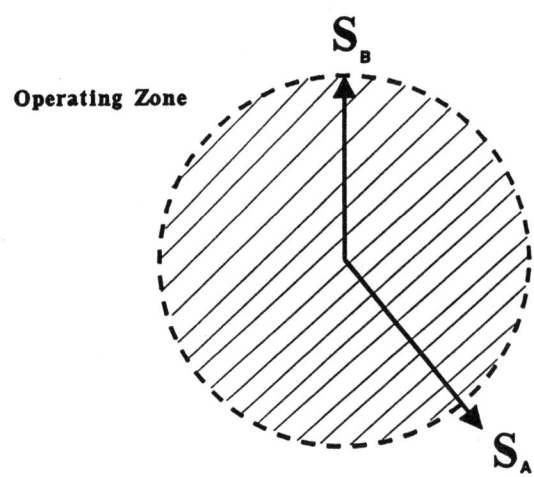

Figure 12B-4   Magnitude comparator operating characteristics.

be related to those to a phase comparator, S_1 and S_2, so that the outputs of both comparators are equivalent. The relationships between (S_A and S_B) and (S_1 and S_2) are

$$S_A = S_1 + S_2 \qquad (B-5)$$
$$S_B = S_1 - S_2 \qquad (B-6)$$

Equations (B-5) and (B-6) imply that a phase comparator with inputs S_1 and S_2 will provide the same output as a magnitude comparator if the inputs S_A and S_B have the values shown above. Refer to Figure 12B-5.

It is also true that if an equivalent phase comparator is to be derived from a magnitude comparator, the

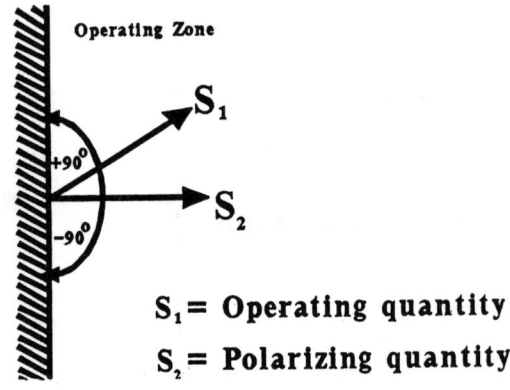

S₁ = Operating quantity

S₂ = Polarizing quantity

Figure 12B-2   Phase comparator operating characteristics.

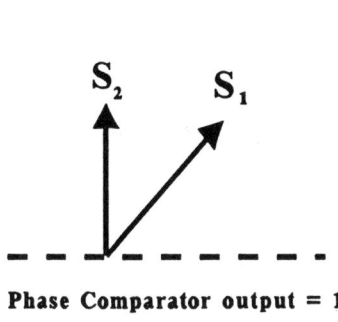

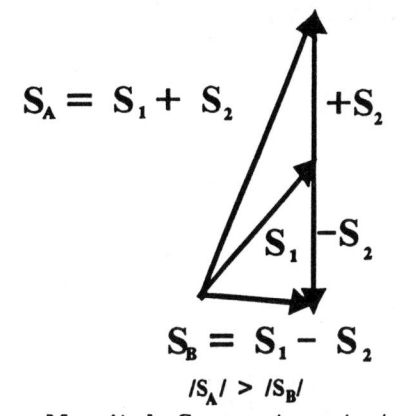

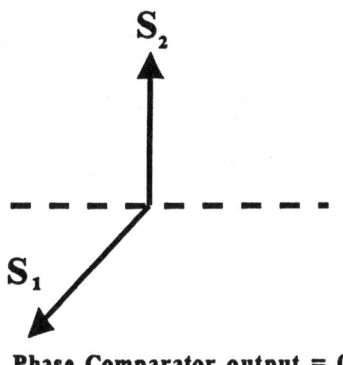

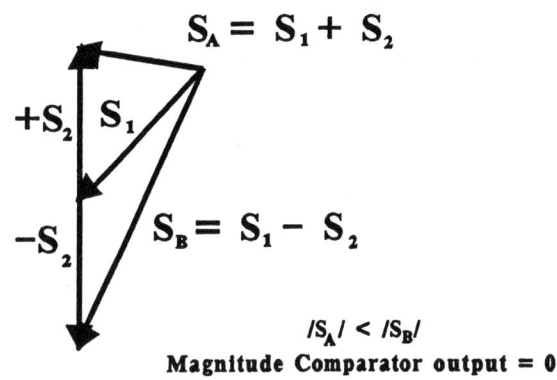

Figure 12B-5 Equivalent operation of a magnitude and a phase comparator.

following relationships are equivalent:

$$S_1 = \frac{(S_A + S_B)}{2} \tag{B-7}$$

$$S_2 = \frac{(S_A - S_B)}{2} \tag{B-8}$$

B.1.2 Generalized Use of Phase Comparators

Phase comparators are used widely in distance relay designs. The input phasors are generally a combination of voltages and currents. From these inputs, the ratio V/I, or impedance, is proportional to the distance to the fault, and it is indeed the quantity of interest.

Most distance relays do not measure $Z = V/I$ directly, but the operating characteristic of the R-X diagram is derived from the characteristics of the comparator.

In general, the inputs to a phase comparator will have the following format:

$$S_1 = k_1 V + k_2 I \tag{B-9}$$

$$S_2 = k_3 V + k_4 I \tag{B-10}$$

where V and I are the voltage and current of interest to derive Z, or the unit characteristic of the R-X diagram. Constants k_1, k_2, k_3, and k_4 are design constants that may be complex and introduce a phase shift.

In this section, a general procedure to derive the impedance characteristic of the comparator on the $Z = V/I$ plane will be developed. The procedure will be used later to derive a variety of distance unit characteristics used in ABB relays.

Using Eq. (B-1), (B-9), and (B-10) for a phase comparator, we get

$$\frac{S_1}{S_2} = M e^{\pm j90°}$$

$$= \frac{k_1 V + k_2 I}{k_3 V + k_4 I}$$

$$= \frac{k_1}{k_3} \frac{(V/I) + (k_2/k_1)}{(V/I) + (k_4/k_3)}$$

$$= \frac{k_1}{k_3} \frac{Z + (k_2/k_1)}{Z + (k_4/k_3)} \tag{B-11}$$

In most applications, k_1 and k_3 are real numbers; therefore, Eq. (B-11) can be simplified, as shown in

Eq. (B-12), for any value of M_1:

$$\frac{Z - a}{Z - b} = M_1 e^{\pm j90°} \qquad (B-12)$$

where

$$a = -\frac{k_2}{k_1} \qquad (B-13)$$

$$b = -\frac{k_4}{k_3} \qquad (B-14)$$

The quantities **a** and **b** are vectors and do have the same units as Z, i.e., they are impedances as well. In general, **a** and **b** will be sufficient to define the operating characteristics of the unit.

Equation (B-12) defines the operating characteristic of the phase comparator. If a number of impedances Z could be found on the $Z = V/I$ plane that satisfy Eq. (B-12), an impedance locus can be determined, defining the operating and nonoperating regions of the comparator in the R-X diagram.

Vectors **a** and **b** are fixed, but can be thought as reference points in the R-X diagram.

B.1.3 Generalized Use of Magnitude Comparators

Conceptually, magnitude comparators, as well as the general procedure to derive the comparator characteristic on the R-X diagram, are simpler than those for a phase comparator.

If the inputs S_A and S_B are expressed in terms of impedances, either the power-system or relay setting impedances, then by using Eq. (B-3), with $C = 1$, the operating characteristic of the comparator in the R-X diagram can be defined.

B.2 Basic Application Example of a Phase Comparator

The basic steps outlined above will be illustrated in this section. For this purpose, let V and I be the voltage and current input to the relay and the inputs to the phase comparator be

$$S_1 = V - Z_c I \qquad (B-15)$$

$$S_2 = V \qquad (B-16)$$

where Z_c is the relay setting. From Eq. (B-9) and (B-10), it follows that $k_1 = 1$, $k_2 = -Z_c$, $k_3 = 1$, and

$k_4 = 0$. Using Eqs. (B-13) and (B-14), we get

$$a = Z_c \qquad (B-17)$$

$$b = 0 \qquad (B-18)$$

For the purpose of illustration only, different constraints will be analyzed.

1. For $-90° < \angle S_1 - \angle S_2 < +90°$,

$$\frac{Z - a}{Z - b} = M_1 e^{\pm j90°} \qquad (B-19)$$

Referring to Figure 12B-6, we see that an infinite number of impedance vectors could be found that satisfies Eq. (B-19). The requirement is that the projection of the vector from **a** to Z, $(Z - a)$, satisfy the $\pm 90°$ angle difference to the vector from **b** to Z, $(Z - b)$, and its projection. In Figure 12B-6, two Z vectors have been projected that satisfy the angle requirement. Notice that the $-90°$ requirement is met to the right of Z_c, and the $+90°$ requirement to the left. A characteristic angle of $\pm 90°$ and the S_1 and S_2 inputs define the popular mho impedance locus. Notice that the plot is symmetric.

2. A characteristic angle of less than 90° (Fig. 12B-7) distorts the mho circle and modifies it to a characteristic that is called "tomato" due to its similarity to the fruit.

The characteristic is really composed of two circles: one for $\angle S_1 - \angle S_2 = +a$, and the other for $\angle S_1 - \angle S_2 = -a$. The centers of both are displaced away from and perpendicular to the middle of Z_c. The characteristic in Figure 12B-7 is for $a = 45°$.

3. A characteristic angle greater than 90° (Fig. B-8) distorts the mho circle and modifies it to a "lens."

Again, the characteristic is composed of two circles: one for $\angle S_1 - \angle S_2 = +a$, and the other for $\angle S_1 - \angle S_2 = -a$. The centers of the circles are on a line perpendicular to the middle of Z_c. The characteristic in Figure B-8 is for $a = 135°$.

The conditions discussed above are practical and have been used for many purposes in relaying. Although a simple description of the procedure has been presented where $a = Z_c$ and $b = 0$, these constants can take on different values.

A practical phase comparator that when modified can provide the characteristics just discussed is illustrated in Figure 12B-9.

Basically, the timer T determines the characteristic. For 60-Hz systems, 4.16 msec equals 90° of the power-system cycle. Therefore, a coincidence of at least 4.16 msec determines the familiar mho characteristic.

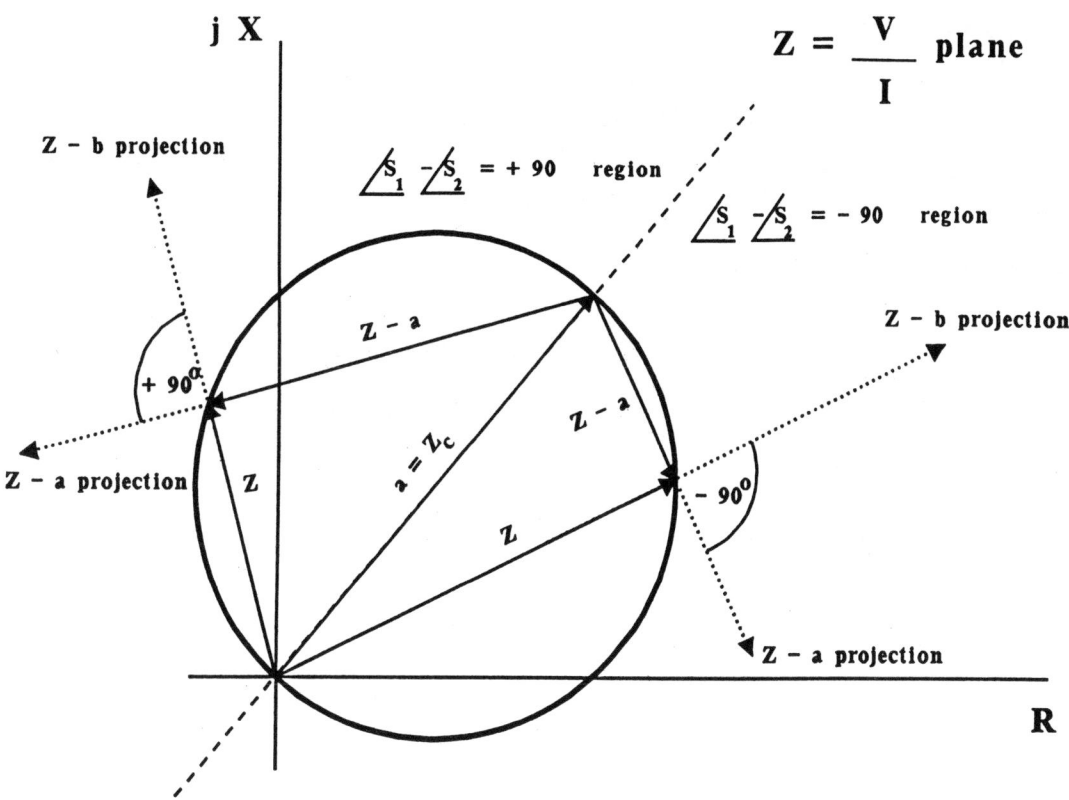

Figure 12B-6   R-X plot for $\angle S_1 - \angle S_2 = 90°$.

B.3   Basic Application Example of a Magnitude Comparator

Let V and I be the voltage and current input to the relay, and the inputs to the magnitude comparator be

$$S_A = 2V - IZ_c \qquad \text{(B-20)}$$
$$S_B = -IZ_c \qquad \text{(B-21)}$$

where Z_c is the relay setting.

For a magnitude comparator, the characteristic is determined by

$$\frac{S_A}{S_B} = e^{j\theta}$$
$$\frac{2V - IZ_c}{-IZ_c} = e^{j\theta}$$

or

$$\frac{V}{I} = Z = \frac{(Z_c - Z_c e^{j\theta})}{2} \qquad \text{(B-22)}$$

The characteristics of this magnitude comparator are determined by Eq. (B-22) and plotted in the R-X diagram of Figure 12B-10. The diagram illustrates the center of $Z_c/2$ and radius of $Z_c/2$. That defines the characteristic mho circle.

B.4   Practical Comparator Applications in Distance Relaying

The above discussion has helped us in understanding the use of comparators for distance relaying. The R-X diagrams above corresponded to the Z = V/I plane, or the impedance seen by the relay due to the voltage and current inputs. It is generally accepted that the R-X plots are representative of the positive sequence line impedance to the fault (Z_{11}). Most distance units operate on a combination of voltages and currents from the power system and have a definite purpose in operating for certain types of faults only. Distribution factors and load flow that influence the different relaying units will be reviewed, and the response of impedance units to different types of faults described.

Distance relays are used in the detection of phase faults (phase-to-phase and three-phase) and ground

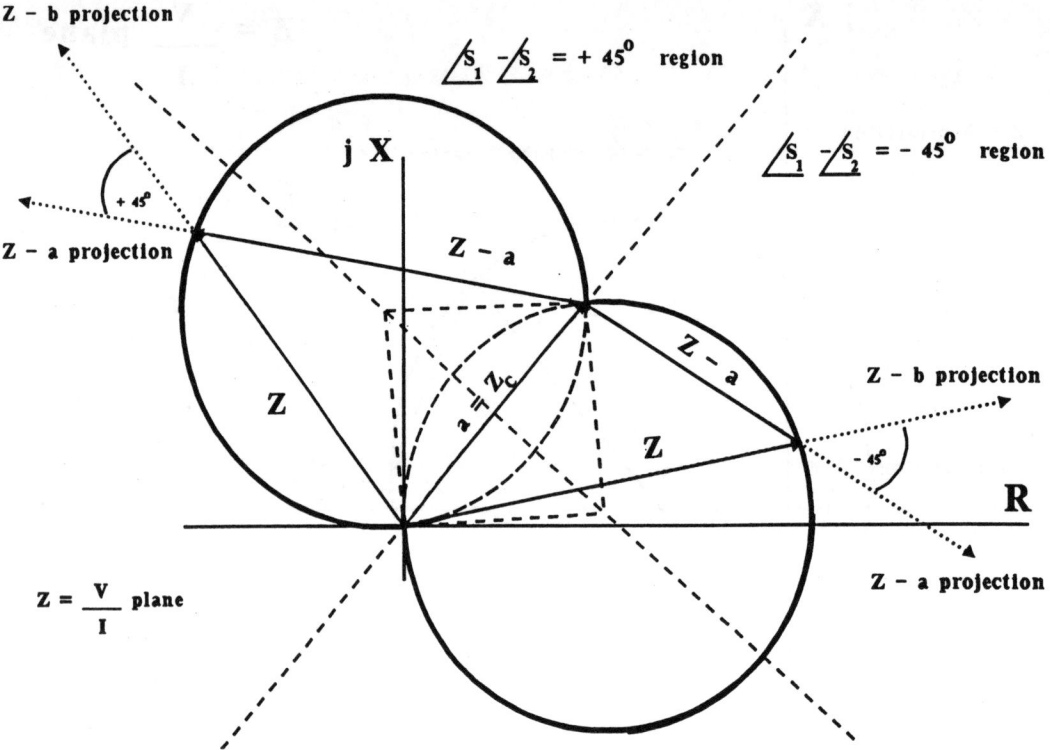

Figure 12B-7   R-X plot for $\angle S_1 - \angle S_2 = 45°$

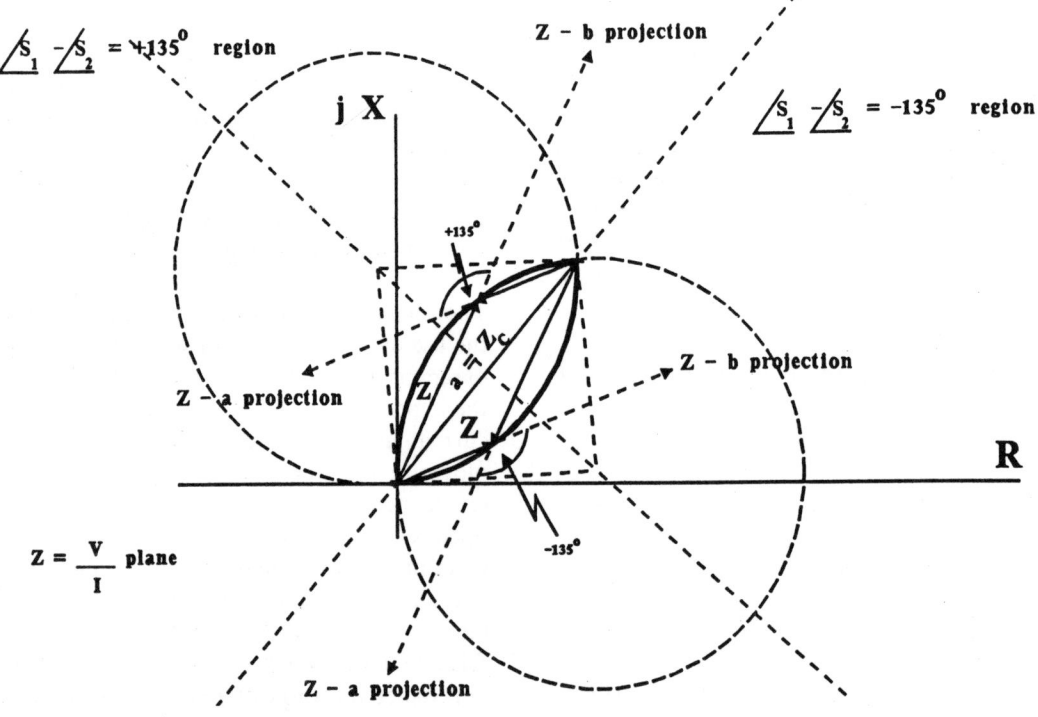

Figure 12B-8   R-sX plot for $\angle S_1 - \angle S_2 = 135°$

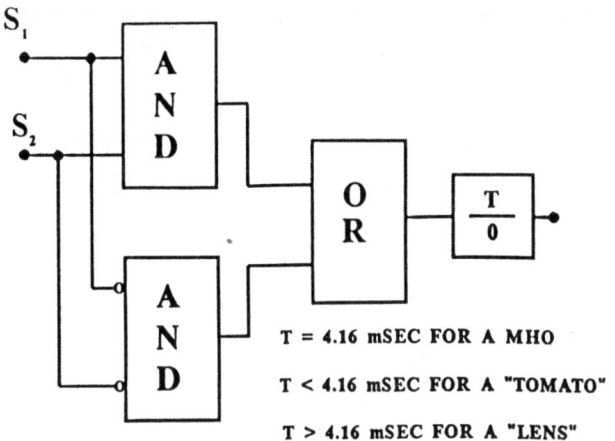

$$T = 4.16 \text{ mSEC FOR A MHO}$$

$$T < 4.16 \text{ mSEC FOR A "TOMATO"}$$

$$T > 4.16 \text{ mSEC FOR A "LENS"}$$

Figure 12B-9 A basic phase comparator.

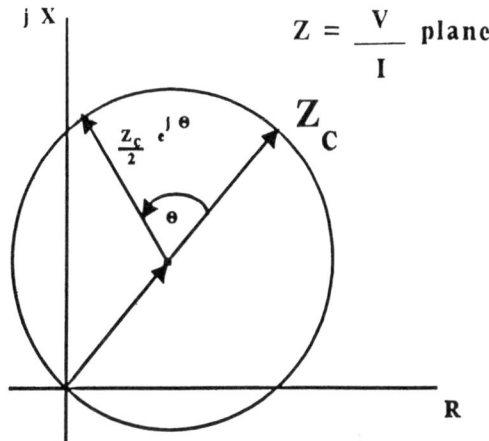

Figure 12B-10 Mho unit derived from a magnitude comparator.

faults (single-phase-to-ground). Detecting phase-to-phase-to-ground faults is achieved by the operation of the phase and/or ground units, depending on the zero sequence impedance (Z_0) of the system. Another use of comparators is for the familiar blinder lines for out-of-step detection.

B.4.1 List of Symbols to Be Used

Throughout the remainder of this section, the following list of symbols and abbreviations will be used:

V_s, V_s'	Source voltages
V_A, V_B, V_C	Line-to-ground voltages at the relay location
V_{Am}, V_{Bm}, V_{Cm}	Prefault line-to-ground voltages at the relay location (memory voltage)
V_{A1}, V_{A2}, V_{A0}	Phase A positive, negative, and zero sequence voltages at the relay location
V_{B1}, V_{B2}, V_{B0}	Phase B positive, negative, and zero sequence voltages at the relay location
V_{C1}, V_{C2}, V_{C0}	Phase C positive, negative, and zero sequence voltages at the relay location
V_{F1}, V_{F2}, V_{F0}	Phase A positive, negative, and zero sequence voltages at the fault location
I_A, I_B, I_C	Line currents at the relay location in the tripping direction
I_{A1}, I_{A2}, I_{A0}	Phase A positive, negative, and zero sequence currents at the relay location in the tripping direction
I_F	Total fault current
I_{F3ph}	Three-phase fault current
I_L	Prefault load flow
Z_{1s}, Z_{0s}	Positive and zero sequence source impedances
Z_{1s}', Z_{0s}'	Positive and zero sequence source impedances
Z_{1s}'', Z_{0s}''	Positive and zero sequence source and line impedances
Z_{1l}, Z_{0l}	Positive and zero sequence line impedances
Z_{1l}', Z_{0l}'	Positive and zero sequence line impedances
Z_c	Relay impedance-reach setting
Z_{1c}, Z_{0c}	Positive and zero sequence relay impedance-reach settings
Z_{1cF}, Z_{0cF}	Forward-positive and zero sequence relay impedance-reach settings
Z_{1cR}, Z_{0cR}	Reverse-positive and zero sequence relay impedance-reach settings
X_{1c}, X_{0c}	Positive and zero sequence relay reactance-reach settings
Z_{1l}'', Z_{0l}''	Positive and zero sequence reverse-looking impedance
ZR_s	Zero sequence to positive sequence source impedance ratio Z_{0s}/Z_{1s}
ZR_s''	Zero sequence to positive sequence source impedance ratio Z_{0s}''/Z_{1s}''
ZR_l	Zero sequence to positive sequence line impedance ratio Z_{0l}/Z_{1l}
ZR_c	Zero sequence to positive sequence reach ratio Z_{0c}/Z_{1c}

ZR_{cF}	Zero sequence to positive sequence line forward-reach ratio Z_{0cF}/Z_{1cF}
ZR_{cR}	Zero sequence to positive sequence line forward-reach ratio Z_{0cR}/Z_{1cR}
XR_c	Zero sequence to positive sequence reactance-reach ratio X_{0c}/X_{1c}
ZR_1''	Zero sequence to positive sequence impedance ratio Z_{01}''/Z_{11}''
PANG	Positive sequence line impedance angle
RT	Blinder setting
S_1, S_2	Phase comparator inputs
S_A, S_B	Magnitude comparator inputs
K_1	Positive sequence current distribution factor
K_2	Negative sequence current distribution factor
K_0	Zero sequence current distribution factor
k_1, k_2, k_3, k_4	General constants of a phase comparator equations
a	Complex operator, e^{+j120}
a, b	Reference vectors

B.4.2   Phase-to-Phase Unit

The operating characteristic of this unit can be better understood and derived using the magnitude comparator concept. It is primarily used to detect phase-to-phase faults, for which Figure 12B-11 applies, for a radial system. It has been used in relaying systems such as the KDAR (KD, KD-4, KD-10, KD-41, KD-11, etc.), Uniflex (LKD), LDAR (LZM, LDM, LDMS, LZ, etc.), and MDAR ($\phi\phi$ unit). The inputs to the magnitude comparator are

$$S_A = V_{A2} - I_{A2}Z_c \tag{B-23}$$

$$S_B = V_{A1} - I_{A1}Z_c \tag{B-24}$$

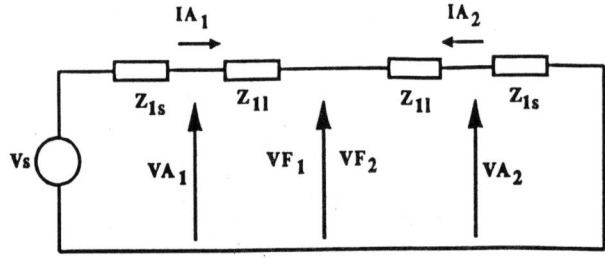

Figure 12B-11   Sequence network for a BC fault.

S_A is composed of negative sequence quantities only, and S_B positive sequence quantities only, for which this unit is not responsive to load or out-of-step conditions.

The magnitude comparator operating characteristic is defined by Eq. (B-3)

$$\frac{S_A}{S_B} = \frac{V_{A2} - I_{A2}Z_c}{V_{A1} - I_{A1}Z_c} = e^{j\theta} \tag{B-25}$$

Using Figure 12B-11, we obtain

$$\frac{S_A}{S_B} = \frac{I_{A1}Z_{1s} - (-I_{A1})Z_c}{[I_{A1}(2Z_{11} + Z_{1s})] - I_{A1}Z_c} = e^{j\theta}$$

or, in the positive sequence R-X plane, the unit's characteristic equation is

$$Z_{11} = \frac{(Z_c - Z_{1s})}{2} + \frac{(Z_c + Z_{1s})}{2}e^{-j\theta} \tag{B-26}$$

Equation (B-26) defines the locus of impedances (Z_{11}) for the magnitude comparator to change its output. Figure B-12 illustrates the R-X diagram of the unit on the positive sequence impedance (Z_{11}) plane.

B.4.3   Three-Phase Unit

This unit can be better understood by using the magnitude comparator concept. Its main purpose is to detect three-phase faults; therefore, the simple

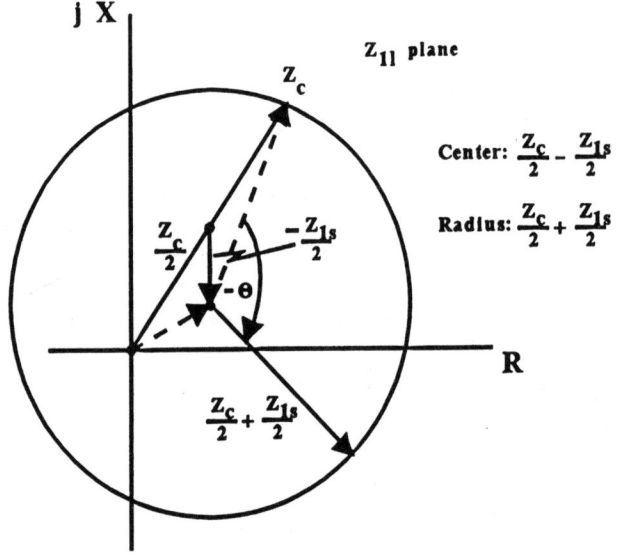

Figure 12B-12   Phase-to-phase unit operating characteristic on the Z_{11} plane.

sequence network connection of Figure 12B-13 applies. It has been used in the KDAR (KD, KD-4, KD-10, KD-41, KD-11, etc.) and Uniflex (LKD) relaying systems. The inputs to the magnitude comparator are

$$S_A = -\left(\frac{Z_c}{2}\right)I_{A1} \qquad (B\text{-}27)$$

$$S_B = V_{A1} - \left(\frac{Z_c}{2}\right)I_{A1} \qquad (B\text{-}28)$$

Since for a three-phase fault all quantities are positive sequence, the operating characteristic of the unit is defined by

$$\frac{S_A}{S_B} = \frac{-(Z_c/2)I_{A1}}{V_{A1} - (Z_c/2)I_{A1}} = e^{j\theta}$$

$$= \frac{-(Z_c/2)I_{A1}}{Z_{11}I_{A1} - (Z_c/2)I_{A1}} = e^{j\theta}$$

Then,

$$Z_{11} = \left(\frac{Z_c}{2}\right) - \left(\frac{Z_c}{2}\right)e^{-j\theta} \qquad (B\text{-}29)$$

Equation (B-29) defines the locus of impedances (Z_{11}) for the magnitude comparator to operate. Figure 12B-14 illustrates the R-X diagram of the unit on the positive sequence impedance (Z_{11}) plane.

B.4.4   Ground Units

The implementation of phase-to-ground units has been the most difficult. The impedance characteristics of these units have been derived for phase-A-to-ground faults and, again, the R-X diagrams are determined on the positive sequence impedance (Z_{11}) plane.

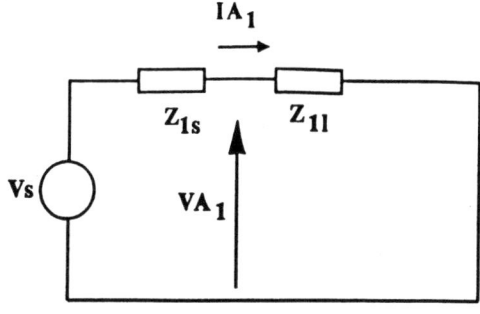

Figure 12B-13   Sequence network for a three-phase fault.

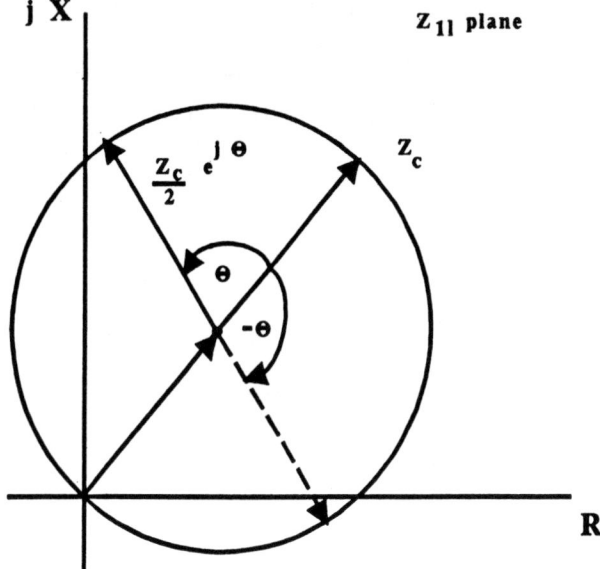

Figure 12B-14   Three-phase unit operating characteristic on the Z_{11} plane.

For the purpose of generality, since a few ground units will be analyzed, Figure 12B-15 defines the quantities of interest for a radial system.

The units to be analyzed in more than one way require a positive sequence impedance setting and zero sequence impedance setting, although most of the time, this requirement is not evident at first glance.

It is also worth emphasizing that the R-X plots to be derived are based on the positive sequence line impedance

$$Z_{11} = \frac{V_{A1} - V_{E1}}{I_{A1}}$$

Type SDG, SDGU, and LDG Ground Units

The principle has been used in the SDGU family (SDG and SDGU) and Uniflex (LDG) relays. It can be better understood and its characteristics are defined in a magnitude comparator.

With the quantities defined in Figure 12B-15 for a phase-A-to-ground fault, the inputs to the magnitude comparator are

$$S_A = V_{A0} - I_{A0}Z_{0c} \qquad (B\text{-}30)$$

$$S_B = (V_{A1} + V_{A2}) - (I_{A1} + I_{A2})Z_{1c} \qquad (B\text{-}31)$$

For a magnitude comparator, the characteristics are

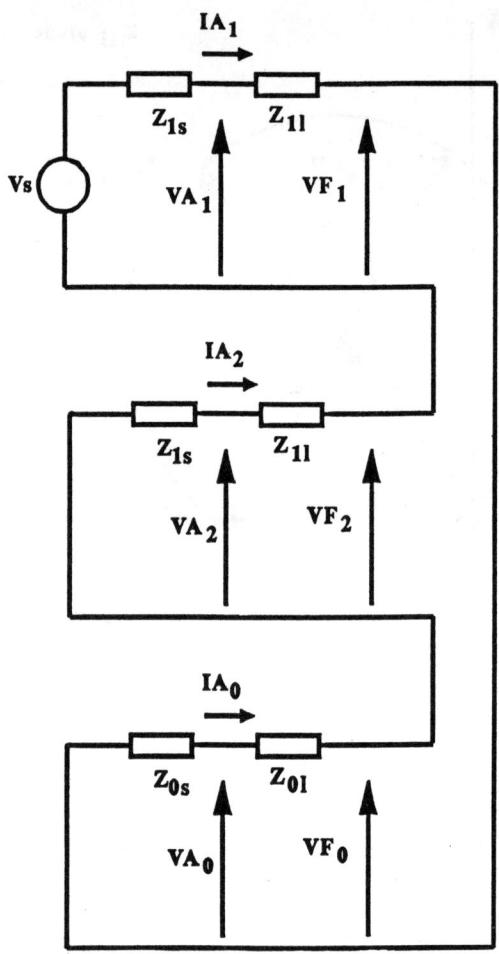

Figure 12B-15 Sequence network for a phase-A-to-ground fault.

defined by

$$\frac{S_A}{S_B} = \frac{V_{A0} - I_{A0}Z_{0c}}{(V_{A1} + V_{A2}) - (I_{A1} + I_{A2})Z_{1c}}$$
$$= e^{jr} \qquad (B-32)$$

From Figure 12B-15, the following relationships are true:

$$V_{A1} = V_S - I_{A1}Z_{1s}$$
$$= I_{A1}(2Z_{11} + Z_{1s} + Z_{0s} + Z_{01}) \qquad (B-33)$$
$$V_{A2} = -I_{A1}Z_{1s} \qquad (B-34)$$

Therefore, simplifying Eq. (B-32), we obtain

$$\frac{S_A}{S_B} = \frac{-I_{A1}Z_{0s} - I_{A0}Z_{0c}}{[I_{A1}(2Z_{11} + Z_{0s} + Z_{01})] - 2I_{A1}Z_{1c}}$$
$$= e^{jr} \qquad (B-35)$$

The relay settings are chosen, such that

$$ZR_1 = ZR_c = \frac{Z_{0c}}{Z_{1c}} = \frac{Z_{01}}{Z_{11}} \qquad (B-36)$$

Therefore, Eq. (B-35) can be simplified to

$$Z_{11} = \frac{2Z_{1c} - Z_{0s}}{(2 + ZR_1)} - \frac{(Z_{0s} + ZR_c Z_{1c})}{(2 + ZR_1)}$$
$$= e^{-jr} \qquad (B-37)$$

which defines the impedance characteristic of the comparator on the Z_{11} plane. Figure 12B-16 illustrates the characteristic. Notice that the reach Z_{1c} is fixed for $r = 180°$.

Quadrature-Polarized Ground Unit

This unit has been successfully used in the LDAR (LZM, LDMS, etc.) and MDAR (Z1G, Z2G, Z3G, and PLTG) relaying systems. The operation of this unit takes advantage of the presence of unfaulted voltages as the reference or polarizing quantity. The quadrature-polarized unit is dependent on system parameters, as will be found next, and has proven to be a reliable and sensitive unit:

$$S_1 = V_A - \left(I_A + \frac{Z_{0c} - Z_{1c}}{Z_{1c}} I_{A0}\right) Z_{1c} \qquad (B-38)$$

$$S_2 = j(V_C - V_B) \qquad (B-39)$$

These two expressions need to be modified. For this

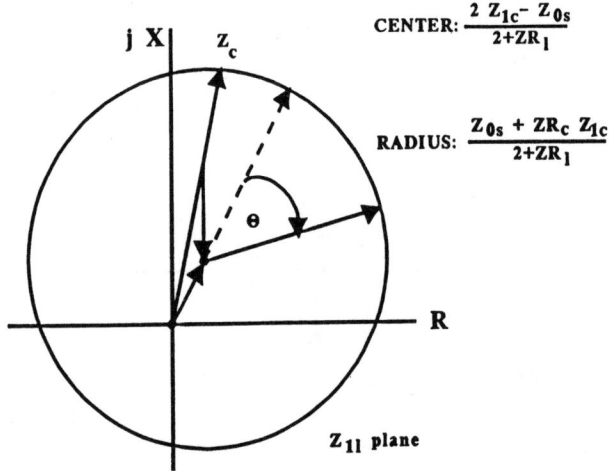

Figure 12B-16 Characteristics of type SDG, SDGU, and LDG units.

purpose, refer to Figure 12B-15. Then

$$
\begin{aligned}
S_1 &= (V_{A1} + V_{A2} + V_{A0}) \\
&\quad - I_{A1}\left(3 + \frac{Z_{0c} - Z_{1c}}{Z_{1c}}\right)Z_{1c} \\
&= V_{A1} + (-I_{A1}Z_{1s}) + (-I_{A1}Z_{0s}) \\
&\quad - I_{A1}(2Z_{1c} + Z_{0c}) + (V_{F1} - V_{F1}) \\
&= (V_{A1} - V_{F1}) - I_{A1}(2Z_{1c} + Z_{0c} \\
&\quad + Z_{1s} + Z_{0s}) + (-V_{F2} - V_{F0}) \\
&= (V_{A1} - V_{F1}) - I_{A1}(2Z_{1c} + Z_{0c} + Z_{1s} \\
&\quad + Z_{0s}) + I_{A1}(Z_{1s} + Z_{11} + Z_{0s} + Z_{01}) \\
&= (V_{A1} - V_{F1}) \\
&\quad - I_{A1}(2Z_{1c} + Z_{0c} - Z_{11} - Z_{01}) \\
&= (V_{A1} - V_{F1}) - I_{A1}[Z_{1c}(2 + ZR_c) \\
&\quad - Z_{11} - Z_{01}]
\end{aligned} \tag{B-40}
$$

Working on the other input, we get

$$
\begin{aligned}
S_2 &= j[(V_{C1} + V_{C2} + V_{C0} - (V_{B1} + V_{B2} + V_{B0})] \\
&= j[V_{A1}(a - a^2) - V_{A2}(a - a^2)] \\
&= -\sqrt{3}(V_{A1} - V_{A2}) \\
&= -\sqrt{3}[V_{A1} - (-I_{A1}Z_{1s}) - V_{F1} + V_{F1}] \\
&= -\sqrt{3}[(V_{A1} - V_{F1}) + I_{A1}Z_{1s} \\
&\quad + I_{A1}(Z_{1s} + Z_{11} + Z_{0s} + Z_{01})] \\
&= -\sqrt{3}[(V_{A1} - V_{F1}) \\
&\quad + I_{A1}(2Z_{1s} + Z_{0s} + Z_{11} + Z_{01})] \\
&= -\sqrt{3}[(V_{A1} - V_{F1}) \\
&\quad + I_{A1}(Z_{11} + Z_{01} + Z_{1s}(2 + ZR_s)]
\end{aligned} \tag{B-41}
$$

For the phase comparator to operate, using Eqs. (B-40) and (B-41), we obtain

$$
\begin{aligned}
\frac{S_1}{S_2} &= \frac{\frac{(V_{A1}-V_{F1})}{I_{A1}} - [Z_{1c}(2 + ZR_c) - Z_{11} - Z_{01}]}{-\sqrt{3}\left\{\frac{(V_{A1}-V_{A1})}{I_{A1}} + [Z_{1s}(2 + ZR_s) + Z_{11} + Z_{01}]\right\}} \\
&\quad \frac{Z_{11} - [Z_{1c}(2 + ZR_c) - Z_{11} - Z_{01}]}{-\sqrt{3}\{Z_{11} + [Z_{1s}(2 + ZR_s) + Z_{11} + Z_{01}]\}}
\end{aligned} \tag{B-42}
$$

Therefore, Eq. (B-42) can be expressed as

$$
\frac{S_1}{S_2} = \frac{Z_{11}(2 + ZR_1) - Z_{1c}(2 + ZR_c)}{-\sqrt{3}Z_{11}(2 + ZR_1) - \sqrt{3}Z_{1s}(2 + ZR_s)} \tag{B-43}
$$

If we follow the general procedure for a phase

comparator, the general constants are

$$
\begin{aligned}
k_1 &= (2 + ZR_1) & k_2 &= -(2 + ZR_c)Z_{1c} \\
k_3 &= \sqrt{3}(2 + ZR_1) & k_4 &= -\sqrt{3}Z_{1s}(2 + ZR_s)
\end{aligned}
$$

The relay settings are chosen such that

$$
ZR_1 = ZR_c = \frac{Z_{01}}{Z_{11}} = \frac{Z_{0c}}{Z_{1c}}
$$

Therefore, the reference vectors in the R-X diagram are

$$
a = Z_{1c} \tag{B-44}
$$

$$
b = -Z_{1s}\frac{(2 + ZR_s)}{(2 + ZR_1)} \tag{B-45}
$$

The reference vectors are identified in Figure 12B-17, and the locus of impedance vectors (Z_{11}) is found using the techniques discussed for a phase comparator previously. A characteristic angle of $\pm 90°$ is used for the phase comparator to obtain the circular characteristic.

The influence of the source impedance is evident in the **b** vector. The unit can accommodate more fault resistance for short lines, in which the source impedance is large compared to the impedance of the protected line.

Self-Polarized Mho Ground Unit

This unit has been implemented in the REL 350 (Z2G, Z3G) relay system to clearly define forward and reverse reach. This type of unit utilizes faulted phase quantities for the operating quantity and the restraint

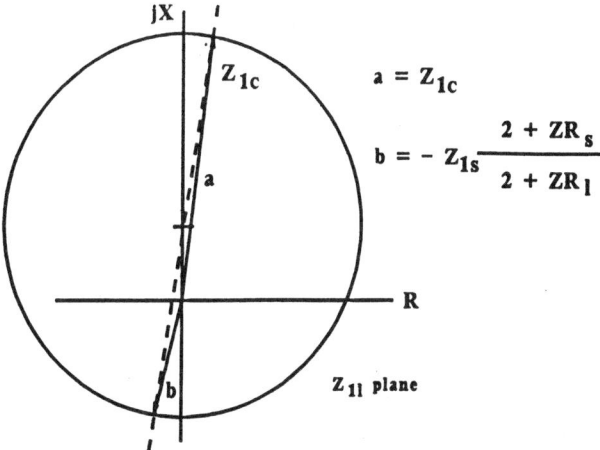

Figure 12B-17 Forward impedance characteristic of the quadrature polarized ground unit on the Z_{11} plane.

quantity for which it is called a self-polarized mho unit. This unit is better understood using a phase comparator that has the following inputs:

$$S_1 = V_A - \left(I_A + \frac{Z_{0cF} - Z_{1cF}}{Z_{1cF}} I_{A0} \right) Z_{1cF} \qquad (B\text{-}46)$$

$$S_2 = -\left[V_A + \left(I_A + \frac{Z_{0cR} - Z_{1cR}}{Z_{1cR}} I_{A0} \right) Z_{1cR} \right] \qquad (B\text{-}47)$$

These two expressions need to be modified. Referring to Figure 12B-15, we obtain

$$\begin{aligned} S_1 &= (V_{A1} + V_{A2} + V_{A0}) \\ &\quad - I_{A1} \left(3 + \frac{Z_{0cF} - Z_{1cF}}{Z_{1cF}} \right) Z_{1cF} \\ &= V_{A1} + (-I_{A1} Z_{1s}) + (-I_{A1} Z_{0s}) \\ &\quad - I_{A1}(2Z_{1cF} + Z_{0cF}) - V_{F1} + V_{F1} \\ &= (V_{A1} - V_{F1} \\ &\quad - I_{A1}(2Z_{1cF} + Z_{0cF} - Z_{1L} - Z_{0L}) \end{aligned} \qquad (B\text{-}48)$$

and

$$\begin{aligned} -S_2 &= (V_{A1} + V_{A2} + V_{A0}) \\ &\quad + I_{A1}(2Z_{1cR} + Z_{0cR}) + V_{F1} - V_{F1} \\ &= (V_{A1} - V_{F1}) + (-I_{A1} Z_{1s}) \\ &\quad + (-I_{A1} Z_{0s}) + I_{A1}(2Z_{1cR} + Z_{0cR} + Z_{1s} \\ &\quad + Z_{1L} + Z_{0s} + Z_{01}) \\ &= (V_{A1} - V_{F1}) \\ &\quad + I_{A1}(Z_{11} + Z_{01} + 2Z_{1cR} + Z_{0cR}) \end{aligned} \qquad (B\text{-}49)$$

For this phase comparator to produce an output, with the use of Eqs. (B-48) and (B-49), the characteristic is defined by

$$\frac{S_1}{S_2} = \frac{\frac{(V_{A1} - V_{F1})}{I_{A1}} - (2Z_{1cF} + Z_{0cF} - Z_{11} - Z_{01})}{-\frac{(V_{A1} - V_{F1})}{I_{A1}} - (Z_{11} + Z_{01} + 2Z_{1cR} + Z_{0cR})} \qquad (B\text{-}50)$$

Therefore, Eq. (B-50) can be reduced to

$$\begin{aligned} \frac{S_1}{S_2} &= \frac{Z_{11}(2 + ZR_1) - Z_{1cF}(2 + ZR_{cF})}{-[Z_{11}(2 + ZR_1) + Z_{1cR}(2 + ZR_{cR})]} \\ &= M_1 e^{\pm j 90^\circ} \end{aligned} \qquad (B\text{-}51)$$

Equation (B-51) has the general format for a phase comparator and to plot its characteristic on the Z_{11} plane.

For this phase comparator, we get

$$k_1 = (2 + ZR_1) \qquad k_2 = -Z_{1cF}(2 + ZR_{cF})$$
$$k_3 = -(2 + ZR_1) \qquad k_4 = -Z_{1cR}(2 + ZR_{cR})$$

The relay settings are chosen such that

$$\begin{aligned} ZR_1 = ZR_{cF} = ZR_{cR} &= \frac{Z_{01}}{Z_{11}} \\ &= \frac{Z_{0cF}}{Z_{1cF}} = \frac{Z_{0cR}}{Z_{1cR}} \end{aligned} \qquad (B\text{-}52)$$

Therefore, the reference vectors defining the characteristic of the unit on the Z_{11} plane are

$$\mathbf{a} = Z_{1cF} \qquad (B\text{-}53)$$
$$\mathbf{b} = -Z_{1cR} \qquad (B\text{-}54)$$

The **a** and **b** vectors are identified in Figure 12B-18 and the locus of impedance vectors, Z_{11}, that satisfy Eq. (B-51) is illustrated using the techniques discussed before.

Notice that the Z_{1cR} setting defines explicitly the characteristics of the unit and the source impedance has no influence. In general, any self-polarized unit will not be dependent on the source impedance; on the other hand, as seen before, any unit that uses a combination of the unfaulted phases will be dependent on the source impedance.

Reactance Ground Unit

This unit has been implemented in the type KDXG reactance ground relay. It is better understood with a

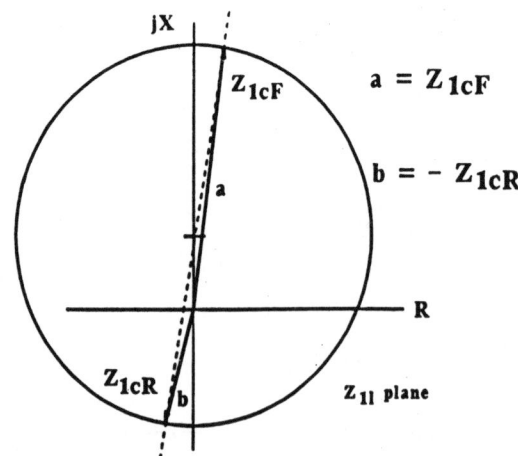

Figure 12B-18 Forward impedance characteristic of the self-polarized mho ground unit.

phase comparator that has the following inputs:

$$S_1 = V_A - \left(I_A + \frac{X_{0c} - X_{1c}}{X_{1c}} I_{A0}\right) jX_{1c} \quad \text{(B-55)}$$

$$S_2 = j\left(I_A + \frac{X_{0c} - X_{1c}}{X_{1c}} I_{A0}\right) X_{1c} \quad \text{(B-56)}$$

These expressions need to be simplified. Referring to Figure 12B-15, for a phase-A-to-ground fault, we get

$$S_1 = V_{A1} + V_{A2} + V_{A0}$$
$$- I_{A1}\left(3 + \frac{X_{0c} - X_{1c}}{X_{1c}}\right) jX_{1c}$$
$$= V_{A1} + (-I_{A1}Z_{1s}) + -(I_{A1}Z_{0s})$$
$$- jI_{A1}(2X_{1c} + X_{0c}) + V_{F1} - V_{F1}$$
$$= (V_{A1} - V_{F1}) - I_{A1}[j(2X_{1C} + X_{0C})$$
$$- Z_{1L} - Z_{0L}] \quad \text{(B-57)}$$

and

$$S_2 = j(2X_{1c} + X_{0c})I_{A1} \quad \text{(B-58)}$$

Equations (B-57) and (B-58) determine the impedance characteristic of the comparator. It follows from the procedure discussed above that

$$\frac{S_1}{S_2} = \frac{Z_{11} - (j2X_{1c} + jX_{0c} - Z_{11} - Z_{01})}{j(2X_{1c} + X_{0c})} \quad \text{(B-59)}$$

$$\frac{S_1}{S_2} = \frac{Z_{11}(2 + ZR_1) - jX_{1C}(2 + XR_c)}{jX_{1C}(2 + XR_c)}$$
$$= M_1 e^{\pm j90°} \quad \text{(B-60)}$$

The settings of the relay are made such that

$$ZR_1 = \frac{Z_{01}}{Z_{11}} = \frac{X_{0c}}{X_{1c}}$$

Finding the general constants for a phase comparator in Eq. (B-60), we obtain

$$k_1 = (2 + ZR_1) \qquad k_2 = jX_{1c}(2 + XR_c)$$
$$k_3 = 0 \qquad k_4 = jX_{1c}(2 + XR_c)$$

It follows that

$$a = jX_{1c} \quad \text{(B-61)}$$

$$b = < \pm e^{j90°} \text{ (infinity at 90°)} \quad \text{(B-62)}$$

Equation (B-62) might not be a rigorous mathematical expression, but it identifies the location of one of our reference vectors. It can be thought, therefore, that $(Z - b)$ is always perpendicular to the reactance line and the phase comparator characteristic angle require-

ment every time met. Figure 12B-19 illustrates the construction and characteristics of the reactance unit. It should be mentioned that this unit was designed to cover more ground fault resistance; however, it could not work by itself since it is nondirectional and would even operate for load. Therefore, it had to be supervised by some other unit, as will be illustrated later.

B.4.5 Blinder Units

Blinders are impedance elements that are used for out-of-step relaying and also supervising impedance units on load current encroachment. These elements have been employed in REL-300 and REL-350 systems. Implementation is similar to that for the reactance unit, and the phase comparator approach makes it easier to understand. A typical implementation in a phase comparator has inputs

$$S_1 = -V_A + I_A R_T e^{j(PANG-90°)} \quad \text{(B-63)}$$
$$S_2 = +I_A e^{j(PANG-90°)} \quad \text{(B-64)}$$

Since blinders are used for out-of-step conditions, it monitors three-phase conditions only; the above quantities are all positive sequence line current and voltage inputs.

Following the procedure yields

$$\frac{S_1}{S_2} = \frac{Z_{11} - R_T e^{j(PANG-90°)}}{-e^{j(PANG-90°)}} = M_1 e^{\pm j90°} \quad \text{(B-65)}$$

Equation (B-65) appears in the standard format so far

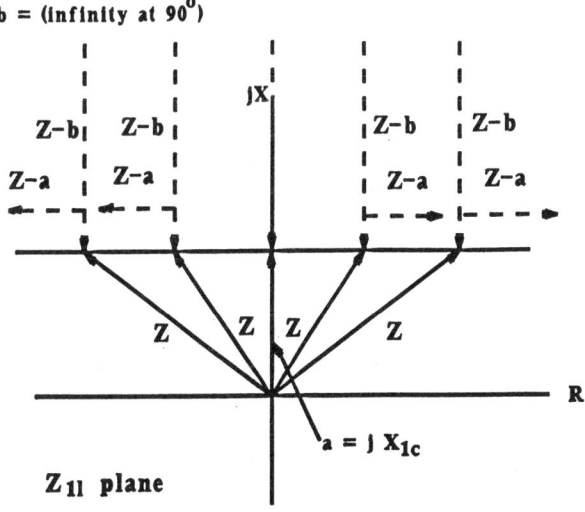

Figure 12B-19 Reactance unit characteristics.

used to derive phase comparator characteristics. Therefore,

$$k_1 = 1 \qquad k_2 = -R_Te^{j(PANG-90°)} \qquad \text{(B-66)}$$

$$k_3 = 0 \qquad k_4 = -e^{j(PANG-90°)} \qquad \text{(B-67)}$$

$$\mathbf{a} = R_Te^{j(PANG-90°)}$$

$$\mathbf{b} = -<e^{j(PANG-90°)} = <e^{j(PANG+90°)}$$

Equation (B-67) may not be a rigorous mathematical expression, but it provides a reference to infinity for our **b** vector. This way, the magnitude does not matter, just the angle to infinity (PANG + 90). If we refer to Figure 12B-20, (**Z** − **b**) can be thought of as always perpendicular to the line impedance Z_{11} that the unit is being applied to.

B.5 Reverse Characteristics of an Impedance Unit

In the previous section, the forward-looking characteristics of several impedance units were analyzed. It should be stressed that in the above R-X plots, the fact that the impedance locus may go through the third and fourth quadrant does not imply that the unit is nondirectional or operate for a reverse fault. The previous plots are all forward-looking.

To investigate the directionality of an impedance unit, the reverse-looking characteristics need to be found. The idea is to find the locus of reverse

impedances (Z_{11}'') for which the unit will operate. For an impedance unit to be secure and directional, the reverse-looking impedance (Z_{11}'') locus should lie on the third quadrant (negative R and negative X).

For the purpose of studying the reverse characteristics of the different units derived, the circuit in Figure 12B-21 is assumed. The source impedance, denoted Z_{1s}'', is the composite of the line being protected and the source impedance "looking" forward. The reverse-looking impedance, Z_{11}'', is the impedance in the reverse direction and the purpose of this study is to find the locus of impedances Z_{11}'' for which the units will operate.

This circuit is again a radial circuit so that the influence of different distribution factors can be disregarded. The directions of the currents are different from before. The plots now will be done on the reverse-looking line impedance Z_{11}''.

B.5.1 Phase-to-Phase Unit

Following the same approach used before, for a phase-to-phase fault, Figure 12B-22 describes the connection of sequence networks for a phase-to-phase fault.

The direction of the currents in Figure 12B-22 indicates the proper direction of the sequence components that the unit sees. The inputs to the magnitude comparator are again

$$S_A = V_{A2} - I_{A2}Z_c \qquad \text{(B-68)}$$

$$S_B = V_{A1} - I_{A1}Z_c \qquad \text{(B-69)}$$

and the impedance characteristic is defined by

$$\frac{S_A}{S_B} = \frac{V_{A2} - I_{A2}Z_c}{V_{A1} - I_{A1}Z_c} = e^{j\theta} \qquad \text{(B-70)}$$

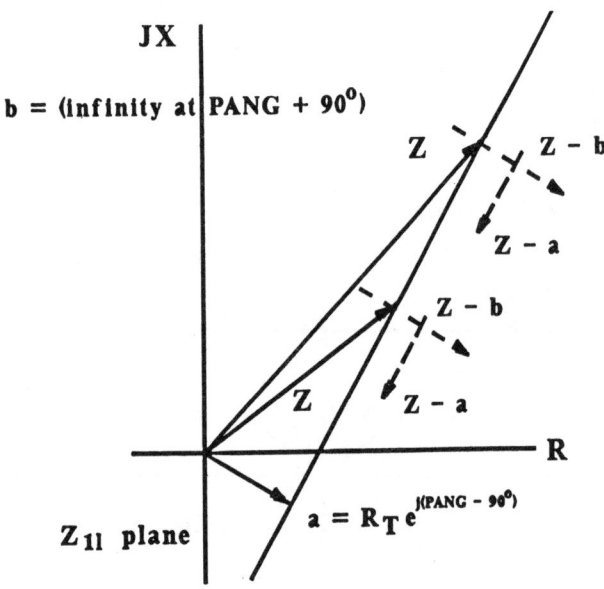

Figure 12B-20 Typical blinder unit characteristics.

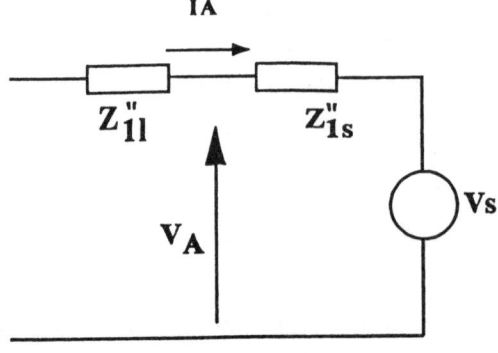

Figure 12B-21 Radial system for studying the reverse characteristic of distance units.

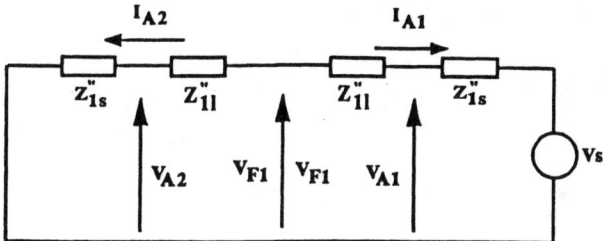

Figure 12B-22 Network connection for a phase-to-phase fault. The direction of the currents indicate the actual component currents seen by the distance units.

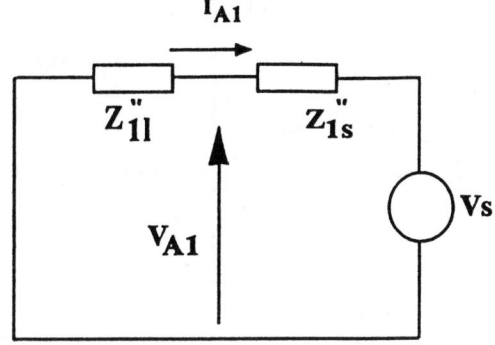

Figure 12B-24 Network for a three-phase fault.

and substituting voltages from the figure yields

$$\frac{S_A}{S_B} = \frac{-(I_{A1}Z''_{1s}) - (-I_{A1}Z_c)}{-I_{A1}(2Z''_{1l} + Z''_{1s}) - I_{A1}Z_c} = e^{j\theta} \qquad (B-71)$$

or on the positive sequence R-X plane, the characteristic is

$$Z''_{1l} = -\frac{1}{2}(Z''_{1s} + Z_c) + \frac{1}{2}(Z''_{1s} - Z_c)e^{-j\theta} \qquad (B-72)$$

Figure 12B-23 illustrates the impedance characteristic of the unit on the R-X diagram of Z''_{1l}, the *reverse-direction impedance*.

B.5.2 Three-Phase Unit

Figure 12B-24 illustrates a three-phase fault in the reverse direction. The voltage and current (V_{A1} and I_{A1}) directions are the references that the distance unit

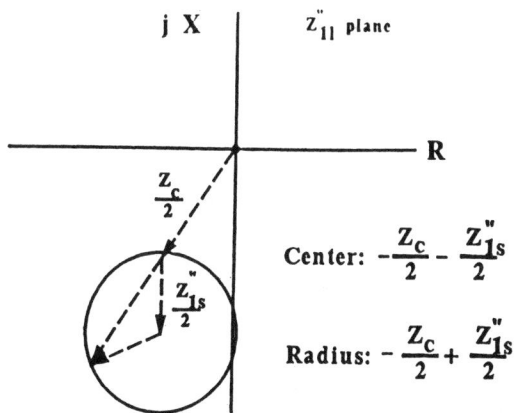

Center: $-\dfrac{Z_c}{2} - \dfrac{Z''_{1s}}{2}$

Radius: $-\dfrac{Z_c}{2} + \dfrac{Z''_{1s}}{2}$

Figure 12B-23 Phase-to-phase unit characteristic on the reverse direction.

"sees." The inputs to the magnitude comparator are

$$S_A = -\frac{Z_c}{2}I_{A1} \qquad (B-73)$$

$$S_B = V_{A1} - \frac{Z_c}{2}I_{A1} \qquad (B-74)$$

Therefore, the characteristic of the device is defined by

$$\frac{S_A}{S_B} = \frac{-\frac{Z_c}{2}I_{A1}}{V_{A1} - \frac{Z_c}{2}I_{A1}} = e^{j\theta} \qquad (B-75)$$

Substituting voltages with currents, we get

$$\frac{S_A}{S_B} = \frac{-\frac{Z_c}{2}I_{A1}}{(-Z''_{1l}I_{A1}) - \frac{Z_c}{2}I_{A1}} = e^{j\theta} \qquad (B-76)$$

The operating characteristic of the unit is defined by

$$Z''_{1l} = -\frac{Z_c}{2} + \frac{Z_c}{2}e^{-j\theta} \qquad (B-77)$$

Figure 12B-25 illustrates the characteristic of the unit on the reverse positive sequence impedance Z''_{1l} plane.

B.5.3 Ground Units

To develop the reverse-looking characteristic of the different ground units, Figure 12B-26 will be used. It illustrates the connection of the sequence networks for an A-to-ground fault. The directions of the currents are those that the relay actually "sees."

SDGU Impedance Unit

The inputs to the magnitude comparator are:

$$S_A = V_{A0} - I_{A0}Z_{0c} \qquad (B-78)$$
$$S_B = (V_{A1} + V_{A2}) - (I_{A1} + I_{A2})Z_{1c} \qquad (B-79)$$

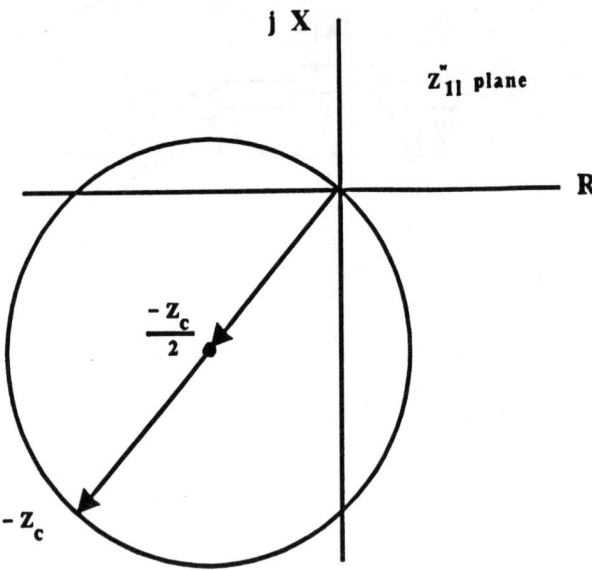

Figure 12B-25 Reverse characteristic of the three-phase unit.

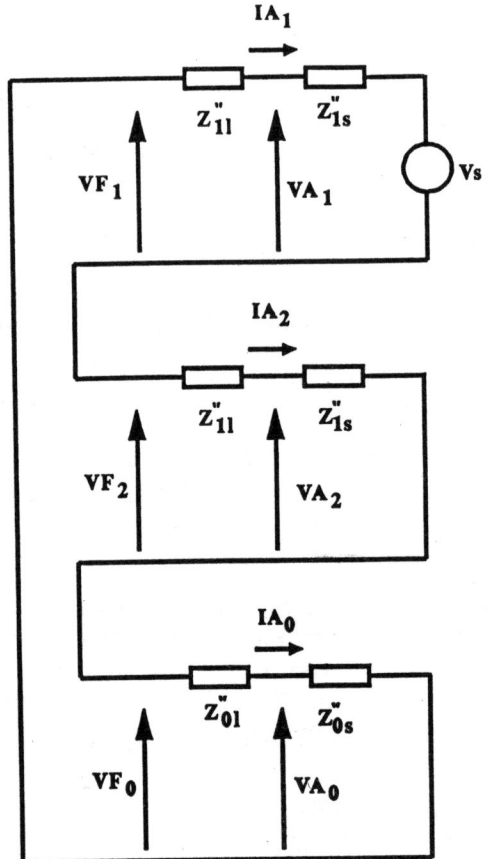

Figure 12B-26 Sequence network connection for a reverse A-to-ground fault.

The characteristic is defined by

$$\frac{S_A}{S_B} = \frac{V_{A0} - I_{A0}Z_{0c}}{(V_{A1} + V_{A2}) - (I_{A1} + I_{A2})Z_{1c}}$$
$$= e^{jr} \qquad (B\text{-}80)$$

making the appropriate substitutions from Figure 12B-31

$$\frac{S_A}{S_B} = \frac{(I_{A0}Z_{0s}'') - I_{A0}Z_{0c}}{-I_{A1}(2Z_{11}'' + Z_{01}'' + Z_{0s}'') - (2I_{A0}Z_{1c})} = e^{j\theta}$$

or

$$\frac{S_A}{S_B} = \frac{-Z_{0s}'' + Z_{0c}}{2Z_{11}'' + Z_{01}'' + Z_{0s}'' + 2Z_{1c}} = e^{j\theta} \qquad (B\text{-}81)$$

Then, the equation that defines the characteristic on the Z_{11}'' plane is

$$Z_{11}'' = \frac{-(2Z_{1c} + Z_{0s}'')}{(2 + ZR_1'')} + \frac{Z_{1c}ZR_c - Z_{0s}''}{(2 + ZR_1'')}e^{j\theta} \qquad (B\text{-}82)$$

This characteristic is illustrated in Figure 12B-27.

Quadrature-Polarized Ground Unit

The inputs to the phase comparator are

$$S_1 = V_A - \left(I_A + I_{A0}\frac{Z_{0c} - Z_{1c}}{Z_{1c}}\right)Z_{1c} \qquad (B\text{-}83)$$

$$S_2 = j(V_C - V_B) \qquad (B\text{-}84)$$

The directions of the currents and voltages are those shown in Figure 12B-26. The above equations can be

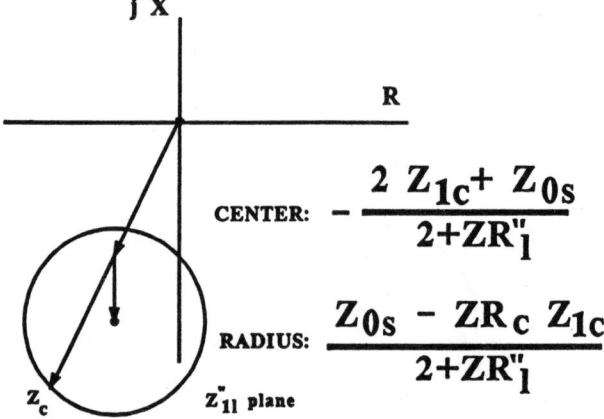

Figure 12B-27 Reverse characteristic of the SDGU relay.

modified to

$$S_1 = (V_{A1} + V_{A2} + V_{A0}) - I_{A1} \times \left(3 + \frac{Z_{0c} + Z_{1c}}{Z_{1c}}\right) Z_{1c}$$

$$= V_{A1} + (I_{A2} Z_{1s}'') + (I_{A0} Z_{0s}'') - I_{A1}(2Z_{1c} + Z_{0c})$$
$$+ (V_{F1} - V_{F1})$$

$$= (V_{A1} - V_{F1}) + I_{A1}(Z_{1s}'' + Z_{0s}'' - 2Z_{1c} - Z_{0c} - Z_{11}''$$
$$- Z_{01}'' + Z_{1s}'' + Z_{0s}'')$$

or

$$\frac{S_1}{I_{A1}} = -Z_{11}'' - Z_{01}'' - 2Z_{1c} - Z_{0c}$$

$$= -Z_{11}''(2 + ZR_1'') - Z_{1c}(2 + ZR_c) \qquad \text{(B-85)}$$

Also, working on S_2, we get

$$S_2 = j(V_{C1} + V_{C2} + V_{C0} - V_{B1} - V_{B2} - V_{B0})$$

$$= j[V_{A1}(a - a^2) - V_{A2}(a - a^2)]$$

$$= -\sqrt{3}(V_{A1} - V_{A2})$$

$$= -\sqrt{3}[V_{A1} - (I_{A1} Z_{1s}'') - V_{F1} + V_{F1}]$$

$$= -\sqrt{3}[(V_{A1} - V_{F1}) - I_{A1} Z_{1s}'' + V_{F1}]$$

or

$$\frac{S_2}{I_{A1}} = -\sqrt{3}[-Z_{11}'' - Z_{1s}'' - (Z_{11}'' + Z_{1s}'' + Z_{01}'' - Z_{0s}'')]$$

$$= -\sqrt{3}[-2Z_{11}'' - 2Z_{1s}'' - Z_{01}'' - Z_{0s}'']$$

$$= -\sqrt{3}[Z_{11}''(2 + ZR_1'') + Z_{1s}''(2 + ZR_s'')] \qquad \text{(B-86)}$$

The operating characteristic is defined by

$$\frac{S_1}{S_2} = \frac{-Z_{11}''(2 + ZR_1'') - Z_{1c}(2 + ZR_c)}{-\sqrt{3}[Z_{11}''(2 + ZR_1'') + Z_{1s}''(2 + ZR_s'')]}$$

$$= M_1 e^{\pm j90°} \qquad \text{(B-87)}$$

If we follow the procedure for a phase comparator, the general constants are

$$k_1 = -(2 + ZR_1'')$$
$$k_2 = -Z_{1c}(2 + ZR_c)$$
$$k_3 = \sqrt{3}(2 + ZR_1'')$$
$$k_4 = \sqrt{3}(2 + ZR_s'')Z_{1s}'' \qquad \text{(B-88)}$$

The reference vectors are therefore

$$a = -\frac{(2 + ZR_c)}{(2 + ZR_1'')} Z_{1c}$$

and

$$b = -\frac{(2 + ZR_s'')}{(2 + ZR_1'')} Z_{1s} \qquad \text{(B-89)}$$

Figure 12B-28 illustrates the reverse characteristics of the quadrature-polarized relay.

Self-Polarized Ground Unit

The inputs to the phase comparator are

$$S_1 = V_A - \left(I_A + \frac{Z_{0cF} - Z_{1cF}}{Z_{1cF}} I_{A0}\right) Z_{1cF} \qquad \text{(B-90)}$$

$$S_2 = -\left[V_A + \left(I_A + \frac{Z_{0cR} - Z_{1cR}}{Z_{0cR}} I_{A0}\right) Z_{1cR}\right] \qquad \text{(B-91)}$$

Modifying the two expressions using Figure 12B-26 and following the same procedure as for the quadrature-polarized distance unit, we obtain

$$\frac{S_1}{I_{A1}} = -Z_{11}''(2 + ZR_1'') - Z_{1cF}(2 + ZR_{cF}) \qquad \text{(B-92)}$$

and

$$\frac{S_2}{I_{A1}} = -Z_{11}''(2 + ZR_1'') + Z_{1cR}(2 + ZR_{cR}) \qquad \text{(B-93)}$$

Figure 12B-29 illustrates the reverse characteristics of this unit. It is assumed that $ZR_{cF} = ZR_{cR} = ZR_1''$.

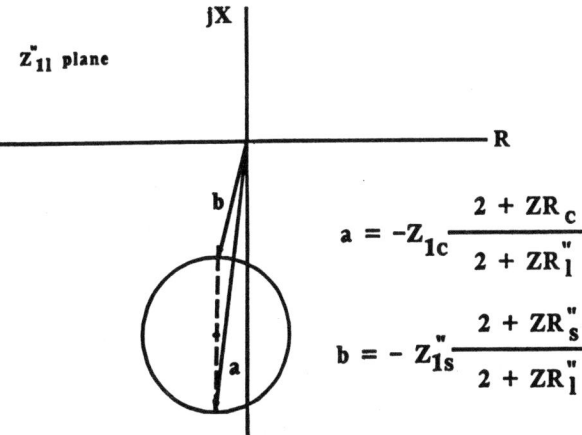

Figure 12B-28 Reverse characteristics of the quadrature polarized relay.

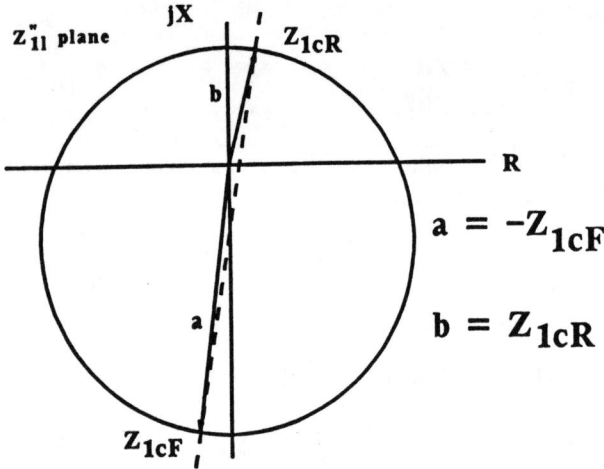

Figure 12B-29   Reverse characteristics of the nondirectional ground unit.

B.6   Response of Distance Units to Different Types of Faults

Distance units are designed specifically to operate for certain faults; a phase A ground distance unit should respond to phase-A-to-ground faults. However, all distance units do have a characteristic for other types of faults. A phase A ground distance unit will have a response to phase-B-to-ground, phase-C-to-ground, phase-to-phase, etc., since the unit will essentially still be receiving the input quantities from the power system. It needs to be remarked that, in general, distance units responding to other faults will have a shorter reach for other types of faults.

In the next sections, the text will concentrate on the phase-to-phase, quadrature-polarized ground unit and self-polarized ground unit only.

B.6.1   Phase-to-Phase Unit

Response to Other Phase-to-Phase Faults

In Section 4.2, the response-to a BC phase-to-phase fault was investigated. If the same approach is used for other phase-to-phase faults, the analysis will yield the same result: For all phase-to-phase faults, this unit has the characteristic shown in Figure 12B-12 in this appendix. This means that we can use only one phase-to-phase unit to detect all types of phase-to-phase faults. There is no need for three units for all phase-to-phase faults.

Response to a Three-Phase Fault.

The inputs to the magnitude comparator are

$$S_A = V_{A2} - I_{A2}Z_c \tag{B-94}$$

$$S_B = V_{A1} - I_{A1}Z_c \tag{B-95}$$

for a three-phase fault $S_A = 0$. Therefore, this unit is not responsive to three-phase faults and it is not affected by normal load flow at all. This is an excellent characteristic that makes this unit unique.

Response to Phase-to-Ground Faults.

The unit shows the same response for all forward phase-to-ground faults. If a figure similar to 12B-15 is used and Eqs. (B-23) and (B-24) (S_A and S_B) are applied, the locus of impedances on the Z_{11} plane is defined by the equation

$$Z_{11} = \frac{Z_c - Z_{1s}(1 + ZR_s)}{(2 + ZR_1)} - \frac{(Z_{1s} + Z_c)e^{-jr}}{(2 + ZR_1)} \tag{B-96}$$

The characteristic for the response of the phase-to-phase unit to all phase-to-ground faults is shown in Figure 12B-30.

B.6.2   Quadrature-Polarized Ground-Distance Unit

In all microprocessor-based relays, there will be three ground units to detect all the ground fault types. The units are constantly receiving the inputs from the power system and they do have a definite response to

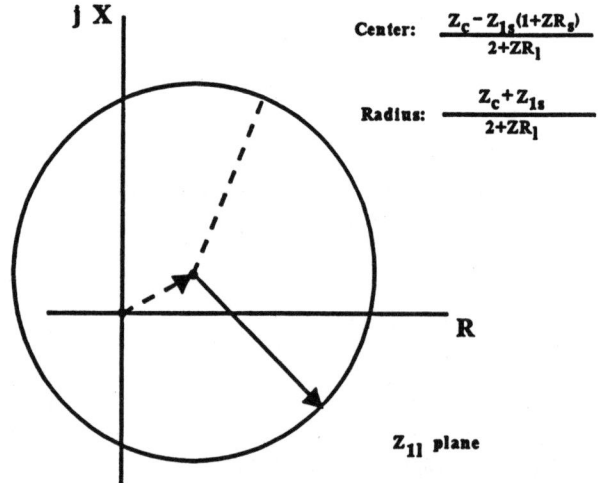

Figure 12B-30   Response of the phase-to-phase unit to all phase-to-ground faults.

other types of faults. In this sense, the basic approach used thus far will still help us in the analysis of the response of the phase A ground-distance unit to all other types of faults.

Response to Other Phase-to-Ground Faults

If other phase-to-ground faults are applied to the phase A unit, it will tend to have a response mainly due to the I_{A0} term in S_1. For analysis, a connection similar to Figure 12B-15 should be used, but always by considering the phase shifts (a or a^2) introduced when using other phases as a reference. In this sense, the reference **a** and **b** vectors can be calculated.

For a phase-B-to-ground fault, we have

$$\mathbf{a} = -Z_{1s}\frac{(1-a)+(1-a^2)ZR_s}{(2+ZR_l)}$$
$$+ Z_{1c}\frac{(1+a)+a^2ZR_c}{(2+ZR_l)} \qquad \text{(B-97)}$$

and

$$\mathbf{b} = -Z_{1s}\frac{(1+a)+ZR_s}{(2+ZR_l)} \qquad \text{(B-98)}$$

For a phase-C-to-ground fault, we have

$$\mathbf{a} = -Z_{1s}\frac{(1-a^2)+(1-a)ZR_s}{(2+ZR_l)}$$
$$+ Z_{1c}\frac{(1+a^2)+aZR_c}{(2+ZR_l)} \qquad \text{(B-99)}$$

and

$$\mathbf{b} = -Z_{1s}\frac{(1+a^2)+ZR_s}{(2+ZR_l)} \qquad \text{(B-100)}$$

The above reference vectors for a phase comparator define the characteristics of the phase A ground unit for phase-B-to-ground and phase-C-to-ground faults. Figure 12B-31 illustrates the characteristics of the phase A unit for all phase-to-ground faults.

Response to a Three-Phase Fault.

For a three-phase fault, the actual inputs to the phase comparator for the phase A unit are

$$S_1 = V_A - I_A Z_{1c} \qquad \text{(B-101)}$$
$$S_2 = j(V_C - V_B) \qquad \text{(B-102)}$$

If we use Figure 12B-13 and the general procedure for

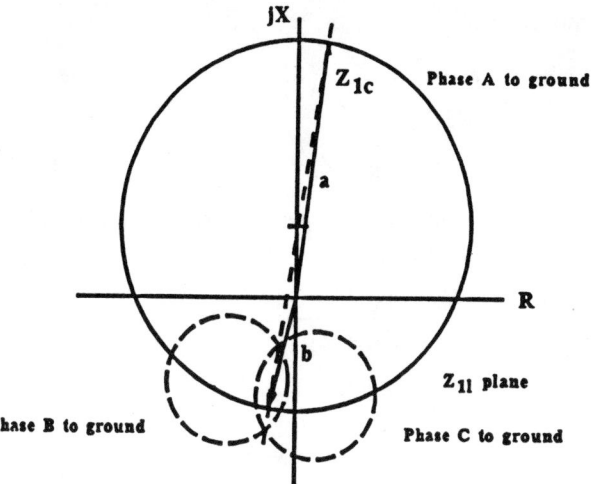

Figure 12B-31 Characteristic for all phase-to-ground faults.

a phase comparator, the reference vectors are

$$\mathbf{a} = Z_{1c} \qquad \text{(B-103)}$$
$$\mathbf{b} = 0 \qquad \text{(B-104)}$$

Notice that the above equations imply that the reach for a three-phase fault for the phase A unit is the same as its setting for phase-to-ground faults. This implies that this same unit can be used to detect three-phase faults. In general, the operation of all three-phase-to-ground units indicates a three-phase fault. It is unfortunate, however, that the unit itself is not dependent on the source impedance, and a fault right at the bus will not be detected because there is no reference quantity, $S_2 = 0$. To make up for the lack of polarizing quantity for a bus fault, REL-300 uses the prefault voltage as a reference. This concept is called "memory voltage," and it is used to increase the coverage of the unit. This concept is used in many other designs to avoid the voltage inversion problem in lines with series-capacitor compensation that was described in the transmission-line protection chapter. Hence, memory voltage increases the coverage and is used to protect series-capacitor lines.

If memory voltage is employed, then the inputs to the phase comparator are

$$S_1 = V_A - (I_A)Z_{1c} \qquad \text{(B-105)}$$
$$S_2 = j(V_{Cm} - V_{Bm}) \qquad \text{(B-106)}$$

S_1 needs no simplification. However, S_2 can be

modified to

$$S_2 = j(a - a^2)V_{Am}$$
$$= -\sqrt{3}(V_s)$$
$$= -\sqrt{3}I_{A1}(Z_{1l} + Z_{1s}) \tag{B-107}$$

Therefore, following the general procedure for a phase comparator, we get

$$\frac{S_1}{S_2} = \frac{V_A - I_{A1}Z_{1c}}{-\sqrt{3}I_{A1}(Z_{1l} + Z_{1s})} = M_1 e^{\pm j90°} \tag{B-108}$$

or

$$\frac{S_1}{S_2} = \frac{Z_{1l} - Z_{1c}}{-\sqrt{3}(Z_{1l} + Z_{1s})} = M_1 e^{\pm j90°}$$

The reference vectors **a** and **b** are

$$\mathbf{a} = Z_{1c} \tag{B-109}$$
$$\mathbf{b} = -Z_{1s} \tag{B-110}$$

Therefore, if the unit has memory voltage, the forward reach will not be modified, but the coverage of the phase A ground-distance unit for three-phase faults will be dependent on the source impedance. It is evident that now bus faults can be detected easily. Figure 12B-32 illustrates the response of the quadrature-polarized ground unit to forward three-phase faults with and without "memory" action.

Response to Phase-to-Phase Faults

It can be found that by using a figure similar to 12B-11 and applying the general procedure for a phase comparator to the S_1 and S_2 inputs in this unit, the reference vectors are as follows. For a BC fault,

$$\mathbf{a} = -Z_{1s} \tag{B-111}$$
$$\mathbf{b} = 0 \tag{B-112}$$

For a CA fault,

$$\mathbf{a} = -\left[\frac{Z_{1s}(a + 1)}{2} + \frac{(a - 1)Z_{1c}}{2}\right] \tag{B-113}$$

$$\mathbf{b} = -\left[\frac{Z_{s1}(1 - a)}{2}\right] \tag{B-114}$$

For a AB fault,

$$\mathbf{a} = -\left[\frac{Z_{1s}(a^2 + 1)}{2} + \frac{(a^2 - 1)Z_{1c}}{2}\right] \tag{B-115}$$

$$\mathbf{b} = -\left[\frac{Z_{s1}(1 - a^2)}{2}\right] \tag{B-116}$$

The response of the quadrature-polarized phase A unit to all forward phase-to-phase faults is shown in Figure 12B-33.

B.6.3 Self-Polarized Ground Unit

Response to Other Phase-to-Ground Faults

For analysis, a connection similar to Figure 12B-15 should be used, but always by considering the phase shifts (a or a^2) introduced when using other phases as a reference. In this sense, the reference **a** and **b** vectors

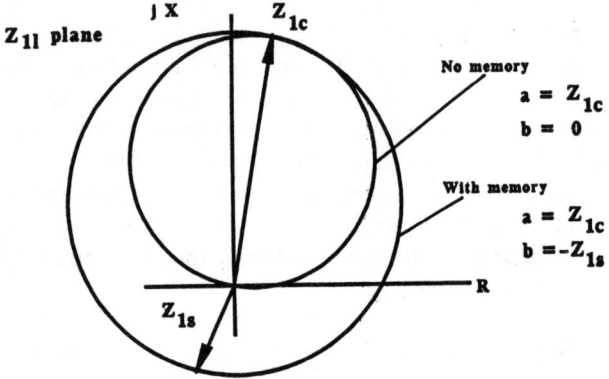

Figure 12B-32 Forward characteristic for three-phase faults. With and without "memory" action.

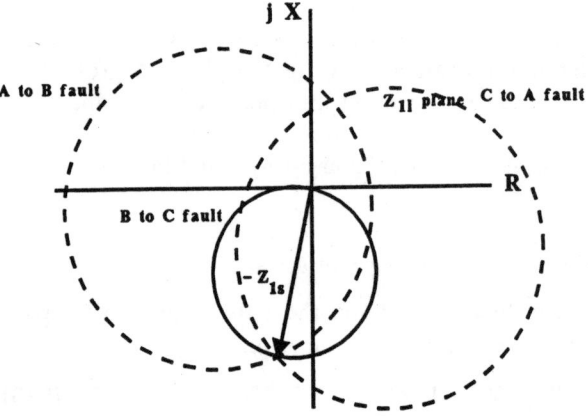

Figure 12B-33 Characteristics for forward phase-to-phase faults.

can be calculated. For a phase-B-to-ground fault,

$$\mathbf{a} = -Z_{1s}\frac{(1-a)+(1-a^2)ZR_s}{(2+ZR_1)}$$
$$+ Z_{1cF}\frac{(1+a)+a^2ZR_{cF}}{(2+ZR_1)} \tag{B-117}$$

$$\mathbf{b} = -Z_{1s}\frac{(1-a)+(1-a^2)ZR_s}{(2+ZR_1)}$$
$$- Z_{1cR}\frac{(1+a)+a^2ZR_{cR}}{(2+ZR_1)} \tag{B-118}$$

For a phase-C-to-ground fault,

$$\mathbf{a} = -Z_{1s}\frac{(1-a^2)+(1-a)ZR_s}{(2+ZR_1)}$$
$$+ Z_{1cF}\frac{(1+a^2)+aZR_{cF}}{(2+ZR_1)} \tag{B-119}$$

$$\mathbf{b} = -Z_{1s}\frac{(1-a^2)+(1-a)ZR_s}{(2+ZR_1)}$$
$$- Z_{1cR}\frac{(1+a^2)+aZR_{cR}}{(2+ZR_1)} \tag{B-120}$$

The above reference vectors for a phase comparator define the characteristics of the phase A ground unit for phase-B-to-ground and phase-C-to-ground faults. Figure 12B-34 illustrates the characteristics of the phase A unit for all phase-to-ground faults.

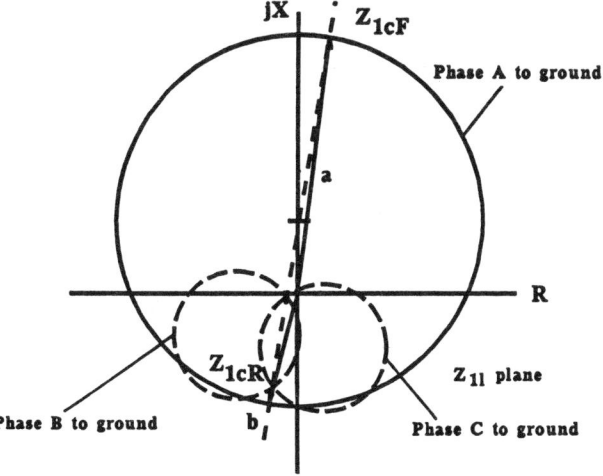

Figure 12B-34 Characteristics for all phase-to-ground faults.

Response to a Three-Phase Fault

For a three-phase fault, the actual inputs to the phase comparator for the phase A unit are

$$S_1 = V_A - I_{A1}Z_{1cF} \tag{B-121}$$
$$S_2 = -(V_A + I_{A1}Z_{1cR}) \tag{B-122}$$

If we use Figure 12B-13 and the general procedure for a phase comparator, the reference vectors are

$$\mathbf{a} = Z_{1cF} \tag{B-123}$$
$$\mathbf{b} = Z_{1cR} \tag{B-124}$$

This means that the unit's characteristic for a three-phase fault will be the same as for phase-to-ground faults (Fig. 12B-18). In REL 350 the operation of all the three-phase-to-ground units indicates the occurrence of a three-phase fault.

Response to Phase-to-Phase Faults

It can be found that by using a figure similar to 12B-11 and applying the general procedure for a phase comparator to the S_1 and S_2 inputs in this unit, the reference vectors are as follows. For a BC fault,

$$\mathbf{a} = -Z_{1s} \tag{B-125}$$
$$\mathbf{b} = -Z_{1s} \tag{B-126}$$

For a CA fault,

$$\mathbf{a} = -\left[\frac{Z_{1s}(1+a)}{2} + \frac{(a-1)Z_{1cF}}{2}\right] \tag{B-127}$$

$$\mathbf{b} = -\left[\frac{Z_{1s}(1+a)}{2} + \frac{(a-1)Z_{1cR}}{2}\right] \tag{B-128}$$

For a AB fault,

$$\mathbf{a} = -\left[\frac{Z_{1s}(1+a^2)}{2} + \frac{(a^2-1)Z_{1cF}}{2}\right] \tag{B-129}$$

$$\mathbf{b} = -\left[\frac{Z_{1s}(1+a^2)}{2} + \frac{(a^2-1)Z_{1cR}}{2}\right] \tag{B-130}$$

The response of the self-polarized phase A ground unit to all forward phase-to-phase faults is shown in Figure 12B-35.

B.6.4 Double Phase-to-Ground Faults

It is noticeable that the double line-to-ground faults have been avoided thus far. The reason is that the analysis of such faults using the general procedure explained above and maybe other procedures becomes too complicated. However, in general, it is accepted

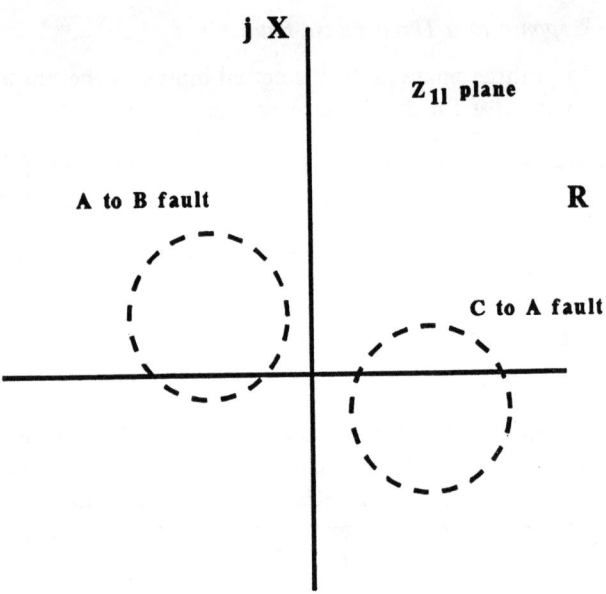

Figure 12B-35 Characteristics for all forward phase-to-phase faults.

that the analysis of the performance of distance units can be made only for phase-to-ground, phase-to-phase, and three-phase faults. Phase-to-phase-to-ground fault detection is achieved by the operation of the units detecting phase-to-phase and three-phase faults.

Figure 12B-36 shows the sequence connection for a phase-to-phase-to-ground fault. Notice that the zero sequence impedance is present in the sequence con-

nection. The detection of a phase-to-phase-to-ground fault will be done by the phase-to-phase unit if the zero sequence impedance is large. This means that a phase-to-phase-to-ground fault approaches a phase-to-phase fault if the zero sequence impedance is large. This fact is evident in Figure 12B-36, and it should be detected by the phase-to-phase unit.

On the other hand, if the zero sequence impedance is small, then the phase-to-phase-to-ground fault approaches a three-phase fault, and the three-phase fault detection units should operate for phase-to-phase faults.

In many years of operation, considering the phase-to-phase-to-ground fault as a variation of the phase-to-phase or three-phase fault has never been a problem.

B.7   The Influence of Current Distribution Factors and Load Flow

In all the above analyses, a radial system was assumed for the purpose of simplicity. In real life, systems are not necessarily radial and they are carrying load. It is the purpose of this section to describe the role that distribution factors and load current have in the operating characteristics of the impedance units.

Figure 12B-37 illustrates the definition of distribution factors for all the sequence networks. Distribution factors are complex per unit factors of the total sequence current at the fault. In the following

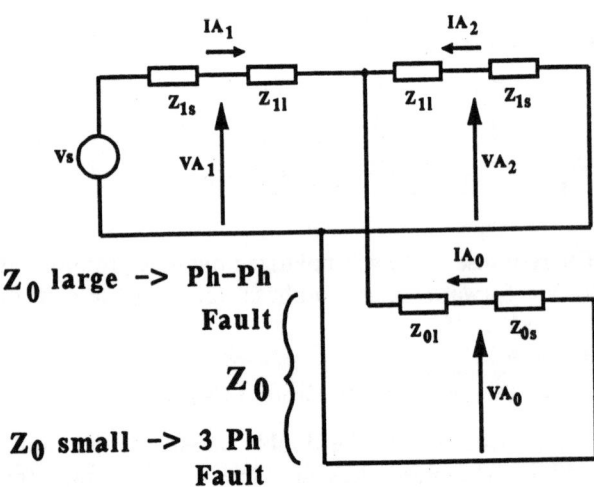

Figure 12B-36 Double phase-to-ground sequence network connection.

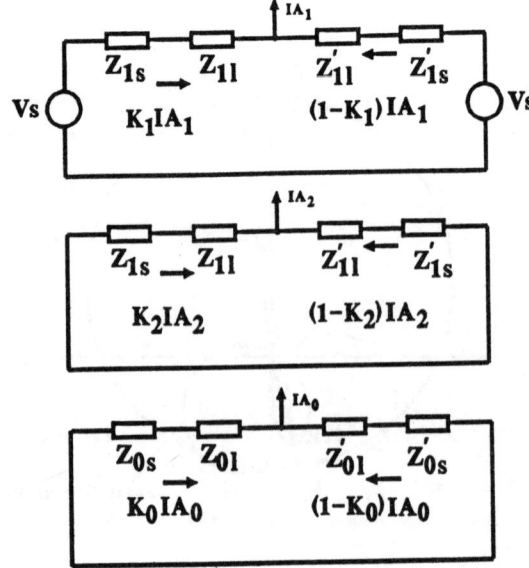

Figure 12B-37 Distribution factors in all the networks.

paragraphs, distribution factors will be used to describe the operating characteristics of the phase-to-phase unit, quadrature-polarized ground unit, and self-polarized ground unit; however, a note of caution should be given. The analysis is a simple academic curiosity, and although it accurately describes the general case, the actual impedance characteristic of the unit will be difficult to visualize since distribution factors (K_0, K_1, K_2) will change, depending on the location of the fault in the line.

From Figure 12B-37, we see that the following is true:

$$K_1 = \frac{V_s(Z'_{1s} + Z'_{11})}{V_s(Z'_{1s} + Z'_{11}) + V'_s(Z_{1s} + Z_{11})} \tag{B-131}$$

$$K_2 = \frac{(Z'_{1s} + Z'_{11})}{(Z'_{1s} + Z'_{11}) + (Z_{1s} + Z_{11})} \tag{B-132}$$

$$K_0 = \frac{(Z'_{0s} + Z'_{01})}{(Z'_{0s} + Z'_{01}) + (Z_{0s} + Z_{01})} \tag{B-133}$$

Notice that if there is no load flow, $V_s = V'_s$, then $K_1 = K_2$. It can be also shown that

$$\left(1 - \frac{K_2}{K_1}\right) = \frac{I_L}{I_{F3ph}} \tag{B-134}$$

B.7.1   Phase-to-Phase Unit

Refer to Figure 12B-38 for the analysis of this unit. The inputs to the phase comparator are still

$$S_A = V_{A2} - K_2 I_{A2} Z_c$$
$$= K_2 I_{A1} Z_{1s} + K_2 I_{A1} Z_c \tag{B-135}$$
$$S_B = V_{A1} - K_1 I_{A1} Z_c$$
$$= (K_1 + K_2) I_{A1} Z_{11} + K_2 Z_{1s} I_{A1} \tag{B-136}$$
$$- K_1 Z_c I_{A1}$$

Using the general procedure for a magnitude comparator, we can show the locus of impedances Z_{11} that defines the impedance characteristics for this unit to be

$$Z_{11} = \frac{Z_c - Z_{1s}(1 - I_L/I_{F3ph})}{(2 - I_L/I_{F3ph})}$$
$$+ \frac{(Z_{s1} + Z_c)(1 - I_L/I_{F3ph})}{(2 - I_L/I_{F3ph})} e^{-jr} \tag{B-137}$$

Notice that the unit's reach is still fixed at Z_c and its center will move depending on the load flow. Its characteristic is not dependent on the distribution factors, but the prefault load flow. The characteristic is plotted in Figure 12B-39.

B.7.2   Quadrature-Polarized Ground Unit

Reference will be made to Figure 12B-40 that shows the sequence network connection for a phase-to-ground fault. Distribution factors for all the networks make the inputs to the phase comparator

$$S_1 = V_A - \left(I_A + K_0 I_{A0} \frac{Z_{0c} - Z_{1c}}{Z_{1c}}\right) Z_{1c}$$

and reduce to

$$\frac{S_1}{K_1 I_{A1}} = Z_{11}\left(2 + \frac{K_0}{K_1} ZR_1 - \frac{I_L}{I_{F3ph}}\right)$$
$$- Z_{1c}\left(2 + \frac{K_0}{K_1} ZR_c - \frac{I_L}{I_{F3ph}}\right) \tag{B-138}$$

Also, working on S_2, we get

$$S_2 = j(V_C - V_B)$$

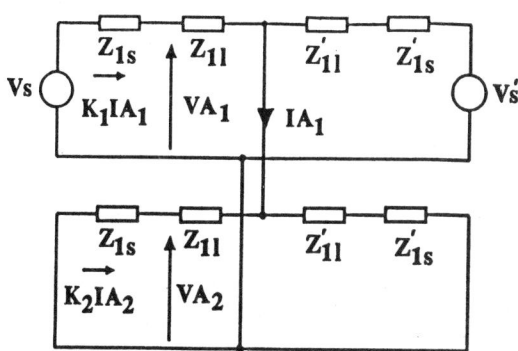

Figure 12B-38   General case of a phase-to-phase fault.

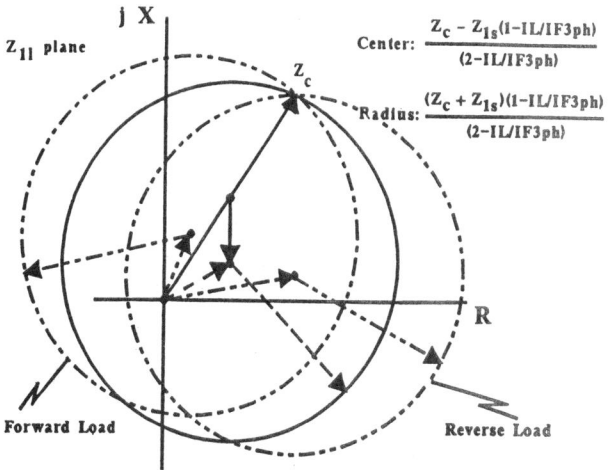

Figure 12B-39   Characteristic for a phase-to-phase fault.

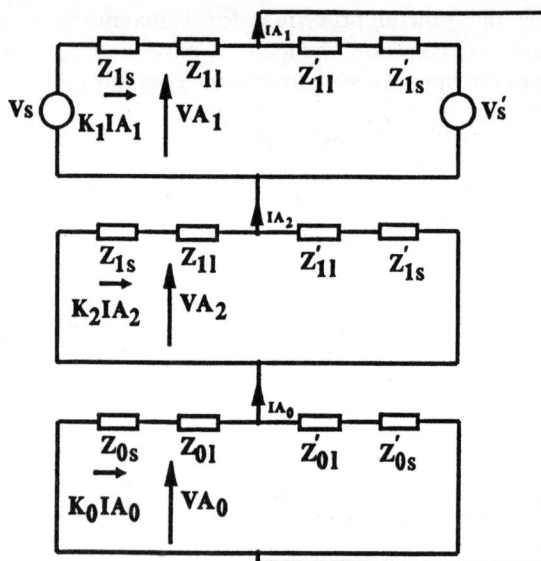

Figure 12B-40 Sequence connection for a general phase-to-ground fault.

that reduces to

$$\frac{S_2}{K_1 I_{A1}} = -\sqrt{3}\left[Z_{11}\left(2 + \frac{K_0}{K_1}ZR_1 - \frac{I_L}{I_{F3ph}}\right) \right.$$
$$\left. + Z_{1s}\left(2 + \frac{K_0}{K_1}ZR_s - 2\frac{I_L}{I_{F3ph}}\right)\right] \qquad (B\text{-}139)$$

If the settings of the unit are made such that $ZR_c = ZR_1$, the **a** and **b** reference vectors for a phase comparator are

$$\mathbf{a} = Z_{1c} \qquad (B\text{-}140)$$

$$\mathbf{b} = -Z_{1s}\left[\frac{\left(2 + \frac{K_0}{K_1}ZR_s - 2\frac{I_L}{I_{F3ph}}\right)}{\left(2 + \frac{K_0}{K_1}ZR_1 - \frac{I_L}{I_{F3ph}}\right)} \right] \qquad (B\text{-}141)$$

The characteristic is illustrated in Figure 12B-41 and shows that the reach of the unit Z_{1c} remains unchanged. It is dependent on the current distribution factors of the power system and the prefault load flow.

B.7.3 Self-Polarized Ground Unit

Reference will be made to Figure 12B-41 that shows the sequence network connection for a phase-to-ground fault. Distribution factors for all the networks make the inputs to the phase comparator

$$S_1 = V_A - \left(I_A + K_0 I_{A0}\frac{Z_{0cF} - Z_{1cF}}{Z_{1cF}} \right)Z_{1cF}$$

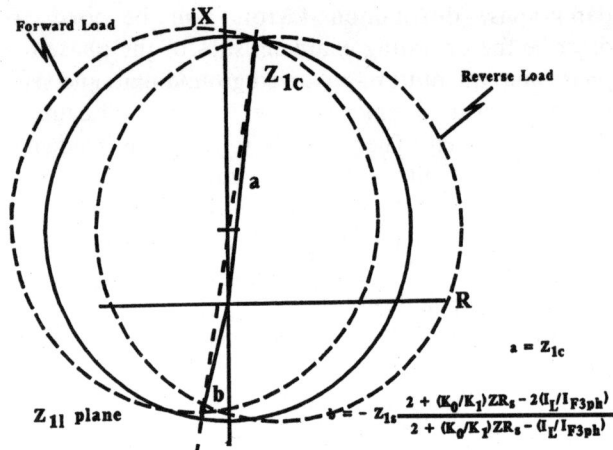

Figure 12B-41 Characteristic for phase-to-ground faults.

and reduce to

$$\frac{S_1}{K_1 I_{A1}} = Z_{11}\left(2 + \frac{K_0}{K_1}ZR_1 - \frac{I_L}{I_{F3ph}}\right)$$
$$- Z_{1cF}\left(2 + \frac{K_0}{K_1}ZR_c - \frac{I_L}{I_{F3ph}}\right) \qquad (B\text{-}142)$$

Also,

$$-S_2 = V_A + \left(I_A + K_0 I_{A0}\frac{Z_{0cR} - Z_{1cR}}{Z_{1cR}} \right)Z_{1cR}$$

and it reduces to

$$\frac{S_2}{K_1 I_{A1}} = -Z_{11}\left(2 + \frac{K_0}{K_1}ZR_1 - \frac{I_L}{I_{F3ph}}\right)$$
$$- Z_{1cR}\left(2 + \frac{K_0}{K_1}ZR_c - \frac{I_L}{I_{F3ph}}\right) \qquad (B\text{-}143)$$

If the settings of the unit are made such that $ZR_{cF} = ZR_{cR} = ZR_1$, the **a** and **b** reference vectors for a phase comparator are

$$\mathbf{a} = Z_{1cF} \qquad (B\text{-}144)$$
$$\mathbf{b} = -Z_{1cR} \qquad (B\text{-}145)$$

The characteristic is the same as that in Figure 12B-18 and does not depend on load flow, or distribution factors. This is true, in general, for units polarized with the same faulted phase; the characteristic in the R-X diagram is static and does not depend on any parameter but the settings of the unit.

B.8   Derived Characteristics

The discussion above showed the different characteristics that can be obtained with either a phase or magnitude comparator. The characteristics of the units derived are smooth and simple.

Designs have been made, however, in which different characteristics, many comparators, have been used to achieve characteristics that are not necessarily smooth. This is done for many purposes, like load restriction for heavily loaded long lines, and to try to accommodate more fault resistance, although actual apparent impedance is not simply a resistance component.

It is not the scope of this section to discuss the advantages and disadvantages of impedance characteristics. Therefore, a simple description of how these characteristics are obtained will be given through examples.

Refer to Figure 12B-42a. It shows an application with mixed reactance and blinder units. To derive this characteristic, four units are required. The reactance unit X determines the X reach, the resistance unit R determines the R reach, and blinders A and B provide the necessary directionality of the unit. For the relay to operate, the four units should pick up.

In Figure 12B-42c, two comparators are used to achieve the characteristic. In this case, any of the units, offset mho or lens, will provide the output of the entire distance unit. This characteristic has been used in heavily loaded lines for the starting impedance unit.

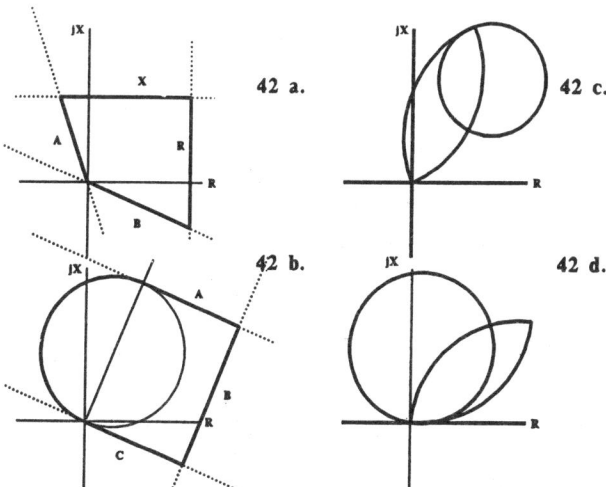

Figure 12B-42   Composite impedance characteristics.

Figure 12B-42b illustrates an intentional increase of the resistance reach for a mho unit. In this case, only the mho unit operates for the $+90°$ coincidence requirement, and the output is left to the operation of all three blinders for the $-90°$ coincidence requirement.

Figure 12B-42d illustrates the use of a mho and lens unit to increase the resistance coverage of the mho element. The angle of Z_c in both units is different and the output taken from the operation of any of the two comparators.

B.9   Apparent Impedance

The term *apparent impedance* is used very commonly to describe the resultant impedance when the voltages and currents entering the relay are used to calculate the impedance to the fault. This expression is confusing if one confuses it with the "impedance characteristics" that have been described in the previous paragraphs.

In this section, we define apparent impedance as the loop impedance using the voltages and currents that the relay receives. In this sense, the following are the apparent impedances for different fault loops. For a three-phase fault,

$$Z_{app} = \frac{V_A}{I_A} \qquad (B\text{-}146)$$

For a phase-to-phase fault,

$$Z_{app} = \frac{V_B - V_C}{I_B - I_C} \qquad (B\text{-}147)$$

For a phase-to-ground fault,

$$Z_{app} = \frac{V_A}{I_A + \frac{Z_{0c} - Z_{1c}}{Z_{1c}} I_{A0}} \qquad (B\text{-}148)$$

If the apparent impedance Z_{app} is within the operating characteristics, those studied above, then the unit will operate. Notice that the operating characteristics of the unit only depend on the setting of the unit and, in some cases, the source impedance.

For the phase loops, three-phase and phase-to-phase, mutual and fault resistance have little influence, and the actual apparent impedance Z_{app} is the same as the impedance to the fault.

The influence of mutuals and fault resistance is more severe in the phase-to-ground loop. Fault resistance and zero sequence mutual effects from other lines tend to make the apparent impedance Z_{app} different from the actual impedance to the fault. Taking into consideration these factors, we know that the apparent impedance to the fault with fault resistance and zero sequence mutual effects is

$$Z_{app} = Z_{1l} + \frac{3I_{A0}R_f + I_{0E}Z_{0m}}{I_A + \frac{Z_{0c} - Z_{1c}}{Z_{1c}}I_{A0}} \qquad (B\text{-}149)$$

where

R_f = fault resistance
I_{0E} = zero sequence current in the parallel line
Z_{0m} = zero sequence mutual between parallel lines

and the rest of the symbols are as described before.

It should be noticed that apparent impedance will change with fault resistance and mutual effects, and if Z_{app} enters the operating characteristics of the different ground units described, the relay may under- or overreach.

B.10 Summary

The purpose of this appendix is to illustrate the different properties of distance elements found in ABB relays. In microprocessor technology, several of these units can be found and the detection of faults in power systems is, in general, a combination of the operation of all the phase and ground units. In some designs, such as MDAR (REL-300) all the distance units are supervised by a phase selector and/or forward directional unit. This limits the interaction of the several units that MDAR has available. In conclusion, although distance elements have been designed with a definite type of fault detection, they will respond to other types of faults as well.

It is an objective of this appendix, as well, to clarify the meaning of forward and reverse. These concepts are totally opposite to each other. It is not proper to draw the forward-looking characteristics (Z_{1l}) plane on the reverse-looking characteristics (Z_{1l}'') plane. It is hoped that the reader can now make such a distinction. They are totally different concepts.

The influence of current distribution factors and load is simply an academic exercise since these factors change as the fault moves within the protected line. To find rigorously the operating characteristic with distribution factors and load flow would be an iterative process. However, the development illustrates the dynamic performance of the different units.

The term *apparent impedance* has also been defined. It corresponds to the fault-loop evaluation, and if the apparent impedance falls inside the operating characteristics of the unit, the unit will operate. Apparent impedance for ground faults is affected by fault resistance and zero sequence mutual effects.

APPENDIX C: INFEED EFFECT ON GROUND-DISTANCE RELAYS

Because the current of the reference and/or operating quantities for ground-distance relays are different from those for phase-distance relays, therefore, the descriptions and results provided in Section 4.5 for phase-distance relays on infeed effect cannot be directly used for ground-distance relay application. The general solution for infeed effect on ground-distance relays will be described as follows.

C.1 Infeed Effect on Type KDXG, LDAR, and MDAR Ground-Distance Relays

Refer to Figure 12C-1.

$$\begin{aligned}
-V_{A1} &+ 2I_{A1G}Z_{1L} \\
&+ 2(I_{A1G} + I_{A1H})Z_{1H} \\
&- V_{A0} + I_{A0G}Z_{0L} + (I_{A0G} \\
&+ I_{A0H})Z_{0H} - V_{A2} = 0 \qquad (C\text{-}1)
\end{aligned}$$

$$V_{AG} = V_{A1} + V_{A2} + V_{A0}$$
$$\text{(voltage at relay location)} \qquad (C\text{-}2)$$

From Eqs. (C-1) and (C-2), we get

$$\begin{aligned}
V_{AG} &= Z_{1L}\left[2I_{A1G} + I_{A0}\left(\frac{Z_{0L}}{Z_{1L}}\right)\right] \\
&+ Z_{1H}\left[2(I_{A1G} + I_{A1H})\right. \\
&+ \left.(I_{A0G} + I_{A0H})\left(\frac{Z_{0L}}{Z_{1L}}\right)\right] \\
&= Z_{1L}\left[I_G + \left(\frac{Z_{0L} - Z_{1L}}{Z_{1L}}\right)I_{A0G}\right] \\
&+ Z_{1H}\left[(I_G + I_H)\right. \\
&+ \left.(I_{A0G} + I_{A0H})\left(\frac{Z_{0L} - Z_{1L}}{Z_{1L}}\right)\right] \qquad (C\text{-}3)
\end{aligned}$$

$$I_R = I_G + \left(\frac{Z_{0L} - Z_{1L}}{Z_{1L}}\right)I_{A0G} \qquad (C\text{-}4)$$

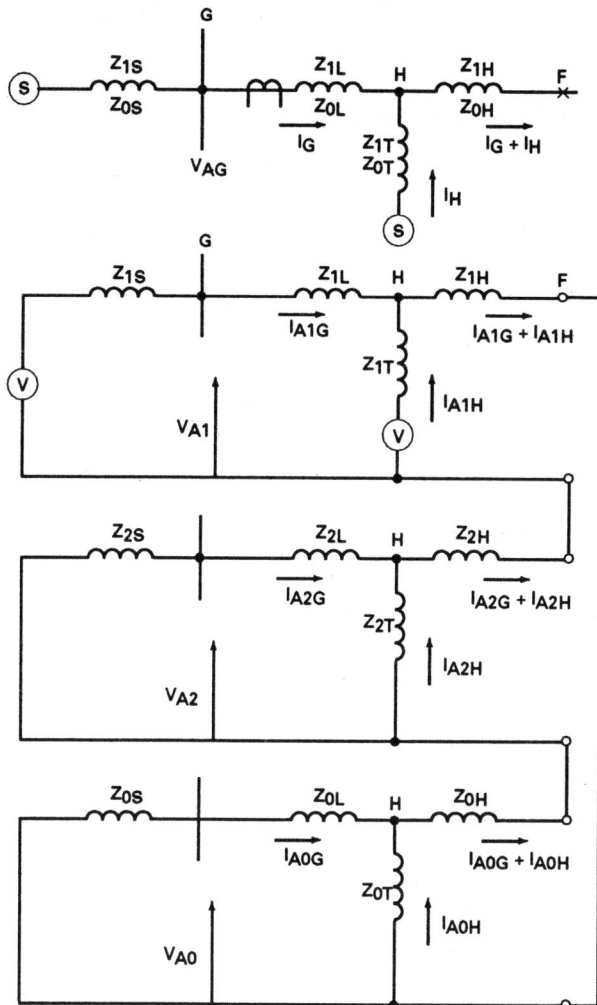

Figure 12C-1   Infeed effect on ground distance relays.

C.2   Infeed Effect on Type SDG and LDG Ground-Distance Relays

Refer again to Figure C-1.

$$-V_{A1} + 2I_{A1G}Z_{1L} + 2(I_{A1G} + I_{A1H})Z_{1H}$$
$$- V_{A0} + I_{A0G}Z_{0L}$$
$$+ (I_{A0G} + I_{A0H})Z_{0H} - V_{A2} = 0 \qquad (C\text{-}1)$$

$$V_{AG} = V_{A1} + V_{A2} + V_{A0}$$
$$\text{(voltage at relay location)} \qquad (C\text{-}2)$$

$$V_{AG} = Z_{1L}\left[2I_{A1G} + I_{A0}\left(\frac{Z_{0L}}{Z_{1L}}\right)\right]$$
$$+ Z_{1H}\left[2(I_{A1G} + I_{A1H})\right.$$
$$\left. + (I_{A0G} + I_{A0H})\left(\frac{Z_{0L}}{Z_{1L}}\right)\right]$$
$$= Z_{1L}\left[I_G + \left(\frac{Z_{0L} - Z_{1L}}{Z_{1L}}\right)I_{A0G}\right]$$
$$+ Z_{1H}\left[(I_G + I_H)\right.$$
$$\left. + (I_{A0G} + I_{A0H})\left(\frac{Z_{0L} - Z_{1L}}{Z_{1L}}\right)\right] \qquad (C\text{-}3)$$

LDG balances when

$$V_{X1} + V_{X2} + V_{X0} = 0 \qquad (C\text{-}6)$$

where V_{X1}, V_{X2}, and V_{X0} are relay-compensated voltages, and

$$V_{X1} = V_{A1} - I_{A1G}Z_{1C}$$
$$V_{X2} = V_{A2} - I_{A1G}Z_{1C}$$
$$V_{X0} = V_{A0} - I_{A0G}Z_{0C} = V_{A0} - I_{A0G}pZ_{1C},$$
$$p = \frac{Z_{0C}}{Z_{1C}} = \frac{Z_{0L}}{Z_{1L}}$$

or

$$V_{A1} = V_{X1} + I_{A1G}Z_{1C}$$
$$V_{A2} = V_{X2} + I_{A1G}Z_{1C}$$
$$V_{A0} = V_{X0} + I_{A0G}pZ_{1C} \qquad (C\text{-}7)$$

From Eqs. (C-3), (C-6), and (C-7), we obtain

$$2V_{X1} + 2I_{A1G}Z_{1C} + V_{X0} + I_{A0G}pZ_{1C}$$
$$= Z_{1L}\left[I_G + I_{A0G}\left(\frac{Z_{0L} - Z_{1L}}{Z_{1L}}\right)\right]$$
$$+ Z_{1H}\left[(I_G + I_H) + (I_{A0G} + I_{A0H})\left(\frac{Z_{0L} - Z_{1L}}{Z_{1L}}\right)\right]$$

Equations (C-3) and (C-4) yield

$$Z_{APP} = \frac{V_{AG}}{I_R} = \frac{V_{AG}}{I_G + \left(\frac{Z_{0L} - Z_{1L}}{Z_{1L}}\right)I_{A0G}}$$

$$= \frac{Z_{1L}\left[I_G + \left(\frac{Z_{0L} - Z_{1L}}{Z_{1L}}\right)I_{A0G}\right] + Z_{1H}\left[(I_G + I_H) + (I_{A0G} + I_{A0H})\left(\frac{Z_{0L} - Z_{1L}}{Z_{1L}}\right)\right]}{I_G + \left(\frac{Z_{0L} - Z_{1L}}{Z_{1L}}\right)I_{A0G}}$$

$$= Z_{1L} + Z_{1H}\left[\frac{(I_G + I_H) + (I_{A0G} + I_{A0H})\left(\frac{Z_{0L} - Z_{1L}}{Z_{1L}}\right)}{I_G + \left(\frac{Z_{0L} - Z_{1L}}{Z_{1L}}\right)I_{A0G}}\right]$$

$$= Z_{1L} + Z_{1H}\left[1 + \frac{I_H + I_{A0H}\left(\frac{Z_{0L} - Z_{1L}}{Z_{1L}}\right)}{I_G + I_{A0G}\left(\frac{Z_{0L} - Z_{1L}}{Z_{1L}}\right)}\right]$$

$$= (Z_{1L} + Z_{1H}) + Z_{1H}\left[\frac{I_H + I_{A0H}\left(\frac{Z_{0L} - Z_{1L}}{Z_{1L}}\right)}{I_G + I_{A0G}\left(\frac{Z_{0L} - Z_{1L}}{Z_{1L}}\right)}\right] \qquad (C\text{-}5)$$

$$= Z_{1L}\left[I_G + I_{A0G}\left(\frac{Z_{0L}-Z_{1L}}{Z_{1L}}\right)\right]$$
$$+ Z_{1H}\left[(I_G + I_H) + (I_{A0G}+I_{A0H})\left(\frac{Z_{0L}-Z_{1L}}{Z_{1L}}\right)\right]$$

$$Z_{1C}\left[I_G + I_{A0G}\left(\frac{Z_{0L}-Z_{1L}}{Z_{1L}}\right)\right]$$
$$= Z_{1L}\left[I_G + I_{A0G}\left(\frac{Z_{0L}-Z_{1L}}{Z_{1L}}\right)\right]$$
$$+ Z_{1H}\left[(I_G + I_H) + (I_{A0G}+I_{A0H})\left(\frac{Z_{0L}-Z_{1L}}{Z_{1L}}\right)\right]$$

$$Z_{1C} = Z_{1L} + Z_{1H}\left[\frac{(I_G+I_H)+(I_{A0G}+I_{A0H})\left(\frac{Z_{0L}-Z_{1L}}{Z_{1L}}\right)}{I_G+I_{A0G}\left(\frac{Z_{0L}-Z_{1L}}{Z_{1L}}\right)}\right]$$

$$Z_{1C} = Z_{1L} + Z_{1H}\left[\frac{I_G+I_{A0G}\left(\frac{Z_{0L}-Z_{1L}}{Z_{1L}}\right)+I_H+I_{A0H}\left(\frac{Z_{0L}-Z_{1L}}{Z_{1L}}\right)}{I_G+I_{A0G}\left(\frac{Z_{0L}-Z_{1L}}{Z_{1L}}\right)}\right]$$

$$Z_{1C} = (Z_{1L}+Z_{1H})$$
$$+ Z_{1H}\left[\frac{I_H+I_{A0H}\left(\frac{Z_{0L}-Z_{1L}}{Z_{1L}}\right)}{I_G+I_{A0G}\left(\frac{Z_{0L}-Z_{1L}}{Z_{1L}}\right)}\right] \tag{C-8}$$

It is important to note that Eqs. (C-5) and (C-8) are the same.

APPENDIX D: COORDINATION IN MULTIPLE-LOOP SYSTEMS

Each relay in a multiple-loop system is coordinated as described in Section 2.2.2. Each loop must be coordinated within itself and with adjacent relays in other loops. This technique is illustrated through the following example.

D.1 System Information

A typical 23-kV loop system has three-phase bus fault currents as shown in Figure 12D-1. The fault locations are marked with an X. Each fault is indicated with a circled number, followed by the total maximum and minimum (with bracket) fault current in amperes at 23 kV. The fault currents that are contributed from each circuit are also shown.

The forward/reverse maximum and minimum currents through the breakers and relays can be calculated from Figure 12D-1 and are shown in Figure 12D-2. Load currents for the 5- and 10-MVA transformers are 125.5 and 251.0 A, respectively. This means that the maximum load current flow on each line section could be 125.5 A in either the forward or reverse direction. Therefore, all overcurrent units should be set higher

than $(2 \times 125.5) = 250$ primary A, or 6.25 secondary A for a 200/5 ct application.

D.2 Relay Type Selection

D.2.1 Selection of Directional Time-Overcurrent Relay

Refer to step 1 in Section 3.6.2. Table 12D-1 is prepared for determining the need of directional time-overcurrent units. It shows that either a directional or nondirectional relay could be applied for D and E. If, however, the relay must provide backup protection beyond the remote bus, a directional relay should be used. Using a directional relay would also mean avoiding future problems caused by system changes.

D.2.2 Application of Instantaneous-Trip Relay

Follow step 2 in Section 3.6.2. Table 12D-2 is prepared for determining the possibility of instantaneous-trip

Table 12D-1 Selection of Directional Time-Overcurrent Relay

Breaker location	I_{RM}	I_{RL}	4 (I_{RM} or I_{RL})	$\geqq I_{FN}$	Directional required
A	250	125	1000	750	Yes
B	1000	125	4000	210	Yes
C	1100	125	4400	180	Yes
D	200	125	800	800	Yes/no
E	200	125	800	800	Yes/no
F	1100	125	4400	180	Yes
G	1000	125	4000	210	Yes
H	250	125	1000	750	Yes

I_{RM} = contributed maximum reverse fault current.
I_{RL} = contributed maximum reverse load current.
I_{FN} = contributed minimum forward bus fault current.

Table 12D-2 Application of Instantaneous-Trip Relay

Breaker location	I_{CM}	$\geqq 2.5\ I_{FM}$	I_{FM}	Instantaneous-trip unit applicable
A	2750	2500	1000	Yes
B	1100	500	250	Yes
C	1000	500	200	Yes
D	3200	2750	1100	Yes
E	3200	2750	1100	Yes
F	1000	500	200	Yes
G	1100	625	250	Yes
H	2750	2500	1000	Yes

I_{CM} = contributed maximum close-in fault current.
I_{FM} = contributed maximum forward bus fault current.

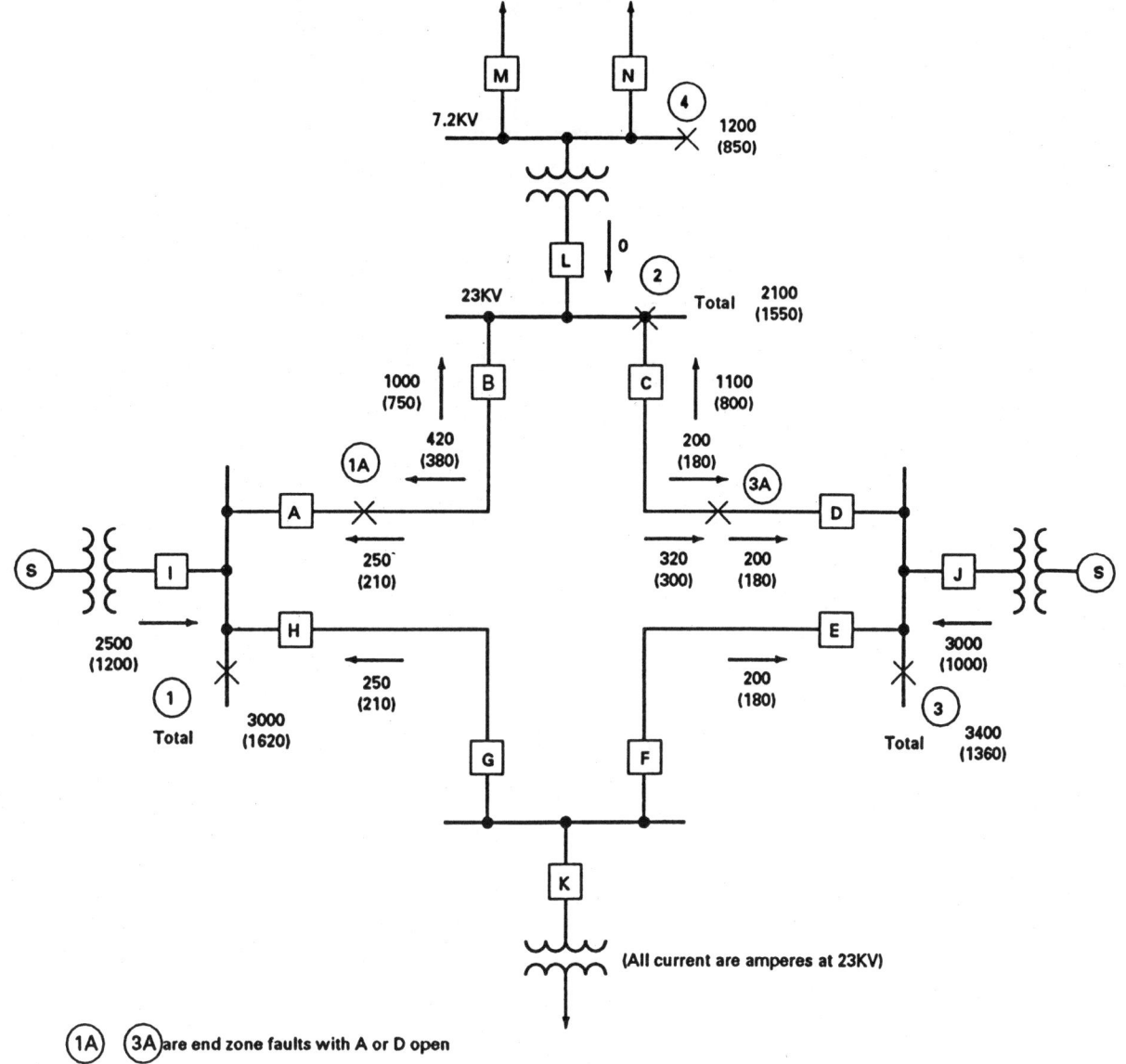

Figure 12D-1 Example of relay application and settings for loop system.

unit application. It indicates that the instantaneous-trip unit can be set to underreach if applied to all these breakers.

D.2.3 Requires Directional Instantaneous-Trip Relay

Follow step 3 of Section 3.6.2. Table 12D-3 is prepared for determining the need of directionally controlled instantaneous-trip overcurrent units.

D.2.4 Summary of Type of Relay Selection

Table 12D-4 is a summary of relay selection for each breaker.

D.3 Relay Setting and Coordination

D.3.1 Coordination Paths

As suggested in Section 5.2.1, a good starting point for the coordination process is at the largest sources.

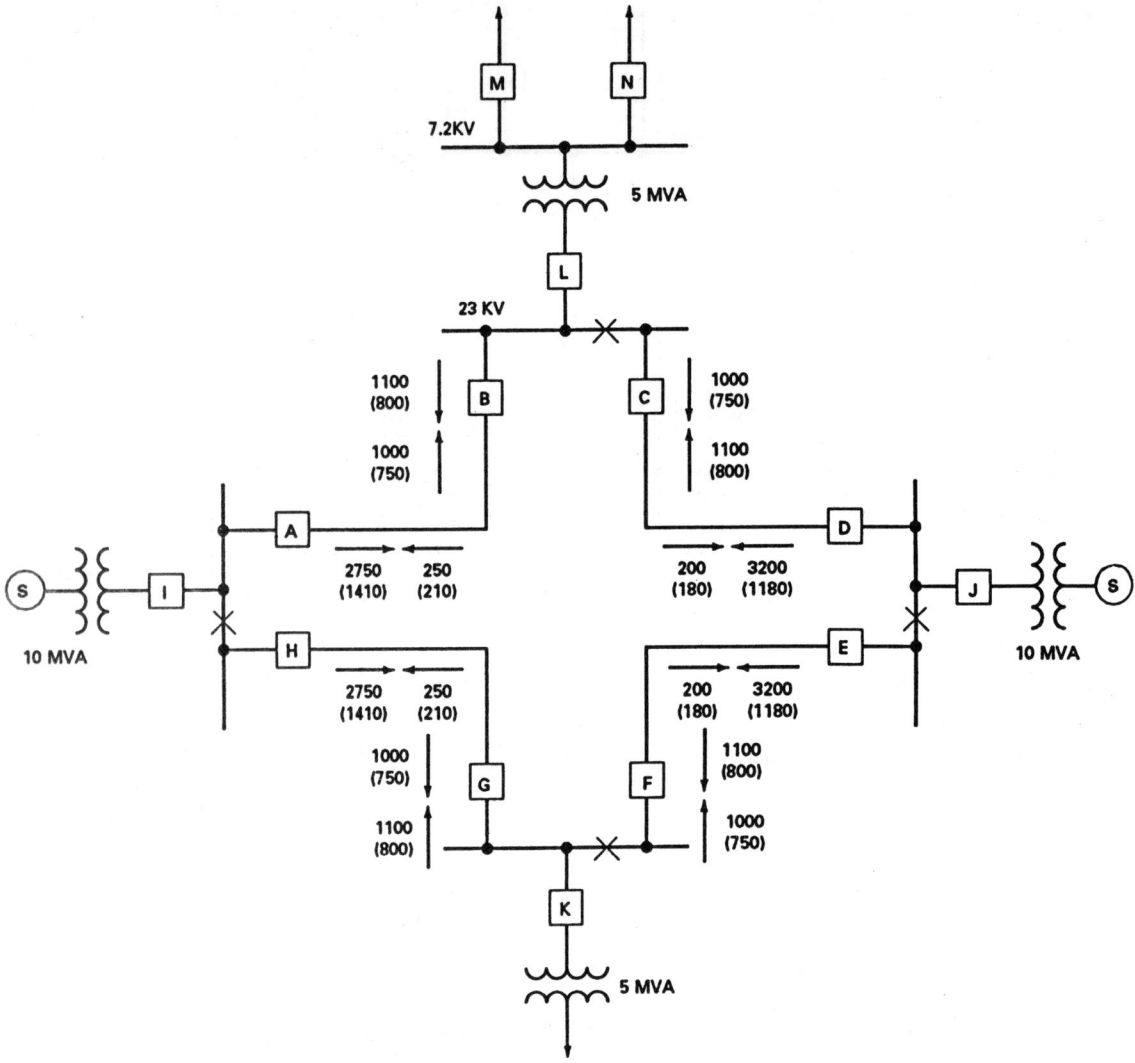

Figure 12D-2   Maximum and minimum fault currents through each breaker.

Table 12D-3   Requiring Directional Instantaneous-Trip Relay

Breaker location	I_{RM}	$\geqq I_{FM}$	Required directionally controlled IT unit
A	250	1000	No
B	1000	250	Yes
C	1100	200	Yes
D	200	1100	No
E	200	1100	No
F	1100	200	Yes
G	1000	250	Yes
H	250	1000	No

I_{RM} = contributed maximum reverse fault current.
I_{FM} = contributed maximum forward bus fault current.

Therefore, the coordination paths for the system as shown in Figure 12D-1 can be attempted as below:

1. Coordinate B with H and I; D with B and L; F with D and J; and H with K and F.
2. Coordinate G with A and I; E with G and K; C with E and J; and A with C and L.

The first step is to define the protections on I and J, set relays L and K based on the known data from relays M and N, and then make coordinations for loops a and b in turn.

D.3.2   Protections on I and J

Assume that locations I and J provide transformer (either HU or CA-26 relays) and bus (either KAB or

Table 12D-4 Type of Relay Selection

Breaker location	Time-overcurrent unit needing directional control	Instantaneous-trip unit can be set to underreach	Instantaneous-trip unit requiring directional control	Type of relay selected
A	Yes	Yes	No	CR with IIT
B	Yes	Yes	Yes	IRV
C	Yes	Yes	Yes	IRV
D	Yes	Yes	No	CR with IIT
E	Yes	Yes	No	CR with IIT
F	Yes	Yes	Yes	IRV
G	Yes	Yes	Yes	IRV
H	Yes	Yes	No	CR with IIT

CA-16 relays) differential protections. Their operations should therefore be high-speed.

D.3.3 Protections on M and N

Assume that the phase relays M and N are type HiLo CO-8 (refer to Fig. 12D-3 for time curves), with 1- to 12-A taps, and with a 6- to 144-A IIT unit. For preventing operation on cold-load inrush, tap 10 and a time dial of 2 have been predetermined.

In this case, the equivalent fault current at 23 kV for the time-overcurrent unit at a 10-A setting would be

$$10 \times \frac{200}{5} \times \frac{7.2}{23} = 125A \quad \text{(at 23 kV)}$$

Then the following data can be obtained from the time curve in Figure 12D-3 for a time dial of 2:

Multiples of tap setting	Primary current at 23 kV	Time to close contacts (sec)
1.0	125	
2.0	250	4.00
2.5	312	2.38
3.0	375	1.65
4.0	500	1.05
5.0	625	0.78
7.0	875	0.58
10.0	1250	0.46
16.0	2000	0.36

Also, assume that the IIT unit setting of 18 A has been selected after analysis of the 7.2-kV circuits. The equivalent fault current at 23 kV for a nondirectional

IIT unit at an 18-A setting would be

$$18 \times \frac{200}{5} \times \frac{7.2}{23} = 225A \quad \text{(at 23 kV)}$$

Plot this information in Figure 12D-4 as curves M and IIT. They will be used for coordination with relays L.

D.3.4 Setting for Relay L (and K) to Coordinate with Relays M and N

Since these relays energize a transformer bank, its minimum pickup setting should be approximately twice the full load, and it should be coordinated with relay M based on the maximum through-fault current.

The recommended ct ratio for relay L is 200% of the transformer self-rating, which is $2 \times 125 = 250$, i.e., select a ct ratio of 250/5. Therefore, the time-overcurrent unit should be set above

$$2 \times 125 \frac{5}{250} = 5 A$$

Tap 5 is selected. The maximum through fault current is

$$1200 \left(\frac{5}{250} \right) = 24 A$$

which is $24/5 = 4.8$ times the tap setting. The operating time of N at 1200 A, from Figure 12D-4, is 0.47 sec. Then set relay L for 0.47 plus 0.3 (one step of CTI), or 0.77 sec. Using the curve from Figure 12D-3 for a 4.8 multiple and 0.77 sec, we select time dial 2, which gives a time of 0.83 sec.

Alternatively, the multiple and time requirements can be specified and the time dial set by actual current in the relay. This latter method would yield a time dial of approximately 1.75, instead of 2. Using tap 5 (250 A primary) and time dial 2, the curve points for the time-overcurrent unit are as follows:

Multiples of tap setting	Primary current at 23 kV	Time to close contacts (sec)
1.0	250	
2.0	500	4.00
2.5	625	2.38
3.0	750	1.65
4.0	1000	1.05
5.0	1250	0.78
7.0	1750	0.58
10.0	2500	0.46

Plot these points in Figure 12D-5 as curve L and copy curve M from Figure 12D-4 to illustrate the setting of relay L to coordinate with relays M and N.

The IIT instantaneous-trip unit for relay L should be set at the highest of the following: (1) about four to six times the transformer self-rating; (2) higher than the inrush current; or (3) 1.1 to 1.3 times the maximum through fault. Condition (3) would be the highest for this example. For example, use a factor of 1.3 for the setting calculation; then $1.3 \times 1200 = 1560$ A (at 23 kV), or $1560/50 = 31.2$ A (relay side). Set the IIT unit at 32 A (or $32 \times 50 = 1600$ A at 23 kV). Relay K (on the bottom line of Figure 12D-2) is assumed to have the same setting as L.

D.3.5 Setting for Relay B to Coordinate with Relays H and I

Based on the assumption in Section D.3.2 that location I provides transformer and bus differential protection, their operation should be high-speed. However, the setting for H is unknown and cannot be determined until all other relays in the loop are set and coordinated.

However, since the relay H is at the end of the B, D, F, H loop, at least it will have 4 (CTI) = 1.2 sec for the setting. Consider the benefit of IIT application; at this point, 1 sec for line faults out of H can be tried. The critical fault for B is (1) maximum, giving 250 A primary.

The maximum forward-direction load flow current will be 125.6 A if breakers I and F are opened. The overcurrent unit at B should be set to 251.2 A for this load current. However, the maximum end-zone fault current is only 250 A. Therefore, from the point of view of protection, operating the system with both breakers I and F opened at the same time is not allowed. Based on this conclusion, it can be assumed that load will not flow through B to A, and since the

relay is directional, then there is no need to set the overcurrent unit at twice the maximum load current. It can be set based on continuous rating. The maximum continuous load is $125.6/40 = 3.14$ A (relay side). The continuous rating of any HiLo CO tap is higher than this value. Therefore, choosing a 1-A tap should be no problem. Tap 1 corresponds to 1×40, or 40 A primary. The critical fault multiple then is $250/40 = 6.26$, at which value the operating time (CTI plus 1) should be 0.3 plus 1, or 1.3 sec. From Figure 12D-3, for a 6.26 multiple and 1.3 sec, the time dial is approximately 4. A time dial of 1 may be used, or the value determined by a test. Using tap 1 (40 A primary) and a time dial of 4, we can arrive at the curve points for Figure 12D-6:

Multiples of tap setting	Primary current at 23 kV	Time to close contacts (sec)
1.0	40	
3.0	120	3.40
4.0	160	2.10
5.0	200	1.58
10.0	400	0.95
14.0	560	0.80
20.0	800	0.70

The instantaneous-trip unit of IRV-8 is a cylinder element. It can be set for 1.10×250, or 275 A primary (6.88 A secondary). A setting of 7 or 280 A primary will provide high-speed, sequential tripping. After A opens, the fault current increases from 420 to 380 A, which is above the instantaneous unit pickup of 280 A.

D.3.6 Setting for Relay D to Coordinate with Relays B and L

Since relay L operates instantaneously for the maximum fault (2), the critical fault is (2) minimum. Critical fault current is 1550 A for relay L and 800 A for relay D. Relay D is set for a minimum pickup of 2×125, or 250 A primary (6.25 A secondary). Tap 7 is selected to give a minimum pickup of 7×40, or 280 A. The critical fault multiple is 800/280, or 2.85 for relay D. The operating time of relay L for the critical fault value of 1550 A is 0.65 sec at the $1550/280 = 5.52$ multiple. The set point is 0.65 plus 0.30, or 0.95 sec at 800 A. From Figure 12D-3, for a 2.85 multiple and 0.95 sec, the time dial is approximately 1.10. The 1.25 value may be used, or the time dial determined by a

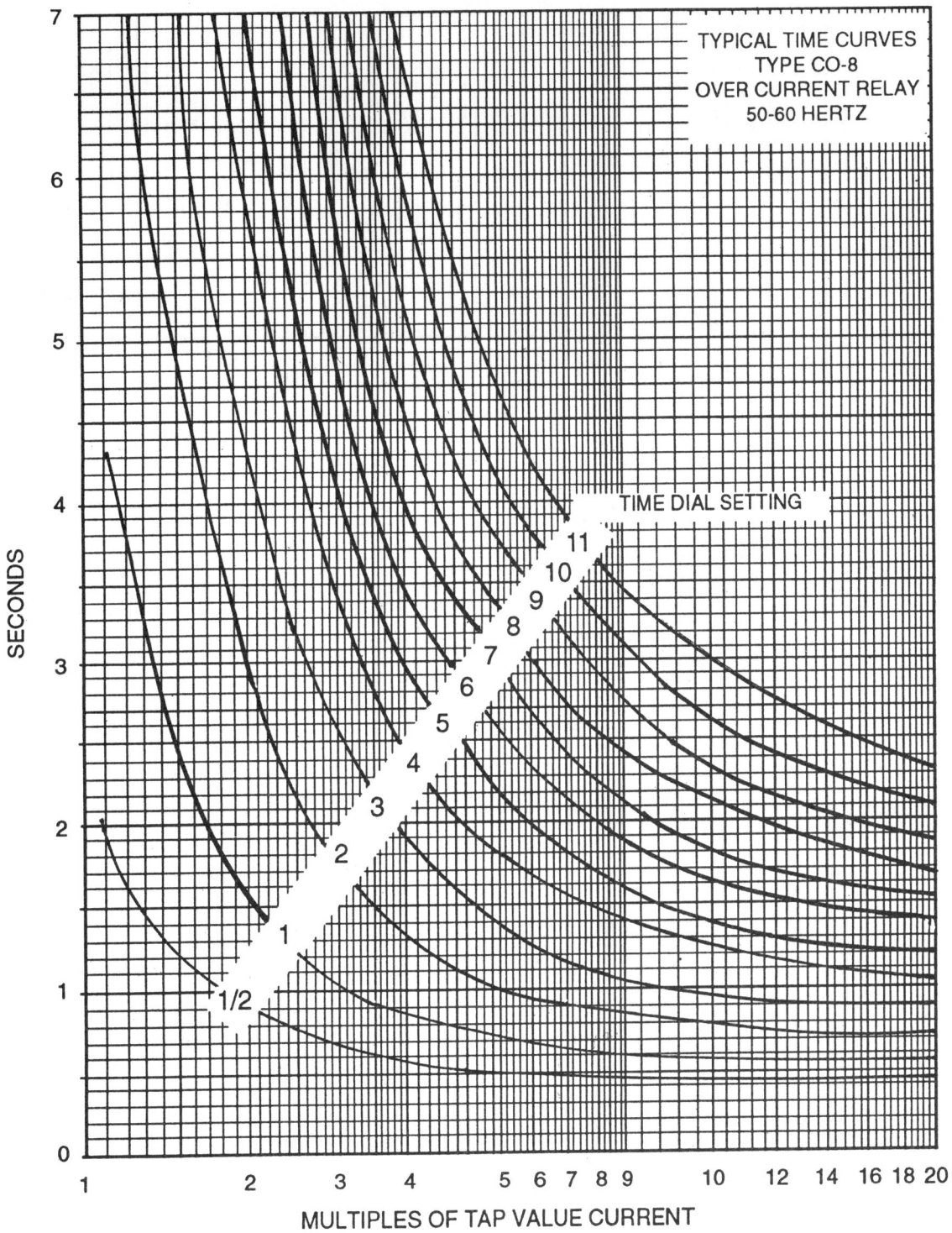

Figure 12D-3 Typical time curve of the type CO-8 or IRV-8 relay.

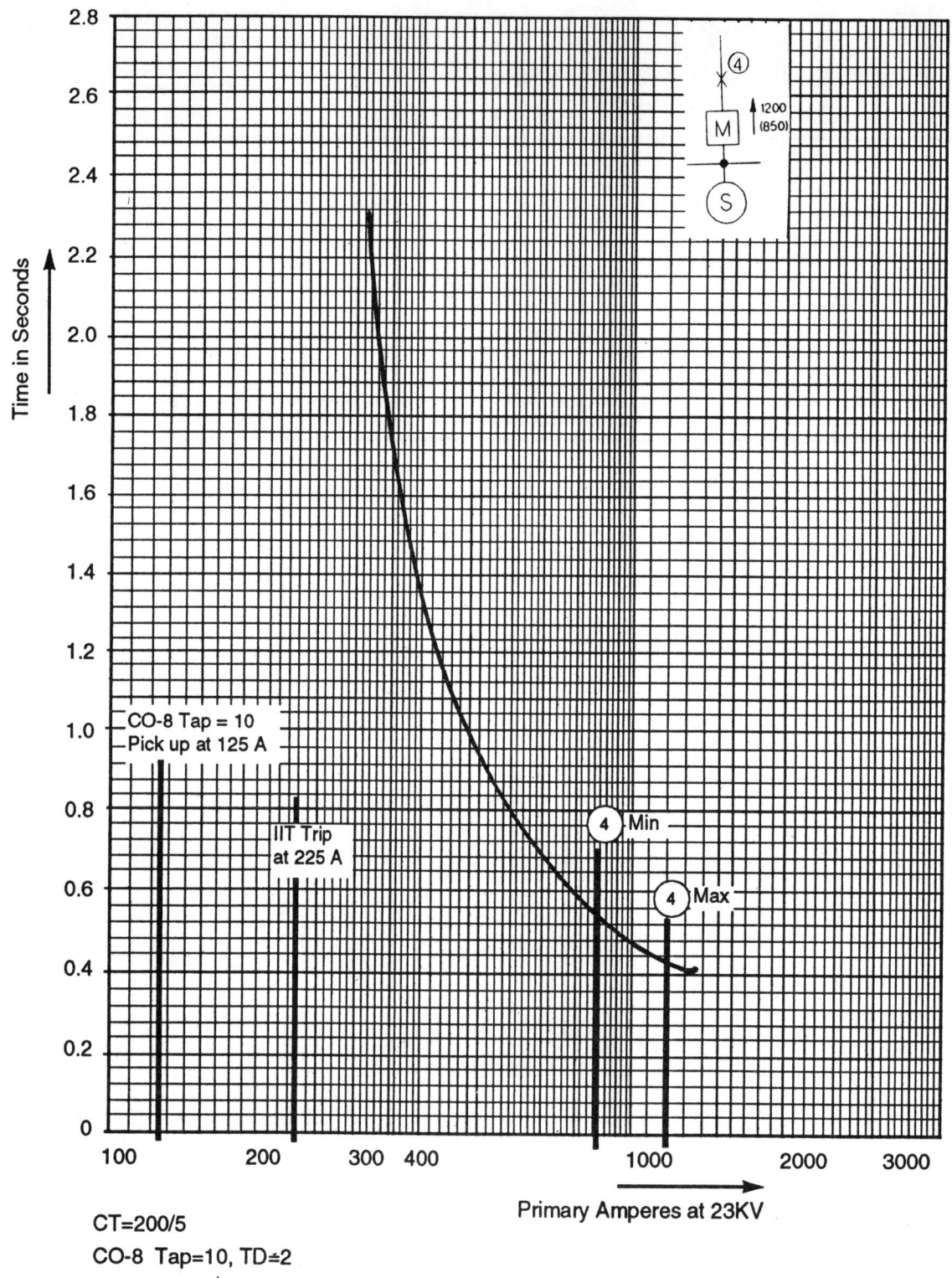

CT=200/5

CO-8 Tap=10, TD≐2

Pick up at 10(200/5) (7.2/23)=125 A

IIT Pick up at 10(200/5) (7.2/23)=125 A

Figure 12D-4   Relay M settings.

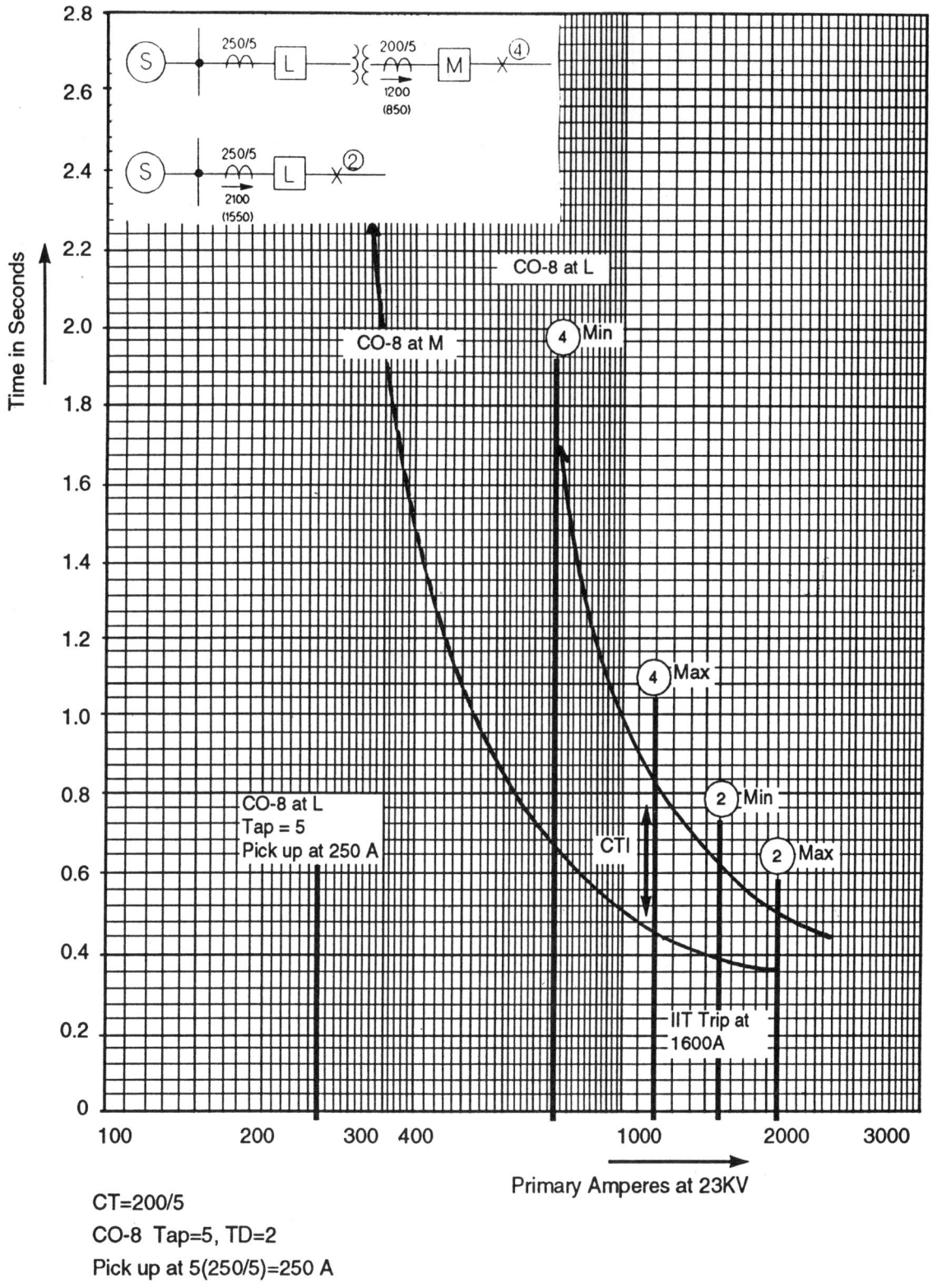

Figure 12D-5   Relay L settings.

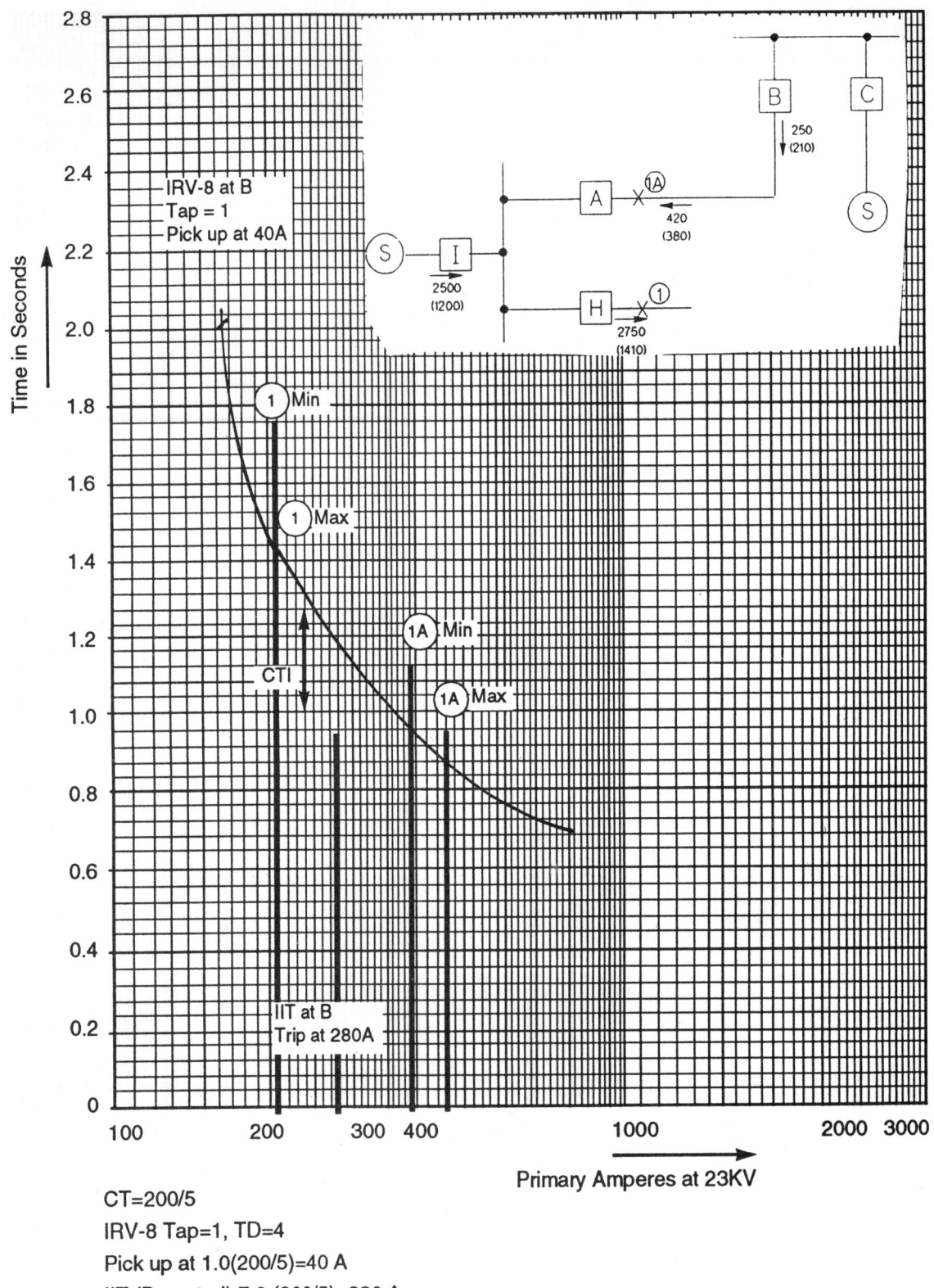

Figure 12D-6 Relay B settings.

CT=200/5

IRV-8 Tap=1, TD=4

Pick up at 1.0(200/5)=40 A

IIT (D-control) 7.0 (200/5)=280 A

test. Using tap 7 (280 A primary) and a time dial of 1.25, we can find the curve points for Figure 12D-7:

Multiples of tap setting	Primary current at 23 kV	Time to close contacts (sec)
1.0	280	
2.0	560	2.50
2.5	700	1.55
3.0	840	1.00
4.0	1120	0.600
5.0	1400	0.45
7.0	1860	0.35

Finally, the IIT of CR-8 is set to 1.3 × 1100, or 1430 A (36 A secondary).

D.3.7 Setting for Relay F to Coordinate with Relays D and J

Based on the assumption in Section E.3.2 that location J provides transformer and bus differential protection, their operations should be high-speed.

The maximum forward-direction load flow current will be 125.6 A if breakers J and B are opened. The overcurrent unit at F should be set to 251.2 A for this load current. However, the maximum end-zone fault current is only 200 A. Therefore, from the point of view of protection, operating the system on both breakers J and B opened at the same time is not allowed. Based on this conclusion, it can be assumed that load will not flow through F to E, and since the relay is directional, then there is no need to set the overcurrent unit at twice the maximum load current. It can be set based on a continuous rating. The maximum continuous load is $125.6/40 = 3.14$ A (relay side). The continuous rating of any HiLo CO tap is higher than this value. Therefore, choosing a 1-A tap should be no problem. Since the IIT unit of the D relay will operate for fault (3) maximum, the critical fault for relay F is fault (3) minimum. The overcurrent unit of the D relay will operate in 0.55 sec (1180/ $280 = 4.22$ multiple).

For relay F, tap 1 corresponds to 1 × 40, or 40 A primary. The critical fault multiple is 180/40 or 4.5, at which value the operating time should be 0.55 plus 0.3, or 0.85 sec. From Figure 12D-3, for a 4.5 multiple and 0.85 sec, the time dial is approximately 2. Using tap 1 (40 A primary) and a time dial of 2, we can find the curve points for Figure 12D-8:

Multiples of tap setting	Primary current at 23 kV	Time to close contacts (sec)
1.0	40	
2.5	100	2.38
3.0	120	1.65
4.0	160	1.05
5.0	200	0.78
7.0	280	0.58
10.0	400	0.46
16.0	640	0.36

The instantaneous-trip unit of the IRV-8 is a cylinder element. It can be set for 1.10 × 200, or 220 A primary (5.5 A secondary). A 6-A setting or 240 A primary current will provide high-speed, sequential tripping. After E opens with a fault near E on line FE, the fault current increases at F to 300 A which is above the instantaneous unit pickup of 220 A.

D.3.8 Setting for Relay H to Coordinate with Relays K and F

Since relay K operates instantaneously for the maximum fault (2), the critical fault is (2) minimum. Critical fault current is 1550 A for relay K and 750 A for relay H. Relay H is set for a minimum pickup of 2 × 125, or 250 A primary (6.25 A secondary) load current. Tap 7 is selected to give a minimum pickup of 7 × 40, or 280 A.

The critical fault multiple is 750/280, or 2.68 for relay H. The operating time of relay K for the critical fault value of 1550 A is 0.70 sec (1550/280, or a 5.52 multiple). The set point for relay H should be 0.70 plus 0.30, or 1 sec at 750 A. From Figure 12D-3, for a 2.68 multiple and 1 sec, the time dial is 1. Using tap 7 (280 A primary) and a time dial of 1, we can find the curve points for Figure 12D-9:

Multiples of tap setting	Primary current at 23 kV	Time to close contacts (sec)
2.0	560	2.1
2.5	700	1.2
3.0	840	0.8
4.0	1120	0.5
5.0	1400	0.38
7.0	1960	0.30

The instantaneous-trip unit of the IRV-8 is a cylinder element. It can be set for 1.10 × 1000, or

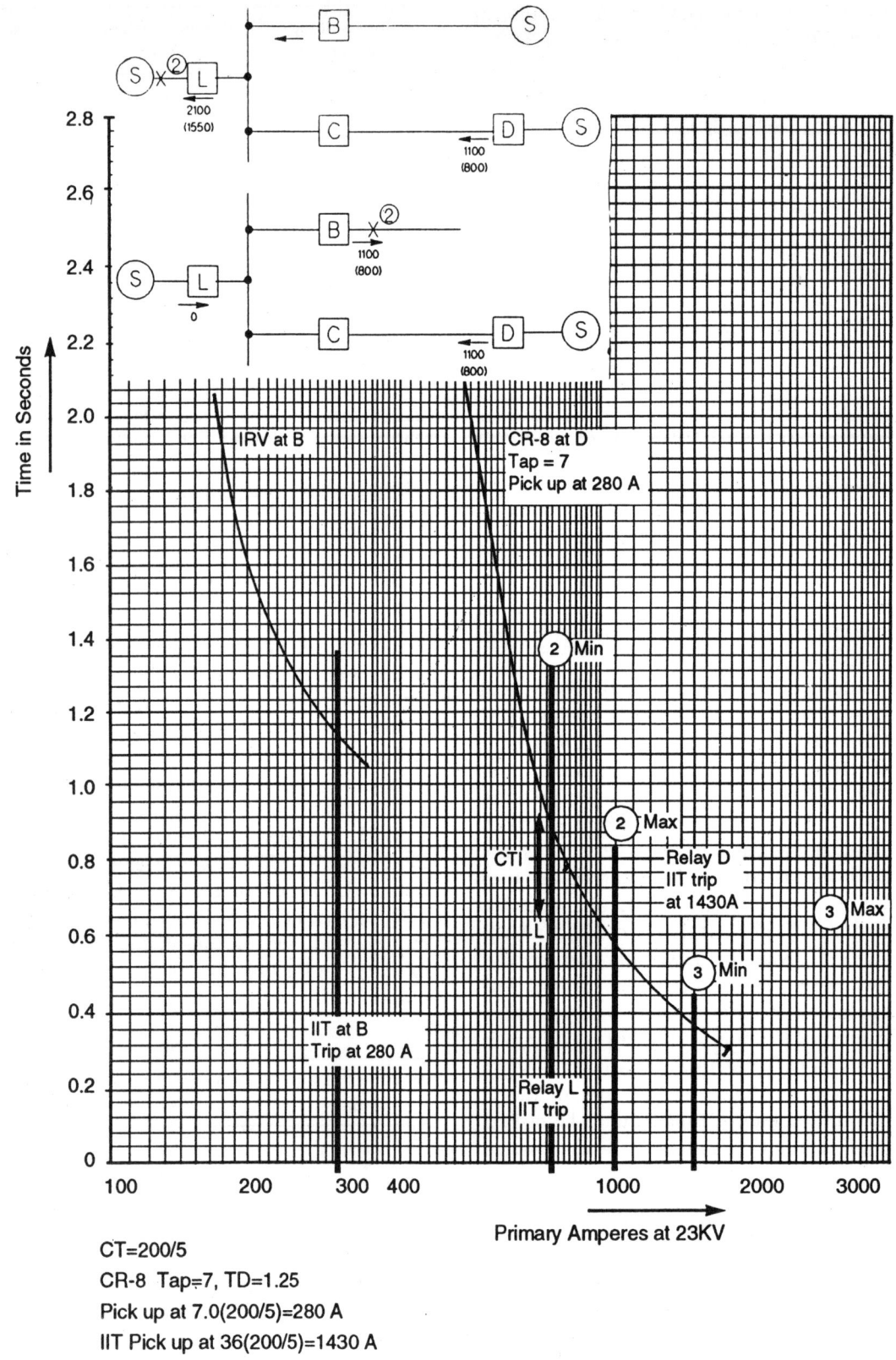

CT=200/5

CR-8 Tap=7, TD=1.25

Pick up at 7.0(200/5)=280 A

IIT Pick up at 36(200/5)=1430 A

Figure 12D-7   Relay D settings.

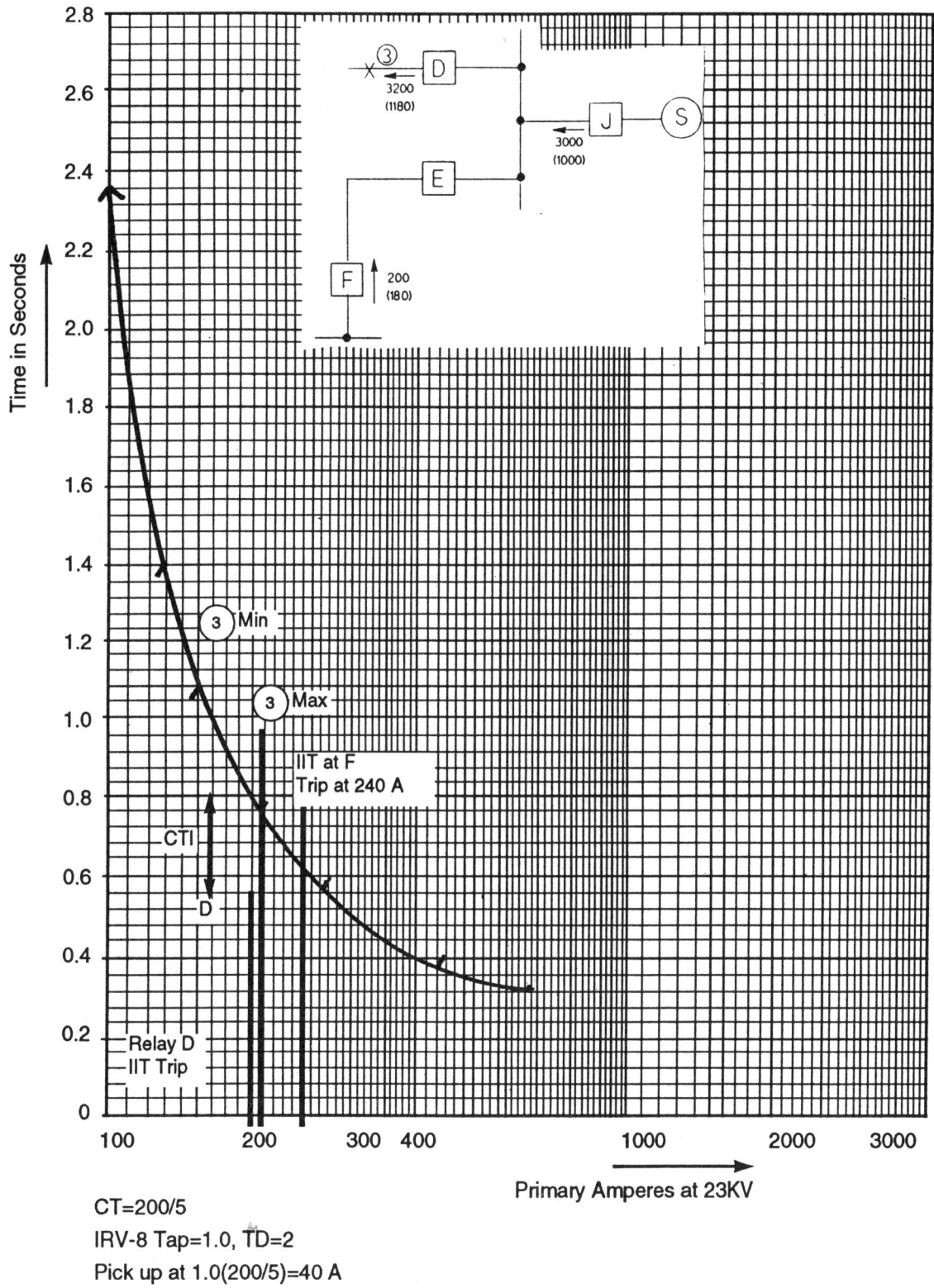

CT=200/5

IRV-8 Tap=1.0, TD=2

Pick up at 1.0(200/5)=40 A

IIT Pick up at 6(200/5)=240 A

Figure 12D-8 Relay F settings.

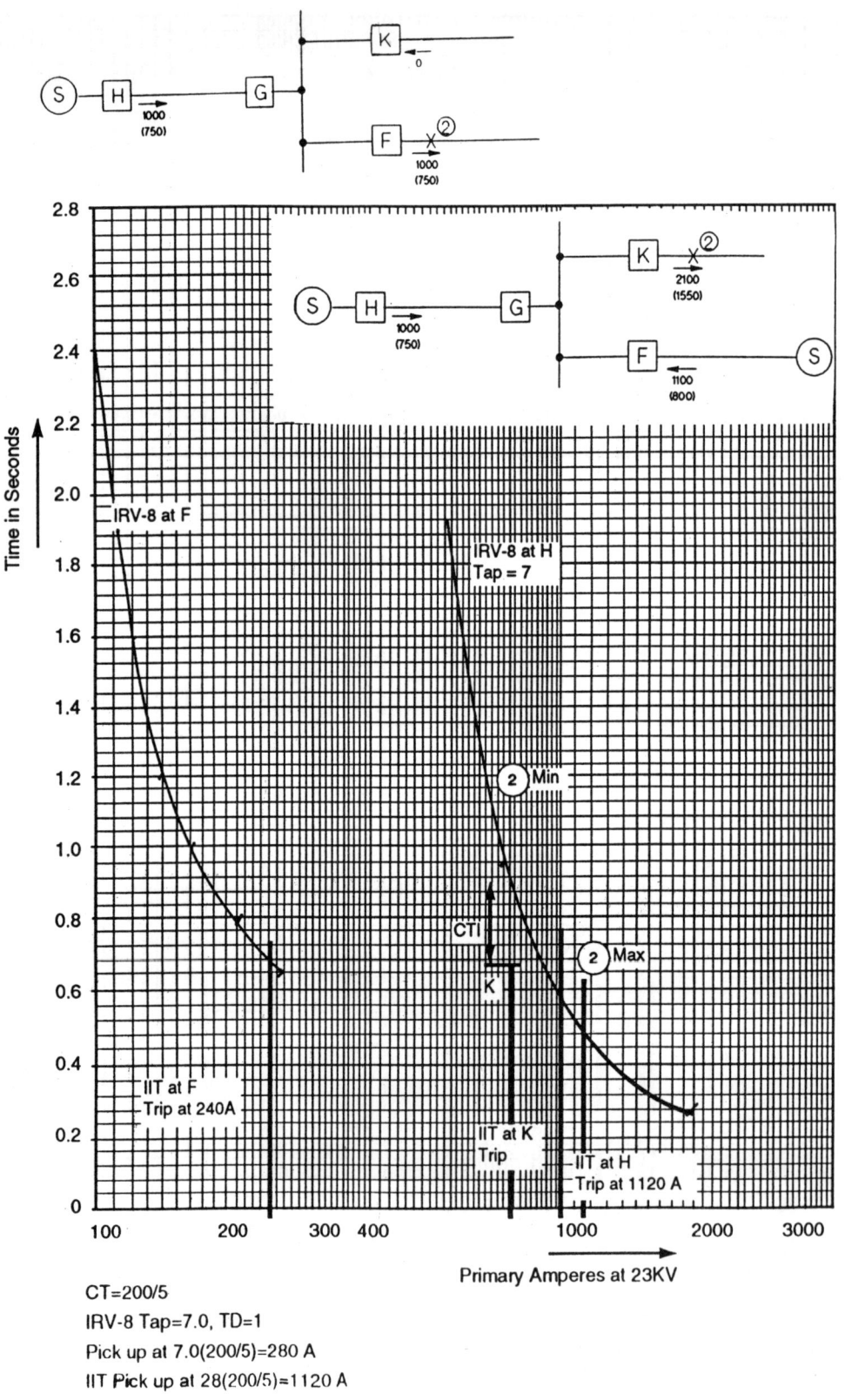

Figure 12D-9   Relay H settings.

1100 A primary (27.5 A secondary). Settings are rounded off to 28.0 A (1120 A primary).

D.3.9   Check Relay Settings on Coordination with Relay B

The instantaneous unit setting of relay H is 1120 A primary. This means that relay H will operate for line faults (2) under both maximum (2750 A) and minimum (1410 A) conditions. On tap 7, with a time dial of 1, the H relay time unit operates for line faults (2) in 0.25 sec (maximum fault condition) or 0.38 sec (minimum fault condition). Both values are well within the 1-sec time used in the coordination of B (Sec. D.3.5).

It should be noted that the major factor slowing down the H relay setting is the speed of relay K (Fig. 12D-9). A better speed may be obtained for the loop if the performance of relay K (and L) is improved.

D.3.10   Settings and Coordinations for Relays G, E, C, and A

For this particular example, from Figure 12D-1, the system configuration and fault currents are correspondingly symmetrical between A, B, C, D and H, G, F, E. Therefore, the settings for relays G, E, C, A should be correspondingly the same as those for relays B, D, F, and H.

D.3.11   Summary

This example has been used to describe an application of inverse-time directional overcurrent relays. Though electromechanical relays have been referred to, identical functions and curve shapes are available in solid-state and microprocessor versions.

13

Backup Protection

Revised by: **E. D. PRICE**

1 INTRODUCTION

Backup relaying, which provides necessary redundancy in protective systems, is defined in the IEEE Standard Dictionary as "protection that operates independently of specified components in the primary protective system and that is intended to operate if the primary protection fails or is temporarily out of service."

Backup protection includes remote backup, local backup, and breaker-failure relaying. Breaker failure is defined as a failure of the breaker to open or interrupt current when a trip signal is received.

Backup protection for equipment such as generators, buses, and transformers usually duplicates primary protection and is arranged to trip the same breakers. In the event of a breaker failure, some remote line protection would isolate the fault.

In the past, backup protection for transmission lines was provided by extending primary protection to line sections beyond the remote bus. This remote backup is defined in the IEEE Dictionary as "backup protection in which the protection is at a station or stations other than that which has the primary protection."

With the advent of EHV and increased concern about both service continuity and possible breaker failures, local backup, including breaker-failure protection, has become common.

2 REMOTE VS. LOCAL BACKUP

2.1 Remote Backup

Circuit breakers occasionally do fail to interrupt or trip for various reasons, and the remote terminal relays and breakers may be able to provide backup for a failed breaker. A remote backup system is shown in Figure 13-1. The relays protecting line GH at bus G for breaker 1 must also overreach and protect all other lines extending from the remote bus H. That is, the relays at G must operate selectively for faults on lines HR, HS, and HT if the relays or breaker 2, 3, or 4, respectively, fail to clear the associated line fault. Thus, the relays on breaker 1 provide primary protection for line GH, as well as backup protection for lines HR, HS, and HT. Backup systems use time discrimination to detect faults in the remote line sections.

In designing remote backup systems, their selectivity, sensitivity, speed, and application must be considered.

2.1.1 Selectivity

Remote backup provides poor selectivity. It interrupts all tapped loads on the unfaulted line sections. Opening breaker 1 at G for faults on a remote section may interrupt service to tapped loads on line GH unnecessarily. For a breaker failure at bus H, all lines feeding fault power through the defective breaker must

323

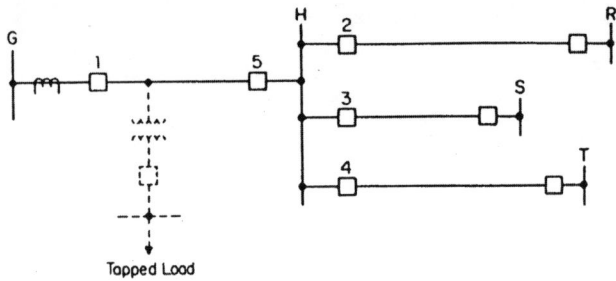

Note:
The Infeed Effect at Bus H Reduces The Current Magnitude and
The Reach of Distance Relays at Breaker 1, Bus G for Faults on
Lines HR, HS or HT.

Figure 13-1 Remote backup at bus G, breaker 1 requires these relays to selectively operate for faults on lines HR, HS, and HT under all operating conditions.

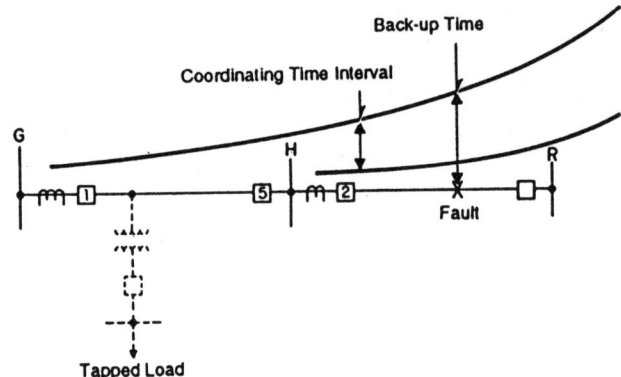

Figure 13-2 Remote backup at bus G, breaker 1 for line HR with inverse time overcurrent relays.

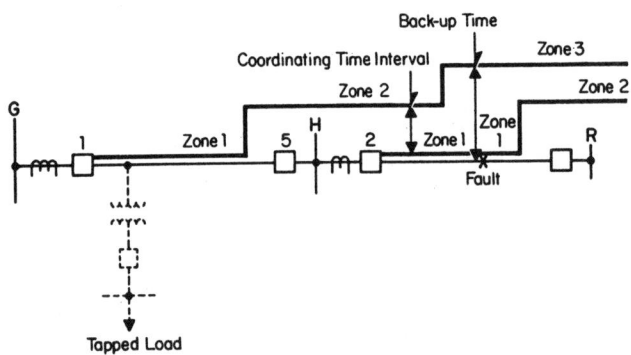

Figure 13-3 Remote backup at bus G, breaker 1 for line HR with distance relays.

be opened at their remote ends with remote backup. Such a scheme interrupts all loads on the lines, as well as on bus H. Thus, if breaker 2 fails to open for a line HR fault, breakers at G, S, and T must be opened with remote backup.

2.1.2 Sensitivity

Remote backup provides poor sensitivity. Relays at bus G may not "see" faults near buses R, S, or T. For the scheme shown in Figure 13-1, the fault infeed effect at bus H for faults near R, S, or T tends to reduce the current magnitude and "reach" of distance relays at breaker 1 at bus G. In these cases, relays at other remote terminals will have to trip first, redistributing the fault currents and increasing the effective reach of the relays at bus G. This can result in sequential tripping.

2.1.3 Speed

Remote backup must be relatively slow to give the primary relays in the remote line time to clear their fault (coordinating time interval in Figs.13-2 and 13-3). As the coordinating time interval is typically 0.3 sec, backup times greater than 20 cycles are common. If sequential tripping is necessary, as indicated in the preceding paragraph, the fault-clearing time for the breaker for a remote backup must be further increased.

2.1.4 Application

The application and setting of relays for remote backup require an understanding of fault levels under all possible operating conditions.

2.2. Local Backup and Breaker Failure

There are various reasons for a circuit breaker to fail to interrupt or trip. The need for remote backup, local backup, or breaker-failure relaying depends on the consequences of such failure.

2.2.1 Local Backup

Unlike remote line protection, local backup is applied at the local station. If the primary relays fail, local backup relays will trip the local breakers. If the local breaker fails, either the primary or backup relays will initiate the breaker-failure protection to trip other breakers adjacent to the failed breaker. Although local backup protection has many advantages and is widely used, it does not automatically eliminate the need for remote backup.

Local backup and breaker-failure protection are characterized by fault detection and initiation of tripping at the local terminal. For example, if a fault on line HR (Fig. 13-1) is not properly cleared by the primary protection system because of a failure in any part of the system other than the circuit breaker, the local secondary relaying system will detect the fault and trip breaker 2. If the fault on line HR is not properly cleared because of a failure of breaker 2, then the primary and/or secondary protective relays will initiate local breaker-failure backup to open breakers 3, 4, and 5 at bus H. Figure 13-4, a block diagram, illustrates the difference between the remote backup, local backup, and breaker-failure schemes.

A protection system involves a number of elements, including protective relays; ac current transformers and wiring; ac voltage transformers, devices, and wiring; dc supply and wiring; circuit breakers or other disconnection means; and a communication channel with pilot relaying. Ideally, a backup protection system should duplicate all these elements to provide total redundancy. In practice, all elements except the circuit breaker can be and frequently are duplicated in a variety of combinations, depending on the degree of protection required. The circuit breakers are not and cannot be duplicated. Many modern breakers have two independent trip coils, and breaker-failure protection provides a duplicate function.

The maximum practical redundancy is separating two protection systems as shown in Figure 13-5. The common element is the circuit breaker; even so, separate trip coils are shown. If a common station battery is used, separate fused leads from the battery are used for the two protection systems. Quite frequently, only one voltage source is used, and

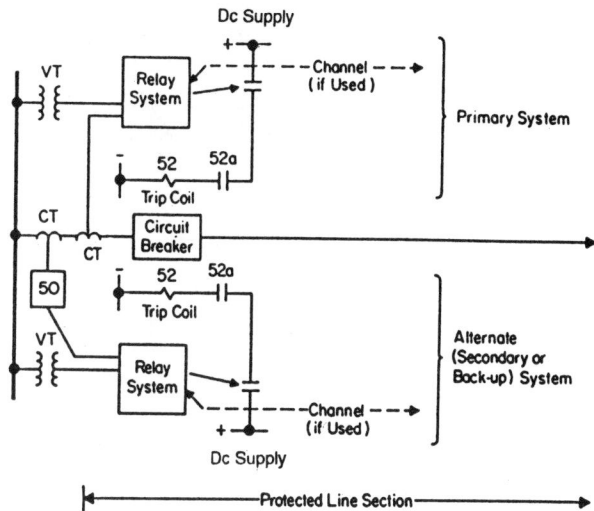

Figure 13-5 Transmission line protection with maximum practical redundancy.

separate fuses leads are run from the voltage transformer to the relays.

Instantaneous relays can be considered an independent protective system. Because these relays do not fully cover the line section, however, a remote end-zone fault would require an additional protection system. Local backup is usually applied only on lines equipped with a primary pilot system backed up by a second pilot system or nonpilot backup relays, or both.

2.2.2 Basic Schematic and Operation

Figures 13-6 and 13-7 illustrate basic local backup and breaker-failure schemes for electromechanical and logic-based relays, respectively. The operating principle is the same for either scheme. In Figure 13-6, operation of one of the protective relay systems trips its associated breaker and energizes the breaker-failure initiation relays 62X and/or 62Y, which are auxiliary SG, MG-6, or AR types of relays. Contacts on 62X and 62Y operate the timer 62BF only if the instantaneous overcurrent relay 50 indicates that current is continuing to flow. This continued current flow, indicating a failure to clear the fault, causes the timer to energize the multicontact 86BF lockout relay, tripping all the adjacent breakers. The 86BF lockout relay may also be used to block reclosing, to stop the carrier with blocking-type pilot relaying so that the remote end can trip, if possible, and to initiate transfer-trip.

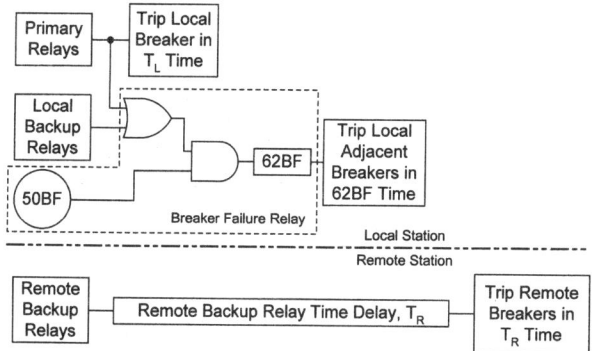

Figure 13-4 Block diagram illustrating the differences between remote backup, local backup, and breaker failure schemes.

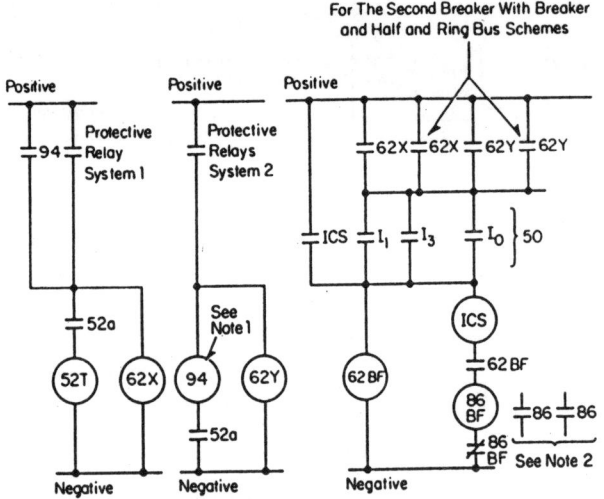

Notes:

1) 94 Function is Auxiliary Tripping Relay Which may not be Required With Dual Trip Coils.
2) Multi-Contacts to Trip Local Circuit Breakers around Failed Circuit and Initiate Remote Trip as Required to Isolate Fault, Block Reclosing, etc.

Figure 13-6 Simplified dc protection schematic for breaker failure and local backup protection. (Alternating current circuit per Figures 13-5, 13-13, or 13-14.)

Timer 62BF should be energized with device 50 (Fig. 13-6) rather than through the breaker auxiliary contacts 52a, since these contacts may be open, while the contacts in a damaged breaker are closed. Alternatively, breaker 52a contact can be used together with the 50 relay (Fig. 13-7).

Since transformer faults may not provide sufficient current to operate device 50, a transformer differential relay auxiliary contact 86T can be used instead,

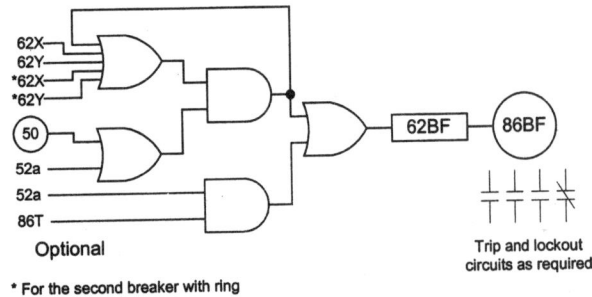

* For the second breaker with ring and 1-1/2 breaker schemes

Figure 13-7 Simplified logic diagram for logic-based breaker failure and local backup protection. (Alternating current circuit per Figures 13-5, 13-13, or 13-14.)

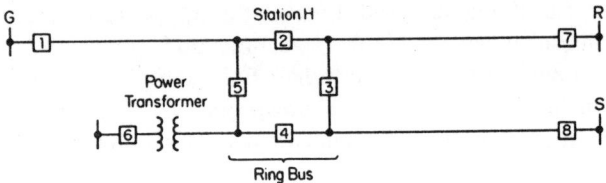

Figure 13-8 Remote backup required with breaker failure at station H where ring bus or breaker and a halfschemes are used.

supervised by the breaker 52a contact to operate timer 62BF (Fig. 13-8).

The three overcurrent units (50) are normally connected in phases A and C and to ground, but can be connected in all three phases if necessary. The phase units must be set below minimum fault current and can be set below load current, if necessary. A setting above load current is preferred in the interest of security if the minimum phase-fault current substantially exceeds the maximum load current. Setting above maximum load helps to prevent undesirable tripping during testing.

2.3 Applications Requiring Remote Backup with Breaker-Failure Protection

When ring buses or breaker-and-a-half schemes are used, breaker-failure protection does not necessarily eliminate the need for remote backup: As shown in Figure 13-8, a fault on line HR, for example, requires tripping both breakers 2 and 3 at station H. If breaker 2 fails to clear the fault, breaker failure would initiate the tripping of breaker 5, but would leave line GH still connected to the fault. The breaker-failure protection for breaker 2 frequently initiates the transfer-tripping of breaker 1 at station G. If transfer-trip is not applied or is not operative, however, remote backup at breaker 1 is still required to clear the fault.

Because of the infeed effect and high apparent impedances, remote backup from the remote stations may be difficult, if not impossible, to achieve when all lines are in service. Opening the breakers around the failed breaker will, however, remove the infeed effect and permit remote backup coverage. If, for example, breaker 2 fails for a fault on line HR (Fig. 13-8), line protection will open breaker 3 and breaker-failure protection will open breaker 5 to remove all infeeds around station H, except that from line GH.

Breaker-failure protection would trip both breakers 5 and 6 upon the failure of breaker 4 for a line HS fault

(Fig. 13-8). Similarly, the failure of breaker 5 in Figure 13-8 for a line GH fault would result in tripping both breakers 4 and 6. All other breaker failure conditions of breakers 4 or 5 would require remote backup, at S or G, respectively, or transfer-trip.

3 BREAKER-FAILURE RELAYING APPLICATIONS

Breaker-failure relaying should not be considered a substitute for good system design and equipment maintenance.

Breaker-failure protection should be as fast as possible without tripping unnecessarily. This criterion is particularly important in EHV lines, where stability is critical. Here, breaker-failure timer settings of 100 to 250 msec (6.5 to 15 cycles on a 60-Hz basis) are used.

These critical operations require dual high-speed solid-state or microprocessor pilot and breaker-failure protections. With electromechanical-type relays, breaker-failure timer settings of around 250 msec or more are practical.

In applying breaker-failure protection, it is recommended that

One breaker-failure circuit per breaker be applied, regardless of the bus configuration.
All adjacent breakers should be tripped by breaker-failure protection, regardless of fault location.
In all cases, all breakers tripped by the breaker-failure scheme should be locked out.
A remote breaker must also be tripped by either its own relays or transfer-trip initiated by local breaker-failure protection. Remote backup clearing of the breakers may be preferred over the direct-transfer trip.
One timer per bus or one timer per breaker may be used.

The latter is recommended, since it provides maximum isolation and flexibility, even though it does involve additional timers. These methods will be illustrated for various bus arrangements.

Circuit-breaker auxiliary switches should not be used to indicate whether or not a circuit breaker is carrying current unless there is no other way to accomplish this.

The fact that an auxiliary switch has operated is not sufficient proof that a circuit breaker has interrupted a fault. The auxiliary switch may be opened because

1. Its operating linkage is broken.
2. It is out of adjustment.
3. The breaker mechanism has operated but the main breaker contacts have failed to interrupt the current.

When protective relays are being tested, the breaker-failure scheme should be properly blocked or isolated to prevent misoperation.

3.1 Single-Line/Single-Breaker Buses

A typical single-line/single-breaker bus is shown in Figure 13-9. Figures 13-10 and 13-11 illustrate the dc schematics or logic diagrams for breaker-failure/local backup protection using one timer per bus section.

Figure 13-12 shows the schematic for one timer per breaker. The two methods, one timer per bus (method 1) and one timer per breaker (method 2), have the following differences:

Method 1 is less costly than method 2, since fewer timers are required.

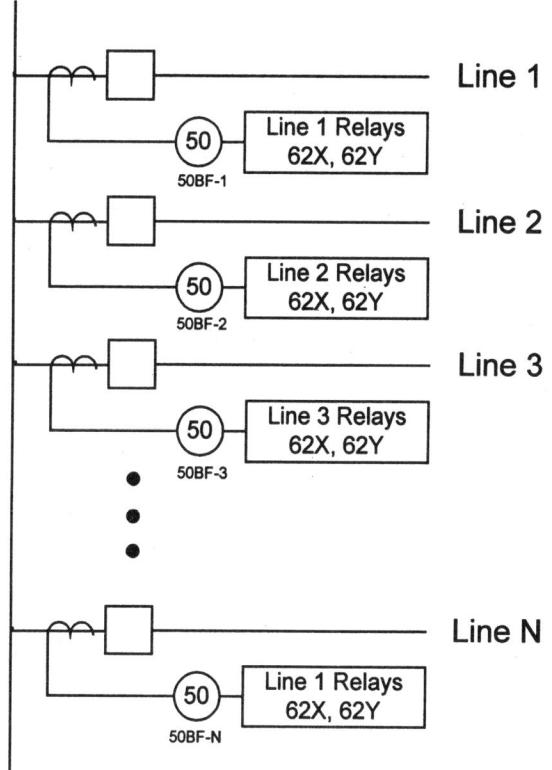

Figure 13-9 A single-line single-breaker bus. (The breaker failure local backup protective scheme is shown in Figures 13-10 and 13-11.)

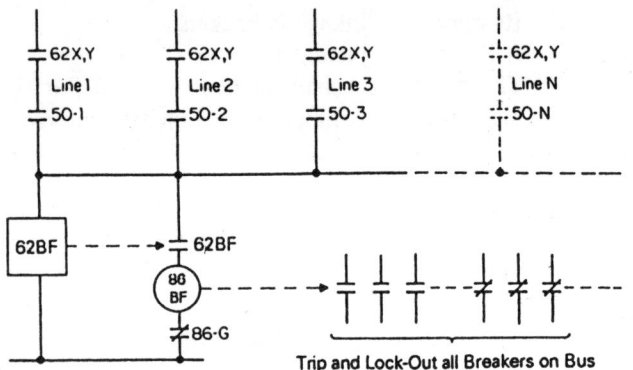

Legend:

62X, Y – Breaker Failure Intiating Auxiliaries Operated by the Line Relaying.
50 – Overcurrent Relay Responsive to Current Flowing Through The Individual Breaker.
62BF – Breaker Failure Timer.
86BF – Lock-out Relay. (can be common with 86B)

Figure 13-10 Typical simplified dc schematic for breaker failure local backup protection using a common timer for a single-line single-breaker bus of Figure 13-9.

Transfer-trip of the remote breaker is easier with method 2. With method 1, the common timer cannot distinguish which breaker has failed.

An evolving fault may cause incorrect breaker-failure operation with method 1. If a line 1 fault evolves to line 2, with sequential operation of 62X, 62Y, and 50 contacts, the common timer circuit may be enrgized long enough to operate and trip all breakers, even though both line 1 and 2 breakers trip normally. With method 2, each timer is deenergized as soon as the associated line fault is cleared.

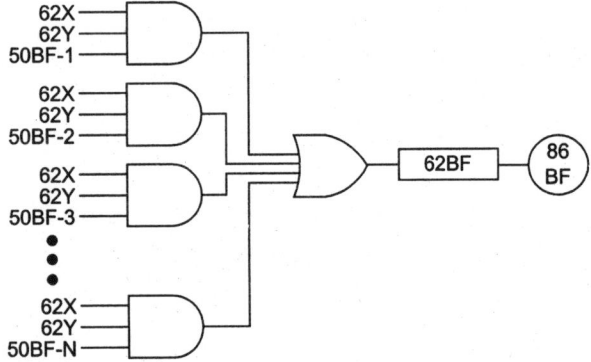

Figure 13-11 Typical simplified logic diagram for breaker failure local backup using a common timer for a single-line single-breaker bus of Figure 13-9.

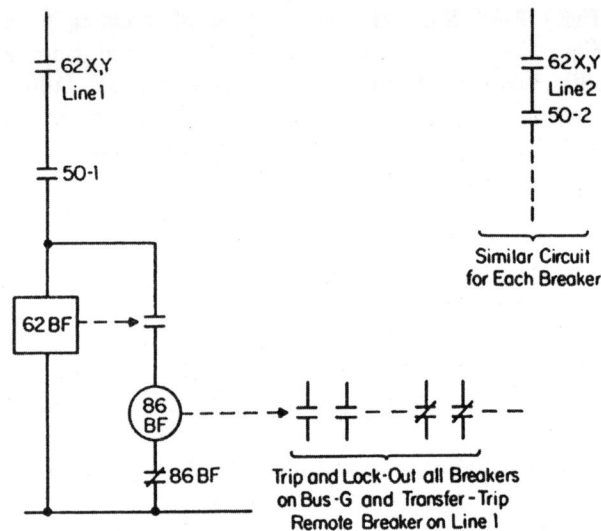

Figure 13-12 Typical simplified dc schematic for breaker failure local backup protection using one timer per breaker for the single-line single-breaker bus of Figure 13-9. (The solid-state logic diagram would be as shown in Figure 13-7 for each breaker.)

The common timer approach of method 1 requires that the timer be set for the slowest breaker interrupting time. Method 2 permits the separate timers to be set for the interrupting times of the individual breakers.

3.2 Breaker-and-a-Half and Ring Buses

Typical breaker-and-a-half and ring buses are shown in Figures 13-13 and 13-14. These arrangements require tripping two breakers and paralleling the current transformers for each line, as shown. A current detector (50) is provided for each breaker. One timer per breaker is recommended for all these bus configurations. Breaker-failure protection systems are shown in Figures 13-6 and 13-7. Another set of 62X, 62Y auxiliaries (shown dotted) must be added for the second breaker.

The breaker-failure/local backup circuits are the same for each breaker, except for the application of the 86 relay contacts. The 86BF relay operations are outlined in Table 13-1 for Figure 13-13 and in Table 13-2 for Figure 13-14. Neither table includes reclosing lookout, which may be desired.

It will be noted from Tables 13-1 and 13-2 that all adjacent breakers are tripped, regardless of fault

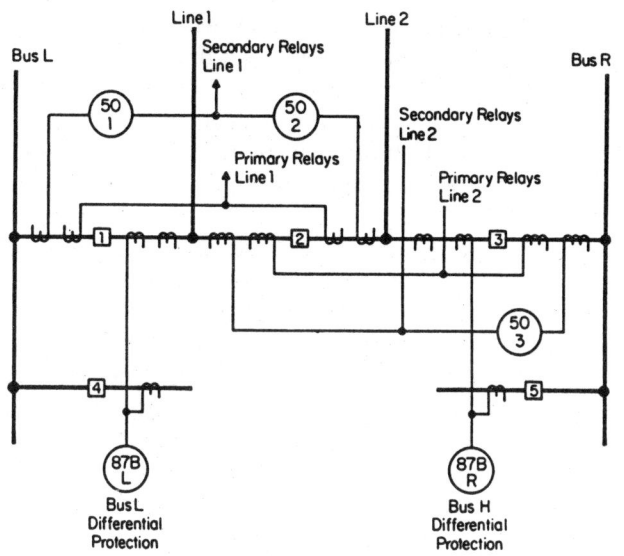

Figure 13-13 Single-line diagram of a breaker and a half bus.

location. For simplicity and reliability, breakers that are already tripped will be retripped. Assume, for example, that a fault occurs on line 1 of Figure 13-14. The protective relays for line 1 will attempt to trip breakers 1 and 4 and the remote end of line 1. Assume

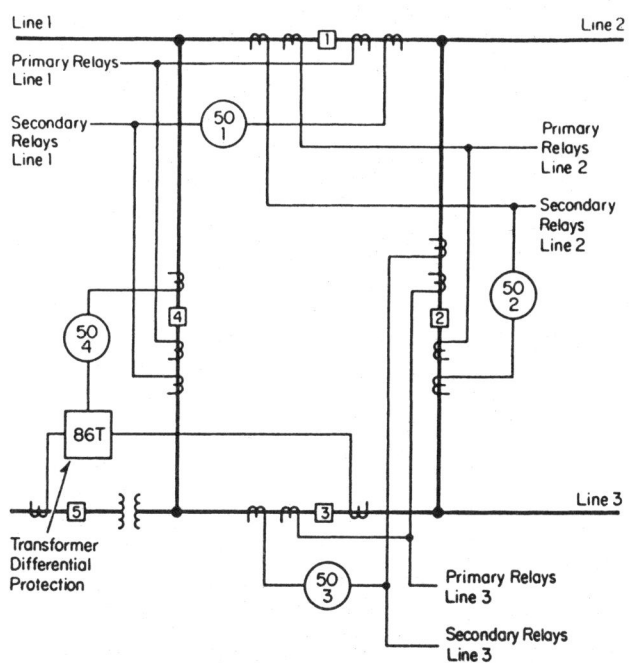

Figure 13-14 Single-line diagram of a ring bus.

Table 13-1 Breaker-Failure Operations for Breaker-and-a-Half Bus (Fig. 13-13)[a]

For local backup or breaker failure no.	86 Relay operations
1	Trip 2 and all other bus breakers, such as 4, etc., on bus L. Transfer-trip line 1.
2	Trip 1 and 3. Transfer-trip lines 1 and 2.
3	Trip 2 and all other bus breakers, such as 5, etc., on bus R. Transfer-trip line 2.

[a]See schematics for Figures 13-6 or 13-7.

Table 13-2 Breaker-Failure Operations for Ring Bus (Figure 13-14)[a]

For local backup or breaker failure no.	86 Relay operations
1	Trip 2 and 4. Transfer-trip line 1 and 2.
2	Trip 1 and 3. Transfer-trip lines 2 and 3.
3	Trip 2, 4, and 5. Transfer-trip line 3.
4	Trip 1, 3, and 5. Transfer-trip line 1.

[a] See schematics for Figures 13-6 or 13-7.

breaker 1 fails to clear, but breaker 4 and the remote line 1 breaker do open. Then breaker failure need not retrip 4 and transfer-trip line 1, as shown in Table 13-2. Similarly, for a fault on line 2, it may be unnecessary to retrip breaker 2 and transfer-trip line 2.

It is simpler, however, to trip all breakers involved, and it also allows reclosing to be blocked throughout. This practice also provides symmetrical protection around the bus.

4 TRADITIONAL BREAKER-FAILURE SCHEME

4.1 Timing Characteristics of the Traditional Breaker-Failure Scheme

Figures 13-6 and 13-7 show the approach of traditional breaker-failure schemes. The operating principle of this approach as illustrated in Figure 13-15 is that the breaker-failure timer 62BF is started by the operations

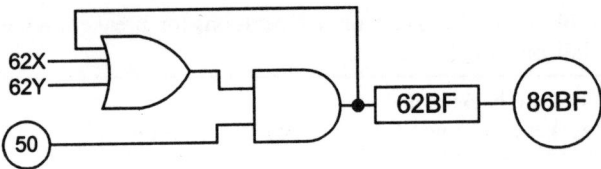

Figure 13-15 Block diagram of the traditional breaker failure scheme.

of the fault detector 50 and the breaker failure initiates signal 62X or 62Y.

A time chart for the traditional breaker-failure scheme is given in Figure 13-16. The shaded margin time provides security and should accommodate the following variables:

1. *Excessive breaker interrupting time.* According to ANSI Standard C37.04, the circuit-breaker interrupting time may be one cycle longer for three-cycle breakers at currents below 25% of the maximum rating. Also, the interrupting time may be longer on close-open duty.

2. *Inconsistency in BFI times.* These are minimized by static and microprocessor breaker-failure initiation. However, the wide time range associated with electromechanical BFI is primarily a function of the variation in dc voltage. For example, the BFI AR contact output in the SRU relay or ARM module (Uniflex) has an operating time of 2 to 5 msec. The BFI telephone relay in the SRU relay has an operating time of 8 to 16 msec. The SG relay 62X/62Y in the KC-4/TD-5 breaker-failure scheme has an operating time of 16 to 50 msec. The breaker-failure total clearing time, as shown in Figure 13-16, assumes that the pickup of device 50 is equal to the protective relay plus the BFI time. Either 50 or BFI signal delaying will delay starting the 62BF timer (Fig. 13-15) and the total backup breaker-failure clearing time will be longer and the margin increased.

3. *Overtravel of the 62BF timer after the fault detector reset.* Static timers have less than 1 msec of overtravel. In microprocessor relays the overtravel really depends on the logic and operation design. The overtravel is usually zero, but the worst case expected would be the computation interval time (the time between logic calculations).

4. *Inconsistency in 62BF timer.* The static timer, as contained in the TD-5, TD-50, and SBFU, has a repeatability (including variations of ambient temperature and voltage supply) of $\pm 5\%$, which is $\pm 5\%$ msec for a timer setting of 100 msec (six cycles on a 60-Hz

base). Microprocessor timers are generally consistent to within the computation interval time.

5. *62BF timer setting error.* Includes human error, instrumentation error, and potentiometer resolution. The static timer may be set within 2 ms. The microprocessor timer can generally be set to within the resolution of the logic computation interval time.

6. *A safety factor.* Because of the widespread harmful effects of a false 86BF operation, it is recommended that a generous safety factor be incorporated in the margin time. The degree of safety required is a direct function of the confidence level of the total protective system. Typical values range from two to six cycles (60-Hz base). A two-cycle safety factor appears to be adequate, with three cycles being a widely used total "margin."

4.2 Traditional Breaker-Failure Relay Characteristics

Four types of relays are used in the traditional breaker-failure scheme: current-detector relays, timer relays, auxiliary relays, and multitrip auxiliary relays.

4.2.1 Fault-Current Detector (50BF)

Fault detectors that have high dropout and whose dropout time is minimally affected by ct saturation and dc decay in the secondary circuit should be considered. Examples of this type are cylinder unit relays and static and microprocessor relays with suitable filtering. Hinged armature and plunger-type relays that can have a significant dropout delay should be considered carefully before use. If such devices are used, their dropout time could be ascertained under the worst conditions and this time should be considered when setting the breaker-failure timer. Determining the worst condition may be difficult.

A type KC-4 relay can be used for the fault-current detector in an electromechanical breaker-failure scheme. It is a high-speed unit with a 98% or greater ratio of dropout to pickup. The KC-4 relay has three-cylinder-type overcurrent units in an FT case. It is available in 0.5- to 2.0-, 1- to 4-, 2- to 8-, 4- to 16-, 10- to 40-, and 20- to 80-A ranges. The pickup is approximately 19 to 24 msec at four times pickup current, and 12 to 16 msec at eight times pickup. Dropout times are on the order of 20 msec after the current decreases to 0. Various combinations of ranges for phase and ground applications are available. Circuit shield types 50H and 50B are static relays with similar characteristics.

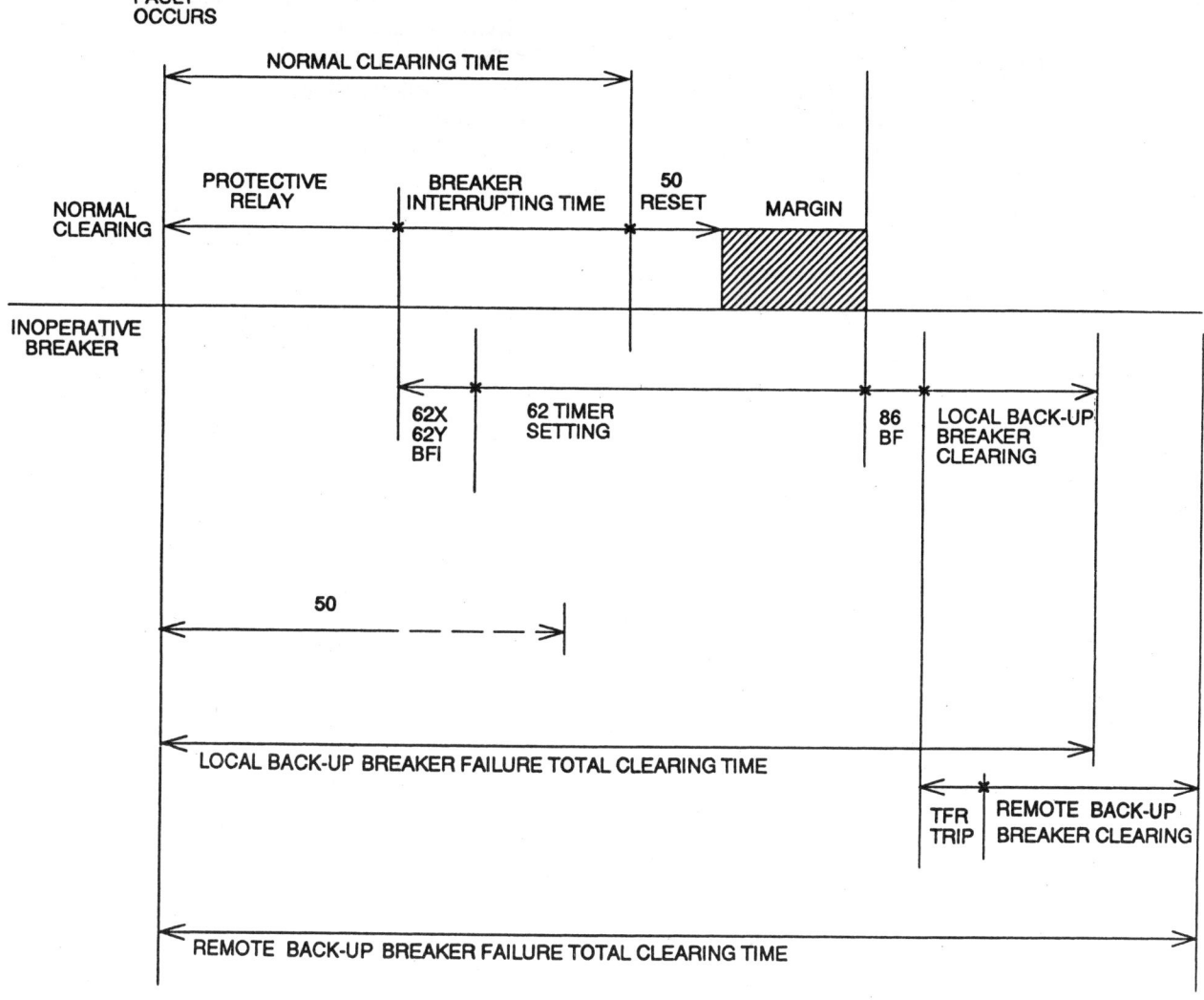

Figure 13-16 Time chart of the traditional breaker failure scheme (time not to scale).

4.2.2 Timer (62BF)

A direct-current static timer, type TD-5, is used in the electromechanical scheme with contact input, as shown in Figure 13-6. The range, for the breaker-failure applications is 0.05 to 0.4 sec. Static timers, types 62B and 62T are generally used with the static fault detectors.

4.2.3 Auxiliary Relay (62X, 62Y, or BFI)

Operating times are 16 to 50 msec for the SG and MG relays, and 2 to 3 msec for the AR relays.

4.2.4 Multitrip and Lockout Auxiliary Relay (86BF)

Either the electroswitch type WL relay or type LOR can be used; their operating times are approximately 16 msec.

4.3 Microprocessor Relays

The microprocessor breaker-failure function is generally implemented in line relays in two levels. The first level provides only a breaker-failure initiate signal (BFI, 62X, 62Y) simultaneously with the breaker trip.

In this case there is no 62X or 62Y operating time to consider.

The second level is where the complete breaker-failure tripping function is included in the line relay. Quite often special design considerations—filtering and algorithms—are implemented to assure fast and secure dropout of the 50-fault detector. These times will be specified and generally range from one-half to one cycle. It is generally safe to assume that the maximum dropout time will be one cycle plus the computation interval time. That is, if the status of the 50-fault detector is computed every one-quarter cycle, then the maximum dropout time will be 1.25 cycles. This assumes the appropriate dc filtering that is expected in microprocessor relays.

The setting resolution of the 62BF timer will also be defined by the computation interval time of the 50-fault detector and the frequency with which the timer is compared with the fault detector.

5 AN IMPROVED BREAKER-FAILURE SCHEME

5.1 Problems in the Traditional Breaker-Failure Scheme

5.1.1 Fault-Detector Reset Time Problem

Referring to Figure 13-16, we see that the reset time of device 50 is a critical characteristic in the performance of the breaker-failure system. It will affect the margin value, as well as the 62BF timer setting. Device 50 resetting stops the timer; with 62X/62Y seal-in, it is the only function that stops the timer. It should be noted that the longer the reset time of device 50, the longer the 62BF time has to be set. Consequently, it may be difficult to set the timer securely and still avoid system instability.

To prevent the device 50 reset time variables from affecting the margin, it is recommended that the 62BF time delay be determined for maximum device 50 reset time. Any faster 50 reset will then merely add to the safety margin.

The reset time of the overcurrent fault detector is affected by several factors:

The reset times are longer when the current after interruption is nonzero. Certain types of circuit breakers are equipped with arcing contacts and shunting resistors. When the breaker main contact interrupts fault current, the current does not drop immediately to 0, but to a level determined by the shunting resistor. It falls to 0 when the arcing contacts open. The reset time of device 50 on such applications may be longer.
The fault-current level at which the unit is energized prior to interruption.
The setting of the unit.
Current transformer decay. When breaker current is interrupted at current zero, the current transformer magnetizing current will not be zero. This current would then have to decay through the connected relays, possibly increasing their reset time.

The reset time of a solid-state overcurrent relay is faster than an electromechanical one.

5.1.2 Fault-Detector Setting Problem

Figure 13-17 illustrates a problem in the traditional breaker-failure scheme for breaker A on a ring bus application where maximum load exceeds minimum fault levels. Because IF1 > IF2, it may be difficult to set I < (IF2)/2 in the traditional single-fault-detector scheme. One solution to this problem (Fig. 13-17) is to use a special two-fault-detector breaker-failure scheme. The low set with 86T (from the transformer differential scheme) is for a fault at F2, whereas the high set with 62X (from the line protection) is for fault F1.

Figure 13-18 illustrates another problem in the traditional breaker-failure scheme for breaker B in a breaker-and-a-half system application. If breaker B fails for a traditional breaker-failure scheme, the 62BF timer for breaker B may not start (because initial = 10% may be lower than the fault-detector setting) until breaker A opens (because sequential = 100%). This condition results in unduly long backup clearing time. One solution to this problem is to use a special

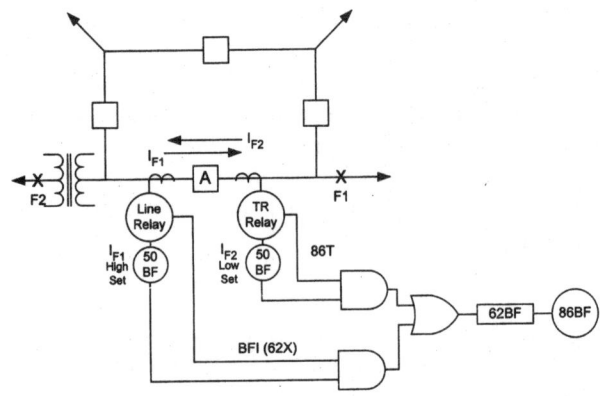

Figure 13-17 Traditional BF scheme with two timers.

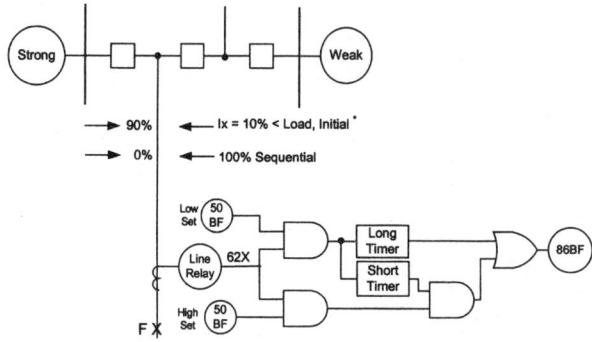

Figure 13-18 Traditional BF with two timers and two fault detectors.

breaker-failure scheme as shown in Figure 13-18. It includes two timers and two overcurrent units. However, the shorter timer setting or lower overcurrent unit setting may affect the security of the traditional breaker-failure scheme.

These problems can be solved by applying the type SBF-1 breaker-failure relay. It uses an improved concept as described in Section 5.2.

5.1.3 Transformer Breaker-Failure Considerations

A low-level fault internal to the transformer may cause the sudden pressure relay (SPR) to operate. The current level may be well below a "low set" value that the overcurrent relay could detect. For this case, an 86T contact in series with the circuit breaker's 52a (logically 86T AND 52a) as shown in Figure 13-19 must be used to assert and start the timer. In this case if

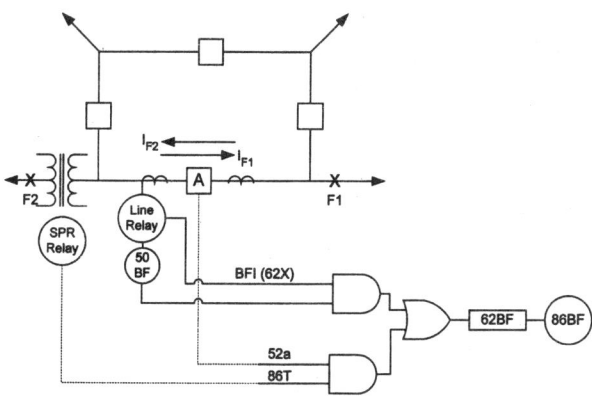

Figure 13-19 Transformer BF considerations.

tripping occurs by the SPR and the breaker fails to trip, breaker failure tripping occurs. The drawback is that if the breaker trips successfully and the 52a contact does not open, breaker failure clearing still takes place.

5.2 The Improved Breaker-Failure Scheme

Many approaches have been devised for improving the traditional breaker-failure scheme such as using a shorter reset time of device 50 (for example, using a static relay), or employing separate timers for different levels of fault current. These approaches will still be affected by the reset time of the device 50 and still present difficulties in determining the reset time, especially when information on the shunting resistor current is not available.

Figure 13-20 shows the basic concept of the improved approach. The unique feature of this approach is that the fault detector operates only after the 62BF times out. By that time, the breaker-failure condition has been confirmed; consequently, the reset time of the fault detector will have no effect on the scheme. Figure 13-21 is the time chart for the improved breaker-failure scheme.

The improved approach has many advantages over the traditional ones:

1. The fault detector 50 will not operate before the 62BF is timed out; therefore, it will never operate when clearing normally and device 50 reset time is not a consideration.

2. It permits shorter margin and shorter overall clearing times and will wield a net saving of one to two cycles over the traditional approach. This can be illustrated as below (compare Fig. 13-16 and 13-21): Net saving in clearing time = maximum 50 reset time, which represents about one cycle for static units and two cycles for electromechanical units.

3. The overcurrent-fault-detector unit never operates when clearing normally so it can be set

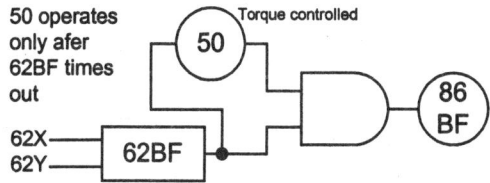

Figure 13-20 Logic diagram of the SBF-1 relay.

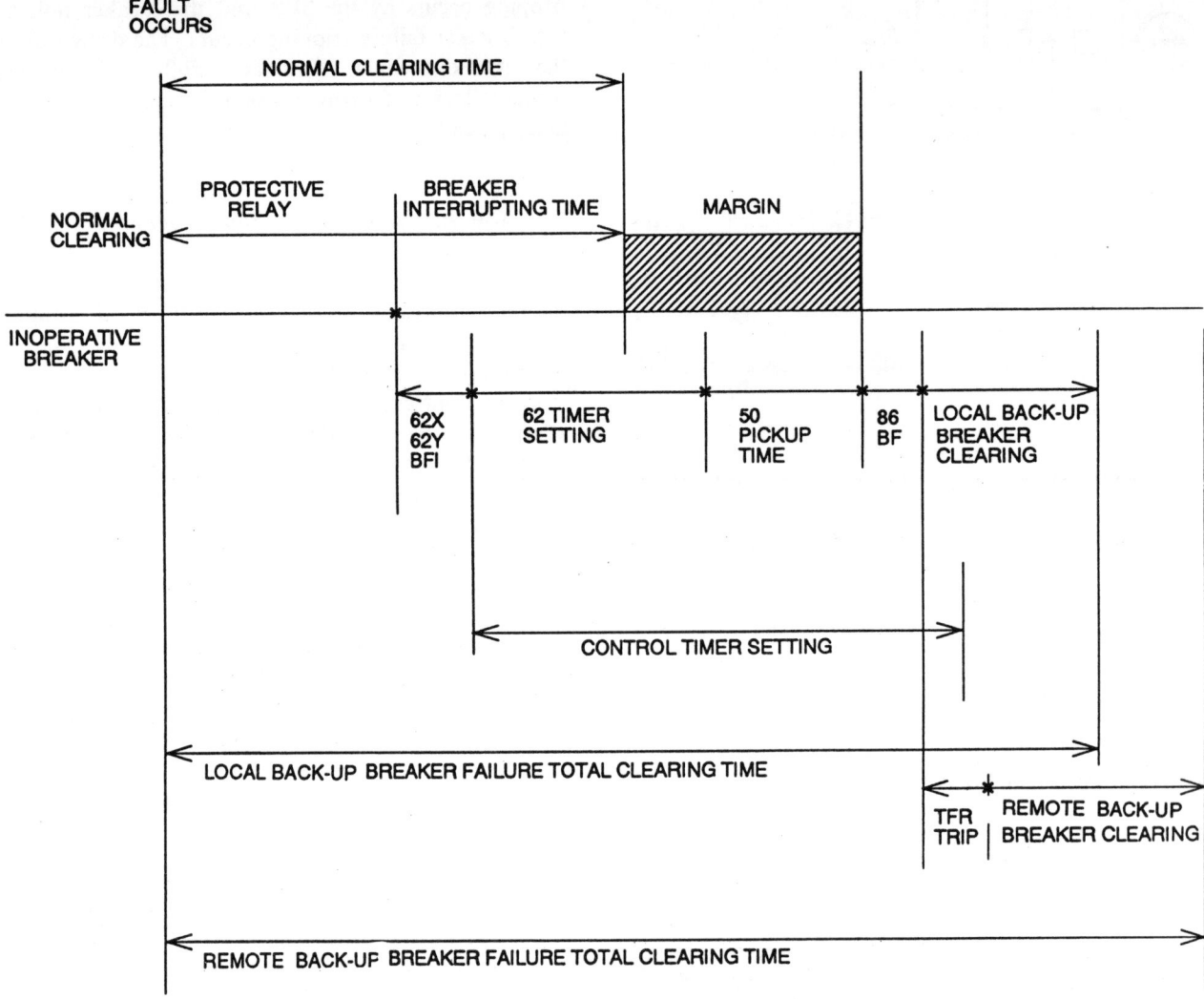

Figure 13-21 Time chart of the improved breaker failure scheme.

lower than load current, if necessary (for example, application problems as described in Section 5.1.2).

Figure 13-16	Figure 13-21
Total clearing time = Protective relay time + breaker interrupting time + maximum 50 reset time + margin + 86BF time	Total clearing time = Protective relay time + breaker interrupting time + margin + 86BF time

5.3 Type SBF-1 Relay

The type SBF-1 breaker-failure relay uses the improved concept for breaker-failure detection, as described in Section 5.2. The relay, as shown in Figure 13-22, consists of phase- and ground-fault detectors, 62BF timer, and a "control" timer. The control timer is for resetting the SBF-1 relay. In applying the SBF-1 relay:

The fault detectors can be set below the load current.

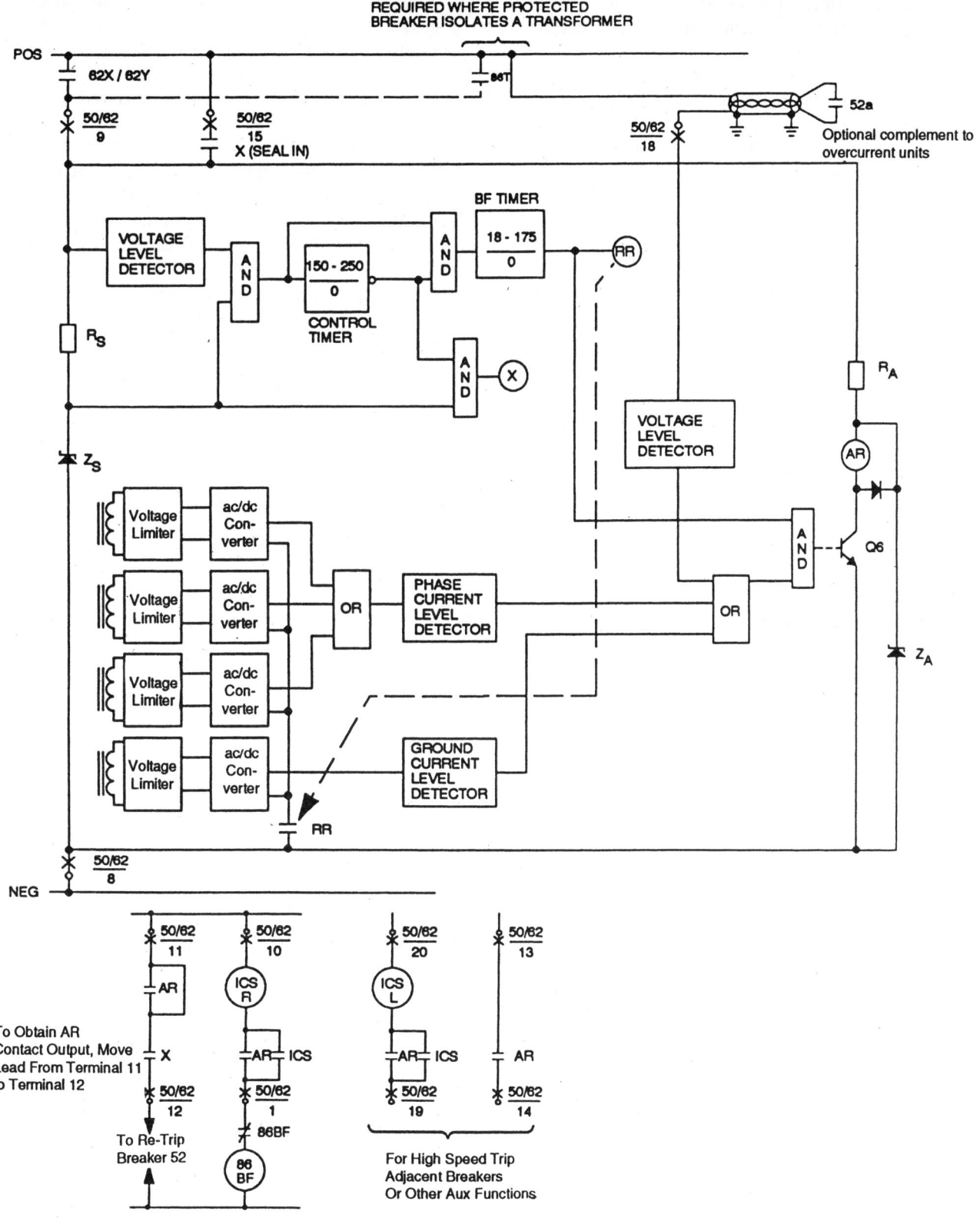

Figure 13-22 Type SBF-1 relay.

The 62BF timer should be set longer than the breaker normal clearing time, with a margin of two to three cycles.

The control timer should be set at least 32 msec longer than the 62BF time setting to provide long enough time for the SBF-1 relay to trip the breakers. The breakdown of this 32 msec is as follows (refer to Fig. 13-22):

Maximum pickup of		
RR relay	1 to 3 ms
50 unit	3 to 8 ms
AR relay	2 to 5 ms
Total	16 ms
86BF relay	16 ms
Total	32 ms

6 OPEN CONDUCTOR AND BREAKER POLE DISAGREEMENT PROTECTION

At voltages of 345 kV and above, the physical size of the operating components and the phase spacing requirements of power circuit breakers have led to the use of an independent operating mechanism for each. Also, with the trend toward larger and larger turbine-generator sizes, system stability criteria have often dictated the use of independent pole-operated breakers at voltages below 345 kV to obviate the three-phase fault, three-pole "hung" breaker fault condition. Rarely will more than one fail to interrupt fault current. Therefore, serious faults (those involving two or three phases) will always be downgraded if not entirely cleared, reducing the danger of system instability.

Field experiences have shown it necessary to consider the consequences of the unsymmetrical operation of such a breaker. Electrical or mechanical failures have left one phase open when the others are closed, and vice versa. Since pole disagreement may occur under a no-fault condition, the breaker-failure initiation circuit of the conventional breaker-failure scheme may not be energized. Therefore, the conventional breaker-failure relay scheme will not respond if the failure occurs under a nonfault condition.

Figure 13-23 shows logic that can be used to detect pole disagreement under a no-fault condition. This logic has an output whenever one or more phases carries current above the I_H setting, while one or more phases carries current below the I_L setting. The current inputs I_L, I_H, and I_M must be capable of measuring current accurately in the 1.0 to 100.0 mA range to distinguish between no current and charging current. This usually requires special current inputs to the relay.

For breaker-and-a-half or ring-bus configuration, with low-magnitude load current flowing through the bus, odd phase current combinations result from unequal phase impedances and multiple paths for current flow through the bus; unequal phase currents flowing through an individual breaker can give the false appearance of pole disagreement. This logic contains a zero sequence voltage comparison circuit (Fig. 13-23) that allows this low-current difference to be ignored, while permitting tripping when a hazardous pole disagreement actually exits.

Simpler methods may be employed that utilize conventional ct inputs that are only reliable with load flow. Typical minimum current levels that can be accurately measured range from 0.25 to 0.5 A (250 to 500 mA). These applications involve the measurement of zero or negative sequence current through the poles

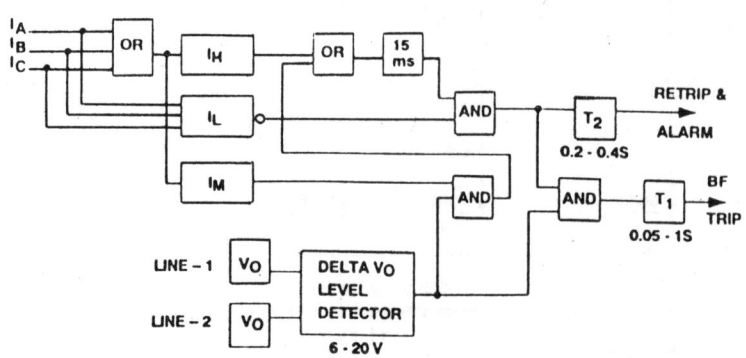

Figure 13-23 Pole disagreement logic example.

of the breaker, and may be used to trip or alarm. The sequence current measured must distinguish between that which is produced by normal unbalances and that produced by an open breaker pole. A long time delay is applied to ride through maximum fault clearing times.

7 SPECIAL BREAKER-FAILURE SCHEME FOR SINGLE-POLE TRIP SYSTEM APPLICATION

Special arrangements should be considered for a breaker-failure scheme in a single-pole trip application.

The problem is that the unequal load current during single phasing may pick up the ground overcurrent I_o unit (the phase overcurrent units may pick up also) in the conventional breaker-failure scheme (Fig. 13-24) and trip the scheme falsely. One special arrangement, as shown in Figure 13-24, has been employed successfully in the field for years for solving this problem. In this special arrangement, no ground overcurrent is used in the scheme. Each phase overcurrent unit is supervised by its individual breaker-failure initiate contact 62Z/A, B, and C. Blocking diodes TRB are used for blocking the sneak path during single phasing.

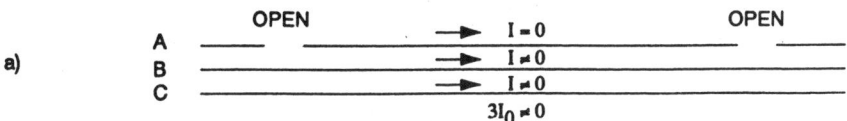

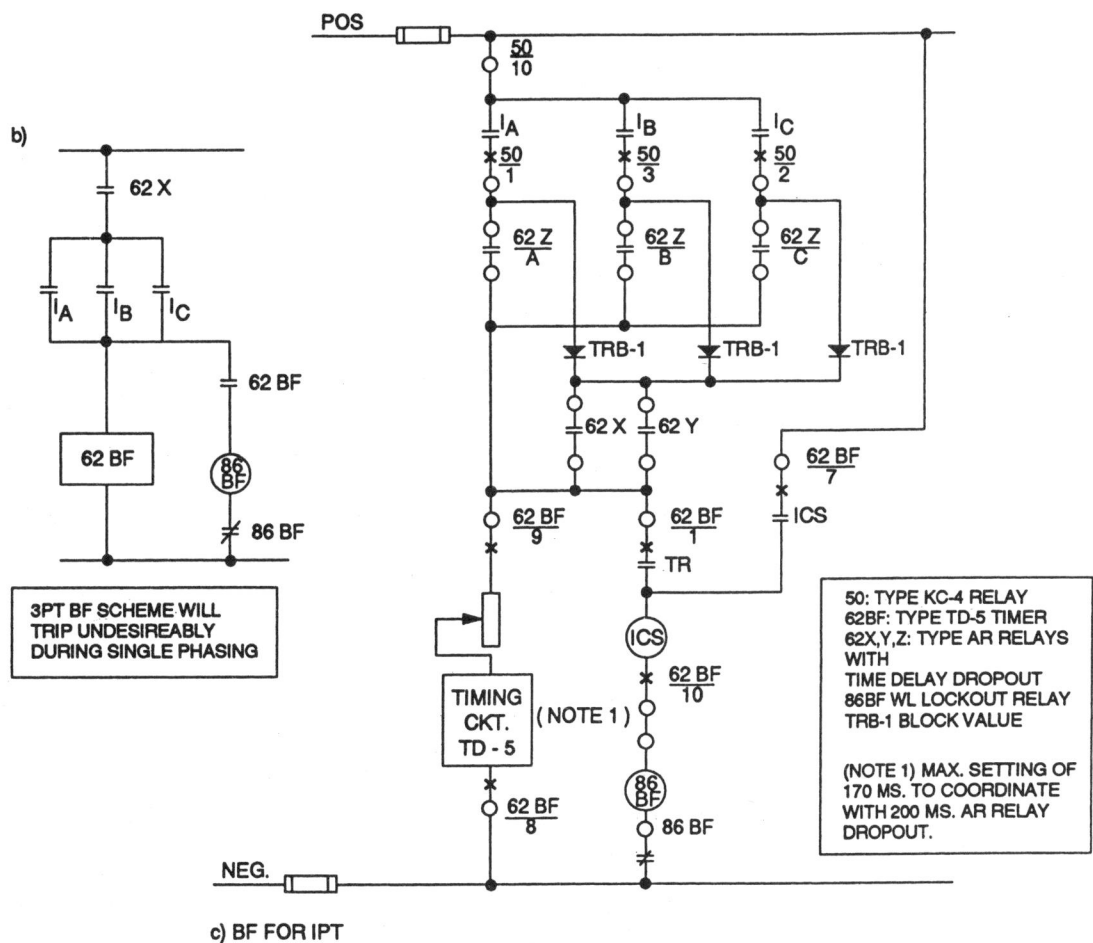

Figure 13-24 Special consideration on breaker failure scheme for single-pole trip application.

14

System Stability and Out-of-Step Relaying

W. A. ELMORE

1 INTRODUCTION

Since relaying systems must function properly during system swings, it is necessary to understand the effects of these disturbances on relay performance. Swings are the oscillations of synchronous machines with respect to other synchronous machines. They are caused by changes in load, switching, and faults. A swing does not necessarily indicate system instability. In some cases, however, the swing is severe enough to cause synchronous machines to go out of step. Before examining the influence of system swings on relay performance, three factors must be considered: steady-state stability, transient stability, and relay quantities encountered during swings.

A system is stable when it is able to develop restorative forces in excess of the disruptive forces to which it is subjected. Disruptive forces are produced by such factors as faults, loss of excitation, or switching of lines, transformers, or generation. Restorative forces are produced by current flow, voltage increase, or impedance reduction produced by fault removal and/or line switching.

2 STEADY-STATE STABILITY

The fundamentals of power transmission and stability are more easily understood if both system resistance, excluding the load impedances, and machine saliency are neglected. In this case, the power P transmitted over circuits connecting two portions of the systems is

given by the following equation:

$$P = \frac{V_S V_R}{X} \sin \phi \qquad (14\text{-}1)$$

where

V_S and V_R = sending- and receiving-end voltages, respectively

X = reactance between V_S and V_R

ϕ = angle by which V_S leads V_R

See Figure 14-1.

If system resistance is not neglected, different equations apply for the sending- and receiving-end power. The variables, however, are essentially the same. If phase-to-phase voltages are used, Eq. (14-1) yields three-phase power. For this discussion, V_S and V_R are taken as per unit quantities, and Eq. (14-1) gives per unit power. If V_S, V_R, and X are held constant in Eq. (14-1), the power flow is changed by varying the angle ϕ.

As the load increases at the receiving end, synchronous machines are momentarily slowed down, and the machine-rotor inertia meets the increased load requirements. That is, an increase in load results in a small reduction of system frequency until there is a change in mechanical input via the governor or manual action.

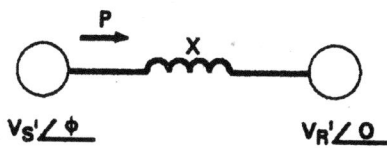

Figure 14-1 Simple equivalent system.

To restore system frequency, the mechanical input to the machines must be increased. This input must be greater than the steady-state load requirements, since the machines must be accelerated to a new and larger angle. When the new angle ϕ is reached, the mechanical input will exceed the load requirements by the amount required to accelerate the machines. The mechanical input must then be reduced to maintain frequency and required power transfer. Any load change therefore results in swings or oscillations as the system adjusts to the changes. Steady-state stability is the ability of the system to adjust to gradual load changes.

The extreme unstable condition occurs when ϕ is equal to 90°. At this point, increased load conditions could only be met by increasing V_S or V_R. An increase in ϕ would cause a reduction, rather than an increase, in power transfer.

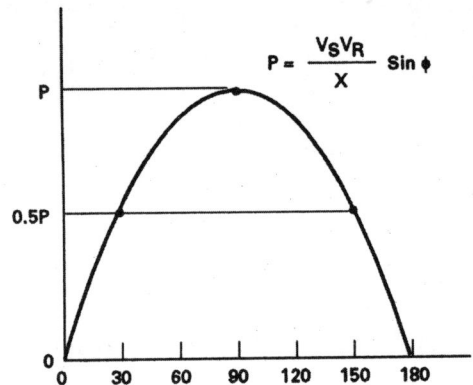

Figure 14-2   Power transfer curve.

3   TRANSIENT STABILITY

Transient stability is the ability of the system to properly adjust (remain in synchronism) to sudden large changes. Again, if we ignore system resistance and machine saliency, the power transmitted during the transient interval P is given by the following equation:

$$P = \frac{V'_S V'_R}{X'} \sin \phi \qquad (14\text{-}2)$$

In Eq. (14–2), P is three-phase MW if V'_S and V'_R are expressed in kV (phase to phase) and X' in ohms per phase, where

V'_S = voltage behind transient reactance at the sending end

V'_R = voltage behind transient reactance at the receiving end

X' = reactance between V'_S and V'_R, including transient reactances of the machines

ϕ = angle by which V'_S leads V'_R

Figure 14-2 shows a power-transfer curve. Note that the peak value is inversely proportional to the total reactance X. Figure 14-3 describes how X is influenced by the presence of a fault, as well as the type of fault.

It is recognized that the quantity to be inserted as X_F in the fault representation is dependent on the type of fault, whether 3ϕ, $\phi\phi$G, $\phi\phi$, or ϕG. For the 3ϕ fault, X_F is 0. For the $\phi\phi$G fault, X_F is X_2 and X_0 in parallel. The positive sequence network is retained

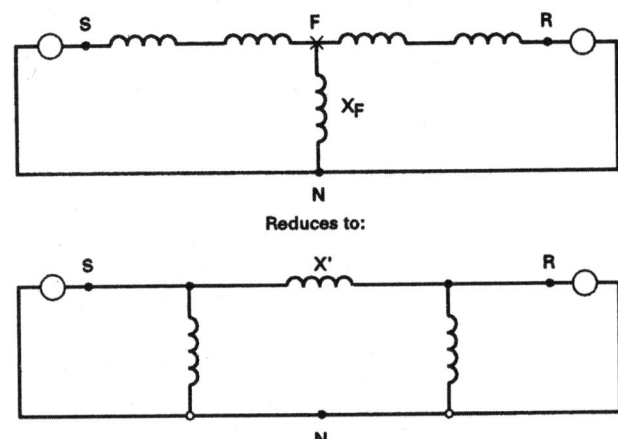

Figure 14-3   Effect on X' of fault.

intact without reduction, whereas X_2 and X_0 are the thevenin impedances in the negative and zero sequence networks, respectively. X_F for the $\phi\phi$ fault is simply X_2. For the ϕG fault, X_F is X_2 and X_0 in series.

Based on the normal relationship found between X_1, X_2, and X_0, the relative order of severity of the various types of faults produces the relative power-transfer curves shown in Figure 14-4. The reduction of the "T," of which X_F is one leg, to an equivalent "pi" produces X' which is the transfer impedance that, with the voltage V_S and V_R, defines the peak value of the power-transfer curve.

Figure 14-5 describes a representative case of two parallel lines with generation at each end. The power transfer curves show that a fault causes an immediate drop in transmitted power from bus S to bus R because of the fact that the "torque angle" m cannot change instantaneously, but X' does. This change in power

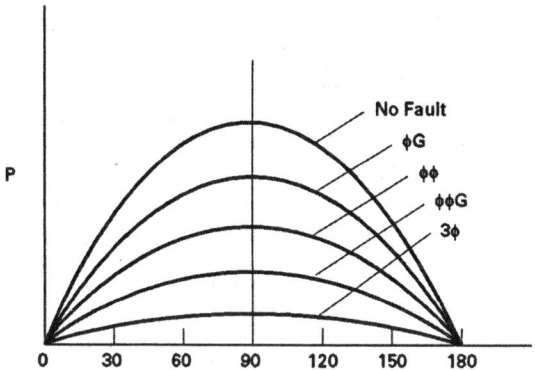

Figure 14-4 Relative order of severity of fault types.

quantity DE. Power DE is the difference between mechanical input and electrical output that, therefore, produces acceleration of the rotating mass. With losses ignored in this example, quantity DE is also the difference between electrical power input and mechanic

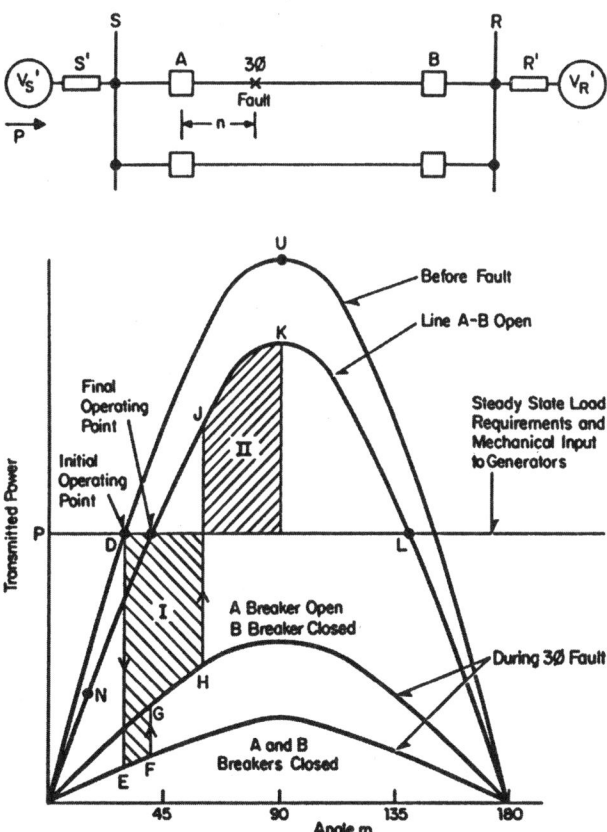

Figure 14-5 Power transfer curves before, during, and after line-to-line fault.

power output at the receiving end, causing deceleration there. Both of these effects increase m.

When the fault occurs (Fig. 14-5), the transmitted power is reduced to point E because of the increased effective X', and the swing begins along E-F. At point F, breaker A opens, and the transmitted power increases to G. The swing then continues along G-H.

When the fault is cleared at H by breaker B opening, the sending-end rotor kinetic energy has increased by an amount proportional to area I, since the mechanical input has exceeded the transmitted electrical power. As the fault is cleared, the transmitted power increases to J, exceeding the mechanical input and causing the sending-end machines to decelerate and the receiving-end machines to accelerate. Since the velocity of a rotating mass cannot be changed instantly, the swing continues to K, at which point the additional sending-end rotor energy, resulting from the fault, is completely absorbed (area II equals area I).

The velocity of the sending-end mass, with respect to the receiving-end mass, is 0 at K. At K, the electrical output of the sending end exceeds the mechanical input; therefore, the swing reverses, reaching a point N. At N, the swing reverses again. Voltage regulator and governor action, as well as system resistance, will dampen the oscillation, until the final operating point is reached.

If the initial swing went to point L and the sending-end generators still had excess rotor energy (area II smaller than area I), the swing would continue in the same direction. After L was passed, the mechanical input of the sending-end generators would again exceed the electrical output, and the swing would be accelerated, resulting in instability with the machines operating out of synchronism with each other. After this, only system separation and resynchronizing of the machines could restore normal system operation.

4 RELAY QUANTITIES DURING SWINGS

With the fault cleared and the system operating out of step, extreme variations in voltage and current will occur throughout the system. Figure 14-6 provides insight into this phenomenon. A simple system is represented with the relay of interest shown at some intermediate point between the two sources. The relay voltage V_R is established by the angle between V_S and V_U (each assumed here to maintain their predisturbance values) and the distribution of impedances (Z_S, Z_L, and Z_U) between these two sources. V_R will assume some position between them. As V_S, the accelerating source,

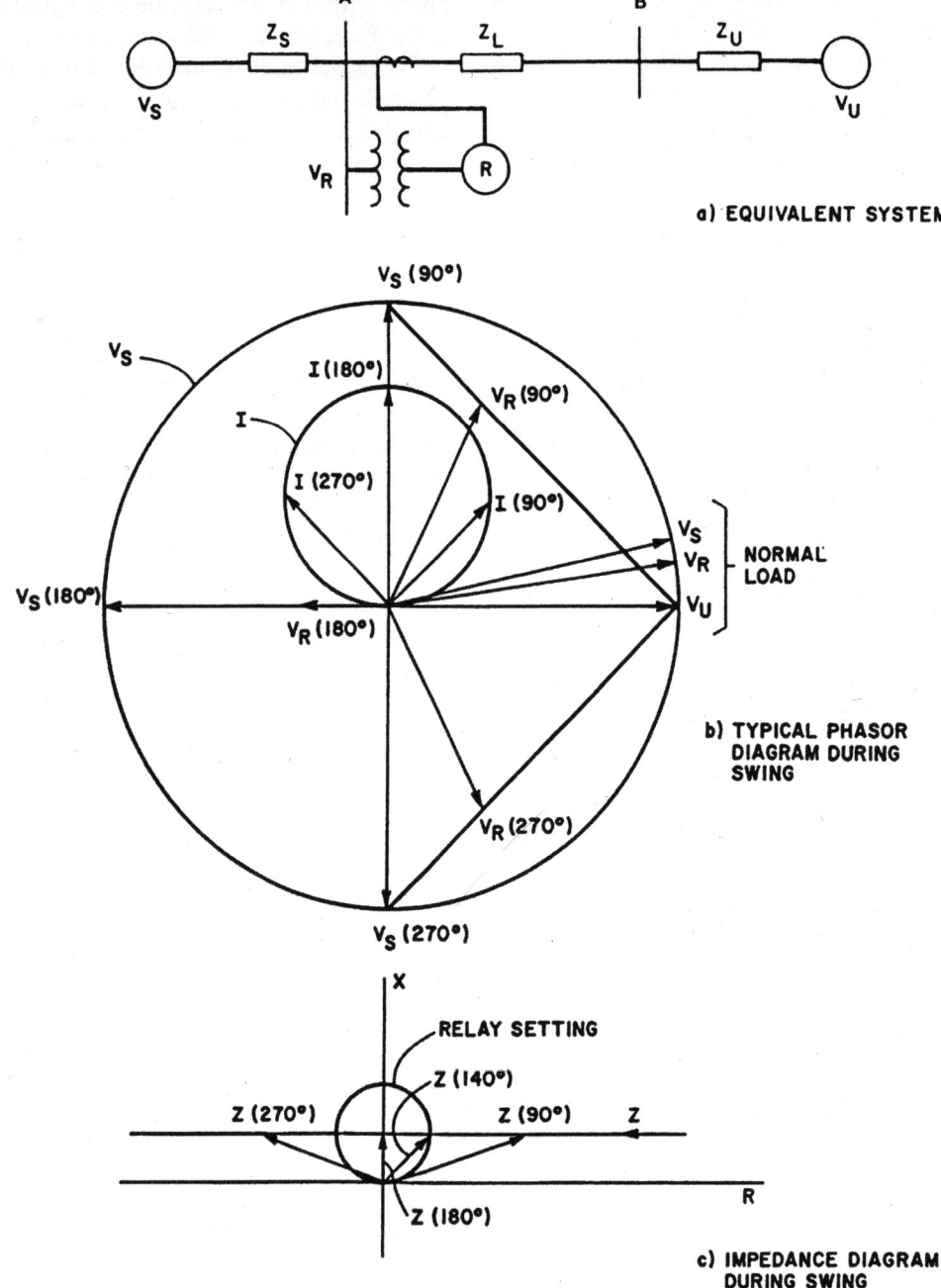

a) EQUIVALENT SYSTEM

b) TYPICAL PHASOR DIAGRAM DURING SWING

c) IMPEDANCE DIAGRAM DURING SWING

Figure 14-6 Relay quantities for OS condition. (Shown for 90° impedance angles.)

moves in phase position with respect to V_U, the apparent impedance viewed by the relay (V_R/I_R) will change with time, producing a trajectory on an R-X (resistance-reactance) diagram such as that of Figure 14-6c.

With the depressed voltage and large current, this swing condition (long after the fault has been success-

fully cleared) appears to be a three-phase fault. The location of this apparent fault is at the electrical center of the system. If this apparent impedance enters the operating area of a distance relay, it will operate.

If V_R and V_U are not supported at full value, their ratio is influential in determining the locus on the R-X diagram as the swing progresses. Also, if impedances are not pure reactance values as is assumed in Figure 14-6, the effect can be determined using a similar simple diagram.

Interaction between machines in complex networks can only be determined by a large-scale digital study. For distance relaying evaluation, actual impedance (using line current and relay voltage) values must be determined with respect to time for each pertinent relay location, and each chosen switching condition, fault location, and type.

5 EFFECT OF OUT-OF-STEP CONDITIONS

5.1 Distance Relays

A distance relay (21) responsive to three-phase faults will operate if an out-of-step (OS) condition produces a swing locus that falls within its operating area (Fig. 14-7).

When swing ohms enter the operating area of a zone 1 relay with a circular characteristic, there is a 90° angle between the voltages at the points along the line angle identified by the relay reach. In Figure 14-7, for example, there must be a 90° phase displacement between the voltage phasors at relay location A and a point at 90% of the line length for a 90% reach zone 1 phase relay to operate on a swing. The effective generator voltages will be displaced substantially more than 90° (angle m in Fig. 14-7). The likelihood of a system attaining a stable operating condition after

such a swing is virtually nil. In general, zone 1 swing-trips occur only on unrecoverable swings.

Some form of blinding is required to screen over-reaching distance relays against tripping on severe swings from which recovery is possible. Operating independently, a phase-distance relay (21) will initiate tripping when the angle between the two system voltages is very large and increasing (Fig. 14-7).

Figure 14-8a shows that zone 1 tripping is highly dependent on the locus of the swing ohms and therefore the distance to the electrical center for the case involved.

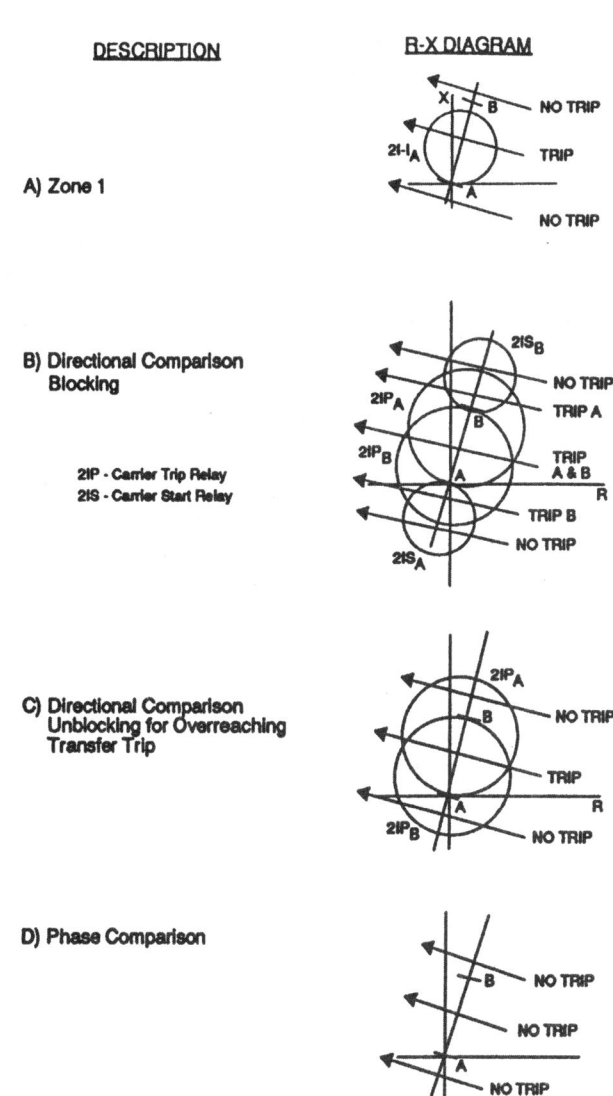

Figure 14-8 Effect of OS swings on various line relaying systems.

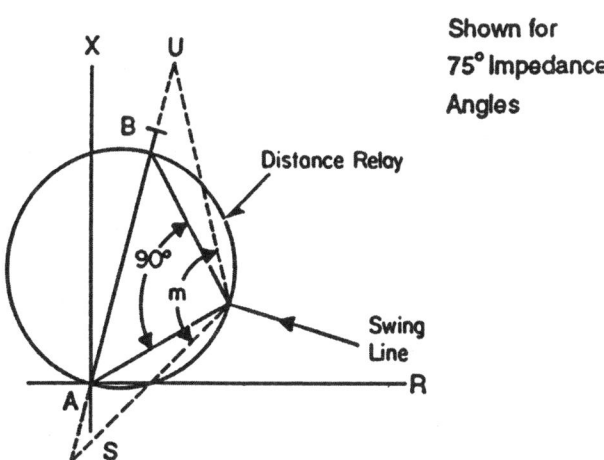

Figure 14-7 Source angle relationship for swing trip on the system of Figure 14-6.

5.2   Directional Comparison Systems

Swing ohms entering a protected line area will produce simultaneous high-speed tripping at the two terminals. Only swings entering external to the line area will block tripping, and then only if local 21S operates and/or the remote 21P does not. For the permissive overreaching transfer-trip or the unblock system, tripping occurs on OS conditions only if the 21P at both terminals operate. If tripping is to be avoided for cases where operation takes place, out-of-step blocking must be used.

5.3   Phase-Comparison or Pilot-Wire Systems

Phase-comparison or pilot-wire schemes are solely current-responsive, and since swings produce a through-current condition, tripping does not occur.

5.4   Underreaching Transfer-Trip Schemes

Out-of-step swings entering the circle of either zone 1 relay will cause tripping at both terminals when using the underreaching transfer-trip (whether direct or permissive) scheme.

5.5   Circuit Breakers

With the two system segments 180° apart at the instant of interruption, a theoretical undamped recovery voltage of four times normal is possible. Figure 14-9 describes this phenomenon with the breaker at the electrical center of the system. At current 0, where interruption takes place, the voltage on each side of the breaker must settle at a new value. In the process of getting there, overshoot takes place as a result of the presence of inductance and distributed capacitance in the system. Recovery voltage is the voltage across the breaker contacts following current interruption.

Figure 14-10 shows that this identical phenomenon occurs even though the breaker is as far away from the electrical center as possible. The extremely large transient recovery voltage still appears. If the circuit breaker has insufficient dielectric strength to withstand this voltage, reignitions will continue until a more favorable angle is reached. To interrupt at all, a breaker must be capable of attempting interruption, possibly for several seconds, at each current 0. If the breaker cannot perform such interruptions, tripping

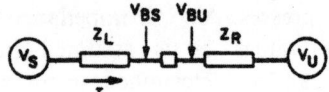

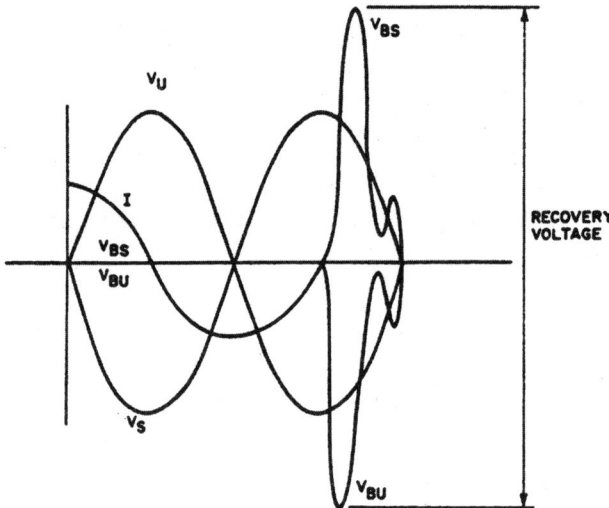

Figure 14-9 Maximum recovery voltage on OS trip with breaker at electrical center.

must be initiated at a favorable angle, preferably just before the two sources are in phase.

5.6   Overcurrent Relays

Figure 14-6 can be used to illustrate the conditions encountered by phase-overcurrent units during swings. Assume, for example, that an instantaneous overcurrent unit set for 2.5 times full load were used in a line connecting V_s and V_u, and that $Z_S + Z_L + Z_U$ equals 0.765 per unit on the full-load base. During an OS condition, the instantaneous unit would operate because the current reaches at least 2.61 (2/0.765) times full load when V_U lags by 180°. Swings during stable conditions will also result in higher than normal currents, although currents will be considerably less than during an OS condition.

5.7   Reclosing

When a fault persists after reclosing, the stability of the system will probably be jeopardized. On the other hand, system stability is greatly improved if the fault is

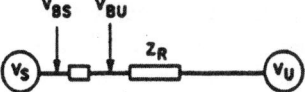

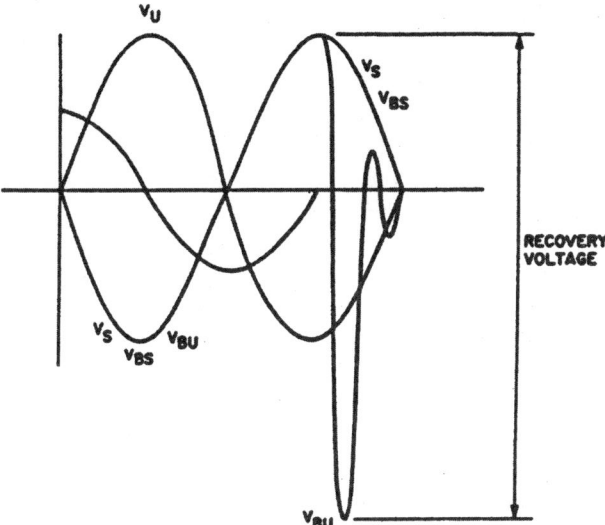

Figure 14-10 Maximum recovery voltage on OS trip with breaker at source.

temporary and does not reignite following reclosure. Three-phase faults tend to be permanent more often than other faults. They also have the most severe effect on stability. For this reason, reclosing is often blocked for three-phase faults and for OS conditions, but allowed for all others.

There appears to be no advantage to the high-speed reclosing of both terminals following an OS trip. Reclosing one terminal, or preferably blocking tripping at one terminal, will facilitate system restoration. The second terminal can be reclosed under synchronism-check relay supervision.

6 OUT-OF-STEP RELAYING

Ideally, fault relays should clear faults fast enough to maintain stability. Also, they should not operate on swings from which the system can recover. If a system does go out of step, it should be split by circuit breakers opening at a few preselected locations, in such a way that generation and load on each side of the split are reasonably balanced. The system should not be split so most of the generation is separated from the major system load.

In Figure 14-11, breaker A is in a poor location for splitting system (1) from (2), since it would dump one unit of load on system (2), which only generates 0.4 unit of power. Splitting the system using breakers D or E would offer a more tolerable generation/load balance. In this scheme, system (1) need only increase its generation from 0.6 to 0.66 to maintain frequency.

Figure 14-11 also illustrates that OS tripping is desirable at some points, but should be blocked at others. This selective tripping/blocking philosophy is basic to the intelligent application of OS relaying.

6.1 Generator Out-of-Step Relaying

Generator per unit reactances have steadily increased over the years, and inertia constants have decreased as machine ratings have increased. This, in turn, has reduced critical clearing times and increased the need for the OS relaying for generators.

Loss-of-field relays, equipped with directional units and undervoltage supervision, may provide a measure of OS protection for generators. Viewed from the terminals of a large modern machine, the ohms will, in general, fall within the machine or unit transformer when the machine is out of step with the system. If the swing ohms fall within the machine for the system shown in Figure 14-11, the KLF (or KLF-1) loss-of-field relay (40) will operate if swing ohms stay inside the characteristic circle for 0.25 sec.

If a loss-of-field relay is used for OS sensing, the timer must not time out for stable swings. It must operate, however, for field failure before damage (or further damage) can occur, and it must recognize the fastest realistic swing rate. Generally, all these time constraints can be satisfied. Figure 14-12 shows a typical stable swing locus following a severe three-phase fault.

If swing ohms pass through the unit transformer, OS detection may not be possible with either a loss-of-

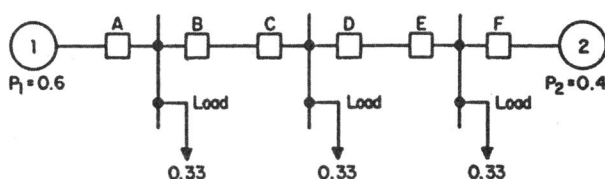

Figure 14-11 Generation and load distribution.

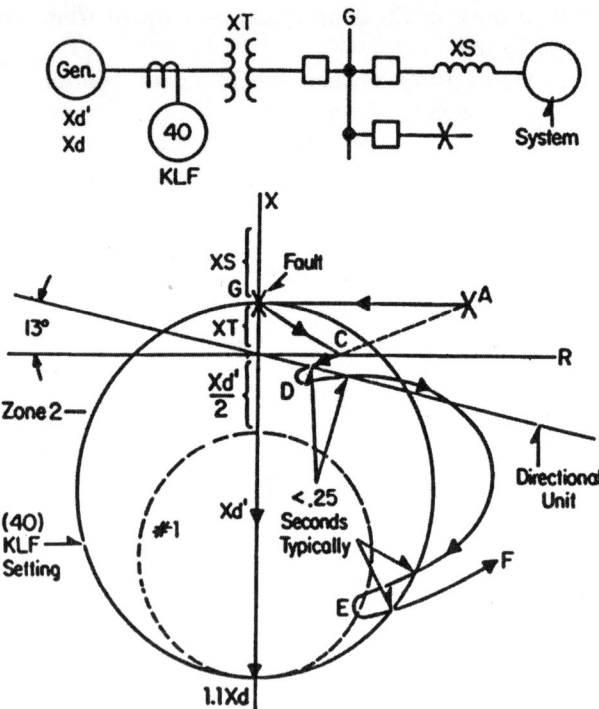

Figure 14-12 Stable swing following clearing of nearby three-phase fault with the KLF relays (40)

field relay or simple distance relay. Moving the directional unit characteristic output to point G on Figure 14-12 would substantially increase the possibility of a false trip on a stable swing, such as GCD. Alternatively, the time criterion could be increased to the point where the stable swing would not trigger relay operation, but then the fastest out-of-step swing may not be recognized.

When the smaller characteristic is used as described by the dotted circle of Figure 14-12, a higher degree of security is achieved. However, it occurs at the expense of making out-of-step recognition impossible, as well as losing the ability to match accurately, as described in Chapter 8, the loss-of-field relay characteristic, steady-state stability, and minimum excitation unit limits for the machine.

In general, devices applied for other functions (fault detection, loss of field, thermal protection, etc.) are unsuitable for detecting OS conditions. The use of relaying, tailored explicitly for OS detection, is the only dependable approach, unless extensive studies demonstrate that other devices will suffice or instability is improbable.

This should not be taken to mean that generator tripping is encouraged when OS conditions develop,

but rather that OS detection may be easiest at the generating plant. Out-of-step separation would then be accomplished by transfer-tripping, or other suitable means, to maintain a generation/load match, as described above.

6.2 Transmission-Line Out-of-Step Relaying

The prime criterion in OS tripping is to maintain a generation/load match in the islands created. If such a match were perfect, no large load shifts and load dropping would be required. Also, little or no generation would be dropped. To even approximate this ideal would, in all probability, require trip-blocking at some locations and trip initiation at others.

Distance-relay operation on OS conditions tends to occur at locations where the relay reach settings are longest. There are two reasons for this phenomenon. First, the minimum system voltage during an OS condition tends to occur in the high-impedance segments of the system. Second, distance relays with long reach settings, such as those on long lines, cover a larger area on the R-X diagram and therefore are more likely to respond to swing conditions. Out-of-step tripping at long-line terminals is not necessarily conducive to ideal system splitting.

7 PHILOSOPHIES OF OUT-OF-STEP RELAYING

Certain fundamental objectives should influence the design of protection systems:

1. Block tripping at all locations for stable swings.
2. Ensure separation for every OS condition.
3. Effect separation at points that will leave a satisfactory load/generation balance in each separated area. Loads should not be interrupted.
4. Block tripping or automatically reclose at *one end* of any line that trips because of an OS condition.
5. Initiate tripping while the systems are less than 120° out of phase and the angle is closing in order to minimize breaker stress.
6. Minimize the possibility of an OS condition occurring by

   (a) Using high-speed relaying
   (b) Using a high-speed excitation system

(c) Employing loss-of-field relays to remove a unit that is drawing excessive reactive power from the system

(d) Providing sufficient transmission capacity

(e) Tripping generators upon the loss of critical transmission lines

(f) Applying generator braking resistors or inserting series capacitors for critical faults

(g) Applying fast valving techniques

(h) Using an independent mechanism for each breaker pole to downgrade faults from three-phase to phase-to-phase

Although easily stated, these objectives are not so readily achieved, particularly item 3 above.

7.1 Utility Practice

Utility practice consists of a combination of:

1. Allowing the line-protection relays to initiate OS line tripping
2. Allowing the loss-of-field relays to initiate OS generator tripping (when the nature of the loss-of-field relay allows it)
3. Restricting relay-trip sensitivity at the higher power factors
4. Blocking tripping on OS
5. Blocking reclosing on OS
6. Initiating tripping using relays designed for OS tripping

There is no industry standard for protection-system design; however, once the difficult functional decisions of "what" and "where" are made, there is reasonable consistency in the "how."

8 TYPES OF OUT-OF-STEP SCHEMES

Some typical systems used in OS relaying are described here.

8.1 Concentric Circle Scheme

A concentric circle scheme for OS detection on terminal A is shown in Figure 14-13. Customarily, an OS relay with a characteristic as shown for 68 is added to a transmission-line relaying system and surrounds an over-reaching element, such as 21P. Because of rotating apparatus inertia, significant time is required for the torque angle to advance and the swing locus to

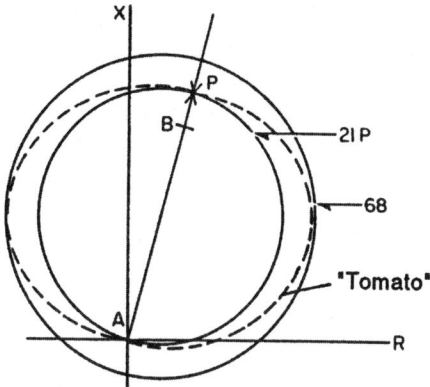

Figure 14-13 The concentric circle scheme for out-of-step detection.

pass from 68 to 21P. For a fault within the 21P reach, however, both elements operate essentially simultaneously. The relaying logic senses the sequence, identifies swing or fault, and initiates the appropriate action.

The dotted arcs of circles described as a "tomato" characteristic represent a popular analog implementation of the "outer" characteristic.

This scheme is appropriate for an OS trip-blocking function or reclose-blocking. It is not appropriate for OS tripping, however, unless additional logic is added. If an external phase fault occurs close to the balance point of 21P, as at P in Figure 14-13, for example, the relay will respond slowly because of its low energy level. Device 68, on the other hand, has an appreciably higher energy level and operates faster than does 21P. Removing remote terminal infeed on external fault clearing following breaker failure can also produce sequential operation of 68 and 21P. If the time in the sensing logic is shorter than this operating-time difference, a fault at P would be incorrectly identified as an OS condition and would cause a false trip at A. If the time is increased to avoid this situation, rapid OS swings would not produce OS tripping. Adding a third concentric circle would allow better perceptive segregation of swings and faults, but could introduce load ohm involvement. OS tripping can be achieved with the two-concentric-circle scheme if the 68/21P *operating* sequence is checked and the *resetting* sequence is correct.

For the three-terminal applications, infeed can adversely affect OS blocking relays that use the concentric circle scheme if the sequential clearing of a three-phase fault can occur. The reach-shortening effect of the third terminal infeed can cause an internal line-end fault to operate 68, but not 21P. Clearing the infeed may then allow 21P to operate. This sequence could cause undesirable OS trip-blocking at one terminal. Out-of-step blocking then should not be used in a three-terminal application unless remote terminal coverage can be obtained with maximum infeed.

8.2 Blinder Scheme

One form of a blinder relay has an operating characteristic that parallels the transmission-line plot in an R-X diagram (Fig. 14-14). A single-blinder relay 21B, gives the two linear characteristics shown.

An obvious application of this device is for limiting the coverage of a distance relay in the load area. This is a side benefit of the application of blinder OS relays.

A single-blinder relay plus auxiliary logic can be used for OS tripping. Its use, however, is limited to only those applications where OS trip-blocking of phase-distance relays is not required, since swings passing through the line section (on an R-X plot) will cause operation of the line relays. A single blinder cannot distinguish between a fault and an OS condition until the resetting sequence is confirmed. Such a scheme delays OS tripping until the swing is well past the 180° position and is returning to an in-phase condition.

The single-blinder OS package is recommended for causing a system splitting to occur through the tripping of a line that is protected by a phase-comparison relay.

Another application is OS tripping on swings passing beyond the reach of the line relays (on an R-X plot). Finally, this relay scheme is particularly well-suited for generator OS trip applications.

The two-blinder scheme (Fig. 14-15) senses OS conditions by observing the operating sequence of the outer and inner blinders. A fault produces essentially simultaneous operation; an OS condition causes the outer blinder to operate first, followed by operation of the inner blinder. The two-blinder scheme allows the trip-area restriction of distance relays, OS trip-blocking, OS reclose-blocking, or OS tripping, regardless of normal load-flow direction.

9 RELAYS FOR OUT-OF-STEP SYSTEMS

9.1 Electromechanical Types

9.1.1 KS-3 (68) OS Blocking Scheme

Figure 14-16 shows the configuration for the type KS-3 out-of-step blocking scheme. If Z_{OS} operates approximately 60 msec or more ahead of 21-2, the OS relay operates to block all or selected tripping. The OS relay also blocks reclosing when some elements in the system, such as zone 1 or time trips, are allowed to operate during OS conditions.

For faults, 21-2 trip contacts or the D_0 and I_0 contacts of the ground relay close to short out the OS unit coil, blocking pickup of the OS unit. This scheme

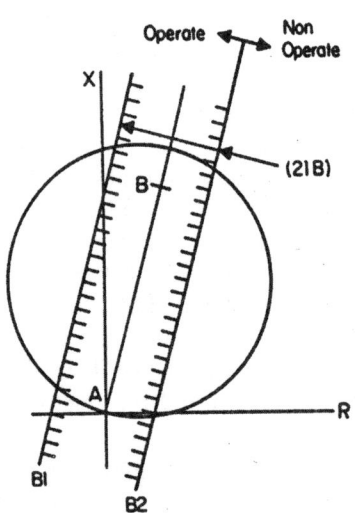

Figure 14-14 The single blinder scheme for out-of-step detection.

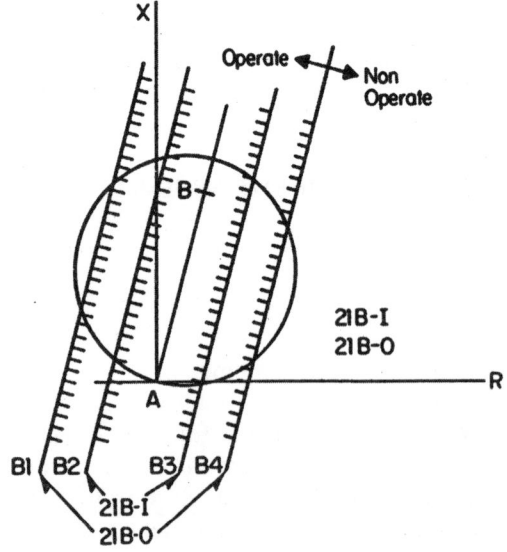

Figure 14-15 The two blinder scheme for out-of-step detection. (I = inner blinder, O = outer blinder.)

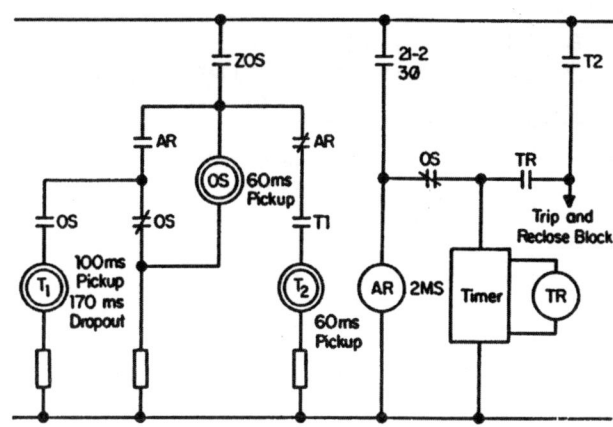

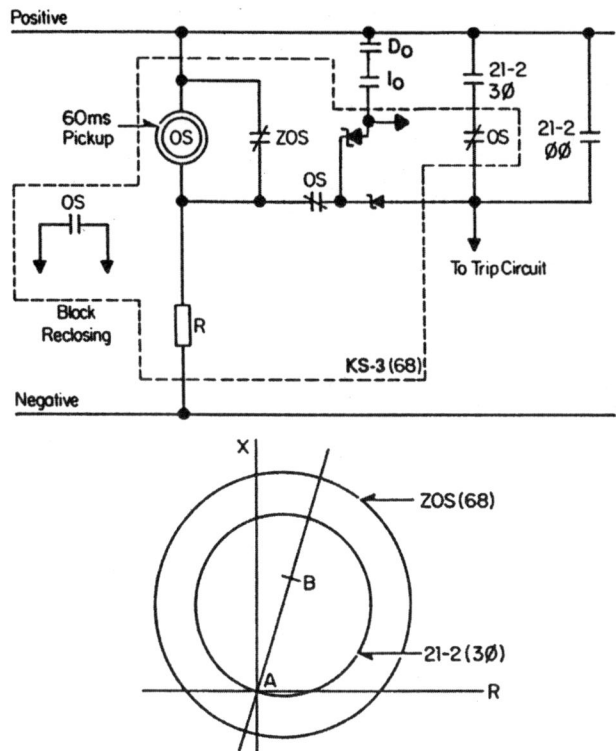

Figure 14-16 The type KS-3 out-of-step blocking scheme.

Figure 14-17 The KST out-of-step tripping scheme.

is recommended for short-to moderate-length lines. It should not be used on long lines, where the load might operate the Z_{OS} unit.

9.1.2 KST (68) OS Tripping Scheme

Figure 14-17 illustrates the KST scheme of OS tripping. After sensing an OS condition in the same way as the KS relay, telephone relays T1 and T2 add two requirements: The 21-2 relay (for example, the KD-10 phase-distance relay) must operate for 100 msec, and 21-2 resets 60 msec or more ahead of Z_{OS}. On a swing, Z_{OS} operates first to energize OS. If 21-2 does not operate before 60 msec, OS operates. Then when 21-2 operates, the AR relay, T1, is energized. If both Z_{OS} and 21-2 remain closed for 100 msec, T1 operates. As the swing moves out, 21-2 resets first, deenergizes AR, and permits the energization of T2 through AR back contacts if Z_{OS} is still closed. If Z_{OS} does not reset for 60 msec, T2 operates to trip and block reclosure as shown.

A fault that operates Z_{OS} and 21-2 together (or within approximately 60 msec) will have no effect because the short around the OS coil will be

maintained. The zone 2 timer would time out and trip in the scheme shown.

9.2 Solid-State Types

9.2.1 SDBU-1 (21B), SI-T (50), ARS (94), OS Tripping Scheme

For the single-blinder OS tripping scheme (Fig. 14-18), swings from the right to left cause B_1 to operate, B_2 to operate, B_1 to reset, and then B_2 to reset. It is of no consequence whether B_1 is initially operated by load and B_2 does not subsequently reset.

Device 50 (SI-T relay) is sensitively set and operates at a current level above maximum zero power factor interchange, line charging, or transformer-magnetizing current. The device operates when a swing begins and prohibits load pickup trip.

Thus, AND 2 operates when B_1 and 50 operate with B_2 reset to identify the swing origin in the positive F region. After 4 msec, the feedback circuit holds the upper input AND 2. AND 4 has an output when the swing moves between B_1 and B_2 to operate both blinders. If AND 4 output persists for 20 msec and the swing moves across B_1 to reset it, AND 6 has an output. An output from AR occurs 20 msec later for tripping and reclose block.

Swings originating to the left of B_1 traveling left to right produce identical action through AND 1, AND 3, and AND 5. The restricted trip feature prevents tripping on recoverable swings. B_1 and B_2 may be used to supervise the tripping of a phase-distance relay.

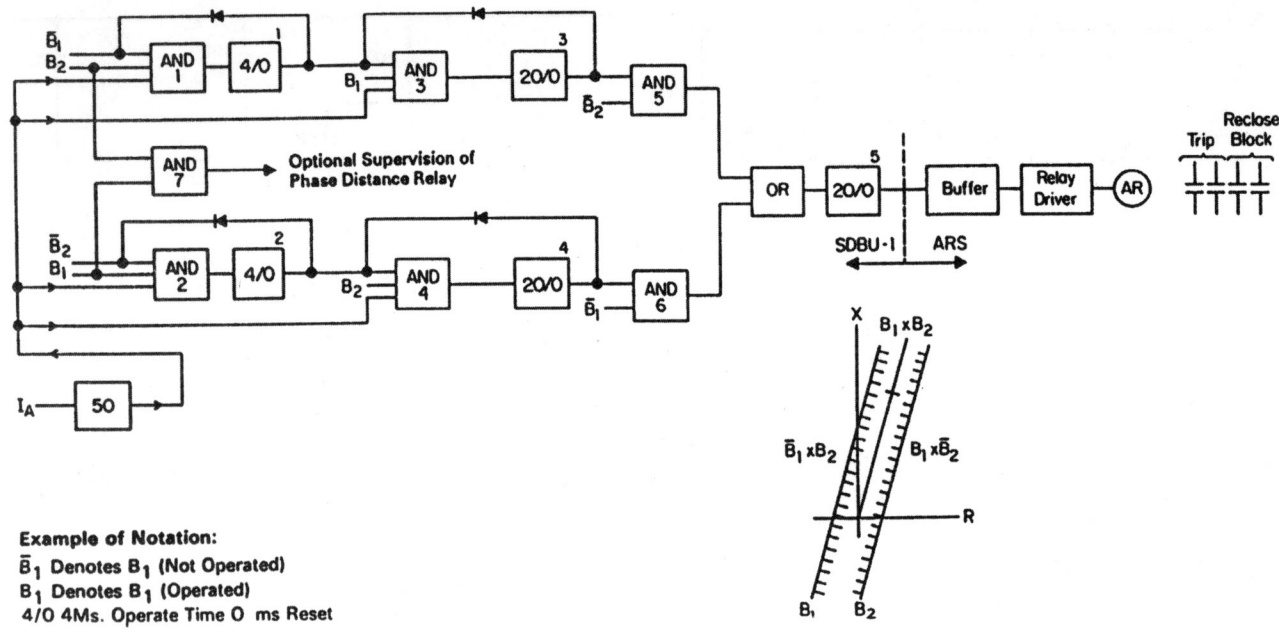

Example of Notation:
$\bar{B}_1$ Denotes B_1 (Not Operated)
B_1 Denotes B_1 (Operated)
4/0 4Ms. Operate Time 0 ms Reset

Figure 14-18   The SDBU-1 out-of-step tripping scheme.

Out-of-step block of reclosing is not available with this complement, unless OS trip is used.

The above scheme is recommended for generator OS sensing, because its logic requires that swing ohms emerge from the side of the relay characteristic opposite to that from which it entered. That is, there must be a reversal of power flow as viewed from the machine terminals, and the reversal must occur during high current. These two conditions will not be satisfied unless the machine is out of step with the system. A low-current reversal can, however, occur during motoring.

This scheme, or its equivalent, supervised by an over-current or distance relay, is the most secure available for generator OS tripping. A severe but stable swing, such as shown in Figure 14-12, cannot cause misoperation, regardless of the timing involved in the transient.

9.2.2   Lens Scheme

One significant variation of the blinder scheme uses a lens and single-blinder line characteristic as described in Figure 14-19 for out-of-step tripping. It is also equipped with a reactance-type characteristic to restrict the reach of the system to the desired extent. Various areas are established (inside the lens, right of the blinder, left of the blinder, above the reactance line, below the reactance line) and, by the addition of

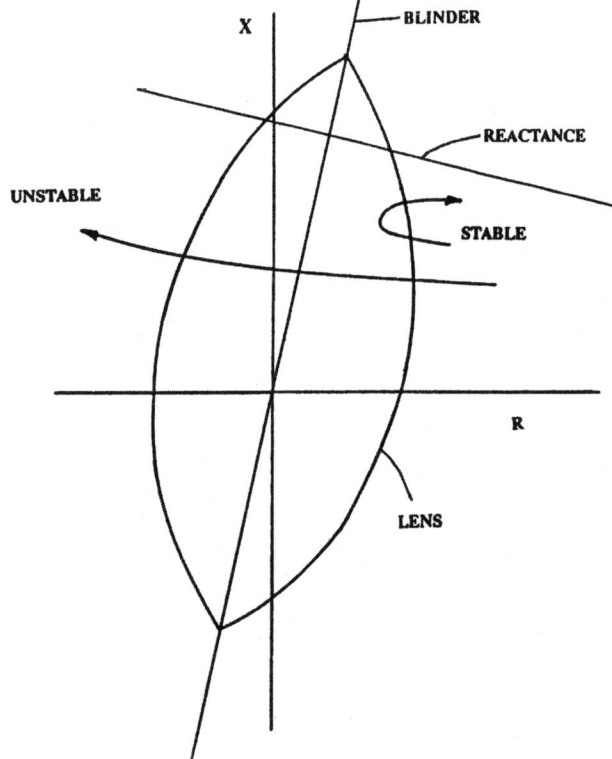

Figure 14-19   Lens scheme for out-of-step tripping GZX-104.

sequence and timing logic, can determine the origin and termination of swings and the time involved in passage, thus identifying out-of-step conditions.

9.2.3 Double-Blinder OS Tripping and Blocking Scheme

Figure 14-20 contains the logic for out-of-step sensing used in MDAR 2.5. 21BO and 21BI are the outer and inner blinder units, respectively. $3\phi F$ is an input from a phase selector identifying the fact that no unbalanced fault exists (not ϕG, not $\phi\phi$, not $\phi\phi G$). I_{AL} signifies that current above a preselected level exists.

The logic differentiates between a swing and fault through the sequence of operation of 21BO and 21BI. If it is a swing, the optional blocking (shown as switches OSB1, OSB2, and OSB3) of each of three distance units is obtained. The inner blinder supervises (permits) tripping of each of the distance units to avoid any possibility of load trip not resulting from a swing condition.

Logic is included to permit delayed OS tripping "on the way out" to minimize the possibility of breaker damage during tripping or "on the way in" to minimize the possibility of thermal damage to a transmission line. The terms "way in" and "way out" refer to the trajectory of the ohmic value seen by the

relay and the point at which tripping is initiated (on the way into the inner-blinder operating region or on the way out of the operating region of the outer blinder).

10 SELECTION OF AN OUT-OF-STEP RELAY SYSTEM

The key ingredients in out-of-step relaying are identification and strategy. Identification is possible using any of the schemes described, though a higher order of security is mandatory for those schemes to be used for OS tripping. Also, OS identification is dependent on suitable voltage and current relationships being present at the point of application of the relaying system to clearly recognize an OS condition when it occurs. Strategy relates to action required once an OS condition has been identified. The choices are OSB (out-of-step block of tripping), OST (out-of-step trip), and OSBR (out-of-step block of reclosing).

The electrical center is not a fixed point in a system. Indeed, several electrical centers may be present for a given swing condition. Further, the electrical center moves as the number of generators and switching conditions vary. For the particular unstable case under consideration, the ohmic value (and angle) manifested at the relay location must be known with respect to

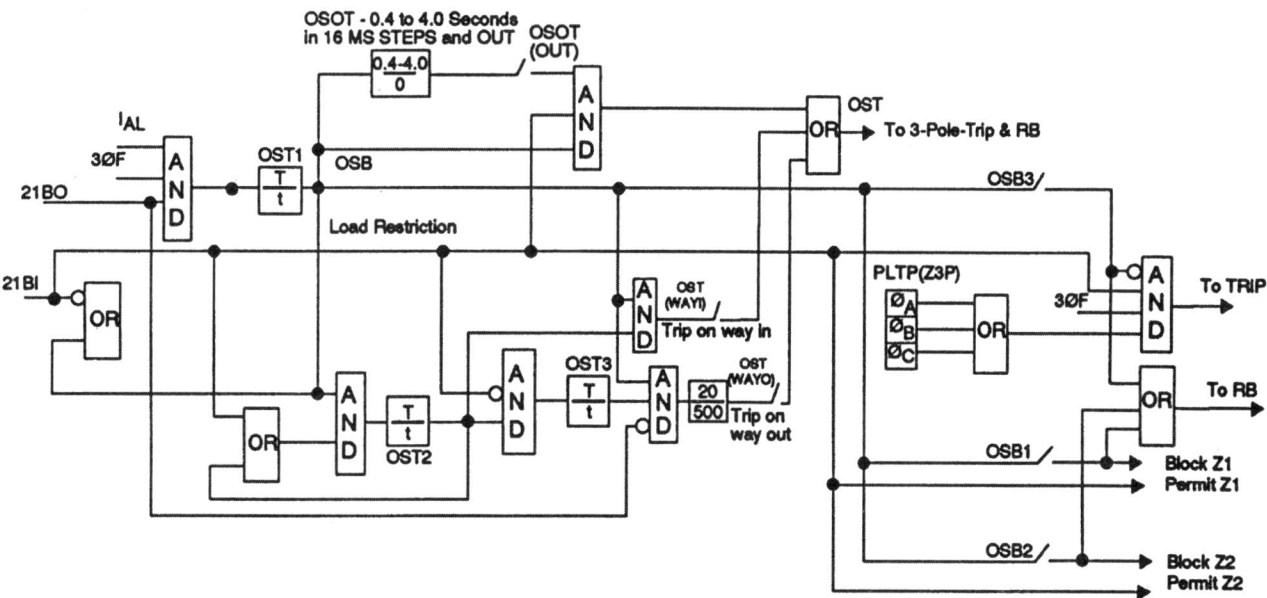

Figure 14-20 Out-of-step trip logic used in MDAR 2.5.

time to assure that proper recognition is possible. In general, a system study is required for this.

Out-of-step blocking is less critical than OS tripping in terms of identification of the swing and security. OS blocking must respond only when the blocked device responds. Swings producing ohmic values beyond the reach of, say, a supervised zone 2 phase-distance relay need not be recognized by the OS detection scheme. Also, a fault appearing on the R-X diagram between the characteristics of an OS relay and a phase-distance relay that is used with it can falsely identify an OS condition with impunity, whereas OS tripping under the same circumstances could not be tolerated.

Stated simply, (1) OS blocking should be applied when swing-produced trips are possible but are intolerable, and (2) OS tripping should be applied when tripping will not naturally occur and tripping must take place for an optimum generation-load match following separation of the two system segments that were out of synchronism. Even when "natural tripping" occurs (the distance relays applied for other functions operate for the OS condition), OS sensing is required to block reclosing. Reclosing following a legitimate OS trip is futile and should not be attempted.

Blinders aid natural tripping by providing a screening effect against undesired tripping in response to large power swings that are stable. This is particularly useful in long-line applications that require large impedance settings on distance relays, making them more susceptible to tripping on load.

Out-of-step relaying is as much art as science, requiring a realistic appraisal of what is possible, what is probable, and what is certain. It may, in particular areas of a complex system, be less expensive to universally apply out-of-step blocking rather than to deduce where it is needed.

15

Voltage Stability

L. WANG

1 INTRODUCTION

Power system instability may occur under two kinds of disturbances: (1) gradual variations of system conditions such as load, and (2) drastic changes of system conditions such as faults and loss of important lines or generators. According to the reasons causing power system instabilities, they can be classified as small-disturbance (SD) voltage instability, SD angle instability, large-disturbance (LD) voltage instability, and LD angle instability or transient instability. While Chapter 14 covers transient instability, this chapter mainly focuses on voltage instability.

1.1 Small-Disturbance Instability

During normal steady-state conditions, power system voltage magnitudes, voltage angles, and frequency are constant. The power system is operating at stable equilibrium points. A power system equilibrium point is composed of a set of independent voltages and currents.

As power system operating conditions change, such as the variations of load conditions or system configurations, the system equilibria will change accordingly. Under small variations, the transition from one equilibrium point to the other can be assumed to be instantaneous. At certain conditions, the number of system equilibria will change, which is called *bifurcation*. For example, both the maximum power point in the power-angle curve and the nose point in the Q-V curve are where the number of

equilibria unite, and both are bifurcation points. At the bifurcation point, a system usually loses its stability. The resulting unstable condition is small-disturbance instability.

For a stable operating condition, it is helpful to know how stable that condition is. This translates to the amount of power a power system can further support from a steady-state condition without losing stability. The study of SD instability is concerned with the stable degree of a stable equilibrium point. A static system model will be sufficient for the study of SD instability.

Small-disturbance instability can be further classified as SD angle instability and SD voltage instability based on the reasons initiating the bifurcation or unification of system equilibria. If a bifurcation occurs because of generator angles increasing to their limits, the instability is called SD angle instability. If a bifurcation occurs because of bus voltages decreasing to their limits, the instability is called SD voltage instability. SD angle and voltage instabilities are explained in the following two examples.

1.1.1 SD Angle Instability

The mechanism of angle instability can be illustrated by considering a single-machine infinite-bus (SMIB) system (Fig. 15-1). Figure 15-2 shows the system power-angle curve, which is the trajectory of the system equilibrium points with the mechanical power input being the variable. Superscripts "s" and "u" represent state variables at the stable and unstable equilibria, respectively. For a specific mechanical

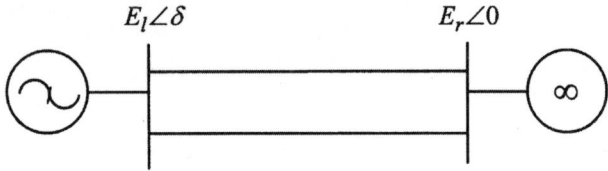

Figure 15-1 Single-machine infinite-bus system.

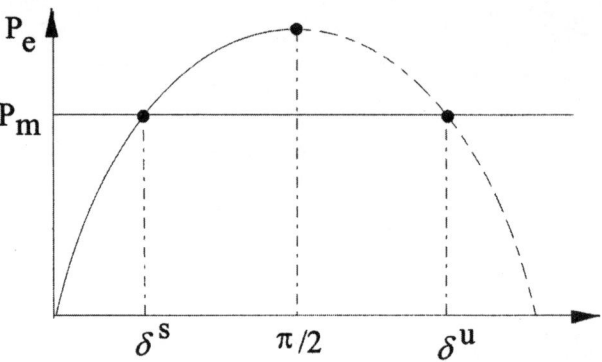

Figure 15-2 Power-angle curve of a SMIB system.

power input, there exist two equilibrium points. One is stable, and the other is unstable. In Figure 15-2, the continuous line in the power-angle curve represents stable equilibria, and the dashed line in the power-angle curve represents unstable equilibria.

As the mechanical power is increased gradually, the stable and unstable equilibria will move closer. Finally, the stable and unstable equilibria will join together and the system reaches the bifurcation or maximum loading point. At this point, the system will lose the stable operation condition. This instability phenomenon is the small-disturbance angle instability because the unstable condition results from a generator angle reaching its limit (90° for this specific situation).

1.1.2 SD Voltage Instability

Analogous to angle instability, the mechanism of SD voltage instability can be illustrated by considering a generator connected to PQ load through a double-circuit transmission line, as shown in Figure 15-3. Figure 15-4 shows the system Q-V curve. The Q-V curve can be obtained by increasing the load while keeping the source voltage constant. The Q-V curve represents the trajectory of stable and unstable equilibrium points. The nose point in the Q-V curve represents the maximum power transfer, which occurs at a load impedance corresponding to the Thevenin impedance as viewed from the load bus.

For any load conditions, there exist stable and unstable equilibrium points. The solid line in the Q-V curve represents the trajectory of stable equilibrium points, and the dashed line in the Q-V curve represents the trajectory of unstable equilibrium points. A power system can never operate at unstable equilibrium points, thus unstable equilibrium points only have analytical importance and do not have practical meaning. With the increase of load, the stable and unstable points will become closer and eventually will join together at the nose point of the Q-V curve. The nose point corresponds to the bifurcation point.

Unlike the situation discussed in the SMIB system, the occurrence of the bifurcation here is caused by the

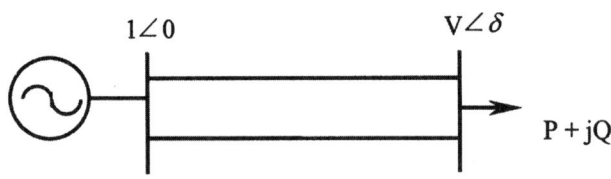

Figure 15-3 Two-bus power system.

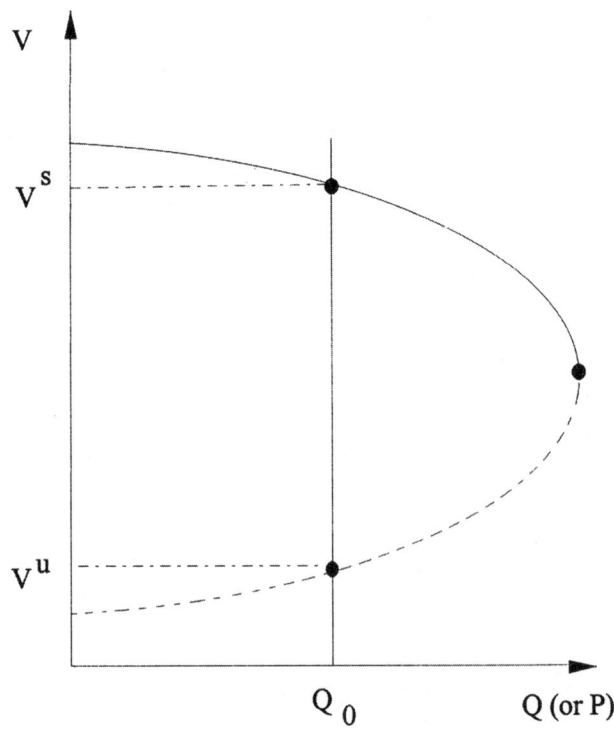

Figure 15-4 Q-V curve of a two-bus system.

voltage decreasing, and thus the system's inability to support further load increases. Therefore, this instability is SD voltage instability. At the bifurcation point, the voltage phase angle may be much less than its 90° limit.

1.2 Large-Disturbance Instability

Large disturbances move a system condition far away from a stable equilibrium. After the disturbances, if a system can recover to its stable equilibrium, it is stable; otherwise instability will occur. These kinds of instability are called LD instability.

The study of LD instability is concerned with the system ability to return to an equilibrium following large disturbances. Based on the dynamic system theory, for any stable equilibrium, there exists a region around it; if the system's initial transient states fall in this region, the system will recover to the stable equilibrium. If the initial states are outside of this region, the system will be unstable. This region is defined as the domain of attraction of the stable equilibrium.

Finding the domain of attraction is a problem only pertinent to nonlinear systems because linear systems, if they operate at steady states, will remain stable after large disturbances irrespective of the strength of the disturbances. In other words, the domains of attraction of linear systems are the whole state space.

Figure 15-5 shows the concept of linear system stability and Figure 15-6 shows nonlinear system stability. In Figure 15-5a, the ball is at its stable equilibrium; in Figure 15-5b, the ball is away from its stable equilibrium but will return to it eventually irrespective of the initial positions of the ball. For

nonlinear system, the situations are quite different, as shown in Figure 15-6. Depending on the disturbances or its initial positions, the ball may not return to its predisturbance stable equilibrium and may never return to any stable equilibrium. Only when the ball falls within the domain of attraction of a stable equilibrium will it become stable and return to that equilibrium eventually.

Under large disturbances such as short circuits or loss of important lines or generators, a power system cannot be described by a linear model. This is mainly because the real and reactive power terms include sine functions and multiplications of voltages. The domain of attraction of its equilibrium point is not the whole state space but equilibrium dependent. To determine the domain of attraction, a dynamic system model will be required.

Similar to SD instability, LD instability may be classified as LD angle instability (or transient instability) and LD Voltage instability. The concepts of transient instability and LD voltage instability will be explained using the example systems shown in Figures 15-1 and 15-3, respectively.

1.2.1 LD Angle Instability (or Transient Instability)

Figure 15-7 shows the transient responses of the SMIB system following two large disturbances, which set the initial states of the postfault system at $(\delta, \omega) = (\delta_1, 0)$ and $(\delta, \omega) = (\delta_2, 0)$. Angle δ represents the angle difference between the two generators, and angular velocity ω represents the angular speed difference between the two generators. For simplicity of illustrations, we assume the angular velocity differences are zero at the two initial states. For the initial state $(\delta_1, 0)$,

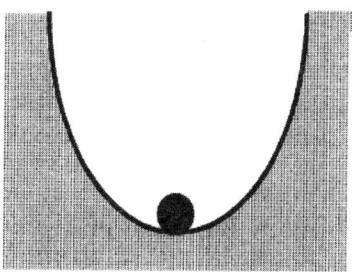

(a) Steady-state

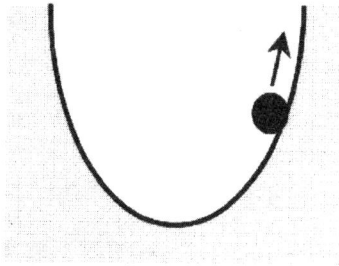

(b) Stable condition

Figure 15-5 Linear system stability: one stable equilibrium.

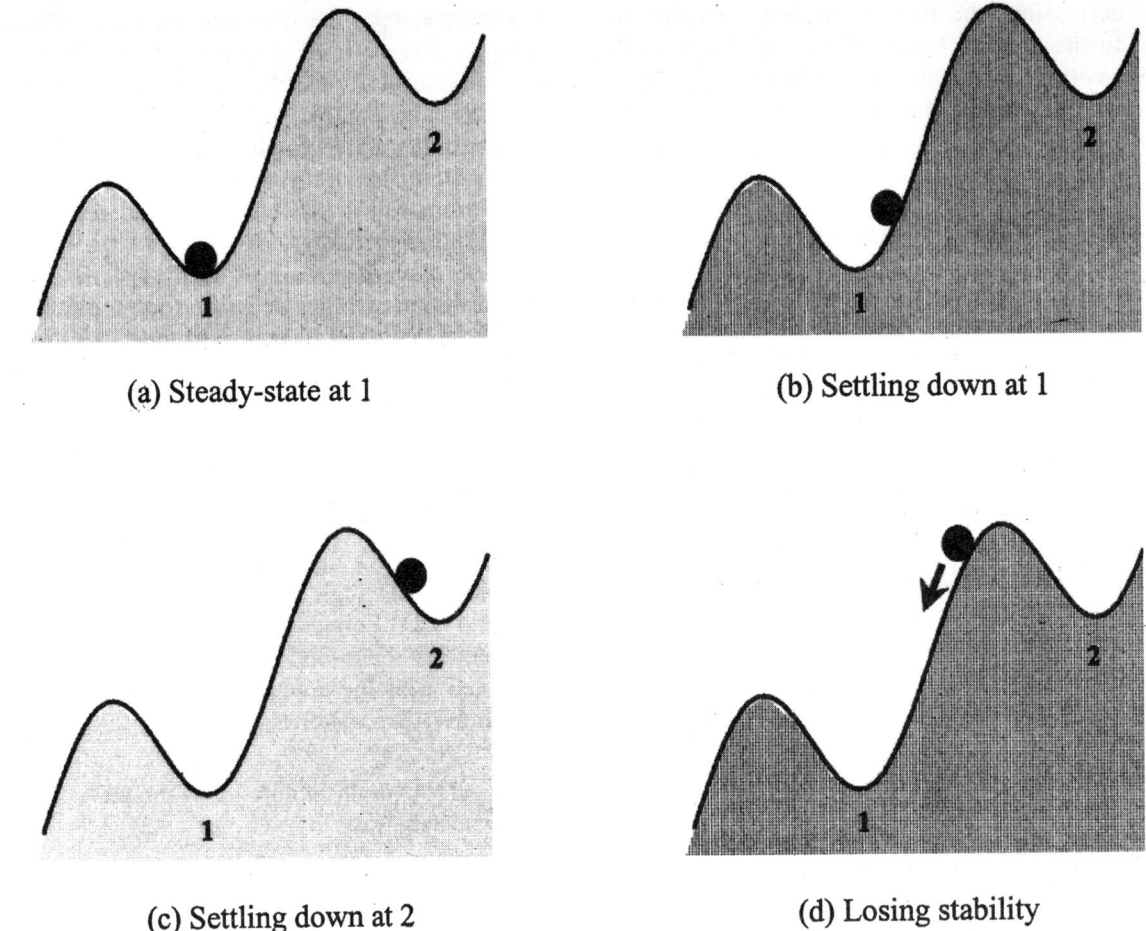

(a) Steady-state at 1   (b) Settling down at 1

(c) Settling down at 2   (d) Losing stability

Figure 15-6   Nonlinear system stability: multiple stable equilibria.

the postdisturbance system is stable and finally set down at $(\delta^s, 0)$, while for the initial state $(\delta_2, 0)$, the postdisturbance system is unstable. The state $(\delta_1, 0)$ is within the domain of attraction of $(\delta^s, 0)$, but $(\delta_2, 0)$ is not. The resulting instability is the LD angle instability or transient instability.

1.2.2   LD Voltage Instability

Shown in Figure 15-8 are the transient responses of the two-bus system illustrated in Figure 15-3 following two large disturbances. The disturbances set the initial states of the postfault system at $(V, \dot{V}) = (V_1, 0)$ and $(V, \dot{V}) = (V_2, 0)$, where V and $\dot{V}$ represent the load bus voltage and the rate of change of the bus voltage, respectively. For simplicity of illustrations, we assume the voltage change rates are zero at the two initial states. The postdisturbance initial state $(V, \dot{V}) = (V_1, 0)$ is within the domain of attraction of $(V^s, 0)$,

but the state $(V_2, 0)$ is not within this region. The unstable condition originating from the initial state $(V_2, 0)$ is the LD voltage instability because the domain of attraction is affected by the bifurcation point, which is mainly caused by a decrease in voltage.

Three mechanisms play a vital role in voltage collapses; tap changer dynamics, load dynamics (particularly those involving large percentages of induction motors), and generator field excitation dynamics. These mechanisms are critical in determining whether a system becomes unstable following a fault condition.

1.3   Voltage Instability Incidents

Many voltage instability incidents have been reported over the past years. Following are some examples:

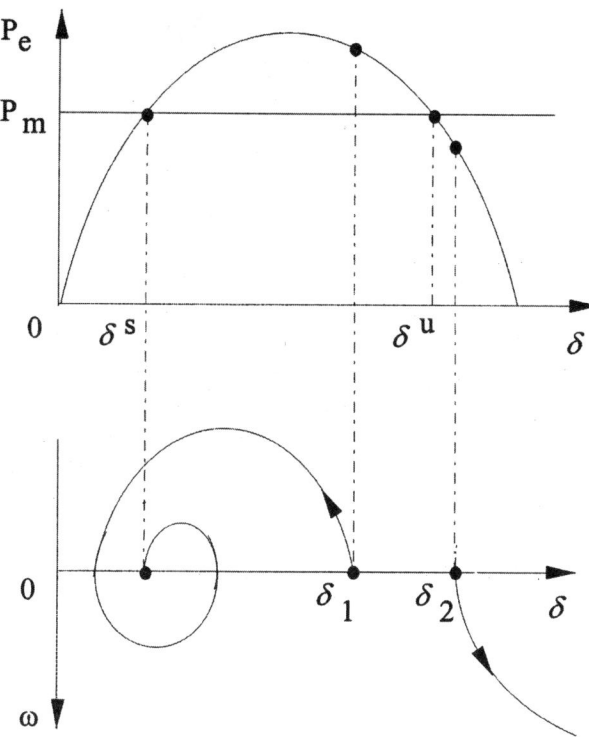

Figure 15-7 Transient responses of a SMIB system.

French system: December 19, 1978; January 12, 1987

Swedish system: December 27, 1983

Japanese system: July 23, 1987

New York system: September 22, 1970; July 13, 1977

Florida system: December 28, 1982; May 17, 1985

TVA system: August 2, 1987

Among these incidents, some were caused by gradual change of loads, while others were caused by loss of important lines or generators.

2 VOLTAGE INSTABILITY INDICES

Voltage stability studies have attracted great attention in recent years, and various indices have been developed for SD voltage instability detection. Some indices require the knowledge of system state variables (bus voltage magnitudes and phase angles) only at a current operating condition, and others require state variables at stressed operating conditions. The former needs less computation time, but the latter can provide power margins to voltage collapse, which is very useful to system operators. This section discusses typical methods in each category.

2.1 Indices Based on Current Operating Condition

Corresponding to a combination of load demands and generator outputs, stable and unstable equilibria can

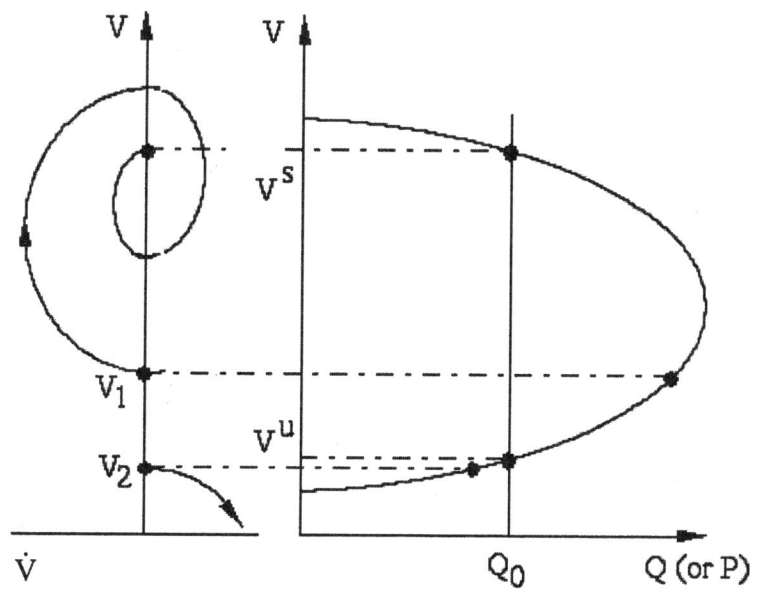

Figure 15-8 Transient responses of a two-bus system.

be found by solving load flow equations. A stable equilibrium is the condition at which the power system operates, while an unstable equilibrium is just a mathematical solution of power flow equations. A power system cannot operate at an unstable equilibrium, but the unstable equilibrium provides a helpful concept in defining voltage stability indices.

Voltage instability indices based solely on stable equilibrium and based on both stable and unstable equilibria have been proposed. The indices requiring the knowledge of unstable equilibrium conditions take more computational time because the procedure of finding the correct unstable equilibrium is more complicated than that of finding a stable one. In this section, we limit our discussions to the method which only requires state variables (voltage magnitudes and phase angles) at a stable equilibrium.

2.1.1   Q Angle Method

Simply speaking, voltage instability (at least for a lot of cases) is due to too much power being transmitted to the grid. In increasing power to the grid, a point will be reached beyond which the grid cannot sustain. By looking at the power flow solutions mathematically at this point, the load flow Jacobian matrix becomes singular. The load flow Jacobian matrix reflects the sensitivity of power flow to voltage variations. Whenever the load flow Jacobian matrix is at or close to singular condition, even a very small increase in power demand or power supply will cause the power system to move away from its stable operation conditions.

Detecting the singularity of a load flow Jacobian matrix is a widely used method for detecting voltage collapse. The proximity of a Jacobian matrix to singularity indicates how close an operating condition is to the voltage collapse point.

To explain the Q angle method, let's start from a simple two-bus system as shown in Figure 15-9. The system may represent a Thevenin equivalent of a system as seen from a load. In this system, a load with real and reactive power, P and Q, is supplied by an infinite system via a single lossless transmission line. The series reactance and shunt admittance of the transmission line are denoted by jX and jB_1, respectively. The source voltage V_1 is kept constant at $1.0\angle 0$, and the load bus voltage is $V_2\angle\delta$.

This system represents a simple load flow problem: given power demands, find voltages. To solve for the voltage magnitude V_2 and voltage angle δ at the load

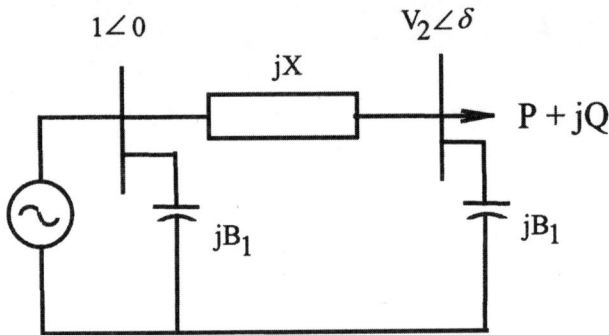

Figure 15-9   One-line diagram of a two-bus test system.

bus, the load flow equations are expressed by

$$-\frac{V_1 V_2}{X}\sin\delta = P \tag{15-1a}$$

$$\frac{V_1 V_2}{X}\cos\delta - \left(\frac{1}{X} - B_1\right)V_2^2 = Q \tag{15-1b}$$

In the equations load active power P and reactive power Q are knowns; only V_2 and δ are unknowns. Solving Eq. (15.1) will provide bus voltage. This is typically done by using the load flow Jacobian matrix, which represents the sensitivity of power flow to voltage variations. By taking partial derivatives of active power P and reactive power Q with respect to voltage magnitude V_2 and phase angle δ, the load flow Jacobian matrix for this simple system is found as

$$J = \begin{bmatrix} \frac{\partial P}{\partial V_2} & \frac{\partial P}{\partial \delta} \\ \frac{\partial Q}{\partial V_2} & \frac{\partial Q}{\partial \delta} \end{bmatrix}$$

On the other hand, with V_2 and δ as variables and with active power P and reactive power Q as constants, Eq. (15.1a) or (15.1b) can be viewed as representing a plane curve. The intersection points of the two plane curves gives the bus voltage magnitude V_2 and voltage phase angle δ.

Figure 15-10 shows the real and reactive power curves, where the active load P is fixed (1.0 p.u.) and the reactive power Q takes three different values. Since P is fixed, the load flow solutions vary along the constant P curve as Q is changed. When the reactive power demand is increased, the load bus voltage magnitude decreases and its angle (absolute value) increases. It is also found that when the reactive load becomes heavy, the angle between the gradient vector ∇Q and the tangent vector of the constant P curve at the feasible load flow solution point (higher voltage magnitude) will increase. For ease of discussion, we define α_1 as the angle between the gradient vector ∇Q and the tangent

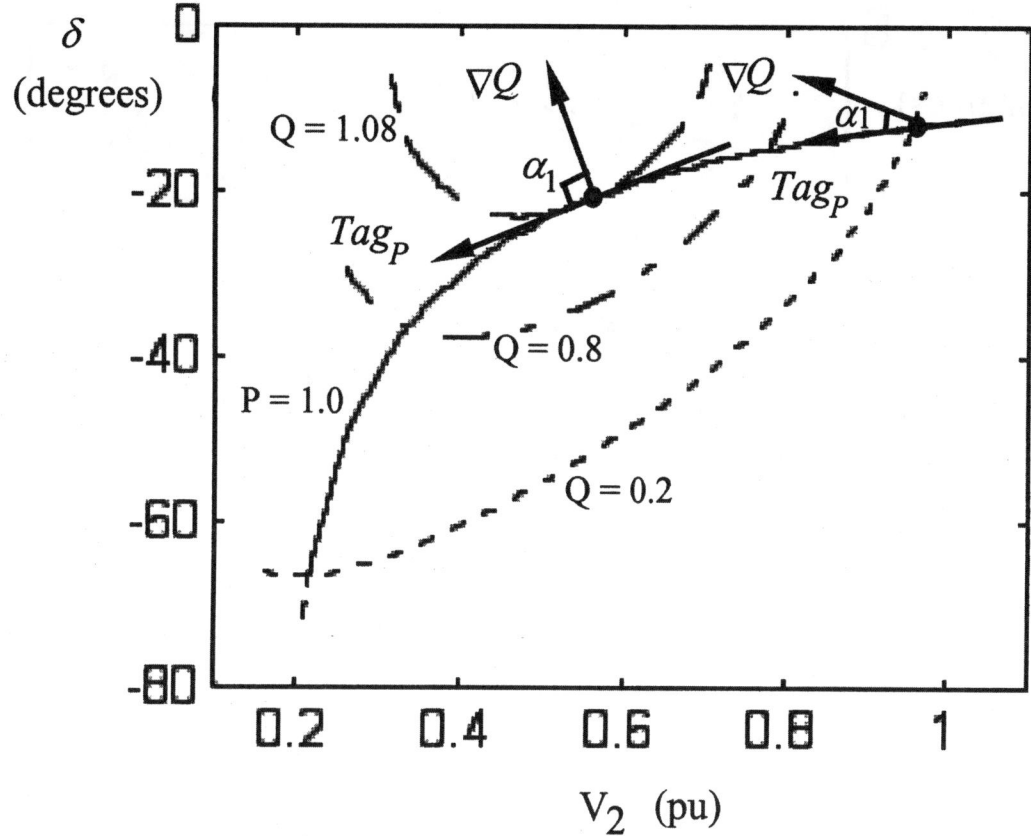

Figure 15-10   Real and reactive power curves in the state space for the two-bus system (P is fixed).

vector of the constant P curve at the feasible load flow solution. When $Q = 1.08$ p.u., the two curves are tangent to each other and the angle α_1 equals $90°$.

The tangent point is the bifurcation point or maximum load point, where two solution points are coincident. Beyond his point, voltage collapse occurs. Thus, the angle α_1, which is the Q angle, can indicate how close an operation condition is to voltage collapse.

Figure 15-11 shows the curves for the case where the reactive load Q is fixed (1.0 p.u.) and the active power P takes different values. The load flow solutions are along the constant Q curve. Similar observations as seen in Figure 15-10 can also be obtained from Figure 15-11.

2.1.2   Application to Multibus Systems

The method described above can be extended to general multibus power systems. Consider a power system, and let n be the total number of buses minus the swing bus. Allow the $(n+1)$th bus to be the reference bus, and assume $m+1$ to be the number of generator buses. At the system buses, the real and

reactive power balance equations for the power system may be expressed by

$$\sum_{j=1}^{n+1} V_i V_j Y_{ij} \cos(\delta_i - \delta_j - \theta_{ij}) - p_i(V_i) = P_i$$
$$(i = 1, \ldots, n) \qquad\qquad (15\text{-}2a)$$

$$\sum_{j=1}^{n+1} V_i V_j Y_{ij} \sin(\delta_i - \delta_j - \theta_{ij}) - q_i(V_i) = Q_i$$
$$(i = 1, \ldots, n - m) \qquad\qquad (15\text{-}2b)$$

where

$\quad (P_i, Q_i) =$ constant part of the net power entering bus i
$(p_i(V_i), q_i(V_i)) =$ voltage dependent part of the net power entering bus i
$\quad\quad V_i \angle \delta_i =$ ith bus complex voltage
$\quad\quad Y_{ij} \angle \theta_{ij} =$ (i, j)th element of the network admittance matrix

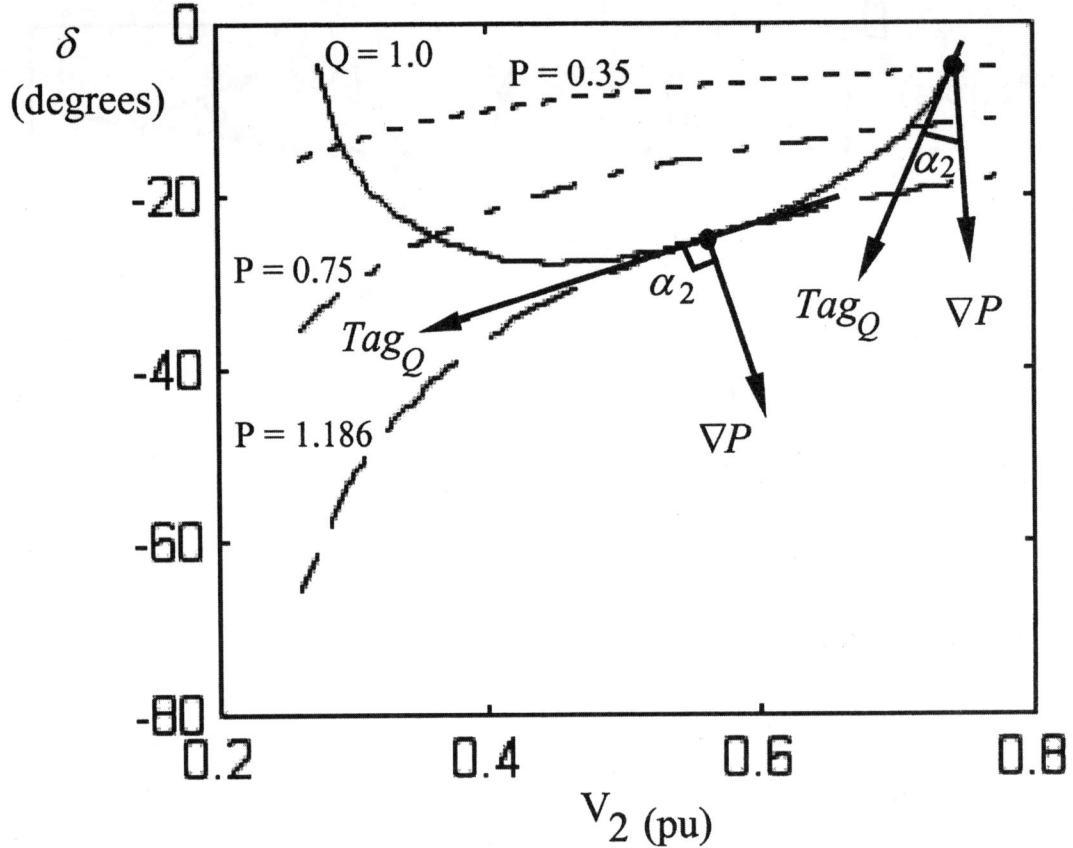

Figure 15-11   Real and reactive power curves in the state space for the two-bus system (Q is fixed).

Based on the assumptions on the number of busses, load flow Eq. (15.2) includes $2n - m$ algebraic equations with $2n - m$ unknowns. From a geometric point of view, every equation in (15.2) represents a space surface, and the remaining $(2n - m) - 1$ simultaneous equations represent a space curve in a $(2n - m)$-dimensional space. The intersection points between the space surface and the space curve give the voltage magnitudes and phase angles corresponding to a specified load level.

Figure 15-12 illustrates the relative movement between the space surface and the space curve as the injection power in the space surface equation is increased from Q to Q'. Also shown in the figure are the gradient vector of the space surface and the tangent vector of the space curve calculated at their intersection point.

The angles corresponding to the gradient vector of the space surfaces expressed by the active power balance equations are defined as P angles; and in a similar way, the angles corresponding to the gradient vector of the space surfaces expressed by the reactive power balance equations are defined as Q angles.

It can be seen that at the voltage collapse point both Q angles and P angles are equal to 90°. Consequently, the closeness of these angles to 90° is an indicator for the voltage instability detection.

2.2   Indices Based on Stressed System Conditions

The indices based on a current operating condition usually take less time to compute and thus are suitable for online application. But, a MVA margin of an operating condition to voltage collapse is sometimes required. To compute the margin, a load and generation variation pattern need be assumed. Along the assumed direction, the system is stressed until the voltage collapse point. In a stressed condition, especially near the voltage collapse point, the conventional iterative power flow may have difficulty in converging; thus the continuation power flow method has to be used. The solutions from the continuation method can be displayed in the form of P-V or Q-V curves.

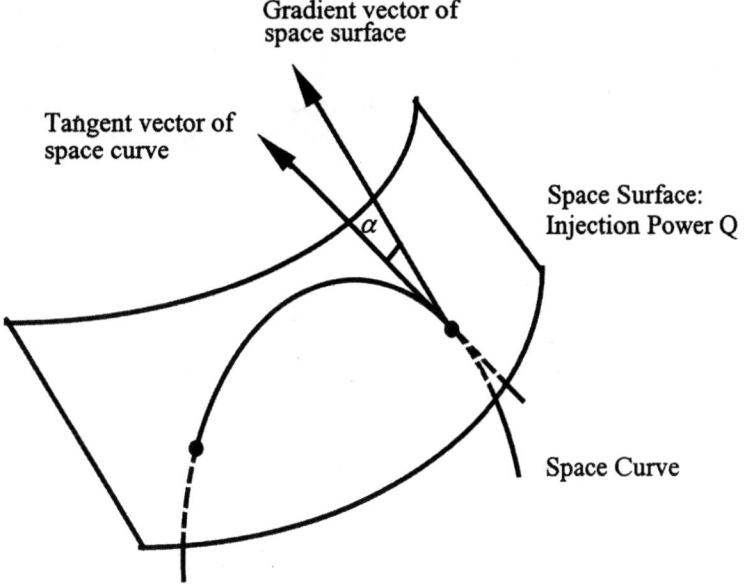

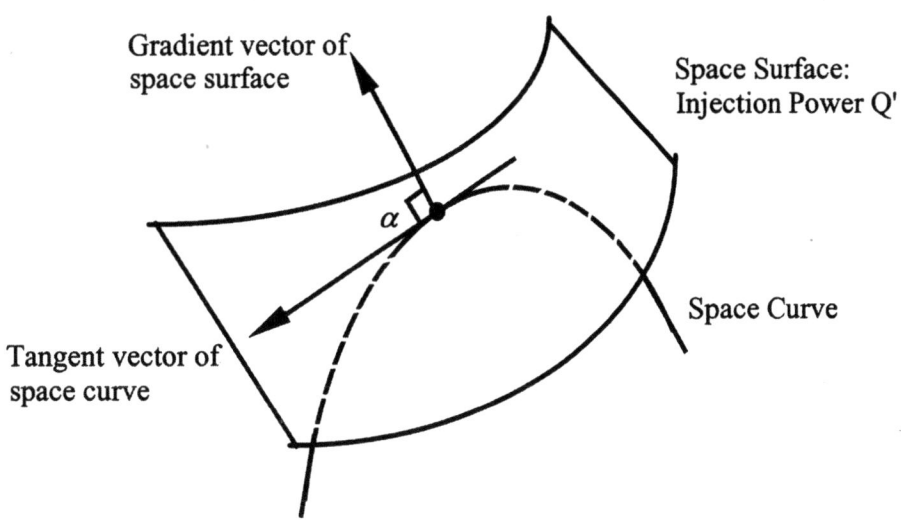

●: Load flow solution candidates

Figure 15-12   Relative movement between the space surface and the space curve, $Q < Q'$.

P-V curves and Q-V curves are widely used by electric utilities to determine voltage instability. On both P-V and Q-V curves, the nose points correspond to the voltage collapse point. These curves can be generated by increasing the load demands at a single bus or at the buses of a specific area according to a predetermined load variation pattern. The continuation load flow method may be utilized to find the low-voltage solutions required in plotting these curves.

Once the P-V and Q-V curves are drawn, system voltage stability can be evaluated based on the distance between a current operating point and the noise points of these curves. The drawback of the method is the heavy computations required to execute the large number of load flows needed to create the curves.

2.3 Summary

This section reviewed the methods for small-disturbance voltage instability detection. The methods based on a current operating condition are computationally fast, while the methods based on stressed system conditions require more computational time. Although the latter can calculate a MW margin to voltage collapse point, a future load variation is required in the calculation.

3 VOLTAGE INSTABILITY PROTECTION

While the voltage instability indices described in the last section are academically sound and useful for system planning and operations, they have to be simplified for protective relaying applications because they require extensive computations and system-wide information. These indices provide excellent guidance in designing more practical voltage instability protection tools. This section focuses on industry practices in protection of voltage instability problems.

The critical issue in voltage instability protection is to identify and predict the occurrence of voltage instability problems. In most cases, this is difficult to do without access to system-wide voltages through communications. Compromises have to be made when detecting voltage instability using localized voltages.

3.1 Reactive Power Control

Voltage instability is mainly caused by system inability to provide sufficient reactive power. Control of the reactive power sources is one of the effective methods in preventing voltage collapses.

Capacitors can deliver reactive power to the power system and keep system voltage within tolerable limits. To ensure that capacitor bank control actions are activated only during voltage collapse conditions, not during fault conditions, capacitor banks switching is sometimes supervised by zero sequence overvoltage relays or overcurrent relays. This purpose can also be achieved by delaying switching actions, provided that faults can be cleared within a short period of time, say, 0.5 sec. These schemes have been applied by a utility to subtransmission systems at 161 kV and distribution systems at 46 kV.

For transmission systems over 230 kV, fast capacitor switching is required due to the potential impact of prolonged low-voltage conditions. Similar to subtransmission systems, capacitor switching is activated only after faults have been cleared. Programmable-logic-controller (PLC)-based high-speed capacitor control schemes have been used for fast capacitor switching. Low voltages at all three phases are the trigger conditions, and a time delay of 0.2 sec is used to avoid switching during fault conditions.

Other than mechanically switched capacitors, there are other reactive power sources which can prevent voltage collapses. The reactive sources include static var compensators (SVCs), static condensers (STATCONs), synchronous condensers, and generators. They are usually controlled by solid-state switches so they provide varied reactive power in a continuous manner. When the reactive power sources reach their maximum limits and voltages at main buses are still in a dangerous state, other remedial means such as load shedding will have to be used.

3.2 Load Tap Changer Blocking Schemes

Load tap changers (LTCs) provide an economic means to help keep voltages at satisfactory levels. During normal conditions (other than voltage collapses), LTCs will adjust the tap to a lower level whenever the controlled voltage is higher than a desired upper limit; similarly LTCs will adjust the tap to a higher level whenever the controlled voltage is lower than the desired lower limit.

Under low-voltage conditions that are a result of possible voltage collapses, the normal LTCs operations cannot help but deteriorate the situations. This is true especially when load reactive power demand is very sensitive to voltage levels. Under such a situation, the voltage rise by LTCs will cause a significant increase in reactive power demand and thus widen the gap between reactive power supply and demand. Therefore, it is recommended that LTCs be blocked from trying to raise voltages whenever there are indications that there is an impending system voltage collapse.

3.3 Load Shedding

Load shedding causes forced power interruptions to customers and results in high costs to power suppliers and consumers, therefore undervoltage load shedding is usually the last option in preventing voltage collapses. It is only applied when all other options have been taken but still without success.

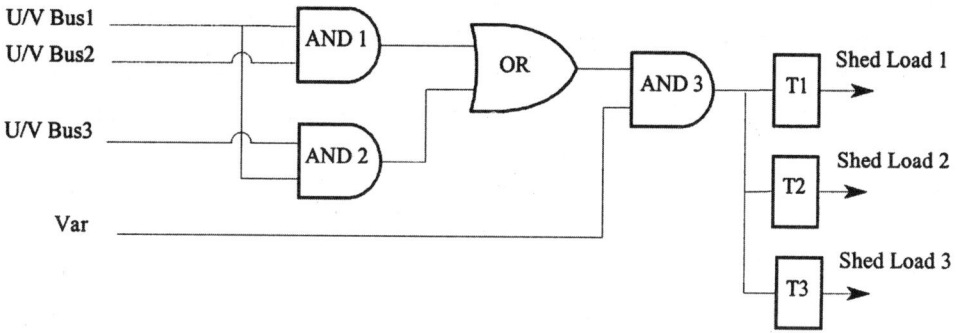

Figure 15-13   Undervoltage load-shedding. (From IEEE Report, reference 2.)

3.3.1   General Considerations

Undervoltage load shedding is implemented in stages and a time-delayed fashion. In general, following are the factors that need to be considered when implementing undervoltage load shedding schemes:

Determination of the amount of load to be shed
Determination of time steps of load shedding
Selection of load to be shed
Determination of voltage levels to shed loads

The selection of appropriate time steps and the amount of load to be shed at each step is important in effective undervoltage load shedding. The proper time step depends on system load characteristics, generation source, and rate of change of voltage. The amount of load being shed at each stage and the time delay associated with each stage are usually predetermined based on extensive voltage stability studies. It is not allowed to shed more load than is necessary because of unnecessary electricity interruptions. On the other hand, insufficient load shedding does not prevent voltage collapses or alleviate the system conditions. Shedding load in stages with proper time delay can solve this dilemma. Some utilities in the United States have elected to shed 15% of peak load in three steps during an undervoltage condition:

1.   5% of area load at voltage 10% below the lowest normal voltage with 3.5 sec time delay
2.   5% of area load at voltage 8% below the lowest normal voltage with 5 sec time delay
3.   5% of area load at voltage 8% below the lowest normal voltage with 8 sec time delay

All kinds of load, including industrial, commercial, and residential, play an important role during voltage collapse conditions. They are normally close to their peak values at voltage collapse, so shedding any of those loads would be helpful to alleviate voltage conditions.

Constant power loads such as motors are the most harmful to voltage conditions during voltage collapse, these loads may be considered first before shedding other types of loads. Also, power factors should be considered when determining what loads should be shed because it is the reactive power shortage that causes voltage drops and voltage collapse. Thus, shedding loads with lower power factor will be more effective in maintaining system voltages.

When considering tripping lines, choose the ones which are less critical to maintain system integrity. Radial lines or lines with heavy tapped loads but light through loads are better choices than other lines would be. Regarding the voltage level for initiating load shedding, it is a common practice to set the level in the range of 0.85 to 0.97 p.u. of the lowest normal voltage.

3.3.2   A Centralized Load Shedding Scheme

Undervoltage load shedding can be implemented either based on local measurements or on measurements from several substations. This section shows a centralized undervoltage load shedding scheme. In the scheme, the load shedding decision is made based on voltage measurements and reactive power measurements from three substations, two at 230 kV and one at 500 kV.

Figure 15-13 shows the implementation, where staged load shedding is initiated when voltages in more than one key bus are lower than settings and the output of reactive power sources is near their maximum Mvar capacity. Three stages are shown in the figure with three time delay constants and three load blocks. In the example, voltage setting is equal to 97% of the lowest normal voltage; each load shedding block is equal to 250 MW; and the first time delay T1 is 10 sec, T2 and T3 are 2 sec.

16

Reclosing and Synchronizing

Revised by: **S. WARD**

1 INTRODUCTION

The large majority of overhead line faults are transient and can be cleared by momentarily deenergizing the line. In fact, utility reports show that less than 10% of all faults are permanent. It is, therefore, feasible to improve service continuity by automatically reclosing the breaker after fault relay operation. For example, automatic reclosing greatly improves service in radial distribution circuits, where service continuity is directly affected by circuit interruption. High-speed reclosing on tie lines, if successful, also assists in maintaining stability. Reclosing is generally not used for cables as cable faults are seldom transient. It may, however, be applied on circuits with both overhead line and cable sections, especially if the overhead line/cable ratio is high.

This chapter will describe the application and operation of reclosing and synchronizing relays.

Microprocessor relays such as feeder protections DPU2000R and REF, distance protections REL, and breaker terminal REB551 all offer optional built-in reclosing and synchronism check devices. These protections implement many of the features and functions discussed below for the individual relay units, and the application of reclosing and synchronism check follow the same guide lines as required for a stand-alone unit.

2 RECLOSING PRECAUTIONS

Automatic reclosing can improve continuity of service and increase the availability of transmission lines, but certain precautions must be followed:

1. A generator must never be connected to a system under conditions that will produce a thermal or mechanical impact that will deprive it of life. The angle of voltages across a breaker in the vicinity of a generator is an inadequate measure of the possible hazard associated with closing that breaker. The sudden change in power in the generator following closure is, however, a key indicator. Steady-state switching may be appreciably less severe than reclosing following a fault because of the transient forces that are established by the original fault condition. A change in power, following steady-state switching, of 0.5 per unit will produce negligible loss of life.

The change in power following reclosure may substantially exceed this level. For this reason, reclosing breakers near a generating plant should be eliminated, restricted in number, be time-delayed beyond roughly 10 sec, or be carefully controlled. Modern reclosing devices offer fault-selective reclosing, so that reclosing may be allowed for the less severe single-phase-to-ground faults, while reclosing is blocked for the more infrequent two- and three-phase faults.

2. When a transformer is subjected to a substantial "through-fault," severe forces are developed within and between windings, which produces motion. Repetitive motion can produce failure of the winding. Cases have been reported of prolonged life of transformers that are subjected to shorter and/or fewer severe faults. Minimizing the number of reclosures can provide benefit in this regard. The protection system should also be designed so that a *transformer* fault, on a line-transformer section, does not result in reclosing.

3.   Bus faults are not common, but when they do occur they are generally severe and produce high fault current. For this condition, it is not desirable to have the breakers reclose automatically and a trip from bus protection relays therefore leads to lock-out.

3   RECLOSING SYSTEM CONSIDERATIONS

3.1   One-Shot vs. Multiple-Shot Reclosing Relays

The desired attributes of a reclosing system vary widely with user requirements. In an area with a high level of lightning incidence, most transmission line breakers will be successfully reclosed on the first try. Here, the small additional percentage of successful reclosures afforded by multiple operations does not warrant the additional breaker operations. Single-shot reclosing relays (those which produce only one reclosure until reset) are entirely justified.

Subtransmission circuit-reclosing practices also vary widely, depending on the requirements of the loads supplied. If there are motors or generators in the system, the first reclosure may be time-delayed. Most often, two or three reclosures are used for subtransmission circuits operating radially, and only one or two reclosures for tie circuits. Approximately half the utilities use some form of circuit checking before time-delayed reclosure to assure that either synchronism exists or one circuit is dead.

Multiple-shot reclosing relays are warranted on distribution circuits with significant tree exposure, where an unsuccessful reclosure would generally mean a customer outage. A typical utility experience on distribution feeders in an area with a large number of annual thunderstorm days is as follows:

Number of successful reclosures	%
Immediate	83.25
Second (15 to 45 sec)	10.05
Third (120 sec)	1.42
Total successful	94.72
Lockouts	5.28

The data show that increasing the number of reclosures does improve service continuity, but the incremental benefit of each additional reclosure is less than for the preceding one.

3.2   Selective Reclosing

The speed of tripping is a significant factor in the success of a reclosure on a transmission circuit. The faster the clearing, the less fault damage and/or degree of arc ionization, the less the shock to the system on reclosure, and the greater the likelihood of reenergization without subsequent tripping. The probability of successful reclosing then is improved if reclosing occurs only after a high-speed pilot trip. By allowing only pilot tripping to initiate high-speed reclosing, maximum success can be assured for single-shot reclosing, and many unsuccessful reclosures can be avoided. Such a system eliminates the high probability of unsuccessful reclosure on nonpilot trips, particularly for end-zone faults in which clearing occurs sequentially and the deenergized time is short. Some modern reclosing devices (e.g., REB551) include a feature to prolong the dead time when the communication link is out of service, thus increasing the likelihood of a successful reclosure.

3.3   Deionizing Times for Three-Pole Reclosing

When reclosing at high speed, the dead time required to deionize the fault arc must be considered. Based on a study of 40 years of operating experience, minimum dead times can reasonably be represented by a straight line, using the following equation:

$$t = 10.5 + \frac{kV}{34.5} \text{cycles} \qquad (16\text{-}1)$$

where kV is rated line-to-line voltage. On a 345-kV system, for example, this formula would give an approximate required dead time of 20.5 cycles.

If the inherent minimum reclosing time of the breaker involved can produce a shorter dead time than indicated, a reclosing delay must be incorporated. In the interests of placing the device in the protected environment of the control house and having a more accurate timing device, the time delay should be in the reclosing relay rather than the breaker.

Single-pole tripping and reclosing requires longer dead time because of the fact that the two phases that remain energized tend to keep the arc conducting longer.

3.4   Synchronism Check

A synchronism-check relay is an element in the reclosing system that senses that the voltages on the two sides of a breaker are in exact synchronism. (An automatic synchronizer, on the other hand, initiates closure at an optimum point when the two system segments are not in precise synchronism, but have a small beat frequency across the breaker contacts.) The setting for most synchronism-check relays is based on the angular difference between the two voltages and designed to minimize the shock to the system when the breaker closes. The angular difference between the voltages does not, however, determine the transient to which the system will be subjected upon closure. Rather, the shock to the system is related to the voltage across the breaker contacts (the "phasing voltage"). The phasing voltage is the critical quantity in determining whether or not a breaker is allowed to close. Therefore, more advanced synchronism-check devices have been developed that also have settings for voltage difference, phase angle difference, and frequency difference.

3.5 Live-Line/Dead-Bus, Live-Bus/Dead-Line Control

Live-line/dead-bus, live-bus/dead-line (LLDB/LBDL) control is frequently introduced in a reclosing system for a transmission or subtransmission circuit. This scheme allows the breaker to be closed when the circuit on one side is energized at full voltage and the circuit on the other side is dead. The synchronism-check and LLDB/LBDL controls are complementary. The synchronism-check unit allows closure when the two voltages are high but in synchronism; the voltage control allows closure when one voltage is normal and the other is very low, preferably 0.

Generally, synchronism check is performed on a single-phase basis, i.e., by comparing one phase voltage on each side of the breaker. For LLDB/LBDL control, however, it might be advantageous to check all three phases. This prevents out-of-synchronism closing of the breaker in case of fuse failure. If check is made in one phase only and the fuse were blown in this phase, the line (or bus) would be seen as "dead" and allow LLDB/LBDL closing.

3.6 Instantaneous-Trip Lockout

On distribution systems, where coordination with fuses is necessary, the fuse is often protected on the first trip with an instantaneous tripping element on the substation breaker relay followed by a reclose attempt. This control is removed after the first trip, allowing the fuse

to blow and preventing a second breaker trip if the fault occurs beyond the fuse. The result is a combination of minimum outage area for permanent lateral faults and reclosing for temporary faults.

It should be recognized that voltage-sensitive loads on unfaulted lateral circuits may be more seriously affected by a fault when this "fuse-saving" strategy is used. A fault on a fused lateral produces a voltage dip throughout the circuit, and the severity of the dip is dependent on the proximity to, and nature of, the fault. Since complete loss of voltage to the load on the unfaulted laterals occurs when the substation breaker is opened, some deterioration in service occurs using this principle, in exchange for the benefit of not having to replace a fuse for a temporary fault.

3.7 Intermediate Lockout

Unattended substations that are not equipped with supervisory control can be controlled more effectively by a reclosing scheme that locks out on a permanent fault before exhausting all its reclosing shots. With an attended (or supervisory controlled) substation at the other terminal of the line, a manual reclosure can be attempted after lockout, whenever the operator judges that the fault no longer exists. If manual closure is successful, a synchronism-check relay will operate—in conjunction with the reclosing relay in the intermediate lockout condition at the unattended station—to restore the second-line terminal to service. This reclosing scheme is very effective for multiterminal lines in which several unattended stations without supervisory control are disconnected for line faults.

Intermediate lockout is achieved in REL301/512 by a "hold cycle" input; in MRC2 by the "pause" input, and in DPU2000R by the "ARCI" input. These relays all perform the same function as described above; when the input is asserted, operation "freezes" and when deasserted, the reclosing cycle resumes.

3.8 Compatibility with Supervisory Control

All reclosing systems should incorporate some provision allowing circuit breakers to be tripped manually or by supervisory control without inadvertent reclosing action. Such a provision is inherent in any reclosing system that has an "initiate" function. In other reclosing systems, lockout must be accomplished by other means, such as a breaker control-switch slip-contact or the temporary removal of control voltage to the reclosing relay.

3.8.1 Reset Time

Reclosing relays include a reset timer that starts when the final reclosing attempt has been made. If the breaker opens before this timer has expired, the relay goes to lockout and further reclosing is blocked until a manual close has been made.

3.8.2 Follow-Breaker Function

The follow-breaker function is used to prevent sustained pumping action of a circuit breaker as a result of a permanent fault, or if the breaker is closed by any other device than the designed reclosing relay action. If during the open interval time the relay sees the breaker close, then the relay will step forward in its program and begin the reset timer. Should the breaker open before the reset time expires, the timing for the next open interval in the reclosing sequence will begin.

3.8.3 Close Fail Time

Generally, the reclosing relay monitors the breaker 52b contact and will keep its close pulse asserted until the breaker has closed. However, should the breaker fail to close, a "close-fail-timer" will deenergize the close contact when the set time has expired.

3.8.4 Manual Close

A manual close input to the reclosing/synchronism check device can be used to supervise closing by synchronism check or LLDB/LBDL control. In addition, the close input is used to ensure that the reclosing relay will be in lockout state for the duration of the reset time.

3.9 Inhibit Control

In some applications, reclosing should be inhibited until further action takes place in other devices. For example, if a dual-breaker scheme is used, and one of the breakers fails while clearing a fault, a transfer-trip signal would be sent to the remote terminal to clear the fault contribution from that source. Reclosing of the remote breaker would be prevented until the transfer-trip signal was removed. (There must also be assurance that removal of the transfer-trip signal is not a result of channel failure.) With this additional logic in the reclosing relay, the remote breaker can be reclosed reliably, simply by resetting the local breaker-failure lockout relay after the faulty breaker has been properly isolated.

3.9.1 Drive to Lockout

When a "drive-to-lockout" input is energized, the reclosing relay will go into lockout from any point in the sequence. The relay will stay in lockout until the input is removed and the breaker is closed manually or by supervisory control. Upon removal, the recloser will go through its reset sequence and return to "ready" state.

3.10 Breaker Supervision Functions

Microprocessor reclosing relays provide a number of breaker operation functions to prevent excessive breaker wear and aid in breaker maintenance scheduling. The number of maximum allowable reclosures per selected time period and recovery time can be set. In addition, counters for cumulative operations will provide alarm when breaker service is needed.

3.11 Factors Governing Application of Reclosing

The factors governing the application of reclosing are summarized below:

1. For instantaneous reclosing, the protective relay contacts must open in less than the breaker reclose time. This presents no problem with high-speed relays but on some slow-speed relays, it may be necessary to reduce the contact follow, or "wipe."
2. The breaker latch check (LC) and, when applicable, the low-pressure switch (LPC) should be used to avoid operating the breaker if the mechanism is not prepared to accept closing energy or gas pressure is inadequate.
3. The breaker should be derated according to the breaker standards.
4. For instantaneous reclosing, arc deionizing time must be considered (see Sec. 3.3).

4 CONSIDERATIONS FOR APPLICATIONS OF INSTANTANEOUS RECLOSING

The applications of instantaneous reclosing fall into three categories:

1. Feeders with no-fault-power back-feed and minimum motor load

2. Single ties to industrial plants equipped with local generation

3. Lines with sources at both ends

4.1 Feeders with No-Fault-Power Back-Feed and Minimum Motor Load

Instantaneous reclosing can be applied to these feeders, but care must be taken to avoid reclosing into motors that are still rotating, since their internal voltage may be out of phase with the system voltage.

4.2 Single Ties to Industrial Plants with Local Generation

Since these circuits must be opened at the plant before reclosing, instantaneous reclosing at the utility end is not practicable unless instantaneous tripping of the plant tie or local generator is assured. Without this instantaneous tripping, the local generation, even when quite small, can maintain the arc for line faults and negate successful reclosure. Reclosing may also occur out of synchronism.

4.3 Lines with Sources at Both Ends

Simultaneous tripping of the line is necessary in instantaneous reclosing that invariably requires some form of pilot relaying. Both ends of the line can be instantaneously reclosed *only* if there are sufficient ties between the terminals or sufficient inertia in both systems to ensure that the two ends will not go out of synchronism during the dead time. In the absence of sufficient ties (or inertia), one end can be reclosed instantaneously by LBDL reclosing, with the other then closed manually or by a synchronism check.

Instantaneous reclosing of both ends of a line without any checking is widely practiced for high-voltage transmission lines in the United States. These lines have multiple parallel circuits and are protected with pilot relaying. This reclosing practice is usually suspended when the pilot relaying is out of service, since zone 1 phase and ground or ground instantaneous relaying do not provide 100% instantaneous line protection. Sequential tripping and instantaneous unsupervised reclosing will produce unsuccessful reclosing.

5 RECLOSING RELAYS AND THEIR OPERATION

5.1 Review of Breaker Operation

Knowledge of the operation of the breaker and its auxiliary contacts is essential to understanding how reclosing relays function. The time sequence of events occurring within the breaker and its auxiliary contacts during a typical instantaneous reclosing cycle is shown in Figure 16-1. The auxiliary contacts are actuated directly by breaker main contact travel (a and b) or by the operating mechanism travel (aa and bb).

5.2 Single-Shot Reclosing Relays

A typical breaker-control scheme is shown in Figure 16-2. Contact 79 provides the reclosing intelligence. In some varieties of reclosing relays, this contact is pre-closed, waiting only for the 52bb contact to close to initiate "closing." This 52bb contact closes

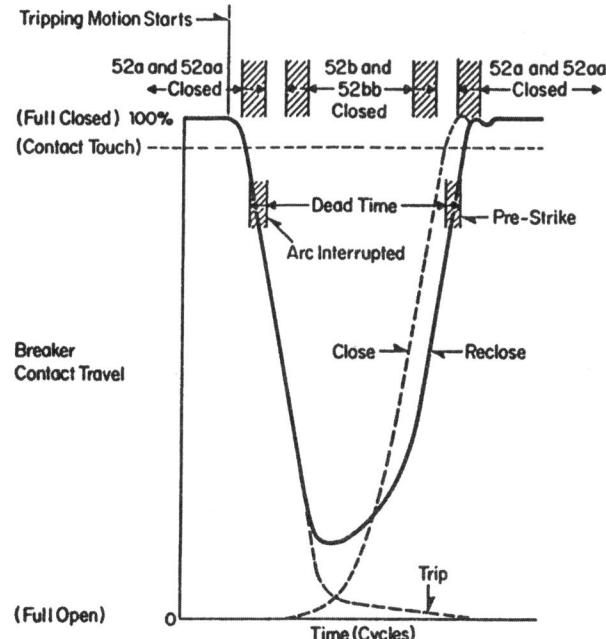

Notes:
1) 52a and 52aa - Closed When Breaker Is Closed.
   52b and 52bb - Closed When Breaker Is Open.
2) 52aa Is a fast opening contact when the breaker opens.
   52bb Is a fast closing contact when the breaker opens.
3) Cross-Hatched Areas Indicate Variable or Adjustable Range.

Figure 16-1 Typical circuit breaker instantaneous reclosing cycle.

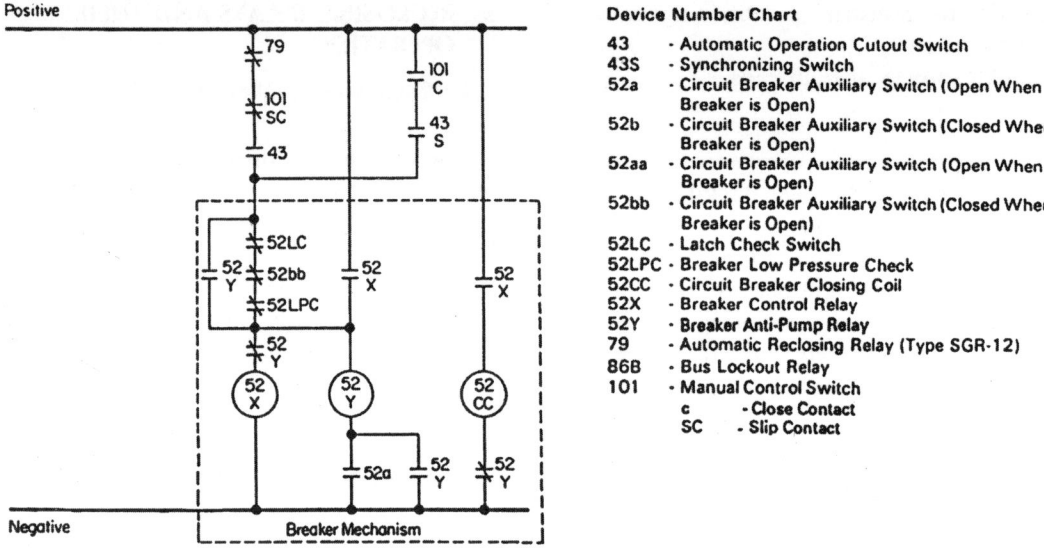

Figure 16-2 External schematic of the SGR-12 reclosing relay.

as the breaker moves to the open position in response to protective relay action (or for a "manual" trip). The closing of the breaker is indicated to the reclosing relay by the opening of the 52b switch. When it opens, the reclosing relay becomes locked out. If reclosing is successful, the 52b stays open, causing an internal timer to reset the reclosing relay, allowing the sequence to be repeated at a later time. If reclosing is unsuccessful (fault still present), the 52b opens before timing is complete, and the reclosing relay stays locked out.

For the manual trip, 101/SC opens and stays open as a result of the movement of the control switch handle by an operator to the trip position. This contact is closed in "close" and closed "after close." It is open in "trip" and open "after trip." It "slips" and is therefore called a slip contact. No reclosing is desired following a manual trip, and the slip contact supervises this.

The breaker-control scheme of Figure 16-2 is only one of a number of schemes used, but the fundamentals of automatic reclosing are similar.

5.2.1 Solid-State Single-Shot Reclosing Relay

The logic diagram for the SGR-51 is shown in Figure 16-3. It requires only a 52b contact to indicate breaker status. The single-shot function provides an output during the closing stroke of the breaker. It has a short-duration output immediately following a 1 input. The output then reverts to 0, regardless of whether the input 1 is short or continuous.

With the SGR-51 reset and the breaker closed, the 52b switch is open, and a continuous 1 exists at the single-shot input with a steady-state 0 output. For this condition, the flip-flop outputs are as shown. With the negated input to the upper amplifier, relay CR is continuously energized, providing a closed contact CR in the breaker close circuit.

As the breaker is tripped by protective relays, the only open contact in this breaker close circuit, 52bb, closes (Fig. 16-3). As the breaker moves, 52b closes to produce a 0 input to the single shot. Then, as the breaker recloses, 52b opens, putting a 1 on the single-shot input. A short 1 output follows to operate the flip-flop. The upper output of the flip-flop changes to 1, putting a 1 on the negated amplifier input and deenergizing CR relay. This opens contact CR in the close circuit. The lower output of the flip-flop changes to 0, operating the amber lamp to indicate a lockout.

If the breaker stays closed (52b open), the two 1 inputs to AND permit an output to the reset timer. If this condition continues for the reset interval (adjustable from 3 to 30 sec), the lower input (c) to the flip-flop is energized. This resets the flip-flop, turns off the amber light, and energizes the CR relay, making it ready for the next automatic reclose operation.

If the breaker retrips before the reset timer times out, the closing of 52b removes the 1 from AND to stop the timer and prevent the reset. Further action is blocked until the breaker is closed manually.

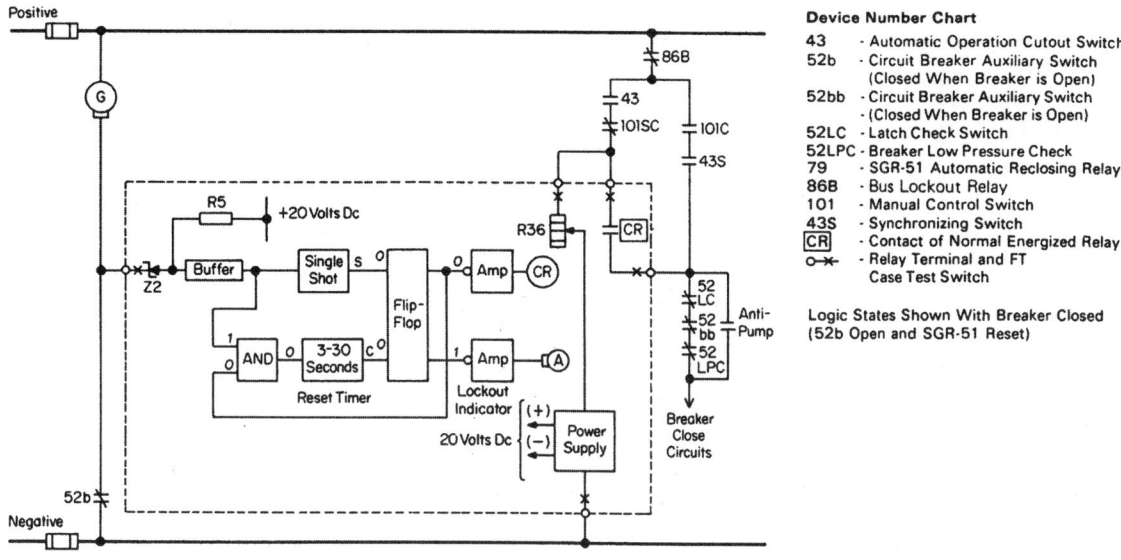

Figure 16-3 Logic diagram of the SGR-51 reclosing relay.

5.2.2 Solid-State SGR-52 Relay Operation

SGR-52 logic, shown in Figure 16-4, is similar to that for SGR-51 but has reclose-initiate and reclose-block functions. Also, the closing contact CR is not preclosed. To initiate reclosing, CR must be energized via the reclose-initiate circuit. This input is under the control of logic that identifies pilot tripping has occurred or some other consideration that assures the success of high-speed unsupervised reclosing. The reclose-initiate circuit includes a 100/0-msec timer that allows time for the 52b contact to operate, even if only a momentary closure of the reclose-initiate contact occurs.

When a trip for which reclosing is desired (such as a pilot trip) occurs, the reclose-initiate auxiliary contact closes to provide a 1 input to OR-1. The output deenergizes the 100/0 timer and places a 0 on the input of OR-2. This puts a 0 on the negated AND-1 input. As the breaker opens and 52b closes, the middle negated input to the AND-1 goes to 0. If a lockout has not occurred, the three-input AND-1 is satisfied to initiate the reclosing timer (0 to 2 or 2 to 20 sec). The output of the timer satisfies AND-2 to operate CR and initiate reclosing.

The 52b switch may close some time after the breaker actually interrupts the flow of fault current. If the reclose timer is set on 0, the CR relay will be energized at the same time as the 52b switch closure. This produces only a slight delay in reclosing relative to the SGR-51 with its preclosed CR contacts. For EHV applications where intentional delay is required, the delay is provided by the 0- to 2-sec timer. When longer delays are required, 2- to 20- or 6- to 60-sec timers are available.

Reclose blocking is necessary for functions such as out-of-step tripping and breaker-failure transfer-tripping. The reclose block input to OR-3 operates a timer to input OR-2. This output of OR-2 overrides the initiate input and blocks reclosing by removing one of the AND-1 inputs, even though the pilot system may be calling for a reclosure. To avoid critical resetting between initiate and block inputs, the 0.2/60-msec timer provides a continuing block signal for 60 msec, after removal of the blocking input.

The lockout and reset operation occurs in the same manner as described above for the SGR-51 relay.

5.3 Multishot Reclosing Relays

5.3.1 Electromechanical RC Relay

This relay provides multishot (up to 6) reclosures through the use of drum-operated switches that allow various reclosing strategies to be elected. Timing is controlled by a small synchronous motor that drives the drum. Instantaneous-trip lockout, multishot selectable time interval reclosing, intermediate lockout, and initiated first-shot instantaneous or time-delayed reclosures are possible using this device.

Since resetting following successful reclosure is dependent on full drum travel, a minute or more of

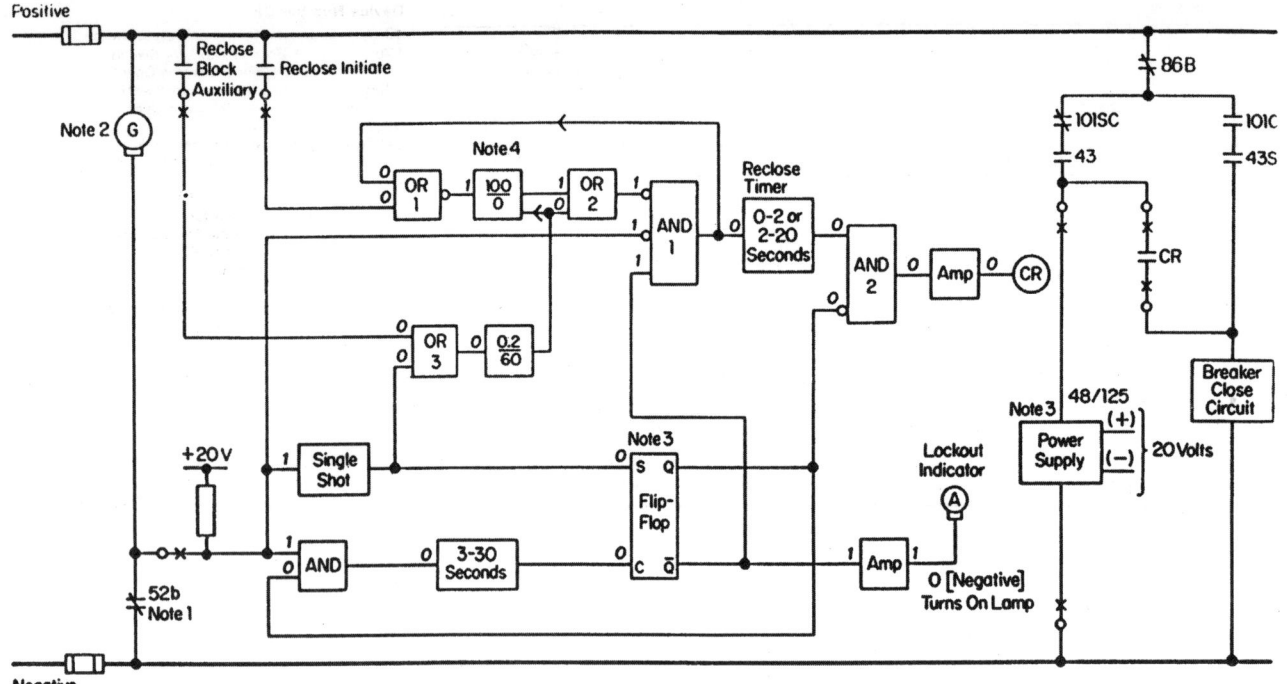

Figure 16-4 Logic diagram of the SGR-52 reclosing relay.

Device Number Chart

43   - Recloser Cutout Switch
43S - Synchronizing Switch
52a - Breaker Auxiliary Switch (Open When Breaker is Open)
52b - Breaker Auxiliary Switch (Open When Breaker is Open)
79   - SGR-52 Relay
86B - Bus Lockout Relay
101C - Breaker Control Switch Close
101SC - Breaker Control Switch Slip Close
o—x— - Relay Terminal and FT Case Test Switch

Notes:

1. Logic States Shown With Breaker Closed, 52b Open and SGR-52 Reset.
2. Light Not Required. Only a Circuit to Negative Through 52b is Required.
3. Upon Energization of Power Supply, Relay Locks out Until Reset.
4. Output from 0.2/60 Causes Fast Timeout of 100/0 Timer.

time may be required to accomplish this. Faults occurring during this resetting period will produce lockout. Fast resetting of solid-state and microprocessor relays overcomes this problem.

5.3.2   Solid-State 79M Relay Operation

The 79M circuit-shield reclosing relay provides up to three shots. Each shot has its own reclose time and reset time settings and the first shot can be set to instantaneous. The 79M close contact remains closed until 52b opens. The instantaneous cutout contact opens on a selected trip count and closes on reset or lockout.

A tap changer cutout contact is provided that opens on first trip and closes on reset or lockout.

5.3.3   Microprocessor-Based Reclosing Relays

Several varieties of microprocessor-based relays are available. Functionally, they are similar to their electromechanical and solid-state counterparts, but in addition, possess the qualities one expects from microprocessor relays: self-testing, "watchdog" timing, failure alarm, etc.

One example the, MRC-2, provides the following:

1. Instantaneous-trip enable (from selected states of the sequence)
2. Drive to lockout (from external contact)
3. Load tap changer lockout (to avoid stepping during reclosing)
4. In-progress output (to identify remotely that the reclosing relay is sequencing)

plus independent alarming of

1. Lockout
2. Failed reclose
3. Alarm (processing failure)

5.3.4 Reclosing in Feeder and Distance Protection Terminals

The feeder protection DPU2000R and distance protections REL301/302, REL512, and REL521/531 all have optional reclosing functions. A summary of the reclosing relay implementation in these protections is given here.

Reclosing in Feeder Protection DPU2000R

DPU2000R provides three overcurrent steps that can be coordinated with up to four-shot reclosing. Each current element can be enabled or disabled at each step. For instance, instantaneous-trip lockout can be employed as described in Section 3.6.

79 CUTOUT TIMER. The 79 cutout timer (79-CO) function provides for the detection of low-level intermittent faults prior to the resetting of the reclose sequence as shown in Figure 16-5. At the end of the selected cutout time period, all overcurrent functions are reenabled based on the 79-1 settings. In fuse-saving applications involving downstream fuses, the 50P and 50N instantaneous functions are set below the fuse curve to detect faults on tapped laterals. These functions are blocked after the first trip in the reclose sequence. The 51P and 51N time overcurrent functions are set above the fuse curve. This results in the

upstream protection being less sensitive to an intermittent or low-level fault during the subsequent reclose operations. If the reset time is too short, the reclosing relay may reset before the fault is detected again. If the reset time is too long, the intermittent or low-level fault is not cleared fast enough by the upstream protective device. In schemes using discrete reclosing relays, blocked instantaneous overcurrent functions are placed in service only after the reclosing relay has reset. However, the 79-CO function in DPU2000R reenables the instantaneous functions at the end of the selected cutout period. Set the time for the 79-CO function according to how long it takes a downstream fuse or other protective relay to clear downstream faults. The typical time setting is between 10 and 15 sec. If an intermittent or low-level fault exists, it will be detected at the end of the 79-CO cutout time period, and the DPU2000R will trip and continue through the reclose sequence until the fault is permanently cleared or lockout is reached. The 79-CO function allows the reset time to be set beyond 60 sec without jeopardizing sensitivity to intermittent or low-level faults.

79 V VOLTAGE BLOCK FUNCTION. The DPU2000R 79 V voltage block function blocks reclosing when one or more of the input voltages is below the 79 V voltage setting. When the input voltage is restored within the 79 V time delay setting, the recloser operation is unblocked and the open time will begin. If the voltage is not restored within the 79 V timer setting, the recloser will proceed to lockout. This function is useful in preventing a feeder breaker reclosure when the bus voltage is low because an upstream device has tripped.

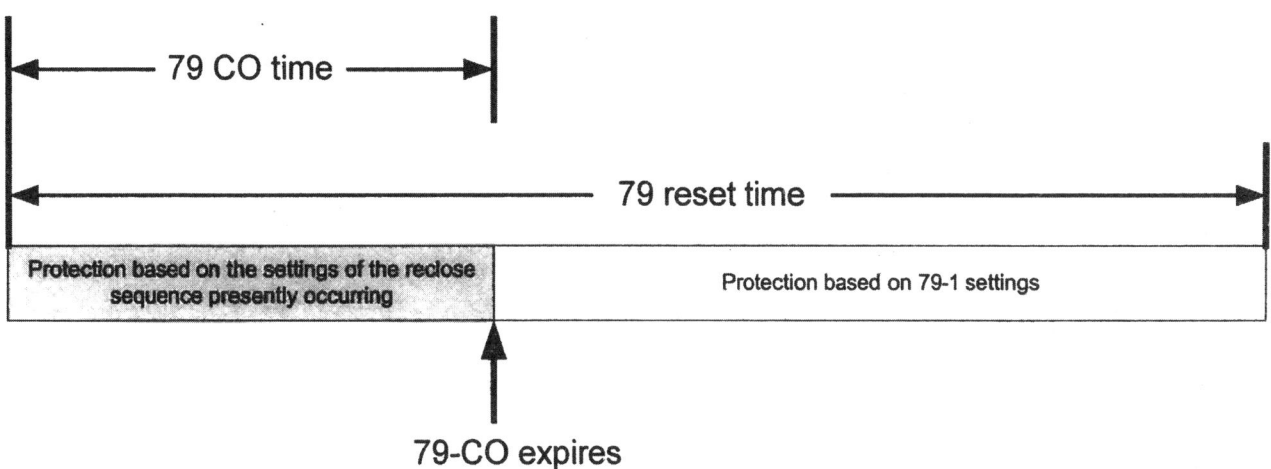

Figure 16-5 Cutout time.

When voltage is restored following successful reclosing of the upstream breaker, the reclosing cycle will proceed.

REL512 Reclosing Coordination for Ring Bus Applications

When applying the REL 512 reclosing to a ring bus configuration, it is important to understand that the recloser module performs the reclosing function using its own I/O, independent of the protection function. The only exceptions are the reclose initiate and block signals developed in the protection which are passed internally to the reclose module. All other inputs, including bus and line system ac voltages and binary (status) dc voltage, are external inputs connected via termination blocks associated with the reclosing module. Thus, this system allows the protection functions to trip two breakers and the reclosing function to control one breaker. This method of protection and reclosing control is shown in Figures 16-6 and 16-7. For simplicity, all REL 512 protection functions are identified by 21 and reclosing functions by 79.

There are separate voltage and breaker auxiliary contact connections made to the protection and reclosing modules of the relay. For ring bus applications, both breakers common to the protected line should have their auxiliary contacts appropriately connected to the protection module via programmable inputs (middle row of terminal blocks). When the protection operates both breakers trip and provide their change of state information to the protection.

For reclosing, one of the line's breakers is designated as the lead breaker and the other is the follow breaker. Only the lead breaker's auxiliary contacts are connected to the reclosing module. It is recommended to use the 52b. Connecting the 52a is optional.

The protected line's voltage is connected to the protection. This voltage is also connected to recloser module as the *line* voltage. It will be used to control the lead breaker. Furthermore, this voltage is also connected to the reclosing module of the adjacent relay that shares control of the follow breaker. It will be connected as the *bus* voltage there.

The REL 512 pilot tripping for a fault on its protected line will provide a reclose initiate signal

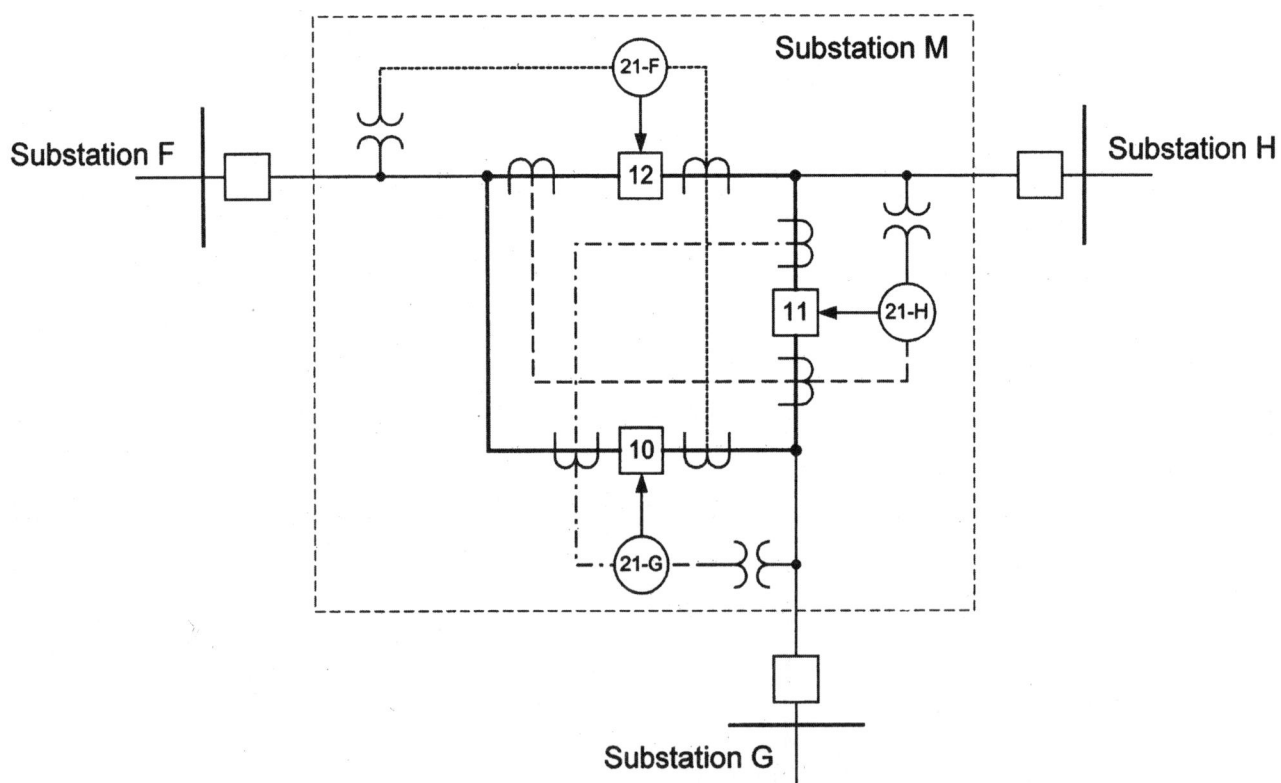

Figure 16-6 REL-512 protection control of ring bus circuit breakers.

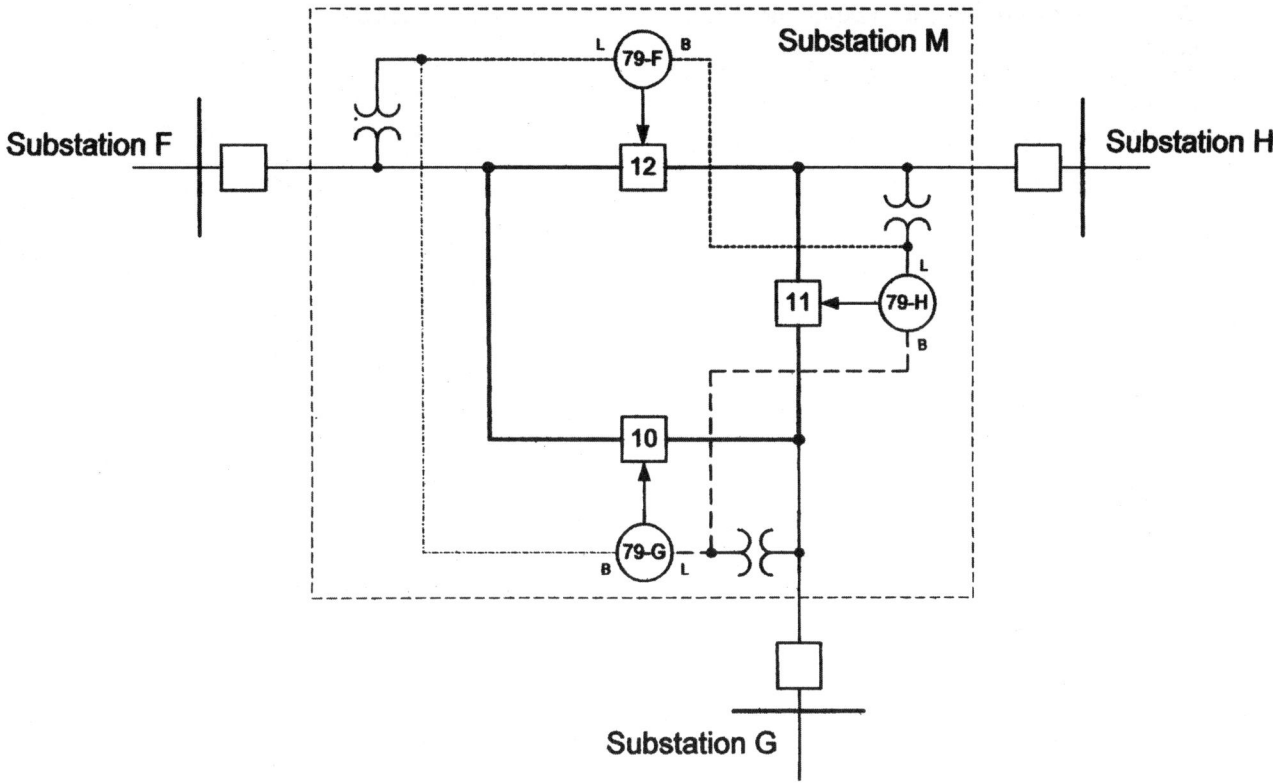

Figure 16-7   REL-512 reclosing control of ring bus circuit breakers.

internally to the recloser module within its chassis. It also will provide an Rl signal for the reclose module of the adjacent relay that controls the follow breaker.

If the *lead* breaker fails to successfully reclose it will advance to "lockout." When in this state the reclosing module will need to lock out the other breaker associated with this line. This situation requires the use of the reclose module lockout output to drive the adjacent recloser controlling the *follow* breaker to "lockout" via its input "drive to lockout."

A failed reclose will be the result of either exceeding the maximum wait time waiting for the appropriate voltage supervision to be satisfied or by reclosing into a fault. In the latter condition a "reclose-block" signal will be generated. This signal will be passed internally to the recloser in the same chassis if the reclose shot number has the reclose-block enabled. The reclose-block signal can also be mapped to a protection programmable contact and used to lock out the adjacent reclosing function.

The number of reclosures is set to 2. The successful reclose reset time delay should be set to 10 sec. This time is arbitrary, but should be coordinated with the shot #2 dead time (reset time + margin > shot #2 dead time).

The reclose shot #1 is set up for high-speed reclosing as required. Generally the reclose will be set up with no voltage supervision or HBDL supervision. If the remote end has closed first and voltage has been restored, then synchronism checking should also be enabled if HBDL is enabled. Otherwise reclose will not be attempted if the line voltage has been restored.

The reclose shot #2 is required to provide the sync-check reclose. The reclose dead time is 12 sec. The maximum wait time should be set longer than the shot #1 maximum wait time. Reclose block is enabled. This will allow immediate blocking and bypassing of shot #2 of the lead breaker for CIFT and advance the recloser to "lockout."

A successful reclose sequence follows:

1. A fault occurs on line M-H.
2. 21-H trips breakers 11 and 12 and sends internal RI to 79-H and external RI to 79-F.
3. 79-H starts shot #1 0.5 sec. dead time count. 79-F bypassed shot #1 (RI #2 input starts at

shot #2) and starts shot #2 12.0 sec. dead time
count.

4. 79-H recloses breaker 11 after 0.5 seconds of
   dead time on HBDL.
   Recloses on synchronism check if line voltage
   has been restored from remote end.
5. 79-H resets 10 seconds after reclose (this
   assumes HBDL was immediately satisfied).
6. 79-F recloses breaker 12.0 seconds on synchronism check.
7. 79-F resets 10 seconds later.

The following is a failed reclose sequence:

1. A fault occurs on line M-H.
2. 21-H trips breakers 11 and 12 and sends
   internal Rl to 79-H and external Rl to 79-F.
3. 79-H starts shot #1 0.5-sec dead time count.
   79-F bypassed shot #1 (RI #2 input starts at
   shot #2) and starts shot #2 12.0-sec dead time
   count.
4. 79-H recloses breaker 11 after 0.5 sec of dead
   time on HBDL into fault.
5. 21-H trips on CIFT and sends internal block
   reclose to 79-H.
6. 79-H advances to lockout and sends external
   signal to 79-F to advance to lockout.

Breaker Terminal REB551 in Breaker-and-a-Half Applications

The microprocessor breaker terminal REB551 is a
complete protection package for functions associated
with a circuit breaker and may in addition to reclosing
and synchronizing functions also include breaker
failure protection. The recloser function of this
terminal (which is also implemented in distance relay
terminals REL521/531 and line differential terminals
REL551/561) is briefly described here.

In breaker-and-a-half applications, the breaker
control functions, such as reclosing, are associated
with the breaker rather than with the line protection.
In breaker-and-a-half substations, there are two
breakers that trip for a line fault. The reclosing of
the breaker should be made sequentially in order to
minimize breaker wear. First one breaker (MASTER)
recloses and if it does not trip again (the fault was
transient), the second breaker (SLAVE) is allowed to
close. In case the fault is permanent, the MASTER
breaker trips again, and closing of the SLAVE breaker
is blocked.

Figure 16-8 shows the operation for a transient
fault. The sequence is as follows:

1. Both circuit breakers, CB1 and CB2, are tripped
   for the line fault.

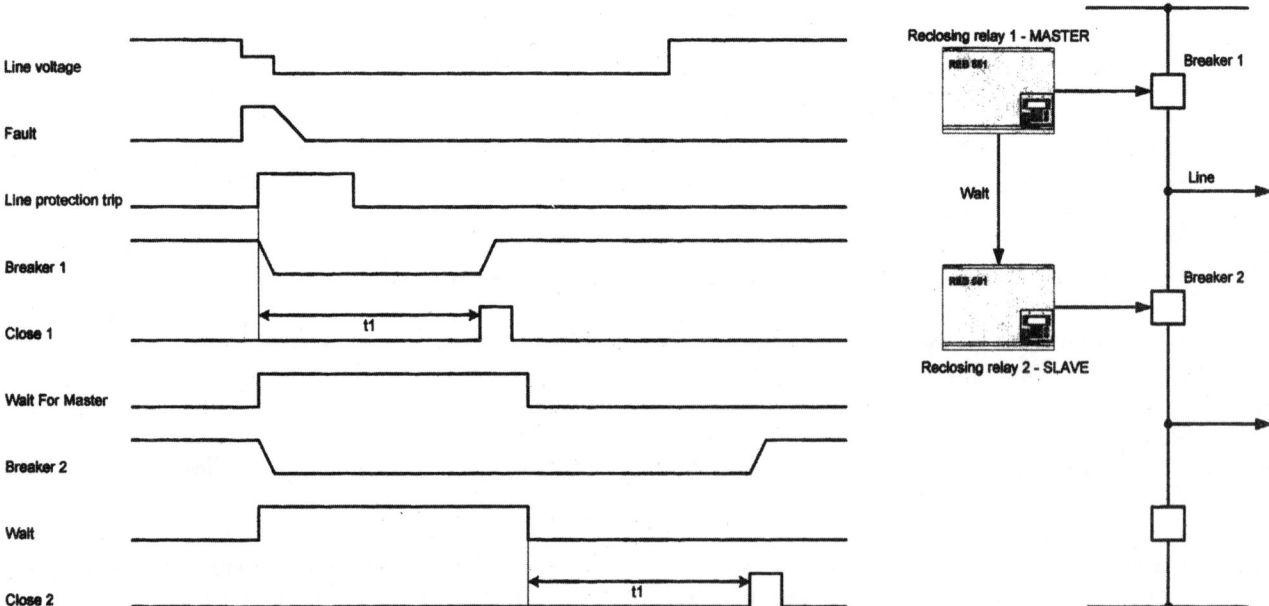

Figure 16-8 Sequential reclosing in breaker-and-a-half application.

2. The MASTER recloser starts its dead time and issues a blocking signal, WF MASTER (wait for MASTER), to the SLAVE recloser.
3. The MASTER recloses its breaker, CB1, and removes the blocking signal.
4. The SLAVE starts its dead time and recloses its breaker CB2.

For a permanent fault, CB1 will trip again and both the MASTER and SLAVE recloser locks out, provided that a single shot reclosing was set. For multiple-shot reclosing, CB1 will continue its cycle and CB2 will be allowed to reclose only when CB1 has been closed successfully for 1 sec.

6 SYNCHRONISM CHECK

Synchronism-check relays verify that the voltages on the two sides of a breaker are approximately the same in magnitude and phase in order to ensure that a minimum impact occurs to the power system when a circuit breaker is closed. The synchronism-check relay supervises automatic or manual closing of a circuit breaker. One type, the CVX, is shown in Figure 16-9 and its connections are typical. The 52b contact ensures that the CVX contact is open immediately after tripping and that a conscious determination of synchronism is made prior to closure.

The driving force for current flow following breaker contact closure is the voltage appearing across the open contacts just prior to closing. The difference between the voltages on the two sides of the breaker is called the *phasing voltage*. The phasing voltage can be controlled by several methods; the conventional *angular* and the more precise *phasing voltage* methods.

6.1 Phasing Voltage Synchronism Check Characteristic

The phasing voltage method provided by CVX, and also the circuit shield type 59S/V, form the characteristics shown in Figures 16-10 and 16-11. The line-to-ground voltage (bus voltage) will provide a reference, and a comparison is made with an input voltage which is supplied to the relay from the other side of the breaker that is to be controlled (line voltage).

The phasing voltage method verifies that the voltage difference between the two voltages, bus voltage and line voltage, is less than the set limit determined by the angular adjustment. The normal angular adjustment is 20°, which may be increased to 60° if such a wide closure angle will not disrupt the system. Figure 16-10 describes the relationship required in order for closing of the circuit breaker to be permitted. The center of the circle is established by the reference voltage phasor (bus voltage). The radius of the circle is determined by the voltage difference setting. Operation results when the synchronizing voltage (line voltage) falls inside the circle. A "close" output is produced in response to a reclosing-directed synchronism check or a manual close command. The 52b contact assures that synchronism which exists prior to tripping is not allowed

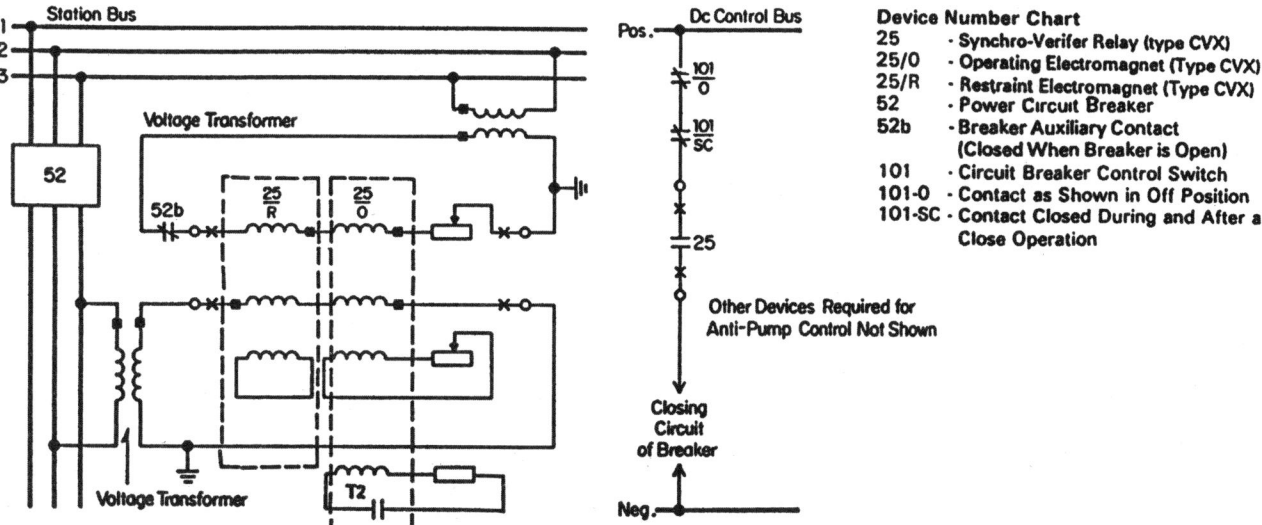

Figure 16-9 Synchronism check relay schematic diagram.

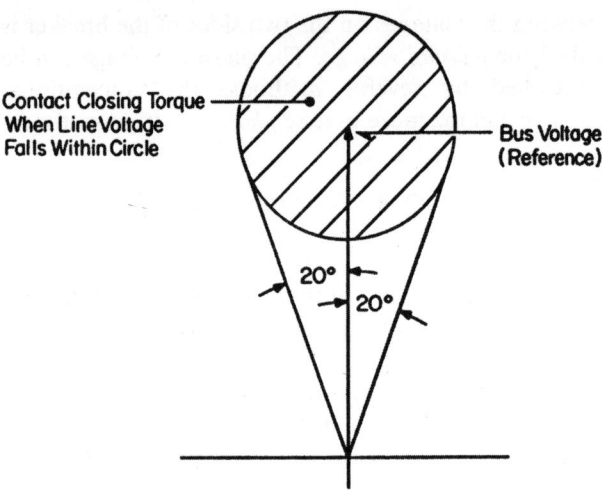

Figure 16-10 Phasing voltage synchro-check relay closing characteristic.

to falsely indicate that synchronism exists following tripping.

Usually in conventional synchronism-check relaying, a relatively long time measurement is used to insure that the voltages across the open breaker are in synchronism. However, this long time delay, which may be 10 sec or more, is undesirable if both ends of the line are being reclosed at high speed. If the time measurement is shortened, a faster synchronism-check measurement can be made, but this may result in reclosing for a nonsynchronous condition with slip frequencies that are higher than desired for proper reclosing. A slip cut-off frequency function can provide

a high-speed synchronization determination when voltages are in synchronism without the risk of reclosing if high slip frequencies are actually present. Figure 16-12 plots the maximum frequency across the open breaker that will allow the CVX contact to close. For example, at a 20° setting of angle of closure and a slip frequency of 0.0167 (the value that gives a synchronscope pointer movement at the rate of a clock's second hand) would require a time-dial setting on the CVX of 2 or less.

6.2 Angular Synchronism Check Characteristic

The angular method is the one most commonly used in microprocessor relays. The line-to-ground voltage (V_{BUS}) will provide a reference, and a comparison is made with an input voltage (V_{SYNC}) which is supplied to the relay from the other side of the breaker that is to be controlled. It verifies that the angular difference (θ_{SYNC}) is within the acceptable range of set angle and verifies both voltage magnitudes are in the acceptable band between V_{MAX} and V_{LIVE}. The resulting synchronism region looks like an automobile windshield

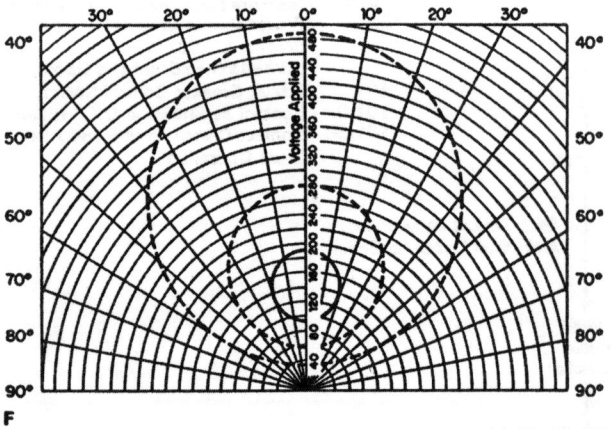

Figure 16-11 Typical voltage angle characteristics of CVX for various angle settings. (Rated voltage on one circuit.)

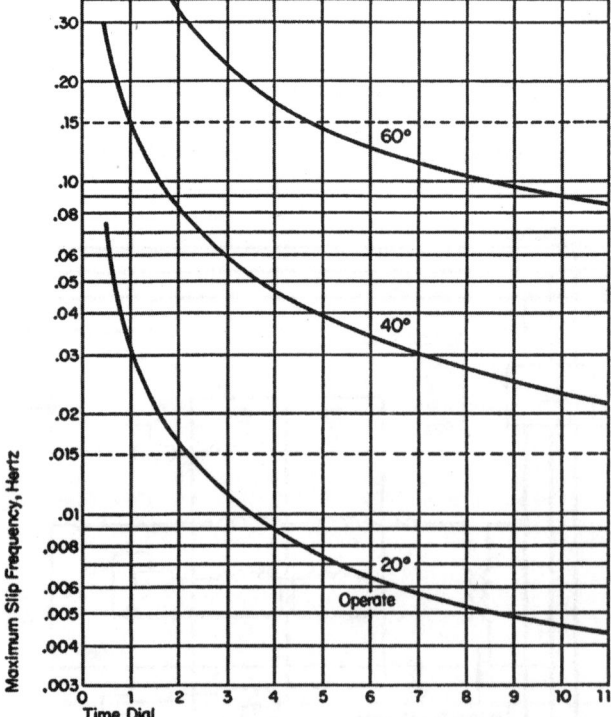

Figure 16-12 Approximate maximum slip frequency for which operation occurs. (Rated voltage on both sides.)

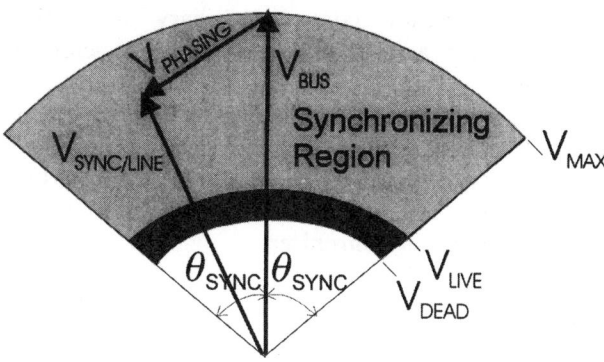

Figure 16-13 Angular synchronism check characteristic.

wiper stroke with V_{BUS} at the center as shown in Figure 16-13. Operation results when the synchronizing voltage ($V_{SYNC/LINE}$) falls inside the synchronizing region and the slip frequency is below the set value of SLIP FREQ.

The synchronism-check function is intended to assure that the two segments of the power system that are to be interconnected by the closure of a breaker are in synchronism with a standing angle. It is not intended to provide an automatic synchronizing function for two systems that are operating at different frequencies. The relays contain no provision for energizing the closing coil at the precise angle that would ensure zero voltage across the breaker contacts at the instant of closure.

7 DEAD-LINE OR DEAD-BUS RECLOSING

At some locations, reclosing is initiated when the bus is hot and the line dead, or vice versa. The synchronism-check function will not provide an output to permit reclosing if no voltage is present on one or both sides of the open breaker. With the possibility that either side of the breaker, bus or line, may be deenergized (dead), hot-busdead-line, dead-bushot-line, or dead-busdead-line checks are necessary. In this case single- or three-phase voltages may be checked to establish clearly that the voltages are as they appear, and not possibly erroneous because of such things as fuse failure. This arrangement may be used, for example, to energize one end of a transmission line if the bus is hot. Reclosing at the other end of the line could be supervised by a synchro-check relay. The CVX relay contains elements for the synchronism-check, LLDB, and LBDL controls. Figure 16-14 illustrates its basic

operation for controlling automatic breaker closure. Manual breaker closures by 101C can also be supervised as shown.

To avoid any possibility of pumping, synchronism-check closures should always be made under control of a reclosing relay or similar device (see Fig. 16-14). The synchro-verifier relay does not include an antipump provision, nor will the antipump scheme in the breaker reliably prevent pumping when closing into a permanent fault.

8 AUTOMATIC SYNCHRONIZING

A synchronizing system can be used at unattended or attended locations for automatic synchronizing or supervision of manual synchronizing. The heart of this system is the synchronizer, which compares the voltage on two sides of an open breaker and energizes the breaker close coil under the following conditions:

1. If the frequency difference is below a preset amount

Partial Schematic

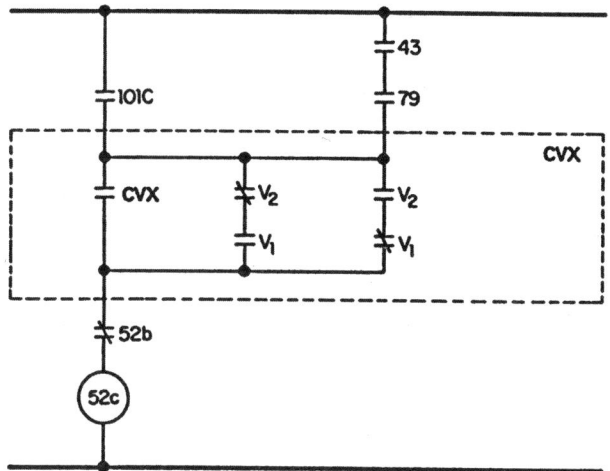

Device Number Chart

52c - Close Coil of Circuit Breaker
43 - Automatic/Manual Switch
79 - Reclosing Relay
101C - Control Switch Contact (Closed in "Close")
V_1 - Voltage Unit for One Side of Breaker
V_2 - Voltage Unit for Other Side of Breaker

Figure 16-14 Typcial synchronism check control of breaker closing using CVX-1.

2. At such a phase angle that the breaker contacts close when the systems are in phase

The automatic synchronizer Synchrotact 5 can be equipped with the following options:

1. An acceptor to restrict the range of voltages at which breaker closing in initiated

2. A governor control auxiliary to initiate angular adjustments needed for synchronizing

3. A regulator control auxiliary to initiate voltage-level adjustments needed for synchronizing

4. A synchronism-check device to avoid possible closure because of component failure in the automatic synchronizer

17

Load-Shedding and Frequency Relaying

Revised by: **W. A. ELMORE**

1 INTRODUCTION

When a power system is in stable operation at normal frequency, the total mechanical power input from the prime movers to the generators is equal to the sum of all the connected loads, plus all real power losses in the system. Any significant upset of this balance causes a frequency change. The huge rotating masses of turbine-generator rotors act as repositories of kinetic energy: When there is insufficient mechanical power input to the system, the rotors slow down, supplying energy to the system. Conversely, when excess mechanical power is input, they speed up, absorbing energy. Any change in speed causes a proportional frequency variation.

Unit governors sense small changes in speed resulting from gradual load changes. These governors adjust the mechanical input power to the generating units in order to maintain normal frequency operation. Sudden and large changes in generation capacity through the loss of a generator or key intertie can produce a severe generation and load imbalance, resulting in a rapid frequency decline. If the governors and boilers cannot respond quickly enough, the system may collapse. Rapid, selective, and temporary dropping of loads can make recovery possible, avoid prolonged system outage, and restore customer service with minimum delay.

2 RATE OF FREQUENCY DECLINE

Before designing a relay scheme for system overload protection, it is necessary to estimate variations in frequency during disturbances. Figure 17-1 shows a system S that consists of two interconnected subsystems, S_1 and S_2. For all of S, the following relationship must hold true for constant-frequency operation:

$$\text{Generation} = \text{loads} + \text{losses} \qquad (17\text{-}1)$$

There can, however, be more generation than load in S_1 and more load than generation in S_2, with the difference being transferred by an intertie as shown. If the total loads and losses are equal to the total mechanical power input, there will be no change in generator speed or frequency with time.

If, however, the tie is suddenly lost as a result of a permanent fault, the kinetic energy in the S_1 generators must increase to absorb the excess power input; that is, the generators must speed up. Conversely, the S_2 generators must slow down. The classical expression for the initial rate of change of frequency is

$$\frac{df}{dt} = -\frac{\Delta P}{2H} \qquad (17\text{-}2)$$

where

$\dfrac{df}{dt}$ = per unit initial rate of change of frequency

Δp = decelerating power in per unit of connected kVA

H = inertia constant, $\dfrac{MW - sec}{MVA}$ or $\dfrac{KW - sec}{kVA}$

In this text, all H constants and ΔP values are related to a kVA rather than kW base.

The inertia constant (H) is defined as the ratio of the moment of inertia of a generator's rotating components to the unit capacity. It is the kinetic energy in

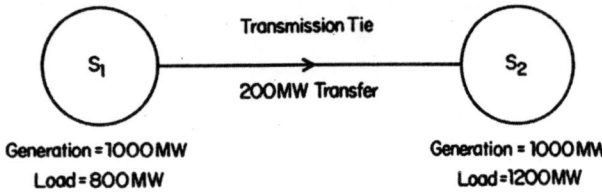

Figure 17-1   Interconnected system S.

these components at the rated speed. For example, a turbine-generator rated at 100 MVA, with an inertia constant of 4, has a kinetic energy of 400 MW-sec, or 400 MJ, in its rotor when spinning at rated speed. If both power output and load were constant with declining frequency and speed, the generator could supply its full load (with p = 1) for 4 sec, with no power input to the turbine, before the rotor would come to a complete halt. The inertia constant H for an individual unit is available from the manufacturer or may be calculated from

$$H = \frac{0.231 \, WR^2 \, RPM^2 \, 10^{-6}}{kVA} \qquad (17\text{-}3)$$

For a system, a composite value is calculated as follows:

$$H_{system} = \frac{H_1 MVA_1 + H_2 MVA_2 + \ldots + H_n MVA_n}{MVA_1 + MVA_2 + \ldots + MVA_n} \qquad (17\text{-}4)$$

where subscripts 1, 2, . . . , n refer to individual generating units.

The larger the inertia constant, the slower the frequency decline for a given overload. Older water-wheel generators, with their massive rotors, have inertia constants as large as 10. Newer turbine-generator units, however, may have inertia constants of only 2 or 3, since the trend is toward larger outputs with smaller rotor masses. Power systems are becoming more prone to serious frequency disturbances for given amounts of sudden load change.

An example is shown in Figure 17-1. System S_2 in the figure has a net load of 1200 MW and total generation of 1000 MW. The tie must carry 200 MW from S_1 to S_2. The inertia constant for S_2 is 4, and the power-factor rating of the machines there is 0.85. If the tie is suddenly lost, the initial rate of frequency drop in system S_2 is calculated as follows:

$$\Delta P = \frac{\text{tie load lost}}{\text{kVA of } S_2} = \frac{200}{1000/0.85}$$

$$= 0.17 \text{ per unit}$$

Then using Eq. (17-2), we obtain

$$\frac{df}{dt} = \frac{0.17}{2(4)} = -0.0213 \text{ per unit}$$

$$\frac{df}{dt} = -0.0213 \times 60 = -1.275 \text{ Hz/sec}$$

The negative rate of change indicates frequency drop.

As the frequency drops, experience has shown that load power also decreases. A frequently used relationship is that 1% frequency drop produces 2% automatic load reduction. 1% frequency drop corresponds to 0.6 Hz. The resulting load change applies to the entire load in the system segment under consideration, not to the ΔP, and in the example above would correspond to 0.02 (1200) = 24 kW. This reduces the generation deficiency to (200 − 24) = 176 kW, slowing the decay. A frequency reduction of 8.35% would reduce the load by 16.7% or 0.167 (1200) = 200 kW. However, at this frequency (55 Hz), generating-plant station auxiliaries would probably collapse and stable operation would not be possible. Motor-driven auxiliaries will slow down, reducing generator output. Safety margins in generator and motor cooling and bearing lubrication systems may become dangerously small.

The simple diagram of Figure 17-2 may be used to illustrate a principle. Loss of generator 2 imposes an increased load on generator 1. The load will continue to be supplied, but at the expense of decreasing speed of the rotating mass. The initial MW overload on the "remaining" generation is exactly equal to the "lost" MW generation. The load reduction which is influenced by the frequency reduction is related to the original total load $(1 + \Delta P)$. The new load following frequency reduction is 1.0, the original load on G1. A new stable operating frequency is reached:

$$f_f = f_0 \left[1 - \frac{\Delta P}{d(1 + \Delta P)} \right] \qquad (17\text{-}5)$$

where

f_f = new stable operating frequency in hertz
f_0 = rated frequency in hertz
ΔP = per unit load reduction (per unit based on remaining generation)
d = per unit change in load per unit change in frequency

The d factor may vary from 1/2 to 7, depending on the mix of loads, although typically most utilities assume d = 2 (that is, a 2% decrease in load for each 1% decrease in frequency). An exact value of d can be

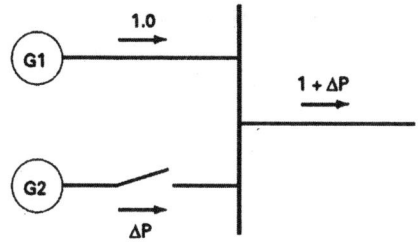

Open Switch at $t = 0$

$$\Delta L = \frac{\Delta P}{1 + \Delta P}$$

$$d = \frac{\Delta L}{\Delta f}$$

$$\Delta f = \frac{\Delta L}{d}$$

$$f_f = f_0 (1 - \Delta f)$$

$$f_f = f_0 \left[1 - \frac{\Delta P}{d(1 + \Delta P)} \right]$$

Figure 17-2 Example of natural load-shedding.

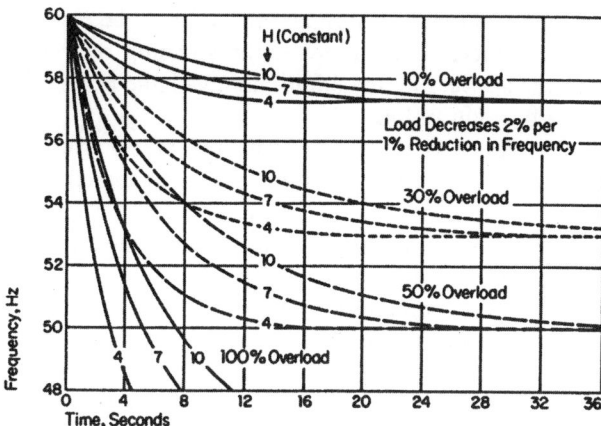

Figure 17-4 Behavior of frequency during overload.

Most 60-Hz plants will operate down to 55 Hz on a temporary basis, but under no circumstances should long-blade turbines be permitted to operate loaded on a steady basis at a frequency below 59.5 Hz or below 58.5 Hz for shorter-blade machines. The last rows of low-pressure blades in steam and gas turbines are tuned to be free of resonance when operating at the rated speed. Off-normal operation can produce failure in a matter of minutes.

Governor response will work to correct the deficiency (or excess) in system speed. However, time delay is involved in reestablishing a new stable relationship in boilers, water flow, etc. To avoid operation at reduced frequency, it is necessary to shed load (trip circuit breakers to disconnect load from its source of power).

determined only by observing the variation of load with frequency on the system under consideration.

Figure 17-3 shows a typical frequency decline resulting from loss of generation. Figure 17-4 describes the behavior of system frequency for several combinations of inertia constant and percent overload using $d = 2.0$.

3 LOAD-SHEDDING

For gradual increases in load, or sudden but mild overloads, unit governors will sense speed change and increase power input to the generator. Extra load is handled by using *spinning reserve*, the unused capacity of all generators operating and synchronized to the system. If all generators are operating at maximum capacity, the spinning reserve is 0, and the governors may be powerless to relieve overloads.

In any case, the rapid frequency plunges that accompany severe overloads require impossibly fast governor and boiler response. To halt such a drop, it is necessary to intentionally and automatically disconnect a portion of the load equal to or greater than the overload. After the decline has been arrested and the

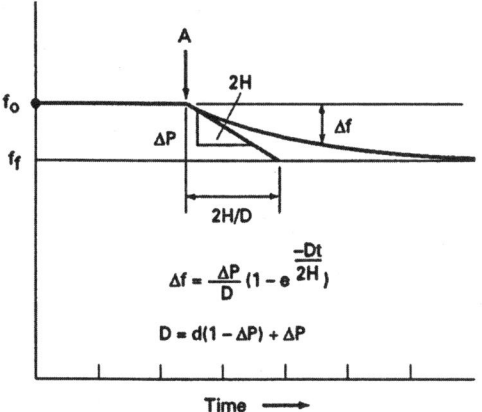

Figure 17-3 Frequency change response to moderate overload (all per unit).

frequency returns to normal, the load may be restored in small increments, allowing the spinning reserve to become active and any additional available generators to be brought online.

Frequency is a reliable indicator of an overload condition. Frequency-sensitive relays can therefore be used to disconnect load automatically. Such an arrangement is referred to as a load-shedding or load-saving scheme and is designed to reserve system integrity and minimize outages. Although utilities generally avoid intentionally interrupting service, it is sometimes necessary to do so in order to avert a major system collapse. In general, noncritical loads, usually residential, can be interrupted for short periods, minimizing the impact of the disturbance on service.

Automatic load-shedding, based on underfrequency, is necessary since sudden, moderate-to-severe overloads can plunge a system into a hazardous state much faster than an operator can react. Underfrequency relays are usually installed at distribution substations, where selected loads can be disconnected.

The object of load-shedding is to balance load and generation. Since the amount of overload is not readily measured at the instant of a disturbance, the load is shed a block at a time until the frequency stabilizes. This is accomplished by using several groups of frequency relays, each controlling its own block of load and each set to a successively lower frequency. The first line of frequency relays is set just below the normal operating frequency range, usually 59.4 to 59.7 Hz. When the frequency drops below this level, these relays will drop a significant percentage of system load. If this load drop is sufficient, the frequency will stabilize or actually increase again. If this first load drop is not sufficient, the frequency will continue to drop, but at a slower rate, until the frequency range of the second line of relays is reached. At this point, a second block of load is shed. This process will continue until the overload is relieved or all the frequency relays have operated. An alternative scheme is to set a number of relays at the same frequency or close frequencies and use different tripping time delays.

Techniques for developing schemes and calculating settings are described in Section 5.

4   FREQUENCY RELAYS

Many different types of frequency relays have been used over the years. The induction-disk relay that was the forerunner of all frequency relays has faded into disuse in favor of more accurate devices. Three general classes of frequency relays are being applied: the induction-cylinder relay, the digital relay, and the microprocessor relay.

4.1   KF Induction-Cylinder Underfrequency Relay

This relay is fast and sufficiently accurate for most applications. The principle of operation of the relay is described in Chapter 3 and is based on a circuit in which the phase angle changes as frequency changes. The phase relationship between the current in this circuit and a reference current produces torque that changes direction when the set-point frequency is reached. With this relay, as with all frequency relays, precaution must be taken in the design of the device to make certain that the phase shift associated with phenomena such as faults, which appear as an extreme change of frequency due to the sudden change of phase angle of the supply voltage, do not produce misoperation. In the case of the induction-cylinder relay, a time delay of at least six cycles is required. The tripping characteristics for the KF induction-cylinder relay are shown in Figure 17-5. This type of plot is useful for predicting the frequency at which tripping will occur during frequency declines. It reflects the fact that the frequency will continue to drop after the relay-setting frequency is crossed, and during the time the relay is operating. As a result, the actual contact closure frequency will be somewhat below the set value.

The "cycles-of-delay" parameter associated with each curve in the family is the intentional time-delay setting after the cylinder unit closes its contacts. The plot shown in Figure 17-5 includes the inherent cylinder operating time. An examination of the six-cycle delay curve with a 10-Hz/sec frequency decline indicates that trip contact will close when the frequency is 2 Hz below the setting. Total operating time following the crossing of the set-point frequency will thus be $2/10 \times 60$ (base) = 12 cycles. In other words, the cylinder unit operates in six cycles and the timer adds a six-cycle delay. Figure 17-6 shows an example of the actual trip frequency that results from the necessary operating time of the induction cylinder and the required security delays. The relay in this case was set for 58.9 Hz and was unable to produce benefit until the frequency had decayed to 57 Hz. This is an exaggerated case. Frequency decay, in general, will not be so severe and it will benefit from the automatic load-shedding resulting from frequency decay that is not shown here.

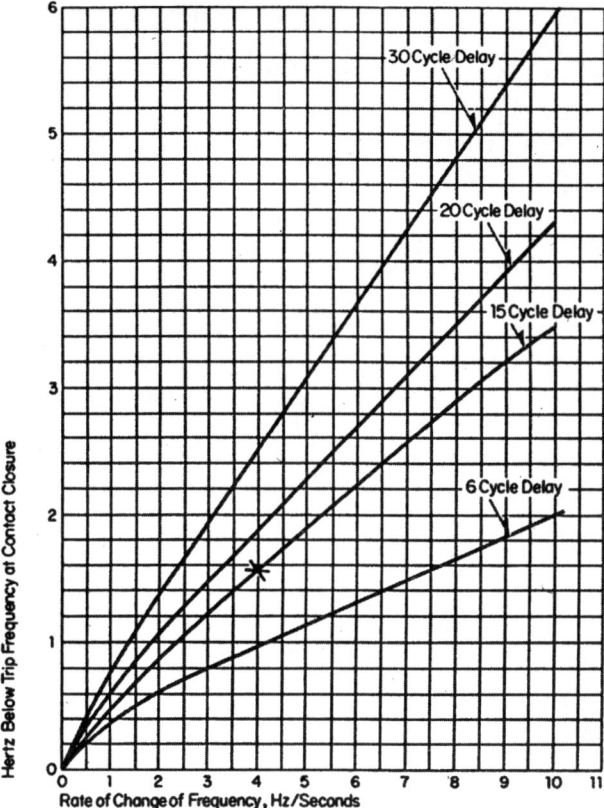

Figure 17-5 Characteristics of 60-Hz induction cylinder (KF) frequency relay.

4.2 Digital Frequency Relays

Digital relays in general utilize a multimegahertz counter. Zero crossing of voltage is detected, and a counter starts and continues counting until the next voltage zero or in some relays until the next positive-going zero crossing. The count accumulated is indicative of the period of the waveform and thus the frequency is identified.

Accuracies of 0.005 Hz are realizable utilizing this concept. Security is achieved by several expedients such as requiring that an abnormal count occur in three consecutive periods. One cycle of adequate frequency will cause the relay to reset, requiring it to begin the "count of 3" again. The sudden phase shift caused by a fault such as a "b to c" fault for a relay that is sensing "a to b" voltage will be sensed as a short period once but will not repeat.

4.3 Microprocessor-Based Frequency Relay

The principle applied in the microprocessor relay is the same as that in a digital relay, but additional sophistication is included. All the self-checking provisions and examination of various failure modes are constantly achieved and alarm and lockout are an inherent part of these relays.

Multiple set points are common among digital and microprocessor frequency relays. Some may be used for over- or underfrequency applications and some include a "restore" function. The restore function may be set at a frequency level to indicate that the power system has recovered and is now able to accommodate the reapplication of the load that was shed. In general, a long time delay is required in the restore function to assure that pumping will not result. Often, an external timer is required.

5 FORMULATING A LOAD-SHEDDING SCHEME

Several procedures and criteria must be considered when designing load-shedding schemes for specific systems. These include:

1. Maximum anticipated overload
2. Number of load-shedding steps
3. Size of the load shed at each step
4. Frequency settings
5. Time delay
6. Location of the frequency relays

5.1 Maximum Anticipated Overload

Underfrequency relays should be able to shed a load equal to the maximum anticipated overload. Logically, there is no reason to limit load-shedding to any percentage of load. Indeed, it is preferable to shed 100% of load, preserving interconnections and keeping generating units on line and synchronized, than to allow the system to collapse with customers still

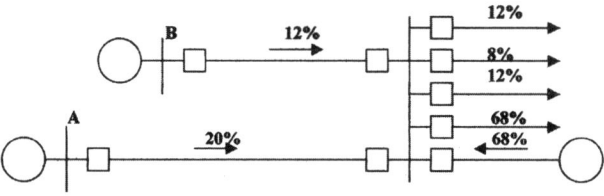

Figure 17-6 Load shedding example.

connected. Even if 100% of the load is shed, service can be restored rapidly; if the system collapses, a prolonged outage would result. For this reason, it is necessary to evaluate the cost of the load-shedding scheme in light of the probability that an overload of a given severity can occur.

The system should be studied with respect to the overload that would result from the unexpected loss of key generating units, transmission ties, and buses. Stability studies can help identify areas that, if separated or islanded from the rest of the system, would have a severe generation deficiency. These areas will need more comprehensive load-shedding.

The load-reduction factor d should also be considered, since it will reduce the overload once the frequency has dropped. If spinning reserve, or additional generation capacity equal to the overload compensated for by d, is not available shortly after the disturbance, it will be impossible to bring back the system to rated frequency. This will mean that an islanded system cannot be resynchronized, and interconnections to neighboring utilities cannot be reclosed. (It should also be remembered that the turbine-generators must not be operated for extended periods below rated speed.)

The load-reduction factor d is rarely known exactly and may vary with time. To design a conservative scheme, which will tend to shed enough load for system recovery to normal frequency, it is safest to assume that d equals 0.

5.2   Number of Load-Shedding Steps

The simplest load-shedding scheme is one in which the predetermined percentage of the load is shed at once when a group of relays senses a frequency drop. Although this scheme will arrest any anticipated frequency decline, it will often disconnect far more customers than necessary. A refinement then would be to use two groups of relays, one operating at a lower frequency than the other, and each shedding half the predetermined load. The higher-set relays would trip first, halting the frequency decline as long as the overload were half or less of the worst-case value. For more severe overloads, the frequency would continue to drop, although at a slower rate, until the second group of relays operated to shed the other half of the expendable load.

The number of load-shedding steps can be increased virtually without limit. With a great many steps, the system can shed load in small increments until the decline stops; almost no excess load need be shed. Such a scheme may, however, inhibit system recovery. As noted below, it may also be difficult to coordinate so many steps.

Most utilities use between two and five load-shedding steps, with three being the most common.

5.3   Size of the Load Shed at Each Step

When possible, the size of the load-shedding steps should be related to expected percentage overloads. When a study of the system configuration, or a stability study, reveals that there is a relatively high probability of losing certain generating units or transmission lines, the load-shedding blocks should be sized accordingly. Sizing can be determined as in the following example.

Assume that a power system has a generating plant A at a remote location, this plant is tied to the rest of the system by long lines, and the system also is connected by a transmission tie B to a neighboring utility (Fig. 17-6). Assume in addition that A carries up to 20% of system load and B up to 12%. Stability studies show that certain faults or disturbances may result in loss of synchronism between A and the system, so that its transmission ties must be opened. Furthermore, problems on the neighboring utility system may necessitate the tripping of B. It is important for such a system to implement a load-shedding scheme that will preserve the remaining system if A and B are lost. It is logical, therefore, to use three load-shedding steps to handle overload

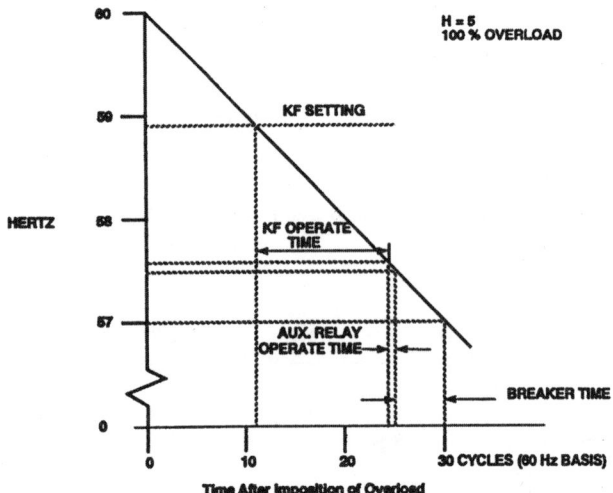

Figure 17-7   Frequency vs. time on 100% overload.

resulting from (1) loss of A, (2) loss of B, or (3) loss of both A and B simultaneously. The overloads, in order of increasing probability and seriousness, are listed in Table 17-1.

The following load-shedding steps are implemented to handle each situation in succession:

Step 1   Shed 12% of total load (12% of total).
Step 2   Shed an additional 8% of remaining load (20% of total).
Step 3   Shed an additional 12% of remaining load (32% of total).

$$\% \text{ overload} = \frac{\text{load} - \text{power input}}{\text{power input}} \times 100 \quad (17\text{-}6)$$

Note that each step sheds only enough load to handle the next, more serious contingency. Each step should be evenly spread over the system by dropping loads at diverse locations.

If the system under consideration is large, there may be many possible combinations of events to consider, each causing only a small percentage overload in itself. In this case, a number of overload situations may be lumped together and handled in one step. Conversely, it may be sufficient to shed a percentage of the overload in a few equal steps. To implement the shedding evenly and at the distribution level, such a system would require a large number of frequency relays distributed over the system. With so many relays, there is no cost penalty in using smaller steps (five, for example) to more closely balance generation and load, provided that all the steps can be coordinated.

5.4   Frequency Settings

The frequency at which each step will shed load depends on the system's normal operating frequency range, the operating speed and accuracy of the

Table 17-1   Overloads Resulting From the Loss of A, B, or Both

Event	Percent of generation lost	Percent overload from Eq. (17-6)
Loss of interconnection B	12	13.6
Loss of generator A	20	25
Loss of both A and B simultaneously	32	47

frequency relays, and the number of load-shedding steps.

The frequency of the first step should be just below the normal operating frequency band of the system, allowing for variation in the tripping frequency of the relay. The stable, solid-state type 81 relays or microprocessor relays may be set from 55 to 59.9 Hz within 0.01 Hz of the lowest expected normal-frequency excursions to trip at the first indication of trouble. For induction-cylinder electromechanical relays, the highest frequency setting should be approximately 0.1 to 0.2 Hz below the system's lowest normal operating frequency. Whatever type of relay is used, the frequency should be selected to avoid shedding for minor disturbances from which the system can recover on its own.

The remaining load-shedding steps may be selected as follows:

1. Based on the best estimate of ΔP, calculate df/dt using Eq. (17-2). Employing relay tripping curves, calculate the actual frequency at which load will be shed by the first-step relays for the most severe expected overload. (See Fig. 17-6 for guidance.)
2. Set the second-step relays just below this frequency, allowing a margin that will tolerate any expected frequency drift for *both* sets of relays.
3. Calculate the actual frequency at which the second load-shedding step will occur. The rate of frequency decline by the second-step relays can be calculated as that resulting from the most severe expected overload *minus* the load shed in the first step.
4. Again, allowing a margin for relay drift, set the third-step relays below the lowest second-step shedding frequency.
5. Repeat the calculations until settings are obtained for all steps. Determine the system's lowest frequency value before the final load block is interrupted for the worst-case overload. This value should not be below the system's low-frequency operating limit.

Continuing with the example given in Section 5.3 above, we may calculate frequency settings as follows. Assume the first load-shedding frequency is 59.5 Hz and that $H = 4$ (based on remaining generator MVA). From Table 17-1 the worst expected overload is 47% (47% of remaining generator MVA). This will cause a frequency decline of approximately 3.5 Hz/sec. The set point will be reached in 0.143 sec. By applying MDF relays with an effective 60-msec time delay and distribution breakers with an interrupting time of five cycles, the additional time to produce trip is 0.143 sec. The frequency at the time of interruption of load would be 59 Hz.

With the first block of load chosen as 12%, there is at this point a reduction in the overload. Since ΔP must be based on the kVA of the remaining generation (68%) and the 12% is of the original total load, the load is $(100 - 12)/68 = 1.294$. ΔP is then 29.4%. The rate of frequency decline becomes $60 \times 0.294/(2 \times 4) = 2.20$ Hz/sec.

A setting of 58.9 is usable for the next shedding step. This will be reached, with a 2.2 Hz/sec decay rate, in $(59 - 58.9)/2.2 = 0.045$ sec. The relay delay and circuit-breaker opening time are again 0.143 sec. The total time for separating load is then 0.188 sec. The frequency would have decayed to 58.58 Hz by this time. A load corresponding to 8% of the original total load is shed at this point.

The load now shed is 20% of the original total. The load on the remaining generation is $(100 - 20)/68 = 1.176$ for a ΔP of 17.6%. The decay rate becomes $60(0.176)/(2 \times 4) = 1.32$ Hz/sec. A setting of 58.5 Hz is realizable for the next stage. The setting will be crossed in $(58.58 - 58.5)/1.32 = 0.06$ sec. As calculated before, the total time in this interval is 0.203 sec and the frequency falls 0.268 Hz for a shedding level of 58.3 Hz for dropping another 12% of the original load.

With 32% generation loss initially and 32% load shed, the worst-case condition is handled with no frequency excursion below 58.3 Hz. Complete recovery occurs. With any lower level of generation loss, recovery will occur with less frequency drop in each stage and fewer levels of underfrequency settings being reached. Also with inherent load-shedding, the frequency decay would have been somewhat slower.

5.5 Time Delay

The above examples illustrate an important rule for load-shedding schemes: Use the minimum possible time delay consistent with relay security. The less the delay, the more easily the scheme can cope with severe overloads. All unnecessary interposing auxiliary devices should be avoided.

Naturally, there are exceptions when an extra time delay may be needed. One such case is a frequency relay connected to a potential supply from a bus that supplies induction-motor loads (Fig. 17-8). If the line breakers 1 and 2 trip and interrupt current to the motors, they will slow down rapidly. Because of

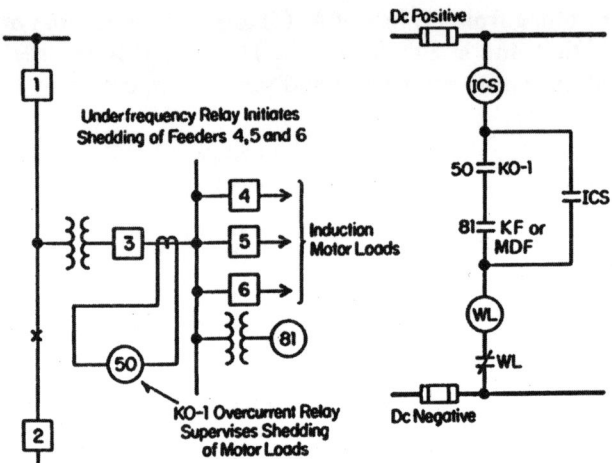

Figure 17-8 Underfrequency relay with induction motor load.

trapped, decaying flux, the motors will excite the bus with ac potential of declining frequency, possibly for 0.5 sec or longer. As a result, the load-shedding frequency relay will trip unnecessarily and lock out the feeder breakers 4, 5, and 6. When remote breakers 1 and 2 reclose, service to the motors will not be restored and breakers 4, 5, and 6 must be reclosed manually. This situation has caused frequency relays with up to 30 cycles of delay to trip.

An intentional time delay long enough to ride through the residual voltage collapse is quite often too long to be consistent with load-shedding requirements. A more effective method is to supervise the underfrequency relay using the overcurrent relay (50) connected to the source current transformer, as shown in Figure 17-8. The frequency relay (81) will trip breakers 4, 5, and 6 and shed load only when significant load current is flowing into the bus.

Consideration should be given to the desirability of allowing the motors to remain on the line as the reclosing takes place. This may produce a severe physical impact on any motor that has a back-voltage that is 180° out of phase with the reapplied system voltage. The main breaker (breaker 3 in Fig. 17-8) is often tripped to avoid just such an unfavorable reenergization.

5.6 Location of the Frequency Relays

In large systems, the load-shedding relays should be spread throughout the system to avoid heavy power flows and undesirable islanding. Load-shedding in one concentrated area, for example, can cause heavy power flow over transmission lines from the area where the load was shed to areas of excess load. Because of the original disturbance, these lines may already be operating at high emergency levels, and the uneven load-shedding may cause thermal overload or system instability.

Concentrated loss of generation in certain areas of the system will also result in frequency dispersion; that is, the frequency in the overloaded areas will drop faster than elsewhere. The difference in frequencies naturally produces rapidly increasing torque angles on the transmission lines, which may cause the system to go out of step. Fortunately, load-shedding relays in the area of greatest frequency decline will trip first. This action alleviates the uneven loading, helps to bring back the system to uniform frequency, and avoids the impending loss of synchronism. It is clearly important, however, to install some extra load-shedding capability in any portion of the system that is prone to concentrated overload.

Finally, load-shedding priorities must be established. The nature of the loads shed can usually be controlled only by tripping feeders at the distribution level. The implication is that frequency relays will be installed in many distribution substations and will control relatively small blocks of load.

6 SPECIAL CONSIDERATIONS FOR INDUSTRIAL SYSTEMS

Load-shedding programs are recommended for industrial power systems. Frequency relaying is highly desirable for those systems in which loads are supplied either exclusively by local generation or a combination of local generators and utility ties. Power must often be maintained to certain essential processes to avoid danger of personal injury, equipment damage, product loss, or process disruption.

For local generators, the same type of single- or multiple-step frequency-based load-shedding program can be applied as that described for utility systems. Special precautions may be necessary, however, to accommodate the relatively small number of power sources, each of which can supply a considerable part of the total load. This type of scheme can produce different, more serious disturbances.

For example, the scheme shown in Figure 17-9a is for a plant that generates about half its own power requirements, the balance being supplied through a utility tie. If this tie is lost, the local generators will be

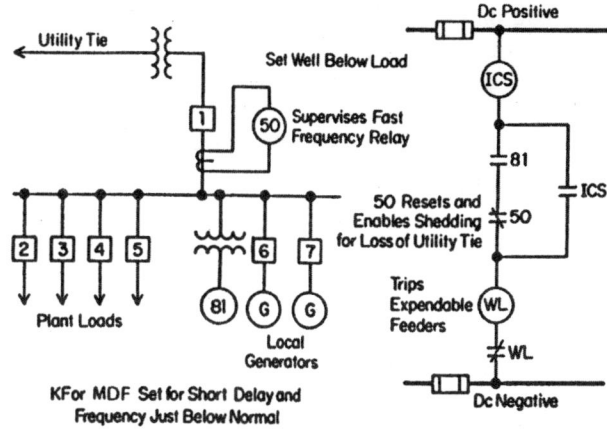

(a) Fast Shedding Scheme for Loss of External Source

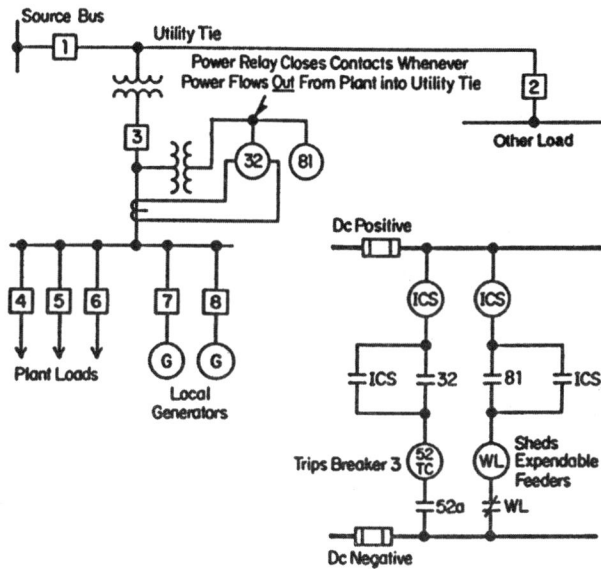

(b) Tripping Tie Breaker for Power Outflow to External Loads

Figure 17-9 Control of industrial plant load-shedding.

100% overloaded, and the rate of frequency drop will be 7.5 Hz/sec (if we assume that $H = 4$). Clearly, this plant needs an exceptionally fast load-shedding scheme that can drop a large percentage of low-priority load with minimum delay. There may be no time for multiple steps. At the same time, the need for speed and sensitivity is often incompatible with security requirements for milder disturbances.

One solution, shown in Figure 17-9a, is to set the frequency relay at a high, near-normal frequency and with minimum delay. Its tripping circuit is supervised by an undercurrent relay, whose contact closes if feeder service from the utility is lost. The large, sudden overload on the local generators is relieved by tripping low-priority plant loads.

An example of an interruptible load would be for a pulp mill where the chipper operation is not a part of the continuous papermaking process and, therefore, its temporary interruption would not interfere with (and indeed might save) the process. Temporary interruption of nonvital lighting in an industrial plant may also be considered for overload relief.

If the tie is in service and the utility suffers a frequency disturbance, the undercurrent relay will prevent load-shedding in the plant, while the utility sheds its overload in lower-priority areas.

The scheme of Figure 17-9b trips breaker 3 for any outfeed of power from the plant and then sheds load as required. A variation of this scheme uses the same complement of relays, but uses the "watt" relay to supervise the 81 trip. This prevents plant load-shedding for cases where the plant is supporting the utility system and the frequency is low.

Figure 17-10 describes a load-shedding scheme which may be used in a plant in which there is no local generation. Simple time-delayed overcurrent relays are used to drop expendable load when one transformer is out of service and the other becomes overloaded. It is assumed that, with proper planning, the maximum load will not exceed the capability of both transformers, and load shedding is required only when one transformer is out of service.

7 RESTORING SERVICE

In general, the reclosing of feeders that have been tripped for load-shedding is left to the discretion of system or station operators. Frequency relays can be used, however, either to supervise restoration or restore loads automatically.

The following considerations apply to any restoration of service, whether manual or automatic:

1. Frequency should be allowed to return to normal before any load is restored. Reclosing feeders when the frequency is still recovering may plunge the system back into crisis and will certainly prevent reunification of islands. Resetting of load-shedding frequency relays cannot be used for the supervision of restoration.

2. Once the frequency has returned to normal, all serviceable interconnections must be allowed to resynchronize and reclose. Unifying an islanded system as much as possible generally facilitates service restoration.

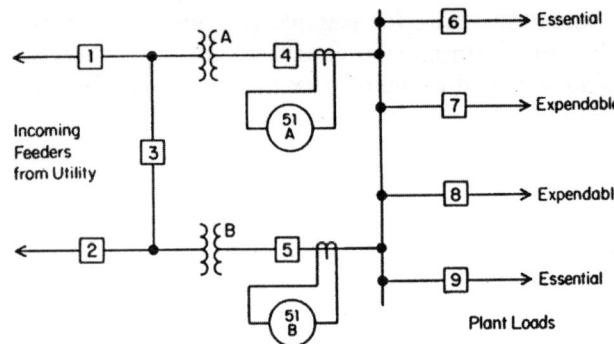

Notes:
1) Relays Set to Pickup for Transformer Overload

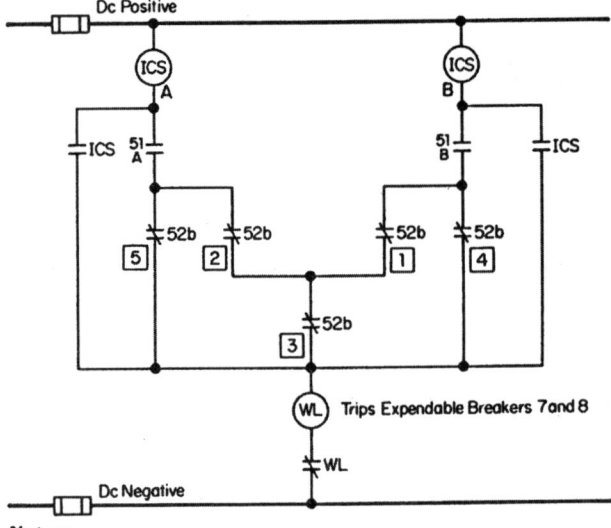

Notes:
1) 52b- Breaker Auxiliary Contact Which is Closed When Breaker is Open.
2) Circuit Permits Shedding When One Transformer is Overloaded and The Other is Disconnected from The System on The High Side, Low Side or Both.

(a) Protection against Transformer Overload.

Figure 17-10 Load-shedding schemes for industrial plants with no local generation.

3. Load should be restored in very small blocks. Reconnecting an entire shedding-step load at once, even at normal system frequency, can cause an overload. Not only may its size exceed spinning reserve, but also high currents resulting from cold load pickup can temporarily cause a severe overload. Reconnecting small blocks of load will cause only small frequency dips, which can be handled by the governors.

More small blocks may be reconnected until most or all of spinning reserve is active. At this point, no further load should be added until additional generating capacity is available. Restoring excessive load may cause the frequency to settle below-normal system frequency, making further reclosing of interconnections impossible.

4. If a significant loss of generation occurs in a concentrated area of the system, transmission lines into that area may be heavily loaded just to supply essential loads. In this case, the imbalance should not be increased by restoring expendable loads.

If frequency relays are used for automatic restoration, as they sometimes are at unattended installations, they should have a frequency setting of the normal system frequency. The load should be restored in blocks of 1 to 2% of system load, and restoration should be sequenced by time delay. After the initial system recovery to normal system frequency, there should be a delay of 30 sec to several minutes, implemented automatically with a timer or manually via supervisory control. This delay allows for resynchronizing of islands, reclosing of interconnections, and starting of peaking generators when available. The first block of load may then be restored; the frequency will dip and return to the normal system frequency. The next block should also incorporate several seconds of delay to permit frequency stabilization.

Each successive block should use a slightly longer time delay than the previous one. Thus, the second-block relays will time out before the third and reclose next. The frequency will reestablish at the normal system frequency, and the third block will time out and reclose. This process will continue until all blocks are restored or the spinning reserve is exhausted.

When restoring "cold" loads, it may be necessary to temporarily disable the instantaneous-overcurrent fault protection to prevent the initial current surge from retripping the feeder.

Restoring load for the example used in Sections 5.3 and 5.4 can be described as follows. Assume that 10% of generation is lost, causing an 11% overload. The first-step relays shed a block equal to 12% of system load. The block shed consists of groups of distribution feeders located at six different unattended substations, each equipped with an underfrequency relay. A second set of six frequency relays (or the same set of relays equipped with a "restore" function) used in the overfrequency mode automatically restores service. All relays are set at the normal system frequency and reclose feeders, one substation at a time, using external timers for sequencing and delay. The initial delay is 45 sec after the frequency returns to normal; subsequent delays are as follows:

Substation no.	External delay (sec)
1	10
2	12
3	14
4	16
5	18
6	20

Figure 17-11 shows the behavior of frequency over time for this shedding and restoration action.

8 OTHER FREQUENCY RELAY APPLICATIONS

1. Overfrequency relays are often applied to generators. These relays protect against overspeed during startup or when the unit is suddenly separated from the system with little or no load. Relay contacts either sound an alarm or remove power input to the turbine.

2. Underfrequency relays, with long external time delays, may also be connected to generating units to protect against turbine-blade damage resulting from prolonged full-load underspeed operation. If the overload exceeds the capability of the load-shedding scheme, these generator relays will isolate the unit (with some load, if possible) to keep it in operation and avoid blade fatigue. When load is small or absent and vibration is minimal, a supervisory overcurrent relay or manual switch can be used to prevent tripping during startup.

3. Underfrequency relays can also be used to sense disturbances and intentionally split systems by opening ties. System splitting will not alleviate overload, but it may allow more orderly and discriminating load-shedding action following the split. Neighboring utilities should agree to uniform load-shedding programs; otherwise, the utility that sheds the most load may find itself relieving overloads for disturbances on other systems. In such cases, the shedding utility could use underfrequency relays to trip the interconnections (possibly in conjunction with reverse-power relays). This scheme would eliminate the possibility of a utility aggravating its own loading problems by adding those of a neighboring utility without adequate load-shedding.

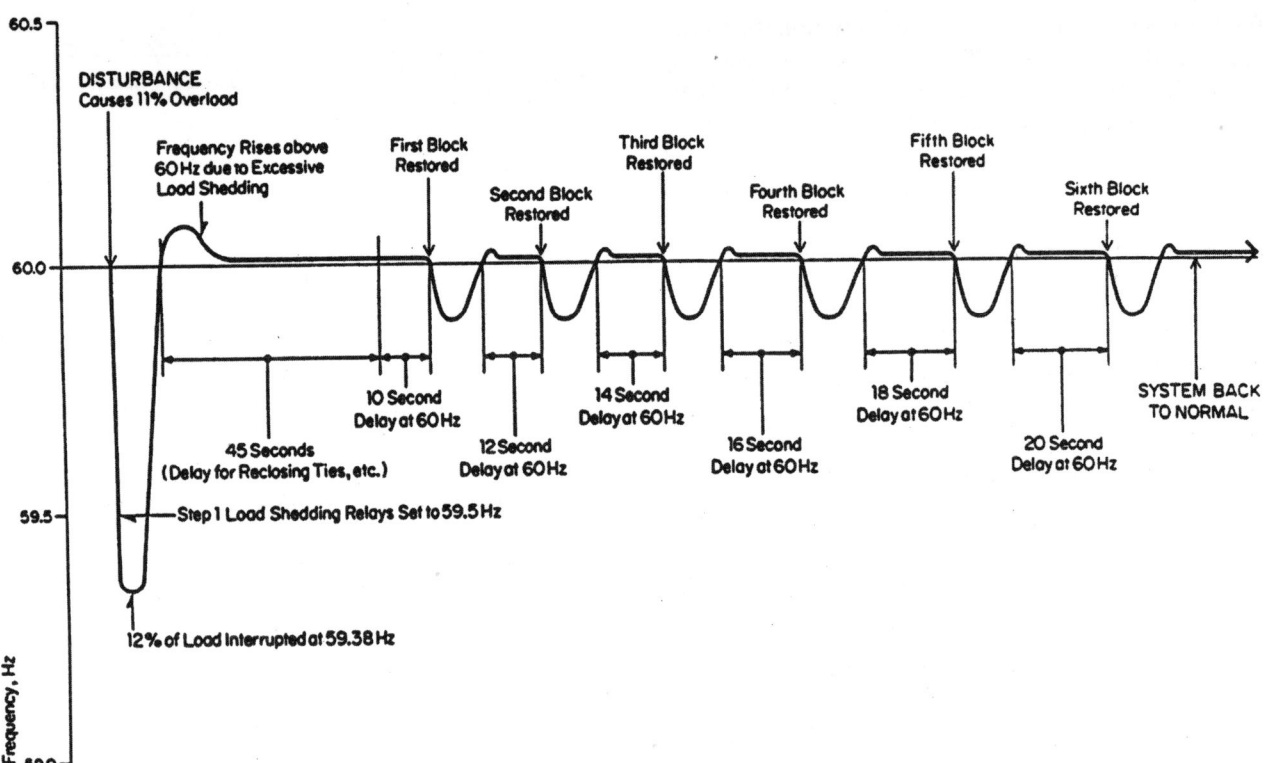

Figure 17-11   Behavior of frequency during automatic load shedding and restoration. (Example for a 60-Hz system.)

4. Another application of underfrequency relays causes the load-tap changer on a transformer to step in the direction to lower the load voltage. With a predominately *resistive load*, this will reduce the load approximately as the square of the voltage. A load of 20 MW, for example, will be reduced, theoretically, to 16.2 MW if the voltage is reduced by 10%. By integrating this procedure with other load-shedding strategies, the seriousness of reduced-frequency operation can be reduced. Reduction of voltage by this expedient on *induction motor loads* may have little effect because an induction motor will adjust its speed to a lower level and maintain essentially a constant kVA. In fact, the power consumed by a motor may actually increase in response to a small reduction in voltage. However, in the case we are considering here, this will be compensated to a degree by a decrease in power resulting from a decrease in frequency. The results of the combination of frequency and voltage reduction are not nearly

so predictable as for a predominately resistive load.

5. Another application in which an underfrequency relay finds use is for the supervision of a generator's offline status. Cross-compound turbines have a high-pressure unit and a low-pressure unit, each of which has its own generator. The two generators are synchronized to each other at reduced speed and are brought up to rated speed together. They may operate at reduced speed for a substantial amount of time. An underfrequency relay is often used to sense that the generators are not yet connected to the power system. Any stator overcurrent during this period of reduced-frequency operation is indicative of the fact that the generators have been inadvertently connected to the power system, either through a generator breaker or through the station service bus. Immediate tripping is mandatory if damage is to be minimized or possibly avoided altogether as a result of this incorrect action.

6. Protection can be inserted by a frequency relay at the appropriate level. Some relays are responsive only at rated frequency and are frustrated in their assignment if prolonged operation is possible at odd frequencies. Some generator ground relays fall into this category. Additional instantaneous relays are applied to provide offline fault protection. They should be removed from service once the machines are synchronized to the system. An overfrequency relay (set for perhaps 55 Hz) may be used to do this. Another problem exists for cross-compound machines that are not yet synchronized to the power system. Reactive interchange between units operating at reduced frequency can undesirably produce operation of a loss-of-field relay. To avoid this, an overfrequency relay should be used to activate this protection at rated frequency.

Bibliography

BASIC FUNDAMENTALS

1. Anderson, P. M. Analysis of Faulted Power Systems. The Iowa State University Press, 1981.
2. Bewley, L. V. Traveling Waves on Transmission Systems. Dover Publications, New York, 1963.
3. Blackburn, J. L. Protective Relaying Principles and Applications. Marcel Dekker, New York, 1987.
4. Blackburn, J. L. Symmetrical Components for Power Systems Engineering. Marcel Dekker, New York, 1993.
5. Greenwood, A. Electrical Transients in Power Systems. John Wiley & Sons, New York, 1971.
6. IEEE/PES Power System Relaying Committee. "IEEE Standard Relays and Relay Systems Associated with Electric Power Apparatus," ANSI/IEEE C37.90-1989.
7. Peterson, H. A. Transients in Power Systems. Dover Publications, New York, 1966.
8. Wagner, C. F., and Evans, R. D. Symmetrical Components. Robert E. Krieger Publishing, Malabar, FL, 1982.

INSTRUMENT TRANSFORMERS

1. IEEE/PES Power System Relaying Committee. "Transient Response of Current Transformers," IEEE New York, 1976, 76-CH, 1130-4 PWR.
2. IEEE/PES Power System Relaying Committee. "Guide for the Grounding of Instrument Transformer Secondary Circuits and Cases," ANSI/IEEE C57.13.3-1983.
3. IEEE/PES Power System Relaying Committee. "Potential Transformer Application on Unit Connected Generators," IEEE Transactions on Power Apparatus and Systems, Vol. PAS-91, Jan/Feb, 1972 pp. 24–28.
4. IEEE/PES Power System Relaying Committee. "IEEE Guide for the Application of Current Transformers Used for Protective Relaying Purposes," Standard No.: C37.110-1996.
5. Wright, A. Current Transformers, Their Transient and Steady State Performance. Chapman and Hall, 1968.

MICROPROCESSOR RELAYING

1. Phadke, A. G., and Thorp, J. S. Computer Relaying for Power Systems. John Wiley & Sons, New York, 1988.
2. Sachdev, M. S. (Coordinator). "IEEE PSRC Tutorial, Advancements in Microprocessor Based Protection and Communication," 97TP120-0.
3. Morrison, I. F. "Prospects of On-Line Computer Control in Transmission Systems and Substations," IE Australia, Vol. EE3, No. 2, Sept. 1967, pp. 234–236.
4. Mann, B. J. "Real Time Computer Calculation of the Impedance of a Faulted Single-Phase Line," IE Australia, Vol. EE4, March 1969, pp. 26–28.
5. Rockefeller, G. D. "Fault Protection with a Digital Computer," IEEE Transactions, Vol. PAS-88, No. 4, April 1969, pp. 438–464.
6. Rockefeller, G. D., and Udren, E. A. "High-Speed Distance Relaying Using a Digital Computer, Part II: Test Results," IEEE Transactions, Vol. 91, No. 3, May/June, 1972, pp. 1244–1258.
7. Girgis, A. A., and Brown, R. G., "Application of Kalman Filtering in Computer Relaying," IEEE Transactions on Power Apparatus and Systems, Vol. 100, No. 7, July 1981, pp. 3387–3397.
8. Schweitzer, E. O. "Unified Protection, Monitoring and Control of Overhead Transmission Lines Achieves Performance and Economy," 35th Relay Conference, Georgia Tech, Atlanta, 1985.

GROUNDING SYSTEM

1. Electrical Transmission and Distribution Book. Westinghouse Electric Corp., Trafford, Pennsylvania, 1964.
2. IEEE. "Recommended Practice for Grounding of Industrial and Commercial Power Systems," IEEE Std. 142-1982

GENERATOR PROTECTION

1. IEEE/PES Power System Relaying Committee. "IEEE Guide for Abnormal Frequency Protection for Power Generating Plants," ANSI/IEEE C37.106-1987.
2. IEEE/PES Power System Relaying Committee. "IEEE Guide for AC Generator Protection," ANSI/IEEE C37, 1995.
3. IEEE/PES Power System Relaying Committee. "Guide for Generator ground Protection," ANSI/IEEE C37.101-1985.
4. IEEE/PES Power System Relaying Committee. "Inadvertent Energizing Protection of Synchronous Generators," IEEE Transactions on Power Delivery, April 1989, p. 965.

MOTOR PROTECTION

1. "AC Motor Protection Guide, Industrial and Commercial Power System Application Series," PRSC-2E Feb. 1992, ABB Allentown.
2. Elmore, W. A., and Kramer, C. A. "Complete Motor Protection by Microprocessor Relay," RPL 87-1, May 1987, ABB Coral Springs, FL.
3. IEEE/PES Power System Relaying Committee. "IEEE Guide for AC Motor Protection," ANSI/IEEE C37.96-2000.
4. "Some Thoughts on Large Motor Protection," RP1 76-2A Sept. 1978, ABB Coral Springs, FL.
5. Zocholl, S. E. "Motor Analysis and Thermal Protection," IEEE/PES Winter Meeting, Feb. 1990, IEEE Paper 90 WM 247-7 PWRD.

TRANSFORMER PROTECTION

1. Elmore, W. A. "Ways to Assure Improper Operation of Transformer Differential Relays," Texas A&M Relay Conference, College Station, TX, April 15, 1991.
2. IEEE/PES Power System Relaying Committee. "IEEE Guide for Protective Relay Applications to Power Transformers," ANSI/IEEE C37.91-2000.
3. Sonnemann, W. K., Wagner, C. L., and Rockefeller, G. D. "Magnetizing Inrush Phenomenon in Transformer Banks," AIEE Transactions Part III, Power Apparatus and Systems, pp. 884–892.
4. "Transformer Protection Guide, Industrial and Commercial Power System Applications Series," PRSC-3E, June 1991, ABB Coral Springs, FL.

BUS PROTECTION

1. "Bus Protection Guide Industrial and Commercial Power System Applications Series," PRSC-9 Nov. 1989, ABB Coral Springs, FL.
2. Forford, T., and Linders, J. R., "A Half-Cycle Bus Differential Relay and Its Application," IEEE Transactions on PAS, Vol. PAS-93, July/Aug. 1974, pp. 1110–1120.
3. IEEE/PES Power System Relaying Committee. "IEEE Guide for Protective Relay Application to Power System Buses," ANSI/IEEE C37.97-1979.

TRANSMISSION LINE PROTECTION

1. Elmore, W. A. Pilot Protective Relaying. Marcel Dekker, 2000.
2. Andersson, Finn, and Elmore, W. A., "Overview of Series-Compensated Line Protection Philosophies," Western Protective Relay Conference 1990, Washington State University, Pullman, WA.
3. Cook, V. Analysis of Distance Protection. John Wiley & Sons, New York, 1985.
4. Crockett, J. M., and Elmore, W. A., "Performance of Phase to Phase Distance Units," Western Protective Relaying Conference 1986, Washington State University, Pullman, WA.
5. Elmore, W. A. "Modern Transmission Line Relaying Variations," Georgia Tech Relaying Conference, Atlanta, May 3–5, 1989.
6. Elmore, W. A. "Zero Sequence Mutual Effects on Ground Distance Relays and Fault Locations," Texas A&M Relay Conference, College Station, TX, April 1992.
7. IEEE/PES Power System Relaying Committee. "Protection Aspects of Multi-Terminal Lines," Special Publication 79TH0056-2-PWR.
8. Lewis, W. A., and Tippett, L. S., "Fundamental Basis for Distance Relaying on a 3-Phase System," AIEE Transaction, Vol. PAS-66, 1947.
9. Martilla, R. J. "Directional Characteristics of Distance Relay Mho Elements Part I," IEEE Transactions on PAS, Vol. PAS-100, Jan. 1981, pp. 96–102.
10. Sonnemann, W. K., and Lensner, H. W. "Compensator Distance Relaying I—General Principles of Operation," AIEE Transactions on PAS, Vol. 77, Part III, pp. 372–382.

11. Sun, S. C., and Ray, R. E., "A Current Differential Relay System Using Fiber Optics Communications," IEEE Transactions on Power Apparatus and Systems, Vol. PAS-102, No. 2, Feb. 1983, pp. 410–419.
12. Udren, E. A., and Cease, T. W. "Transmission Line Protection with Magneto-Optic Current Transducers and Microprocessor-Based Relays," Texas A&M Relay Conference, College Station, TX, April 13–15, 1992.
13. Udren, E. A., and Li, H. J. "Transmission Line Relaying Conference," 1987, Washington State University, Pullman, WA.
14. Crockett, J. M, and Elmore, W. A. "Performance of Phase-to-Phase Distance Units," Conference for Protective Relay Engineers, Texas A&M, College Station, TX, April 13–15, 1987.
15. Gilchrist, G. B., Rockefeller, G. D., and Udren, E. A. "High Speed Distance Relaying Using a Digital Computer," IEEE Transactions, Vol. PAS 91, May/June 1972, pp. 1235–1258.
16. Calero F. "Development of a Numerical Comparator for Protective Relaying: Part I," IEEE Transactions on Power Delivery, Vol. 11, No. 3, July 1996, pp. 1266–1273.
17. Elmore, W. A., Calero, F., and Yang, L., "Evolution of Distance Relaying Principles," Conference for Protective Relay Engineers, Texas A&M, College Station, TX, April 3–5, 1995.
18. Wang, L., and Price, E. "High Speed Microprocessor Distance Relaying for Transmission Lines," Western Protective Relaying Conference, Spoken, WA, October 20–22, 1998.
19. Elmore, W. A., and Price, E. "Ground Relaying Fundamentals," 53rd Texas A&M Conference for Protective Relay Engineers, College Station, TX, April 11–13, 2000.
20. IEEE/PES Power System Relaying Committee. "Guide for Protective Relay Applications to Transmission Lines," IEEE Std. C37.113-1999.
21. Novosel, D., Phadke, A., Saha, M. M., and Lindahl, S. "Problems and Solutions for Microprocessor Protection of Series Compensated Lines," Sixth International Conference on Developments in Power System Protection, The University of Nottingham, UK, March 25–27, 1997.

STABILITY AND OUT-OF-STEP RELAYING

1. "Fundamentals of Out-of-Step Relaying," RPL 79-1C, Nov. 1991, ABB Coral Springs, FL.
2. IEEE/PES Power System Relaying Committee. "Out-of-Step Relaying for Generators—Working Group Report," IEEE Transactions on Power Apparatus and Systems, Vol. PAS-96, No. 5, Sept./Oct. 1977, pp. 1556–1564.
3. Taylor, C. W., et al. "A New Out-of-Step Relay with Rate of Change of Apparent Resistant Augmentation," IEEE Transactions on Power Apparatus and Systems, Vol. PAS-102, No. 3, March 1983 pp. 631–639.
4. Haner, J. M., Laughlin, T. D., and Taylor, C. W., "Experience with the R-Dot Out-of-Step Relay," IEEE Transactions on Power Systems, Vol. 1, No. 2, April 1986 pp. 35–39.
5. Ohura, Y., et al. "A Predictive Out-of-Step Protection System Based on Observation of the Phase Difference Between Substations," IEEE Transactions on Power Delivery, Vol. 5, No. 4, November 1990 pp. 1695–1704.
6. Morioka, Y., et al. "System Separation Equipment to Minimize Power System Instability Using Generator's Angular-Velocity Measurements," IEEE Transactions on Power Delivery, Vol. 8, No. 3, July 1993 pp. 941–947.
7. Roemish, W. R., and Wall, E. T. "A New Synchronous Generator Out-of-Step Relay Scheme, Part I: Abbreviated Version," IEEE Transactions on Power Apparatus and Systems," Vol. PAS-104, No. 3, March 1985 pp. 563–571.

VOLTAGE STABILITY

1. IEEE Special Publication 90TH0358-2-PWR, "Voltage Stability of Power Systems: Concepts, Analytical Tools, and Industry Experience," Prepared by the IEEE Working Group on Voltage Stability, 1990.
2. IEEE/PES Power System Relaying Committee, "System Protection and Voltage Stability," 93THO 596-7 PWR, June 1993.
3. Taylor, C. W., "Power System Voltage Stability," McGraw-Hill, New York, 1994.
4. IEEE Special Publication 93TH0620-5-PWR, "Suggested Techniques for Voltage Stability Analysis," Prepared by the IEEE Working Group on Voltage Stability, 1993.
5. IEEE/PES Power System Relaying Committee, "Voltage Collapse Mitigation," 24th Annual Western Protective Relay Conference, Spokane, WA, October 21–23, 1997.
6. Wang, L., and Girgis, A. A., "On-Line Detection of Power System Small Disturbance Voltage Instability," IEEE Trans. on Power Systems, Aug. 1996, pp. 1304–1313.
7. Vu, K., Begovic, M. M., Novosel, D., and Saha, M. M., "Use of Local Measurements to Estimate Voltage-Stability Margin," IEEE Trans. on Power Systems, Aug. 1999.
8. Shuh, H. M., and Cowan, J. R. "Undervoltage Load Shedding an Ultimate Application for Voltage Collapse," Georgia Tech Protective Relaying Conference, April 29–May 1, 1992.
9. Karlsson, D. "System Protection Schemes in Power Network Based on New Principles," 27th Annual Western Protective Relay Conference, Spokane, WA, October 21–23, 2000.

RECLOSING

1. IEEE/PES T&D Committee Working Group Report. "Arc Deionization Times on High Speed Three Pole Reclosing," IEEE Transactions on Power Apparatus and Systems, Vol. S82, 1963, pp. 236–252.
2. Sturton, A. B. "One, Two and Three-Phase Automatic Reclosing of 230 kV and 345 kV Lines," IEEE Transactions on Power Apparatus and Systems, Vol. S82, 1963, pp. 304–318.

LOAD-SHEDDING AND FREQUENCY RELAYING

1. Dalziel, C. F., and Steinback, E. W. "Underfrequency Protection of Power Systems for System Relief," AIEE Transactions on Power Apparatus and Systems, Part III-B, Vol. 78, pp. 1227–1238.
2. Lokay, H. E., and Burtynk, V. "Application of Underfrequency Relays for Automatic Load Shedding," IEEE Transactions on Power Apparatus and Systems, Vol. PAS-87, No. 3, March 1968, pp. 776–783.
3. Smaha, D. W., Roland, C. R., and Pope, J. W. "Load Shedding Coordination With Turbine-Generator Underfrequency Protection on the Southern Electric System," Georgia Tech Relaying Conference, Atlanta, 1980.

Index

About the Editor

WALTER A. ELMORE is Consulting Engineer, Blue Ridge, Virginia, and retired from the Relay Division of the ABB Power T & D Company, Inc., Coral Springs, Florida. The author or coauthor of more than 100 professional publications including *Pilot Protective Relaying* (Marcel Dekker, Inc.), Mr. Elmore is a Life Fellow of the Institute of Electrical and Electronics Engineers (IEEE) and a member of the National Academy of Engineering, and holds six patents. He is a recipient of the IEEE's Gold Medal for Engineering Excellence (1989) and the Power System Relaying Committee Award for Distinguished Service (1989). A registered Professional Engineer in Florida, Mr. Elmore received the B.S. degree (1949) in electrical engineering from the University of Tennessee, Knoxville